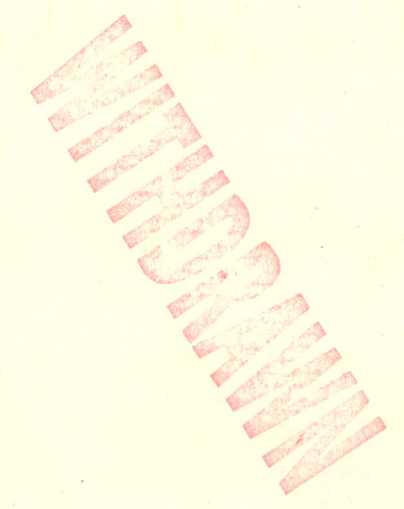

Preparative Organic Photochemistry

Preparative Organic Photochemistry

By

Alexander Schönberg

Technische Universität Berlin

In cooperation with

Günther Otto Schenck · Otto-Albrecht Neumüller

Abteilung Strahlenchemie
Max-Planck-Institut für Kohlenforschung Mülheim/Ruhr

Second completely revised edition of
"Präparative Organische Photochemie"
by A. Schönberg with a contribution by G. O. Schenck

Springer-Verlag New York Inc. 1968

All rights reserved. No part of this book may be translated or reproduced in any form without written permission
from Springer-Verlag. © by Springer-Verlag Berlin · Heidelberg 1968
Printed in Germany · Library of Congress Catalog Card Number 67-16134
Satz: Werk- und Feindruckerei Dr. Alexander Krebs, Weinheim/Bergstr. und Bad Homburg v. d. H.
Druck: Offsetdruckerei Julius Beltz, Weinheim/Bergstr.

The use of general descriptive names, trade names, trade marks, etc. in this publication, even if the former are not
especially identified, is not to be taken as a sign that such names, as understood by the Trade Marks and Merchandise
Marks Act, may accordingly be used freely by anyone

Title-No. 1382

Preface to the First Edition

(Abridged and translated)

Organic photochemistry may be divided into three parts: theory which is the province of the physical chemist; instrumentation which requires the skill of both physicist and engineer; and preparation which falls within the sphere of the organic chemist.

At one time the same person could cover all three fields without too much difficulty, but this has now become virtually impossible because the disciplines involved have expanded in both breadth and depth; it is therefore timely to have a separate treatment of preparative organic photochemistry.

There appears to be no review of the main photochemical reactions which includes the advances made in recent years available to the organic chemist working in the preparative field. An exception is the excellent "Photochemical Reactions" by C. R. MASSON, V. BOEKELHEIDE and W. A. NOYES JR., published in 1956, which gives a brief review of the reactions which are important in preparative organic photochemistry. The present monograph on the other hand seeks to provide a detailed survey for the chemist; the author does not set out to discuss every photochemical reaction in the field of organic chemistry but he does include in addition to those of current interest in the preparative field some which are likely to be of interest in the future and which result in single end-products of known composition. The photochemical synthesis of highly polymerized products falls outside the scope of the work.

The author hopes his book will not only show chemists what has been achieved already but will also stimulate them to make greater use of photochemical reactions than heretofore. This is not possible, however, unless the basic principles of theory and instrumentation are thoroughly known. The author wishes to thank Prof. G. O. SCHENCK for his contribution, "General remarks on carrying out photochemical reactions", which supplies this knowledge.

As regards future work, it should be noted that it is not sufficient merely to publish the results of experiments carried out in the light but that the results of equivalent experiments in the dark must also be given.

Older publications, in particular, often describe reactions which occurred under irradiation but lack data on parallel experiments where light was excluded, so that the material presented does not allow us to decide whether the light was indispensable in such cases or whether it merely accelerated the reaction.

The author does not wish to conclude without thanking his Egyptian colleagues, and in particular Professor AHMED MUSTAFA, for their assistance with his photochemical work.

Cairo, June 1957 A. Schönberg
University and National Research Centre

Preface to the Second Edition

The same guiding principles laid down for the first edition have been followed in the second. Since the first edition came out, our knowledge in the field of preparative organic chemistry has grown in both range and depth to an amazing extent. Most of the material dealt with in this edition was unknown when the first edition was compiled, so that the present edition may be said to be an essentially new book.

The indexes have been greatly enlarged and new lists of relevant monographs and review articles have been compiled. As regards nomenclature, the present edition follows the IUPAC Rules. References have been included up to the end of 1965, and a few results from the authors' own laboratories have been inserted, although they were not at that time available in print.

It is the intention of the publishers and the authors to bring the work up-to-date by the periodical issue of supplementary volumes. In a field which is growing so fast it seems best not to await a possible third edition before summarizing the results.

Acknowledgements: Our sincere thanks are due to F. SCHALLER for the compilation of data pertaining to irradiation technique, to C. MAUCY (IBM Düsseldorf) for writing the Compound Index program, to R. BERNHARDT and V. KUBE (Zentralstelle für Maschinelle Dokumentation, Frankfurt, Direktor K. SCHNEIDER) for the KWOC program and to Mrs. M. LEHMANN for proofreading. We are grateful to the publishers for their interest and encouragement.

Berlin	A. Schönberg
Mülheim/Ruhr	G. O. Schenck
	O.-A. Neumüller
August 1967	

Contents

Chapter 1

Photoisomerization of unsaturated systems proceeding with the formation of four-membered homocyclic rings

1. Formation of cyclobutane derivatives by intramolecular addition between non-conjugated olefinic linkages (CIAMICIAN-addition) 1
2. Photochemical valence tautomerization of 1,3-dienes 4
   a) Homoannular 1,3-dienes 4
   b) Aromatic compounds 9
   c) Transoid 1,3-dienes 10
3. Intramolecular dianthracene formation 11

References 12

Chapter 2

Photoisomerization of dienes and trienes not leading to the formation of cyclobutane derivatives

1. Photoisomerization of open chain 1,3-dienes to 1,2-dienes 14
2. Photoisomerization of 1,3-cyclohexadienes 14
   a) Cleavage to 1,3,5-hexatrienes 14
   b) Rearrangement to bicyclo[3.1.0]hex-2-enes 16
3. Photoisomerization of 1,5-cyclooctadienes 18
4. Photoisomerization of 1,3,5-hexatrienes to 1,3-cyclohexadienes 18
5. Photoisomerization of an acene involving loss of aromaticity 20

References 21

Chapter 3

Photoisomerization of aldehydes and ketones not leading to the formation of oxygen heterocycles

1. Photoisomerization of α,β-unsaturated carbonyl compounds to β,γ-unsaturated or cyclopropyl isomers 22
2. Photoenolization of aromatic ketones 24
3. Photoisomerization of saturated ketones via photoelimination 25

Contents

4. Photoisomerization of α,β-unsaturated carbonyl compounds via cyclization . 27
   a) Cyclization of citral 27
   b) Rearrangement of eucarvone 27
   c) Rearrangement of verbenone 28
5. Photoisomerization of cross-conjugated cyclohexadienones 28
   a) Simple 2,5-cyclohexadien-1-ones 29
   b) Condensed 2,5-cyclohexadien-1-ones 29
References 32

Chapter 4

Various photoisomerizations of ketones, esters and halides

1. Isomerization of ketones to tertiary alcohols by intramolecular cyclization . 34
   a) Cyclobutanols from saturated cyclic and acyclic ketones 34
   b) 2-Hydroxycyclobutanones from 1,2-diketones 35
   c) Cyclobutanols in the steroid series 36
   d) Cyclohexenols from non-conjugated ketones 37
2. Photorearrangement of enol and dienol esters 37
3. Isomerizations with migration of bromine atoms 39
4. Light-induced von Auwers rearrangement 40
References 40

Chapter 5

Photoisomerizations involving formation and transformation of five or six membered heterocyclic oxygen compounds

1. Photochemical formation of heterocyclic oxygen compounds 41
   a) Formation of furan derivatives by intramolecular cyclization 41
   b) Formation of pyran derivatives by intramolecular cyclization 41
   c) Formation of heterocyclic oxygen compounds by isomerization of cyclic ketones 43
2. Photoisomerization of heterocyclic oxygen compounds to open chain compounds 44
   a) Photoisomerization of furan derivatives 44
   b) Photoisomerization of cyclic acetals to esters 45
3. Photochemical rearrangements among heterocyclic oxygen compounds . . 46
   a) Isomerization of a 4H-pyran to a 2H-pyran 46
   b) Isomerization of epidioxides 46
References 47

Chapter 6

Photoisomerizations involving nitrogen compounds

1. Photoisomerizations with rupture of nitrogen-oxygen bonds 48
   a) Amides from aldoximes 48
   b) Anilides from nitrones 48
   c) Oxaziridines from nitrones 50

d) Unsaturated lactams from heterocyclic N-oxides 51
   e) 2-Hydroxyazobenzenes from azoxybenzenes 53
2. Photoisomerization with rupture of nitrogen-halogen bonds 53
3. Photoisomerizations with rupture of nitrogen-carbon bonds 54
   a) Rearrangement of a pyrimidine 54
   b) Rearrangement of pyrazoles 54
References 55

Chapter 7

Photochemical stereoisomerization

1. Photochemical cis-trans isomerizations at double bonds 56
   a) Compounds containing isolated C=C bonds 57
   b) Dienes and polyenes 58
   c) Indigoid systems 61
   d) Azo compounds 62
   e) Compounds containing C=N bonds 63
2. Photochemical cis-trans isomerizations of cyclopropane and cyclobutane derivatives 64
3. Photoepimerizations 65
   a) Epimerization at carbon atoms 65
   b) Epimerization at sulfur atoms 66
References 67

Chapter 8

Photodimerization with formation of cyclobutane derivatives (Cyclodimerization)

1. Stilbene and related compounds 70
2. Unsaturated hydrocarbons 73
   a) Acyclic dienes 74
   b) Cyclic olefins 74
   c) Cyclic dienes 75
3. α,β-Unsaturated ketones 75
   a) Acyclic α,β-unsaturated ketones 76
   b) Cyclic α,β-unsaturated ketones 77
   c) 1,4-Quinones 79
4. α,β-Unsaturated acids and related compounds 82
   a) Derivatives of maleic and fumaric acids 82
   b) Cinnamic acids 83
5. Coumarin and isocoumarin 85
   a) Coumarin 85
   b) Isocoumarin 87
   c) Furocoumarins 87
6. 2,6-Dimethyl-4-pyrone 89
7. Tetraphenylbutatriene 89

8. Unsaturated sulfur compounds 90
   a) Benzo[b]thiophene 1,1-dioxide 90
   b) 2-Nitrobenzo-1,4-dithiin 90
9. Thymine and related pyrimidine derivatives 91
10. β-Lumicolchicine 92
11. Acetylene compounds 92
References 94

Chapter 9

Photodimerizations involving formation of eight-membered rings

1. Carbocyclic aromatic compounds 97
   a) Naphthalene derivatives 97
   b) Anthracene and derivatives 97
   c) Higher condensed aromatic compounds 99
2. Heterocyclic nitrogen compounds 99
   a) Pyridine derivatives 99
   b) Condensed nitrogen heterocyclic compounds 101
3. Heterocyclic oxygen compounds 102
References 103

Chapter 10

Various photodimerizations involving aldehydes, halides and thiocarbonyl compounds

1. Photodimerization of aliphatic compounds 105
   a) Butyraldehyde 105
   b) Dimerization of halides 105
   c) Dimerization of thiocarbonyl compounds 106
2. Photodimerizations of aromatic aldehydes 107
References 108

Chapter 11

Cycloaddition of alkenes or alkynes to other alkenes or aromatic nuclei leading to the formation of four-membered rings. Photolyses involving retro-cycloaddition

1. Addition of maleic anhydride to benzene and related aromatic compounds including furan and thiophene 109
2. Addition of maleic acid derivatives to alkenes 110
3. Addition of 2-methyl-2-butene to benzonitrile 111
4. Photochemical cycloaddition of α,β-unsaturated ketones to alkenes . . . 112
   a) 2-Cyclohexenone and isobutylene 112
   b) Cyclopentenone and cyclopentene 112
   c) 2,4-Pentanedione or dimedone and cyclohexene 113

5. Photochemical cycloaddition reactions of acetylene compounds 114
   a) Addition to benzene and derivatives 114
   b) Addition to cyclopentenone 115
   c) Addition to dimethyl cyclobutene-1,2-dicarboxylate 115
   d) Addition of diphenyl acetylene to naphthalene 115
6. Photolyses involving retro-cycloaddition 116
   a) Photolysis of dehydronorcamphor 116
   b) Photochemical synthesis of bullvalene via retro-cycloaddition 116
References 117

Chapter 12

Photochemical cycloaddition of 1,2-quinones, 1,2-diketones and 1,2,3-triketones to multiple bonds

Photochemical cycloaddition of 1,2-quinones, 1,2-diketones and 1,2,3-triketones to multiple bonds 118
References 125

Chapter 13

Photochemical cyclization of aromatic compounds via elimination of hydrogen and/or halogen atoms. Formation of carbocycles

1. Formation of five-membered homocycles. A fluorene derivative from triphenylmethyl 126
2. Formation of six-membered homocycles 127
   a) Phenanthrenes from stilbenes 127
   b) Fused aromatic compounds from o-dibenzylidene compounds . . . 129
   c) Phenanthroperylenediones, dibenzoperylenediones and analogous compounds from less condensed aromatic precursors 131
   d) Dehydrocyclization of some fused heterocyclic hydrocarbons 134
References 137

Chapter 14

Photochemical dehydrocyclization of aromatic compounds via elimination of hydrogen atoms. Formation of heterocycles

1. Formation of five-membered heterocycles. Carbazoles from diphenylamines . 138
2. Formation of six-membered heterocycles 138
   a) Benzocinnolines from azobenzenes 138
   b) Phenanthridines from Schiff's bases 141
   c) Phenanthridizinium salts from styrylpyridinium salts 142
   d) Benzo[c]tetrazolo[2,3-a]cinnolinium salts from triphenyltetrazolium salts . 143
References 144

Contents   XIII

Chapter 15

Photochemical dehydrodimerization

1. Dehydrogenation by oxygen 146
2. Dehydrogenation by carbonyl compounds 147
3. Dehydrogenation by dyes 148
4. Dianthracene from dihydroanthracene 149
References 150

Chapter 16

Photochemical dehydrogenation

1. Quinones as dehydrogenating agents 151
2. 1,2-Disulfides as dehydrogenating agents 152
3. Photosensitized dehydrogenation using dyes 153
References 154

Chapter 17

Photochemical additions to carbon-carbon multiple bonds not resulting in ring formation

1. Water 155
2. Hydrogen peroxide (MILAS reaction) 157
3. Hydrogen bromide 157
4. Nitrosyl chloride 159
5. Alcohols, ethers, and tert. butyl hypochlorite 159
    a) Alcohols 159
    b) Ethers 160
    c) Tert. butyl hypochlorite 162
6. Sulfur compounds 162
    a) Hydrogen sulfide, thiols and thiocarboxylic acids 162
    b) Sulfenyl chlorides 164
    c) Sulfonyl chlorides 165
7. Ammonia, amines and formamide 165
    a) Ammonia and amines 165
    b) Formamides 166
8. Dimethylmaleic anhydride 168
9. Aldehydes and ketones 169
    a) Aldehydes 169
    b) Ketones 169
10. Aliphatic polyhalides 170
    a) Polyhalides not containing fluorine 171
    b) Polyfluoroalkyl iodides 173
    c) Photoaddition of polyhalides to conjugated systems . . . 175

11. Organophosphorus compounds 176
   a) Phosphines 176
   b) Phosphonates 177
12. Organosilicon compounds 178
   a) Trichlorosilane 178
   b) Organosilicon compounds 179
13. Organogermanium compounds 179
References 180

Chapter 18
Photochemical addition reactions of 1,4- and 1,2-quinones with alkylbenzenes or with ethers

1. Addition of chloranil to hydrocarbons 182
2. Addition of phenanthrenequinone to hydrocarbons 182
3. Addition of phenanthrenequinone or tetrachloro-o-quinone to ethers . . 184
References 185

Chapter 19
Photochemical additions of aldehydes to quinones, quinone imines and quinone oximes

1. Addition of aldehydes to 1,2-quinones 186
2. Addition of aldehydes to 1,4-quinones 188
3. Addition of aldehydes to quinone imines and quinone oximes . . 190
   a) Quinone imines 190
   b) Quinone oximes 191
References 192

Chapter 20
Photoreductions with the aid of alcohols, ethers and other hydrogen donors

1. Photoreductions of C=C bonds 193
2. Photoreductions of C=O bonds. Formation of benzhydrols . . 194
3. Photoreductions of C=N bonds 195
4. Photoreductions of gem. chloronitroso compounds. Formation of oximes . 197
References 197

Chapter 21
Formation of carbinols by photochemical addition of ketones and aldehydes to methylene groups

1. Addition of ketones 198
2. Addition of aldehydes 201
References 202

Chapter 22

Photochemical formation and photolysis of 1,2-ethanediols

1. Formation of 1,2-ethanediols by the addition of alcohols to ketones . . . 203
2. Formation of 1,2-ethanediols via reductive dimerization 203
   a) Aldehydes 203
   b) Monoketones 204
   c) 1,2,3-Triketones 209
   d) α-Ketocarboxylic acids and o-acylbenzoic acids 210
3. Photochemical cleavage of 1,2-ethanediols by carbonyl compounds . . . 211

References 213

Chapter 23

Photochemistry of deoxybenzoin derivatives

Photochemistry of deoxybenzoin derivatives 215
References 217

Chapter 24

Photochemical decarbonylation

1. Decarbonylation of ketones 218
   a) Saturated cyclic ketones 218
   b) Unsaturated cyclic ketones 220
   c) Aromatic ketones 220
2. Decarbonylation of a ketene 222
3. Decarbonylation of aldehydes 222
4. Decarbonylation of S-acyl xanthates 223

References 224

Chapter 25

Photochemical formation and reactions of carboxylic acids and their derivatives

1. Formation of aliphatic carboxylic acids by the action of oxygen and water on alkyl halides 225
2. Formation of carboxylic acids by photolysis of cyclic ketones 225
   a) Saturated ketones 225
   b) Unsaturated ketones 227
3. Formation of acid derivatives 230
   a) Amides 230
   b) Lactones 231
   c) Acyl chlorides 233

XVI Contents

4. Photochemical reactions of esters involving the acyloxy groups 234
   a) Rearrangements of esters 234
   b) Reductive elimination of an acetoxy group 235
   c) Light-induced FRIES rearrangement 236
5. Further formation modes of esters 238
   a) Formation of α-ketocarboxylates from α-keto acetals 238
   b) Formation of esters from 2-butene-1,4-diones 238
   c) Formation of an ester from a diflavylene compound 239
6. Formation of fluorenecarboxylic acids by photolysis of fluoranthenols . . 239
References 240

Chapter 26

Photochemical reactions with N-halogenated amines

1. Photochemical replacement of chlorine in N-chloroamines by hydrogen . . 242
2. The light-induced HOFMANN-LÖFFLER reaction 242
   a) Synthesis of pyrrolidines 242
   b) Synthesis of bridged nitrogen compounds 244
   c) Synthesis of conanines 244
   d) Synthesis of pyrrolizidines 245
3. Formation of chloroalkylamines from N-chloroamines 246
   a) N-Butyl-4-chlorobutylamine 246
   b) Molecular rearrangements of steroidal N-chloroamines 246
References 247

Chapter 27

Photochemical transformations of organic nitrites

1. Photochemical reactions of nitrites involving fission of oxygen-nitrogen bonds.
   The BARTON reaction 248
   a) Simple aliphatic and alicyclic nitrites 248
   b) Epimerization in nitrite photolysis 249
   c) Steroidal nitrites 250
2. Photochemical reactions of nitrites involving fission of carbon-carbon bonds . 251
   a) Syntheses of nitrosoalkanes 251
   b) Hydroxamic acids 252
   c) Fragmentation reactions in the steroid series 253
References 254

Chapter 28

Photochemical dealkylation of nitrogen compounds

1. Photolysis of N-alkyl and N-aralkyl amines 255
2. Photolysis of N-alkyl nitrogen heterocycles 257
References 258

Chapter 29

Photochemical introduction of cyano and nitroso groups

1. Formation of nitriles 260
2. Formation of geminal chloronitroso and dimeric nitroso compounds . . 261
3. Formation of diphenylfuroxan via α-chloro-α-nitrosotoluene 262
4. Formation of oximes from hydrocarbons through photolysis of NOCl or NO/Cl_2 mixtures 262
   a) Photolysis of NOCl 263
   b) Photolysis of NOCl in the presence of HCl or NO 263
   c) Photolysis of NO, Cl_2 and HCl 263
References 265

Chapter 30

Photochemical transformations of unsaturated nitro compounds

1. Photolysis reactions of unsaturated nitro compounds 266
2. Photochemical conversion of aromatic nitro compounds to nitroso compounds 267
3. Photoreduction of an aromatic nitro compound to an aniline derivative . . 270
4. Photocyclization reactions of aromatic nitro compounds 271
References 273

Chapter 31

Light-induced reactions of diazoalkanes, diazirines and related compounds

1. Photoaddition reactions of carbenes to unsaturated systems resulting in ring formation 275
   a) Formation of cyclopropane compounds 275
   b) Formation of oxide rings 279
   c) Formation of aziridine rings 280
   d) Formation of cyclopropene compounds 281
2. Photoaddition reactions of carbenes to saturated compounds resulting in insertion into sigma bonds 281
   a) Insertion into C—C, C—O and C-halogen bonds 281
   b) Insertion into C—H bonds 283
   c) Insertion into O—H bonds 286
   d) Insertion into N—H bonds 287
3. Photodimerization reactions of carbenes 288
4. Miscellaneous photochemical reactions of carbenes 290
   a) Isomerization reactions with formation of olefins 290
   b) Univalent hydrogenation with subsequent dimerization 290
   c) Addition to oxygen 291
5. Photolysis of diazirines 291
6. Addition of diazomethane to olefins with formation of a pyrazoline . . 292
References 292

Chapter 32

Photochemical syntheses with diazoketones, quinone diazides and iminoquinone diazides

1. Acyclic mono-diazoketones 294
   a) Conversion to ketenes and α,β-unsaturated ketones 294
   b) Photolysis in the presence of water or alcohol 295
   c) Photolysis in the presence of N-methylaniline and ethanethiol 297
   d) Photolysis in the presence of azo compounds 298
   e) Photolysis in the presence of azomethines 299
   f) Replacement of the diazo group by hydrogen 300
   g) Intramolecular addition of a carbene 300
2. Acyclic bis-diazoketones 301
3. Cyclic α-diazoketones 301
   a) Ring contraction of five-membered rings 302
   b) Ring contraction of six-membered ring systems 303
   c) Ring contraction of o-quinone diazides (Süs reaction) 304
   d) Photolysis of cyclic α-diazoketones not leading to ring contraction . . . 306
4. Azo dyes from quinone diazides 309
5. p-Quinone diazides and p-iminoquinone diazides 310
References 312

Chapter 33

Photochemical syntheses with diazonium salts and diazosulfonates

1. Reductive deamination of diazonium salts 313
2. Replacement of the diazonium group by halogen or the hydroxy group . 314
3. Photolysis of diazonium salts as a method of cyclization 315
4. Change in reactivity of aryl diazosulfonates 316
5. Photolysis of a 1,2,3-thiadiazine S,S-dioxide 316
References 317

Chapter 34

Synthetic applications of light-induced reactions of azides

1. Photolysis of alkyl azides 319
2. Photolysis of aryl azides 320
   a) Carbazoles and 4-phenylbenzofuroxan from 2-azidobiphenyl derivatives . 320
   b) Photochemical conversion of aryl azides to azo compounds 321
3. Photolysis of acyl azides 321
   a) The light-induced CURTIUS rearrangement 321
   b) Formation of lactams 322
   c) Formation of amides 324
4. Photochemical reaction of ethyl azidoformate 324
   a) Reaction with cyclic hydrocarbons 324
   b) Reaction with alcohols 325

Contents XIX

5. Photochemical syntheses with sulfonic acid azides 326
   a) Photolysis in methanol 326
   b) Photolysis in sulfoxides 326
   c) Photolysis in sulfides 327
References 327

Chapter 35
Photolysis of pyrazolines, pyrazoles, azo compounds, 1,2,3-thiadiazoles, and p-benzoquinone diimine N,N'-dioxides

1. Photolysis of pyrazolines 328
2. Photolysis of pyrazoles 329
3. Photolysis of diaroyl azo compounds 330
4. Photolysis of 1,2,3-thiadiazoles 331
5. Photolysis of p-benzoquinone diimine N,N'-dioxides 332
References 333

Chapter 36
Miscellaneous light-induced reactions of organic nitrogen compounds

1. Incorporation of C_1 or C_2 fragments by the photochemical reaction of various nitrogen compounds with alcohols 334
   a) Benzo[f]quinolines from SCHIFF's bases 334
   b) Imidazolidines from diamines or from SCHIFF s bases 335
   c) Oxazolidines from SCHIFF's bases 335
2. Photolysis of oxadiazolinones 337
3. Light-induced abnormal benzidine rearrangement 337
4. Light-induced condensations involving primary amines and aldehydes . . 338
5. Aromatic nitriles by photochemical cleavage of aromatic aldazines . . 339
References 339

Chapter 37
Photohalogenation

1. Photohalogenation. Scope of the reaction 341
2. Photochlorination 344
   a) Chlorination of benzene in the presence of iodine 344
   b) Chlorination of benzene in the presence of maleic anhydride . . 345
   c) Replacement of the sulfonyl chloride group by chlorine 346
   d) Replacement of the nitroso group by chlorine 347
   e) Replacement of alkyl groups by chlorine 347
   f) Solvent effects on the site of attack by chlorine 348
   g) Chlorination of a cyclic trisulfone 349
3. Photobromination 349
   a) Migration of alkyl groups during bromination 349
   b) Bromination in the presence of oxygen 350
   c) Bromination in the allylic position with the aid of bromine . . 351
   d) Bromination with chlorine-bromine mixtures 351
   e) Orienting effects in the photobromination of alkyl bromides . . 351
   f) Bromination of aryl selenocyanates 353

4. Photoiodination 353
5. Photohalogenation with the aid of inorganic and organic halides 354
   a) Experiments with iodine chloride 354
   b) Chlorination with sulfuryl chloride 354
   c) Chlorination with trichloromethanesulfonyl chloride 355
   d) Chlorination with trichloromethanesulfenyl chloride 356
   e) Chlorination and bromination with the aid of N-chloro- and N-bromo-succinimide 356
   f) Bromination with the aid of dibromodimethylhydantoin 359
References 360

Chapter 38

Photochemical conversions of organic halides

1. Replacement of bromine by hydrogen, chlorine or ^{82}Br 362
   a) Replacement of bromine by hydrogen 362
   b) Replacement of bromine by chlorine or ^{82}Br 362
2. Replacement of iodine in iodides by hydrogen, nitric oxide or chlorine . . 363
   a) Experiments with aliphatic iodides 363
   b) Chlorobenzene from iodobenzene by photochemical decomposition of iodine chloride 365
3. Deiodination of aliphatic iodides 365
4. Debromination of 1,1-diaryl-2-bromoethylenes 366
5. Formation of organomercury compounds by the action of mercury on alkyl iodides 366
6. Preparation of hexaarylethanes by the action of triarylmethyl halides on tri-arylmethanes 367
7. Photolysis of aromatic iodo compounds 368
   a) Iodobenzene and related substances 368
   b) Formation of benzyne on photolysis of 1,2-diiodobenzene or (2-iodophenyl)-mercury iodide 368
8. Photolysis of alkyl hypoiodites, acyl hypoiodites and N-iodoamides . . . 369
   a) Photolysis of alkyl hypoiodites. Preparation of ethers 369
   b) Photolysis of acyl hypoiodites. Replacement of the carboxyl group by iodine 370
   c) Photolysis of N-iodoamides. Preparation of lactones 370
References 371

Chapter 39

Photochemical formation of hydroperoxides and peroxides

1. Replacement of hydrogen by the hydroperoxide group 373
   a) Unsensitized photooxidation of unsaturated compounds 373
   b) Photosensitized oxidation of olefins 375
   c) Photosensitized oxidation of secondary alcohols 379
   d) Photooxidation of ethers 379
   e) Photooxidation of nitrogen compounds 380
   f) Photooxidation of a phenol 382
2. Transannular peroxides from cyclic 1,3-dienes 382
   a) Epidioxides from alicyclic 1,3-dienes 383
   b) Epidioxides from fused 1,3-cyclohexadienes 385

c) Epidioxides from acenes 389
d) Photooxidation of carbo- and heterocyclic cyclopentadienes 394
3. Photochemical formation of bis-aralkyl and bis-acyl peroxides 397
   a) Formation of six-membered cyclic peroxides 397
   b) Formation of open chain peroxides 398
   c) Formation of acyl peroxides 399
4. Miscellaneous photochemical oxidation reactions 400
   a) Oxidative cleavage of C—C bonds 401
   b) Photooxidation of Curare alkaloids 401
References 402

Chapter 40

Photochemical formation and transformations of epoxides

1. Photochemical formation of epoxides 407
2. Photoisomerization of epoxyketones 408
   a) Acyclic α,β-epoxyketones 408
   b) Acyclic β,γ-epoxyketones 409
   c) Cyclic α,β-epoxyketones 409
   d) Steroid α,β-epoxyketones 412
References 413

Chapter 41

Photochemical formation of four membered rings with one oxygen atom (PATERNÒ-BÜCHI reaction)

1. Formation of oxetanes 414
   a) Cycloaddition of aldehydes or ketones to olefins 414
   b) Intramolecular cycloaddition leading to oxetanes 417
   c) Cycloaddition of p-quinones to olefins leading to spirooxetanes . . . 418
   d) Cycloaddition of 1,2-dicarbonyl compounds to olefins leading to α-keto-oxtanes 419
   e) Formation of oxetanes bearing functional groups 420
2. Formation of oxetes as intermediates in the cycloaddition of carbonyl compounds to acetylenes 423
3. Cycloaddition of p-benzoquinone to conjugated dienes not resulting in formation of oxetanes 424
References 424

Chapter 42

Photochemical formation and reactions of furans

1. Photoisomerization of quinoid compounds to furan derivatives 426
2. Photochemical reactions of furans with oxygen 427
   a) Simple furan derivatives 427
   b) Fused furan systems 429
   c) Aryl furans 430
   d) [2.2](2,5)Furanophane 432
3. Photoaddition of methanol to a furan derivative 432
References 433

Chapter 43

Photochemical formation and transformations of organic sulfur compounds

1. Photochemical syntheses using SO_2, SO_2Cl_2 and SCl_2 434
   - a) Sulfochlorination 434
   - b) Sulfenylchlorination 437
   - c) Sulfoxidation 438
   - d) Cyclic sulfates from sulfur dioxide and o-quinones 439
2. Photochemical formation and transformations of sulfides 439
   - a) Thioethers from di- and trisulfides 439
   - b) Photolysis of 9,9'-bis-(phenylthio)-9,9'-bifluorene 440
   - c) Insertion reaction of mercury with disulfides 441
   - d) Conversion of a disulfide to a sulfenyl chloride 441
   - e) Action of mercaptoacetic acid on benzo[a]pyrene 442
   - f) Photolysis of a dixanthate 442
3. Photochemical thiocyanation 443
4. Photooxidation 444
   - a) Sulfoxides 444
   - b) Thiourea 444
   - c) Conversion of thioketones into ketones 445
   - d) Co-oxidation of thiols and olefins by oxygen 446
5. Miscellaneous photochemical reactions of sulfur compounds 447
   - a) Photolysis of sulfones 447
   - b) Photolysis of unsaturated sultones 448
   - c) Photolysis of thiobenzophenone in the presence of olefins 448
   - d) Photochemical formation of a dipyridyl sulfide from a 1,4-dihydropyridine-thione 449
   - e) Photochemical reactions involving extrusion of sulfur from dithietanes and thiiranes 450
   - f) Photochemical aryl migration in arylthiophenes 451
- References 451

Chapter 44

Photochemical reactions of organophosphorus and organoarsenic compounds

1. Organophosphorus compounds 453
   - a) Light-induced addition of dialkyl phosphonates to quinones 453
   - b) Photochemical reactions of trialkyl phosphites 454
   - c) Photolysis of triarylphosphines 455
   - d) Photochemical synthesis of phosphonium salts 455
   - e) Photolysis of tetraarylphosphonium salts 457
2. Organoarsenic compounds 457
- References 458

Chapter 45

Photochemical formation and reactions of organometallic compounds

1. Light-induced formation of organometallic carbonyl compounds from metal carbonyls and organic compounds 459
2. Photochemical reactions of organometallic carbonyl compounds 462
   - a) Substitution reactions with electron donors 462

Contents XXIII

b) Photochemical decarbonylation of organometallic acyl derivatives to alkyl derivatives 464
3. Photochemical reactions of organotin compounds 465
   a) Light-induced reaction of hexamethylditin with trifluoroiodomethane . 465
   b) Synthesis of 1,2,3-triphenylazulene by photolysis of an organotin compound 465
   c) Photosensitized oxygenation of organotin compounds 466
4. Photochemical reactions of organomercury compounds 466
   a) Photolysis reactions with formation of arylmercury halides 467
   b) Photolysis reactions with formation of mercury 467
5. Miscellaneous photochemical reactions of organometallic compounds . . 468
   a) Light-induced formation of a GRIGNARD reagent from an aliphatic bromide 468
   b) Photodimerization of a metal-complexed olefin 468
   c) Photolysis of aromatic lithium compounds 469
   d) Photolysis of iron pentacarbonyl in nitrobenzene solution 469
References 470

Chapter 46

Light sources and light filters in preparative organic photochemistry

G. O. SCHENCK

1. Light sources 472
   a) Low-pressure sodium vapor lamps 474
   b) High-pressure sodium vapor lamps 475
   c) Low-pressure mercury lamps 475
   d) High-pressure mercury lamps 475
   e) Super-high-pressure mercury lamps 476
   f) Super-high-pressure xenon-mercury lamps 477
   g) High-pressure xenon lamps 477
   h) Sunlight 477
   i) Incandescent lamps 477
   j) Halogen incandescent lamps 478
   k) Fluorescent tubes 478
   l) Vortex-stabilized plasma lamps 478
   m) Low-pressure gas discharge lamps for the far UV 478
   n) Flash lamps 479
   o) Lasers 479
   Tables 479
2. Light filters 490
   a) Solid filter materials 490
   b) Liquid filters 491
   c) Interference filters 492
   d) Reflection interference filters 493
References 494

A selective bibliography on photochemistry 495
Author Index 501
Reaction Index 523
Sensitizer Index 559
Compound Index 560

Chapter 1

Photoisomerization of unsaturated systems proceeding with the formation of four-membered homocyclic rings

1. Formation of cyclobutane derivatives by intramolecular addition between nonconjugated olefinic linkages (CIAMICIAN-addition)

Investigation in this field began in 1908 with an observation by CIAMICIAN and SILBER [1] who exposed to sunlight a dilute alcoholic solution of carvone (1) and obtained carvonecamphor (2).

It is very remarkable that at that time the correct formula for the reaction product 2 was already proposed by the Italian workers, though it was later criticized [2]. However, it has been confirmed by the investigations of BÜCHI [3] and of MEINWALD [4], who found that 2 in its turn may, in alcohol or aqueous dioxane, undergo photolysis to acid derivatives, unless the irradiation of 1 was carried out under suitable light conditions by using, for example, a black light fluorescent lamp. Under these circumstances yields of 35% could be obtained within 23 days.

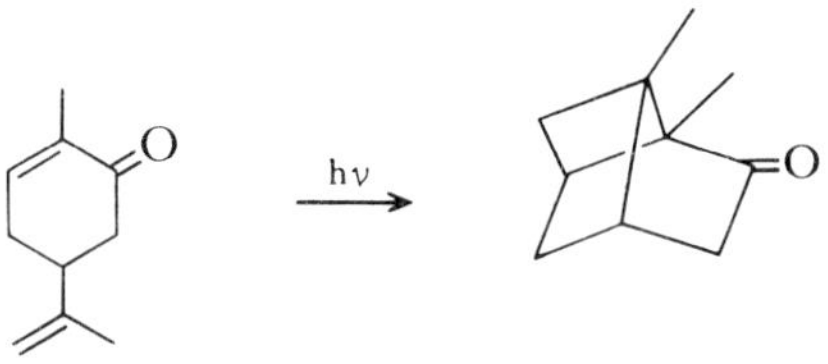

Although it might be supposed that the discovery of the reaction 1 → 2 would have occasioned a search for similar reactions on a broader basis, this did not happen and the great significance of CIAMICIAN's discovery has only been realized in recent times. Table 1 though not exhaustive gives a review of such reactions; the feature common to all the reaction products is the photochemical formation of a cyclobutane ring. In honor of CIAMICIAN it is suggested that this intramolecular reaction be denoted as "CIAMICIAN-addition".

References, pp. 12—13

Table 1

diolefin		cyclobutane	ref.
3	hv →	**4**	[5]
5: R = H **7**: R = COOH	hv, ±Sens →	**6**: R = H **8**: R = COOH	[6—8] [9]
9: X = H **11**: X = Cl	hv →	**10**: X = H **12**: X = Cl	[10] [11]
13	hv, Sens →	**14**	[12]
15	hv →	**16**	[13, 14]
17	hv →	**18**	[15, 16]
19	hv →	**20**	[17]

References, pp. 12—13

Names of the compounds shown are, myrcene (3), β-pinene (4), tricyclo[5.2.1.0^{2,6}]-deca-4,8-dien-3-one (9), pentacyclo[5.3.0.0^{2,6}.0^{3,9}.0^{5,8}]decan-4-one (10), 2.4.5.6-tetrachlorotricyclo[5.2.1.0^{2,6}]deca-4,8-dien-3-one (11), 2.3.5.6-tetrachloropentacyclo-[5.3.0.0^{2,6}.0^{3,9}.0^{5,8}]decan-4-one (12), tricyclo[6.2.1.0^{2,7}]undeca-4,9-diene-3,6-dione (15), pentacyclo[6.2.1.0^{2,7}.0^{4,10}.0^{5,9}]undecane-3,6-dione (16), isodrin (17), 1,2,3,3,4,11-hexachlorohexacyclo[5.4.1.0^{2,6}.0^{4,11}.0^{5,9}.0^{10,12}]dodecane (18), 2,4,8,10-tetramethyl-3,9-dioxatricyclo[6.4.0.0^{2,7}]dodeca-4,10-diene-6,12-dione (19) and 2,4,8,10-tetramethyl-3,9-dioxapentacyclo[6.4.0.0^{2,7}0^{4,11}.0^{5,10}]dodecane-6,12-dione (20). Names lacking here will be mentioned in the text.

The following requirements are regarded as indispensible for the occurrence of the CIAMICIAN-addition [18]: (a) an olefinic linkage capable of being activated by the light used and (b) a nearby, but not necessarily activated, double bond. Activation may be brought about either by a conjugated carbonyl group or by substitution with heavy atoms, e.g., chlorine (cf. the formation of 18). Otherwise addition of sensitizers is necessary to effect the intramolecular cycloaddition.

CIAMICIAN-addition may lead to strained bridged systems not easily available by means of dark reactions. In this way, tetracyclo[2.2.1.0^{2,6}.0^{3,5}]-heptane (quadricyclene, 6) was synthesized from bicyclo[2.2.1]hepta-2,5-diene (norbornadiene, 5) either in the absence [6] or in the presence of sensitizers [7,8], and the diacid 8 was obtained from 7 [9]. In a series of cases, CIAMICIAN-addition leads to the formation of cage compounds. Examples are the production of 10, 12, or 16 from the dicyclopentadienones 9 and 11 [10, 11] or from the DIELS-ALDER adduct 15 of cyclopentadiene and p-benzoquinone [13, 14]. The photoisomerization of the insecticide isodrin (17) to the cage isomer 18 [15, 16] is remarkable since 17 is carbonyl-free.

The majority of the reaction products listed in table — were obtained without sensitizers. On irradiation of endo-dicyclopentadiene (13) in cyclohexane, SCHENCK and STEINMETZ [12] observed that no pentacyclo-[5.3.0.0^{2,6}.0^{3,9}.0^{5,8}]decane (14) was formed. However, 14 was produced on irradiation of 13 in acetone which acts as sensitizer. Compound 14, which is related to cubane, is stable above 400°.

Carvonecamphor (1,2-dimethyltricyclo[3.3.0.0^{2,7}]octan-3-one, 2) [4]. A solution of 50 g of (−)-carvone in 2500 ml of 95% ethanol in a 3-l. round-bottomed Pyrex flask was irradiated with continuous stirring for 23 days. The Pyrex vessel was suspended in the center of a metal cylinder lined with eight F 15 T 8/BL Sylvania black-light lamps. At the end of the irradiation the solution was brought to pH 7 with a little aqueous ammonia, 100 ml of benzene was added, and the solvent was removed by distillation through a Podbielniak column at reduced pressure. The residue was distilled at 13 mm Hg. to give 21.4 g of a slush which was nearly pure 2 as indicated by g.l.p.c. analysis, and a second fraction 8.4 g of a liquid mixture of roughly equal parts of 2, carvone (1) and an ester. For the further purification the original paper should be consulted. Finally, 17.3 g (35%) of carvonecamphor, m.p. 99−103°, were obtained.

Pentacyclo[5.3.0.0^{2,6}.0^{3,9}.0^{5,8}]decane (14) [12]. A nitrogen-flushed solution of 10 g of endo-dicyclopentadiene (13) in 150 ml of acetone was irradiated for 18 hrs. with a water-cooled quartz mercury immersion lamp Philips HPK 125 W. The pale-yellow oil which remained after distilling off the acetone under normal pressure was chromatographed on silica gel. After elution with cyclohexane and distilling this off under slightly reduced pressure, 6.2 g of colorless **14** remained, which had m.p. 134—136° after recrystallization from methanol.

Tetracyclo[2.2.1.0^{2,6}.0^{3,5}]heptane-2,3-dicarboxylic acid (8) [*9*]. A solution of 3 g of bicyclo[2.2.1]heptadiene-2,3-dicarboxylic acid (**7**) in 200 ml of anhydrous ether was placed in a Vycor flask equipped with a magnetic stirrer and a reflux condenser. The solution was irradiated for 8 hrs. with a General Electric AH-4 mercury lamp. During this period 1.41 g (47%) of **8** crystallized from the solution as it was formed. Concentration of the ethereal solution gave impure material from which additional **8** could be crystallized from dry acetone.

Pentacyclo[6.2.1.0^{2,7}.0^{4,10}.0^{5,9}]undecane-3,6-dione (16) [*14*]. The Diels-Alder adduct **15** (19 g) was irradiated in 300 ml of ethyl acetate for 6 hours. The almost colorless solution was evaporated under reduced pressure and the product (90%) which crystallized was filtered off, washed and sublimed (120°/14 mm) to give pure **16**, m.p. 245°.

2. Photochemical valence tautomerization of 1,3-dienes

a) Homoannular 1,3-dienes

The term "valence tautomerization" has been advanced for isomerizations in which no atoms or groups shift [*19, 20*]. It is of historical interest that the expression valence tautomerism was coined by Wieland [*21*]. He introduced this concept in a work on the radical type reaction modes of p-benzquinones and illustrated it with the following formulae.

The study of valence tautomerism has received great stimulation from organic photochemistry since numerous novel valence tautomer systems have thus become available. Some simple examples of photochemical valence tautomerizations, which may formally be regarded as intramolecular Diels-Alder reactions [*22*], will be found in table 2 (comp. also Crowley [*5*]). For reviews on the subject of strained valence tautomers of 1,3-dienes the reader is referred to, inter alia, Vogel [*23*], Hammond and Turro [*24*] and Dauben [*25*].

The names of the compounds given in table 2 are, unless mentioned in the text, 1,3-cycloheptadiene (**23**), bicyclo[3.2.0]hept-6-ene (**24**), 1,4-diphenyl-1,3-cycloheptadiene (**25**), 1,5-diphenylbicyclo[3.2.0]hept-6-ene (**26**), 1,3-cyclooctadiene (**27**), bicyclo[4.2.0]oct-7-ene (**28**), 1,3,5-cycloheptatriene (**29**), bicyclo[3.2.0]hepta-2,6-diene (**30**), 5-methoxy-2,4-cycloheptadien-1-ol (**33**), 5-methoxybicyclo[3.2.0]hept-6-en-2-ol (**34**), 4-methoxy-3,5-cycloheptadien-1-ol (**35**), 6-methoxybicyclo[3.2.0]hept-6-en-3-ol (**36**), 5-methoxy-2,4-cycloheptadien-1-one (**37**), 5-methoxybicyclo[3.2.0]hept-6-en-2-one (**38**), 4-methoxy-2,4,6-cycloheptatrien-1-one (γ-tropolone methyl ether, **39**), 5-methoxybicyclo[3.2.0]hepta-3,6-dien-2-one (**40**), eucarvone (**41**) and 1,4,4-trimethylbicyclo[3.2.0]hept-6-en-2-one (**42**).

References, pp. 12—13

Table 2

cisoid 1,3-diene	cyclobutene	ref.
21	**22**	[26]
23: R=H **25**: R=C$_6$H$_5$	**24**: R=H **26**: R=C$_6$H$_5$	[27, 28] [29]
27	**28**	[30]
29	**30**	[27]
31	**32**	[31]
33	**34**	[28]
35	**36**	[28]

Table 2 (continued)

cisoid 1,3-diene	cyclobutene	ref.
37	38	[28]
39	40	[32]
41	42	[33]
43: R = H 45: R = CH₃	44: R = H 46: R = CH₃	[34, 35] [34]

The valence isomerizations of cyclic 1,3-dienes usually proceed without sensitization. Thus, ultraviolet irradiation (Hanovia lamp, Vycor filter) of 1,2-dihydrophthalic anhydride (21) in ether resulted in formation of bicyclo[2.2.0]hex-5-ene-2,3-dicarboxylic anhydride (22) [26]. The preparative significance of this reaction is that it opens a way to "Dewar benzene" (47). When 22 was in pyridine solution treated with lead tetraacetate, a solution of bicyclo[2.2.0]hexa-2,5-diene (47) in pyridine distilled over, representing a 20 % yield.

The cyclobutene derivatives obtained by photochemical valence tautomerization are frequently thermally unstable and on heating give the respective starting materials. Thus 5-methoxy-2,4-cycloheptadien-1-one (37) is formed from 38 by heating at 60° [28].

CHAPMAN, to whom we owe much of our knowledge on photochemical valence tautomerization of cyclic 1,3-dienes, has also investigated the case of 1,3,5-cyclooctatriene (**31**). Irradiation in ether with a quartz mercury lamp gave a 41 % yield of 3 volatile compounds in the ratio 38:14:48 [*31*]. These were, in the order given, bicyclo[4.2.0]octa-2,7-diene (**32**), tricyclo-[3.3.0.0^{2,8}]oct-3-ene (**48**) and 1,5-cyclooctadiene (**49**).

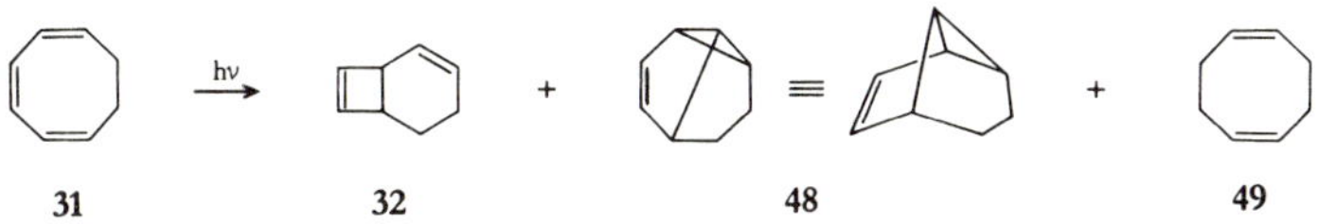

        31   32   48   49

Irradiation [*34, 35*] of ether solutions of 2,3-dihydro-3,5,7-trimethyl-1H-azepin-2-one (**43**) gives after removal of the ether and sublimation of the product a single bicyclic valence tautomer which has been shown to be the endo isomer of 1,4,6-trimethyl-2-azabicyclo[3.2.0]hept-6-en-3-one (**44**). Pyrolysis (430°) of **44** gives the starting diene-lactam **43** in 50 % yield.

Analogously irradiation of **45** [*34*] gives a single photoisomer (**46**). The stereoselectivity of the photoisomerization of the dihydroazepinones is remarkable.

Valence tautomers of natural products have received much attention in recent years. Irradiation of levopimaric acid (**50**) [*36*] in absolute ethanol was found to give crystalline photolevopimaric acid (**51**). Heating the latter for 40 minutes at 120° gave rise to 67 % of starting material **50**.

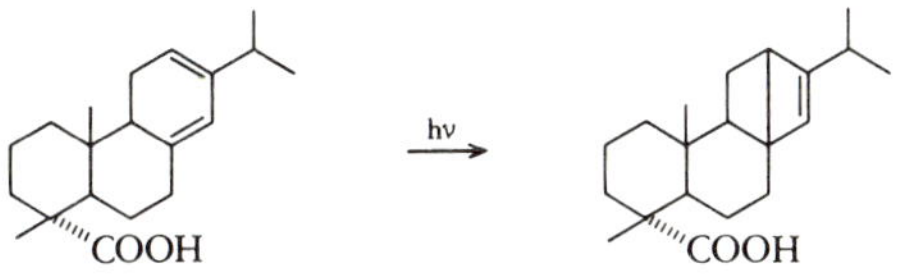

      50   51

The photoisomerization of pyrocalciferol (9α-lumisterol, **52**) and isopyrocalciferol (9β-ergosterol, **54**) [*37, 38*] gives in each case one isomer, which were termed photopyrocalciferol and photoisopyrocalciferol, and

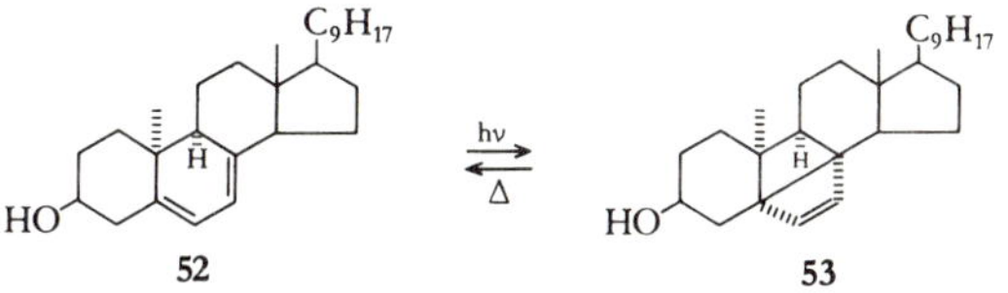

      52   53

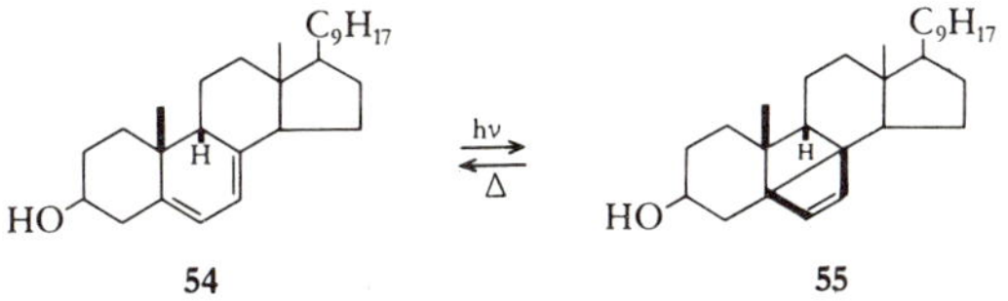

      54   55

for which valence tautomeric structures (cf. **53**, though no stereochemistry was considered) were proposed by WINDAUS et al. already in 1940 [*37*]. Later, DAUBEN and FONKEN [*38*] were able to substantiate these assumptions. Hence, photopyrocalciferol is **53**, and photoisopyrocalciferol is **55**.

Valence tautomerizations of 5,7-dienes in the ergosterol series occurs only with the syn-isomers **52** (9α,10α) and **54** (9β,10β) while the anti-isomers ergosterol (9α, 10β) and lumisterol (9β,10α) respond to irradiation by ring opening (comp. chapter 2).

Although it was known since the middle of the previous century [*39, 40*] that colchicine (**56**), an alkaloid constituent of meadow saffron, was a photolabile compound, a long time was to pass before its photochemical properties were investigated closely [*41, 42*]. GREWE and WULF [*41*] obtained three substances isomeric with colchicine on insolation of dilute aqueous solutions of the alkaloid. These were called α-, β- and γ-lumicolchicine; the latter two are present together in meadow saffron [*42*], possibly formed by a photochemical process within the plant.

The gross structures of the β- and γ-lumicolchicines were established by FORBES [*43*], and final confirmation was adduced by GARDNER, CHAPMAN and their co-workers [*44, 45*]. β-Lumicolchicine is **57**, and γ-lumicolchicine is **58** and both are valence tautomers of **56**. For α-lumicolchicine, for which a dimeric structure was deduced [*46*], see chapter 8.

The photoisomers undergo a variety of photochemical and thermal interconversions, the exact nature of which was disclosed by SCHENCK and co-workers [*47, 48*] and which are represented below:

$$\text{colchicine (56)} \underset{\Delta}{\overset{h\nu}{\rightleftarrows}} \begin{array}{c} \gamma\text{-lumicolchicine (58)} \\ h\nu \updownarrow \quad \downarrow \Delta \\ \beta\text{-lumicolchicine (57)} \\ \Delta \uparrow \quad h\nu \updownarrow h\nu \\ \alpha\text{-lumicolchicine} \end{array}$$

Isocolchicine (**59**) underwent photochemical valence tautomerization, too. When irradiated in water it turned over to lumiisocolchicine (**60**) [*49*] while with methanol as solvent [*50*] a methanol adduct (**61**) was obtained in addition to **60**. The methanol adduct was correlated with γ-lumicolchicine, thus proving the stereochemistry depicted in **61** [*50*].

References, pp. 12—13

Bicyclo[3.2.0]hept-6-ene (24) [*28*]. 1,3-Cycloheptadiene (23) (7.0 g) was placed in a 40×200 mm cylindrical quartz tube fitted with an internal cooling condenser and drying tube, and absolute ether (160 ml) was pipetted into the reaction vessel. After 54 hrs. irradiation with a General Electric UA-3 mercury arc lamp, distillation of the solution through a column packed with glass helices gave 4.1 g (58%) of **24**, b.p.$_{745}$ 96°.

5-Methoxybicyclo[3.2.0]hepta-3,6-diene-2-one (40) [*32*]. A solution of 5 g of γ-tropolone methyl ether (39) in one liter of distilled water was placed in a Pyrex vessel with internal cooling and irradiated with a General Electric UA-3 lamp for 24 hrs. at a distance of 15 cm. The solution became turbid and a red solid precipitated. The solution was extracted with three portions of methylene chloride, the extract dried and evaporated. Fractional distillation of the residue gave, besides 3 g of starting material **39**, 1.2 g of **40**, b.p.$_{0.1}$ 34°.

1,4,6-Trimethyl-2-azabicyclo[3.2.0]hept-6-en-3-one (44) [*34*]. A solution of **43** (3 g) in anhydrous ether (1800 ml) was flushed with nitrogen and then irradiated with a type A Hanovia mercury arc lamp encased in a water-cooled quartz immersion well. After 1 hour the 252 mµ maximum characteristic of **43** had completely disappeared, and the irradiation was stopped. Evaporation of the ether gave a thick, brown residue which was distilled (72–74° at 0.7 mm) to give crystalline **44**. Sublimation (50–80° at 0.005 mm) of the distillate gave pure **44**, m.p. 67–69° (2.1 g, 70%).

Photolevopimaric acid (51) [*36*]. A solution of 0.906 g of levopimaric acid (50) in 30 ml of absolute ethanol was charged to a fused quartz actinometer cell 75 mm in diameter, 10 mm thick, of 32 ml capacity and with a bottle mouth opening at the top. The vessel was irradiated with unfiltered radiation from a 100 Watt Hanovia 30600 quartz high-pressure mercury lamp at a distance of 30 cm. After 85 hrs. irradiation water was added short of turbidity and the solution allowed to stand. The product precipitated to give 0.73 g (81%) of crude material, which after recrystallization from aqueous ethanol gave pure **51**, m.p. 112–114° (0.42 g, 46%).

β- and γ-Lumicolchicine (57 and 58) [*48*]. Colchicine (56) (9.0 g) was dissolved in 150 ml of water in an argon atmosphere. This solution was irradiated with a mercury immersion lamp through solidex whilst magnetically stirred. Crystalline material precipitated during the irradiation which was filtered off and recrystallized to give 500 mg (5.6%) of α-lumicolchicine. The filtrates were irradiated as above for a total of 19 days when the solvent was evaporated in vacuo and the residue chromatographed over silica gel. Gradient elution with ether and chloroform afforded 3.99 g of pure **57**, m.p. 182–185° and 1.378 g (15.3%) of pure **58**, m.p. 268–272°, besides crude material which on repeated chromatography gave another 110 mg of **57** (total yield 45.6%). When the irradiation was carried out in methanol instead of water, β-lumicolchicine was obtained in 81.3% yield and had m.p. 209–212°.

b) Aromatic compounds

As non-planar structures, i.e., as valence tautomers of the benzenoid systems, bicyclo[2.2.0]hexa-2,5-dienes (cf. **47**) are capable of independent existence (a summary of such reactions has been given by VAN TAMELEN [*51*]).

1,2,5-Tri-tert.butylbicyclo[2.2.0]hexa-2,5-diene (**63**) is a remarkably stable compound obtained by irradiation of 1,2,4-tri-tert.butylbenzene (**62**) [*52*]. The repulsion between the bulky tert.butyl groups, which cannot be planar in the benzenoid form, is held responsible for the photorearrangement **62** → **63**. This repulsion also prevents the back isomerization at room temperature; hence, **63** may be distilled in high vacuum (10^{-7} mm Hg) at room temperature. Aromatization is accomplished on heating to 200° for 15 minutes.

Recently it was shown [*53*] that **63** is prone to further photoisomerization, a "prismane" (**64**) arising in a reaction reminiscent of the quadricyclene (**6**) synthesis from norbornadiene (**5**) [*6—8*]. 1,2,5-Tri-tert.butyltetracyclo[2.2.0.0^{2,6}.0^{3,5}]hexane (**64**) was of comparable stability at room temperature as was **63**. Complete separation of the isomers was not possible.

1,2,5-Tri-tert.butylbicyclo[2.2.0]hexa-2,5-diene (63) [*52*]. Irradiation by means of a Hanovia Type L ultraviolet lamp (Vycor filter) of **62** in ether solution gave **63**, which was separated from other products by thin layer chromatography (silica gel/cyclohexane). The hydrocarbon **63**, a liquid at room temperature but solid below 0°, was distilled in high vacuum at room temperature.

c) Transoid 1,3-dienes

Only a few examples of valence tautomerizations of transoid 1,3-dienes are known [*54*]. Here the valence isomerization of 1,3-butadiene, which leads to bicyclobutane (**65**) and cyclobutene [*55*], is referred to.

The formation of 3,5:4,6-dicyclo-5α-cholestane (**67**) from cholesta-3,5-diene (**66**) has been observed by DAUBEN and WILLEY [*56*]. The structure of **67** (cf. also [*54*]) is based, inter alia, on NMR and IR spectra and on chemical properties of the bicyclobutane compound: on treatment with ethanol in the dark, **67** yields 6β-ethoxy-3,5-cyclo-5α-cholestane (**68**) and 3α-ethoxymethyl-A-norcholest-5-ene (**69**), which may also be obtained directly upon irradiation of **66** in ethanol.

References, pp. 12—13

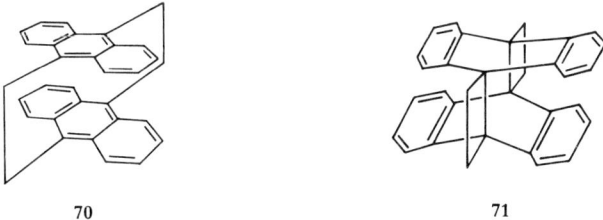

3. Intramolecular dianthracene formation

Exposure of the orange, crystalline [2.2](9,10)anthracenophane (**70**) to sunlight rapidly transformed it into a colorless, crystalline isomer **71** [*57*].

The ready formation of the photoisomer **71** can be attributed to a proximity effect of the two anthracene nuclei. Reversion to compound **70** occurs when the photoisomer is kept in the dark at room temperature or when heated, the transformation rate increasing with temperature. The regenerated **70** was shown to be a polymorph of the original; the initial and regenerated forms have been designated α and β.

[2.2](9,10)Anthracenophane photoisomer (**71**) [*57*]. The orange crystals (α- or β-form) of [2.2](9,10)anthracenophane (tetrabenzo[2.2]paracyclophane, **70**) were exposed to sunlight either in air or in a vacuum. Rapid conversion to colorless, monoclinic plates of **71** occurred; these reverted to the β-form of **70** slowly at room temperature or rapidly at 100°. Recrystallization of **71** in chloroform gave the α-form of **70**.

References

[1] G. CIAMICIAN and P. SILBER: Ber. dtsch. chem. Ges. **41**, 1928 (1908).
[2] E. SERNAGIOTTO: Gazz. Chim. Ital. **48**, 52 (1918).
[3] G. BÜCHI and I. M. GOLDMAN: J. Amer. chem. Soc. **79**, 4741 (1957).
[4] J. MEINWALD and R. A. SCHNEIDER: J. Amer. chem. Soc. **87**, 5218 (1965).
[5] K. J. CROWLEY: Tetrahedron **21**, 1001 (1965).
[6] W. G. DAUBEN and R. L. CARGILL: Tetrahedron **15**, 197 (1961).
[7] G. S. HAMMOND, N. J. TURRO and A. FISCHER: J. Amer. chem. Soc. **83**, 4674 (1961).
[8] G. S. HAMMOND, P. WYATT, C. D. DE BOER and N. J. TURRO: J. Amer. chem. Soc. **86**, 2532 (1964).
[9] S. J. CRISTOL and R. L. SNELL: J. Amer. chem. Soc. **80**, 1950 (1958).
[10] R. C. COOKSON, J. HUDEC and R. O. WILLIAMS: Tetrahedron Letters **22**, 29 (1960).
[11] P. YATES and P. EATON: Tetrahedron **12**, 13 (1961).
[12] G. O. SCHENCK and R. STEINMETZ: Chem. Ber. **96**, 520 (1963).
[13] R. C. COOKSON, E. CRUNDWELL and J. HUDEC: Chem. and Ind. **1958**, 1003.
[14] R. C. COOKSON, E. CRUNDWELL, R. R. HILL and J. HUDEC: J. Chem. Soc. **1964**, 3062.
[15] R. C. COOKSON and E. CRUNDWELL: Chem. and Ind. **1958**, 1004.
[16] C. W. BIRD, R. C. COOKSON and E. CRUNDWELL: J. Chem. Soc. **1961**, 4809.
[17] P. YATES and M. J. JORGENSON: J. Amer. chem. Soc. **85**, 2956 (1963).
[18] P. DE MAYO in: Adv. Org. Chem., Ed. by R. A. RAPHAEL, E. C. TAYLOR and H. WYNBERG, Vol. **2**, pp 367−425; esp. p. 390. New York: Interscience 1960.
[19] J. W. BAKER: Tautomerism. London: Routledge 1934; esp. p. 38, 201−226.
[20] A. C. COPE, A. C. HAVEN, JR., F. L. RAMP and E. R. TRUMBULL: J. Amer. chem. Soc. **74**, 4867 (1952).
[21] H. WIELAND: Ber. dtsch. chem. Ges. **53**, 1313 (1920).
[22] K. ALDER and G. JACOBS: Chem. Ber. **86**, 1528 (1953).
[23] E. VOGEL: Angew. Chem. **74**, 829 (1962).
[24] G. S. HAMMOND and N. J. TURRO: Science **142**, 1541 (1963).
[25] W. G. DAUBEN: Chem. Weekblad **60**, 381 (1964).
[26] E. E. VAN TAMELEN and S. P. PAPPAS: J. Amer. chem. Soc. **85**, 3297 (1963).
[27] W. G. DAUBEN and R. L. CARGILL: Tetrahedron **12**, 186 (1961).
[28] O. L. CHAPMAN, D. J. PASTO, G. W. BORDEN and A. A. GRISWOLD: J. Amer. chem. Soc. **84**, 1220 (1962).
[29] P. COURTOT: Ann. Chim. [13] **8**, 197 (1963).
[30] W. G. DAUBEN and R. L. CARGILL: J. Org. Chem. **27**, 1910 (1962).
[31] O. L. CHAPMAN, G. W. BORDEN, R. W. KING and B. WINKLER: J. Amer. chem. Soc. **86**, 2660 (1964).
[32] O. L. CHAPMAN and D. J. PASTO: J. Amer. chem. Soc. **82**, 3642 (1960).
[33] G. BÜCHI and E. M. BURGESS: J. Amer. chem. Soc. **82**, 4333 (1960).
[34] O. L. CHAPMAN and E. D. HOGANSON: J. Amer. chem. Soc. **86**, 498 (1964).
[35] L. A. PAQUETTE: J. Amer. chem. Soc. **86**, 500 (1964).
[36] W. H. SCHULLER, R. N. MOORE, J. E. HAWKINS and R. V. LAWRENCE: J. Org. Chem. **27**, 1178 (1962).
[37] A. WINDAUS, J. DIMROTH and W. BREYWISCH: Liebigs Ann. Chem. **543**, 240 (1940).
[38] W. G. DAUBEN and G. J. FONKEN: J. Amer. chem. Soc. **81**, 4060 (1959).
[39] M. HÜBLER: Arch. Pharm. **171**, 193 (1865).
[40] H. STRUVE: Z. Anal. Chemie **12**, 164 (1873).
[41] R. GREWE and W. WULF: Chem. Ber. **84**, 621 (1951).
[42] F. ŠANTAVÝ: Coll. Czech. Chem. Comm. **16**, 665 (1951).
[43] E. J. FORBES: J. Chem. Soc. **1955**, 3864.

[44] P. D. GARDNER, R. L. BRANDON and G. R. HAYNES: J. Amer. chem. Soc. **79**, 6334 (1957).
[45] O. L. CHAPMAN, H. G. SMITH and R. W. KING: J. Amer. chem. Soc. **85**, 803 (1963).
[46] O. L. CHAPMAN, H. G. SMITH and R. W. KING: J. Amer. chem. Soc. **85**, 806 (1963).
[47] G. O. SCHENCK, H. J. KUHN and O.-A. NEUMÜLLER: Tetrahedron Letters **1961**, 12.
[48] H. J. KUHN, O.-A. NEUMÜLLER and G. O. SCHENCK: Forschungsber. Land Nordrhein-Westfalen **1624**; esp. p. 71, 158. Köln: Westdeutscher Verlag 1966.
[49] O. L. CHAPMAN, H. G. SMITH and P. A. BARKS: J. Amer. chem. Soc. **85**, 3171 (1963).
[50] W. G. DAUBEN and D. A. COX: J. Amer. chem. Soc. **85**, 2130 (1963).
[51] E. E. VAN TAMELEN: Angew. Chem. **77**, 759 (1965); internat. ed. **4**, 738 (1965).
[52] E. E. VAN TAMELEN and S. P. PAPPAS: J. Amer. chem. Soc. **84**, 3789 (1962).
[53] K. E. WILZBACH and L. KAPLAN: J. Amer. chem. Soc. **87**, 4004 (1965).
[54] W. G. DAUBEN and W. T. WIPKE: Pure Appl. Chem. **9**, 539 (1964).
[55] I. HALLER and R. SRINIVASAN: J. Chem. Phys. **40**, 1992 (1964).
[56] W. G. DAUBEN and F. G. WILLEY: Tetrahedron Letters **1962**, 893.
[57] J. H. GOLDEN: J. Chem. Soc. **1961**, 3741.

Chapter 2

Photoisomerization of dienes and trienes not leading to the formation of cyclobutane derivatives

1. Photoisomerization of open chain 1,3-dienes to 1,2-dienes

The vapor phase photoisomerization of 1,3-butadiene to 1,2-butadiene has been reported by SRINIVASAN [1]. Many other products arise at the same time thus impairing the preparative value of this reaction. Analogously, 1,3,5-hexatriene gave, among other products, 1,2,4-hexatriene [2]. The photoisomerization of sorbic acid to 3,4-hexadienoic acid has been described elsewhere (comp. chapter 3).

2. Photoisomerization of 1,3-cyclohexadienes

a) Cleavage to 1,3,5-hexatrienes

1,3-Cyclohexadienes may be regarded [3] as ring systems containing 2n ring members and (n-1) conjugated double bonds, when n = 3. Such systems, on illumination may undergo bond cleavage reactions yielding open chain compounds with n conjugated double bonds, valence tautomerizations (comp. chapter 1), "bond-switching" processes and dimerization reactions.

As an example for a cleavage reaction we may refer to the photochemical conversion of 1,3-cyclohexadiene (**1**) to 1,3,5-hexatriene (**2**) [2—4] and of α-phellandrene (**3**) to 3,7-dimethyl-1,3,5-octatrienes (cf. **4**) [4] besides higher boiling material (see page 17).

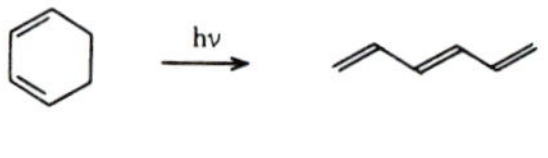

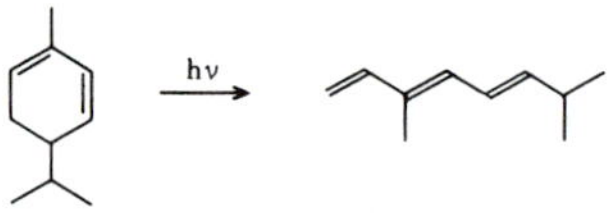

Similar results were obtained with substituted 1,3-cyclohexadienes. Thus, irradiation [5] of trans-3,5-cyclohexadiene-1,2-dicarboxylic acid (**5**) in ether solution with a low-pressure mercury lamp gave a 30 % yield of

References, p. 21

2,4,6-octatrienedioic acid (**6**), and ethyl 1,3-cyclohexadiene-1-carboxylate (**7**) was photoisomerized [6] to ethyl 2-methylene-3,5-hexadienoate (**8**) and further to ethyl 2-methyl-2,4,5-hexatrienoate (**9**) and a valence tautomer, ethyl bicyclo[3.1.0]hex-2-ene-3-carboxylate (**10**).

The most prominent example of photochemical cleavage reactions of 1,3-cyclohexadienes is the photoisomerization of ergosterol (**11**) [7]. Calciferol (vitamin D$_2$, **13**) had for long been regarded as the final product of photoisomerization of ergosterol, until in 1948 VELLUZ and co-workers [8, 9] succeeded in isolating an intermediate, precalciferol, in the **11** → **13** transformation. This isomer was found [8, 9] (comp. also [10]) to be thermally equilibrated with calciferol; its structure was elucidated independently by VELLUZ and HAVINGA [11, 12] to be that of 9,10-secoergosta-5(10),6,8,22-tetraen-3β-ol (**12**, only one geometrical isomer shown).

The photochemical and thermal interconversions of ergosterol (**11**), lumisterol (**14**), precalciferol (**12**), calciferol (**13**), tachysterol (**15**), and suprasterol-II (**16**) [13, 14] are illustrated in the annexed scheme. Resolution of these most intriguing and complex interconversions is mainly due to the work of HAVINGA and his co-workers [15—18].

The transformation of a triterpenoid 1,3-cyclohexadiene to a conjugated triene is illustrated by the photoisomerization [*19*] of methyl 3β-acetoxyursa-9(11),12-dien-28-oate (**17**) which yields methyl 3β-acetoxyursa-8(26),9(11),12-trien-28-oate (**19**) via methyl 3β-acetoxyursa-8,11,13-trien-28-oate (**18**).

Methyl 3β-acetoxyursa-8(26),9(11),12-trien-28-oate (19) [*19*]. The ester **17** (1.0 g) in 400 ml of ethanol was irradiated in a Pyrex flask with a bare mercury arc lamp (125 W), the heat of the lamp keeping the solution at reflux. All irradiations were conducted under oxygen-free nitrogen. The progress of the photolysis was determined by ultraviolet spectroscopy, and when the 283 mµ band had disappeared, the mixture was refluxed for 2 hours. The combined ethanol solutions of two such irradiations were concentrated in vacuo to about 25 ml and left at room temperature for 12 hours, when 250 mg of **17** precipitated. The filtrate of this, left at 0° for 12 hrs., gave prisms, m.p. 144—147°. Filtration and removal of the residual solvent in vacuo afforded a yellow gum which was chromatographed over alumina. Elution with 1 : 1 light petroleum-benzene gave material which was combined with the above crystals (1.53 g). Recrystallization from ether-methanol afforded then 57% of **19**, m.p. 152—153°.

b) Rearrangement to bicyclo[3.1.0]hex-2-enes

PRINZBACH et al. [*20, 21*] were able to show that dimethyl 1,3-cyclohexadiene-1,4-dicarboxylate (**20**) isomerized photochemically to dimethyl bicyclo-[3.1.0]hex-2-ene-1,5-dicarboxylate (**23**) in 60—70 % yield. The isomerization was visualized as involving dimethyl 2,5-dimethylenehex-3-enedioate (**21**) and dimethyl bicyclo[3.1.0]hex-3-ene-1,3-dicarboxylate (**22**) as intermediates.

References, p. 21

An analogous photochemical recyclization reaction was observed by MEINWALD et al. [*22*] who established that prolonged irradiation of α-phellandrene (**3**) did not give rise to 5-isopropyl-2-methylbicyclo[2.1.1]-hex-2-ene (**24**) as formulated earlier [*23*], but led to the formation of 6-isopropyl-2-methylbicyclo[3.1.0]hex-2-ene (**25**).

When 1,2,3,4,5-pentaphenyl-1,3-cyclohexadiene (**26**) was irradiated in benzene in the absence of oxygen, a high yield of 1,2,3,5,6-pentaphenylbicyclo[3.1.0]hex-2-ene (**27**) was produced [*24*]. The same sequence of photoreactions may plausibly be assumed in both the **3** → **25** and the **26** → **27** conversions.

Another bicyclo[3.1.0]hex-2-ene formation from a 1,3-cyclohexadiene has been mentioned above in the ergosterol (**11**) to suprasterol-II (**16**) transformation. Again open chain trienes are involved in this sequence of photochemical and thermal steps.

Dimethyl bicyclo[3.1.0]hex-2-ene-1,5-dicarboxylate (23) [*21*]. Dimethyl ester **20** (980 mg) was dissolved in 300 ml of ether and the solution purged with purified nitrogen. The mixture was irradiated with a water-cooled Quarzlampengesellschaft Q 81 mercury high-pressure lamp through Uviol glass, external cooling maintaining the temperature at −15°. After 9 hrs. irradiation the solvent was removed, and the oily residue was distilled at 80−90°/0.0001 mm Hg to give colorless **23**.

6-Isopropyl-2-methylbicyclo[3.1.0]hex-2-ene (25, [22]) [*23*]. A 3% ethereal solution of α-phellandrene (**3**) (80%) was irradiated with light above 250 mμ until it showed no absorption in this region. Distillation gave as principal photoproduct (45%) the isomer **25**, b.p.$_{18}$ 55°.

1,2,3,5,6-Pentaphenylbicyclo[3.1.0]hex-2-ene (27) [*24*]. A solution of 4.25 g of **26** in 1.1 liter of distilled benzene was purged with prepurified nitrogen for 24 hours. The solution was irradiated in a Hanovia immersion apparatus equipped with a Vycor filter (a 2 mm thick, cylindrical tube) for 65 minutes. The course of the isomerization was followed spectroscopically. After removal of the solvent, the yellow solid was chromatographed over alumina (250 g). Elution with hexane-benzene (9:1) gave 3.97 g (93%) of **27**, which was recrystallized from benzene-hexane to afford colorless prisms, m.p. 163.1−163.6°.

3. Photoisomerization of 1,5-cyclooctadienes

1,5-Cyclooctadiene (**28**) by a mercury-sensitized gas phase photolysis yielded bicyclo[5.1.0]oct-3-ene (**29**) and tricyclo[3.3.0.0^{2,6}]octane (**30**), though in trace amounts [*25*], while direct irradiation of **28** in the presence of cuprous chloride yielded 30 % of **30**. The mechanism of the reaction leading to **30** in the presence of CuCl was unraveled to involve free radicals [*26*].

Irradiation [*27*] of 5,6-diphenyldibenzo[a,e]cyclooctatetraene (**31**) in methylcyclohexane at 0° produced a 36 % yield of 5,11-diphenyldibenzo-[a,e]cyclooctatetraene (**33**), m.p. 183—184°. The transformation **31** → **33** could also be brought about by heating **31** to temperatures of 140—200°, the yield climbing to 96 % in this case. The reaction is believed to proceed via the "twisted" intermediate **32** which is analogous to **30**. Reversion of the bond-switching reaction could then lead either **31** or **33**.

Tricyclo[3.3.0.0^{2,6}]octane (30) *(a)* [*25*]. A round-bottom quartz flask of 2-liter capacity was charged with 3.5 g of 1,5-cyclooctadiene (**28**) and 0.53 g of mercury. It was evacuated on a vacuum line to the vapor pressure of the diene (ca. 1 mm), detached from the line and irradiated for 5 hrs. with 254 mμ light from sixteen General Electric G8T5 lamps surrounding the quartz vessel. After irradiation, the flask was connected to the vacuum line, and the volatile material (0.9 g) was distilled off. The distillate consisted of unreacted **28** and the two reaction products, which together amounted to 200 mg. Pure samples of **29** and **30** were obtained by gas chromatographic separation.

(b) [*26*]. A solution of 2 g of **28** in 380 ml of ether was saturated with CuCl. The saturated solution was irradiated for 24 hrs. with a Hanovia 450 W high-pressure mercury vapor lamp through a Vycor filter. After removal of the ether by distillation through a 20-cm glass helix packed column the residue was fractionated to give 0.79 g (39.5%) of **30**, b.p.$_{26}$ 43°.

4. Photoisomerization of 1,3,5-hexatrienes to 1,3-cyclohexadienes

The photochemistry of 1,3,5-hexatrienes is closely connected to that of their cyclic isomers, 1,3-cyclohexadienes, from which they may be photochemically generated. Thus, irradiation of 1,3,5-hexatriene (**2**) was reported [*2*] to give, among other products (comp. p. 14), 1,3-cyclo-

References, p. 21

hexadiene (**1**), and 2,4,6-octatriene (**34**) was reported [*28*] to photocyclize to trans-5,6-dimethyl-1,3-cyclohexadiene (**35**).

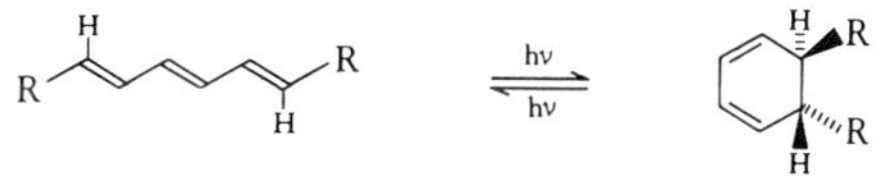

2: R = H
34: R = CH₃

1: R = H
35: R = CH₃

Alloocimene (**36**) when analogously irradiated, in ether solution [*28*] isomerized to α-pyronene (**37**), and in 5 % hexane solution [*29*] to 3,4,6,6-tetramethylbicyclo[3.1.0]hex-2-ene (**38**).

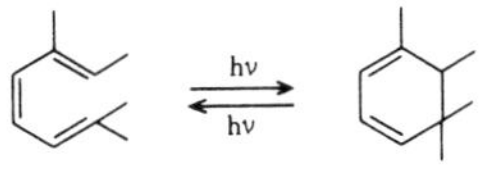

36   37 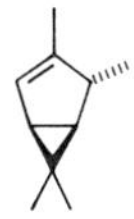   38

Further examples of photochemical recyclization are constituted by the conversion of precalciferol (**12**) to ergosterol (**11**) and lumisterol (**14**) (see above).

The stilbene (**41**) → 4a,4b-dihydrophenanthrene (**42**) photoisomerization does also belong to the same type of cyclizations though usually the cyclic isomers evade isolation by undergoing rapid oxidation with formation of the corresponding phenanthrenes. These reactions, which have been widely studied by MALLORY and co-workers (comp.[*30*]) have been dealt with in more detail in chapter 13.

Suitable ortho substituents at the stilbene chromophore may render the 4a,4b-dihydrophenanthrene isomers accessible. Thus, 2,2',4,4',6,6'-hexamethylstilbene (**39**) is on irradiation of its methylcyclohexane solution with UV light transformed into the red isomer, 4a,4b-dihydro-1,3,4a,4b,6,8-hexamethylphenanthrene (**40**) [*31*]. The cyclization is reverted by irradiation with visible light.

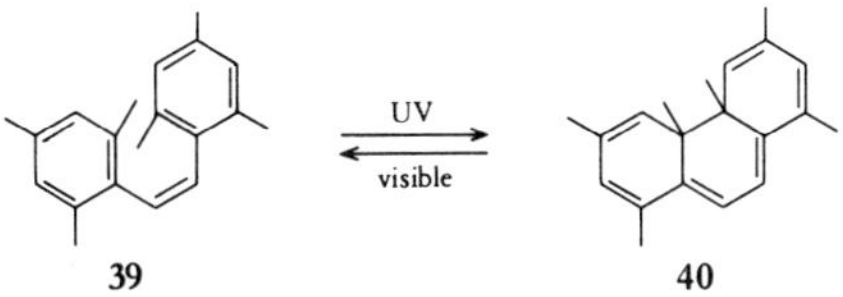

39   40

SARGENT and TIMMONS [*32*] have reported on the photoisomerization of trans-α,α'-dicyanostilbene (**43**) and of 2,3-diphenylmaleimide (**46**) in the absence of oxygen. The compounds isolated, **45** and 9,10-dihydro-9,10-phenanthrenedicarboximide (**47**) are secondary isomerization products of the primary cyclized isomers, e.g., 9,10-dicyano-4a,4b-dihydrophenanthrene (**44**).

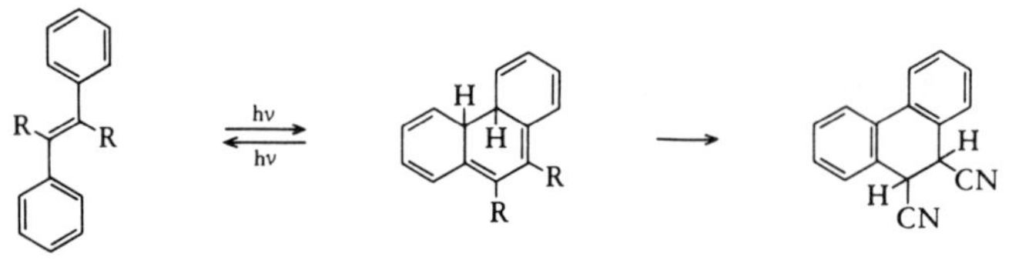

41: R = H
43: R = CN

42: R = H
44: R = CN

45

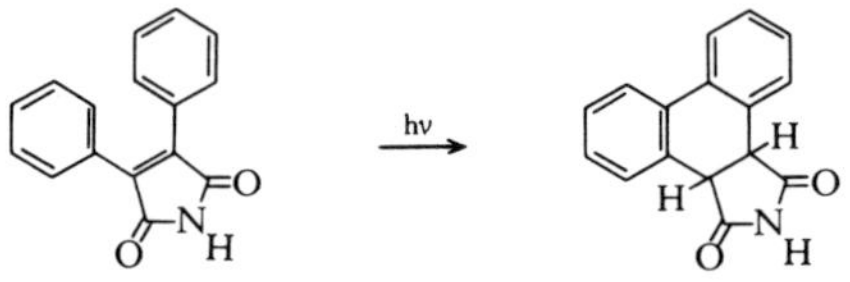

46

47

An interesting variation in the open chain triene → 1,3-cyclohexadiene photoisomerization was found recently by BLATTMANN et al. [*33*]. Trans-10b,10c-dihydro-10b,10c-dimethylpyrene (**48**) is converted by visible light to a valence tautomer of the metacyclophane type, viz., 8,16-dimethyl-[2.2]metacyclophane-1,9-diene (**49**). Spontaneous recyclization occurs in the dark.

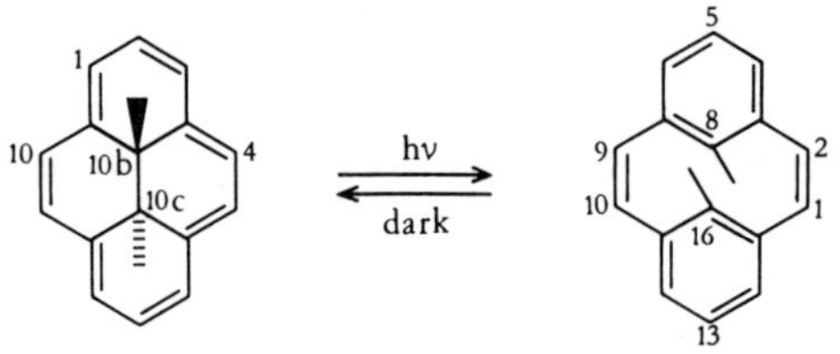

48

49

9,10-Dicyano-9,10-dihydrophenanthrene (45) [*32*]. Trans-α,α'-dicyanostilbene (**43**) (5.22 g) in 275 ml of purified benzene was degassed at −70° under high vacuum. The solution was then irradiated under vacuum for 15.75 hrs. with a Hanovia 400 W medium-pressure mercury lamp surrounded by a water-cooled Pyrex jacket. The product crystallized from the solution. After filtration and removal of the solvent the residue was recrystallized from acetone to give a total of 3.23 g (85%) of **45**, m.p. 199−204°.

8,16-Dimethyl[2.2]metacyclophane-1,9-diene (49) [*33*]. A solution of 30 mg of **48** in 1 l of pentane was subjected to irradiation using four ordinary Mazda lamps (a total of 500 W) for 6 hours. After concentration of the solution, a spectral inspection showed the presence of about 50% of the isomer **49** which did not absorb above 300 mμ. With other solvents the proportion of **49** in the irradiation mixture may be as high as 95%.

5. Photoisomerization of an acene involving loss of aromaticity

The conversion of a methyl group to a methylene group under the influence of irradiation has been described by CLAR [*34*] when it was shown that on illumination under exclusion of air the red-violet solution of

References, p. 21

6-methylpentacene (**50**) was very rapidly decolorized. Decoloration was believed to be due to the formation of 6,13-dihydro-6-methylenepentacene (**51**).

References

[1] R. Srinivasan: J. Amer. chem. Soc. **82**, 5063 (1960).
[2] R. Srinivasan: J. Amer. chem. Soc. **83**, 2806 (1961).
[3] D. H. R. Barton: Helv. Chim. Acta **42**, 2604 (1959).
[4] R. J. de Kock, N. G. Minnaard and E. Havinga: Rec. Trav. Chim. **79**, 922 (1960).
[5] P. Courtot and J.-M. Robert: Bull. Soc. Chim. France **1965**, 3362.
[6] H. Prinzbach and E. Druckrey: Tetrahedron Letters **1965**, 2959.
[7] H. H. Inhoffen and K. Irmscher: in: Fortschr. Chem. Org. Naturstoffe, ed. by L. Zechmeister, Vol. **17**, pp. 70–123; esp. p. 93. Wien: Springer 1959.
[8] L. Velluz, A. Petit and G. Amiard: Bull. Soc. Chim. France **1948**, 1115.
[9] L. Velluz, G. Amiard and A. Petit: Bull. Soc. Chim. France **1949**, 501.
[10] A. Akhtar and C. J. Gibbons: J. Chem. Soc **1965**, 5964.
[11] L. Velluz, G. Amiard and B. Goffinet: Bull. Soc. Chim. France **1955**, 1341.
[12] A. L. Koevoet, A. Verloop and E. Havinga: Rec. Trav. Chim. **74**, 788 (1955).
[13] W. G. Dauben, I. Bell, T. W. Hutton, G. F. Laws, A. Rheiner, jr. and H. Urscheler: J. Amer. chem. Soc. **80**, 4116 (1958).
[14] W. G. Dauben and P. Baumann: Tetrahedron Letters **1961**, 565.
[15] E. Havinga, R. J. de Kock and M. P. Rappoldt: Tetrahedron **11**, 276 (1960).
[16] E. Havinga and J. L. M. A. Schlatmann: Tetrahedron **16**, 146 (1961).
[17] R. J. de Kock, G. van der Kuip, A. Verloop and E. Havinga: Rec. Trav. Chim. **80**, 20 (1961).
[18] G. M. Sanders and E. Havinga: Rec. Trav. Chim. **83**, 665 (1964).
[19] R. L. Autrey, D. H. R. Barton, A. K. Ganguly and W. H. Reusch: J. Chem. Soc. **1961**, 3313.
[20] H. Prinzbach and J. H. Hartenstein: Angew. Chem. **75**, 639 (1963); internat. ed. **2**, 477 (1963).
[21] H. Prinzbach, H. Hagemann, J. H. Hartenstein and R. Kitzing: Chem. Ber. **98**, 2201 (1965).
[22] J. Meinwald, A. Eckell and K. L. Erickson: J. Amer. chem. Soc. **87**, 3532 (1965).
[23] K. J. Crowley: J. Amer. chem. Soc. **86**, 5692 (1964).
[24] G. R. Evanega, W. Bergmann and J. English, jr.: J. Org. Chem. **27**, 13 (1962).
[25] R. Srinivasan: J. Amer. chem. Soc. **86**, 3318 (1964).
[26] J. E. Baldwin and R. H. Greeley: J. Amer. chem. Soc. **87**, 4514 (1965).
[27] M. Stiles and U. Burckhardt: J. Amer. chem. Soc. **86**, 3396 (1964).
[28] G. J. Fonken: Tetrahedron Letters **1962**, 549.
[29] K. J. Crowley: Tetrahedron Letters **1965**, 2863.
[30] F. B. Mallory, C. S. Wood and J. T. Gordon: J. Amer. chem. Soc. **86**, 3094 (1964).
[31] K. A. Muszkat, D. Gegiou and E. Fischer: Chem. Comm. **1965**, 447.
[32] M. V. Sargent and C. J. Timmons: J. Chem. Soc. **1964**, 5544.
[33] H.-R. Blattmann, D. Meuche, E. Heilbronner, R. J. Molyneux and V. Boekelheide: J. Amer. chem. Soc. **87**, 130 (1965).
[34] E. Clar: Chem. Ber. **82**, 495 (1949).

Chapter 3

Photoisomerization of aldehydes and ketones not leading to the formation of oxygen heterocycles

1. Photoisomerization of α,β-unsaturated carbonyl compounds to β,γ-unsaturated or cyclopropyl isomers

Shift of a double bond from conjugation to non-conjugation under the influence of irradiation is a frequently observed feature. The major part of these isomerizations can be visualized as proceeding via photoenolization (comp. the transformation of 5-methyl-3-hexen-2-one (**1**) into 5-methyl-4-hexen-2-one (**2**) [*1*]) but there are cases where another reaction path seems to be followed (comp. the conversion **14** → **15**). It should, however, be noted that photoenols are usually not isolable.

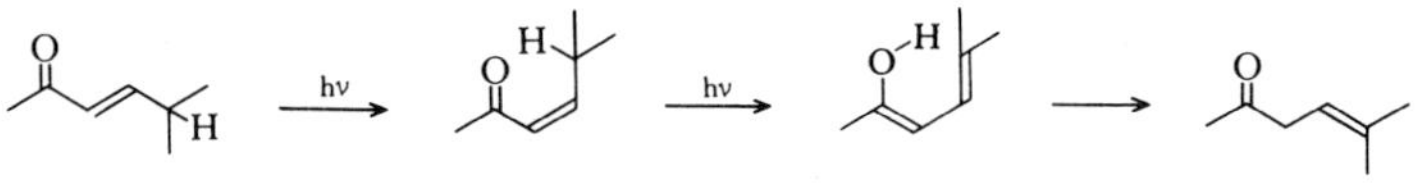

The photoisomerization of 4,5,5-trimethyl-3-hexen-2-one (**3**) and of 1-acetylcyclohexene (**5**) to 4-tert. butyl-4-penten-2-one (**4**) and 3-acetylcyclohexene (**6**) was reported by LEVINA et al. [*2*]. This report, however, could not be substantiated [*1*, *3*]. In contrast, it was shown that **3** was converted in 55 % yield to 1-(1,2,2-trimethylcyclopropyl)-2-propanone (**7**) [*3*].

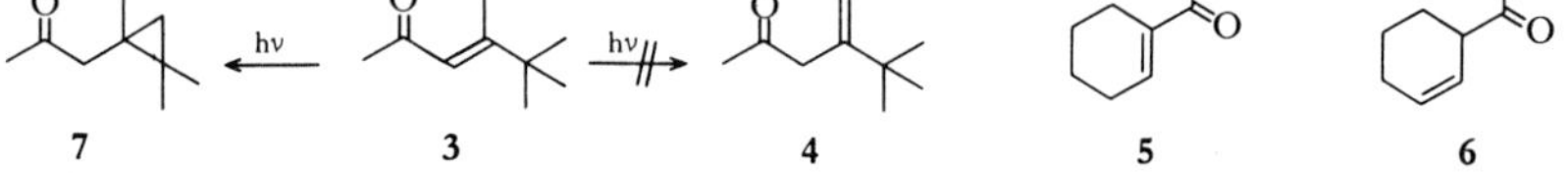

A similar photoenolization path as elaborated by YANG [*1*, *3*] seems to be followed in the case of irradiation of sorbic acid (**8**), reported by CROWLEY [*4*]. When illuminated in dilute ethereal solution, sorbic acid (**8**) undergoes first a trans→cis isomerization and then a migration of hydrogen to form 3,4-hexadienoic acid (**9**). It was assumed [*4*] that the free carboxy group was a prerequisite for the isomerization to proceed.

References, pp. 32—33

This assumption was proved unjustified when the irradiation of methyl sorbate was carried out in benzene solution in the presence of benzophenone as sensitizer [5]: the allene still did form, though dimerization to cyclobutane type dimers was competing.

According to the investigations of Lutz et al. [6] certain β-aroylacrylic acids may be photoisomerized in either of two ways. For example, trans-3-benzoyl-2-methylacrylic acid (10), on brief exposure to sunlight, underwent a trans-cis isomerization, the cis-acid 11 being in a ring-chain tautomeric equilibrium with 12 (comp. chapter 7).

By longer lasting irradiation, however, the photoequilibrium 10 ⇌ 11 is deprived of 10, which irreversibly isomerizes to 3-benzoyl-2-methylenepropionic acid (13) [6].

Shift of a double bond from the α,β- to the β,γ-position is also prominent in the photochemistry of natural products. Photoisomerizations of this type in the ionone series will be dealt with in another context (comp. chapter 5), but a remarkable example out of the steroid class will be discussed here.

10α-Testosterone (14) was irradiated in tert. butanol with the full light from a mercury high-pressure lamp [7]. From the resultant mixture 22% of unchanged starting material as well as 21% of a new β,γ-unsaturated isomer 15 could be separated chromatographically. The latter on treatment with boiling potassium carbonate reverted to 14 quantitatively. No dimerization of 14 was reported [7]. Testosterone (16), on the other hand, gave a cyclopropylketone isomer, 17β-hydroxy-1β,5-cyclo-5β,10α-androstan-2-one (17) besides other products [8].

5-Methyl-4-hexen-2-one (2) [*1*]. Ketone **1** (10 g) was irradiated in 450 ml of ether for 12 hrs. with a Hanovia 450 W lamp employing a Pyrex filter. At this time the complete conversion of **1** to **2** was indicated by IR and UV spectroscopy. Distillation afforded 7.5 g (75%) of **2**.

1-(1,2,2-Trimethylcyclopropyl)-2-propanone (7) [*3*]. Irradiation of **3** (1.1 g) in 400 ml of ether as above gave, after 31 hours, 55% of **7** which was isolated by vapor phase chromatography besides approximately 10% of starting material.

3,4-Hexadienoic acid (9) [*4*]. A 3% solution of sorbic acid (**8**) in dry ether was irradiated with a 450 W Hanovia high-pressure mercury lamp in a double-walled water-cooled Vycor glass immersion well. More than 1 hr. irradiation was required per gram of **8**, and when almost no starting material remained, fractional distillation (b.p.$_{0.4}$ 63°) gave 20% of the acid **9**, m.p. 23°. The yield of **9** was considerably increased and the irradiation time reduced by the addition of 1% of formic acid.

Cis-3-benzoyl-2-methylacrylic acid (12) [*6*]. A solution of trans-acid **10** (1 g) in 50 ml dry ether was placed in the sunlight until the yellow color disappeared (2 days). The solution was taken to dryness and the residue crystallized twice from hot benzene. The cyclic form **12** was obtained in a yield of 0.9 g; m.p. 89—92°.

3-Benzoyl-2-methylenepropionic acid (13) [*6*]. Ethereal solutions of **10** or **12** were insolated (1 week or 4 days, respectively) and **13** was obtained in 40—50% yield with m.p. 152—154° (from dilute alcohol).

17β-Hydroxy-10α-androst-5-en-3-one (15) [*7*]. A solution of 950 mg of **14** in 500 ml of tert. butanol was irradiated for 72 hrs. by a Quarzlampengesellschaft Q 81 high-pressure burner in a concentrically arranged water-cooled quartz finger at room temperature. After evaporation of the solvent in vacuo the oily residue was chromatographed on 24 g neutral alumina. Benzene eluted 197 mg of **15**, which after two recrystallizations from acetone-petroleum ether had m.p. 173—174°.

2. Photoenolization of aromatic ketones

While benzophenone on irradiation in the presence of hydrogen donors is readily reduced (comp. chapter 22), 2-alkylbenzophenones are not. For example, 2-benzylbenzophenone (**18**) is extremely stable towards ultraviolet radiation: the compound was recovered unchanged after prolonged irradiation in alcohol solution and no pinacol formation could be detected

18: R = C$_6$H$_5$
20: R = H

19: R = C$_6$H$_5$
21: R = H

22: R = H

23

24

References, pp. 32—33

[*9*]. 2-Methylbenzophenone (**20**) behaved similarly. It is believed that **18** or **20** in benzene undergo an intramolecular photoenolization, and the photoenols **19** and **21** thus formed revert back to **18** or **20** in a dark reaction.

The enol **21** could not be isolated; its existence was proved by the smooth reaction with dimethyl acetylenedicarboxylate to give the adduct **22** in excellent yield [*9*]. Various alkylsubstituted derivatives of **22** have subsequently been synthesized by PFAU et al. [*10*] who also found that 2,6-dimethylbenzophenone (**23**) and 2-methylacetophenone (**24**) resisted photoenolization.

The reversible photoenolization reaction of the **20** → **21** type was further investigated by PORTER, ULLMAN and their co-workers [*11*—*14*]. Interestingly enough, under carefully chosen conditions the intermediate photoenol **21** underwent a reversible photocyclization reaction [*12*]. The resultant 4a,10-dihydro-9-anthrol **25** was, on admission of oxygen, converted to 9-anthrone (**26**).

An analogous sequence consisting of photoenolization, photocyclization and oxidation, gave a high yield of 11-hydroxy-6-phenyl-12H-benzo[b]xanthen-12-one (**28**) from 3-benzoyl-2-benzylchromone (**27**) [*14*]. Experimental details on this synthesis will be found in chapter 13.

Dimethyl 1,4-dihydro-1-hydroxy-1-phenyl-2,3-naphthalenedicarboxylate (22) [*9*]. An equimolar solution of **20** and dimethyl acetylenedicarboxylate in benzene was irradiated with a Hanovia S 200 source at 15—20° for 24 hours. After the solvent was removed, the residue crystallized and no appreciable amount of polymeric material was formed. The residue was recrystallized from benzene-cyclohexane to give a 85% yield of **22**, m.p. 112°.

3. Photoisomerization of saturated ketones via photoelimination

Among photoreactions involving intramolecular hydrogen transfer, the NORRISH "type II" elimination process [*15*] (for the definition see [*47*]) has been fairly extensively studied. With saturated cyclic ketones this fragmentation becomes an isomerization process.

It should, however, be noted that in several cases the open chain isomers obtained photochemically may also be envisaged as arising through a disproportionation reaction of biradicals generated in the first instance or via even other routes. Thus, d,l-camphor (**29**) yields α-campholenaldehyde (**30**) as the normal product of α-cleavage together with 4-acetyl-3,3,4-trimethylcyclopentene (**31**) [*16, 17*].

The "type II" cleavage is also observed in the photoisomerization of 20-oxopregnanes, leading to 13,17-secosteroids. The formation of 3,3-ethylenedioxy-13,17-secopregna-5,13(18)-dien-20-one (**33**) from 3,3-ethylenedioxypregn-5-en-20-one (**32**) [*18*] has been described in chapter 4.

Some reactions which can not be pictured as involving a photoenolization step will for practical reasons also be mentioned here: the photoisomerization of cyclohexanone to 5-hexenal (**34**) [*19, 20*] and of menthone (**35**) to 3,7-dimethyl-5-octenal (**36**) [*21*]. The same primary process may also lead to ketenes and thence to saturated acids (see chapter 25).

A recent application of this photocleavage was reported by IRIARTE et al. [*22*]. When 3,3:20,20-bis-(ethylenedioxy)-C-nor-5α-pregnan-11-one (**37**) was irradiated in dilute solution, rapid and quantitative formation of the unsaturated aldehyde **38** occurred. This behavior contrasts with that of the corresponding ketone with six-membered ring C (comp. chapter 4).

References, pp. 32—33

3,7-Dimethyl-5-octenal (36) [*21*]. Homogenous solutions of 260 g of menthone (**35**), 975 ml alcohol and 520 ml water were exposed to sunlight during several months in summer. The oily layer which formed during the insolation was separated after neutralization of the aqueous phase. The oil was then dissolved in three volumes of ether and the solution shaken with a concentrated bisulfite solution. The crystals obtained were decomposed with potassium carbonate and the aldehyde extracted with ether. Evaporation of the solvent gave 7% of the aldehyde **36**, which after repetition of the bisulfite purification procedure had b.p. 195°.

3,3:20,20-Bis-(ethylenedioxy)-C-nor-11,13-seco-5α-pregn-13-en-11-al (38) [*22*]. The ketone **37** (750 mg) was dissolved in 250 ml of ethanol which was saturated with potassium carbonate. This solution was irradiated with a Quarzlampengesellschaft Q 81 mercury high-pressure lamp for 30 mins. at 20°. The light source, being surrounded by a water-cooled quartz jacket, was immersed into the liquid which was stirred magnetically under nitrogen. Evaporation of the solvent and chromatography over alumina gave 635 mg of **38**, m.p. 122.5—123.5° (from methylene chloride-petroleum ether).

4. Photoisomerization of α,β-unsaturated carbonyl compounds via cyclization

a) Cyclization of citral

When a cyclohexane solution of citral (**39**) was irradiated [*23*] (cf. [*24*]) the UV absorption at 238 mμ faded indicating that the conjugation was being destroyed. The product was mainly a mixture of two isomers roughly in the proportion of 2 : 1. The more abundant isomer proved to be 2-isopropenyl-5-methyl-1-cyclopentanecarboxaldehyde (**40**) and the minor product was 1,6,6-trimethylbicyclo[2.1.1]hexane-5-carboxaldehyde (**41**). The latter product arises from **39** through intramolecular cycloaddition of the carvone → carvonecamphor type (comp. chapter 1 for CIAMICIAN-addition).

2-Isopropenyl-5-methyl-1-cyclopentanecarboxaldehyde (40) [*23*]. A 1 : 1 mixture of the cis- and trans-isomers of citral (**39**) (200 g) was dissolved in 400 ml of cyclohexane and the solution irradiated with two 125 W mercury arc lamps through a Pyrex filter for 15 days. The solvent was evaporated under reduced pressure and the residual oil stirred for 4 hrs. with a solution of sodium bisulfite (850 g) and NaHCO₃ (320 g) in 2 liters of water. After removal of the citral-bisulfite adduct by filtration the filtrate was extracted with light petroleum and dried over sodium sulfate. The solvent was evaporated, giving a liquid (yield 50—60 g) which was fractionally distilled. The first fraction (35 g) was a 1 : 1 azeotrope of **40** and **41**, and the second fraction (20 g) was **40**, b.p.$_{15}$ 85°.

b) Rearrangement of eucarvone

The photochemical valence tautomerization of eucarvone (**42**) to 1,4,4-trimethylbicyclo[3.2.0]hept-6-en-2-one (**43**) was described by BÜCHI and BURGESS [*25*] (comp. chapter 1). A photoisomer of different structure (**44**) was detected more recently on irradiation of **42** in aqueous acetic acid [*26*]. 1,5,5-Trimethylbicyclo[2.2.1]hept-2-en-7-one (**44**) arises directly from **42**, and the presence of acid is a prerequisite for the occurrence of **44** among the photoisomerization products [*26*].

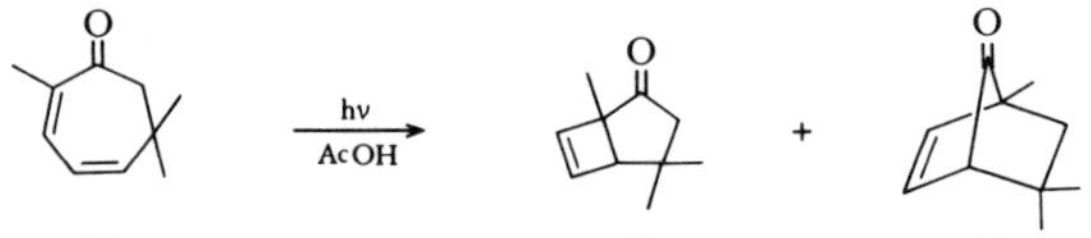

The formation **44** from **42** represents a novel rearrangement of a cycloheptane derivative into the bicyclo[2.2.1]heptane system and the mechanism of the reaction has found some interest.

1,5,5-Trimethylbicyclo[2.2.1]hept-2-en-7-one (44) [*26*]. Eucarvone (26 g) in 1 liter of acetic acid-water (4 : 6) in a closed Pyrex tube was irradiated in sunlight for 4 weeks. The solution was poured into water and extracted with ether. After being washed with aqueous Na_2CO_3 and drying, the ethereal solution was evaporated and distilled through a helix-packed column to give 12.6 g of a mixture of photoproducts containing no eucarvone. The mixture in light petroleum was chromatographed on alumina. Light petroleum-ether (99.5 : 0.5) eluted 1.73 g of **44**, b. p.$_{12}$ 68—70°. Further quantities of **44** were eluted with increasing amounts of ether in light petroleum. With pure ether 1.81 g of **43** was obtained, b.p.$_7$ 65—67.5°.

c) Rearrangement of verbenone

Irradiation of the α,β-unsaturated ketone verbenone (**45**) in cyclohexane or ether solution gave 47 % of the β,γ-isomer chrysanthenone (**46**) [*27*].

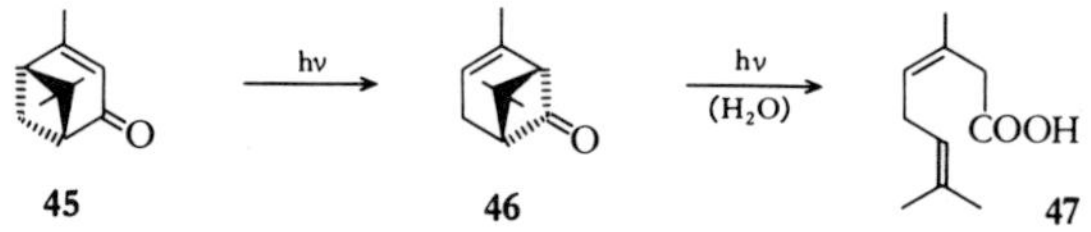

Different results were obtained on irradiation of **45** or **46** in aqueous ether [*27*]: both isomers gave 3,7-dimethyl-3,6-octadienoic acid (**47**). The formation of the latter was interpreted as involving a ketene intermediate. For similar reactions see chapter 25.

Chrysanthenone (46) [*27*]. Verbenone (**45**) (1.25 g) in 125 ml of cyclohexane was irradiated with a Hanovia UVS 500 mercury arc lamp under reflux in a quartz flask for 3 hours. After evaporation of the solvent the residue was taken up in light petroleum, filtered and adsorbed on a column of alumina. Elution with light petroleum gave 0.58 g. (47%) of **46**, which after a second chromatography in light petroleum on a column of charcoalcelite (1 : 1) had b.p.$_{12}$ 88—89°. Further elution of the above reaction mixture with light petroleum-ether (4 : 1) afforded 0.41 g of unchanged starting ketone **45**.

5. Photoisomerization of cross-conjugated cyclohexadienones

The photochemistry of cross-conjugated dienones has received much attention in recent years, and several reviews on this complex matter have appeared [*28—30*]. Only a few examples out of many will be referred to here.

References, pp. 32—33

a) Simple 2,5-cyclohexadien-1-ones

The photolysis [*31*] of 2,6-di-tert. butyl-4-hydroxy-4-phenyl-2,5-cyclohexadien-1-one (**48**) was reported to give 4-benzoyl-2,5-di-tert. butyl-2-cyclopenten-1-one (**49**) in ca. 20% yield.

When 1-hydroxy-4-oxo-α,α-diphenyl-2,5-cyclohexadien-1-acetic acid lactone (**50**) is irradiated in the solid state or if its benzene solution is boiled in sunlight, the result is rearrangement to the γ-lactone **51** [*32*]. This reaction does not take place in the dark at room temperature, while at higher temperatures decarboxylation occurs with formation of 4-(α-phenylbenzylidene)-2,5-cyclohexadien-1-one (**52**) [*32*].

4-Benzoyl-2,5-di-tert. butyl-2-cyclopenten-1-one (49) [*31*]. A 0.5% solution of **48** in moist dioxane was flushed with nitrogen for 30 mins. and irradiated in Pyrex with a Hanovia Utility lamp at 22 ± 2° until most of **48** had been consumed. The solvent was removed by vacuum distillation at about 30° and the viscous residue treated with hexane. Recrystallization from hexane-chloroform gave pure **49**, m. p. 157—158°, the yields varying somewhat (10—25%) depending on work up.

(2,5-Dihydroxyphenyl)-diphenylacetic acid γ-lactone (51) [*32*]. (*a*). A thin layer of **50** is irradiated for 2 months when rearrangement takes place to the lactone **51**, which forms large colorless crystals on recrystallization from methanol or acetic acid; m.p. 196°.

(*b*) 2 g of **50** were refluxed in 50 ml of benzene under passage of CO_2 for 36 hrs. in sunlight. The benzene was evaporated and the residue taken up in carbon disulfide. γ-Lactone **51** remained as insoluble residue, and the diphenylquinomethane **52**, m.p. 167°, was obtainable from the CS_2 mother liquor.

b) Condensed 2,5-cyclohexadien-1-ones

In the isomerizations to be described, cross-conjugated dienones rearrange to α,β-unsaturated ketones in which the carbonyl group is conjugated to a cyclopropyl ring. Secondary photoproducts comprise linearly conjugated dienones, unsaturated spiroketones, and phenols. A great many compounds, particularly in the terpene and steroid fields, have thus become available albeit frequently in very low yield. Since a full discussion of the formation mode of these photoisomers would be outside the scope of this book only a few representative examples will be given. For leading references to the work of, inter alia, BARTON, JEGER, KROPP, SCHAFFNER and their co-workers the reader should consult the reviews quoted above [*28—30*] and SCHAFFNER [*33*].

One of the dienones the photochemistry of which was studied very extensively, is santonin (**53**), an anthelmintic sesquiterpene lactone contained in various species of artemisia. Photochemical experiments with santonin were carried out comparatively early but were mainly directed toward the elucidation of the structures of photosantonic acid (comp. chapter 25), isophotosantonic lactone, and photosantoninic acid; the early literature has been covered by SIMONSEN and BARTON [*34*].

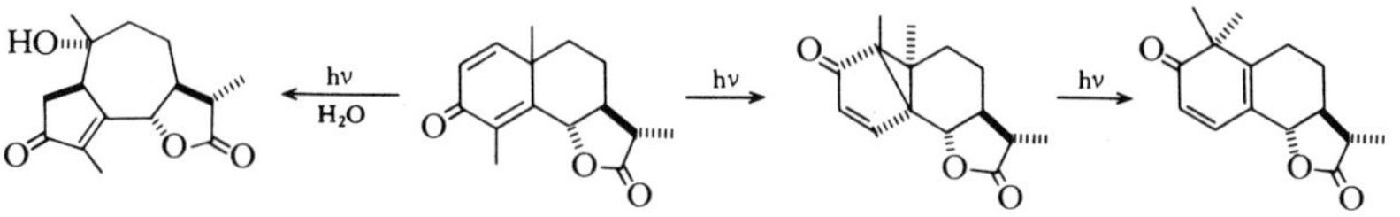

  56   53   54   55

Lumisantonin (**54**), which is formed on irradiation of santonin in ethanol, was obtained independently by a series of investigators [*35, 36*]. Benzophenone was reported to sensitize the rearrangement **53** → **54** [*37*].

More detailed investigations [*37, 38*] revealed that lumisantonin (**54**) on further irradiation furnished the linearly conjugated isomer **55** which turned over to photosantonic acid when irradiated in the presence of water (see page 229). Isophotosantonic lactone (**56**) was obtained [*39*] on illumination of santonin (**53**) in aqueous acetic acid. Although this conversion represents a gross addition of a water molecule the simultaneously proceeding skeletal rearrangement of the santonin molecule warrants inclusion of this reaction here.

The following scheme illustrates the sequence of intriguing though complex photoisomerizations of cross-conjugated cyclohexadienones as elaborated by KROPP [*40, 41*]. Yields of the different products vary depending on wavelength of the incident light, on the nature of the solvent and on the irradiation time.

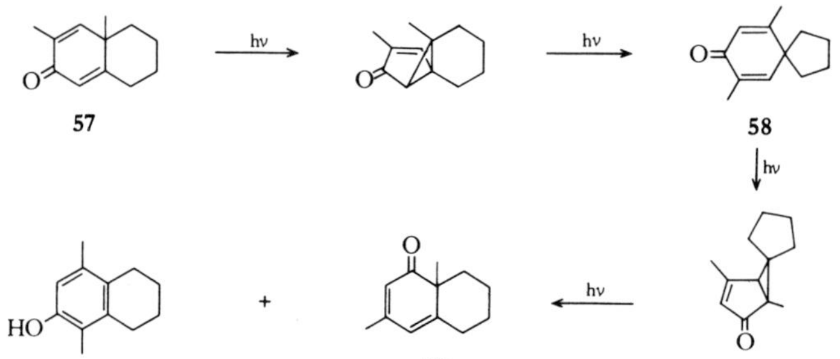

  57   58

                                  59

References, pp. 32—33

The interesting results obtained on irradiation of santonin stimulated investigations on the photochemistry of steroid 1,4-dienones. Thus, BARTON and TAYLOR [*42*] irradiated 21-acetoxy-17α-hydroxyandrosta-1,4-diene-3,11,20-trione (prednisone acetate, **60**) in ethanol and obtained lumiprednisone acetate (**61**) together with minor amounts of 21-acetoxy-17α-hydroxy-5-methyl-19-norpregna-1(10),3-diene-2,11,20-trione (neoprednisone acetate, **62**).

As in the case of santonin, irradiation in the presence of aqueous acetic acid gave a A-nor-B-homo compound **63**, while in dioxane 21-acetoxy-1,17α-dihydroxy-4-methyl-19-norpregna-1,3,5(10)-triene-11,20-dione (**64**) was the major product [*42*].

Methyl substitution on the dienone chromophore affects the photoisomerization path in a dramatical way as becomes evident from the work of the Zurich school [*43—46*]. While 17β-acetoxyandrosta-1,4-dien-3-one (**65**) on irradiation in dioxane solution yields a mixture of 9 isomeric compounds among which are ketones and phenols [*43, 44*] and 17β-acetoxy-2-methylandrosta-1,4-dien-3-one gives 5 products [*45*], the introduction of a 4-methyl substituent (cf. **67** and **69**, respectively) alters the reactivity in such a way as to favor the production of a single photoproduct in each case [*46*]. The structures deduced for these isomers were (in the case of irradiation of the free alcohols),17β-hydroxy-1-methyl-1β,5-cyclo-5β,10α-androst-3-en-2-one (**68**) and the respective 3-methyl derivative (**70**).

65: $R_1 = R_2 = H$
66: $R_1 = CH_3$, $R_2 = H$
67: $R_1 = H$, $R_2 = CH_3$
69: $R_1 = R_2 = CH_3$

68: $R_1 = H$, $R_2 = CH_3$
70: $R_1 = R_2 = CH_3$

Lumisantonin (54) [*36*]. Santonin (**53**) (1 g) was dissolved in 100 ml of absolute dioxane in a concentric quartz vessel equipped with a cold finger. The strongly agitated solution was irradiated for 1 hr. with a Philips Biosol 250 W mercury high-pressure lamp at a distance of 10 cm. The solvent was removed in vacuo and the residue chromatographed over neutral alumina. Benzene eluted 417 mg of crystals which after three crystallizations from acetone-hexane gave 335 mg of pure **54**, m.p. 156—157°.

Isophotosantonic lactone (56) [*39*]. A solution of 4 g of **53** in 110 ml of a mixture of acetic acid and water (4 : 5) was placed in a quartz flask under reflux and irradiated by a bare mercury arc lamp (125 W) for about 7 hrs. until the rotation fell to about 2°. The solvent was removed under reduced pressure leaving a gum; on treatment with $NaHCO_3$ this was separated into neutral and acidic fractions. The neutral fraction (3.2 g) was chromatographed on silica gel (110 g), and isophotosantonic lactone (**56**) (1.2 g) was obtained on elution with ether-acetone (1 : 2); m.p. 165—167° (from ethyl acetate-light petroleum).

6,7,8,8a-Tetrahydro-3,8a-dimethyl-1(5H)-naphthalenone (59) [*40*]. A solution containing 201 mg of 5,6,7,8-tetrahydro-3,4a-dimethyl-2(4aH)-naphthalenone (**57**) in 100 ml of dioxane was irradiated for 5 hrs. with a Quarzlampengesellschaft NK 6/20 low-pressure mercury lamp. The solution was stirred with a stream of nitrogen during the irradiation. Removal of the solvent on a rotary evaporator and chromatography of the residue over silica gel (6 g) gave, on elution with 480 ml of benzene and 240 ml of ether-benzene (1 : 99) mixture, 99 mg (50%) of dienone **59** as a colorless oil. Analogous irradiation of 6,9-dimethylspiro[4.5]deca-6,9-dien-8-one (**58**) afforded **59** in 63% yield [*41*].

Lumiprednisone acetate (61) [*42*]. Prednisone acetate (**60**) (3.4 g) in 350 ml of ethanol was irradiated in a quartz flask at room temperature until the rotation fell to ca. 0° (about 20 hrs.). Removal of the solvent under reduced pressure afforded a yellow gum which was fractionally crystallized from ethyl acetate-light petroleum to constant rotation, giving as the less soluble product lumiprednisone acetate (**61**), m. p. 224—226° (from methanol); yield 0.8 g. Starting material constituted the more soluble fraction (540 mg), and 7 mg of neoprednisone acetate (**62**), m.p. 230—233°, were obtained on chromatography of the mother liquors.

17β-Hydroxy-1-methyl-1β,5-cyclo-5β,10α-androst-3-en-2-one (68) [*46*]. 17β-Hydroxy-4-methylandrosta-1,4-dien-3-one (**67**, OH instead of OAc) (295 mg) was dissolved in 50 ml of dioxane and irradiated for 55 mins. in a cylindrical quartz vessel fitted with a reflux condenser or with a cold finger. The light source was a Philips Biosol 250 W high-pressure mercury burner 7—8 cm from the reaction vessel. During the irradiation the solution was stirred vigorously by means of a magnetic stirrer. The solvent was evaporated in vacuo and the residue chromatographed on a column of 20 g neutral alumina. Benzene-ether (9 : 1) mixture eluted a total of 190 mg crystals of **68**, which after recrystallization from ether-hexane had m.p. 196.5—198.5°.

References

[*1*] N. C. YANG and M. J. JORGENSON: Tetrahedron Letters **1964**, 1203.

[*2*] R. YA. LEVINA, V. N. KOSTIN and P. A. GEMBITSKII: Zhur. Obshchei Khim. **29**, 2456 (1959); J. Gen. Chem. USSR, **29**, 2421 (1959).

[*3*] M. J. JORGENSON and N. C. YANG: J. Amer. chem. Soc. **85**, 1698 (1963).

[*4*] K. J. CROWLEY: J. Amer. chem. Soc. **85**, 1210 (1963).

[*5*] H. P. KAUFMANN and A. K. SEN GUPTA: Liebigs Ann. Chem. **681**, 39 (1965).

[*6*] R. E. LUTZ, P. S. BAILEY, C.-K. DIEN and J. W. RINKER: J. Amer. chem. Soc. **75**, 5039 (1953).

[*7*] H. WEHRLI, R. WENGER, K. SCHAFFNER and O. JEGER: Helv. Chim. Acta **46**, 678 (1963).

[*8*] B. NANN, D. GRAVEL, R. SCHORTA, H. WEHRLI, K. SCHAFFNER and O. JEGER: Helv. Chim. Acta **46**, 2473 (1963).

[9] N. C. Yang and C. Rivas: J. Amer. chem. Soc. **83**, 2213 (1961).
[10] M. Pfau, N. D. Heindel and T. F. Lemke: Comptes rendus **261**, 1017 (1965).
[11] A. Beckett and G. Porter: Trans. Faraday Soc. **59**, 2051 (1963).
[12] E. F. Ullman and K. R. Huffman: Tetrahedron Letters **1965**, 1863.
[13] K. R. Huffman, M. Loy and E. F. Ullman: J. Amer. chem. Soc. **87**, 5417 (1965).
[14] W. A. Henderson, jr. and E. F. Ullman: J. Amer. chem. Soc. **87**, 5424 (1965).
[15] W. Davis, jr. and W. A. Noyes, jr.: J. Amer. chem. Soc. **69**, 2153 (1947).
[16] G. Ciamician and P. Silber: Ber. dtsch. chem. Ges. **43**, 1340 (1910).
[17] R. Srinivasan: J. Amer. chem. Soc. **81**, 2604 (1959).
[18] P. Buchschacher, M. Cereghetti, H. Wehrli, K. Schaffner and O. Jeger: Helv. Chim. Acta **42**, 2122 (1959).
[19] G. Ciamician and P. Silber: Ber. dtsch. chem. Ges. **41**, 1071 (1908).
[20] R. Srinivasan: J. Amer. chem. Soc. **81**, 2601 (1959).
[21] G. Ciamician and P. Silber: Ber. dtsch. chem. Ges. **40**, 2415 (1907).
[22] J. Iriarte, K. Schaffner and O. Jeger: Helv. Chim. Acta **47**, 1255 (1964).
[23] R. C. Cookson, J. Hudec, S. A. Knight and B. R. D. Whitear: Tetrahedron **19**, 1995 (1963).
[24] R. C. Cookson: Pure Appl. Chem. **9**, 575 (1964).
[25] G. Büchi and E. M. Burgess: J. Amer. chem. Soc. **82**, 4333 (1960).
[26] J. J. Hurst and G. H. Whitham: J. Chem. Soc. **1963**, 710.
[27] J. J. Hurst and G. H. Whitham: J. Chem. Soc. **1960**, 2864.
[28] H. E. Zimmerman in: Adv. Photochem., Ed. by W. A. Noyes, jr., G. S. Hammond and J. N. Pitts, jr., Vol. **1**, pp. 183–208. New York: Interscience 1963.
[29] O. L. Chapman in: Adv. Photochem., Ed. by W. A. Noyes, jr., G. S. Hammond and J. N. Pitts, jr., Vol **1**, pp. 323–420. New York: Interscience 1963.
[30] J. Saltiel in: Survey Progr. Chem., Ed. by A. F. Scott, Vol. **2**, pp. 239–328; New York: Academic Press 1964.
[31] E. R. Altwicker and C. D. Cook: J. Org. Chem. **29**, 3087 (1964).
[32] H. Staudinger and S. Bereza: Liebigs Ann. Chem. **380**, 243 (1911).
[33] K. Schaffner in: Fortschr. Chem. Org. Naturstoffe, Ed. by L. Zechmeister, Vol. **22**, pp. 1–114. Wien: Springer 1964.
[34] J. Simonsen and D. H. R. Barton: The Terpenes, Vol. **3**, pp. 249–326, esp. p. 292. Cambridge: University Press 1952.
[35] D. H. R. Barton, P. de Mayo and M. Shafiq: J. Chem. Soc. **1958**, 140.
[36] D. Arigoni, H. Bosshard, H. Bruderer, G. Büchi, O. Jeger and L. J. Krebaum: Helv. Chim. Acta **40**, 1732 (1957).
[37] M. H. Fisch and J. H. Richards: J. Amer. chem. Soc. **85**, 3029 (1963).
[38] O. L. Chapman and L. F. Englert: J. Amer. chem. Soc. **85**, 3028 (1963).
[39] D. H. R. Barton, P. de Mayo and M. Shafiq: J. Chem. Soc. **1957**, 929.
[40] P. J. Kropp: J. Amer. chem. Soc. **86**, 4053 (1964).
[41] P. J. Kropp: Tetrahedron **21**, 2183 (1965).
[42] D. H. R. Barton and W. C. Taylor: J. Chem. Soc. **1958**, 2500.
[43] H. Dutler, H. Bosshard and O. Jeger: Helv. Chim. Acta **40**, 494 (1957).
[44] H. Dutler, C. Ganter, H Ryf, E. C. Utzinger, K. Weinberg, K. Schaffner, D. Arigoni and O. Jeger: Helv. Chim. Acta **45**, 2346 (1962).
[45] C. Ganter, F. Greuter, D. Kägi, K. Schaffner and O. Jeger: Helv. Chim. Acta **47**, 627 (1964).
[46] K. Weinberg, E. C. Utzinger, D. Arigoni and O. Jeger: Helv. Chim. Acta **43**, 236 (1960).
[47] J. N. Pitts, jr., F. Wilkinson and G. S. Hammond in: Adv. Photochem., Ed. by W. A. Noyes, jr., G. S. Hammond and J. N. Pitts, jr., Vol. **1**, pp. 1–21. New York: Interscience 1963.

Chapter 4

Various photoisomerizations of ketones, esters and halides

1. Isomerization of ketones to tertiary alcohols by intramolecular cyclization

a) Cyclobutanols from saturated cyclic and acyclic ketones

Research on photochemical intramolecular cyclization reactions of ketones, leading to the formation of tertiary alcohols, was inaugurated by observations from YANG and collaborators [1, 2] who investigated the photochemistry of cyclodecanone (1) [1] and of 2-alkanones [2].

The formation of 2 can be explained by a transannular interaction between the excited carbonyl group and one of the nearest methylene groups, followed by cyclization [1].

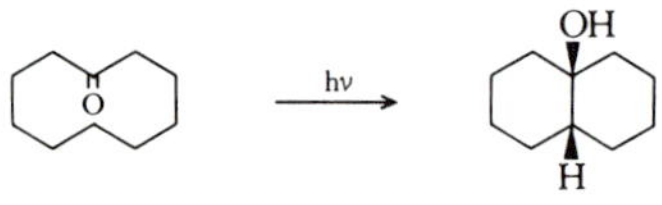

Similar proximity effects in photocyclization reactions of cyclic ketones were observed by SCHULTE-ELTE and OHLOFF [3]. Irradiation of 10% cyclohexane solutions of cyclododecanone (3) gave 64% of cis-bicyclo-[8.2.0]dodecan-1-ol (4) and 11% of its trans-isomer (5) together with minor amounts of cyclododecanol and 3-cyclododecen-1-one.

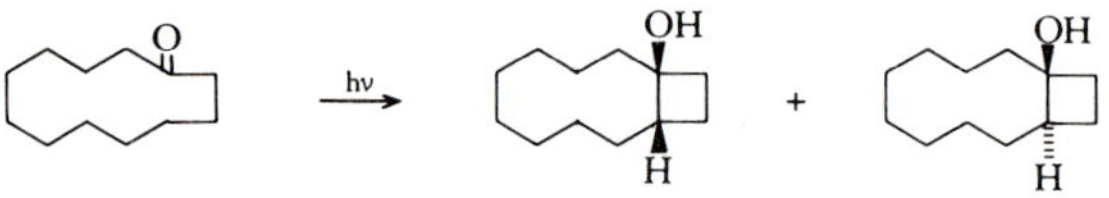

Cyclobutanol formation is now fairly well understood due to the extensive studies of YANG, JEGER and their co-workers (for a summary see [4]). Irradiation of 2-pentanone, 2-octanone and 2-nonanone [2] gave, in 10—20% yield, the corresponding cyclobutanols (cf. the formation of the two [5] 1-methyl-2-propyl-1-cyclobutanols from 2-octanone [2, 5]).

References, p. 40

The formation of 2,3-epoxy-cis-1,2-diphenyl-1-cyclobutanol (**6**) on irradiation of 3,4-epoxy-1,4-diphenyl-1-butanone has been mentioned elsewhere (cf. chapter 40).

cis-Decahydro-4a-naphthol (2) [*1*]. A 5% solution of cyclodecanone in cyclohexane was irradiated in a quartz immersion irradiator [*6*] for 132 hours. The solvent was removed and the mixture distilled in vacuo. Removal of unchanged ketone (30%) with GIRARD reagent gave nonketonic material (35%) which crystallized upon cooling. Sublimation gave pure **2**, m.p. 64—65°.

b) 2-Hydroxycyclobutanones from 1,2-diketones

URRY and TRECKER [*7, 8*] observed the formation of 2-hydroxycyclobutanones on photolysis of 1,2-diketones. Examples are the conversion of 5,6-decanedione (**7**) to 2-butyl-3-ethyl-2-hydroxycyclobutanone (**8**) (89%), of 4,5-octanedione (**10**) to 2-hydroxy-3-methyl-2-propylcyclobutanone (**11**) (92%) [*7*] and of 2,3-pentanedione (**12**) to 2-hydroxy-2-methylcyclobutanone (**13**) [*8*]. The latter dimerizes on standing, 3,8-dimethyl-2,7-dioxatricyclo[6.2.0.0^{3,6}]decane-1,6-diol (**14**) thus becoming easily available.

7: $R_1 = R_2 = C_4H_9$
10: $R_1 = R_2 = C_3H_7$
12: $R_1 = C_2H_5$, $R_2 = CH_3$

8: $R_1 = C_4H_9$, $R_2 = C_2H_5$
11: $R_1 = C_3H_7$, $R_2 = CH_3$
13: $R_1 = CH_3$, $R_2 = H$

The selectivity of the photolyses of the above 1,2-diketones is remarkable since no 1-acylcyclobutanols (cf. **9**) were detected.

Unexpectedly, 1,2-cyclodecanedione (**15**) gave no decalol (see preceding section) but 1-hydroxybicyclo[6.2.0]decan-10-one (**16**) in excellent yield (74%) [*8*]. Cyclooctanone (9%) accompanying the main product was derived from **16** by a second photochemical reaction, together with ketene.

2-Butyl-3-ethyl-2-hydroxycyclobutanone (8) [7]. A solution of 5,6-decanedione (7) (35.1 g) in cyclohexane (277.2 g) in a Pyrex culture flask was exposed to sunlight for 12 hours. Analysis of the reaction mixture by vapor-phase-chromatography indicated only unreacted diketone (11%) and cyclization product (89%). Distillation gave pure **8** (28.3 g), b.p.$_2$ 83°.

1-Hydroxybicyclo[6.2.0]decan-10-one (16) [8]. The 1,2-diketone 15 (15.0 g) was dissolved in 200 ml of benzene and the solution irradiated in a Pyrex flask with a General Electric Sunlamp 275 W a distance of 15 cm. When the yellow color had faded (70 hrs.), the mixture was distilled in vacuo. A fraction containing starting material and cyclooctanone was followed by 4.66 g of **16**, m.p. 49—50° (from ligroin).

c) Cyclobutanols in the steroid series

Proximity conditions for this intramolecular cyclization are most easily met with in the steroid series. Thus, in 20-oxopregnanes the carbonyl group is in such a favorable juxtaposition to the 18-methyl group as to facilitate formation of a 18,20-cyclosteroid (cf. **17 → 18**). Research in this field was mainly done by JEGER, SCHAFFNER and their co-workers [4, 9—13].

Irradiation of 3β-acetoxypregn-5-en-20-one (**17**) [9] in hexane gave crystalline 3β-acetoxy-18,20-cyclopregn-5-en-20-ol (**18**) (33% conversion).

As was found subsequently [5, 10] with related substrates, usually both 20-isomers are formed besides fragmentation products. This is illustrated by the conversion [9, 10] of 3,3-ethylenedioxypregn-5-en-20-one (**19**) to the (20S)- and (20R)-isomers of 3,3-ethylenedioxy-18,20-cyclopregn-5-en-20-ol

References, p. 40

(**20**, 45%, and **21**, 3—4%). Besides, 3,3-ethylenedioxy-13,17-secopregna-5,13(18)dien-20-one (**22**, 24%) and, after prolonged irradiation, a secondary photodegradation product 3,3-ethylenedioxy-17-nor-13,17-secoandrosta-5,13(18),15-triene (**23**, 23%) were obtained. Yields were found to vary with solvent character and duration of irradiation [*10*].

In analogy to the **17** → **18** transformation, 11-oxopregnanes were converted into 11,19-cyclosteroids by irradiation [*11, 13*]. In this manner, 3,3:20,20-bis-(ethylenedioxy)-11β,19-cyclo-5α-pregnan-11α-ol (**25**) was prepared in 61 % yield from 3,3:20,20-bis-(ethylenedioxy)-5α-pregnan-11-one (**24**) [*11*].

The stereochemical requirements to be complied with in this photocyclization reaction were elegantly elaborated also by the Zurich group [*12, 13*]. Reactions **17** → **18** and **24** → **25** open a ready access to compounds with functionalized 18- and 19-methyl groups which are otherwise hardly amenable.

d) Cyclohexenols from non-conjugated ketones

Also 1-alkyl-1-cyclohexenols are accessible by intramolecular photocyclization, though in low yield. Thus, YANG et al. [*14*] obtained, on irradiation of 6-hepten-2-one (**26**), 1-methyl-3-cyclohexen-1-ol (**27**) and 1-methyl-2-vinylcyclobutanol (**28**).

2. Photorearrangement of enol and dienol esters

Acetyl migrations were observed by MAZUR and co-workers who studied the irradiation of enol and dienol acetates [*15, 16*]. Photolysis of 1-propen-2-ol acetate (**29**) with 254 mμ radiation yielded acetylacetone (2,4-pentanedione, **30**), and 1-cyclohexen-1-ol acetate (**31**) yielded 2-acetylcyclohexanone (**32**) as the only isolable product. By contrast, 2-methyl-1-cyclohexen-1-ol acetate (**33**) furnished three transformation products,

2-methylcyclohexanone (**34**), 2-acetyl-2-methylcyclohexanone (**35**), and the dehydrodimer of **34**, viz., 2-methyl-2-(1-methyl-2-oxocyclohex-1-yl)-cyclohexanone (**36**) [*15*]. For related photorearrangements of esters see chapter 25.

29: R = CH$_3$
40: R = C$_6$H$_5$

30: R = CH$_3$
41: R = C$_6$H$_5$

31: R = CH$_3$
42: R = C$_6$H$_5$

32: R = CH$_3$
43: R = C$_6$H$_5$

33   **34**   **35**   **36**

The same reaction, applied to 3,17β-diacetoxyandrosta-3,5-diene (**37**) [*16*], furnished the normal 1,3-diketone, 17β-acetoxy-4-acetylandrost-5-en-3-one (enolized, cf. **38**), together with a vinylog β-diketone, 17β-acetoxy-6β-acetylandrost-4-en-3-one (**39**).

37   **38**   **39**

Enol benzoates on irradiation rearrange in the same way [*17*], e.g., 1-propen-2-ol benzoate (**40**) gave 1-phenyl-1,3-butanedione (**41**). However, in the case of 1-cyclohexen-1-ol benzoate (**42**) no **43** was isolated but only ring cleavage products. **43** was obtainable on pyrolysis of **42**.

2,4-Pentanedione (30) [*15*]. A solution of 1 g of isopropenyl acetate (**29**) in 100 ml of cyclohexane was transferred to an externally cooled tube of 40 mm diameter and ca. 150 ml volume. This solution was irradiated for 45 hrs. with a Quarzlampengesellschaft NT 6/20 low-pressure mercury immersion lamp. The solvent was evaporated and the residue was dissolved in 5 ml of methanol and treated with 5 ml of saturated aqueous cupric acetate solution on a water bath. After being left at the same temperature for 5 mins., the mixture was extracted with ether, and the isolated compound was crystallized from benzene to give 0.2 g of copper chelate of acetylacetone, m.p. 330° (dec.). Decomposition of the ether solution of the copper chelate with 5% HCl gave **30**.

17β-Acetoxy-4-acetylandrost-5-en-3-one (38) [*16*]. A solution of 1.5 g of **37** in 80 ml of cyclohexane was irradiated as above for 7 hours. The solvent was then evaporated under reduced pressure and the oily residue treated with 15 ml of methanol to give 0.4 g (26%) of **38**, m.p. 150—152°.

References, p. 40

3. Isomerizations with migration of bromine atoms

Photoisomerizations of organic halides will also be treated in this context though the overall reaction can usually be divided into homolytic cleavage of a halogen atom and subsequent addition to another site.

According to NESMEYANOV et al. [18] (comp. also [19]) the photochemical isomerization of 2-bromo-3,3,3-trichloropropene (44) proceeds differently in light to that achieved by the action of antimony pentachloride. Whereas the photochemical reaction yields 3-bromo-1,1,2-trichloropropene (45), 2-bromo-1,1,3-trichloropropene (46) is obtained from the dark reaction.

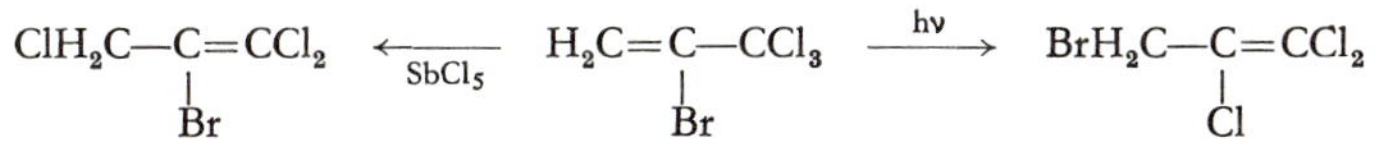

$$ClH_2C-C(Br)=CCl_2 \xleftarrow{SbCl_5} H_2C=C(Br)-CCl_3 \xrightarrow{h\nu} BrH_2C-C(Cl)=CCl_2$$

46   44   45

The photoisomerization of 11a-bromo-6-demethyl-6-deoxytetracycline (47) to 7-bromo-6-demethyl-6-deoxytetracycline (48) was reported by HLAVKA and KRAZINSKI [20]. Replacement of the 11a-bromine atom by hydrogen will be discussed in chapter 38.

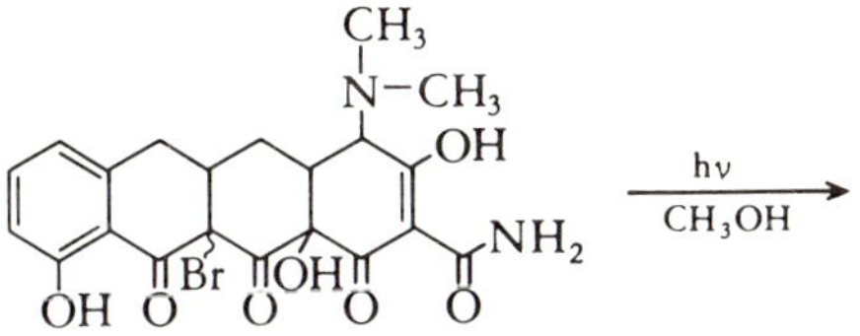

47 $\xrightarrow[CH_3OH]{h\nu}$

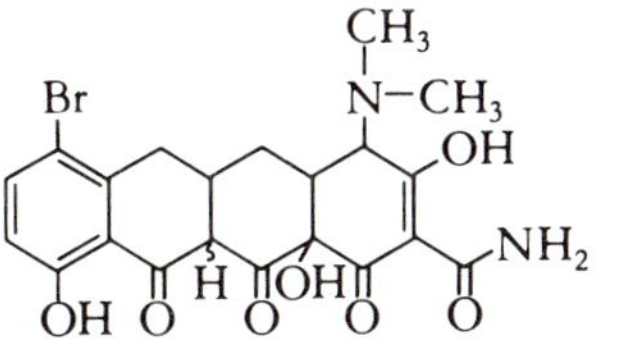

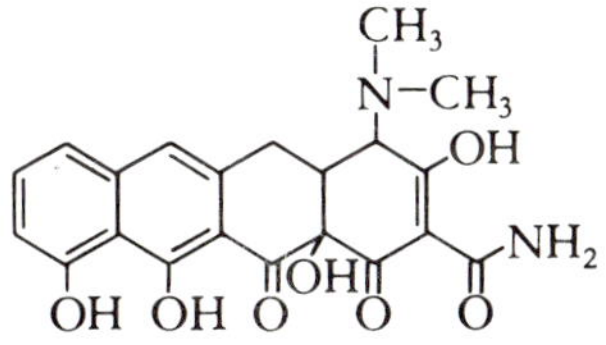

48   +   49

3-Bromo-1,1,2-trichloropropene (45) [18]. 2-Bromo-3,3,3-trichloropropene (44) undergoes quantitative rearrangement to 45 under the action of UV light. Compound 45 has b.p.$_{19}$ 78—79°.

7-Bromo-6-demethyl-6-deoxytetracycline (48) [20]. A solution of 100 mg of the hydrochloride of 47 in 50 ml of methanol was irradiated with a Hanovia 30,600 lamp in a double-walled Hanovia immersion well. After 4 hrs. the reaction mixture was brought to dryness in vacuo. Partition column chromatography, the details of which should be taken from the original paper, gave 10% of anhydro-6-demethyltetracycline (49) and 90% of 48.

4. Light-induced von Auwers rearrangement

Ultraviolet irradiation of a petroleum ether solution of 4-dichloromethyl-4-methyl-2,5-cyclohexadiene-$\Delta^{1,\alpha}$-acetic acid (**50**) yields the rearranged acid **51** [*21*]. It is believed that rearrangement to 3,3-dichloro-2-p-tolylpropionic acid (**51**) proceeds via free radicals with the dichloromethyl radical acting as the chain carrier.

References

[*1*] M. BARNARD and N. C. YANG: Proc. Chem. Soc. **1958**, 302.
[*2*] N. C. YANG and D.-D. H. YANG: J. Amer. chem. Soc. **80**, 2913 (1958).
[*3*] K. H. SCHULTE-ELTE and G. OHLOFF: Chimia **18**, 183 (1964).
[*4*] O. JEGER and K. SCHAFFNER: Chem. Weekblad **60**, 389 (1964).
[*5*] N. C. YANG and D.-D. H. YANG: Tetrahedron Letters **4**, 10 (1960).
[*6*] M. S. KHARASCH and H. N. FRIEDLANDER: J. Org. Chem. **14**, 239 (1949).
[*7*] W. H. URRY and D. J. TRECKER: J. Amer. chem. Soc. **84**, 118 (1962).
[*8*] W. H. URRY, D. J. TRECKER and D. A. WINEY: Tetrahedron Letters **14**, 609 (1962).
[*9*] P. BUCHSCHACHER, M. CEREGHETTI, H. WEHRLI, K. SCHAFFNER and O. JEGER: Helv. Chim. Acta **42**, 2122 (1959).
[*10*] M. CEREGHETTI, H. WEHRLI, K. SCHAFFNER and O. JEGER: Helv. Chim. Acta **43**, 354 (1960).
[*11*] H. WEHRLI, M. S. HELLER, K. SCHAFFNER and O. JEGER: Helv. Chim. Acta **44**, 2162 (1961).
[*12*] M. S. HELLER, H. WEHRLI, K. SCHAFFNER and O. JEGER: Helv. Chim. Acta **45**, 1261 (1962).
[*13*] J. IRIARTE, K. SCHAFFNER and O. JEGER: Helv. Chim. Acta **46**, 1599 (1963).
[*14*] N. C. YANG, A. MORDUCHOWITZ and D.-D. H. YANG: J. Amer. chem. Soc. **85**, 1017 (1963).
[*15*] A. YOGEV, M. GORODETSKY and Y. MAZUR: J. Amer. chem. Soc. **86**, 5208 (1964).
[*16*] M. GORODETSKY and Y. MAZUR: J. Amer. chem. Soc. **86**, 5213 (1964).
[*17*] M. FELDKIMEL-GORODETSKY and Y. MAZUR: Tetrahedron Letters **1963**, 369.
[*18*] A. N. NESMEYANOV, R. KH. FREIDLINA and V. N. KOST: Tetrahedron **1**, 241 (1957).
[*19*] A. N. NESMEYANOV, R. KH. FREIDLINA, V. N. KOST and M. YA. KHORLINA: Tetrahedron **16**, 94 (1961).
[*20*] J. J. HLAVKA and H. M. KRAZINSKI: J. Org. Chem. **28**, 1422 (1963).
[*21*] C. W. BIRD and R. C. COOKSON: J. Org. Chem. **24**, 441 (1959).

Chapter 5

Photoisomerizations involving formation and transformation of five or six membered heterocyclic oxygen compounds

1. Photochemical formation of heterocyclic oxygen compounds

a) Formation of furan derivatives by intramolecular cyclization

The intramolecular cyclization of an acetylenic alcohol to a furan derivative was reported by JONES [1] who irradiated trans-dec-2-ene-4,6,8-triyn-1-ol (1) and observed, besides considerable decomposition, the formation of the cis-alcohol 2 (in photochemical equilibrium with 1) and of the cyclization product 3 which was constituted of a mixture of geometrical isomers. 2-(2,4-Hexadiynylidene)-2,5-dihydrofuran (3) was also obtainable from 2 on treatment with alkali.

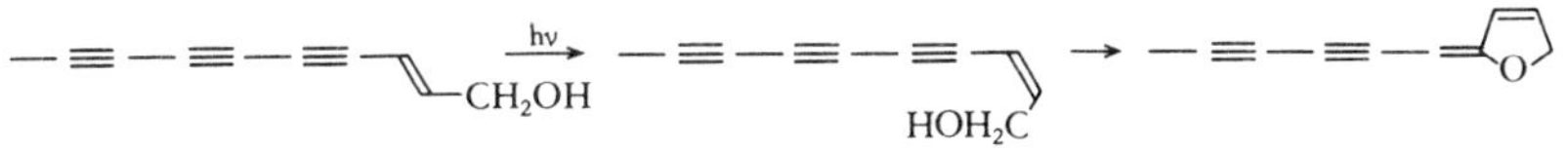

2-(2,4-Hexadiynylidene)-2,5-dihydrofuran (3) [1]. Trans-alcohol 1 (210 mg) was dissolved in 250 ml of ethanol and irradiated for 45 mins. in a quartz water-cooled 500 W Hanovia Photochemical Reactor at 20°. The resulting brown solution was evaporated under nitrogen, and the residue was chromatographed in the dark on alumina (40 g). Elution with light petroleum afforded 20 mg of 3, which on further chromatography was resolved to give 3.2 mg of the α-isomer and 8.2 mg of the β-isomer, m.p. 78°.

b) Formation of pyran derivatives by intramolecular cyclization

While on irradiation of trans-α-ionone in methanol solution the corresponding cis-isomer may be isolated (comp. chapter 7) this is not the case with trans-β-ionone (4). Instead, 6,7,8,8a-tetrahydro-2,5,5,8a-tetramethyl-5H-1-benzopyran (6) is obtained [2, 3] which is in a photochemical equilibrium with 4. Both in the cyclization as well as in the cyclohexadiene type cleavage (comp. chapter 2), cis-β-ionone (5) has to be assumed as the intermediate.

References, p. 47

Irradiation of trans-β-ionone (**4**) also produces a small quantity of a second isomer which was first regarded as the cyclopropane ketone **7** [*2*, *3*] but which was later shown [*4*] to be 4-(6,6-dimethyl-2-methylene-cyclohexylidene)-2-butanone (**8**). This compound arises from cis-β-ionone through a photoenolization process (comp. chapter 3 for related reactions), and several other examples of this isomerization have become known through the work of MOUSSERON et al. [*5*, *6*].

Isomerizations related to that described (cf. **5** → **6**) are encountered in the photochromic spiropyrans [*7*, *8*] which have in recent times found application in photographic science [*9*] and in technical devices for eye protection against intense light [*10*]. Thus, the photoisomerization of 1′,3′,3′-trimethyl-6-nitrospiro[2H-1-benzopyran-2,2′-indoline] (**9**) with UV radiation was reported [*11*] to lead to an open chain ketone (**10**) [*12*]. Irradiation of the latter with visible light again produced the colorless spiropyran **9**.

6,7,8,8a-Tetrahydro-2,5,5,8a-tetramethyl-5H-1-benzopyran (6) [*3*]. A solution of **4** (24 g) in 600 ml of ethanol was irradiated for 48 hrs. by a low intensity mercury arc in a quartz apparatus at 35—40°. Removal of the solvent under reduced pressure yielded a product which was fractionally distilled to give a 22—32% yield of **6**, b.p.$_1$ 81—82°.

β-Ionone (4) [*3*]. A solution of the pyran **6** (22 g) in 650 ml of ethanol was irradiated as above for 48 hrs. at 35°. On fractionation of the products, among other compounds, **4** was obtained as proved by the formation of its semicarbazone, m.p. 147—148°.

Ketone 10 [*11*]. A saturated, colorless solution of **9** in hexane was irradiated whereupon the solution rapidly became dark-blue (almost black) and a dark-purple, crystalline solid precipitated. In contrast to the starting material which melted at 179—180°, this compound partially melted at about 150°, resolidified, changed color from purple to pink (or was said to become colorless at 155° [*12*]) and finally melted at about 175°. On standing in room light the purple solid **10** gradually reverted to a greenish solid, similar in appearance to the starting compound **9**.

References, p. 47

c) Formation of heterocyclic oxygen compounds by isomerization of cyclic ketones

Ultraviolet irradiation [*13*] of a dioxane solution of hecogenin acetate (**11**) gave 3β-acetoxy-12,13-seco-5α,22α,25D-spirost-13-en-12-al (lumihecogenin acetate, **12**), which could be obtained in 80 % yield by interrupting the reaction at the appropriate time. Prolonged irradiation resulted in further isomerization of the latter to 3β-acetoxy-12α,14α-epoxy-5α,22α,25D-spirostan (photohecogenin acetate, **13**) which was also obtained on irradiating **12** [*13*]. This reaction is an intramolecular PATERNÒ-reaction (comp. chapter 41 for oxetane formation).

A photorearrangement producing γ-lactones was reported by RIGAUDY and DERIBLE [*14, 15*]. Irradiation of ethereal solutions of, for example, 2,2-diphenyl-1,3-indandione (**14**) gave 92 % of 3-α-phenylbenzylidenephthalide (**15**) [*14*].

14: $R_1 = R_2 = C_6H_5$
16: $R_1 = CH_3, R_2 = C_6H_5$

15: $R_1 = R_2 = C_6H_5$
17: $R_1 = C_6H_5, R_2 = CH_3$

18: $R_1 = CH_3, R_2 = C_6H_5$

With dissymmetric substituents R_1 and R_2 isomeric phthalides were obtained [*14*]. Thus, 2-methyl-2-phenyl-1,3-indandione (**16**) gave, on irradiation in ether solution, a mixture consisting of 43 % of **17**, 11 % of **18** and 9 % of starting material **16**. This ratio corresponds to the photostationary state between the components since irradiation of either cis-3-α-methylbenzylidenephthalide (**17**) or of trans-3-α-methylbenzylidenephthalide (**18**) led to approximately the same ratio [*14*]. By choice of appropriate filter conditions the photochemical equilibrium could be biased to one or the other direction [*15*].

3β-Acetoxy-12α,14α-epoxy-5α,22α,25 D-spirostan (13) [*13*]. A solution of hecogenin acetate (**11**) (100 g) in dioxane (7 liters) was irradiated in a Hanovia Photochemical Reactor by means of a 500 W medium-pressure mercury lamp. The lamp was surrounded by a double walled quartz thimble which fitted into the reaction flask and through which cooling water circulated. During the irradiation the reaction mixture was stirred and a stream of purified oxygen-free nitrogen was slowly passed through. Irradiation was discontinued after 36 hrs., the dioxane solution filtered through alumina to remove traces of peroxides, and evaporated. The solid residue was recrystallized from methanol to yield 30 g of **13**, m.p. 205–206°. Interruption of the above irradiation after 8 hrs. would have given approximately 80% of the aldehyde **12**, m.p. 143–147° (from methanol).

3-α-Phenylbenzylidenephthalide (15) [*14*]. 2,2-Diphenyl-1,3-indandione (**14**) (200 mg) was dissolved in 200 ml of ether and the solution transferred to a glass vessel with parallel walls. This vessel was contained in another glass trough through which cooling water circulated. Irradiation was done with a Philips SP 500 mercury high-pressure lamp. After 7 hrs., the colorless solution was evaporated in vacuo and the resulting crystals washed with methanol to give 184 mg (92%) of product, m.p. 148–149°. Recrystallization from methanol gave 112 mg of pure **15**, m.p. 150–151° and 54 mg of less pure material.

2. Photoisomerization of heterocyclic oxygen compounds to open chain compounds

a) Photoisomerization of furan derivatives

On irradiation of 2,5-dihydro-2,3,5,5-tetraphenylfuran (**19**) in dilute benzene solution, the bond between oxygen and carbon breaks and, after migration of a phenyl group, 1,2,3,4-tetraphenyl-3-buten-1-one (**20**), m.p. 181.5–183°, is formed in 37% yield [*16*]. It should be mentioned that on treating **19** with acids, 1,2,4,4-tetraphenyl-3-buten-1-one (**21**) is formed without migration of a phenyl group.

References, p. 47

The enol ether type isomer, 2,3-dihydro-2,2,4,5-tetraphenylfuran (**22**), behaved differently by undergoing a valence tautomerization to 1-benzoyl-1,2,2-triphenylcyclopropane (**23**), m.p. 161–162° (yield 35 %) [*16*].

The work of BOYKIN JR. and LUTZ [*16*] has found analogies with simpler systems. When 2,3-dihydro-4,5-dimethylfuran (**24**) was exposed to UV radiation in ether solution, a 74 % yield of 1-acetyl-1-methyl-cyclopropane (**25**) was obtained besides minor quantities of the geometrical isomers of 3-methyl-3-penten-2-one, and 3-methyl-4-penten-2-one [*17*].

Aldehydes were produced on irradiation [*18*] of 2,3-dihydro-3-methyl-2-propenylfuran (**26**) in cyclohexane. From a cis/trans mixture of **26** there were produced only trans-3-methyl-trans-2-propenylcyclopropanecarboxaldehyde (**27**) and cis-3-methyl-trans-2-propenylcyclopropanecarboxaldehyde (**28**) while on pyrolysis of the same starting material all 4 isomers were formed.

b) Photoisomerization of cyclic acetals to esters

ELAD and YOUSSEFYEH [*19*] have reported on the acetone sensitized photochemical conversion of cyclic acetals to the open chain carboxylic esters (cf. table 3).

Table 3 [*19*]

acetal (general formula)		ester (general formula)		yield
2-pentyl-1,3-dioxolane	(**29**)	ethyl hexanoate	(**30**)	36%
2-heptyl-1,3-dioxolane	(**29**)	ethyl octanoate	(**30**)	55%
2-nonyl-1,3-dioxolane	(**29**)	ethyl decanoate	(**30**)	33%
2-benzyl-1,3-dioxolane	(**29**)	ethyl phenylacetate	(**30**)	35%
2-phenethyl-1,3-dioxolane	(**29**)	ethyl hydrocinnamate	(**30**)	30%*
2-heptyl-1,3-dioxane	(**31**)	propyl octanoate	(**32**)	23%
2-phenethyl-1,3-dioxane	(**31**)	propyl hydrocinnamate	(**32**)	14%

* when irradiated through quartz the yield rose to 52%.

Ethyl or propyl carboxylates (30 or 32) [*19*]. The acetals **29** or **31** (3 g) were dissolved in a mixture of 10 ml of acetone and 90 ml of tert. butanol. These solutions were irradiated at room temperature for 20–40 hrs. with a Quarzlampengesellschaft Q 81 high-pressure mercury vapor lamp contained in a Pyrex immersion tube. The esters were isolated by chromatography over alumina, and yields are based on total amount of acetal irradiated. Use of a quartz immersion tube allowed for higher yields as was shown for the case of ethyl hydrocinnamate synthesis (cf. table).

3. Photochemical rearrangements among heterocyclic oxygen compounds

a) Isomerization of a 4H-pyran to a 2H-pyran

If 4-benzyl-2,4,6-triphenyl-4H-pyran (**33**) is irradiated under nitrogen in cyclohexane [*20*], the 2H-isomer **34** is formed, which on treatment with acids or bases yields 1,2,3,5-tetraphenylbenzene (**35**).

2-Benzyl-2,4,6-triphenyl-2H-pyran (34) [*20*]. The 4H-pyran **33** (2.0 g) is dissolved in 200 ml of cyclohexane and irradiated for 2 hrs. by a UV immersion lamp while nitrogen is passed through the solution. The yellow solution is then evaporated and the residue freed from viscous material with a little alcohol. The mixture of **33** and **34** (1.1 g) is separated by fractional crystallization from methanol to give 0.64 g (32%) of **34**, m.p. 125 — 126.5° (from methanol).

b) Isomerization of epidioxides

A German patent [*21*] states that solutions of ascaridole (**36**) in carbon tetrachloride are light sensitive and that the deterioration of such solutions can be prevented by the addition of dyes. These early observations were confirmed and the efficiency of, for example, sudan G as stabilizer evaluated [*22*]. The light sensitivity of **36** was shown to be due to its photoisomerization to isoascaridole (**37**) which is also the product of thermal rearrangement of ascaridole (cf. [*23*]).

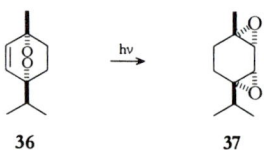

A similar photorearrangement of an epidioxide had already earlier been reported by BERGMANN et al. [*24, 25*], though final proof of the structures of the products involved was adduced only later [*26*]. 2α,5-Epidioxy-5α-cholest-3-ene (**39**) (comp. chapter 39) is quantitatively converted to 4α,5-epoxy-5α-cholestan-2-one (**40**) on exposure of its solutions to sunlight. The latter may also be obtained directly on photosensitized oxygenation of 2,4-cholestadiene (**38**) in sunlight [*24, 25*].

Isoascaridole (37) [*22*]. Ascaridole (**36**) was irradiated for 45 hrs. as a thin layer in an open irradiation vessel by means of a water-cooled Osram HQA 500 mercury high-pressure lamp. Vacuum distillation of 28.7 g of irradiated material gave 11.93 g of **37**.

4α,5-Epoxy-5α-cholestan-2-one (40) [*25*]. One part of the peroxide **39** was dissolved in 50 parts of absolute alcohol and exposed to sunlight in a closed Pyrex flask. After one week the crystals precipitated were filtered off, washed with a little alcohol and then recrystallized from absolute alcohol to give 88% of **40**, m.p. 172°.

References

[*1*] J. B. Jones: J. Chem. Soc. **1963**, 5759.
[*2*] G. Büchi and N. C. Yang: Chem. and Ind. **1955**, 357.
[*3*] G. Büchi and N. C. Yang: J. Amer. chem. Soc. 79, 2318 (1957).
[*4*] P. de Mayo, J. B. Stothers and R. W. Yip: Can. J. Chem. 30, 2135 (1961).
[*5*] M. Mousseron-Canet, M. Mousseron, P. Legendre and J. Wylde: Bull. Soc. Chim. France **1963**, 379.
[*6*] M. Mousseron: Pure Appl. Chem. 9, 481 (1964).
[*7*] R. Exelby and R. Grinter: Chem. Rev. 65, 247 (1965); esp. p. 255.
[*8*] G. H. Brown and W. G. Shaw: Rev. Pure Appl. Chem. 11, 2 (1961).
[*9*] R. E. Miller: USP 3,116,148 of 21. 12. 1959; C. A. 60, 6388 (1964).
[*10*] R. E. Fox: Report AD 261608 of April 1961; Nuclear Sci. Abstr. 16, 13390 (1962).
[*11*] C. A. Heller, D. A. Fine and R. A. Henry: J. Phys. Chem. 65, 1908 (1961).
[*12*] E. Berman: J. Phys. Chem. 66, 2275 (1962).
[*13*] P. Bladon, W. McMeekin and I. A. Williams: J. Chem. Soc. **1963**, 5727.
[*14*] J. Rigaudy and P. Derible: Bull. Soc. Chim. France **1965**, 3047.
[*15*] J. Rigaudy and P. Derible: Bull. Soc. Chim. France **1965**, 3055.
[*16*] D. W. Boykin, jr. and R. E. Lutz: J. Amer. chem. Soc. 86, 5046 (1964).
[*17*] D. E. McGreer, M. G. Vinje and R. S. McDaniel: Can. J. Chem. 43, 1417 (1965).
[*18*] J. Wiemann, N. Thoai and F. Weisbuch: Tetrahedron Letters **1965**, 2983.
[*19*] D. Elad and R. D. Youssefyeh: Tetrahedron Letters **1963**, 2189.
[*20*] K. Dimroth, K. Wolf and H. Kroke: Liebigs Ann. Chem. 678, 183 (1964).
[*21*] S. Deichsel: DRP 620 636 of 22. 10. 1933; C. **1936**, I, 4622.
[*22*] G. O. Schenck, K. G. Kinkel and H.-J. Mertens: Liebigs Ann. Chem. 584, 125 (1953).
[*23*] M. Matic and D. A. Sutton: J. Chem. Soc. **1953**, 349.
[*24*] E. L. Skau and W. Bergmann: J. Org. Chem. 3, 166 (1938).
[*25*] W. Bergmann, F. Hirschmann and E. L. Skau: J. Org. Chem. 4, 29 (1939).
[*26*] R. J. Conca and W. Bergmann: J. Org. Chem. 18, 1104 (1953).

Chapter 6

Photoisomerizations involving nitrogen compounds

1. Photoisomerizations with rupture of nitrogen-oxygen bonds

a) Amides from aldoximes

Irradiation of aryl aldoximes in a variety of solvents leads to the formation of amides [1] besides geometrical isomerization. Syn-benzaldehyde oxime (1), for instance, gives benzamide (2). It was suggested [1] that this BECKMANN type rearrangement might involve oxaziridines as intermediates (see below).

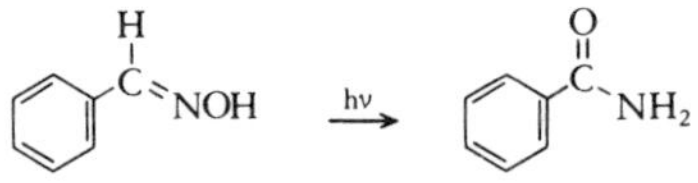

Benzamide (2) [1]. A 0.5% solution of **1** in about 200 ml of acetic acid was irradiated for 3 hrs. at 10°, a quartz immersion apparatus with 80 W Hanovia CH3 lamp being used. 41% of benzamide (2) was isolated whereas yield in the absence of radiation was nil.

b) Anilides from nitrones

Nitrones (cf. **3**) may on irradiation undergo a variety of rearrangements which will be discussed in this and the following sections. A more complete account of nitrone chemistry will be found with HAMER and MACALUSO [2].

One of the early examples of reactions of this type is the rearrangement of α-acridin-9-yl-N-phenylnitrone (**3**) to 9-acridinecarboxanilide (**4**) carried out by CHARDONNENS and HEINRICH [3].

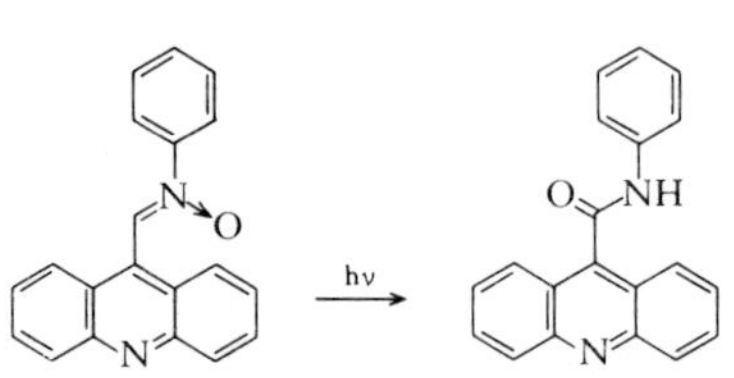

References, p. 55

In this connection, reference is made to the experiments of MIKHAILOV and TER-SARKISYAN [4], involving the irradiation of an ethanolic solution of 12-methylbenz[a]acridine and N,N-dimethyl-4-nitrosoaniline with light from a quartz mercury lamp. The Russian authors obtained a mixture containing α-benz[a]acridin-12-yl-N-(4-dimethylaminophenyl)-nitrone (5) and 4'-dimethylaminobenz[a]acridine-12-carboxanilide (6). It was assumed [4] that the anilide is derived from the nitrone by photochemical rearrangement.

In the light of more recent studies it must be assumed, that oxaziridines (see below) are intervening in the transformation discussed above. Photolysis of α-benzoyl-α,N-diphenylnitrone (7) in anhydrous ether or benzene gave an almost quantitative yield of N-phenyldibenzamide (9) [5]. The oxaziridine 8 was regarded to be the intermediate.

7: R = H₅C₆-CO-   8: R = H₅C₆-CO-   9: R = H₅C₆-CO-   13: R = H
10: R = H   11: R = H   12: R = H

Diaryloxaziridines are very unstable and break down thermally [6—8] to the anilides actually observed. When N,α-diphenylnitrone (10) was irradiated in absolute ethanol [7] a 53 % yield of N,N-diphenylformamide (13) was obtained besides 5 % of benzanilide (12). When acetone was used as solvent, the ratio between the two products was reversed, 12 being the main product (75 %) with 5 % of 13. 2,3-Diphenyloxaziridine (11) could not be isolated.

9-Acridinecarboxanilide (4) [3]. A thin layer of 3 in a watch-glass was moistened with acetone, covered with a second watch-glass and exposed to sunlight for 8 days (Switzerland). Evaporated acetone was replaced with additional quantities. The reaction product was found to be no longer soluble in boiling benzene, and the insoluble residue was crystallized from glacial acetic acid to give 4 as small yellow prisms, m.p. 315°.

N-Phenyldibenzamide (N,N-dibenzoylaniline, 9) [5]. A solution of 3.0 g of nitrone 7 in 1000 ml of anhydrous ether was irradiated with a water-cooled mercury immersion lamp Hanovia Type L-450 W with a Vycor filter. The solution was flushed with nitrogen before and during the illumination. After 24 hrs., 95% of the starting material had disappeared as became evident from thin-layer chromatography. The solution was brought to dryness to give 9, m.p. 158—161°.

c) Oxaziridines from nitrones

In some instances SPLITTER and CALVIN [6, 7] succeeded in isolating the oxaziridines suspected to be intermediates in the nitrone → anilide transformation. Thus, when N-tert. butyl-α-(4-nitrophenyl)-nitrone (14) was irradiated in ethanol, the oxaziridine 15 could be isolated.

In some cases the (dark) reverse reaction was observed. When α-(4-dimethylaminophenyl)-N-(3-nitrophenyl)-nitrone was irradiated at room temperature in ethanol to complete disappearance of the nitrone spectrum, it would re-form nitrone to the extent of 94% within some seconds after irradiation ceased [7].

The photochemical formation of the oxaziridine 17 from 7-chloro-2-methylamino-5-phenyl-3H-1,4-benzodiazepine 4-oxide (16) and the reversal of the process by refluxing 17 in isopropanol was reported by STERNBACH et al. [9].

The phototransformation of 5,5-dimethyl-1-pyrroline 1-oxide (18) to the bicyclic oxaziridine 19 has been observed by BONNETT et al. [10]. Thermal rearrangement of 2,2-dimethyl-6-oxa-1-azabicyclo[3.1.0]hexane (19) did not re-produce the nitrone 18, but, instead, led to 5,5-dimethyl-2-pyrrolidinone (20). 2-Substituted 1-pyrroline 1-oxides were, in contrast to BONNETT's statement [10], also prone to the nitrone → oxaziridine photorearrangement, as was found subsequently [11]. 2,5,5-Trimethyl-1-pyrroline

References, p. 55

1-oxide (**21**) gave, on irradiation, a 28% yield of 2,2,5-trimethyl-6-oxa-1-azabicyclo[3.1.0]hexane (**22**). This finding is of some bearing to the photoisomerization of 2-substituted quinoline oxides (comp. following section).

18: R = H
21: R = CH$_3$

19: R = H
22: R = CH$_3$

20

2-Tert. butyl-3-(4-nitrophenyl)-oxaziridine (**15**) [7]. A solution of 30 mg of **14** in 204 ml of absolute ethanol was irradiated in three portions for 25 mins. each. The solution was contained in a water-cooled glass container 1 cm thick, between two General Electric DxB photospots 50 cm apart. After evaporation of the solvent, 50 ml of petroleum ether was added to the residue and the insoluble material was filtered. The filtrate was evaporated leaving 12 mg (40%) of the oxaziridine **15**, m.p. 56—59°.

8-Chloro-3,9b-dihydro-4-methylamino-9b-phenyloxazirino[2,3-d] [1,4] benzodiazepine (**17**) [9]. A solution of 10 g of **16** in 1000 ml of isopropanol was exposed to diffuse daylight in a borosilicate glass flask. After 12 days' irradiation the solvent was evaporated at low temperature, and the irradiation product was separated from residual **16** by fractional crystallization from a mixture of ether and petroleum ether, in which **17** was considerably more soluble. Finally 65% of **17** were isolated in the form of colorless prisms, m.p. 167—170° (bath preheated to 160°). The material solidified immediately and melted again at 236—236.5°, i.e., the melting point of **16**.

2,2-Dimethyl-6-oxa-1-azabicyclo[3.1.0]hexane (**19**) [10]. Nitrone **18** (1.6 g) in 20 ml of a cyclohexane-ethanol (1:1) mixture was placed 1 cm. from a Hanovia UVS 250 lamp and irradiated for 5 days. Fractionation of the product gave 0.18 g (11%) of crude **19**, b.p.$_{25}$ 55—60°. The distillation residue consisted of nitrone **18** and lactam **20**.

d) Unsaturated lactams from heterocyclic N-oxides

Isomerizations of the kind **18** → **20** proceeding under illumination have, in fact, already earlier been noted in the quinoline and quinoxaline series. Thus, quinoxaline 1,4-dioxide (**23**) in water is converted to 2-quinoxalinol 4-oxide (**24**) by the action of radiation. Under the same conditions, quinoxaline 1-oxide forms 2-quinoxalinol [12].

23

24

Similarly, quinoline 1-oxide (**25**) on irradiation in aqueous solution rearranges to carbostyryl (**26**) in good yield [13]. On irradiation of a solution of **25** in absolute ethanol a dimeric product is formed in addition to **26**. High yields (80—98%) were attained in the photochemical synthesis of all possible methylcarbostyryls [14] from the corresponding methylquinoline 1-oxide precursors.

25: R = H
27: R = CH₃

26: R = H
28: R = CH₃

As expected, quinaldine 1-oxide (**27**) differed in behavior; only 10 % of 3-methylcarbostyryl (**28**) could be obtained by BUCHARDT et al. [*14*] together with an unidentified product. Independent research [*15*] showed that **27** in fact underwent an interesting series of phototransformations which may be visualized as proceeding via the intermediate oxaziridine **29**. 1-Methyl- and 3-methylcarbostyryl (**30** and **28**, respectively) and 1-acetylindole (**31**) were formed in 16 %, 22 %, and 8 % yield, respectively.

Adenine 1-oxide (**32**) was found to be very sensitive to radiation [*16*]. In this case, however, deoxygenation (to adenine, **33**) was competing with rearrangement to isoguanine (**34**). When all ortho-positions were blocked (as with isoguanine 1-oxide, **35**) deoxygenation was the only reaction mode observed, with isoguanine (**34**) being the product [*16*].

2-Quinoxalinol 4-oxide (24) [*12*]. The dioxide **23** (7.5 g) in water (500 ml) was exposed to direct sunlight in a glass vessel. After 2—3 hrs., **24** began to separate out together with some **23** which was re-dissolved by warming. After 2 weeks, the precipitate (3 g) was filtered off and dissolved in dilute NaOH solution. Following treatment with charcoal, **24** was precipitated with acetic acid and recrystallized from glacial acetic acid to give buff prisms, m.p. 274—275°.

Carbostyryl (26) [*13*]. The oxide **25** (501 mg) was dissolved in 45 ml water, transferred to 25 ml Pyrex test tubes and closed with a cork stopper. The tubes were wrapped with aluminium foil above the surface of the reaction mixture to avoid formation of strongly colored oxidation products. The tubes were irradiated with a 600 W sun lamp at a distance of 15 cm (15 hours, 35°). Cooling the reaction mixture caused colorless crystals to separate which were filtered off (276 mg). The mother-liquor on further treatment yielded 50 mg less pure material. After recrystallization from ethanol, **26** had m.p. ca. 300° (subl.).

References, p. 55

e) 2-Hydroxyazobenzenes from azoxybenzenes

The sensitivity of azoxy compounds to light [17, 18] has been investigated further by BADGER and BUTTERY [19]. These authors found, that the photoisomerization, which in the case of azoxybenzene leads to the formation of 2-hydroxyazobenzene, takes place in such a way that the oxygen atom migrates to the non-adjacent ring. The photochemical rearrangement of 4-bromo-O,N,N-azoxybenzene (36) affords 4-bromo-2-hydroxyazobenzene (37), and the rearrangement of 2,2'-azoxynaphthalene (38) produces 2-(2-naphthylazo)-1-naphthol (39) in a similar manner [19].

The solvent has a great influence on the rate of the photochemical rearrangement of azoxy compounds. The rate is low in benzene and relatively high in ethanol. Yields are reported to be poor (5—15 %) even following 30 days' irradiation [19].

The ortho-isomers appear to be formed exclusively. BADGER and BUTTERY [19] point out in this respect that the rearrangement of azoxybenzene and related compounds achieved by using acids, but in the absence of light, leads mainly to the formation of the 4-hydroxyazobenzenes with the ortho-isomers occurring only as by-products. This relationship has been more fully explored by SHEMYAKIN and MAIMIND [20] with the aid of labelled azoxy compounds. The acid catalyzed WALLACH rearrangement yielding the para isomers was formulated as proceeding via oxadiaziridines [20].

2-(2-Naphthylazo)-1-naphthol (39) [*19*]. A solution of **38** in chloroform was insolated for 30 days (Australia). After evaporation of the solvent, the residue was dissolved in benzene and chromatographed over alumina. Unconverted **38** was recovered, together with red-brown needles of **39**, m.p. 168° (from ethanol).

2. Photoisomerization with rupture of nitrogen-halogen bonds

MARTIN and BARTLETT [21] have observed the isomerization 40 → 41 when N-bromosuccinimide (40) was irradiated in carbon tetrachloride. This free-radical chain reaction may also be started by heating the system in the dark in the presence of peroxides. In this case, higher yields of 3-bromopropionyl isocyanate (**41**) are obtained[*21*]. For further photolyses of nitrogen-halogen compounds (HOFMANN-LÖFFLER reaction) see chapter 26.

$$\underset{40}{\text{[cyclic imide-NBr]}} \xrightarrow{h\nu} \underset{41}{Br-CH_2-CH_2-C\underset{N=C=O}{\overset{O}{\diagdown}}}$$

3. Photoisomerizations with rupture of nitrogen-carbon bonds

a) Rearrangement of a pyrimidine

Ultraviolet irradiation (254 mµ) of 4-amino-2,6-dimethylpyrimidine (**42**) results in an intramolecular rearrangement with the formation, either in aqueous or anhydrous medium, of only one photoproduct, 3-amino-2-(1-iminoethyl)-crotonitrile (**44**) in quantitative yield.

$$\underset{42}{\text{pyrimidine}} \xrightarrow{h\nu} [\,\mathbf{43}\,] \rightarrow [\,\ldots\,] \rightarrow \underset{44}{\text{product}}$$

The discoverers of this reaction [*22, 23*] suggest that the transformation **42** → **44** proceeds via isomeride **43** which breaks down to **44** in the manner indicated. This photodegradation is of some bearing to the photochemistry of nucleic acids.

3-Amino-2-(1-iminoethyl)-crotonitrile (44) [*23*]. A 0.01–0.03 M solution of **42** in 0.02 M phosphate buffer (pH 8.5–8.7) was irradiated at 15–20° in the form of a layer of about 1 mm thickness contained between the walls of a 30 W Philips germicidal lamp and a concentric glass cylinder; for further details of the irradiation equipment the original literature should be consulted. The solution was stirred by means of a current of nitrogen. Needle-like crystals of **44**, which had separated out during the course of the irradiation, were removed and the filtrate again irradiated. The yield of **44** usually exceeded 50%, and this could be increased by concentration of the final filtrate. The crude **44** was washed with a little cold water and then recrystallized from methanol to give pale yellow needles, m.p. 206–207° (dec.).

b) Rearrangement of pyrazoles

A novel rearrangement was discovered by TIEFENTHALER et al. [*24*] when pyrazole was transformed to imidazole on irradiation in dioxane. 1H-Indazoles are similarly converted to benzimidazoles as is illustrated for the case of 7-methyl-1H-indazole (**45**). Yields of 7-methylbenzimidazole (**46**) are 9–31 %, depending on reaction conditions.

References, p. 55

45   46   47   48

2-Alkyl-1H-indazoles were later reported [25] to be almost quantitatively isomerized, while the 1-substituted isomers behaved differently, e.g., 1-methyl-1H-indazole (47) rearranged photochemically to 2-N-methylaminobenzonitrile (48) in 33 % yields. This observation may shed some light on the above mentioned, deep-rooted rearrangement of pyrimidine 42 to nitrile 44.

References

[1] J. H. Amin and P. de Mayo: Tetrahedron Letters **1963**, 1585.
[2] J. Hamer and A. Macaluso: Chem. Rev. **64**, 473 (1964).
[3] L. Chardonnens and P. Heinrich: Helv. Chim. Acta **32**, 656 (1949).
[4] B. M. Mikhailov and G. S. Ter-Sarkisyan: Izvest. Akad. Nauk SSSR, Otdel. Khim. Nauk **1954**, 656; Bull. Acad. Sci. USSR, Div. Chem. Sci. **1954**, 559; C. A. **49**, 10953 (1955).
[5] A. Padwa: J. Amer. chem. Soc. **87**, 4365 (1965).
[6] J. S. Splitter and M. Calvin: J. Org. Chem. **23**, 651 (1958).
[7] J. S. Splitter and M. Calvin: J. Org. Chem. **30**, 3427 (1965).
[8] K. Shinzawa and I. Tanaka: J. Phys. Chem. **68**, 1205 (1964).
[9] L. H. Sternbach, B. A. Koechlin and E. Reeder: J. Org. Chem. **27**, 4671 (1962).
[10] R. Bonnett, V. M. Clark and A. Todd: J. Chem. Soc. **1959**, 2102.
[11] L. Kaminsky and M. Lamchen: Chem. Comm. **1965**, 130.
[12] J. K. Landquist: J. Chem. Soc. **1953**, 2830.
[13] O. Buchardt: Acta Chem. Scand. **17**, 1461 (1963).
[14] O. Buchardt, J. Becher and C. Lohse: Acta Chem. Scand. **19**, 1120 (1965).
[15] M. Ishikawa, S. Yamada and C. Kaneko: Chem. Pharm. Bull. **13**, 747 (1965).
[16] B. G. Brown, G. Levin and S. Murphy: Biochemistry **3**, 880 (1964); C. A. **61**, 3334 (1964).
[17] W. M. Cumming and J. K. Steel: J. Chem. Soc. **123**, 2464 (1923).
[18] W. M. Cumming and G. S. Ferrier: J. Chem. Soc. **127**, 2374 (1925).
[19] G. M. Badger and R. G. Buttery: J. Chem. Soc. **1954**, 2243.
[20] M. M. Shemyakin and V. I. Maimind in: Recent Progress in the Chemistry of Natural and Synthetic Colouring Matters and Related Fields, Ed. by T. S. Gore, B. S. Joshi, S. V. Sunthankar and B. D. Tilak, pp. 441–449. New York: Academic Press 1962.
[21] J. C. Martin and P. D. Bartlett: J. Amer. chem. Soc. **79**, 2533 (1957).
[22] K. L. Wierzchowski, D. Shugar and A. R. Katritsky: J. Amer. chem. Soc. **85**, 827 (1963).
[23] K. L. Wierzchowski and D. Shugar: Photochem. Photobiol. **2**, 377 (1963).
[24] H. Tiefenthaler, W. Dörscheln, H. Göth and H. Schmid: Tetrahedron Letters **1964**, 2999.
[25] H. Göth, H. Tiefenthaler and W. Dörscheln: Chimia **19**, 596 (1965).

Chapter 7

Photochemical stereoisomerization

1. Photochemical cis-trans isomerizations at double bonds

The stereoisomerization of olefinic compounds in light belongs to the longest known and best investigated photochemical reactions. These reactions are of great significance for preparative chemistry since they are in many cases the simplest, in a few cases the only, method by which specific stereoisomers may be synthesized.

Photochemical cis-trans isomerizations play a large rôle in the field of natural products, and reference should be made to SCHAFFNER [1] for a summary of literature on this subject.

It has frequently been possible to convert the cis to the trans form, equally the trans to the cis form by photochemical means. Thus, in the case of the reversible interconversion of fumaric to maleic acid, irradiation of either isomer with ultraviolet light establishes an equilibrium between the two forms [2]. Nevertheless, in cases where the absorption spectra of the two isomers are different it is possible to use photochemical stereoisomerization for the selective preparation of one of the isomers; one works with light which will only be absorbed by one of the forms, thus causing its isomerization. The isomer which is formed will not be re-isomerized since it will not absorb the radiation (comp. [3]).

In the presence of sensitizers the photoequilibrium may be shifted to one side or the other. Bromine has long been known as sensitizer; its use is referred to in the formation of fumaric acid described below. Photochemical isomerization of cis- and trans unsaturated compounds, particularly of the stilbenes, has been investigated mainly by HAMMOND and collaborators [4, 5] (comp. also [6, 7]).

According to HAMMOND [5], sensitizers having low excitation energy function as "photocatalysts", i.e., in the presence of these sensitizers the composition of the photostationary state approaches that of thermal equilibrium. It is of great interest that different sensitizers may lead to different photostationary compositions as may be seen from table 4 illustrating the sensitized isomerization of the 1,3-pentadienes (piperylenes) in benzene solution under irradiation for 5—10 hrs. [5].

References, pp. 67—69

Table 4 [5]

sensitizer	initial cis [%]	trans [%] at photoequilibr.
acetophenone	0	55
	100	53
benzophenone	0	57
	100	56
fluoren-9-one	0	69
	100	67
2,3-butanedione	0	77
	100	77
1'-acetonaphthone	0	79
	100	79

a) Compounds containing isolated C=C bonds

Out of the numerous photochemical cis-trans isomerizations which have become known (comp. WYMAN [3]) only a few will be discussed in the sequel. A summary of earlier literature describing photochemical stereoisomerizations of, inter alia, α,β-unsaturated acids will also be found with BACHÉR [8].

Examples shown here comprise stereoisomerization in the presence (cf. **1** → **2** [9], **3** → **4** [10]) and in the absence of sensitizers (cf. **5** → **6** [11, 12], **7** → **8** [13]). The isomerization of maleic (**1**) to fumaric acid (**2**) is also possible without bromine as sensitizer [14] albeit in lower yield.

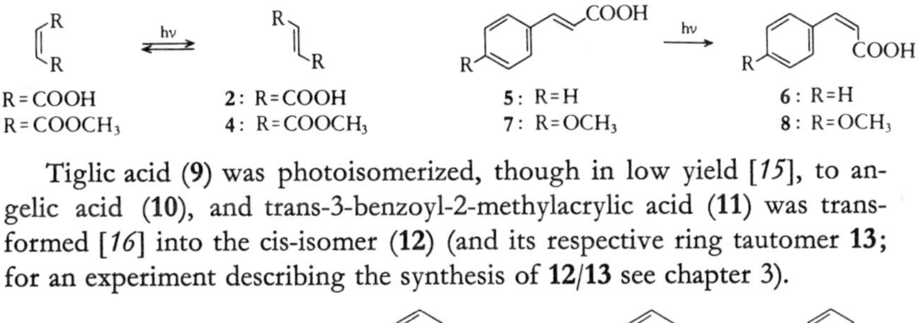

1: R=COOH   2: R=COOH   5: R=H   6: R=H
3: R=COOCH₃   4: R=COOCH₃   7: R=OCH₃   8: R=OCH₃

Tiglic acid (**9**) was photoisomerized, though in low yield [15], to angelic acid (**10**), and trans-3-benzoyl-2-methylacrylic acid (**11**) was transformed [16] into the cis-isomer (**12**) (and its respective ring tautomer **13**; for an experiment describing the synthesis of **12/13** see chapter 3).

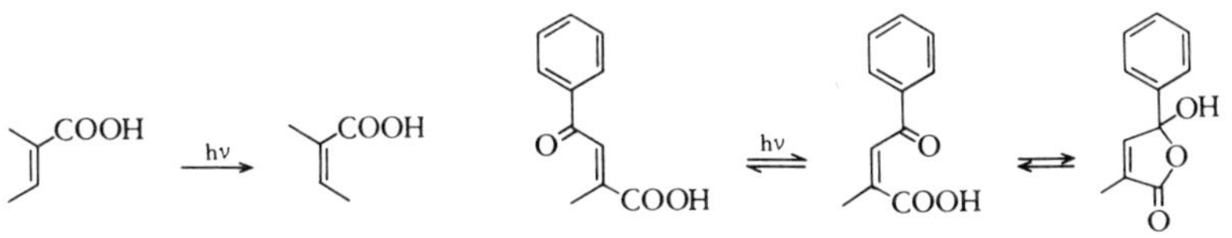

9   10   11   12   13

The photosensitized stereoisomerization of cis,trans,trans-1,5,9-cyclododecatriene (**14**) to the all-trans (**15**) and the cis,cis,trans (**16**) isomers has been described by NOZAKI et al. [17].

14 ⇌ 15 + 16

The composition of the isomerizate was found to depend on the nature of the sensitizer used; table 5 gives the ratios of the isomers being present in benzene solutions of **14** irradiated through Pyrex at 10—15° under nitrogen for 200 hours [*17*].

Table 5 [*17*]

sensitizer	triene recovered	14	15	16
none (before irradiation)		97	3	0
none	95%	93	4	3
acetone	92%	41	12	47
cyclododecanone	100%	65	9	26
acetophenone	96%	52	39	9
triphenylene	100%	97	3	0

Fumaric acid (2) [*9*]. If maleic acid (1) (2 g) is dissolved in water (5 ml) and a little bromine water is added, then on placing the solution in sunlight crystallization of **2** begins after one minute. The quantity which may be obtained amounts to 82—93.5%, according to the intensity of the sunlight. No transformation occurs in the dark.

Dimethyl fumarate (4) [*10*]. If a trace of $HgNO_3$ is dissolved in an aqueous solution of dimethyl maleate (3), together with a small amount of nitric acid to prevent hydrolysis, and this mixture is placed in a quartz vessel in front of a mercury lamp, needles of solid, sparingly soluble **4** separate after a few seconds, while prolonged warming does not produce fumarate.

Cis-cinnamic acid (6) [*12*]. A solution of trans-cinnamic acid (5) (3.5 g) in benzene (100 g) is placed in a Uviol-glass tube of 2—3 cm width and exposed to radiation from a Schott Uviol lamp for 100 hours. The yield of cis-acid amounts to 25—30%.

4-Methoxy-cis-cinnamic acid (8) [*13*]. 4-Methoxy-trans-cinnamic acid (7) (14 g) was dissolved in a solution of sodium carbonate (14 g) in 2.7 liters of water and irradiated for 24 hrs. with a mercury high-pressure immersion lamp (300 W). The solution was deaerated before illumination with a stream of nitrogen. The residual trans acid was precipitated with HCl and filtered off. The cis acid **8** had remained in solution and was extracted from the filtrate with ether. Evaporation of the $CaSO_4$-dried solution gave 10.3 g (74%) of **8**, m.p. 67—68°.

b) Dienes and polyenes

BÜCHI and YANG [*18*] have shown that cis-α-ionone (**18**) is formed on irradiation of trans-α-ionone (**17**). The smell of cis-α-ionone is completely different from the violaceous one of the trans isomer and resembles that of cedarwood. As a consequence of the steric hindrance by the two neighboring methyl groups, **18** can only exist in the s-cis configuration.

References, pp. 67—69

Later on [*19*], a second photoisomer of α-ionone was found, 4-(2,6,6-trimethyl-2-cyclohexenylidene)-2-butanone (**19**) the formation of which was rationalized as proceeding via photoenolization of **17** or **18**.

In order to synthesize simple calciferol analogs, HARRISON and LYTHGOE [*20*] obtained cis-2-(2-cyclohexylideneethylidene)-cyclohexanone (**21**) by photoisomerization of the trans-dienone **20**. For the photochemical interconversions of precalciferol and tachysterol see chapter 2.

As an example of a smoothly proceeding partial rearrangement catalyzed by iodine the photoisomerization of 9-cis,13-cis-3-dehydroretinoic acid (**22**) to 9-cis-3-dehydroretinoic acid (**23**) [*21*] may be cited.

Cis-trans isomerizations of the kind **22** → **23** play an important rôle in the photochemistry of the visual process (comp. WALD [*22*]) and in the chemistry of vitamin A (retinol, **24**) (comp. MOUSSERON [*23*]).

The great ease with which light induces cis-trans rearrangements with polyenes has caused considerable difficulty to chemists working with such compounds, e.g., with the carotenoids; for a review on work done in this field see ZECHMEISTER [*24*]. In a series of cases photoisomerization occurs so readily that in order to obtain products free from isomers, the synthesis has to be carried out as far as possible with exclusion of light. Separation of the isomers is readily achieved by chromatography [*25*].

According to ZECHMEISTER and co-workers [*26, 27*] ordinary 1,8-diphenyl-1,3,5,7-octatetraene is the all-trans compound (**25**), which may be converted by illumination in benzene-hexane solution into the 1-cis (**26**) and the 3-cis (**27**) isomer, the structures of which were confirmed [*28*] by independent synthesis. A third isomer arising in the irradiation of **25** [*27*] was ascribed [*24*] the constitution of 1,5-di-cis-1,8-diphenyl-1,3,5,7-octatetraene (**28**).

All carotenoid solutions are light sensitive [24, 29], but the extent of the change varies considerably and is dependent on the spatial configuration of the substrate and on the wave length of the incident light, visible light being more effective than ultraviolet. Intensive sun irradiation in quartz vessels in an inert gas atmosphere has proved efficient, provided temperatures are kept down to retard thermal reactions. If all-trans carotenoids are irradiated, cis-trans isomerizations and irreversible decomposition take place which are both accompanied by a loss in color intensity. On the other hand, poly-cis compounds behave differently in that the obvious deepening of the color due to steric rearrangements outweighs all other effects.

15-cis-β-carotene (**29**) photoisomerizes so readily [30] that in its synthesis all operations must be carried out either in complete darkness or in red light. If hexane solutions of **29** are exposed to diffuse daylight, an almost quantitative rearrangement into all-trans-β-carotene may be observed within one hour.

Hexane solutions of 15-cis-β-carotene (**29**) treated with iodine remain practically unchanged in the dark during 20 hours; however, if such solutions are exposed to daylight for only 3 mins., a photochemical equilibrium is attained which differs from that of the iodine-free irradiation mixture in that only about 50 % of the all-trans isomer are contained besides other structural isomers [30].

Cis-α-ionone (18) [*18*]. A solution of 25 g of trans-α-ionone (**17**) in 625 ml of ethanol was irradiated for 72 hrs. under nitrogen. The solvent was then evaporated in vacuo and the residue distilled from a HICKMANN flask. The main fractions (b.p.$_{0.05}$ 35—37°) from four such preparations were collected and the material (74 g) distilled through a spinning-band column. Redistillation of fraction 2 (19.1 g) afforded 23 % of **18**, b.p.$_1$ 72—72.5°.

References, pp. 67—69

Cis-2-(2-cyclohexylideneethylidene)-cyclohexanone (21) [20]. The irradiation vessel was constructed from two large Pyrex beakers. The larger outer beaker was cooled in a bath; the inner, smaller beaker contained the lamp, and the space between the beakers was filled with a solution of dienone **20** (400 mg) in 1300 ml of methanol which was stirred with a current of nitrogen. After irradiation with a Mazda No. 19 mercury vapor lamp for 1 hr. at room temperature (ice-bath) the solution was irradiated for a further 2 hrs. at −60° (−80° bath) until the extinction at 309 mµ had dropped by 32%. The product was isolated by dilution with water, extraction with ether-light petroleum (1:1) and rapid chromatography on a wide column (3.5 cm) of alumina (75 g, Woelm grade II neutral). Elution with light petroleum (60−80°) containing 15% (v/v) benzene and evaporation of the first eluate gave an oil (305 mg) which on crystallization from ethanol at −40° gave 230 mg of **21**, m.p. 33°. Any delay during the chromatography led to isomerization.

9-Cis-3-dehydroretinoic acid (23) [21]. The di-cis-acid **22** (1.05 g) was dissolved in 20 ml of isopropyl ether and, after addition of 10 mg of iodine, irradiated with a 60 W lamp. Precipitation of **23** began very soon. After 1 hr. the solution was diluted with ether, washed with 0.1 N sodium thiosulfate solution, water and dried over sodium sulfate. After filtration the solution was evaporated in vacuo at 25° to give 0.8 g of **23**, m.p. 158−160° (ethanol).

3-Cis-1,8-diphenyl-1,3,5,7-octatetraene (27) [27]. A solution of 350 mg of **25** in 500 ml of warm benzene was diluted with hexane to 2500 ml. The solution was distributed in ten Pyrex vessels and placed in strong sunlight for 2 hours. The solution was then adsorbed onto a column (30 × 7.2 cm) composed of Magnesia, lime and Celite in the ratio 1:1:1. Development was carried out with benzene-hexane (2:1) and later with benzene, whereon **26** and **28** passed into the filtrate. Work-up, the details of which should be taken from the original paper, finally gave 17 mg of **27** as pale yellow prisms, m.p. 139−141°. All chromatographic and other operations were carried out in darkness or in dim light.

All-trans-β-carotene [30]. A solution of 2 mg of 15-cis-β-carotene (**29**) in 25 ml of hexane was exposed to daylight. Chromatography over Al_2O_3, elution with cyclohexane-alcohol and evaporation of the eluate in vacuo gave red crystals of all-trans-β-carotene, m.p. 178° (benzene-methanol).

c) Indigoid systems

Attempts to convert indigotin (**30**), which is a trans compound, photochemically to the corresponding cis product have been unsuccessful: the spectrum of indigotin solutions in chloroform does not change on irradiation [31]. Solutions of a number of halogen substituted compounds of the indigo series, e.g., 5,5'-dibromoindigotin, behave similarly. The great photochemical stability of indigotin can be explained by formula **30** illustrating the fixation of the trans form by hydrogen bonds (for a recent discussion see LÜTTKE and KLESSINGER [32]). This explanation is further substantiated by the observation [31, 33] that 1,1'-diacetylindigotin shows phototropic behavior in benzene solution.

WYMAN and BRODE [*34*] reported the existence of a photochemical equilibrium between cis (**32**) and trans-thioindigo (**31**) in inert solvents. The relative concentrations of these isomers were found to depend on the temperature and on the wave length of the light used for irradiation. Light of longer wave lengths promoted the formation of **32** whereas heating in the dark, or irradiation with UV-light had the opposite effect. The separation of the isomers could be achieved by chromatography in the dark. A non-planar configuration was proposed [*35*] for the cis-isomer **32**.

Cis-thioindigo (32) [*34*]. Thioindigo (10 mg) in benzene (50 ml) was irradiated for 20 mins. with yellow light ($\lambda > 520$ mμ). The solution was then poured onto a column packed with silicagel, and elution with benzene (ca. 1 liter) was continued until a narrow maroon zone of **32** at the upper end of the column was clearly separated from a diffuse blue zone of **31**. The column was sucked dry and the dyes extracted from the separated zones. All the operations were carried out in a darkened room illuminated with deep red light of low intensity. Benzene solutions of **32** are yellow-orange while those of **31** are purple-red.

d) Azo compounds

Photochemical cis-trans isomerization is, of course, not confined to compounds incorporating C=C bonds. The conversion of trans-azobenzene into the cis isomer was first accomplished by HARTLEY [*36, 37*]. He irradiated azobenzene solutions with sunlight or UV-light and obtained the cis form by fractional crystallization. In solution, besides the slow thermal isomerization, which leads to the formation of the trans form, a much faster photoisomerization yielding the cis isomer is proceeding. The composition of the equilibrium depends on the nature and the temperature of the solvent. Separation of the isomers is readily brought about by chromatography [*38, 39*]. In this manner, the synthesis of cis-azobenzene (**33**) and some of its derivatives, for example, 4-methyl-cis-azobenzene (**34**) has been described [*38*]. Chromatographic methods were also used in the preparation of cis-2,2'-azopyridine (**35**) [*40*] and the cis-2,2'- or cis-1,2'-azonaph-

33: R=H
34: R=CH$_3$

35

36

37

thalenes (**36** and **37**, respectively) [*41*]; in the latter case it was necessary to carry out the experiments at 0° owing to the great tendency of the cis forms to undergo thermal re-isomerizations. Reference is made here to secondary photoreactions of cis-azo compounds leading to dehydrocyclization (comp. chapter 14).

Cis-azobenzene (33) [*38*]. A solution of trans-azobenzene (1 g) in 50 ml of petroleum ether was irradiated for 30 mins. with a quartz mercury vapor lamp from a distance of 30—38 cm. The solution, which had assumed a red color, was filtered through alumina and washed with 100 ml of petroleum ether. The unchanged trans compound was eluted but **33** remained adsorbed at the upper end of the column. This zone was eluted with 150 ml of petroleum ether containing 1% methanol. The eluate was filtered and dried over sodium sulfate after removal of the methanol by washing with water. The petroleum ether was then evaporated under reduced pressure below 22° to leave crystals which were recrystallized from a little cold petroleum ether to afford pure **33**, m.p. 71°.

e) Compounds containing C=N bonds

Oximes may exist either in the syn (cf. **38**) or the anti (cf. **39**) configuration. Interconversions between these geometrical isomers may in certain cases be accomplished by irradiation [*14*, *42—44*]. Oxime O-ethers also undergo photochemical stereoisomerization. Thus, the 3- and 4-nitrobenzaldehyde O-methyloximes [*43*, *44*] and 4-nitrobenzaldehyde O-benzyloxime [*44*] have been successfully isomerized.

38: R = H   **39**: R = H

Other systems in which syn-anti photoisomerizations have been observed are nitrones and formazans. For example, cis- and trans-α-cyano-α,N-diphenylnitrone (**40** and **41**, respectively) are photochemically interconvertible [*45*] provided irradiation is carried out in the presence of a sensitizer. A photoequilibrium containing ca. 44% of the cis isomer is attainable from either side with eosine as sensitizer. Iodine, however, catalyzes only the cis → trans photoisomerization. By irradiation at 313 mμ both of these forms isomerize to 3-cyano-2,3-diphenyloxaziridine (**42**) (comp. chapter 6 for related reactions).

40   41   42

Formazans (cf. **43**) [*46*] carry two different chromophores which when irradiated with visible light may give rise to 4 isomers,

trans-syn   cis-syn   trans-anti   cis-anti

A sequence of photochemical and thermal stereoisomerizations was held responsible for the interconversions of the yellow and red forms of, e.g., 1,3,5-triphenylformazan (**43**) [*47, 48*]. In the case of 3-ethyl-1,5-diphenylformazan (**44**) HAUSSER et al. [*47*] eventually succeeded in isolating a red as well as a yellow stereoisomer.

43: R = C_6H_5
44: R = C_2H_5

Anti-4-nitrobenzaldehyde oxime (39) [*14*]. A suspension of syn-4-nitrobenzaldehyde oxime (**38**) (2 g) in 40 ml of benzene was contained in a glass tube and exposed to sunlight for 4 days. Precipitation of crystals began very soon while starting material dissolved completely. Filtration afforded light yellow crystals of the anti compound, which after recrystallization from petroleum ether had m.p. 174°. After resolidification the crystals melted at 128° (syn compound **38**).

2. Photochemical cis-trans isomerizations of cyclopropane and cyclobutane derivatives

Irradiation of degassed solutions (benzene, ether, 2-propanol, methanol) of trans-1,2-dibenzoylcyclopropane (**45**) in Pyrex vessels with a General Electric 275 W sunlamp was reported [*49*] to afford cis-1,2-dibenzoylcyclopropane (**46**). This photoisomerization was shown to be reversible, the photostationary mixture containing **45** and **46** in the ratio 2.5:1 [*50*].

45: R = CO–C_6H_5
47: R = C_6H_5

46: R = CO–C_6H_5
48: R = C_6H_5

Introduction of a photosensitizer may alter the reaction path significantly [49]. Thus, the primary product formed on irradiation of **45** in a mixture of 2-propanol and benzene and in the presence of benzophenone was found to be 1,5-diphenyl-1,5-pentanedione. This reduction product is in its turn photolabile and breaks down on further irradiation to acetophenone and a polymer.

Irradiation of 0.1 M benzene or cyclohexane solutions of trans- or cis-1,2-diphenylcyclopropane (**47** and **48**, respectively) in quartz vessels with 254 mμ light (Rayonet reactor with mercury low-pressure lamps, 40°) produced a 0.65 ratio of **47:48** [50]. Prolonged irradiation resulted in formation of a complex mixture of isomers and reduction products.

β- and γ-lumicolchicine (**49** and **50**, respectively; comp. chapter 1) are stereoisomeric compounds, which are UV-spectroscopically indistinguishable. Irradiation of either isomer leads to a photostationary state containing **49** and **50** in equal amounts [51, 52]. The progress of the photoisomerization may easily be followed by polarimetry and thin layer chromatography.

3. Photoepimerizations

a) Epimerization at carbon atoms

According to BUTENANDT et al. [53—55] 17-oxosteroids undergo a photoepimerization at C-13. Thus, androsterone (3α-hydroxy-5α-androstan-17-one, **51**) is isomerized to lumiandrosterone (3α-hydroxy-5α,13α-androstan-17-one, **52**), the latter possessing a ring C/D cis junction [53], and estrone (**53**) is photoisomerized to lumiestrone (3-hydroxy-13α-estra-1,3,5(10)-trien-17-one, **54**) [54, 55].

5 Schönberg, Photochemistry

Later on it was demonstrated [56] that photoepimerization at C-13 was reversible, the photostationary mixture containing the thermodynamically stabler 13α-isomer (54) in great excess, however. Irradiations [56] were carried out in dioxane at 90° with a mercury high-pressure lamp; at room temperature the ratio 53:54 was even smaller.

The photochemical racemization of usnic acid (55), a substance isolated from lichens, was reported by BARTON and QUINKERT [57]. Thermal treatment of 55 also produced racemic usnic acid.

Photoepimerization at a spiranic center was observed by TAUB et al. [58] when optically active dehydrogriseofulvin (56) was irradiated with UV-light in acetonitrile solution.

3α-Hydroxy-5α,13α-androstan-17-one (52) [53]. Androsteron (51) (2.38 g) was dissolved in 60 ml of dioxane and irradiated for 3.5 hrs. with a Heraeus lamp in a quartz vessel under nitrogen. The solvent was evaporated in vacuo and the residue boiled with GIRARD reagent T (2.5 g) in 35 ml of absolute ethanol with 3.2 ml glacial acetic acid (2 hours). The unreacted material (1.6 g) was dissolved in benzene and chromatographed over alumina. Benzene eluted 536 mg (33%) of substance, which was recrystallized from hexane to give pure 52, m.p. 145—146°.

(±)-Usnic acid [57]. (—)-Usnic acid (cf. 55) (723 mg) in anhydrous dioxane (720 ml) was irradiated for 124 hrs. below 55°. The solution was stirred by passage of a stream of dry oxygen-free nitrogen. Removal of the solvent in vacuo gave 628 mg (87%) of racemic usnic acid.

(±)-Dehydrogriseofulvin [58]. A solution of 1.4 g of (—)-dehydrogriseofulvin (cf. 56) in 20 ml acetonitrile in a quartz flask was irradiated with a Hanovia Type 16A13 broad spectrum low-pressure light source at 40°. A colorless crystalline precipitate began to form within 30 minutes. After 60 hrs. the product (918 mg) was filtered and recrystallized from acetonitrile to give (±)-56, m.p. 288—290°.

b) Epimerization at a sulfur atom

Optically active sulfoxides may undergo photochemical racemization as was found by MISLOW et al. [59]. For example, irradiation of (—)-(S)-1-naphthyl p-tolyl sulfoxide (57) in ether solution resulted in complete racemization.

References, pp. 67—69

57   58

Irradiation of a 0.1 M solution of (+)-(R)-methyl p-tolyl sulfoxide (**58**) in ether through Pyrex for 2 hrs. resulted in 5—10 % racemization and no decomposition. On the other hand, irradiation of (+)-(S)-butyl methyl sulfoxide for 10 mins. through Vycor went along with extensive decomposition but no racemization of recovered sulfoxide. Thus it became clear that the benzenesulfinyl chromophore was a prerequisite for the photoracemization to occur [*59*].

Extension of the above results to sulfoxides in the thianthrene series [*59*] led to the discovery of another cis-trans photoisomerization. Thus, when trans-thianthrene 5,10-dioxide (**59**) was irradiated for 2 hrs. through Pyrex in dioxane, complete conversion to the cis isomer (**60**) occurred, whereas analogous treatment of the latter did not produce any change. The same trans → cis isomerization could also be effected thermally.

59   60

(±)-1-Naphthyl p-tolyl sulfoxide [*59*]. An approximately 0.001 M solution of **57** in ether was transferred to a 300 ml Pyrex flask, equipped with a gas capillary at the base throught which nitrogen was introduced. Irradiation was effected with a Hanovia 450 W quartz mercury high-pressure lamp at room temperature, a Pyrex sleeve cutting off wave lengths below 285 mμ. After 1 hour, about 70 % of sulfoxide was recovered by chromatography which was completely racemized.

References

[*1*] K. SCHAFFNER in: Fortschr. Chem. Org. Naturstoffe, Ed. by L. ZECHMEISTER, Vol. **22**, pp. 1—114; esp. p. 61. Wien: Springer 1964.

[*2*] E. WARBURG: Sitz.ber. Preuß. Akad. Wiss. Berlin, Phys.-Math. Kl. **1919**, 960.

[*3*] G. M. WYMAN: Chem. Rev. **55**, 625 (1955).

[*4*] G. S. HAMMOND and J. SALTIEL: J. Amer. chem. Soc. **84**, 4983 (1962).

[*5*] G. S. HAMMOND, J. SALTIEL, A. A. LAMOLA, N. J. TURRO, J. S. BRADSHAW, D. O. COWAN, R. C. COUNSELL, V. VOGT and C. DALTON: J. Amer. chem. Soc. **86**, 3197 (1964).

[*6*] J. SALTIEL in: Survey Progr. Chem., Ed. by A. F. SCOTT, Vol. **2**, pp. 239—328; esp. p. 254. New York: Academic Press 1964.

[7] G. S. HAMMOND and N. J. TURRO: Science **142**, 1541 (1963).
[8] F. BACHÉR in: Handbuch Biol. Arbeitsmeth., Ed. by E. ABDERHALDEN, Abt. I, Teil 2/II, pp. 1339—1968; esp. p. 1817—1860. Berlin: Urban und Schwarzenberg 1929.
[9] J. WISLICENUS: Ber. Verhdl. Kgl. Sächs. Ges. Wiss. [Leipzig] **47**, 489 (1895); Ber. dtsch. chem. Ges. **29**, IV, 1080 (1897).
[10] F. WACHHOLTZ: Z. Physik. Chem. **125**, 1 (1927).
[11] R. STOERMER: Ber. dtsch. chem. Ges. **42**, 4865 (1909).
[12] H. STOBBE and F. K. STEINBERGER: Ber. dtsch. chem. Ges. **55**, 2225 (1922).
[13] J. BREGMAN, K. OSAKI, G. M. J. SCHMIDT and F. I. SONNTAG: J. Chem. Soc. **1964**, 2021.
[14] G. CIAMICIAN and P. SILBER: Ber. dtsch. chem. Ges. **36**, 4266 (1903).
[15] S. W. PELLETIER and W. L. MCLEISH: J. Amer. chem. Soc. **74**, 6292 (1952).
[16] R. E. LUTZ, P. S. BAILEY, C.-K. DIEN and J. W. RINKER: J. Amer. chem. Soc. **75**, 5039 (1953).
[17] H. NOZAKI, Y. NISIKAWA, Y. KAMATANI and R. NOYORI: Tetrahedron Letters **1965**, 2161.
[18] G. BÜCHI and N. C. YANG: Helv. Chim. Acta **38**, 1338 (1955).
[19] M. MOUSSERON-CANET, M. MOUSSERON and P. LEGENDRE: Bull. Soc. Chim. France **1961**, 1509.
[20] I. T. HARRISON and B. LYTHGOE: J. Chem. Soc. **1958**, 837.
[21] U. SCHWIETER, C. VON PLANTA, R. RÜEGG and O. ISLER: Helv. Chim. Acta **45**, 528 (1962).
[22] G. WALD in: Recent progress in photobiology, ed. by E. J. BOWEN, pp. 133—144. Oxford: Blackwell 1965.
[23] M. MOUSSERON: Pure Appl. Chem. **9**, 481 (1964).
[24] L. ZECHMEISTER: Cis-trans isomeric carotenoids, vitamins A, and arylpolyenes. Wien: Springer 1962.
[25] L. ZECHMEISTER and A. L. LEROSEN: Science **95**, 587 (1942).
[26] L. ZECHMEISTER and A. L. LEROSEN: J. Amer. chem. Soc. **64**, 2755 (1942).
[27] L. ZECHMEISTER and J. H. PINCKARD: J. Amer. chem. Soc. **76**, 4144 (1954).
[28] M. AKHTAR, T. A. RICHARDS and B. C. L. WEEDON: J. Chem. Soc. **1959**, 933.
[29] H. H. INHOFFEN and H. SIEMER in: Fortschr. Chem. Org. Naturstoffe, ed. by L. ZECHMEISTER, Vol. 9, pp. 1—40. Wien: Springer 1952.
[30] H. H. INHOFFEN, F. BOHLMANN and G. RUMMERT: Liebigs Ann. Chem. **571**, 75 (1951).
[31] W. R. BRODE, E. G. PEARSON and G. M. WYMAN: J. Amer. chem. Soc. **76**, 1034 (1954).
[32] W. LÜTTKE and M. KLESSINGER: Chem. Ber. **97**, 2342 (1964).
[33] G. M. WYMAN and A. F. ZENHÄUSERN: J. Org. Chem. **30**, 2348 (1965).
[34] G. M. WYMAN and W. R. BRODE: J. Amer. chem. Soc. **73**, 1487 (1951).
[35] D. A. ROGERS, J. D. MARGERUM and G. M. WYMAN: J. Amer. chem. Soc. **79**, 2464 (1957).
[36] G. S. HARTLEY: Nature **140**, 281 (1937).
[37] G. S. HARTLEY: J. Chem. Soc. **1938**, 633.
[38] A. H. COOK: J. Chem. Soc. **1938**, 876.
[39] L. ZECHMEISTER, O. FREHDEN and P. F. JÖRGENSEN: Naturwissenschaften **26**, 495 (1938).
[40] N. CAMPBELL, A. W. HENDERSON and D. TAYLOR: J. Chem. Soc. **1953**, 1281.
[41] M. FRANKEL, R. WOLOVSKY and E. FISCHER: J. Chem. Soc. **1955**, 3441.
[42] R. STOERMER: Ber. dtsch. chem. Ges. **44**, 637 (1911).
[43] O. L. BRADY and F. P. DUNN: J. Chem. Soc. **103**, 1619 (1913).
[44] O. L. BRADY and G. P. MCHUGH: J. Chem. Soc. **125**, 547 (1924).

[45] K. Koyano and I. Tanaka: J. Phys. Chem. **69**, 2545 (1965).
[46] A. W. Nineham: Chem. Rev. **55**, 355 (1955).
[47] I. Hausser, D. Jerchel and R. Kuhn: Chem. Ber. **82**, 515 (1949).
[48] R. Kuhn and H. M. Weitz: Chem. Ber. **86**, 1199 (1953).
[49] G. W. Griffin, E. J. O'Connell and H. A. Hammond: J. Amer. chem. Soc. **85**, 1001 (1963).
[50] G. W. Griffin, J. Covell, R. C. Petterson, R. M. Dodson and G. Klose: J. Amer. chem. Soc. **87**, 1410 (1965).
[51] G. O. Schenck, H. J. Kuhn and O.-A. Neumüller: Tetrahedron Letters **1961**, 12.
[52] H. J. Kuhn, O.-A. Neumüller and G. O. Schenck: Forschungsber. Land Nordrhein-Westfalen 1624; esp. p. 34. Köln: Westdeutscher Verlag 1966.
[53] A. Butenandt and L. Poschmann: Ber. dtsch. chem. Ges. **77**, 394 (1944).
[54] A. Butenandt, A. Wolff and P. Karlson: Ber. dtsch. chem. Ges. **74**, 1308 (1941).
[55] A. Butenandt, W. Friedrich and L. Poschmann: Ber. dtsch chem. Ges. **75**, 1931 (1942).
[56] H. Wehrli and K. Schaffner: Helv. Chim. Acta **45**, 385 (1962).
[57] D. H. R. Barton and G. Quinkert: J. Chem. Soc. **1960**, 1.
[58] D. Taub, C. H. Kuo, H. L. Slates and N. L. Wendler: Tetrahedron **19**, 1 (1963).
[59] K. Mislow, M. Axelrod, D. R. Rayner, H. Gotthardt, L. M. Coyne and G. S. Hammond: J. Amer. chem. Soc. **87**, 4958 (1965).

Chapter 8

Photodimerization with formation of cyclobutane derivatives (Cyclodimerization)

The formation of cyclobutane derivatives from unsaturated compounds is one of the oldest and most investigated reactions in photochemistry. This is explained not only by the large number and variety in type of the compounds which show the "cyclobutane reaction", but also by the fact

$$2 \quad \begin{matrix} \diagdown C \diagup \\ \| \\ \diagup C \diagdown \end{matrix} \quad \rightarrow \quad \begin{matrix} -C-C- \\ | \quad | \\ -C-C- \end{matrix}$$

that the 4-membered ring compounds formed are of great stereochemical interest and have, therefore, been extensively investigated (e.g. in the case of truxillic acids). It should also be emphasized that quite a few natural products (e.g. coumarin) undergo photochemical cyclodimerization reactions. Reactions of this type are, in some cases, of interest to the pharmaceutical chemist, as is demonstrated below with stilbamidine.

In many instances, it has been observed that the monomers can be obtained thermally from the photochemically prepared cyclobutane derivatives. The photodimer of benzo[b]thiophene 1,1-dioxide is an exception to this (cf. p. 90).

Many of the unsaturated acyclic compounds which can be photochemically converted to cyclobutane derivatives are also subject to photochemical cis-trans-isomerization (e.g. stilbene).

1. Stilbene and related compounds

Olefinic compounds of the general formula $Ar\overset{R}{C}=\overset{R'}{C}Ar'$, where Ar represents an aromatic or pyridyl radical, undergo photochemical cyclodimerization reactions with particular ease. In the following only a few outstanding examples will be discussed (cf. table 6).

References, pp. 94—96

Table 6

Stilbene [*1*]
Stilbamidine (1) [*2*]
Acenaphthylene (2) [*3*]
10-Benzylideneanthrone (3) [*4*]
Benzylidenephthalide (4) [*5*]
2,4-Dichloro-3-cyano-6-(β-styryl)-pyridine (5) [*6*]
2-(β-Styryl)-quinoline (6) [*7*]

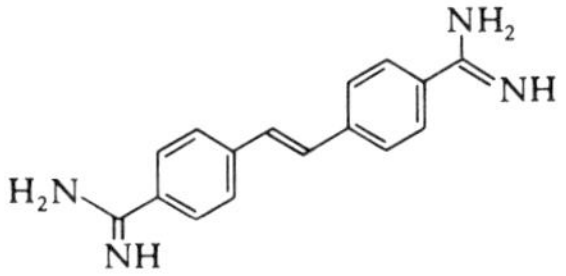

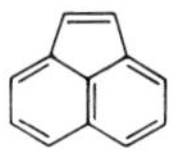

The photodimerization of stilbene was first observed by CIAMICIAN and SILBER [*1*] who exposed a benzene solution of stilbene to sunlight (Bologna) for two years.

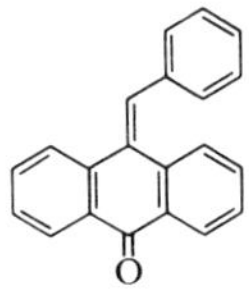

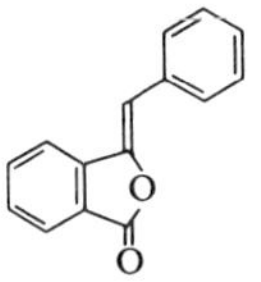

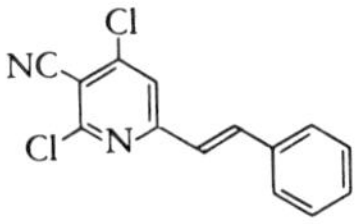

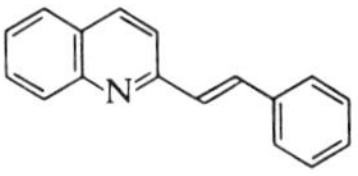

FULTON [*8*] exposed stilbene (1 g) in benzene (10 ml) to not very intense sunlight for six weeks in a sealed glass tube but did not observe any dimerization. Irradiation with a mercury vapor lamp, however, led to formation of the same product as that previously described by CIAMICIAN [*1*]. These results were confirmed by BAKER [*9*] who isolated cis,trans,cis-1,2,3,4-tetraphenylcyclobutane (7) in 20% yield. The stereochemistry of 7 was determined by crystallographic studies [*2, 10*]. SHECHTER [*11*] has obtained with a similar procedure trans,trans,trans-1,2,3,4-tetraphenylcyclobutane (8) in addition to 7, and its structure has been determined from the NMR spectrum.

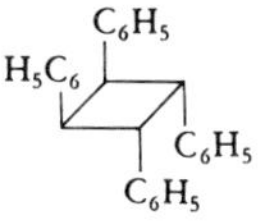

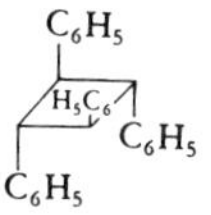

Stilbamidine (trans-4,4'-stilbenedicarboxamidine, 1) is used in the treatment of sleeping sickness.

BOWESMAN [12] observed that the toxicity of stilbamidine increased on standing, and FULTON and YORKE [13] showed that this increase in toxicity was due to photochemical changes, the exact nature of which was established by FULTON [2, 8] and HENRY [14]. Thus trans-stilbamidine (1) in aqueous solution was partly photoisomerized to the cis-isomer and partly photodimerized to cis,trans,cis-1,2,3,4-tetrakis-(4-amidinophenyl)-cyclobutane (9) the stereochemistry of which was ascertained by its degradation to 7.

Investigations on the photodimerization of acenaphthylene (2) were mainly carried out by DZIEWOŃSKI [3, 15] who found that acenaphthylene forms two dimeric products in sunlight for which the names α- and β-heptacyclene were suggested by DZIEWOŃSKI. Later on, X-ray diffraction studies [16] and ozonolysis work [17] showed that the high melting α-heptacyclene was, in fact, trans-6b,6c,12b,12c-tetrahydrocyclobuta[1,2-a: 3,4-a']diacenaphthylene (10) and the lower melting β-heptacyclene the corresponding cis-isomer (11).

The product ratio of the photodimers was shown to be greatly affected by the nature of the solvent used and by the concentration of the solution. Thus it was observed [15], that irradiation of dilute benzene solutions of 2 produced relatively more trans-isomer 10 than when concentrated solutions were used. The dependence of the photodimerization reaction on the nature of the solvent, on the concentration of the substrate, on oxygen being present in the solution etc. was reinvestigated by SCHENCK [18—20]. In the course of these studies it was established that the cis-isomer (11) was exclusively formed when benzene solutions of 2 were irradiated under oxygen, or when 2 was illuminated in hexane solution under argon. Besides, it was shown that the photodimerization was reversible: irradiation of 10 or 11 in solution led to the formation of 2. Since 11 is more soluble it is photolyzed preponderantly and thus the product ratio is shifted in favor of the trans-isomer. The photodimerization reaction was found to be susceptible to sensitization [19, 20].

References, pp. 94—96

2,4-Dichloro-3-cyano-6-(β-styryl)-pyridine dimer [6]. The pyridine derivative (5, 0.3 g) was exposed to diffuse sunlight (Vienna) between glass plates; after only a few hours the surface layer of the yellow compound had been bleached. Irradiation was carried out for several days. Microscopic examination showed decomposition of the original yellow needles to give colorless, opaque crystals. The melting point had risen from 169° to 212—213°. A colorless substance (0.23 g) was obtained on recrystallization from ethanol, m.p. 213—214°, after sintering.

Cis,trans,cis-1,2,3,4-tetraphenylcyclobutane (7) [9]. Stilbene (20 g) in warm benzene (150 ml) contained in an annular Pyrex vessel (internal diam. 5.5 cm) surrounding a Hanovia U.V.S. 500 UV-lamp was irradiated in an atmosphere of nitrogen for 48 hrs. Following evaporation of the solvent, the residue was shaken with ether and filtered. The insoluble fraction was recrystallized four times from ethanol to give 7 (4 g) (m.p. 162—163°). Stilbene (3 g) was obtained from the ether extract.

7 and **trans,trans,trans-1,2,3,4-tetraphenylcyclobutane (8)** [11]. Trans-stilbene (136 g) dissolved in benzene (1000 ml) was irradiated in a quartz flask for 2 months using two external ultraviolet lamps. Part of the solvent was removed and replaced by ether and ethanol. After the mixture had been stored at room temperature for 12—14 days, the solid deposit was filtered and washed with ether. The precipitate (36.5 g, 27% conversion) consisted of a mixture of the photodimers 7 and 8; trans-stilbene was absent. Recrystallization of the solid from ether/ethanol yielded pure 7 (18 g; m.p. 163°), pure 8 (5.0 g; m. p. 150°) and mixed dimers. The mixed dimers were retained and repeatedly treated to obtain pure 7 and 8, ether being a fairly good solvent for carrying out separation.

Cis,trans,cis-1,2,3,4-tetrakis-(4-amidinophenyl)-cyclobutane (9) [8]. 4,4′-Stilbenedicarboxamidine (1) diisethionate (di-2-hydroxyethanesulfonate) (50 g) was exposed to direct sunlight in aqueous solutions (1%) contained in Pyrex glass vessels; each vessel holding 50 ml of solution. On conclusion of the irradiation (1 week), an equal volume of 2 N sulfuric acid was added when, on standing, crystals (33 g) separated out. The solution was concentrated to 1800 ml under reduced pressure below 50° and a further yield of crystals (1.9 g) was obtained on standing. The product was recrystallized from water, in which it dissolves with difficulty, and obtained as short rod-like crystals, $C_{32}H_{32}N_8 \cdot 2H_2SO_4 \cdot 8H_2O$; m.p. 280—290°.

Trans- and cis-6b,6c,12b,12c-tetrahydrocyclobuta[1,2-a:3,4-a′]diacenaphthylene (10, 11) [18]. Acenaphthylene (2.28 g) was dissolved in carbon tetrachloride (150 ml) and the solution carefully flushed with argon to remove traces of oxygen. Exposure of the solution to the filtered light (UV Grün II Filterglas) of a mercury high pressure burner Philips HPK 125 W at 20° caused decoloration of the solution. After 12 hrs. the solvent was distilled in vacuo. Trituration of the residue with 20 ml benzene left 1.77 g (78%) needles of 10, subliming above 300°.

The benzene mother-liquor was evaporated in vacuo, the resulting residue washed with methanol and recrystallized from benzene to yield 0.412 g (18%) of 11, m.p. 230—234°.

2. Unsaturated hydrocarbons

Unsaturated hydrocarbons other than aromatically substituted have not been photodimerized until recently. On behalf of their low absorbancy direct photodimerization of aliphatic and alicyclic olefins is difficult. Usually photosensitizers have to be added in order to effect cyclodimerization. Our knowledge of this process is mainly derived from the investigations of SCHENCK and of HAMMOND and their schools. The results have acquired great significance not only for the theory of photosensitization but they are also of preparative importance. In some instances the composition of the reaction products was shown to depend on the nature of the sensitizer used.

a) Acyclic dienes

When 1,3-butadiene is subjected to irradiation in the presence of a photosensitizer [*21*] trans-1,2-divinylcyclobutane (**12**) is the major product; the cis-isomer (**13**) is formed in small, the third product, 4-vinylcyclohexene (**14**), in varying amounts (43 to 2 %). The composition of the product mixture depends strongly on the sensitizers employed. Isoprene, piperylene, and 2,3-dimethylbutadiene behaved analogously on irradiation.

Trans-1,2-divinylcyclobutane (12) [*21*]. Butadiene (37 g) saturated with 4,4'-bis-(dimethylamino)-benzophenone was added to a constricted 25 × 200 mm Pyrex test tube, frozen, degassed, and sealed under vacuum. The tube was irradiated for 50 hr. with a 450 Watt Hanovia arc provided with a Pyrex housing. The entire system was maintained below 30° by a water bath. The irradiated sample was heated at reflux for 30 min. to convert the cis-1,2-divinylcyclobutane (**13**) to cis,cis-1,5-cyclooctadiene and then fractionated through a spinning-band column. Trans-1,2-divinylcyclobutane (**12**) (19 g) in greater than 95—96% purity was obtained, b.p. 111—113°.

b) Cyclic olefins

Few cycloolefins have been cyclodimerized photochemically. In all cases the presence of a sensitizer was prerequisite. UV-irradiation of cyclopentene in acetone which acts not only as a solvent but also as the absorbing sensitizer, leads to a complex mixture of saturated and unsaturated hydrocarbons along with addition products of sensitizer to substrate [*22*]. Maximum yield in trans-tricyclo[5.3.0.0^{2,6}]decane (**15**) was 55 % of the hydrocarbon fraction. For details of the separation procedure the original literature should be consulted.

Similarly irradiation of norbornene (bicyclo[2.2.1]hept-2-ene, **16**) in acetone solution produced, inter alia, 2 different pentacyclo[6.4.1^{3,6}.1^{9,12}.0.0^{2,7}]-tetradecanes (**17**) [*23*]. For another photochemical synthesis of **17** see chapter 45.

References, pp. 94—96

c) Cyclic dienes

The sensitized photodimerization of cyclopentadiene [21] leads to the formation of endo-dicyclopentadiene (**18**), exo-dicyclopentadiene (**19**) and trans-tricyclo[5.3.0.0^{2,6}]deca-3,9-diene (**20**). These three dimers are obtained in approximately equal amounts, the nature of the sensitizer being without influence in this case.

The photosensitized cyclodimerization of 1,3-cyclohexadiene was first observed by SCHENCK [24]. If benzophenone is used as a sensitizer the dimerization proceeds almost quantitatively and may be conducted without a solvent. Products thus arising are cis,trans,cis- and cis,cis,cis-tricyclo-[6.4.0.0^{2,7}]dodeca-3,11-dienes **21** and **22**, respectively, together with the hitherto unknown exo dimer of the DIELS-ALDER type, exo-tricyclo-[6.2.2.0^{2,7}]dodeca-3,9-diene (**24**) which is to be regarded a true photoproduct. The (thermal) endo dimer (**23**) is not obtained in this way but may be produced exclusively from **21** by heating. Analogously **22** rearranges to **24** at 175°.

With lamps or sunlight quite large-scale reactions may be carried out, e.g., a 125 W mercury high pressure lamp produces about 220 g dimerizate, which is a mixture of isomers **21**, **22** and **24** in the ratio 60:15:25, in 48 hours. Preparative details concerning these dimerizations may be found in a paper by VALENTINE [25].

Photodimers of 1,3-cyclohexadiene [25]. Cyclohexadiene (100 g) and 2'-acetonaphthone (9 g) were dissolved in sufficient isopentane to give 500 ml of solution. Nitrogen was bubbled through the solution for 5 min. and the solution was then irradiated for 24 hr. with a 450 Watt medium pressure mercury arc housed in a Hanovia immersion reactor. The reaction mixture was concentrated by rotary evaporation with aspirator suction, then vacuum distilled to give, in addition to about 6 g of polymeric material, 92 g (92%) of dimeric products, b.p.$_1$ 34—39°. Separation of dimers was effected by preparative gas chromatography.

3. α,β-unsaturated ketones

A large number of α,β-unsaturated ketones have been converted to cyclobutane derivatives by the action of light. This cyclodimerization reaction, which has been reviewed upon by MUSTAFA [26], applies to acyclic (e.g. chalcones) as well as to cyclic unsaturated ketones (e.g. quinones).

a) Acyclic α,β-unsaturated ketones

CIAMICIAN and SILBER [27] found that insolation of dibenzylideneacetone (1,5-diphenylpentadien-3-one) in ethanol produced a dimer of m.p. 125—135°. If the ketone in glacial acetic acid solution is exposed to direct sunlight in the presence of uranyl chloride, the main product is a high melting dimer (m.p. 245°) [28]. Since oxidative degradation led to α-truxillic acid this dimer was ascribed the constitution 25. Its synthesis represents one of the earliest examples of the use of sensitizers in preparative organic photochemistry.

$$2\ C_6H_5-CH{=}CH-CO-CH{=}CH-C_6H_5 \xrightarrow{h\nu}$$

25

According to RECKTENWALD [29] the photodimerization of dibenzylideneacetone in isopropanol/benzene solution proceeds in a substantially different manner. The dimeric product, which is assumed to be identical with that obtained by CIAMICIAN and SILBER [27], melts at 139.5—140°. Oxidation with potassium permanganate yields δ-truxinic acid (27) and the dimer therefore has the structure 26.

26 27

1,3-Dicinnamoyl-2,4-diphenylcyclobutane (25) [28]. Dibenzylideneacetone (5 g) and uranyl chloride hydrate (8 g) dissolved in 100 ml of glacial acetic acid were exposed to direct sunlight. Two days later, a product had already separated out as well-defined crystals. Gentle warming accelerated the reaction but also increased the formation of resin. The crystalline product, except for a sparse residue, dissolved in boiling glacial acetic acid and separated out as colorless needles on cooling. It is almost insoluble in ethanol, ether, acids and alkali and easily soluble in chloroform. It melts at 245°, with partial decomposition, giving a yellow liquid which will not solidify after continued heating and subsequent cooling. If the compound is distilled it decomposes completely, a fraction of the distillate solidifying to give dibenzylideneacetone.

1,2-Dicinnamoyl-3,4-diphenylcyclobutane (26) [29]. Dibenzylideneacetone (20 g), dissolved in thiophene-free benzene (30 ml) and isopropanol (90 ml) was exposed to light from a quartz mercury lamp (Hanovia SH) for 90 hours. A pyrex plate was used as a filter to absorb short-wave radiation.

During irradiation the solution was maintained in an atmosphere of nitrogen and stirred with a magnetic stirrer. The reaction mixture became warmer under irradiation, and all the ketone had dissolved after eight hours. The temperature was then maintained at 25°, by cooling, and after 60 hours a white precipitate began to separate out. When the reaction was complete the solid residue (**26**) was filtered off and washed with cold ether, m. p. 139.5—140°; yield 6.1 g.

b) Cyclic α,β-unsaturated ketones

Cyclopentenone (**28**) furnishes, on exposure to light of wavelength above 300 mµ, an approximately equal mixture of two tricyclic dimers **29** and **30**. This cyclodimerization occurs in solution in a variety of solvents as well as in the pure liquid [*30*].

Irradiation of 3,5-dimethyl-2-cyclohexen-1-one (**31**) produced several dimers, two of which were shown to be cyclobutane derivatives. They were tentatively assigned head-to-head and head-to-tail dimeric structures [*31*].

Piperitone (**32**) on irradiation with a quartz mercury lamp gave rise to a mixture of three crystalline dimers. In a comparative experiment with sunlight, however, only one dimer was formed — in fact the dimer produced in the smallest quantity under UV-irradiation [*32*]. The structures of the dimers one of which is regarded a cyclobutane derivative have not yet been elucidated conclusively.

The photochemical cyclodimerization reaction has been applied to unsaturated steroid ketones as well. Thus testosterone esters (**33**) and androst-4-en-3-one (**34**) when irradiated yield one dimer of cyclobutane structure each [*33*, *34*]. A fair number of structures can be formulated for these dimers but with the aid of IR- and NMR-spectroscopy this number could be reduced to 3, viz. **35**—**37** [*35*].

Dimerization with formation of 4-membered homocyclic rings

35 **36** **37**

An even higher selectivity in product formation was observed when cholesta-4,6-dien-3-one (**38**) was irradiated. Only one dimer out of 20 possible was obtained [*36, 37*]. Interestingly enough the α,β-double bond of one monomer unit combines under the influence of light with the γ,δ-double bond of another **38** to yield a dimer of structures **39** or **40**.

38   **39**   **40**

The dimerization can be reversed both thermally and photochemically; in the latter case a photostationary state is attained which has the composition 31 % **38** and 69 % **39/40** [*37*].

Cis,trans,cis-tricyclo[5.3.0.0^{2,6}]decane-3,8-dione (29) and cis,trans,cis-tricyclo-[5.3.0.0^{2,6}]decane-3,10-dione (30) [*30*]. Cyclopentenone (82 g) was placed in a water-cooled Pyrex chamber and exposed to the radiation of a mercury arc lamp (Hanovia, 450 Watt) through two thicknesses of Pyrex glass. The sample remained at, or slightly below, room temperature and the irradiation was continued for 24 hours. The products were dissolved in methylene chloride and the solution was concentrated to 100 ml before dilution with carbon tetrachloride (300 ml). Further concentration removed most of the methylene chloride, in which the photo-dimers are remarkably soluble. On cooling, the concentrate deposited almost white crystals of **29**. These were washed free of adhering mother-liquor with cold carbon tetrachloride, recrystallized from that solvent and dried overnight in vacuo. Sublimation at 115° (0.5 mm Hg.) yielded material of high purity. In various runs, the yield ranged from 35 to 40 g (43—49%). Analytical material was prepared by recrystallization from methylene chloride/hexane followed by resublimation, m.p. 125—126.5°.

The mother-liquors from the recrystallization of **29** were concentrated under vacuum and then rapidly distilled at 125° (1 mm Hg.). The distillate, now free of the colored polymeric products of the photo-reaction, solidified with ease. After several recrystallizations from hexane, pure **30** (30—37 g, 37—45%) was obtained, m.p. 66—67°.

Androst-4-en-3-one dimer (35/37, R = C$_8$H$_{14}$) [*34*]. A solution of **34** (3 g) in 100 ml hexane was irradiated for 4^1/$_2$ hrs. with a mercury immersion lamp. On con-

References, pp. 94—96

centration of the solution to half its volume 103 mg of a dimer of the pinacol type separated which was filtered off. On further standing the hexane solution deposited 1.157 g of crystals (m.p. 80—95°) which were triturated with methanol. Recrystallization of the insoluble part from ethanol gave shining plates of **35/37**, R = C_8H_{14}, m. p. 289—291°.

Cholesta-4,6-dien-3-one dimer (39/40) [*36*]. 4.5 g of **38** in ethanol solution (200 ml) were irradiated with a mercury high pressure burner (Philips Biosol) at room temperature for 24 hours. At the end of the illumination 3.32 g crystalline **39/40** had precipitated, which after three recrystallizations from benzene/ethanol had m.p. 173—174°.

c) 1,4-Quinones

Irradiation of 1,4-benzoquinone in solution or as crystals, with light from a medium pressure mercury arc or with sunlight, was reported to yield an insoluble, non-sublimable polymer containing hydroxyl groups. No dimers could be detected. Similarly mono-, tri-, and tetramethyl-1,4-benzoquinones were converted to polymers by the action of light [*38*].

More recent investigations on the photochemistry of 1,4-benzoquinone have demonstrated that in fact 1,4-benzoquinone is able to photodimerize. Thus it has been shown by BRYCE-SMITH [*39*] that irradiation in molten maleic anhydride produces very low yields of a dimer which has been attributed the cage structure **41**, though also other interpretations have been offered [*105*].

Dialkyl derivatives of 1,4-benzoquinone differ from the aforementioned quinone derivatives insofar as irradiation produces a variety of dimers, among which are cage compounds, unsaturated cyclobutane derivatives, and oxetanes.

Sunlight converted crystalline 2,5-dimethyl-1,4-benzoquinone into a mixture of two dimers (one yellow, one colorless) in fairly good yield [*38*]. Under a mercury lamp, only a poor yield of the colorless isomer was produced. On the other hand, irradiation of a solution of the quinone in ethyl acetate produced a quantity of the yellow isomer. Above its m. p. (164°), this yellow dimer dissociated into the monomer. The stability to further irradiation of the crystals or of a solution and its thermal stability were taken as proof for structure **42**, but careful reinterpretation of NMR-spectra made clear that the dimer in fact was an oxetane (**43**) [*40*].

The structure of the colorless dimer must be left in doubt. The structures **44** and **45** have to be taken into consideration because pyrolysis of this cage dimer produced only the parent quinone with no trace of the 2,3-dimethyl-1,4-benzoquinone.

Irradiation of 2,6-dimethyl-1,4-benzoquinone [*38*] resulted in the formation of two unsaturated dimers and one saturated dimer of the cage type. One of the unsaturated dimers was shown to possess a cis-configuration of the two enedione moieties since on further irradiation it turned over into the cage compound. The other unsaturated dimer was shown later on [*40*] not to possess the trans-cyclobutane structure anticipated but to be an oxetane derivative.

Exposure of 2,3-dimethyl-1,4-benzoquinone to daylight produced a yellow dimer which readily formed a diacetate. Evidence favoring the enolic structure **46** has been obtained from the NMR-spectrum [*41*]. The ketoform of dimer **46** (viz. **47**) had previously been isolated by COOKSON [*38*] after irradiation of crystalline 2,3-dimethyl-1,4-benzoquinone. **47** was attributed a cis-configuration since on further irradiation it yielded a colorless cage compound (**48**).

The photodimer of thymoquinone (**49**) has been known for more than 80 years [*42*]. According to ZAVARIN [*43*] the formulation **50** is in agreement with the known facts but there is no evidence to decide whether the monomer units are joined head-to-head or head-to-tail (as in **50**). COOKSON [*38*] favors **50** since the dimer is relatively stable to heat and does not yield a saturated isomer on further irradiation.

1,4-Naphthoquinone in sunlight forms the colorless dimer **51** which decomposes at 270° into the monomer. The dimer is insoluble in alkali at room temperature [*44*]. With hot sodium hydroxide in aqueous dioxane, **51** gave an orange solution from which **52** was obtained by acidification [*41*].

51   52

A dimer similar to **51** was obtained from 2-methyl-1,4-naphthoquinone [*45*] but no dimerization occurred on exposure of 2,3-dimethyl-1,4-naphthoquinone (as crystals or in benzene) to sunlight [*44*].

Pentacyclo[6.4.0.0^{2,7}.0^{4,11}.0^{5,10}]dodecane-3,6,9,12-tetrone (41) [*39*]. A melt of 1,4-benzoquinone (15 g) and maleic anhydride (20 g) was irradiated in air, from 15 cm above, at 60—65° with a 500 W medium pressure mercury arc lamp for 1 hour. After solidification, the irradiated mixture was digested with ether. The pale yellow dimer (30 mg) was filtered off. After recrystallization from dimethyl sulfoxide **41** had m.p. 248° (dec.). Ultraviolet irradiation of crystalline or molten 1,4-benzoquinone produced the same dimer but in reduced quantities (ca. 20 mg in 1 hour). (However, see [*105*]).

1,4,4a,8b - Tetrahydro - 5,8 - dihydroxy - 2,3,6,7-tetramethylbiphenylene-1,4-dione (46) [*41*]. 2,3-Dimethyl-1,4-benzoquinone was exposed to daylight and the benzene-soluble fraction of the product was sublimed at 180°/0.01 mm to give the dimer **46** as yellow rhombic plates m.p. 229—229.5° (from benzene/light petroleum).

Cis - 4,5,10,11 - tetramethyltricyclo[6.4.0.0^{2,7}]dodeca - 4,10 - diene - 3,6,9,12-tetrone (47) [*38*]. The crystalline 2,3-dimethyl-1,4-benzoquinone was exposed to sunlight and then extracted with boiling ether. The cream-colored solid was removed and the filtrate evaporated. On chromatography of the residue on silica gel, unchanged quinone passed through in benzene. Benzene containing 10% of ethyl acetate eluted the cis-dimer **47**, which crystallized from ether in pale yellow cubes, m.p. 163—164°. **47** passes into **46** under the influence of alkali.

1,2,7,8-Tetramethylpentacyclo[6.4.0.0^{2,7}.0^{4,11}.0^{5,10}]dodecane-3,6,9,12-tetrone (48) [*38*]. The extract from several days' Soxhlet extraction of the above cream-colored solid yielded more of the dimer **47**. The insoluble white residue of **48** was purified by sublimation at 240°/20 mm. It began to decompose at ca. 270° with formation of a trace of yellow sublimate.

Thymoquinone dimer (50) [*42*]. Pure thymoquinone (**49**) is dissolved in absolute ether in such quantity that a solution of 1.5—2 g thymoquinone suffices to deposit the substance evenly on the inner walls of a round-bottomed flask of 3—4 litre capacity. This is exposed to sunlight and the originally yellow coating gradually becomes an opaque dull white. This conversion takes place during a few days exposure to diffuse daylight. The dimer is separated from the thymoquinone by flushing the flask with a small quantity of ether, in which **49** dissolves with ease, the dimer remaining undissolved. The dimer is purified [*43*] by recrystallization from isooctane/chloroform, m.p. 198—200°.

1,4-Naphthoquinone dimer (51) [*44*]. 1,4-Naphthoquinone (0.5 g) in benzene (12 ml) was exposed to sunlight (mid-November to mid-December, Cairo); colorless crystals began to form after a week. At the end of the experiment these were filtered off, washed with a little cold benzene and recrystallized from excess alcohol, m.p. 244—248° (decomp.), yield 0.2 g. A m.p. of 273—274° was reported for **51** by Bruce [*41*].

4. α,β-unsaturated acids and related compounds

a) Derivatives of maleic and fumaric acids

Photodimerization of maleic and fumaric acid derivatives has been used as a means of synthesizing the isomeric cyclobutanetetracarboxylic acids [46]. According to SCHENCK et al. [47] maleic anhydride and related compounds will dimerize to cyclobutane derivatives particularly well in the presence of sensitizers, e. g. benzophenone. The following table illustrates the scope of the reaction:

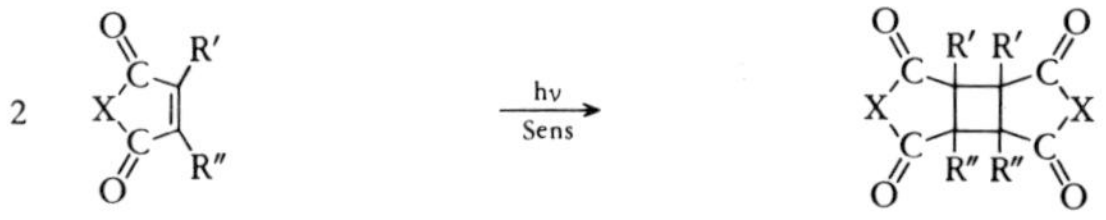

a: $R'=R''=CH_3$, $X=O$
b: $R'=CH_3$, $R''=H$ or vice versa, $X=O$
c: $R'=R''=H$, $X=O$
d: $R'=R''=CH_3$, $X=NH$
e: $R'=R''=H$, $X=N-C_6H_5$
f: $R'=R''=H$, $X=N-C_6H_{11}$

Crystalline dimethyl fumarate cyclodimerized to 1,2,3,4-tetracarbomethoxycyclobutane (**55**) under irradiation, as was found by GRIFFIN et al. [48]. The cis,trans,cis-configuration as in **55** was derived from X-ray crystallographic data for dimethyl fumarate [49]. Definite proof was obtained by direct comparison with an authentic sample of the "chair" form of 1,2,3,4-tetracarbomethoxycyclobutane (**55**) synthesized by CRIEGEE [50] from a cinnamic acid dimer of the same stereochemistry.

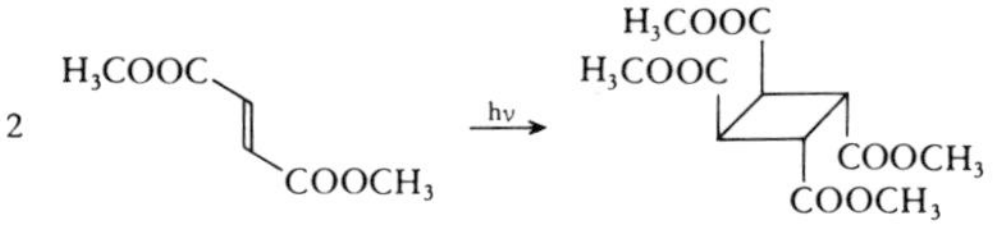

The irradiation of fumaronitrile (**56**) in the crystalline state [51] yielded cis,trans,cis-1,2,3,4-tetracyanocyclobutane (**57**). The stereochemistry was proved by its conversion to **55**, which had previously been prepared by CRIEGEE [50].

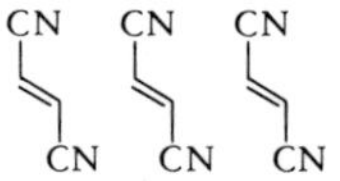

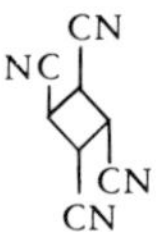

The formation of **57** provides additional evidence that the stereochemistry of photodimerization reactions conducted in the solid state are controlled by crystal lattice factors. In the crystal lattice the molecules of fumaronitrile are arranged as indicated in **56**. This orientation is completely consistent with the formation of a dimer having the structure **57** provided bond formation occurs between nearest neighbor molecules and no major shifts in the relative positions of the atoms take place during the reaction (cf. also COHEN and SCHMIDT [*52*]).

Tetramethylcyclobutanetetracarboxylic acid dianhydride (54a) [*47*]. **53a** (3.15 g) with benzophenone (1.5 g) in benzene (150 ml) was illuminated at 8° for 6 hrs. when **54a** (1.92 g) had separated out. After sublimation (230°/0.3 mm), **54a** melts above 380° with decomposition to **53a**.

0.3 g **54a** separated in 6 hrs. from the same mixture as that above, but without benzophenone.

Illumination was carried out under argon in an apparatus fitted with a water-jacketed glass immersion lamp (Mercury high-pressure lamp Philips HPK 125 W). During illumination a magnetic stirrer was used.

Cis,trans,cis-1,2,3,4-tetracarbomethoxycyclobutane (55) [*48*]. The solvent from a solution of 10 g of dimethyl fumarate in acetone was evaporated in a nitrogen stream while the container, a 500-ml gratuated cylinder, was rotated in a nearly horizontal position. A Westinghouse 15T8 Germicidal Sterilamp (maximum ultraviolet radiation (95%) at 253.7 mµ) was then inserted and irradiation was continued for a period of 1—5 days. During the course of the irradiation the temperature was maintained at 25—30° by immersing the tube in a large water-bath. On completion of the irradiation the dimer was separated from unreacted dimethyl fumarate and polymer by extraction with benzene, and finally recrystallized from the same solvent; m.p. 144—145°, yield 60%.

Cis,trans,cis-1,2,3,4-tetracyanocyclobutane (57) [*51*]. Fumaronitrile was deposited on the inside of a 1-l graduated cylinder and irradiated as in the foregoing experiment. After one week, the brown residue was removed from the tube and extracted with hot ether to remove starting material. The insoluble residue was recrystallized from dry acetonitrile, yielding 2.1 g (68% of converted **56**, 3.5% over-all) of a white crystalline solid, m.p. 250° (dec.).

b) Cinnamic acids

The photodimerization of cinnamic acid ranks among the best-investigated photochemical reactions [*1, 53—56*]. A comprehensive survey on work done until 1929 has been given by BACHÉR [*57*] and STOBBE and BREMER [*58*]. The cyclodimerization reaction is not confined to unsubstituted cinnamic acids; a wide variety of derivatives have been subjected to photochemical polymerization (cf. MUSTAFA [*26*]).

It was formerly accepted that the dimerizations could be represented by the following diagram [*59*]:

$$\begin{array}{ccc} \text{cis-cinnamic acid (solid)} & \rightleftarrows & \text{trans-cinnamic acid (solid)} \\ \updownarrow & & \updownarrow \\ \beta\text{-truxinic acid} & & \alpha\text{-truxillic acid} \end{array}$$

The assumption that trans-cinnamic acid (**58**) exclusively forms α-truxillic acid (**59**) by photodimerization, whilst β-truxinic acid (**61**) is exclusively formed from cis-cinnamic acid (**60**) could not, however, be substantiated. Both the dimorphism and the solid-state dimerization of **58** have been known for many years; nevertheless the precise details of the photoreaction were never agreed upon by the several schools of investigators, and even a more recent study [60] failed to settle the problem.

Ph-CH=CH-COOH   (Ar,COOH)₂ cyclobutane

58   **59**

cis-Ph-CH=CH-COOH   (Ar,COOH)₂ cyclobutane (β)

60   **61**

A re-investigation of the solid-state photodimerization [61] of the two crystal modifications of trans-cinnamic acid (**58**) has revealed that the stable α-trans-cinnamic acid dimerizes to α-truxillic acid (**59**) only; the metastable β-form yields pure β-truxinic acid (**61**) at temperatures below 50°, where the thermal β → α-transformation is slow. At higher temperatures the reaction product will be a mixture of **59** and **61** due to the β → α phase transformation with subsequent dimerization of the latter.

Ring-substituted cinnamic acids have been dimerized as well; depending upon the crystal modifications of the monomers derivatives of the α-truxillic or β-truxinic acid type were formed. A number of monomers occurred in three modifications of which the γ-form proved to be light-stable. X-ray crystallographic studies [62] revealed the following relationship between crystal geometry and photochemical behavior [63] (cf. table 7).

Table 7

packing type	nearest-neighbor relation	double-bond separation	photoproduct
α	centric	3.6—4.1 Å	α-truxillic acid
β	translation	3.9—4.1	β-truxinic acid
γ	translation	4.7—5.1	no reaction

Contrary to earlier assumptions the cis-cinnamic acids do not photodimerize directly; the first stage in the photoreaction of solid cis-cinnamic acids (cf. **60**) is the formation of the trans-acid, and this is followed by the crystallization of the trans-acid into its polymorphic forms which in turn undergo their characteristic photoreactions. Thus the present knowledge of the solid-state photochemistry of cinnamic acids may be summarized as follows [63]:

[cis] $\xrightarrow{h\nu}$ trans $\longrightarrow$ [trans]-α $\xrightarrow{h\nu}$ α-truxillic acid

[trans]-β $\xrightarrow{h\nu}$ β-truxinic acid

[trans]-γ   light-stable

where the species in square brackets indicate a molecule in its crystalline assembly, and the species without brackets a molecule in solid solution or some other 'disordered' state.

Although the photolysis of α-truxillic and β-truxinic acids (**59** and **61**, resp.) has not, formerly, been considered as important preparatively, it is mentioned here on account of its theoretical interest [*64, 65*]. Thus it was reported that α-truxillic acid when irradiated with UV-light in the solid state or in benzene suspension was converted to trans-cinnamic acid (**58**) while β-truxinic acid (**61**) on identical treatment was photolyzed to cis-cinnamic acid (**60**) [*65*].

α-Truxillic acid (59) [*61*]. α-trans-Cinnamic acid was grown in large crystals either by evaporation of a solution of **58** in ether, or from a benzene solution by slow cooling. The material must be checked for homogeneity. Samples are then irradiated in Cellophane-covered Petri-dishes or in a rotating Pyrex cylinder (sunlight, Israel). The powder was turned continuously (automatically) or once or twice daily (manually). The progress of the reaction was followed either by X-ray photography or by UV-spectroscopy. After 15 days the reaction was complete. Dimeric material was separated from residual monomer either by Soxhlet-extraction or by direct recrystallization, yield 74%, m.p. [*60*] 276—280°.

β-Truxinic acid (61) [*61*]. β-trans-Cinnamic acid (β-**58**) can best be prepared from concentrated solutions in ether which are rapidly filtered through a cotton plug previously washed with ether, and are then carefully overlaid with two to three volumes of light petroleum (30—60°), the whole operation being carried out at 0°. Large crystals of β-**58** grow over a period of a few days. This material being checked for homogeneity is irradiated as described above. Work-up after 15 days insolation yielded 3% **59** and 80% **61** besides small amounts of α-**58** when irradiation was carried out at 20°; at 50° the ratio was 46% **59**: 30% **61**. β-Truxinic acid was the sole product of an 18 days' irradiation of β-trans-cinnamic acid, if the reaction vessel consisted of a double walled, water-cooled quartz container; m.p. of the dimer [*60*] 208—210°.

5. Coumarin and isocoumarin

a) Coumarin

The photodimerization of coumarin (**62**) was first observed by CIAMICIAN and SILBER [*1*]. A cyclobutane structure for this coumarin dimer was first suggested by DE JONG [*66*] and this assumption was supported by SCHÖNBERG et al. [*5*], who observed that the coumarin dimer (and related substances, e.g. 3-phenylcoumarin dimer) split into two monomer units (**62**) on heating.

No conclusive evidence for the constitution of the photodimer of coumarin was obtained until 1960, when ANET [*67, 68*] fully elucidated its structure (**63**) as well as that of an isomer (**65**) obtained by lactonization of 2-hydroxycinnamic acid photodimer. The stereochemistry of the 4 possible coumarin dimers (**63—66**) is given in the chart on page 86.

Irradiation of coumarin in ethanol [*1*] or in aqueous suspension [*5*] produced the same dimer having structure **63**. More systematic investigations, however, demonstrated the unsensitized cyclodimerization of coumarin to be more complex in nature [*69*]. Thus irradiation of **62** in a variety of solvents yields dimers **63, 64,** and **66,** of which the latter had previously been overlooked since it forms an eutectic mixture (m.p. 261°) with **64** of the same m.p. as **63.** The relative amounts of the dimers obtained vary with the dielectricity of the solvent chosen.

If a solution of **62** in ethanol or benzene, containing sensitizers as benzophenone or β-carotene is irradiated the trans-head-to-head isomer **64** is predominantly formed together with a small quantity of **65** [*70*].

Coumarin dimers (63, 64 and 66) [*69*]. Coumarin (**62**, 14.6 g) in 200 ml ethylene glycol was irradiated for 60 hrs. at 35—40°. The solvent was evaporated in vacuo and the residue steam-distilled to remove the bulk of non-converted **62**. The dimeric material was dissolved in 120 ml 2N NaOH, the solution weakly acidified with 2N HCl and heated to 80° during 1 hour. The soluble part of this was neutralized with sodium bicarbonate and extracted with ether to remove residual **62**. The aqueous phase after acidification with strong HCl and extraction with ethyl acetate yielded material which upon heating to 120° during 20 mins. left **64** of m.p. 179—180° (from aqueous acetone); yield 1.23 g.

The precipitate of the above separation after boiling with ca. 100 ml benzene and subsequent cooling to 5—10° left 3.33 g **63** undissolved, m.p. 279—280° (hot stage) or 261—262° (capillary). The benzene mother-liquor was chromatographed on Florisil. Benzene/chloroform (4 : 1) eluted **66** of m.p. 204—206°, yield 1.08 g.

Coumarin dimer (64) [*70*]. 29 g **62** with 5 g benzophenone was illuminated under argon in 250 ml benzene at 10—15° for 60 hrs. (water-cooled glass immersion lamp Philips HPK 125 W). **64** (19.35 g) crystallized during the irradiation and a further 8.64 g were obtained from the solution giving a total yield of 27.9 g (96%). Recrystallized from ethanol, benzene or glacial acetic acid and sublimed in high vacuum, **64** had m. p. 176.5° (subl.).

Coumarin dimer (65) [*70*]. On recrystallization of the crude **64** (cf. preceding experiment) from ethanol, 0.45 g of **65** (1.5%) remained undissolved, m.p. 320—325°.

References, pp. 94—96

b) Isocoumarin

The photodimerization of isocoumarin has been little investigated. 3-Phenylisocoumarin (**67**) dimerizes in light, a reaction reversible by heat [*5*].

3-Phenylisocoumarin dimer [*5*]. A solution of **67** (1 g) in dry, thiophene-free benzene was exposed under carbon dioxide in sealed Pyrex tubes to sunlight during 25 days (November, Cairo). The precipitate was filtered off and washed several times with ether. Further yields were obtained from the concentrated filtrate by slow cooling. The photodimer crystallized in colorless crystals from benzene, m.p. 254°; yield almost quantitative.

c) Furocoumarins

Following the observation of WESSELY and DINJAŠKI [*71*] that the furocoumarin pimpinellin (**68**) in contrast to isopimpinellin (**69**), is prone to photodimerization, WESSELY and KOTLAN [*72*] investigated the possible connection between the ease of dimerization of the furocoumarins and their angular (cf. **72**) or linear (cf. **73**) structure. If such a relationship could have been detected, it would have served as a guide in structure elucidation of the naturally occuring furocoumarins, but no such relationship could be observed. In the isomeric pairs isobergapten (**70**) — bergapten (**71**) and angelicin (**72**) — psoralen (**73**) each of the compounds yielded a photodimer [*72, 73*]. Apparently the ease of photodimerization is unrelated to the spatial orientation of the rings. It should be noted, that the dimers could be converted to the corresponding monomers by pyrolysis.

A relationship between photodynamic action of furocoumarins and photodimerization reactions was suspected by MUSAJO et al. [*74*]. More detailed investigations [*73*], however, proved such correlations to be non-existant. Thus the ability to provoke erythemata in skin previously treated with furocoumarins and then exposed to light, decreased in the order psoralen (**73**) ≫ xanthotoxin (**74**) > bergapten (**71**) > angelicin (**72**) whereas the tendency to form photodimers decreases in the order psoralen > angelicin > bergapten > xanthotoxin.

According to SCHÖNBERG and SINA [75,76] xanthotoxin (74) is found in the seeds of Ammi majus L., a plant common in the Nile delta. The powdered seeds have been used for centuries by the Arabs as a remedy for pigmentation; the seeds were eaten and the depigmented areas of the skin then exposed to the sunlight. The areas thus treated showed blistering following pigmentation (but only after irradiation), then gradually assumed the color of the surrounding skin. Nowadays, xanthotoxin itself is used instead of the seeds and is obtained industrially. The chemical and clinical aspects of furocoumarin photosensitization have been discussed at the 1958 Kalamazoo conference on 'Psoralens and Radiant Energy' (cf. [77]).

As for xanthotoxin (74), RODIGHIERO and CAPPELLINA [73] were unable to isolate the photodimer as anticipated on account of spectroscopic changes in irradiated aqueous alcoholic solutions of 74. KRAUCH and FARID [78], by irradiation of xanthotoxin in dioxane solution, obtained a dimer which was ascribed the trans-head-to-head structure 75.

Possibly more detailed studies on the solvent dependence of product formation will reveal the photodimerization of furocoumarins to be as complex in nature as that of the coumarins. Thus pimpinellin (68) on irradiation forms two different dimers depending upon whether the monomer was irradiated in the solid state or in ethyl acetate solution [71].

5-Methoxypsoralen (bergapten) dimer [73]. A solution of bergapten (71, 0.5 g) in acetone (30 ml) was transferred to 3 glass dishes and allowed to evaporate. The resulting layer of crystalline 71 was irradiated with a 500 W quartz mercury lamp at a distance of 50 cm for 2 hours. Sublimation of unaltered 71 at 140°/0.0005 mm left a residue which after repeated recrystallizations from alcohol melted at 242°.

5-Methoxyangelicin (isobergapten) dimer [74]. Finely powdered isobergapten (70) (0.16 g) in a quartz flask filled with nitrogen was exposed to the unfiltered light of a mercury vapor lamp at a distance of 15—20 cm for 16 hours. The powder was shaken and cooled in an air-stream. A weakly-colored yellow product was obtained from which the unchanged 70 could be sublimed by heating to 180° at 0.005 mm. The fraction found to be non-volatile recrystallized from ethyl acetate and yielded pale yellow crystals of the dimer, m.p. 300—320° (dec.).

8-Methoxypsoralen (xanthotoxin) dimer (75) [78]. A solution of xanthotoxin (74) (2.16 g) in 200 ml dioxane was irradiated through a filter for 90 hours. The solvent was then removed and the residue warmed with 50 ml methanol. The insoluble part (410 mg) was recrystallized from glacial acetic acid to give 75, m.p. 303° (subl.).

References, pp. 94—96

6. 2,6-Dimethyl-4-pyrone

The photodimer of 2,6-dimethyl-4-pyrone (76) was first prepared by PATERNÒ [*79*]. GIUA and CIVERA [*80*] proposed structure 77 but according to YATES and JORGENSON [*81, 82*] the dimer has the cage structure 78. The best yields were obtained on irradiation of solid 76, or of aqueous solutions, but dimerization occurred also in concentrated solutions in ethanol, benzene or acetic acid, yields being, however, inferior. 78 did not revert to the monomer upon exposure to the same light source in solution or in the solid state, but depolymerization was effected by warming with dilute acid.

It should be noted that irradiation of 76 in very dilute (0.2 %) aqueous solution led to 78 in addition to a rearrangement product, 4,5-dimethyl-2-furaldehyde (79) [*83*].

2,4,8,10-Tetramethyl-3,9-dioxapentacyclo[6.4.0.0^{2,7}.0^{4,11}.0^{5,10}]dodecane-6,12-dione (78) [*82*]. The pyrone was irradiated in the solid state in thin layers on a large surface with a General Electric 250 Watt Sunlamp. The layers were frequently turned to expose new surfaces to the radiation. After a period of about three weeks the total solid was triturated twice with acetone to dissolve unreacted starting material. The insoluble white solid was filtered off and thoroughly washed with warm acetone. In this way the pure photodimer could be isolated in yields of about 30%. 78 crystallized from dimethylformamide in large, flat needles which sublimed when heated above 200°; m.p. 281—284° (dec.).

7. Tetraphenylbutatriene

Tetraphenylbutatriene (80) was photodimerized as early as 1921 [*84*] but the constitution of the dimer as 81 has only recently been established [*85*].

81 is obtained by sunlamp irradiation or in sunlight. The dimer decomposes to 80 on melting; it does not form adducts with maleic anhydride or tetracyanoethylene.

Tetrakis-(diphenylmethylene)-cyclobutane (81) [*84*]. Finely powdered tetraphenylbutatriene (**80**) was exposed to sunlight on a watch-glass, with continuous mixing during the summer months. It assumed a dirty yellow color. In order to remove unchanged tetraphenylbutatriene, the irradiated powder was first extracted with methyl ethyl ketone in a Soxhlet apparatus and the undissolved residue further extracted with chloroform. The crystals obtained had a strong green fluorescence and were recrystallized from hot chloroform, m.p. 280—281°. UHLER, SHECHTER and TIERS [*85*] report m.p. 290—293°.

8. Unsaturated sulfur compounds

a) Benzo[b]thiophene 1,1-dioxide

The 1,1-dioxide of benzo[b]thiophene (**82**) forms a dimeric product under the action of sunlight [*86*]. This dimer was formulated as **83** or **84** by DAVIES and JAMES [*87*]. The sulfones of 3-bromo-, 3-methyl- and 3,4-dimethylbenzo[b]thiophene behave similarly.

The dimer sublimes at 260—270°/1 mm; on heating in butyl phthalate, however, benzo[b]naphtho[1,2-d]thiophene 7,7-dioxide (**85**) is obtained, which probably arises in a thermal reaction from monomer **82** formed during the pyrolysis.

Benzo[b]thiophene 1,1-dioxide dimer (83 or 84) [*87*]. An almost saturated solution of the sulfone **82** (10 g) in benzene in a quartz flask was exposed to direct sunlight for 20 days. The dimer separated out and was recrystallized from a large quantity of acetone, m.p. 330—331°; yield 5 g.

b) 2-Nitrobenzo-1,4-dithiin

When the red 2-nitrobenzo-1,4-dithiin (**86**) is exposed to light it is converted to the yellow dimer **87** (or **88**) [*88*]. The proposed structure is based on the molecular weight determination and on inspection of its infrared spectrum, which shows the nitro group to have become less conjugated as a consequence of the dimerization.

References, pp. 94—96

5a,11a- or **5a,5b-Dinitrocyclobuta[1,2-b:3,4-b']bis[1,4]benzodithiin (87** or **88)**
[*88*]. **86** (0.40 g) was exposed to sunlight for 15 days during which time the color changed from red to light brown. The resulting brown solid was extracted with several portions of benzene; the insoluble residue was discarded. The combined benzene extract was evaporated and the resulting light tan solid was washed with ethanol (two 1 ml portions) to remove any unchanged starting material. The residue was recrystallized from ethanol giving 0.274 g (69%) of pale yellow needles, m.p. 170.5—172° (dec.).

9. Thymine and related pyrimidine derivatives

BEUKERS and BERENDS [*89*] found that UV-irradiation of thymine (**89**) in a frozen aqueous solution leads to the formation of a dimer. This process is reversible since renewed irradiation of the thawed solution of the reaction product leads to thymine again.

The constitution of the thymine dimer was assumed [*89*] to be **90**, but WULFF and FRAENKEL [*90*] with the aid of NMR spectroscopy showed that evidence in favor of **90** was not compelling. Instead, structure **91** was preferred; final proof was obtained only recently by BLACKBURN and DAVIES [*91*] who found the stereochemistry of the dimer to be cis,cis,cis (**92**).

The formation of **92** is of great biochemical interest; it was suggested [*92, 93*] that the photodimerization is responsible for the photoinactivation of deoxyribonucleic acid (DNA). This opinion has been approved [*94*] as well as rejected [*95*].

Irradiation of pyrimidine derivatives other than **89** and of nucleotides often results in dimer formation, this sometimes competing with addition of water across the 5,6-double bond (cf. p. 155). An excellent review of the whole field has been given by MCLAREN and SHUGAR [*96*]. The mechanism of the dimerization process seems still open to discussion (cf. [*97*]).

Thymine dimer (92) [*89*]. Since thymine shows a very high molecular extinction only dilute solutions can be used in the irradiation procedure (400 mg/liter). After being frozen in large Petri-dishes (35 ml portions) the thymine solution was subjected to UV-light of 254 mμ. The solutions were allowed to thaw and the irradiation product was separated from thymine by repeated extraction with absolute ethanol. In contrast to thymine, the product is almost insoluble in this solvent. The completeness of the extraction was determined by measuring the extinction of the extract. After complete removal of thymine, (no residual absorption at 264 mμ), the residue was recrystallized from water to yield small needles, decomp. 320°.

10. β-Lumicolchicine

Irradiation of colchicine **93** gives three crystalline photoproducts, viz. α-, β- (**94**) and γ-lumicolchicine, which are, in part, photochemically and thermally interconvertible. The exact nature of these consecutive reactions has first been investigated by SCHENCK and collaborators [*98—100*]. These transformations of colchicine and its photoproducts are illustrated on p. 8.

The principal difficulty in the earlier investigations on α-lumicolchicine was its preparation; yields reported varied from 0% to 31%. Choice of appropriate experimental conditions, for example, use of filter solutions, allowed its preparation in yields of up to 58 %. α-Lumicolchicine may be prepared in 38 % yield, when concentrated solutions of **94** are irradiated, the formation being reversible either by heating or by irradiating dilute solutions of α-lumicolchicine. Though for reasons of its low solubility in the usual solvents a dimeric structure for α-lumicolchicine seemed most probable, results of molecular weight determinations were ambiguous and seemed to depend on the method used [*98, 99*].

CHAPMAN and coworkers [*101, 102*] found α-lumicolchicine to be **95**, a cyclobutane dimer of β-lumicolchicine (**94**).

α-Lumicolchicine [*99*]. β-Lumicolchicine (600 mg) in methanol (300 ml) was irradiated in an argon atmosphere with a mercury burner Philips HPK 125 W (water-jacketed immersion lamp, Solidex glass). After 36 mins. irradiation time 230 mg (38.3%) of α-lumicolchicine had precipitated, m.p. 157°.

11. Acetylene compounds

Although the dimers isolated following irradiation of phenylacetylene and diphenylacetylene, respectively, are not cyclobutane derivatives, these photoreactions should be treated here since substituted cyclobutadienes may be regarded as the primary products.

References, pp. 94—96

Irradiation of diphenylacetylene (**96**) in hexane solution [*103*] gave 1,2,3-triphenylazulene (**98**) as well as 1,2,3-triphenylnaphthalene (**99**) which are dimers of **96**. Furthermore hexaphenylbenzene (**100**), a trimer of **96**, was also formed.

BÜCHI [*103*] surmises that tolane (**96**) is dimerized to tetraphenylcyclobutadiene (**97**) (not isolated). By reaction of **97** with diphenylacetylene, hexaphenylbenzene (**100**) could be formed. BÜCHI has also proposed theories, to which reference may be made, to account for the formation of **98** and **99**.

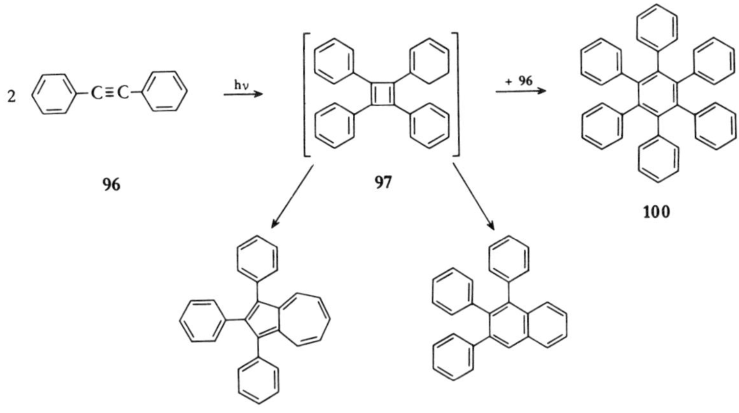

UV-irradiation of phenylacetylene [*104*] in cyclohexane gave 1-phenylnaphthalene (**101**) and 1-phenylazulene (**102**) in the ratio 1 : 5. **101** and **102** were found not to be inconvertible when irradiated separately in cyclohexane. Each compound was free from any detectable amount of nuclear isomers so the dimerization is evidently stereospecific. The following reaction mechanism was suggested [*104*]:

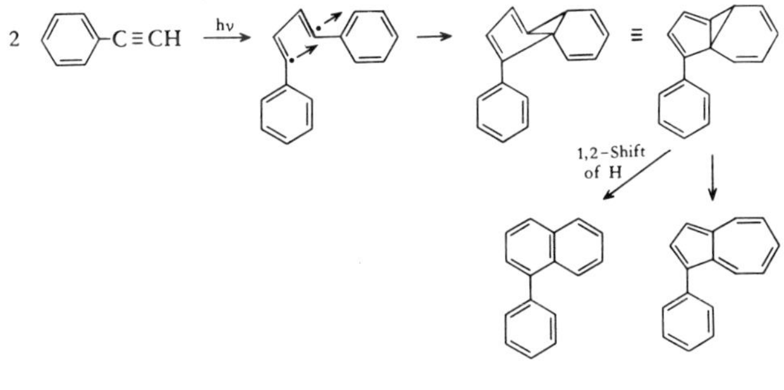

Repetition of the above experiment in benzene solution resulted in concomitant formation of phenylcyclooctatetraene (by cycloaddition to the solvent) and of other unidentified products. The same ratio (1 : 5) cf **101** : **102** was observed as in the cyclohexane experiment.

1,2,3-Triphenylazulene (98) and **1,2,3-triphenylnaphthalene (99)** [*103*]. Diphenylacetylene (**96**) (17 g) dissolved in pure hexane (50 ml) was irradiated with a mercury lamp S 81 (Quarzlampenges.) for 1 week. "Prepurified" nitrogen was bubbled through the solution and the apparatus was cooled to 15—30° using a water bath. The reaction mixture was filtered to remove insoluble material (92 mg) and the filtrate was concentrated to a green solid.

Recrystallization from ethanol (50 ml) removed unchanged starting material and the mother-liquor was chromatographed in hexane solution over a column of "Davison 923" silica gel. Elution with the same solvent gave **96** (16.0 g). Further elution with hexane containing 5% benzene yielded 1,2,3-triphenylnaphthalene (**99**) (172 mg.), m.p. 151—153°. Later fractions contained the blue **98** (151 mg) which after recrystallization from ether-pentane had m.p. 214—216°.

References

[*1*] G. CIAMICIAN and P. SILBER: Ber. dtsch. chem. Ges. **35**, 4128 (1902).
[*2*] J. D. FULTON and J. D. DUNITZ: Nature **160**, 161 (1947).
[*3*] K. DZIEWOŃSKI and G. RAPALSKI: Ber. dtsch. chem. Ges. **45**, 2491 (1912).
[*4*] A. MUSTAFA and A. M. ISLAM: J. Chem. Soc. **1949**, Suppl. 81.
[*5*] A. SCHÖNBERG, N. LATIF, R. MOUBASHER and W. I. AWAD: J. Chem. Soc. **1950**, 374.
[*6*] E. SPÄTH and G. KOLLER: Ber. dtsch. chem. Ges. **58**, 2124 (1925).
[*7*] M. HENZE: Ber. dtsch. chem. Ges. **70**, 1273 (1937).
[*8*] J. D. FULTON: Brit. J. Pharmacol. **3**, 75 (1948).
[*9*] W. BAKER, J. W. HILPERN and J. F. W. McOMIE: J. Chem. Soc. **1961**, 479.
[*10*] J. D. DUNITZ: Acta Cryst. **2**, 1 (1949).
[*11*] H. SHECHTER, W. J. LINK and G. V. D. TIERS: J. Amer. chem. Soc. **85**, 1601 (1963).
[*12*] C. BOWESMAN: Ann. Trop. Med. Parasitol. **34**, 217 (1940).
[*13*] J. D. FULTON and W. YORKE: Ann. Trop. Med. Parasitol. **36**, 134 (1942).
[*14*] A. J. HENRY: J. Chem. Soc. **1946**, 1156.
[*15*] K. DZIEWOŃSKI and C. PASCHALSKI: Ber. dtsch. chem. Ges. **46**, 1986 (1913).
[*16*] J. D. DUNITZ and L. WEISSMAN: Acta Cryst. **2**, 62 (1949).
[*17*] G. W. GRIFFIN and D. F. VEBER: J. Amer. chem. Soc. **82**, 6417 (1960).
[*18*] G. O. SCHENCK, W. HARTMANN and I.-M. HARTMANN: unpublished results.
[*19*] G. O. SCHENCK and R. WOLGAST: Naturwissenschaften **49**, 36 (1962).
[*20*] I.-M. HARTMANN: Dissertation Göttingen 1964.
[*21*] G. S. HAMMOND, N. J. TURRO and R. S. H. LIU: J. Org. Chem. **28**, 3297 (1963).
[*22*] H.-D. SCHARF and F. KORTE: Chem. Ber. **97**, 2425 (1964).
[*23*] D. SCHARF and F. KORTE: Tetrahedron Letters **1963**, 821.
[*24*] G. O. SCHENCK, S.-P. MANNSFELD, G. SCHOMBURG and C. H. KRAUCH: Z. Naturforsch. **19 b**, 18 (1964).
[*25*] D. VALENTINE, N. J. TURRO, JR. and G. S. HAMMOND: J. Amer. chem. Soc. **86**, 5202 (1964).
[*26*] A. MUSTAFA: Chem. Rev. **51**, 1 (1952).
[*27*] G. CIAMICIAN and P. SILBER: Ber. dtsch. chem. Ges. **42**, 1386 (1909).
[*28*] P. PRAETORIUS and F. KORN: Ber. dtsch. chem. Ges. **43**, 2744 (1910).
[*29*] G. W. RECKTENWALD, J. N. PITTS, JR. and R. L. LETSINGER: J. Amer. chem. Soc. **75**, 3028 (1953).

[30] P. E. Eaton: J. Amer. chem. Soc. **84**, 2344 (1962).
[31] W. Treibs: J. prakt. Chem. [2], **138**, 299 (1933).
[32] W. Treibs: Ber. dtsch. chem. Ges. **63**, 2738 (1930).
[33] A. Butenandt and A. Wolff: Ber. dtsch. chem. Ges. **72**, 1121 (1939).
[34] A. Butenandt, L. Karlson-Poschmann, G. Failer, U. Schiedt and E. Biekert: Liebigs Ann. Chem. **575**, 123 (1952).
[35] B. Nann, D. Gravel, R. Schorta, H. Wehrli, K. Schaffner and O. Jeger: Helv. chim. acta **46**, 2473 (1963).
[36] H. P. Throndsen, G. Cainelli, D. Arigoni and O. Jeger: Helv. chim. acta **45**, 2342 (1962).
[37] M. B. Rubin, G. E. Hipps and D. Glover: J. Org. Chem. **29**, 68 (1964).
[38] R. C. Cookson, D. A. Cox and J. Hudec: J. Chem. Soc. **1961**, 4499.
[39] D. Bryce-Smith and A. Gilbert: J. Chem. Soc. **1964**, 2428.
[40] R. C. Cookson, J. J. Frankel and J. Hudec: Chem. Comm. **1965**, 16.
[41] J. M. Bruce: J. Chem. Soc. **1962**, 2782.
[42] C. Liebermann and M. Ilinski: Ber. dtsch. chem. Ges. **18**, 3193 (1885).
[43] E. Zavarin: J. Org. Chem. **23**, 47 (1958).
[44] A. Schönberg, A. Mustafa, M. Z. Barakat, N. Latif, R. Moubasher and A. Mustafa: J. Chem. Soc. **1948**, 2126.
[45] J. Madinaveitia: Rev. Real Acad. Cienc. Fis. Nat. Madrid **31**, 617 (1934); C. A. **29**, 5438 (1935).
[46] G. W. Griffin, J. E. Basinski and A. F. Vellturo: Tetrahedron Letters **3**, 13 (1960).
[47] G. O. Schenck, W. Hartmann, S.-P. Mannsfeld, W. Metzner and C. H. Krauch: Chem. Ber. **95**, 1642 (1962).
[48] G. W. Griffin, A. F. Vellturo and K. Furukawa: J. Amer. chem. Soc. **83**, 2725 (1961).
[49] I. E. Knaggs and K. Lonsdale: J. Chem. Soc. **1942**, 417.
[50] R. Criegee and H. Höver: Chem. Ber. **93**, 2521 (1960).
[51] G. W. Griffin, J. E. Basinski and L. I. Peterson: J. Amer. chem. Soc. **84**, 1012 (1962).
[52] M. D. Cohen and G. M. J. Schmidt: in: Reactivity of Solids, ed. by J. H. de Boer. Amsterdam: Elsevier **1961**, p. 556.
[53] C. N. Riiber: Ber. dtsch. chem. Ges. **35**, 2908 (1902).
[54] A. W. K. de Jong: Verslag. gewone Vergader. Wis. Natuurk. Afdeel. K. N. A. W. **20**, 55 (1911).
[55] H. Stobbe: Ber. dtsch. chem. Ges. **52**, 666 (1919).
[56] R. Stoermer and E. Laage: Ber. dtsch. chem. Ges. **54**, 77 (1921).
[57] F. Bachér, in: Handbuch der biologischen Arbeitsmethoden, ed. by E. Abderhalden, Abt. I, Teil 2/II, pp. 1339—1968; esp. p. 1826. Berlin: Urban und Schwarzenberg 1929.
[58] H. Stobbe and K. Bremer: J. prakt. Chem. **123** (231), 1 (1929).
[59] H. Stobbe and F. K. Steinberger: Ber. dtsch. chem. Ges. **55**, 2225 (1922).
[60] H. I. Bernstein and W. C. Quimby: J. Amer. chem. Soc. **65**, 1845 (1943).
[61] M. D. Cohen, G. M. J. Schmidt and F. I. Sonntag: J. Chem. Soc. **1964**, 2000.
[62] G. M. J. Schmidt: J. Chem. Soc. **1964**, 2014.
[63] J. Bregman, K. Osaki, G. M. J. Schmidt and F. I. Sonntag: J. Chem. Soc. **1964**, 2021.
[64] R. Stoermer and F. Foerster: Ber. dtsch. chem. Ges. **52**, 1255 (1919).
[65] H. Stobbe and A. Lehfeldt: Ber. dtsch. chem. Ges. **58**, 2415 (1925).
[66] A. W. K. de Jong: Rec. Trav. chim. Pays-Bas **43**, 316 (1924).
[67] R. Anet: Chem. and Ind. **1960**, 897.

[68] R. ANET: Can. J. Chem. **40**, 1249 (1962).
[69] C. H. KRAUCH, S. FARID and G. O. SCHENCK: Chem. Ber. **99**, 625 (1966).
[70] G. O. SCHENCK, I. VON WILUCKI and C. H. KRAUCH: Chem. Ber. **95**, 1409 (1962).
[71] F. WESSELY and K. DINJAŠKI: Mh. Chem. **64**, 131 (1934).
[72] F. WESSELY and J. KOTLAN: Mh. Chem. **86**, 430 (1955).
[73] G. RODIGHIERO and V. CAPPELLINA: Gazz. chim. Ital. **91**, 103 (1961).
[74] L. MUSAJO, G. RODIGHIERO and G. CAPORALE: Bull. Soc. Chim. Biol. **36**, 1213 (1954).
[75] A. SCHÖNBERG and A. SINA: Nature **161**, 481 (1948).
[76] A. SCHÖNBERG and A. SINA: J. Amer. chem. Soc. **72**, 4826 (1950).
[77] KALAMAZOO CONFERENCE: J. Invest. Dermatol. **32**, pp. 131—391 (1959).
[78] C. H. KRAUCH and S. FARID: Chem. Ber. **100**, 1685 (1967).
[79] E. PATERNÒ: Gazz. chim. Ital. **44**, I, 151 (1914).
[80] M. GIUA and M. CIVERA: Gazz. chim. Ital. **81**, 875 (1951).
[81] P. YATES and M. J. JORGENSON: J. Amer. chem. Soc. **80**, 6150 (1958).
[82] P. YATES and M. J. JORGENSON: J. Amer. chem. Soc. **85**, 2956 (1963).
[83] P. YATES and I. W. J. STILL: J. Amer. chem. Soc. **85**, 1208 (1963).
[84] K. BRAND: Ber. dtsch. chem. Ges. **54**, 1987 (1921).
[85] R. O. UHLER, H. SHECHTER and G. V. D. TIERS: J. Amer. chem. Soc. **84**, 3397 (1962).
[86] A. MUSTAFA: Nature **175**, 992 (1955).
[87] W. DAVIES and F. C. JAMES: J. Chem. Soc. **1955**, 314.
[88] W. E. PARHAM, P. L. STRIGHT and W. R. HASEK: J. Org. Chem. **24**, 262 (1959).
[89] R. BEUKERS and W. BERENDS: Biochim. Biophys. Acta **41**, 550 (1960).
[90] D. L. WULFF and G. FRAENKEL: Biochim. Biophys. Acta **51**, 323 (1961).
[91] G. M. BLACKBURN and R. J. H. DAVIES: Chem. Comm. **1965**, 215.
[92] A. WACKER, H. DELLWEG and E. LODEMANN: Angew. Chem. **73**, 64 (1961).
[93] A. WACKER, H. DELLWEG, L. TRÄGER, A. KORNHAUSER, E. LODEMANN, G. TÜRCK, R. SELZER, P. CHANDRA and M. ISHIMOTO: Photochem. Photobiol. **3**, 369 (1964).
[94] J. K. SETLOW: Photochem. Photobiol. **3**, 405 (1964).
[95] C. S. RUPERT: Photochem. Photobiol. **3**, 399 (1964).
[96] A. D. MCLAREN and D. SHUGAR: Photochemistry of proteins and nucleic acids. Oxford: Pergamon 1964; esp. p. 184.
[97] S. Y. WANG: Photochem. Photobiol. **3**, 395 (1964).
[98] G. O. SCHENCK, H. J. KUHN and O.-A. NEUMÜLLER, Tetrahedron Letters **1961**, 12.
[99] H. J. KUHN: Dissertation Göttingen 1964.
[100] H. J. KUHN, O.-A. NEUMÜLLER and G. O. SCHENCK: Forschungsber. Land Nordrhein-Westfalen **1624**, Köln: Westd. Verlag 1966.
[101] O. L. CHAPMAN and H. G. SMITH: J. Amer. chem. Soc. **83**, 3914 (1961).
[102] O. L. CHAPMAN, H. G. SMITH and R. W. KING: J. Amer. chem. Soc. **85**, 806 (1963).
[103] G. BÜCHI, C. W. PERRY and E. W. ROBB: J. Org. Chem. **27**, 4106 (1962).
[104] D. BRYCE-SMITH and J. E. LODGE: J. Chem. Soc. **1963**, 695.
[105] E. H. GOLD and D. GINSBURG: J. Chem. Soc. (C) **1967**, 15

Chapter 9

Photodimerizations involving formation of eight-membered rings

1. Carbocyclic aromatic compounds

a) Naphthalene derivatives

Naphthalene has hitherto not been reported to undergo photodimerization but, according to BRADSHAW and HAMMOND [1], 2-methoxynaphthalene (1) dimerizes upon irradiation with ultraviolet light in a variety of solvents. Other naphthalenes carrying methyl, hydroxy, bromo or amino substituents in the 1- or 2-positions resist photodimerization as does also 1-methoxynaphthalene. The photodimer from 1 was regarded to have the structure of a 5,8,9,12-tetrahydro-2,14- or 2,15-dimethyl-1,4[1′,4′]naphthalenonaphthalene (syn or anti, 2 or 3).

2-Methoxynaphthalene dimer (2 or 3) [1]. Two grams of 1 was dissolved in a mixture of 10 ml of benzene and 10 ml of 2-propanol. The mixture was placed in a Pyrex tube, degassed, sealed in vacuo, and irradiated using a Hanovia quartz immersion reactor. The progress of the reaction was followed by observing a slow separation of a white solid. After 2 weeks of continuous irradiation, 900 mg (45%), and after about 2 months 90%, of solid **2** or **3**, m.p. 155—160° (dec.) was isolated. The product reverted to the starting material, **1**, m.p. 71—72°, on heating above the decomposition point.

b) Anthracene and derivatives

Anthracene (4) forms a dimer under the action of sunlight or UV light, a process which belongs to the longest known and most widely studied photochemical reactions; numerous investigations followed the first report by FRITZSCHE [2]. The photodimerization takes place in a variety of solvents among which are benzene, xylene and acetic acid. Photodimerization has also been reported to take place in the solid state [3]. Since the dimeric product is far less soluble than is anthracene, it may be easily isolated. CAPPER and MARSH [4] have used the formation of dianthracene in the purification of phenanthrene and fluorene from anthracene traces. On heating the dimer decomposes to give the monomer.

The structure 5 was proposed by LINEBARGER [5] and ORNDORFF and CAMERON [6] and confirmed by physical means (comp. [7]). The mechanism of the anthracene photodimerization has received much attention; BOWEN and TANNER [8] and LIVINGSTON [9] have produced evidence for a singlet mechanism in the photodimerization, and the rôle of excimers in anthracene photodimerization has been discussed by BIRKS et al. [10].

4: R = H

5: R = H

6
7: R = CHO
8: R = NH$_2$

Many anthracene derivatives with one or two substituents at the 1-, 2-, and 9-positions form photodimers. A survey of such derivatives covering, among others, bromo, chloro, methyl, ethyl, carboxy and formyl substituents, has been given by GREENE et al. [11] and LALANDE and CALAS [12].

9-Monosubstituted anthracenes could give dimers having either cis (head-to-head, cf. 5) or trans (head-to-tail, cf. 6) structures but until now only dimers of the latter type have been encountered. Thus, 9-anthraldehyde dimer, which was earlier [11] supposed to be cis, was later shown to have the trans structure (7) [13, 14]. Electric moment measurements greatly helped in the elucidation of the dimer structures [15].

Compounds resisting photodimerization for steric and/or electronic reasons are 9-anthranol [16], 9-phenylanthracene [17] and 9,10-dimethylanthracene [17]. Photodimerization of 9,10-diphenylanthracene has also not so far been observed. It is possible that the phenyl groups, due to their spatial requirements, prevent the assembly of two molecules or that the photodimer is in fact formed but is so thermally unstable that it rapidly decomposes into the monomer even at room temperature [18]. Since this early proposal several investigations (comp. [12, 19]) have shown, that the thermal stability of the photodimers varies considerably; thus, 9-cyanoanthracene dimer decomposes to the monomer merely in boiling chloroform [12].

References, pp. 103—104

9,10,11,16-Tetrahydro-9,10[9′,10′]anthracenoanthracene (dianthracene, anthracene dimer, 5) [20]. A solution of 2.673 g of anthracene (4) in 150 ml of benzene was irradiated through Solidex with a Philips HPK 125 W mercury high-pressure lamp under argon at 20°. Precipitation of the dimer began after only 10 minutes, and after 5 hrs. irradiation 2.221 g of pure **5** were collected.

9-Anthraldehyde dimer (7) [11]. Exposure of a solution of 4 g of 9-anthraldehyde in 100 ml of glacial acetic acid in a Pyrex flask to sunlight for one day resulted in precipitation of a mixture of white powder and yellow needles (3.6 g). The yellow material (anthraquinone) was removed by washing with benzene, and recrystallization of the residue from benzene afforded colorless crystals of **7**, m.p. 186—187°. After resolidification the material had m.p. 104—105° (9-anthraldehyde).

9-Aminoanthracene dimer (8) [21]. A solution of 9-aminoanthracene in ether was irradiated with the unfiltered radiation from a mercury lamp and without exclusion of oxygen. Within one hour a 60% yield of dimer **8**, m.p. 255°, could be obtained.

c) Higher condensed aromatic compounds

Complex derivatives of anthracene have also been photodimerized, e.g., benz[a]anthracene (**9**) [22] and the carcinogenic 3-methylcholanthrene (**10**) [23]. Structures **11** and **12** have been proposed [24] for tetracene dimer and pentacene dimer, respectively, which have until now only been obtained in solution.

3-Methylcholanthrene dimer [23]. The hydrocarbon **10**, dissolved in benzene, was exposed in a sealed Pyrex tube under CO_2 to the action of sunlight for 30 days (Winter, Cairo). The almost colorless dimer was washed with hot benzene and crystallized from a large volume of xylene; it melted at 295—305°. Heating to 360° (bath temperature) in a carbon dioxide atmosphere converted the dimer to **10**.

2. Heterocyclic nitrogen compounds

a) Pyridine derivatives

2(1H)-Pyridone (**13**) and 1-methyl-2(1H)-pyridone (**15**) are rapidly converted by ultraviolet light to solid dimers, as was found by TAYLOR and PAUDLER [25]. These photodimers have since been shown by two independent groups [26, 27] to be 3,7-diazatricyclo[4.2.2.2^{2,5}]dodeca-9,11-diene-4,8-dione (**14**) and the 3,7-dimethyl analogue (**16**), respectively.

Similar photodimers (cf. **19**) have subsequently been obtained from 2-aminopyridines (cf. **17**), and the two series have been interrelated by hydrolysis of the tetrahydro derivative of **18** to the tetrahydro derivative of **14** [28]. Assignment of the anti-trans configuration to all of the dimers has been made on the basis of detailed study of their NMR spectra.

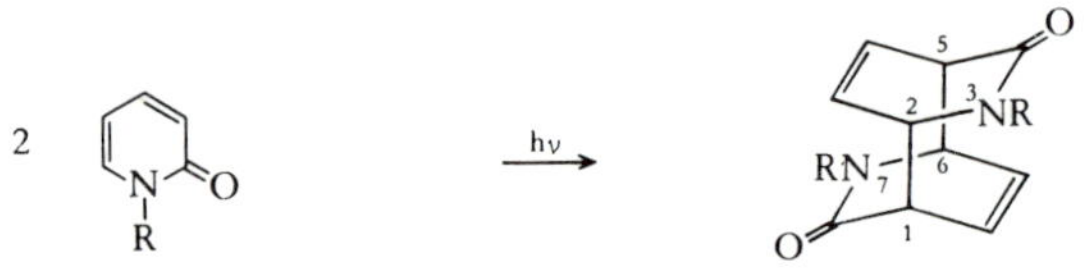

13: R = H
15: R = CH$_3$

14: R = H
16: R = CH$_3$

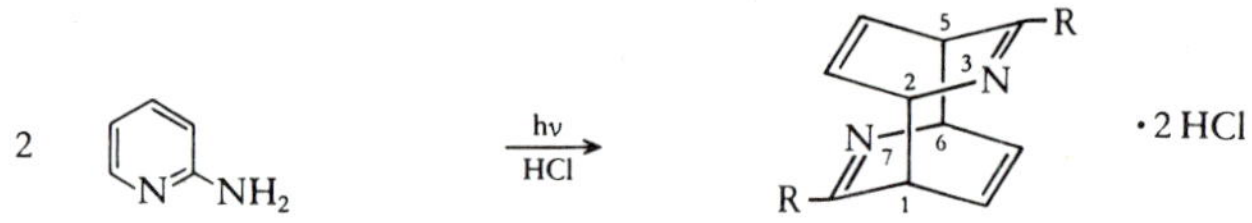

17

18: R = NH$_2$
19

Table 8

monomer	dimer	ref.
2(1H)-pyridone (**13**)	**14**	[25, 28, 29]
1-methyl-2(1H)-pyridone (**15**)	3,7-dimethyl-3,7-diazatricyclo-[4.2.2.2^{2,5}]dodeca-9,11-diene-4,8-dione (**16**)	[25—29]
3-methyl-2(1H)-pyridone	1,5-dimethyl-**14**	[28]
4-methyl-2(1H)-pyridone	9,11-dimethyl-**14**	[28]
1,4-dimethyl-2(1H)-pyridone	3,7,9,11-tetramethyl-**14**	[26, 29]
1,6-dimethyl-2(1H)-pyridone	2,3,6,7-tetramethyl-**14**	[26, 28]
1-(2-hydroxyethyl)-2(1H)-pyridone	3,7-bis-(2-hydroxyethyl)-**14**	[29]
2-aminopyridine (**17**)	4,8-diamino-3,7-diazatricyclo-[4.2.2.2^{2,5}]dodeca-3,7,9,11-tetraene (**18**)	[28]
2-amino-3-methylpyridine	4,8-diamino-1,5-dimethyl-**19**	[28]
2-amino-6-methylpyridine	4,8-diamino-2,6-dimethyl-**19**	[28]
2-amino-5-chloropyridine	4,8-diamino-10,12-dichloro-**19**	[28]
2-benzylaminopyridine	4,8-bis-(benzylamino)-**19**	[28]
3-chloro-2(1H)-pyridone	no dimer	[29]
6-chloro-2(1H)-pyridone	no dimer	[29]
2-amino-5-nitropyridine	no dimer	[28]
2,5-diaminopyridine	no dimer	[28]
2-amino-5-pyridinecarboxylic acid	no dimer	[28]

References, pp. 103—104

A great number of 2-aminopyridine and 2(1H)-pyridone derivatives have thus been photodimerized, and the above table 8 gives some examples (cf. [28, 29]). Also included are compounds which failed to undergo photodimerization on irradiation of their aqueous and aqueous alcoholic (for pyridones) or hydrochloric acid (for aminopyridines) solutions.

3,7-Diazatricyclo[4.2.2.2^{2,5}]dodeca-9,11-diene-4,8-dione (14) [28]. A solution of 69 g of **13** in 100 ml of ethanol was contained in a long, narrow tube designed to fit around a large Hanau low-pressure immersion lamp. After 5 days of irradiation 19.5 g (28%) of the 2(1H)-pyridone dimer had separated at the bottom and sides of the tube and irradiation was discontinued. Recrystallization of solid material from glacial acetic acid gave **14**, m.p. 225.5–227.5°.

In an alternative experiment, a solution of 20 g of **13** in 125 ml of ethanol was irradiated as described below in the preparation of **18** to give, after 3 days, a total yield of 8 g (40%) of **14**.

4,8-Diamino-3,7-diazatricyclo[4.2.2.2^{2,5}]dodeca-3,7,9,11-tetraene (18) [28]. A solution of 30 g of **17** in 40 ml of concentrated HCl contained in a 250-ml Pyrex Erlenmeyer flask was immersed in a 1500 ml Pyrex beaker fitted with an inlet and an outlet for continuous water cooling. The beaker was placed in front of a G.E. AH-6 high-pressure mercury arc. After 2 hrs. irradiation large colorless crystals had formed along the walls of the flask. These were removed in order to allow further passage of light into the solution, and irradiation was continued until the decrease in concentration of the starting material made further illumination unfruitful. In this way, 20 g (66%) of 2-aminopyridine dimer hydrochloride (**18**) was obtained after about 20 hrs. of irradiation. The product was recrystallized from aqueous ethanol to give pure **18**, m.p. 215° (decomposition).

b) Condensed nitrogen heterocyclic compounds

Investigations in this field have been carried out largely by Étienne and his co-workers. Among other compounds, benzo[g]quinoline (**20**) [30], benz[b]acridine (**21**) [31], dibenz[b,h]acridine (**22**) [31], dibenz[a,i]acridine (**23**) [31] and benzo[g]quinazoline (**24**) [32] have been reported to be converted to dimers by irradiation in alcoholic solution. It has, however, been stated that mixtures of dimers were formed, and the structures of these dimers have not yet been determined unambiguously. In this connection it should be borne in mind, that acridine was formerly believed [6, 33] to be converted to a dimer of the dianthracene type; later on it was shown that the photoproduct was in fact 9,9′-biacridan (comp. chapter 20).

Crystalline benzo[b]quinolizinium bromide monohydrate (25), when exposed to radiation from the sun or from a sunlamp, was converted to a higher melting, less soluble salt lacking the yellow color and the characteristic fluorescence of the starting material [34]. A dimeric structure analogous to that of dianthracene was suggested with the arrangement as shown (26) being the most plausible since it allows for maximum separation of the like charges.

Benz[b]acridine dimer [31]. A solution of 64 mg of **21** in 12 ml of ether and under vacuum was exposed to sunlight. After only one hour of illumination, the separation of the dimer could be observed, and after 3 days a precipitate consisting of colorless, sparingly soluble crystals was obtained (48 mg, 75%). Recrystallization from acetic acid gave dimer of m.p. 369—370°. On heating to 260° under vacuum, the compound decomposed to the monomer again. If the illumination was carried out in the presence of air a yield of only 40—43% was obtained.

Benzo[b]quinolizinium dimer dibromide (26) [34]. One gram of the crystalline salt **25** was irradiated with a General Electric sunlamp for about 5.5 hrs. with occasional stirring to ensure complete exposure. As the reaction proceeded, the color of the material changed from yellow to light tan and the crystals disintegrated. Recrystallization from ethanol-ether yielded 0.96 g (93%) of colorless prisms, m.p. 260—263°.

3. Heterocyclic oxygen compounds

Irradiation of 4,6-dimethyl-2H-pyran-2-one (27) in benzene solution through Pyrex leads to the formation of three dimers, namely, syn-trans-2,6,9,11-tetramethyl-3,7-dioxatricyclo[4.2.2.2^{2,5}]dodeca-9,11-diene-4,8-dione (28, m.p. 182—184°), the anti-trans dimer 29, m.p. 245° and 2,6,8,10-tetramethyl-3,11-dioxatricyclo[6.4.0.0^{2,7}]dodeca-5,9-diene-4,12-dione (30, m.p. 152—154°). The photochemical and thermal reactions in this field as elaborated by DE MAYO and YIP [*35*], which finally lead to 1,3,5,7-tetramethyl-1,3,5,7-cyclooctatetraene (**31**), are depicted in the scheme.

GUYOT and CATEL [*36*] found that irradiation of the orange-yellow solutions of 1,3-diphenylisobenzofuran (**32**) resulted in formation of a colorless dimer for which formula **33** was suggested. This structure proposal was later substantiated by COURTOT and SACHS [*37*] on the basis of chemical and spectroscopic evidence.

5,6,11,12-Tetrahydro-5,6,11,12-tetraphenyl-5,12: 6,11-diepoxydibenzo[a,e]cyclooctadiene, 1,3-diphenylisobenzofuran dimer, (33) [*36*]. A saturated alcoholic solution of **32** was exposed to sunlight in the absence of air. The solution lost its strong fluorescence within a few minutes and became colorless. Crystals separated out, the quantity of which increased with time. The crystals were washed with alcohol and dried in vacuo to give **33**, m.p. [*37*] 190—200°. Slow heating of **33** gave **32**, m.p. 125°.

References

[*1*] J. S. BRADSHAW and G. S. HAMMOND: J. Amer. chem. Soc. **85**, 3953 (1963).
[*2*] FRITZSCHE: J. Prakt. Chemie **101**, 333 (1867).
[*3*] R. LUTHER and F. WEIGERT: Z. Physik. Chem. **51**, 297 (1905).
[*4*] N. S. CAPPER and J. K. MARSH: J. Chem. Soc. **1926**, 724.
[*5*] C. E. LINEBARGER: Am. Chem. J. **14**, 597 (1892).
[*6*] W. R. ORNDORFF and F. K. CAMERON: Am. Chem. J. **17**, 658 (1895).
[*7*] C. A. COULSON, L. E. ORGEL, W. TAYLOR and J. WEISS: J. Chem. Soc. **1955**, 2961.
[*8*] E. J. BOWEN and D. W. TANNER: Trans. Faraday Soc. **51**, 475 (1955).
[*9*] R. LIVINGSTON: In: Photochemistry in the liquid and solid states. Ed. by L. J. HEIDT, R. S. LIVINGSTON, E. RABINOWITCH and F. DANIELS, pp. 75—82. New York: Wiley 1960.
[*10*] J. B. BIRKS and J. B. ALADEKOMO: Photochem. Photobiol. **2**, 415 (1963).
[*11*] F. D. GREENE, S. L. MISROCK and J. R. WOLFE, JR.: J. Amer. chem. Soc. **77**, 3852 (1955).
[*12*] R. LALANDE and R. CALAS: Bull. Soc. Chim. France **1960**, 144.
[*13*] D. E. APPLEQUIST, T. L. BROWN, J. P. KLEIMAN and S. T. YOUNG: Chem. and Ind. **1959**, 850.
[*14*] R. CALAS, R. LALANDE, J.-G. FAUGÈRE and F. MOULINES: Bull. Soc. Chim. France **1965**, 119.
[*15*] D. E. APPLEQUIST, E. C. FRIEDRICH and M. T. ROGERS: J. Amer. chem. Soc. **81**, 457 (1959).

[16] J.-G. FAUGÈRE and R. CALAS: Comptes rendus **260**, 585 [1965].
[17] A. WILLEMART: Comptes rendus **205**, 993 [1937].
[18] A. SCHÖNBERG: Trans. Faraday Soc. **32**, 514 [1936].
[19] F. D. GREENE: Bull. Soc. Chim. France **1960**, 1356.
[20] C. H. KRAUCH and D. HESS: unpublished results.
[21] R. LALANDE, H. BOUAS-LAURENT and A. COT: Bull. Soc. Chim. France **1965**, 2695.
[22] A. SCHÖNBERG, A. MUSTAFA, M. Z. BARAKAT, N. LATIF, R. MOUBASHER and A. MUSTAFA: J. Chem. Soc. **1948**, 2126.
[23] A. SCHÖNBERG and A. MUSTAFA: J. Chem. Soc. **1949**, 1039.
[24] J. B. BIRKS, J. H. APPLEYARD and R. POPE: Photochem. Photobiol. **2**, 493 (1963).
[25] E. C. TAYLOR and W. W. PAUDLER: Tetrahedron Letters **25**, 1 [1960].
[26] W. A. AYER, R. HAYATSU, P. DE MAYO, S. T. REID and J. B. STOTHERS: Tetrahedron Letters **1961**, 648.
[27] G. SLOMP, F. A. MACKELLAR and L. A. PAQUETTE: J. Amer. chem. Soc. **83**, 4472 (1961).
[28] E. C. TAYLOR and R. O. KAN: J. Amer. chem. Soc. **85**, 776 (1963).
[29] L. A. PAQUETTE and G. SLOMP: J. Amer. chem. Soc. **85**, 765 (1963).
[30] A. ÉTIENNE: Comptes rendus **218**, 841 (1944).
[31] A. ÉTIENNE and A. STAEHELIN: Bull. Soc. Chim. France **1954**, 748.
[32] A. ÉTIENNE and M. LEGRAND: Comptes rendus **232**, 1223 (1951).
[33] A. KELLMANN: J. Chim. Phys. **57**, 468 (1957).
[34] C. K. BRADSHER, L. E. BEAVERS and J. H. JONES: J. Org. Chem. **22**, 1740 [1957].
[35] P. DE MAYO and R. W. YIP: Proc. Chem. Soc. **1964**, 84.
[36] A. GUYOT and J. CATEL: Bull. Soc. Chim. France [3] **35**, 1124 (1906).
[37] P. COURTOT and D. H. SACHS: Bull. Soc. Chim. France **1965**, 2259.

Chapter 10

Various photodimerizations involving aldehydes, halides and thiocarbonyl compounds

1. Photodimerization of aliphatic compounds

a) Butyraldehyde

When butanal was exposed to 254 mμ radiation, 5-hydroxy-4-octanone (1) was obtained in 58 % yield [1].

$$2\ CH_3-CH_2-CH_2-CHO \xrightarrow{h\nu} CH_3-CH_2-CH_2-\underset{\underset{O}{\|}}{C}-\underset{\underset{OH}{|}}{CH}-CH_2-CH_2-CH_3$$
$$\mathbf{1}$$

b) Dimerization of halides

The photodimerization of trifluoroiodoethylene, as investigated by PARK et al. [2], is known to produce hexafluoro-4,4-diiodo-1-butene (2).

$$2\ F_2C=CFI \xrightarrow{h\nu} F_2C=CF-CF_2-CFI_2$$
$$\mathbf{2}$$

The formation of 2, which is assumed to involve initial homolysis of the C—I bond, is only a special instance of a general reaction which will be discussed in more detail in another context (comp. chapter 38).

HASZELDINE and MATTINSON [3] found that blue trifluoronitrosomethane (3) is converted quantitatively to an orange-red gas of double molecular weight by irradiation. This gas was shown to be N,N-bis-(trifluoromethyl)-O-nitrosohydroxylamine (4). The following free radical chain reaction mechanism was suggested [3]:

$$F_3C-NO\ (3) \xrightarrow{h\nu} F_3C\cdot\ +\ \cdot NO$$
$$F_3C\cdot\ +\ F_3C-NO \longrightarrow (F_3C)_2NO\cdot$$
$$(F_3C)_2NO\cdot\ +\ \cdot NO \longrightarrow (F_3C)_2N-O-NO\ (4)$$
$$\text{or}\ (F_3C)_2NO\cdot\ +\ F_3C-NO \longrightarrow (F_3C)_2N-O-NO\ +\ F_3C\cdot$$

References, p. 108

More recent investigations [4] revealed the reversibility of the 3 → 4 reaction; among further irradiation products isolated [4, 5], bis-(trifluoromethyl)-nitramide (5) will be mentioned here. Presence of air and moisture during irradiation may alter the reaction path drastically as was found by WHITE and PARCELL [6].

$$\begin{matrix} F_3C \\ \diagdown \\ N-O-NO \\ \diagup \\ F_3C \\ \mathbf{4} \end{matrix} \quad \xrightarrow{h\nu} \quad \begin{matrix} F_3C \\ \diagdown \\ N-NO_2 \\ \diagup \\ F_3C \\ \mathbf{5} \end{matrix}$$

N,N-Bis-(trifluoromethyl)-O-nitrosohydroxylamine (4) [3]. Trifluoronitrosomethane (3) (57 mmole) in a sealed quartz tube of 188 ml capacity was exposed to radiation from a Hanovia UV-lamp for 40 hours. The lower part of the tube (5 cm) was covered with black paper in order to protect liquid material from further irradiation. The reaction product was distilled under reduced pressure to yield 3 (18%), carbon dioxide (4%), which was separated from 3 by treatment with caustic soda, and 4 (96%, calculated on the basis of 3 converted). Dimer 4 was a red-brown liquid of b.p. 10° which gave a yellow solid on cooling.

For an improved preparation of 4 utilizing a specially devised 20-liter photochemical reactor see DINWOODIE and HASZELDINE [4].

c) Dimerization of thiocarbonyl compounds

RATHKE [7, 8] observed that the red thiophosgene (6) on exposure to sunlight in the presence of hydrochloric acid was converted to a colorless dimer the correct constitution of which as 7 was established only 60 years later [9]. Final confirmation was adduced on the basis of spectroscopic evidence [10, 11].

$$2 \quad \begin{matrix} Cl \\ \diagdown \\ =S \\ \diagup \\ Cl \end{matrix} \quad \xrightarrow{h\nu} \quad \begin{matrix} Cl \diagdown \diagup S \diagdown \diagup Cl \\ \times\times \\ Cl \diagup \diagdown S \diagup \diagdown Cl \end{matrix}$$

$$\mathbf{6} \qquad\qquad\qquad\qquad \mathbf{7}$$

2,2,4,4-Tetrachloro-1,3-dithietane (7) may be photolyzed to its progenitor; it sublimes on heating and when heated in a closed vessel decomposes to two molecules of thiophosgene [9]. It is of potential industrial interest (cf. [12]).

On exposure to ultraviolet light trifluorothioacetyl fluoride (8) dimerized to a mixture of cis- and trans-2,4-bis-(trifluoromethyl)-2,4-difluoro-1,3-dithietane (9) in 60% yield [13]. Analogously, chlorodifluorothioacetyl fluoride (10) gave dimer 11 in 83% yield.

$$2 \quad \begin{matrix} R \\ \diagdown \\ =S \\ \diagup \\ F \end{matrix} \quad \xrightarrow{h\nu} \quad \begin{matrix} R \diagdown \diagup S \diagdown \diagup F \\ \times\times \\ F \diagup \diagdown S \diagup \diagdown R \end{matrix}$$

8: R = CF₃   9: R = CF₃
10: R = CF₂Cl   11: R = CF₂Cl

References, p. 108

2,2,4,4-Tetrachloro-1,3-dithietane (7) [*9*]. Commercial **6** (15 ml) in a quartz test tube with an internal water-cooling system was irradiated with a Hanau Analytical Quartz Lamp for 20 hours. Large colorless crystals separated from the red liquid, and these were isolated and recrystallized several times from hot ligroin. Sublimation in a RIIBER apparatus at 14 mm Hg gave 5 g of pure **7**, m.p. 119°. When the molten substance was heated to 130°, decomposition occurred giving a red-brown coloration and the smell of thiophosgene.

2,4-Bis-(chlorodifluoromethyl)-2,4-difluoro-1,3-dithietane (11) [*13*]. A solution of 38 g of **10** in 25 ml of dichlorodifluoromethane contained in a quartz reaction vessel was irradiated with ultraviolet light. The light source was a helix-shaped mercury low-pressure lamp which was slipped over the reaction tube so that its radiation impinged primarily upon the liquid portion of the reaction mixture. After 3 hrs. irradiation the solvent was evaporated and the residue distilled to give 31.3 g of **11**, b.p.$_{23}$ 44°. The liquid was a mixture of cis and trans isomers.

2. Photodimerizations of aromatic aldehydes

SCHÖNBERG and MUSTAFA [*14*] found that a benzene solution of phthalaldehyde (**12**), exposed to sunlight for 1 day, deposited colorless crystals of a dimer which was attributed structure **13**. No reaction occurred when the solution was kept in the dark. The aldehyde **13** may, of course, also exist as the ring tautomer **14**.

RIED and WILK [*15*] reported that irradiation of an ethereal solution of 2-nitrosobenzaldehyde (**15**) led to the formation of 2,2′-azodibenzoic acid (**16**).

3-(2-Formyl-α-hydroxybenzyl)-phthalide (13 or 14) [*14*]. The air in a SCHLENK tube, containing 1 g of **12** in 25 ml of dry thiophene-free benzene, was displaced by dry carbon dioxide and the tube sealed. The reaction mixture was insolated (Cairo, July) for one day. The colorless crystals that separated during irradiation had a m.p. about 184° (decomp. with yellow melt). Crystals from xylene contained xylene of crystallization which was readily lost when the sample was dried at 100° for 6 hours. The yield was almost quantitative.

2,2′-Azodibenzoic acid (16) [15]. 2-Nitrosobenzaldehyde (15) (0.1 g) was dissolved in 150 ml of ether and irradiated under cooling with a UV immersion lamp when the solution turned red-brown in color. A dark brown resin deposited eventually, mainly on the surface of the lamp. After evaporation of the ether, the resin was treated with cold benzene. The insoluble residue was repeatedly recrystallized from ethanol and then chromatographed in alcoholic solution on a column of starch. The product thus purified formed yellow needles (8 mg), m.p. 235° (dec.).

References

[1] W. H. URRY and D. J. TRECKER: J. Amer. chem. Soc. **84**, 118 (1962)
[2] J. D. PARK, R. J. SEFFL and J. R. LACHER: J. Amer. chem. Soc. **78**, 59 (1956).
[3] R. N. HASZELDINE and B. J. H. MATTINSON: J. Chem. Soc. **1957**, 1741.
[4] A. H. DINWOODIE and R. N. HASZELDINE: J. Chem. Soc. **1965**, 1675.
[5] J. MASON: J. Chem. Soc. **1963**, 4537.
[6] R. C. WHITE and L. J. PARCELL: J. Phys. Chem. **69**, 4409 (1965).
[7] B. RATHKE: Ann. Chem. Pharm. **167**, 195 (1873).
[8] B. RATHKE: Ber. dtsch. chem. Ges. **21**, 2539 (1888).
[9] A. SCHÖNBERG and A. STEPHENSON: Ber. dtsch. chem. Ges. **66**, 567 (1933).
[10] J. I. JONES, W. KYNASTON and J. L. HALES: J. Chem. Soc. **1957**, 614.
[11] W. K. BUSFIELD, M. J. TAYLOR and E. WHALLEY: Can. J. Chem. **42**, 2107 (1964).
[12] W. H. SHARKEY, W. J. MIDDLETON, H. W. JACOBSON, D. S. ACKER and H. C. WALTER: Chem. Eng. News **41** (38), 46 (1963).
[13] W. J. MIDDLETON, E. G. HOWARD and W. H. SHARKEY: J. Org. Chem. **30**, 1375 (1965).
[14] A. SCHÖNBERG and A. MUSTAFA: J. Amer. chem. Soc. **77**, 5755 (1955).
[15] W. RIED and M. WILK: Liebigs Ann. Chem. **590**, 91 (1954).

Chapter 11

Cycloaddition of alkenes or alkynes to other alkenes or aromatic nuclei leading to the formation of four-membered rings. Photolyses involving retro-cycloaddition

The photodimerisation of unsaturated aliphatic compounds leading to the formation of cyclobutane derivatives has already been considered in chapter 8. Reactions of this type are among the earliest known photochemical reactions. In contrast photoreactions between two different unsaturated molecules are a relatively late discovery but, in recent years, a large number of such reactions have been described. The literature has been reviewed by inter alia, SCHENCK [1] and by COREY [2]. A recent account has been given by STEINMETZ [3].

1. Addition of maleic anhydride to benzene and related aromatic compounds including furan and thiophene

Several independent reports of light-induced addition reactions of maleic anhydride to benzene and derivatives, including phenanthrene, have appeared in recent years.

According to ANGUS [4] benzene and maleic anhydride react in light to form the stable adduct **1**. GROVENSTEIN [5] assigned to the adduct the stereochemical formula **2**. The intermediate (**3**) was supposed to be stabilized by rapid addition of a further molecule of maleic anhydride to the alicyclic diene system giving **1**. No trace of **1** was formed when benzene was heated with maleic anhydride in the dark at 250—300° for 19 hours.

The use of benzophenone as a sensitizer produces a five-fold increase in the yield of **2** in 3 hours. Similar irradiations have produced 2 : 1 adducts from maleic anhydride and toluene, o-xylene and chlorobenzene [*6*].

The tendency of naphthalene to add maleic anhydride in a light-induced reaction is negligible [*7*]. With phenanthrene the addition product **4** has been obtained which decomposes at 300° yielding maleic anhydride and phenanthrene [*8*].

4

The photosensitized cycloaddition of dimethylmaleic anhydride to furan yields **5** [*9*]. The corresponding thiophene adduct, which is easily obtainable in 80 % yield, has been assigned the analogous constitution **6** [*10*].

5   6

Tricyclo[4.2.2.0^{2,5}]dec-9-ene-3,4,7,8,-tetracarboxylic acid dianhydride (2) [*6*]. A solution of maleic anhydride (10 g) and benzophenone (2.0 g) in benzene (150 ml) was irradiated with a mercury burner Philips HPK 125 W in an argon atmosphere. After 3 hrs. 1.5 g of **2**, m.p. 350—355°, had formed.

1 : 1 Adduct from phenanthrene and maleic anhydride (1,2,2a,10b-tetrahydrocyclobuta[l]phenanthrene-1,2-dicarboxylic anhydride) (4) [*8*]. Phenanthrene when irradiated with maleic anhydride in hexane at 30—60° formed an adduct, m.p. 222°, which formed colorless crystals from acetone. The addition can be photosensitized by benzophenone or, more effectively, by benzil.

6,7-Dimethyl-2-oxabicyclo[3.2.0]hept-3-ene-6,7-dicarboxylic anhydride (5) [*9*]. The irradiation was conducted under argon using a mercury Philips HPK 125 W lamp; the mixture was stirred magnetically during irradiation. 48 hours irradiation of a solution of 20 g dimethylmaleic anhydride and 2 g benzophenone in 150 ml furan at 10°, followed by concentration yielded 10.6 g **5**, m.p. 153° after sublimation at 100°/0.1 mm Hg.

2. Addition of maleic acid derivatives to alkenes

The photochemical cycloaddition reaction of maleic anhydride to unsaturated systems was found to be widely applicable. Both components of the reaction system may be varied to some extent. Thus maleic anhydride may be replaced by alkyl maleates, maleimides, or alkylsubstituted maleic anhydrides. Olefins on the other hand may be open-chain or cyclic. An account of work done until 1962 has been given by SCHENCK and STEINMETZ [*1*].

References, p. 117

Photosensitized cycloaddition of dimethylmaleic anhydride (**7**) to acyclic olefins was investigated by SCHENCK and collaborators [*10*]. High yields of cycloadducts were obtained, when **7** was irradiated in chlorinated ethylenes in the presence of benzophenone as sensitizer. The cycloaddition of **7** to 1,1,2-trichloroethylene may be taken as an example:

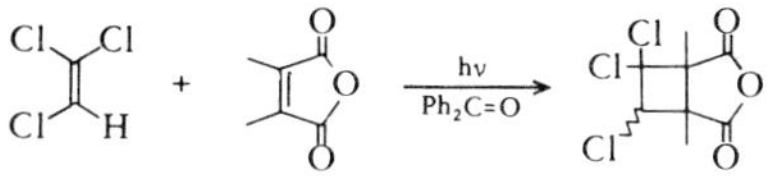

Cyclohexene appears to form charge-transfer complexes with maleic anhydride. Irradiation gives rise to a variety of products among which are three 1 : 1 cycloadducts. Assuming cis junction of the anhydride ring, these may be described as the cis,trans (**9**), cis,cis (**10**) and trans (**11**) isomers of bicyclo[4.2.0]octane-7,8-dicarboxylic anhydride [*11*].

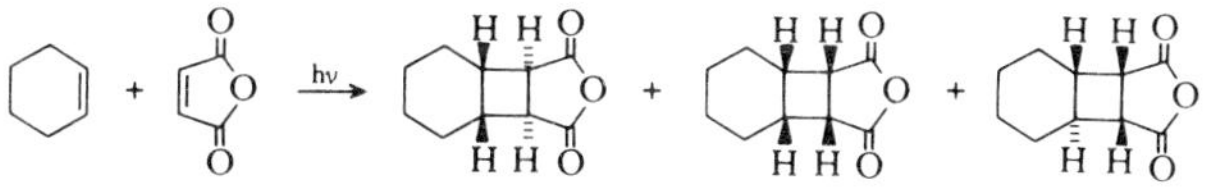

In a similar manner, DE MAYO [*12*] prepared an even greater number of dimethyl esters of stereoisomeric bicyclo[4.2.0]octane-7,8-dicarboxylic acids which were not easily separable from one another. With cyclopentene as addend only three esters with cis junction of the rings were obtained.

3,3,4-Trichloro-1,2-dimethylcyclobutane-1,2-dicarboxylic anhydride (8) [*10*]. 8.0 g **7** and 4.0 g benzophenone were dissolved in 150 ml trichloroethylene and the solution irradiated with a water-cooled mercury immersion lamp at 10°. After 32 hrs. the solution had adopted a brown color, and 70 mg of the dimer of **7** had separated out. The filtrate was freed from trichloroethylene by distillation and the residue digested with ether to furnish 5.6 g of **8**, which after recrystallization from ether melted at 172—173°; conversion 90%. On chromatography of the ether mother-liquor 4.95 g of **7** and 3.2 g of benzophenone were recovered.
An analogous irradiation omitting benzophenone yielded 3.6 g **8**.

3. Addition of 2-methyl-2-butene to benzonitrile

Irradiation of a 1 : 1 mixture of benzonitrile and 2-methyl-2-butene for 7—14 days with a mercury arc yielded a mixture of neutral and basic materials. The neutral fraction yielded a product which crystallized spontaneously when stored at —20°. The 1 : 1 adduct (m. p. 26—30°) which was isolated in 63% yield, was shown to be 1-cyano-7,8,8-trimethylbicyclo-[4.2.0]octa-2,4-diene (**12**) [*13*]. On exposure to ultraviolet light **12** is easily reconverted to the starting materials (for related retro-cycloadditions see p. 116). However, pyrolysis (128°, 38 hrs.) yielded a product which was a mixture of cis- and trans-isomers of 3-cyano-2-methyl-2,4,6,8-decatetraene (**13**).

4. Photochemical cycloaddition of α,β-unsaturated ketones to alkenes

a) 2-Cyclohexenone and isobutylene

COREY [2] discovered that the photochemical cycloaddition of isobutylene and 2-cyclohexenone led to 7,7-dimethylbicyclo[4.2.0]octan-2-one (**14**) as the major product. Both cis and trans isomers of **14** were obtained with the latter predominating considerably even though the trans fused ring system is relatively strained. Further products were the alternative cycloaddition product 8,8-dimethylbicyclo[4.2.0]octan-2-one (**15**), 2- and 3-(β-methylallyl)-cyclohexanone.

Similar photoreactions were carried out with 2-cyclohexenone and, inter alia, benzyl vinyl ether, the major product being a mixture of stereoisomers of 7-benzyloxybicyclo[4.2.0]octan-2-one (**16**). For details of the irradiation and workup procedures the reader is referred to the original literature.

b) Cyclopentenone and cyclopentene

Photochemical cycloaddition reactions starting with cyclopentenone have been investigated by EATON [*14*]. Thus irradiation of the ketone with cyclopentene gave **17** as sole product.

cis,trans,cis-Tricyclo[5.3.0.0^{2,6}]decan-3-one (17) [*14*]. A solution of 2-cyclopentenone (0.064 mole) in cyclopentene (0.64 mole) was illuminated with a Hanovia 450 W mercury lamp equipped with a Pyrex filter. After three hrs. the excess olefin was removed and the residue fractionated through a spinning band column. The main fraction of b.p.$_{0.7}$ 78—80° consisted of **17**, yield 67%.

c) 2,4-Pentanedione or dimedone and cyclohexene

Enolisable β-diketones are also prone to photochemical cycloaddition reactions as was established by DE MAYO. Irradiation [*15*] of 2,4-pentanedione (acetylacetone, **18**) in cyclohexene induced a reaction which could be followed IR-spectroscopically. The product, a 1 : 1 adduct of cyclohexene to **18**, proved to be acetonyl-2-acetylcyclohexane (**20**). Its formation is believed to proceed via cycloaddition of cyclohexene to the enolic form of acetylacetone (**18**) leading to the cyclobutane derivative **19**, which then isomerizes to **20**. Similarly, irradiation of **18** with cyclopentene gave the expected diketone, acetonyl-2-acetylcyclopentane (**21**).

The versatility of this particular photoaddition reaction was further demonstrated by the synthesis of **22** from 5,5-dimethyl-1,3-cyclohexanedione (dimedone) and cyclohexene [*16*]. **23**, which was formed together with the diketone **22**, was found to arise from **22** via photoreduction in cyclohexene.

Acetonyl-2-acetylcyclohexane (20) [*15*]. A solution of the diketone (**18**) (15.02 g) in cyclohexene (135 ml) was irradiated under nitrogen with an 80 W water-cooled immersion lamp for 45 hours. After removal of the solvent the product was fractionated on a 40-in. spinning band column to give the adduct **20**, b.p.$_{1.7}$ 89°; yield 20.2 g (78%).

4,4-Dimethylbicyclo[6.4.0]dodecane-2,6-dione (22) and 11,11-dimethyltricyclo-[7.3.0.0^{2,7}]dodecane-1,9-diol (23) [*16*]. Irradiation of a 0.3—0.5% solution of dimedone in cyclohexene-ether (1 : 3) (water-cooled Hanovia 450 W immersion lamp with Pyrex filter) resulted in formation of **22**, m.p. 96—97.5°, and **23**, m.p. 79—81°.

5. Photochemical cycloaddition reactions of acetylene compounds

a) Addition to benzene and derivatives

The photochemical addition of acetylene compounds (cf. **24**) to benzene was investigated by BRYCE-SMITH and LODGE [*17*]. Whereas the initial 1 : 1 benzene-maleic anhydride adduct undergoes 1,4-addition of a further molecule of anhydride (comp. p. 109) the corresponding adducts of the acetylenes undergo ring opening to give cyclooctatetraene derivatives. In the case of acetylene itself, however, only traces of cyclooctatetraene (**25**) were detected.

24: R=R'=H
26: R=COOCH$_3$, R'=H

25: R=R'=H
27: R=COOCH$_3$, R'=H

According to ATKINSON [*13*] photochemical addition of acetylenes e.g. 3-hexyne and 5-decyne to benzonitrile in methanol solution — irradiation of the pure liquids yielded no adduct — led to the formation of the 1 : 1 adducts, 1-cyano-2,3-diethylcyclooctatetraene (**28**) and 1-cyano-2,3-di-n-butylcyclooctatetraene (**29**) respectively, which are to be regarded as isomerization products of the cyclobutene derivatives initially formed.

28: R=C$_2$H$_5$
29: R=C$_4$H$_9$

Methyl cyclooctatetraenecarboxylate (27) [*18*]. Irradiation with a Hanovia S 500 mercury lamp of methyl propiolate (**26**) (5.0 g) in benzene (140 ml) under nitrogen for 20 hrs. at 53° gave methyl cyclooctatetraenecarboxylate (0.8 g) as a yellow oil b.p.$_{0.4}$ 89°.

1-Cyano-2,3-diethylcyclooctatetraene (28) [*13*]. Benzonitrile (35.9 g, 0.35 moles) and redistilled 3-hexyne (28 g, 0.34 moles) were added to an irradiation vessel, and 400 ml of anhydrous methanol was distilled into the reaction vessel under a slightly positive pressure of nitrogen. A thin layer of polymer forming around the heated section of the lamp had to be removed twice a day. This was most easily done by immersing the lamp, first in conc. sulfuric acid, and then in conc. ammonium hydroxide. After 3 weeks' irradiation, excess acetylene and solvent were removed at 50°/80 mm, and unreacted benzonitrile was distilled from the dark orange residue through a spinning-band column at a bath temperature of 60—70°/1.5 mm. Recovery: 20 g (56%). The thick residue (7.1 g) was chromatographed over 200 g of Florisil. Elution with pentane-benzene (40 : 60) gave a product which was distilled through a microspinning-band column at 62—72°/0.07—0.40 mm. **28** was obtained as a light yellow pleasant-smelling oil, yield 2.43 g (8%).

b) Addition to cyclopentenone

2-Cyclopentenone (**30**) may be added photochemically to 2-butyne [*19*]. Besides the head-to-tail dimer of cyclopentenone (**31**) (comp. p. 77) a mixture of two unsaturated bicyclic ketones (**32** and **33**) was isolated in 79 % yield. **32** and **33** are in photochemical equilibrium with one another in the approximate ratio of 1 : 2, the photoisomerization resembling that of the eucarvone photoisomers (comp. p. 6).

1,7-Dimethylbicyclo[3.2.0]hept-6-en-2-one (32) and 6,7-dimethylbicyclo[3.2.0]-hept-6-en-2-one (33) [*19*]. In a Jena glass apparatus 16.1 g 2-cyclopentenone and 400 ml butyne were irradiated under nitrogen by a mercury high pressure immersion lamp Philips HOQ 400 W for 48 hours. After distilling off the butyne at a bath temperature of 40—50°, 21.2 g (79%) of the mixture of adducts distilled over between 55° and 70°/5 mm Hg. The residue (3.3 g, 20%) consisted of **31**. Fractionation of the distillate on a spinning band column gave 6.8 g (25%) **32**, b.p.$_{12}$ 60° and 14.4 g (54%) **33**, b.p.$_{12}$ 75°.

c) Addition to dimethyl cyclobutene-1,2-dicarboxylate

SEEBACH [*20*] irradiated dimethyl cyclobutene-1,2-dicarboxylate (**34**) in 2-butyne. After 320 hrs. irradiation time the reaction mixture contained 19 % of the dimer of **34** viz. tetramethyl tricyclo[4.2.0.0^{2,5}]octane-1,2,5,6-tetracarboxylate (**36**) and a 50 % yield of the expected addition product, dimethyl 2,3-dimethylbicyclo[2.2.0]hex-2-ene-1,4-dicarboxylate (**35**). With benzophenone as sensitizer in the photoaddition reaction the necessary irradiation time may be shortened, but the resulting solution then adopts a deep yellow color, and the working up is more difficult.

The bicyclohexene derivative **35** easily undergoes isomerization under the influence of heat, the isomerization product being dimethyl 2,3-dimethyl-1,3-cyclohexadiene-1,4-dicarboxylate (**37**) [*20*].

d) Addition of diphenyl acetylene to naphthalene

Whereas acetylene added to benzene in the manner indicated above (cf. the formation of **25**), diphenylacetylene and naphthalene underwent a double cycloaddition reaction, 1,2,3,4-tetrahydro-9,10-diphenyl-1,4,2,3-ethanediylidenenaphthalene (**38**) being the product isolated in 29% yield

[21]. With methyl substituted naphthalenes, e.g., 2,3-dimethylnaphthalene, an improved yield (58%) of the respective adduct (39) was obtained. In this case no cycloaddition of diphenylacetylene to the unsubstituted ring was observed.

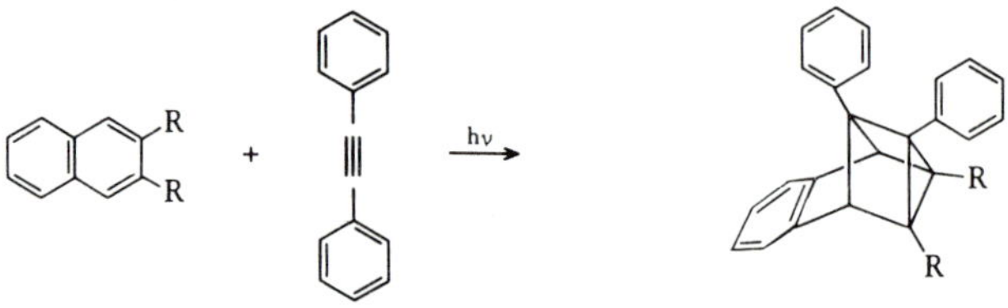

38: R = H
39: R = CH$_3$

1,2,3,4-Tetrahydro-2,3-dimethyl-9,10-diphenyl-1,4,2,3-ethanediylidenenaphthalene (39) [21]. A solution of 6.4 mmole diphenylacetylene and 25.6 mmole 2,3-dimethylnaphthalene in 50 ml of cyclohexane was irradiated for 72 hrs. with a Pyrex-jacketed Philips HPK 125 W lamp to give 58% (based on diphenylacetylene) of 39, m.p. 147°.

6. Photolyses involving retro-cycloaddition

a) Photolysis of dehydronorcamphor

According to SCHENCK and STEINMETZ [22] photolysis of dehydronorcamphor (bicyclo[2.2.1]hept-5-en-2-one, 40) leads to the quantitative formation of cyclopentadiene and ketene. Bicyclo[3.2.0]hept-2-en-7-one (41) was later shown to be an intermediate in this reaction [23], and the photolysis of the latter may be regarded as a retro-cycloaddition. A related reaction, namely, the formation of cyclooctanone and ketene on irradiation of 1-hydroxybicyclo[6.2.0]decan-10-one has briefly been mentioned in chapter 4.

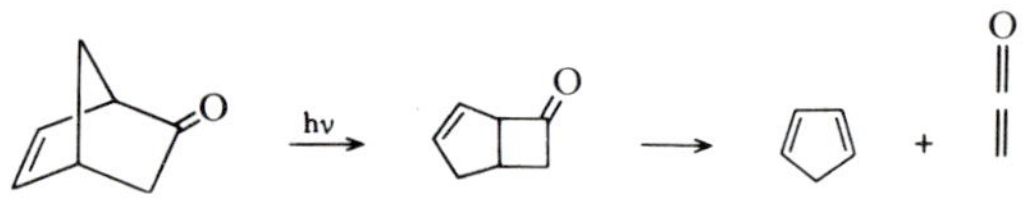

40   41

b) Photochemical synthesis of bullvalene via retro-cycloaddition

Cyclooctatetraene dimer (42) — obtained by heating cyclooctatetraene — on irradiation yields, besides benzene, the hydrocarbon bullvalene (tricyclo[3.3.2.0^{2,8}]deca-3,6,9-triene, 43) [24, 25]. Again, the photolysis of 42 may formally be regarded as a retro-cycloaddition. For the intriguing properties of bullvalene see SCHRÖDER [25, 26].

References, p. 117

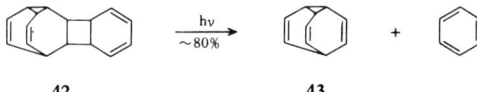

42   43

Bullvalene (43) [24]. Pentacyclo[9.3.2.0^{2,9}.0^{3,8}.0^{10,12}]hexadeca-4,6,13,15-tetraene (cyclooctatetraene dimer (42), m.p. 76°) (22 g) was dissolved in 900 ml of anhydrous ether in a 1-liter three-necked flask fitted with magnetic stirrer, reflux condenser, inlet for pure nitrogen and a quartz-jacketed water-cooled Philips HPK 125 W mercury immersion lamp. After 24 hrs. irradiation the solvent was distilled off through a column and then benzene and residual ether were removed by careful distillation at 20 mm Hg and bath temperature up to 40°. The crystalline, somewhat oily and volatile residue was purified by two sublimations (1—14 mm Hg and bath temperature about 40—60°) and recrystallization from ethanol to give 10.5 g (75%) of pure 43, m.p. 95—96°.

References

[1] G. O. SCHENCK and R. STEINMETZ: Bull. Soc. Chim. Belges **71**, 781 (1962).
[2] E. J. COREY, J. D. BASS, R. LEMAHIEU and R. B. MITRA: J. Amer. chem. Soc. **86**, 5570 (1964).
[3] R. STEINMETZ: Fortschr. chem. Forsch. **7**, 445 (1967).
[4] H. J. F. ANGUS and D. BRYCE-SMITH: Proc. Chem. Soc. **1959**, 326.
[5] E. GROVENSTEIN JR., D. V. RAO and J. W. TAYLOR: J. Amer. chem. Soc. **83**, 1705 (1961).
[6] G. O. SCHENCK and R. STEINMETZ: Tetrahedron Letters **21**, 1 (1960).
[7] G. O. SCHENCK, J. KUHLS, S.-P. MANNSFELD and C. H. KRAUCH: Chem. Ber. **96**, 813 (1963).
[8] D. BRYCE-SMITH and B. VICKERY: Chem. and Ind. **1961**, 429.
[9] G. O. SCHENCK, W. HARTMANN, S.-P. MANNSFELD, W. METZNER and C. H. KRAUCH: Chem. Ber. **95**, 1642 (1962).
[10] G. O. SCHENCK, W. HARTMANN and R. STEINMETZ: Chem. Ber. **96**, 498 (1963).
[11] R. ROBSON, P. W. GRUBB and J. A. BARLTROP: J. Chem. Soc. **1964**, 2153.
[12] P. DE MAYO, S. T. REID and R. W. YIP: Can. J. Chem. **42**, 2828 (1964).
[13] J. G. ATKINSON, D. E. AYER, G. BÜCHI and E. W. ROBB: J. Amer. chem. Soc. **85**, 2257 (1963).
[14] P. E. EATON: J. Amer. chem. Soc. **84**, 2454 (1962).
[15] P. DE MAYO and H. TAKESHITA: Can. J. Chem. **41**, 440 (1963).
[16] H. HIKINO and P. DE MAYO: J. Amer. chem. Soc. **86**, 3582 (1964).
[17] D. BRYCE-SMITH and J. E. LODGE: Proc. Chem. Soc. **1961**, 333.
[18] D. BRYCE-SMITH and J. E. LODGE: J. Chem. Soc. **1963**, 695.
[19] R. CRIEGEE and H. FURRER: Chem. Ber. **97**, 2949 (1964).
[20] D. SEEBACH: Chem. Ber. **97**, 2953 (1964).
[21] W. H. F. SASSE, P. J. COLLIN and G. SUGOWDZ: Tetrahedron Letters **1965**, 3373.
[22] G. O. SCHENCK and R. STEINMETZ: Chem. Ber. **96**, 520 (1963).
[23] D. I. SCHUSTER, M. AXELROD and J. AUERBACH: Tetrahedron Letters **1963**, 1911.
[24] G. SCHRÖDER: Chem. Ber. **97**, 3140 (1964).
[25] G. SCHRÖDER: Cyclooctatetraen, esp. p. 63 Weinheim: Verlag Chemie 1965.
[26] G. SCHRÖDER, J. F. M. OTH and R. MERÉNYI: Angew. Chem. **77**, 774 (1965); internat. ed. **4**, 752 (1965).

Chapter 12

Photochemical cycloaddition of 1,2-quinones, 1,2-diketones and 1,2,3-triketones to multiple bonds

SCHÖNBERG and MUSTAFA [1,2] found that aromatic substituted ethylenes (e.g. styrene, stilbene, triphenylethylene) undergo an addition reaction with phenanthrenequinone (1) in benzene solution under the influence of light:

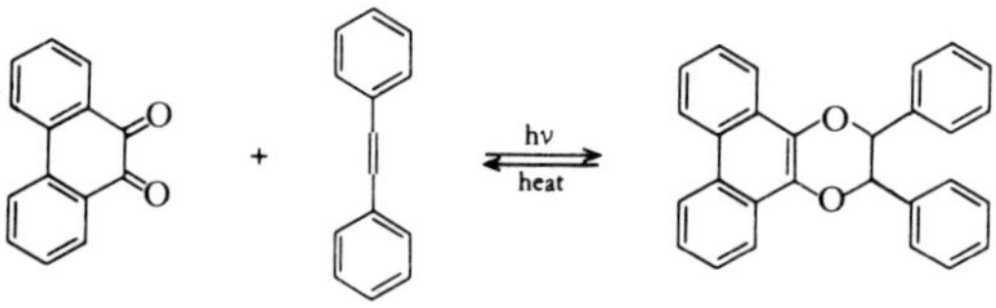

The reaction with stilbene (2) leads to formation of 2,3-dihydro-2,3-diphenylphenanthro[9,10-b][1,4]dioxin (3), the constitution of which was derived from its lack of color and its facile (270°) decomposition to phenanthrenequinone and stilbene as well as from the formation of phenanthrenequinone by the action of concentrated sulfuric acid on 3. Compound 3 was later obtained in a dark process by BUTENANDT et al. [3].

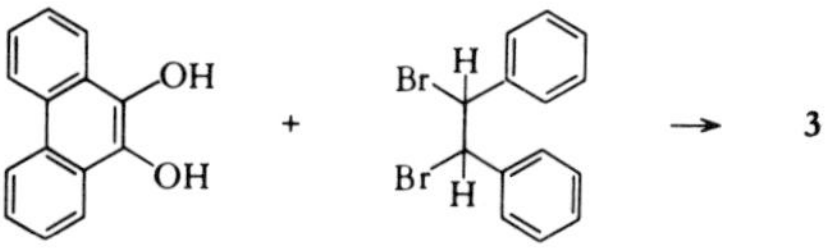

Photochemical reactions proceeding according to the above 1 → 3 conversion will be called "C_4O_2-cycloaddition reactions" in this chapter.

1,2-Dicarbonyl compounds and olefins, which react photochemically yielding dihydro-1,4-dioxin derivatives, are collected in table 9. This list is by no means comprehensive; a complete account has recently been given by PFUNDT and SCHENCK [4]. The similarity between photochemical cycloaddition leading to dihydro-1,4-dioxins and DIELS-ALDER-synthesis has been referred to repeatedly (comp. amongst others [5].

References, p. 125

Cycloaddition of 1,2-dicarbonyl compounds to multiple bonds 119

Table 9

The table lists olefins versus di- and tricarbonyl compounds, the numbers referring to references

1,2-Dicarbonyl compounds Olefinic compounds	9,10-Phenanthrenequinone (1)	5,6-Chrysenedione (4)	Benzo[h]quinoline-5,6-dione (5)	3,4-Dichloro-1,2-naphthoquinone (6)	Tetrachloro-1,2-benzoquinone (7)	Tetrabromo-1,2-benzoquinone (8)	Benzil (9)	1,2,3-Indantrione (10)	1,2,3-Phenalenetrione (11)
Stilbene (2)	2, 3, 6			9	10	11	10	12	13
Chlorostilbene (12)	12								
Indene (13)	8								
Styrene (14)	2				10				
2-Phenyl-2-butene (15)	14								
1,1-Diphenylethylene (16)	2, 15			9			12		
1,1-Diphenyl-1-propene (17)	12								
Triphenylethylene (18)	2		16	9					
2-Styrylnaphthalene (19)	13				10				
2-Stilbazole (2-styrylpyridine) (20)	12								
1,2-Di-(4-pyridyl)-ethylene (21)	17								
2,3-Diphenylacrylophenone (22)	16								
Diphenylketene (23)	18								
3-Ethylidenephthalide (24)	19								
3-Benzylidenephthalide (25)	18			9					
Benzylidenenaphthalide (26)						20			
10-Methyleneanthrone (27)	18								
10-Benzylidenenanthrone (28)	14								
9-Benzylidenexanthene (29)	18								
9-Benzylidenethioxanthene (30)	18								
14-Methylene-14H-dibenzo[a, j]xanthene (31)	19								
Ethyl vinyl ether (32)	8								
Furan (33)	8								
2,5-Diphenylfuran (34)	8								
2,3-Diphenylbenzofuran (35)	14								
Khellinone (36)	10								
2,2-Dimethyl-2H-1-benzopyran (37)	8								
1,3-Diphenyl-1H-2-benzopyran (38)	17								
3-Phenylisocoumarin (39)	8, 17				10				
Spiro[3H-naphtho[2,1-b]-pyran-3,9′-thioxanthene] (40)	19								
Vinyl chloride	21								
3,4,6-Triacetyl-D-glucal	22–24								

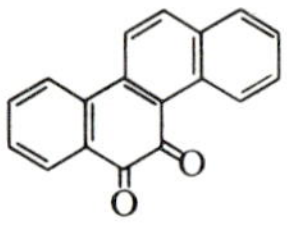

4

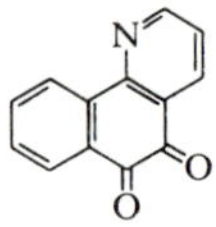

5

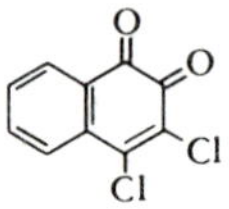

6

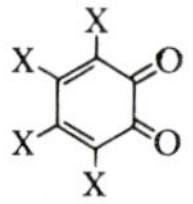

7: X=Cl
8: X=Br

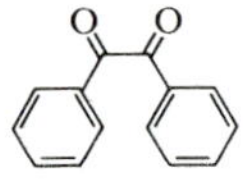

9

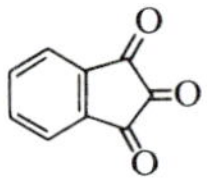

10

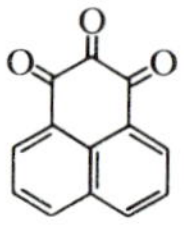

11

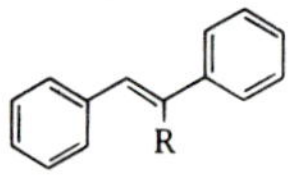

2: R=H
12: R=Cl
22: R=CO–C$_6$H$_5$

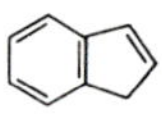

13

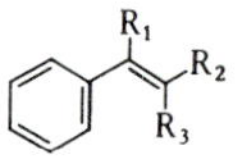

14: R$_1$=R$_2$=R$_3$=H
15: R$_1$=R$_3$=CH$_3$, R$_2$=H.
16: R$_1$=C$_6$H$_5$, R$_2$=R$_3$=H
17: R$_1$=C$_6$H$_5$, R$_2$=CH$_3$, R$_3$=H
18: R$_1$=R$_2$=C$_6$H$_5$, R$_3$=H
19: R$_1$=R$_3$=H, R$_2$=2–naphthyl
20: R$_1$=R$_3$=H, R$_2$=2–pyridyl

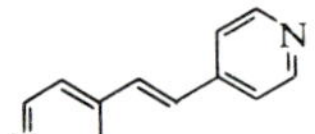

21

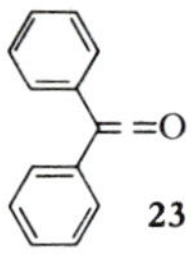

23

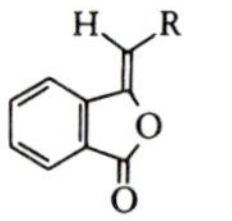

24: R=CH$_3$
25: R=C$_6$H$_5$

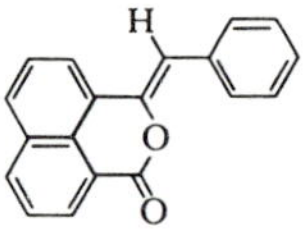

26

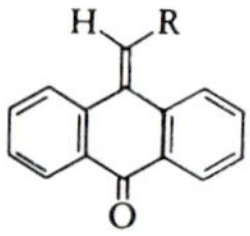

27: R=H
28: R=C$_6$H$_5$

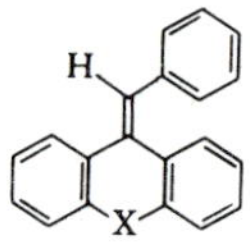

29: X=O
30: X=S

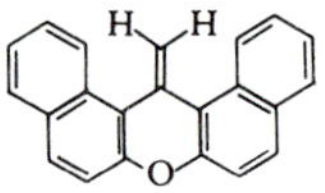

31

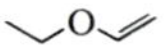

32

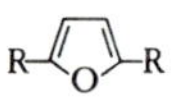

33: R=H
34: R=C$_6$H$_5$

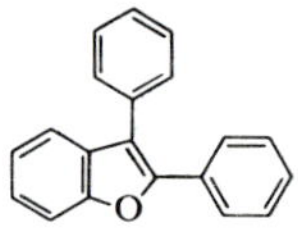

35

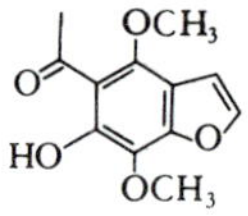

36

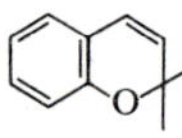

37

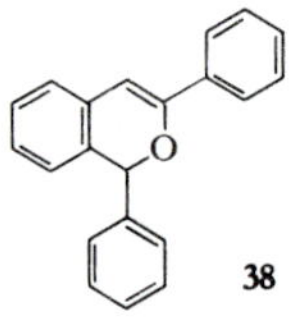

38

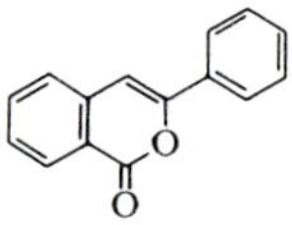

39

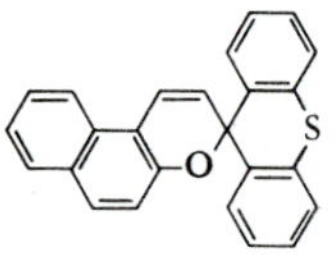

40

The stereochemistry of the addition products has only recently been investigated. Thus starting with either cis- or trans-stilbene cis- and trans-isomers of **3** were obtained [6]. The initial rates of adduct formation were reported to be identical in both cases [7]. It is not out of the question that some of the cycloadducts mentioned in the table are mixtures of stereoisomers.

References, p. 125

Besides, the constitution of several of the products mentioned has not yet been proven unambiguously. A number of products which in the earlier literature were described as C_4O_2-cycloadducts have turned out to be C_3O-cycloadducts (oxetanes) (cf. [8] and chapter 41) and their number may increase in the future. The new formulation as α-ketooxetanes is based mainly on IR and NMR investigations which could not be carried out at the early dates of the syntheses of these compounds.

Most of the C_4O_2-cycloaddition reactions described in the earlier literature have been carried out in sunlight. In several cases it was, however, shown that UV irradiation led to identical products [3, 25]. Among the 1,2-dicarbonyl compounds mentioned in table 9, phenanthrenequinone reacts most readily with formation of dihydro-1,4-dioxins. On the other hand, benzil reacts but poorly [12] and, in some cases, gives oxetanes instead [4, 8].

The mechanism of the reaction will not be dealt with here; SCHENCK [26, 27] assumed that the photochemical cycloaddition involves biradicals to which the olefins add. According to more recent investigations [8, 21], however, the reaction of **1** with olefins seems to be more complicated in nature.

SCHÖNBERG and MUSTAFA [18] found that **1** adds to diphenylketene under the influence of sunlight with formation of 2,3-dihydro-2-oxo-3,3-diphenyl-phenanthro[9,10-b][1,4]dioxin (**41**).

Similar cycloadditions to diphenylketene (**23**) may also be effected, if instead of **23** itself azibenzil is used which in light decomposes into **23** with the liberation of nitrogen. This is visualized below for the formation of 5,6,7,8-tetrachloro-2,3-dihydro-2-oxo-3,3-diphenyl-1,4-benzodioxin (**42**) from tetrachloro-1,2-benzoquinone and azibenzil [28].

A great variety of olefinic compounds has been examined by SCHENCK and co-workers [4, 8, 15, 21, 29] on their tendency to form C_4O_2- and/or C_3O-cycloadducts with 1,2-quinones on irradiation of the components. When indene (**13**) or 3-phenylisocoumarin (**39**) are irradiated with phenanthrenequinone [8], dihydro-1,4-dioxins **43** or **44** are formed together with the α-ketooxetanes. With ethyl vinyl ether (**32**) as olefinic component the product is **45**. Aliphatic and alicyclic olefins such as e.g. 2-methyl-2-butene and bicyclohexylidene undergo photochemical cycloaddition to C_4O_2-adducts (e.g. **46**) as well [15, 29].

The addition compounds which have been obtained by HELFERICH et al. [22—24] are particularly interesting. The photoaddition of phenanthrenequinone to e.g. 3,4,6-triacetyl-D-glucal gives rise to a 1:1 cycloadduct, viz. 1,2-O-9',10'-phenanthrylene-D-glucopyranose 3,4,6-triacetate (**47**). Ozonization of the latter with subsequent saponification leads then to D-glucopyranose. Similarly derivatives of triacetylgalactal [22], diacetylxylal and hexaacetylcellobial [24] have been prepared.

If phenanthrenequinone acts for a long time on chlorostilbene (**12**) in the presence of light, 2,3-diphenylphenanthro[9,10-b][1,4]dioxin (**48**) is obtained, which arises from the intermediate chlorine-containing adduct through loss of HCl [12]. Experiments to obtain **48** by photochemical addition of diphenylacetylene to **1** failed [3].

References, p. 125

When vinyl chloride was similarly added to phenanthrenequinone [*21*], a chlorine free product was isolated which turned out to be formally a 2:1 adduct of phenanthrenequinone to acetylene, viz. **49**.

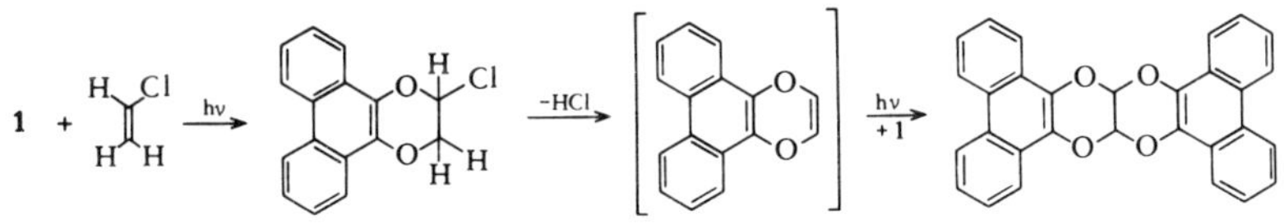

49

Tetrahalogeno-1,2-benzoquinones are able to react with olefins even in darkness at the boiling point of benzene forming C_4O_2-cycloadducts. This has been found independently by SCHÖNBERG [*11*] and by HORNER [*30*]. Thus 5,6,7,8-tetrachloro-2,3-dihydro-2,3-diphenyl-1,4-benzodioxin (**50**) arises from **2** and **7** in boiling benzene.

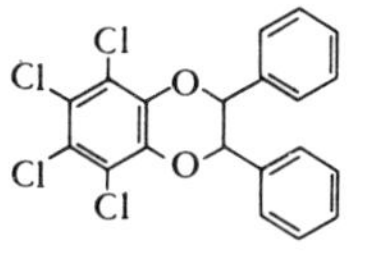

50

It is particularly interesting that the photochemical and thermal reactions of tetrahalogeno-1,2-benzoquinones do not always lead to the same products; since two diene moieties are present in these 1,2-quinones addition may take place either with formation of C_6-cycloadducts or of C_4O_2-cycloadducts. Thus the bright yellow DIELS-ALDER adduct 1,2,3,4-tetrachloro-7-phenylbicyclo[2.2.2]oct-2-ene-5,6-dione (**51**) [*30*] is formed if styrene acts on **7** in boiling benzene, whereas the reaction in sunlight results in nearly colorless 5,6,7,8-tetrachloro-2,3-dihydro-2-phenyl-1,4-benzodioxin (**52**) [*10*].

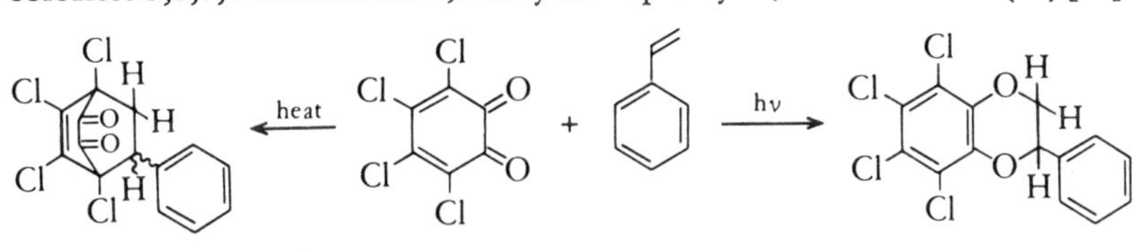

51   **7**   **14**   **52**

2,3-Dihydro-2,3-diphenylphenanthro[9,10-b][1,4]dioxin (3) [*2*]. Phenanthrenequinone (**1**) (1 g) and stilbene (**2**) (0.9 g) in thiophene-free benzene (50 ml) were exposed to sunlight (9 days, August, Cairo), whereupon **1** dissolved completely. After evaporation of the solvent the residue was washed with cold absolute alcohol and then extracted with hot ligroin. On cooling nearly colorless crystals separated which after recrystallization from ligroin melted at about 260°; yield 70%.

Cis- and trans-isomers of **3** [*6*]. 2 g **1** and 4 g of either cis-**2** or trans-**2** are dissolved in 200 ml benzene and the solution irradiated with a mercury immersion lamp (Philips HP 125 W), the water-cooled lamp jacket consisting of Solidex glass. After 6 hrs. the solution is concentrated to about 15 ml, whereupon a part of the trans-cycloadduct precipitates. Chromatography of the filtrate on Al_2O_3 and elution with cyclohexane gives a 1:1 mixture of the unreacted stilbenes. With cyclohexane/benzene 1:1 a blue fluorescing mixture of the isomeric adducts is eluted. Repeated crystallization from benzene/methanol allows a separation of the adducts; cis-3 has m.p. 166—168°, yield 1.52 g; trans-3 has m.p. 257—260°, yield 1.57 g.

9a,14b-Dihydro-10H-indeno[1,2-b]phenanthro[9,10-e][1,4]dioxin (43) [8]. 1.04 g of 1 and 5.8 g of indene (13) were dissolved in 100 ml benzene. This solution was irradiated with a water-cooled mercury immersion lamp (Philips HPK 125 W), the lamp jacket consisting of filterglass WG 1 (Schott). After 1.5 hrs. illumination in an argon atmosphere the solution was evaporated to dryness and the residue digested with 100 ml petroleum ether. 420 mg of yellow crystals, m.p. 174°, remained undissolved which consisted of an α-ketooxetane. The soluble part was brought to dryness and chromatographed on Al_2O_3. Elution with benzene afforded 210 mg (13%) of 43, m.p. 163—165° (from ethanol).

9a,15b-Dihydro-11-oxo-9a-phenyl-11H-[2]benzopyrano[3,4-b]phenanthro[9,10-e][1,4]dioxin (44) [8]. 1.04 g 1 and 2.22 g 3-phenylisocoumarin (39) were dissolved in 100 ml benzene and the solution irradiated as above. After 2 hrs. 1.25 g colorless crystals of 44 had separated. Concentration of the solution to 50 ml and addition of 50 ml of ether gave a second crop, the yield totalling 1.53 g (71%) of 44, m.p. 303° (from acetone). The mother-liquor contained 1.19 g 39, 30 mg 1 and 350 mg of the α-ketooxetane, m.p. 203—206°.

2-Ethoxy-2,3-dihydrophenanthro[9,10-b][1,4]dioxin (45) [8]. A solution of 1.04 g of 1 in 100 ml of ethyl vinyl ether (32) was similarly irradiated during 1.5 hrs. Evaporation to dryness and chromatography of the residue yielded 62% of the cycloadduct 45, which after recrystallization from aqueous methanol melted at 58—62°.

2,3-Dihydro-2,2,3-trimethylphenanthro[9,10-b][1,4]dioxin (46) [29]. 2 g 1 and 4 g 2-methyl-2-butene were introduced together with 200 ml benzene into a water-cooled immersion lamp apparatus of Jena glass and were irradiated with a mercury high-pressure lamp (Philips HP 125 W) under argon. After 6 hrs. the solution was concentrated to about 25 ml and chromatographed on Al_2O_3. Elution with benzene afforded colorless crystals of 46 which, when recrystallized from ethanol, melted at 102°; yield 1.5 g, 56%.

9a,19a-Dihydrophenanthro[9',10':5,6]-1,4-dioxino[2,3-b]phenanthro-[9,10-e]-1,4-dioxin, "phenanthrenequinone-acetylene 2:1 adduct"(49) [21]. 1.04 g 1 were dissolved in 65 ml benzene and the solution saturated with vinyl chloride at a temperature of about —15°. The reaction mixture totalling 110 ml was irradiated at —10° with a mercury immersion lamp through a filterglass GWV (Wertheim) during 1.5 hrs. Evaporation to dryness in vacuo and trituration of the residue with ether left some 49 undissolved. A second crop was obtained on chromatography of the ether soluble part on Florisil and elution with petroleum ether. Recrystallization from xylene gave pure 49, m.p. 348—349°, yield ca. 40%. For details of the work-up procedure and constitution of 3 other products contained in the irradiation mixture the original literature should be consulted.

1,2-O-9',10'-Phenanthrylene-D-glucopyranose 3,4,6-triacetate (11-acetoxymethyl-12,13-diacetoxy-9a,12,13,13a-tetrahydro-11H-phenanthro[9,10-b]pyrano[2,3-e]-[1,4]-dioxin 47) [23]. A suspension of 11.25 g 1 in a solution of 14.7 g 3,4,6-triacetyl-D-glucal in 750 ml benzene is stirred magnetically and irradiated by a watercooled quartz lamp Quarzlampengesellschaft S 700 for 15 hrs. at room temperature. The quinone slowly dissolves. The syrup remaining after evaporation of the solvent in vacuo is triturated while still warm with 150 ml of dry methanol. The addition product precipitates in a crystalline form (yield 12.8 g, 50%) and is obtained by recrystallization from absolute alcohol as pure white feltlike needles of m. p. 209—210°.

References, p. 125

References

[1] A. SCHÖNBERG and A. MUSTAFA: Nature **153**, 195 (1944).
[2] A. SCHÖNBERG and A. MUSTAFA: J. Chem. Soc. **1944**, 387.
[3] A. BUTENANDT, L. KARLSON-POSCHMANN, G. FAILER, U. SCHIEDT and E. BIEKERT: Liebigs Ann. Chem. **575**, 123 (1952).
[4] G. PFUNDT and G. O. SCHENCK, in: 1,4-Cycloaddition Reactions, Ed. by J. HAMER, p. 345–417. New York: Academic Press 1967.
[5] K. ALDER and M. SCHUMACHER, in: Fortschritte der Chemie Organischer Naturstoffe, Ed. by L. ZECHMEISTER, Vol. **10**, p. 1—118, esp. p. 24, Wien etc.: Springer 1953.
[6] G. PFUNDT and S. FARID: Tetrahedron **22**, 2237 (1966).
[7] J. SALTIEL, in: Survey of Progress in Chemistry, Ed. by A. F. SCOTT, Vol. **2**, p. 239—327, esp. p. 279. New York: Academic Press 1964.
[8] C. H. KRAUCH, S. FARID and G. O. SCHENCK: Chem. Ber. **98**, 3102 (1965).
[9] A. MUSTAFA, A. K. MANSOUR and A. F. A. M. SHALABY: J. Amer. chem. Soc. **81**, 3409 (1959).
[10] A. SCHÖNBERG, N. LATIF, R. MOUBASHER and A. SINA: J. Chem. Soc. **1951**, 1364.
[11] A. SCHÖNBERG and N. LATIF: J. Amer. chem. Soc. **72**, 4828 (1950).
[12] A. SCHÖNBERG and A. MUSTAFA: J. Chem. Soc. **1945**, 551.
[13] A. SCHÖNBERG, A. MUSTAFA, M. Z. BARAKAT, N. LATIF, R. MOUBASHER and A. MUSTAFA: J. Chem. Soc. **1948**, 2126.
[14] A. MUSTAFA and A. M. ISLAM: J. Chem. Soc. **1949**, Suppl. 81.
[15] G. O. SCHENCK: Strahlentherapie **115**, 497 (1961).
[16] A. MUSTAFA: J. Chem. Soc. **1949**, Suppl. 83.
[17] A. SCHÖNBERG, N. LATIF, R. MOUBASHER and W. I. AWAD: J. Chem. Soc. **1950**, 374.
[18] A. SCHÖNBERG and A. MUSTAFA: J. Chem. Soc. **1947**, 997.
[19] A. MUSTAFA: J. Chem. Soc. **1951**, 1034.
[20] O. M. ALY, W. I. AWAD and A. M. ISLAM: J. Org. Chem. **23**, 1624 (1958).
[21] S. FARID: Diss. Göttingen 1967.
[22] B. HELFERICH and E. VON GROSS: Chem. Ber. **85**, 531 (1952).
[23] B. HELFERICH, E. N. MULCAHY and H. ZIEGLER: Chem. Ber. **87**, 233 (1954).
[24] B. HELFERICH and M. GINDY: Chem. Ber. **87**, 1488 (1954).
[25] G. O. SCHENCK: Angew. Chem. **64**, 12 (1952).
[26] G. O. SCHENCK: Naturwiss. **40**, 229 (1953).
[27] G. O. SCHENCK and G. A. SCHMIDT-THOMÉE: Liebigs Ann. Chem. **584**, 199.
[28] L. HORNER, E. SPIETSCHKA and A. GROSS: Liebigs Ann. Chem. **573**, 17 (1951).
[29] G. PFUNDT: Diss. Göttingen 1962.
[30] L. HORNER and H. MERZ: Liebigs Ann. Chem. **570**, 89 (1950).

Chapter 13

Photochemical cyclization of aromatic compounds via elimination of hydrogen and/or halogen atoms. Formation of carbocycles

1. Formation of five-membered homocycles. A fluorene derivative from triphenylmethyl

According to SCHMIDLIN and GARCIA-BANUS [1] solutions of triphenylmethyl in benzene are not apparently changed after three months. However, in diffuse daylight complete decolorization of the solution occurs with formation of triphenylmethane and 9,9′-diphenyl-9,9′-bifluorene (**1**).

Light of wavelength 530—400 mμ, which is strongly absorbed by triphenylmethyl, has been shown to be most active in the photochemical transformation of hexaphenylethane [2]. In ultraviolet light, triphenylmethyl is not noticeably oxidized or reduced, irrespective of whether the solution in benzene or solid hexaphenylethane itself is used. The decolorization rate of triphenylmethyl solutions in sunlight is solvent dependent, the velocity increasing in the order: benzene, toluene, m-xylene, hexane [2].

Solutions of tris-(4-biphenylyl)-methyl (**2**), which only exists as a free radical, show no change in their color intensity even after 45 days irradiation with sunlight [1].

9,9'-Diphenyl-9,9'-bifluorene (1) [1]. A benzene solution of triphenylmethyl, prepared from 50 g chlorotriphenylmethane and vacuum-dried silver, was sealed under oxygen-free carbon dioxide. Exposure to diffuse daylight (Winter, Zurich) for 45 days caused the solution to decolorize. The colorless crystals of **1** (2.85 g) formed had m.p. 225°. The filtrate yielded triphenylmethane on evaporation. Total yields from 9.49 g triphenylmethyl were, 3.15 g of **1** and 6.34 g of triphenylmethane.

2. Formation of six-membered homocycles

a) Phenanthrenes from stilbenes

The formation of phenanthrene (**5**) by irradiation of stilbene (**3**) in solution requires a suitable oxidant, such as dissolved molecular oxygen or iodine; on irradiation under a nitrogen atmosphere with careful exclusion of oxygen the only net reaction is cis-trans isomerization with no detectable conversion to phenanthrene [3, 4]. Scope and mechanism of this reaction have found much attention (cf. amongst others [3—9]). The reaction path leading from stilbene to phenanthrene is illustrated below [7, 9]. The intermediate 4a,4b-dihydrophenanthrenes (comp. **4**) could, however, not yet be isolated. With 2,2',4,4',6,6'-hexamethylstilbene (**6**) photocyclization led to the analog **7**, which resisted oxidation by oxygen. Photochemically and thermally 4a,4b-dihydro-1,3,4a,4b,6,8-hexamethylphenanthrene (**7**) reverts to **6** [10]. In certain cases, e.g. with α,α'-dicyanostilbene, isomers of the type **4** compounds were obtained, viz. 9,10-dicyano-9,10-dihydrophenanthrene (**9**) [9].

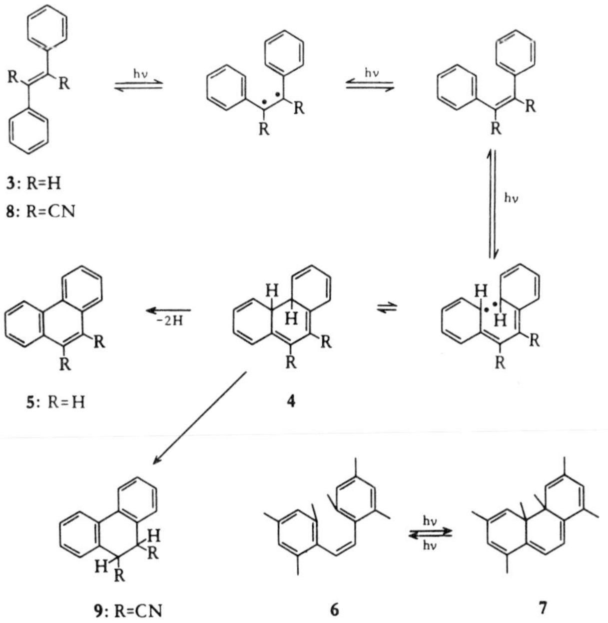

Since its discovery the oxidative photocyclization of stilbenes to phenanthrenes has developed into a valuable synthetic method. 1-, 3- or 9-substituted as well as 9,10-disubstituted phenanthrenes are thus obtainable in 50—90 % yields [6, 9] starting from the readily accessible 2-, 4-, α- or α,α'-substituted stilbenes, respectively, with substituents such as, CH_3, OCH_3, F, Cl, Br, COOH, $COOCH_3$, CN, CF_3 and C_6H_5. Stilbenes with substituents in the 3-position give mixtures of 2- and 4-substituted phenanthrenes. For similar ring closure reactions leading to heterocyclic systems see chapter 14.

The photochemical dehydrocyclization reactions mentioned are most conveniently carried out with iodine as hydrogen acceptor since this oxidant gives faster and cleaner reactions with higher yields than does oxygen. Usually [6] 0.01 mole of the stilbene and 5 mole-% of iodine in 1 l of cyclohexane are irradiated in quartz vessels.

In cases where irradiation of suitably substituted stilbenes fail to produce the desired phenanthrenes another method may be applied with success. When 2-carbomethoxy-α-nitrostilbene (**10**) was irradiated in the presence of dissolved iodine or oxygen no **12** could be isolated [*11*]. On irradiation of 2-carbomethoxy-2'-iodo-α-nitrostilbene (**11**), however, a 45 % yield of the product anticipated, 1-carbomethoxy-10-nitrophenanthrene (**12**), was obtained. In this case a radical reaction path is assumed instead of the dihydrophenanthrene path discussed earlier.

10: R=H
11: R=I

12

Application of the photochemical dehydrocyclization reaction to trans-1-styrylnaphthalene (**13**), carried out under nitrogen in glacial acetic acid and in the presence of ferric chloride, led to the formation of chrysene (**14**) [*8*]. On the other hand, 1,1'-binaphthyl was recovered unchanged after irradiation under the same conditions [*8*].

13   **14**

3-Bromophenanthrene [*6*]. 2.59 g of 4-bromostilbene and 0.127 g of iodine were dissolved in 1 l of cyclohexane and irradiated at 22° for 16 hrs. with a Hanovia water-cooled 19433 Vycor immersion well and a General Electric H 100A4/T mercury lamp. The solution was evaporated to dryness, the residue was taken up in cyclohexane and filtered

References, p. 137

through alumina to remove yellow by-products. Recrystallization from 95% ethanol gave 76% of 3-bromophenanthrene, m.p. 81.2—82.8°.

Chrysene (14) [*8*]. 331 mg of 1-styrylnaphthalene (**13**) and 1.3 g of ferric chloride were dissolved in 150 ml glacial acetic acid and irradiated at 20° in quartz vessels for 7.5 hrs. using a 250 W mercury high pressure lamp Philips Biosol A 10/27. After evaporation of the solvent, the residue was taken up in water and extracted with chloroform. 335 mg. oily yellow crystals were obtained which were chromatographed on neutral alumina. Elution with a mixture of petroleum-ether and benzene (9 : 1) yielded 95 mg (29%) starting material and 124 mg (38%) of chrysene (**14**) which, after two-fold recrystallization from benzene, melted at 242—242.5°.

Another preparation [*4*] gave **14**, m.p. 253.5—254.5°, in 77% yield.

b) Fused aromatic compounds from o-dibenzylidene compounds

According to the investigations of STOBBE [*12*], dibenzylidenesuccinic anhydride (6,7-diphenylfulgide, **15**) may be photochemically dehydrogenated. To this end, a saturated solution of the anhydride in benzene or chloroform, containing some iodine, is exposed to sunlight: the sparingly soluble 1-phenylnaphthalene-2,3-dicarboxylic anhydride (**16**), m.p. 255°, is precipitated from the solution. The reaction principles governing in the stilbene photocyclization are, doubtlessly, at work in the above dehydrocyclization reaction, too.

15: $R_1=R_2=H$
17: $R_1=OCH_3, R_2=H$
19: $R_1=H, R_2=OCH_3$
21: $R_1=H, R_2=CH_3$

16: $R_1=R_2=H$
18: $R_1=OCH_3, R_2=H$
20: $R_1=H, R_2=OCH_3$
22: $R_1=H, R_2=CH_3$

Similar photochemical transformations were carried out by BADDAR et al. [*13*] with the dimethoxy derivatives **17** and **19**. 7-Methyl-1-p-tolylnaphthalene-2,3-dicarboxylic anhydride (**22**) was similarly obtained on irradiation of benzene solutions of bis-(p-methylbenzylidene)-succinic anhydride (6,7-di-p-tolylfulgide, **21**).

The conversion of, for example, **17** to **18** may also be conducted thermally as a dark reaction. Thus the preparation of **20** required either 4 days of insolation or 2 hrs. of heating to 205—210° [*13*].

2-Benzylidene-1-(diphenylmethylene)-indan (**23**), a thermochromic compound, dissolves in organic solvents with a yellow color. When exposed to sunlight in hexane, **23** underwent a dehydrocyclization reaction affording 5,10-diphenyl-11H-benzo[b]fluorene (**26**) [*14*]. CAMPBELL and coworkers [*14*] assume **23** to exist in a biradical form (cf. **24**) in solution. Cyclization is then pictured as free-radical attack of the phenyl ring to give the dihydro intermediate **25** which undergoes dehydrogenation possibly by atmospheric oxygen.

When degassed benzene solutions of 3-benzoyl-2-benzylchromone (**27**) were briefly irradiated with UV-light [*15*] the orange photoenol **28** was formed (for photoenolization see chapter 3). On further irradiation a longer wave length absorption band developed which was ascribed to **29**. The latter compound upon admission of air was converted to 11-hydroxy-6-phenyl-12H-benzo[b]xanthen-12-one (**31**). Heating an oxygen-free solution of **29** resulted in formation of 5a,6-dihydro-11-hydroxy-6-phenyl-12H-benzo[b]xanthen-12-one (**30**) which with oxygen was easily convertible to **31**.

Cyclization with participation of the other phenyl ring was observed to a much lesser extent [*15*]. Thus, 11-phenyl-12H-benzo[b]xanthen-12-one

(**32**), the formation of which can be envisaged by a similar reaction sequence involving, however, the elimination of the elements of water, was obtained in 3.4 % yield on irradiation of **27**.

5-Methoxy-1-(2-methoxyphenyl)-naphthalene-2,3-dicarboxylic anhydride (18) [*13*]. Bis-(2-methoxybenzylidene)-succinic anhydride (6,7-bis-(2-methoxyphenyl)-fulgide, **17**) was dissolved in either benzene or chloroform, containing a trace of iodine, and exposed to sunlight for two days (March, Cairo). The product, which was obtained in nearly quantitative yield, was recrystallized from glacial acetic acid to give straw-yellow, tabular prisms of **18**, m.p. 245—246°. The same product was also accessible by heating **17** at 280—285° for 2 hours.

5,10-Diphenyl-11H-benzo[b]fluorene (26) [*14*]. The hydrocarbon **23** (0.1 g) in 500 ml n-hexane gave a colorless solution after exposure to sunlight for 2 days. Evaporation of the solution gave **26**, m.p. 199°.

11-Hydroxy-6-phenyl-12H-benzo[b]xanthen-12-one (31) [*15*]. The UV- and visible light sources used in the following experiment were a General Electric B-H6 high pressure mercury arc lamp equipped with Corning 9863 and 7740 filters in conjunction with a 500 W Argus 540 slide projector equipped with a Corning 3387 filter.

A solution of 500 mg of **27** in 500 ml of benzene was irradiated for 24 hrs. while dry air was bubbled through. The solvent was evaporated leaving a red-orange residue which was twice recrystallized to give 200 mg of bright orange crystals of **31**, m.p. 261—263°.

The mother liquors of the recrystallization of **31** were evaporated and the residue chromatographed on neutral alumina. Elution with benzene gave 17 mg of **32** and 4.3 mg of a furan compound. Chloroform eluted 220 mg of starting material.

c) Phenanthroperylenediones, dibenzoperylenediones and analogous compounds from less condensed aromatic precursors

$\Delta^{10,10'}$-Bianthrone (**33**) and helianthrone (**34**) are converted to phenanthro[1,10,9,8-opqra]perylene-7,14 dione (**35**) on irradiation of their solutions in organic solvents [*16*]. Similar results were obtained with methoxy derivatives of $\Delta^{10,10'}$-bianthrone [*17*].

The photochemical cyclization of **33** and **34** was more closely investigated by BROCKMANN and MÜHLMANN [*18*]. To account for the hydrogen split off the following equation was established in the case of irradiation of $\Delta^{10,10'}$-bianthrone (**33**) in the absence of oxygen

$$3 \ 33 \xrightarrow{h\nu} 35 + 2 \ 36$$

Irradiation of **33** or **34** in an oxygen atmosphere led to consumption of 2 moles or 1 mole oxygen respectively, hydrogen peroxide being formed. Nitrobenzene or phenanthrenequinone may also be used as oxidants [*18*].

If $\Delta^{10,10'}$-bianthrone be dissolved in acetic anhydride and irradiated for several hours under reflux, a high yield of 10,10'-bi-9-anthrol diacetate (**37**) is obtained [*16, 18*] besides **35**. Similar treatment of helianthrone (**34**) [*18*] leads to formation of dibenzo[a,o]perylene-7,16-diol diacetate (**38**).

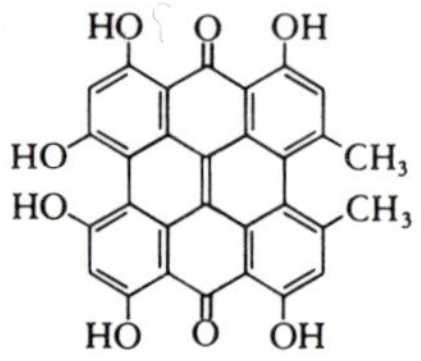

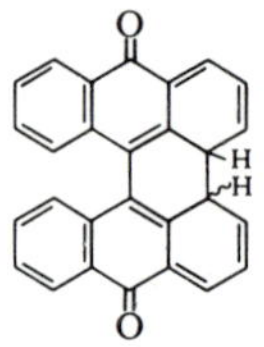

36: R=H
37: R=Ac
38
39
40

Photochemical dehydrocyclization reactions have also enabled the synthesis of hypericin (**39**) and of related compounds. **39** is a photodynamically active dye which has been isolated from plants (for a review on this subject see BROCKMANN [*19*]).

In the foregoing as well as in the following photocyclizations the most plausible formation mechanism involves dihydro intermediates of the kind **40** [*20*], which may be formulated in close analogy to the stilbene to dihydrophenanthrene isomerization.

10-Diphenylmethylene-9-anthrone (**41**) may be transformed to 13-phenyl-8H-dibenz[a,de]anthracen-8-one (**42**) by treatment with aluminium chloride in inert solvents; nevertheless, a better yield and a purer product is obtained if the conversion is carried out photochemically [*21*]. A similar dehydrocyclization permits the synthesis of 16H-benzo[4,5]phenaleno[1,2,3-kl]-

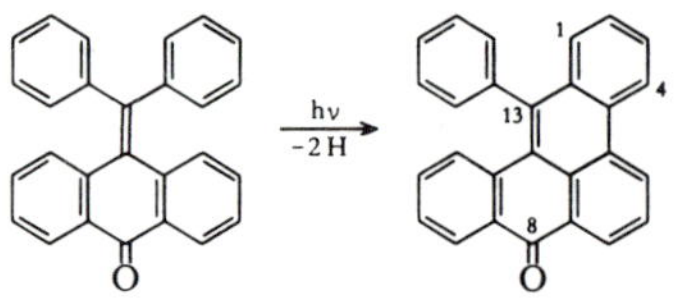

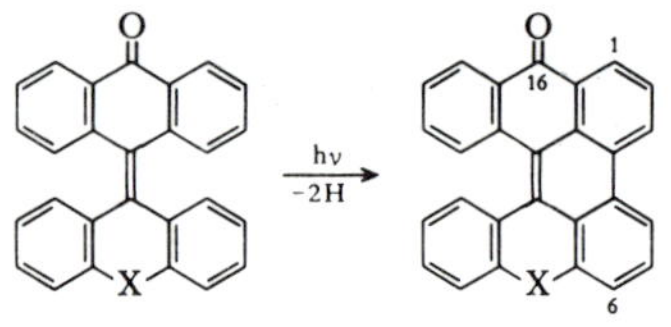

41
42
43: X=O
45: X=S
44: X=O
46: X=S

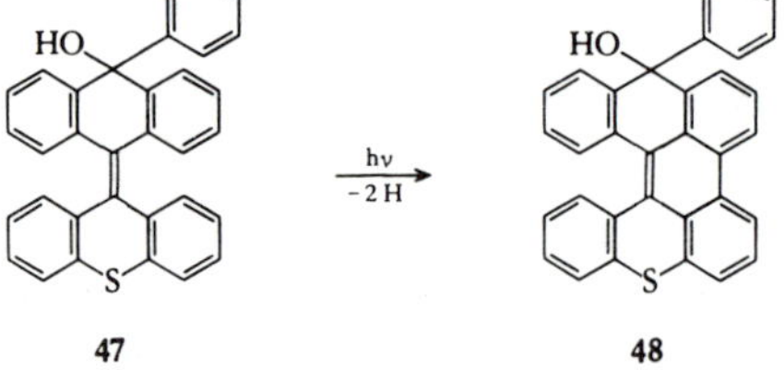

47
48

References, p. 137

xanthen-16-one (**44**) from 10-xanthen-9-ylideneanthrone (**43**) or of 16H-benzo[4,5]phenaleno[1,2,3-kl]thioxanthen-16-one (**46**) from 10-thioxanthen-9-ylideneanthrone (**45**) [*22, 23*]. This dehydrogenation takes place equally well in a CO_2 atmosphere as under oxygen. 9,10-Dihydro-9-phenyl-10-(thioxanthen-9-ylidene)-9-anthrol (**47**) behaves similarly [*23*] with formation of **48**.

Not only hydrogen may be removed photochemically from fused aromatic compounds. The elimination of two chlorine atoms is concerned [*24*] in the preparation of 1,6,8,13-tetrachlorophenanthro[1,10,9,8-opqra]perylene-7,14-dione (**50**) from 1,6,8,11,12,15-hexachlorodibenzo[a,o]perylene-7,16-dione (**49**). The conversion has to be carried out in concentrated sulfuric acid since in benzene solution **49** remains unaffected. **50** is also accessible by irradiation of nitrobenzene solutions of 1,1′,8,8′-tetrachloro-$\Delta^{10,10'}$-bianthrone (**51**) [*25*].

According to SCHÖNBERG and JUNGHANS [*20*] $\Delta^{9,9'}$-bixanthene (**52**) may be photochemically dehydrogenated to benzo[1,2,3-kl:6,5,4-k′l′]dixanthene (**54**). The reaction is visualized as proceeding via the dihydro isomer **53** in analogy to the mechanism established for the formation of phenanthrene from stilbene (see above). **53** is believed to be responsible for the thermochromism observed with **52** [*20*].

Similarly $\Delta^{9,9'}$-bithioxanthene or $\Delta^{9,9'}$-thioxanthylidenexanthene afforded benzo[1,2,3-kl:6,5,4-k′l′]dithioxanthene (**55**) or benzo[1,2,3-kl]thioxantheno[6,5,4-k′l′]xanthene (**56**) [*20*].

Phenanthro[1,10,9,8-opqra]perylene-7,14-dione (35) [*16*]. A yellow saturated solution of 10,10′-bi-9-anthrol (**36**) in glacial acetic acid rapidly turns green on boiling under exposure to sunlight and at the same time crystals are precipitated. On filtering and inspection with a magnifying glass, the fine needles of **35** are readily distinguishable from the compact crystals of starting material **36**. By repeated boiling with glacial acetic acid the two substances may be separated. Yields of higher than 90% may be obtained on prolonged irradiation. In an analogous manner **35** may be prepared from **33** or **34**.

10,10′-Bi-9-anthrol diacetate (37) [*18*]. 150 mg of **33** and 200 mg of anhydrous sodium acetate are dissolved in acetic anhydride and irradiated with a 1000 W lamp at a distance of 20 cm. Oxygen was introduced into the boiling solution until the reaction was complete (3 hrs.). At the end of the reaction, which was signalized by a color change from green to yellow, 63 mg of **35** had precipitated which were filtered off and dried at 180°. The filtrate was concentrated to 15 ml and the residual acetic anhydride destroyed to give 83 mg crystalline **37**, m.p. 272—274°.

13-Phenyl-8H-dibenz[a,de]anthracen-8-one (42) [*21*]. 2 g of **41** are dissolved in xylene in a round-bottomed quartz flask fitted with an aircondenser and exposed to radiation from a Heraeus mercury vapor lamp. The heat of the lamp is sufficient to keep the solution boiling. After 45 hrs. irradiation the xylene solution, which is now light ruby red with a greenish fluorescence, is concentrated to small volume whereupon crystals of **42** separate, m.p. 229° (from glacial acetic acid).

16H-Benzo[4,5]phenaleno[1,2,3-kl]xanthen-16-one (44) [*22*]. 1 g of 10-xanthen-9-ylideneanthrone (**43**) was dissolved in 25 ml dry, thiophene-free benzene and the solution sealed under CO_2 in a Monax glass tube. During the insolation (1 week, June, Cairo) the solution became red with an intense green fluorescence. On evaporation to a volume of 5 ml, red crystals (0.5 g) separated on cooling which were recrystallized from benzene to give **44**, m.p. 245—246°.

1,6,8,13-Tetrachlorophenanthro[1,10,9,8-opqra]perylene-7,14-dione (50) [*24*]. If about 1 g **49** is dissolved in 50 ml concentrated sulfuric acid and this solution is exposed to sunlight for a few days, the evolution of considerable quantities of hydrogen chloride may be observed, as demonstrated by intense fuming, reaction with ammonia, etc. In the course of the irradiation the weak though significant odor of chlorine is also detectable. At the same time the green color disappears little by little and the solution turns blue-red. On addition of water the reaction product is precipitated as a brown amorphous powder. For purification, this is repeatedly boiled with xylene and then recrystallized several times from nitrobenzene to give pure **50** in yellow needles.

Benzo[1,2,3-kl: 6,5,4-k′l′]dixanthene (54) [*20*]. 0.5 g of **52** in 330 ml absolute benzene was irradiated by means of an immersion lamp Hanau 313 with a mercury high pressure burner Q 81 under nitrogen. After 64 hrs. the brown solution which showed a greenish fluorescence was concentrated and filtered through alumina. On evaporating the filtrate and cooling 0.4 g (80%) of **54** was obtained as yellow crystals with a greenish fluorescence, m.p. 245°.

d) Dehydrocyclization of some fused heterocyclic hydrocarbons

DILTHEY and QUINT [*26*] obtained the benzophenanthroxanthylium salts **58** and **63** when the corresponding dibenzoxanthylium salts **57** and **62** (only the carbonium ion structures shown, X usually being ClO_4) were illuminated in aerated glacial acetic acid solutions. Methylene blue or p-benzoquinone were suited as well as hydrogen acceptors. No dehydrogenations took place in the dark under these conditions. Cyclization could, however, also be brought about with the aid of $AlCl_3$.

58 is also obtainable by photochemical elimination of HCl from 14-(2-chlorophenyl)-dibenzo[a,j]xanthylium perchlorate (**59**) or from 14-(2-chlorophenyl)-dibenzo[a,j]xanthene (**60**). The dehydrocyclization of the xanthylium salts was imagined [*26*] to proceed via disproportionation, e.g.

$$2\ 57\ \xrightarrow{h\nu}\ 58\ +\ 61\ +\ HX$$

References, p. 137

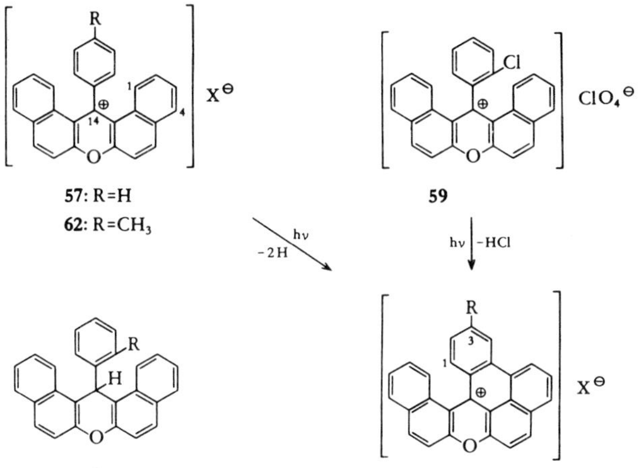

57: R=H
62: R=CH₃

59

60: R=Cl
61: R=H

58: R=H
63: R=CH₃

Thus, 2 moles of the dibenzoxanthylium salt **57** should yield only 1 mole of the benzophenanthroxanthylium salt **58**, if oxidizing agents are absent; the second product, the dibenzoxanthene **61** might then be dehydrogenated in the presence of oxygen or other hydrogen acceptors to a second molecule of **58** via the corresponding dibenzoxanthenol. In the light of more recent research, however, reaction paths requiring formation of dihydro isomers of the kind mentioned earlier (cf. **4**) should rather be drawn into consideration.

The same sort of reaction may also be carried out with analogs of **57** bearing S or NR functions instead of the oxygen hetero atom [27, 28]. Thus, 7,14-diphenyldibenz[a,j]acridinium chloride (**64**) on irradiation in glacial acetic acid gave 10-phenylbenzo[a]phenanthro[1,10,9-jkl]acridinium chloride (**65**).

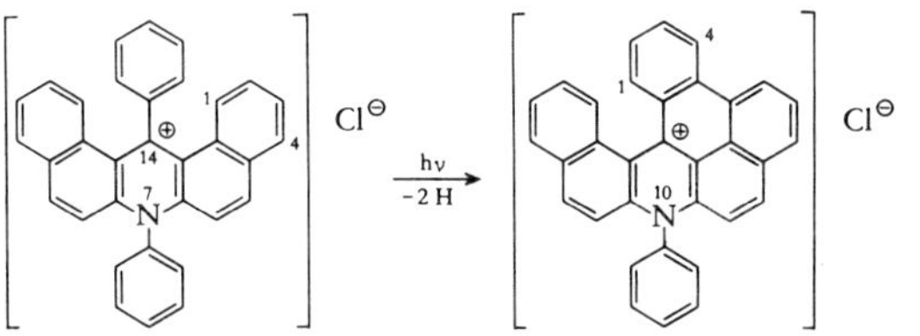

64

65

A reaction similar to the foregoing dehydrocyclization was reported by HUISGEN [29]. 1,3-Diphenylbenzo[f]quinoline (**66**) is stable when irradiated in neutral solution. In an acid medium, however, the rapid formation of a compound is observed, which was formulated as 2-phenylphenanthro-[9,10,1-def]quinoline (**67**).

66 → hv/H⊕, −2H → **67**

Photochemical dehydrocyclizations of sulfur heterocyclic compounds were carried out by CARRUTHERS and STEWART [30]. Thus, irradiation of 3-β-styrylbenzo[b]thiophene (**68**) in boiling hexane in the presence of iodine resulted in smooth conversion to benzo[b]naphtho[1,2-d]thiophene (**69**).

68 → hv, I$_2$ → **69**

Benzo[a]phenanthro[1,10,9-jkl]xanthylium perchlorate (**58**, X = ClO$_4$) [26]. 500 mg of **59** is dissolved in 150 ml boiling glacial acetic acid and exposed to sunlight for 3—4 hrs. while oxygen is passed through the solution. Violet needles separate which are filtered off whilst hot and washed with glacial acetic acid and ether. **58**, X = ClO$_4$, forms violet needles which have a strong bronze surface-lustre; yield 420 mg.

Benzo[a]phenanthro[1,10,9-jkl]xanthylium picrate (**58**, X = C$_6$H$_2$N$_3$O$_7$) [26]. 2 g of **60** are boiled under reflux in 1000 ml glacial acetic acid while oxygen is passed through and the system is irradiated with a mercury vapor lamp. The initial yellow color of the solution is replaced by an intensifying orange-red fluorescence. After 7—8 hrs. the solution is dark red in color while the fluorescence has been almost completely quenched. The solution is concentrated to 120 ml and allowed to cool, when 0.6 g unchanged starting material separates. On shaking with a three-fold quantity of ether, a flocculent blue precipitate is formed which is collected on a filter, washed with ether and dissolved in 100 ml 66% acetic acid. This solution is treated with 5 ml fuming HCl and then added to a hot solution of 0.2 g picric acid in 20 ml hot water. The picrate of **58** crystallizes in blue-black needles or prisms with a bronze lustre.

3-Methylbenzo[a]phenanthro[1,10,9-jkl]xanthylium perchlorate (**63**, X = ClO$_4$) [26]. If a boiling solution of 0.3 g of 14-p-tolyldibenzo[a,j]xanthylium perchlorate (**62**, X = ClO$_4$) in 150 ml glacial acetic acid is exposed to sunlight, precipitation of the dehydrogenated product begins after only a few minutes and is complete in 90 minutes. The mixture is filtered hot and the product washed with warm glacial acetic acid to give 0.1 g of dark brown needles having a greenish-yellow lustre.

2-Phenylphenanthro[9,10,1-def]quinoline (**67**) [29]. 150 mg of **66** is dissolved in 30 ml alcohol and a few drops of concentrated sulfuric acid are added. The colorless solution on exposure to sunlight becomes bright yellow within a few minutes. After 1 hour 170 mg lemon-yellow needles have separated; on further irradiation the filtrate furnishes another 24 mg of the same substance. When the sulfate of the photodehydrogenation product is heated with ammonia and the resulting **67** is recrystallized from much chloroform-alcohol, colorless, felt-like crystals (136 mg) are obtained, m.p. 220°.

References, p. 137

References

[1] J. SCHMIDLIN and A. GARCIA-BANUS: Ber. dtsch. chem. Ges. **45**, 1344 (1912).
[2] S. T. BOWDEN and W. J. JONES: J. Chem. Soc. **1928**, 1149.
[3] F. B. MALLORY, C. S. WOOD, J. T. GORDON, L. C. LINDQUIST and M. L. SAVITZ: J. Amer. chem. Soc. **84**, 4361 (1962).
[4] F. B. MALLORY, C. S. WOOD and J. T. GORDON: J. Amer. chem. Soc. **86**, 3094 (1964).
[5] F. B. MALLORY, J. T. GORDON and C. S. WOOD: J. Amer. chem. Soc. **85**, 828 (1963).
[6] C. S. WOOD and F. B. MALLORY: J. Org. Chem. **29**, 3373 (1964).
[7] W. M. MOORE, D. D. MORGAN and F. R. STERMITZ: J. Amer. chem. Soc. **85**, 829 (1963).
[8] P. HUGELSHOFER, J. KALVODA and K. SCHAFFNER: Helv. chim. acta **43**, 1322 (1960).
[9] M. V. SARGENT and C. J. TIMMONS: J. Chem. Soc. **1964**, 5544.
[10] K. A. MUSZKAT, D. GEGIOU and E. FISCHER: Chem. Comm. **1965**, 447.
[11] S. M. KUPCHAN and H. C. WORMSER: Tetrahedron Letters **1965**, 359.
[12] H. STOBBE: Ber. dtsch. chem. Ges. **40**, 3372 (1907).
[13] F. G. BADDAR, L. S. EL-ASSAL and M. GINDY: J. Chem. Soc. **1948**, 1270.
[14] N. CAMPBELL, P. S. DAVISON and H. G. HELLER: J. Chem. Soc. **1963**, 993.
[15] W. A. HENDERSON, JR. and E. F. ULLMAN: J. Amer. chem. Soc. **87**, 5424 (1965).
[16] H. MEYER, R. BONDY and A. ECKERT: Monatsh. Chem. **33**, 1447 (1912).
[17] G. F. ATTREE and A. G. PERKIN: J. Chem. Soc. **1931**, 144.
[18] H. BROCKMANN and R. MÜHLMANN: Chem. Ber. **82**, 348 (1949).
[19] H. BROCKMANN in: Fortschr. Chem. org. Naturstoffe, ed. by L. ZECHMEISTER, Vol. **14**, 141—185. Wien: Springer 1957.
[20] A. SCHÖNBERG and K. JUNGHANS: Chem. Ber. **98**, 2539 (1965).
[21] E. CLAR and W. MÜLLER: Ber. dtsch. chem. Ges. **63**, 869 (1930).
[22] A. SCHÖNBERG, A. F. A. ISMAIL and W. ASKER: J. Chem. Soc. **1946**, 442.
[23] A. F. A. ISMAIL and Z. M. EL-SHAFEI: J. Chem. Soc. **1957**, 3393.
[24] A. ECKERT: Ber. dtsch. chem. Ges. **58**, 322 (1925).
[25] A. ECKERT and R. TOMASCHEK: Monatsh. Chem. **39**, 839 (1918).
[26] W. DILTHEY and F. QUINT: Ber. dtsch. chem. Ges. **69**, 1575 (1936).
[27] IG Farbenindustrie. Swiss pat. 176 926 of 30. 4. 1934; Chem. Zentr. **1936** I, 648.
[28] IG Farbenindustrie. French pat. 772 781 of 2. 5. 1934; Chem. Zentr. **1936** I, 648.
[29] R. HUISGEN: Liebigs Ann. Chem. **564**, 16 (1949).
[30] W. CARRUTHERS and H. N. M. STEWART: J. Chem. Soc. **1965**, 6221.

Chapter 14

Photochemical dehydrocyclization of aromatic compounds via elimination of hydrogen atoms. Formation of heterocycles

1. Formation of five-membered heterocycles. Carbazoles from diphenylamines

When aerated solutions of N-substituted diphenylamines in organic solvents were irradiated with a mercury lamp, high yields of N-substituted carbazoles were obtained [1]. Two different reaction paths leading to carbazoles were discussed by LINSCHITZ and GRELLMANN [2]. One of these involves 2.

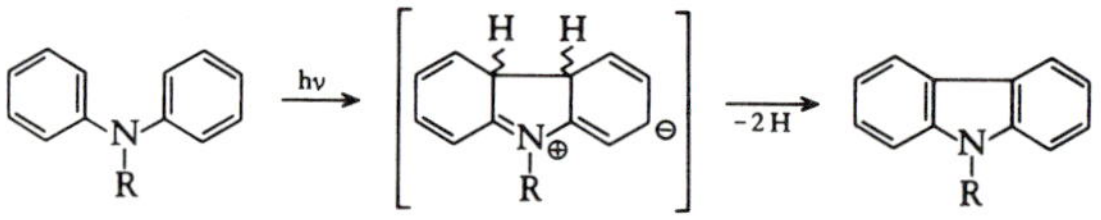

1: R=H
4: R=CH$_3$
6: R=C$_6$H$_5$

2

3: R=H
5: R=CH$_3$
7: R=C$_6$H$_5$

N-Methyldiphenylamine (4) on illumination in aerated hexane solution is thus transformed to 9-methylcarbazole (5), the conversion being 70 % [1]. Analogously triphenylamine (6) gives a 65 % yield of 9-phenylcarbazole (7). Diphenylamine (1) itself is also converted to carbazole (3), but with considerable side reactions.

2. Formation of six-membered heterocycles

a) Benzocinnolines from azobenzenes

In 1960 LEWIS [3] reported on the first example of a photochemical dehydrocyclization process leading to condensed cinnolines. Azobenzene (8) when exposed to light in solutions containing high concentrations of sulfuric acid undergoes oxidative cyclization to benzo[c]cinnoline (9). The same product may also be obtained if the irradiation is carried out in glacial acetic acid in the presence of ferric chloride and under nitrogen [4].

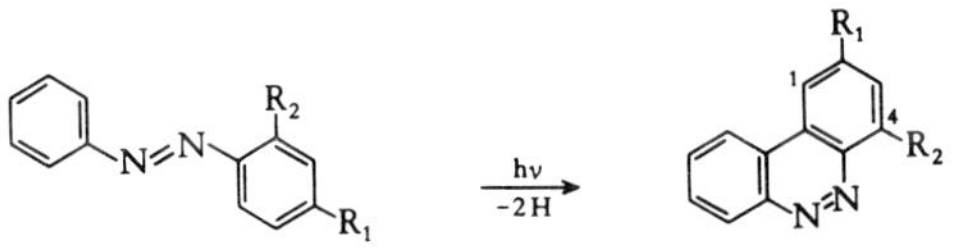

8: $R_1=R_2=H$
10: $R_1=CH_3$, $R_2=H$
12: $R_1=H$, $R_2=CH_3$
20: $R_1=COOH$, $R_2=H$
24: $R_1=NH_2$, $R_2=H$

9: $R_1=R_2=H$
11: $R_1=CH_3$, $R_2=H$
13: $R_1=H$, $R_2=CH_3$
21: $R_1=COOCH_3$, $R_2=H$

The reaction **8 → 9** bears a close resemblance to the stilbene → phenanthrene photodehydrocyclization reaction discussed in chapter 13. Analogous intermediates could be envisaged in the two photoprocesses, but the presence of acid seems to enforce a different reaction pathway (see below).

The above reaction was investigated extensively by BADGER, LEWIS and collaborators [5—8]. Extension to substituted azobenzenes evinced the usefulness of the photocyclization method in the synthesis of substituted cinnolines. Thus, 4-methylazobenzene (**10**) gave 2-methylbenzo[c]cinnoline (**11**) together with an unusual rearrangement product (comp. chapter 36) and 2-methylazobenzene (**12**) afforded 4-methylbenzo[c]cinnoline (**13**) besides benzo[c]cinnoline (**9**). This shows that cyclization can proceed with elimination of a methyl substituent [5].

From 3-methylazobenzene (**14**) both 3-methylbenzo[c]cinnoline (**15**) and 1-methylbenzo[c]cinnoline (**16**) were obtained [5].

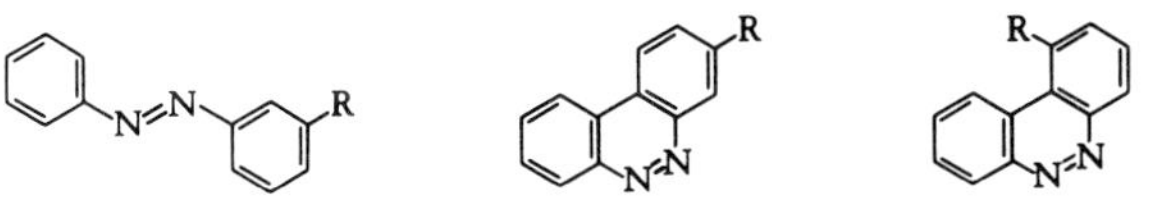

14: R=CH$_3$
22: R=NH$_2$

15: R=CH$_3$
23: R=NH$_2$

16: R=CH$_3$

In all cases benzidines accompanied the benzo[c]cinnolines. Since the former must be derived from hydrazobenzenes the irradiation of azobenzenes was explained [5] to involve a photochemical disproportionation. Accordingly the benzo[c]cinnolines were usually obtained in about 50 % yield.

2,4,6-Trimethylazobenzene (**17**) on irradiation in 22 N sulfuric acid reacted very slowly but more rapidly in 20.5 N sulfuric acid [6]. Product formation proceeded either with ejection or with migration of a methyl group, 2,4-dimethylbenzo[c]cinnoline (**18**) and 1,2,4-trimethylbenzo[c]-cinnoline (**19**) being the respective dehydrocyclization products.

17 → **18** + **19** (hv, −2H)

Scope and limitation of the photocyclization reaction were further explored with azobenzenes substituted in the 2-, 3-, and 4-position with a carboxy group (cf. **20 → 21**), an iodo or a chloro group [6]. Nitro- and acetylazobenzenes gave nitro- and acetylbenzo[c]cinnolines [8]. While 3-aminoazobenzene (**22**) was smoothly converted into 3-aminobenzo[c]cinnoline (**23**) (formation of the isomer being apparently hindered for steric reasons) no cyclization was observed with 4-aminoazobenzene (**24**) [8].

1,1'- and 2,2'-azonaphthalene could not be brought to cyclize, but 1-naphthaleneazobenzene (**25**) did so, producing dibenzo[c,h]cinnoline (**26**) [7].

25 → **26** (hv, −2H)

Benzo[c]cinnoline (9) *(a)* [*3*]. 200 mg of azobenzene (**8**) was dissolved in a mixture of 50 ml ethanol and 150 ml 24 N sulfuric acid and the solution exposed to sunlight for 5 hours. A benzene extract of the neutralized solution was then passed through a column of alumina, elution of the product being effected with a mixture (1:9) of chloroform and benzene. After evaporation of the solvent and crystallization from aqueous ethanol 83 mg of **9**, m.p. 154°, were obtained.

(b) [*4*]. A solution of 160 mg **8** and 1.3 g of ferric chloride in 80 ml glacial acetic acid, contained in a quartz vessel, was irradiated with a Philips Biosol A 10/27 mercury high pressure lamp for 4.5 hrs. under nitrogen. The solvent was evaporated off and the residue chromatographed on neutral alumina. Benzene eluted 65 mg (41%) bright yellow crystals which, after 6-fold reprecipitation from acetone-hexane, melted at 148—148.5°.

Methyl benzo[c]cinnoline-2-carboxylate (21) [*6*]. Azobenzene-4-carboxylic acid (**20**) (2.5 g) was suspended in 22 N sulfuric acid (135 ml) and irradiated in a Pyrex reactor during 360 hours. The reactor consisted of a Philips 125 W mercury-quartz lamp surrounded by a water jacket which in turn was surrounded by a jacket of 150 ml capacity containing the solution to be irradiated. The resulting mixture of solid and solution was diluted with 250 ml water and extracted with benzene (100 ml). This solution gave 0.068 g of unchanged **20**. The acid layer was then partly neutralized. Workup, the details of which should be taken from the original paper, gave, after esterification with methanol, 1.15 g (44%) of crude product which was recrystallized from ethanol to give pure methyl ester **21** as yellow needles, m.p. 185.5°. Hydrolysis gave the free acid, m.p. 363—364°.

3-Aminobenzo[c]cinnoline (23) [*8*]. A solution of 1 g of **22** in 120 ml 98% sulfuric acid was irradiated in the apparatus described above for 108 hours. The mixture was then diluted with ice, basified with sodium hydroxide and repeatedly extracted with benzene.

The benzene solution was evaporated and the residue (0.46 g) chromatographed on alumina. Benzene eluted red-yellow fractions which were evaporated and recrystallized from benzene to afford pure **23**, m.p. 163°.

b) Phenanthridines from Schiff's bases

N-Benzylideneaniline (**27**) in hydrocarbon solutions containing oxidants is easily destroyed by irradiation, the anticipated dehydrocyclization product phenanthridine (**28**) not being formed in detectable quantities [*4*, *9*]. The non-formation of **28** under these conditions was tentatively attributed [*9*] to a low trans-cis isomerization rate shown by **27**. This assumption was supported by the fact that N-diphenylmethyleneaniline (**29**) on analogous treatment gave rise to a dehydrocyclization product, viz. 6-phenylphenanthridine (**30**). This photoconversion did not take place in the absence of oxidants e.g. iodine or oxygen, a result which was interpreted in favor of dihydrophenanthridine being the intermediate [*9*].

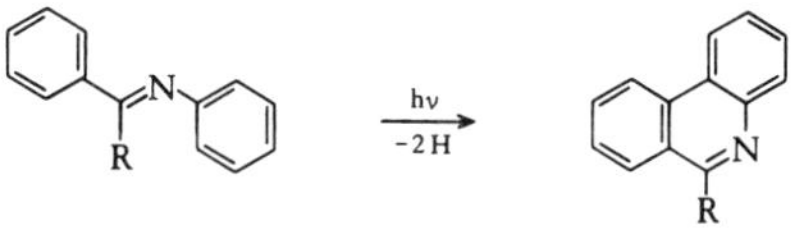

27: R=H
29: R=C_6H_5

28: R=H
30: R=C_6H_5

When the irradiation of N-benzylideneaniline (**27**) itself is carried out in concentrated sulfuric acid [*10*] a mixture of phenanthridine (**28**) and N-phenylbenzylamine is obtained, which points to the participation of the disproportionation mechanism in acid solution.

In these photodehydrocyclization reactions several side reactions may occur which are likely to impair the yields of phenanthridines. Thus

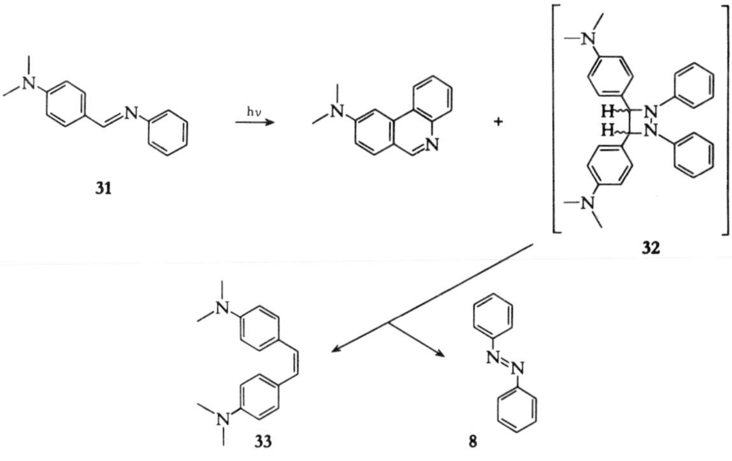

31 **32** **33** **8**

occasionally C$_2$-fragments have been found to be incorporated in products (comp. chapter 36). A rather surprising deviation from the usual reaction path was observed by SEARLES and CLASEN [*11*]. When dilute, ethereal solutions of N-(4-dimethylaminobenzylidene)-aniline (**31**) were irradiated with a mercury lamp [*11*] trans-azobenzene (**8**) and cis-4,4'-bis-(dimethylamino)-stilbene (**33**) were obtained in 35% and 25% yield respectively, besides 15% of 9-dimethylaminophenanthridine. This unusual result indicates the intermediacy of the dimeric compound 3,4-bis-(4-dimethylaminophenyl)-1,2-diphenyldiazetidine (**32**).

The dehydrocyclization of N-(1-naphthylmethylene)-1-naphthylamine (**34**) yielding 40 % of dibenzo[c,i]phenanthridine (**36**) was observed by CAVA and SCHLESSINGER [*12*] when aerobic alcoholic solutions of **34** were irradiated by light of wavelength 280—320 mµ. The Schiff base **34**, being irradiated in benzene solution in the presence of dissolved oxygen afforded a 15% yield of **36**, whereas in a degassed benzene solution no **36** could be isolated. This result was assessed in favor of **35** being an intermediate.

Phenanthridine (28) [*10*]. A solution of 76 mg of **27** in 100 ml of 98% sulfuric acid was irradiated as described above for the preparation of **21**. After 72 hrs. the mixture was diluted with ice and neutralized with sodium hydroxide. Extraction with benzene gave a gum (58 mg) which was chromatographed on alumina. Elution with an ether-benzene (1 : 9) mixture gave, first, N-phenylbenzylamine and then 29 mg of **28**, m.p. 103°.

c) Phenanthridizinium salts from styrylpyridinium salts

Adaptation of the principles elaborated earlier to pyridinium analogs of stilbene proved the photochemical dehydrocyclization method to be very useful for the synthesis of phenanthridizinium salts which are only difficultly accessible by other routes. Irradiation [*13*] of an ethanol solution of 1-styrylpyridinium bromide (**37**) in the presence of iodine afforded after suitable workup a 60 % yield of phenanthridizinium perchlorate (**38**). Similarly 4'-chlorostyrylpyridinium bromide (**39**) furnished 10-chlorophenanthridizinium perchlorate (**40**) in 60 % yield.

37: R=H
39: R=Cl

38: R=H
40: R=Cl

d) Benzo[c]tetrazolo[2,3-a]cinnolinium salts from triphenyltetrazolium salts

2,3,5-Triphenyl-2H-tetrazolium chloride (**41**) and its derivatives are reagents used in the determination of the viability of plant seeds. Aqueous solutions of **41** when exposed to diffuse light [*14*] soon turn red, and the red 1,3,5-triphenylformazan (**43**) precipitates. The filtrate was shown to contain a colorless dehydrogenation product, the constitution of which was established as 2-phenylbenzo[c]tetrazolo[2,3-a]cinnolin-4-ium chloride (**42**) [*14, 15*]. The latter substance may be synthesized almost quantitatively by UV-irradiation of alcoholic solutions of **41** [*15*].

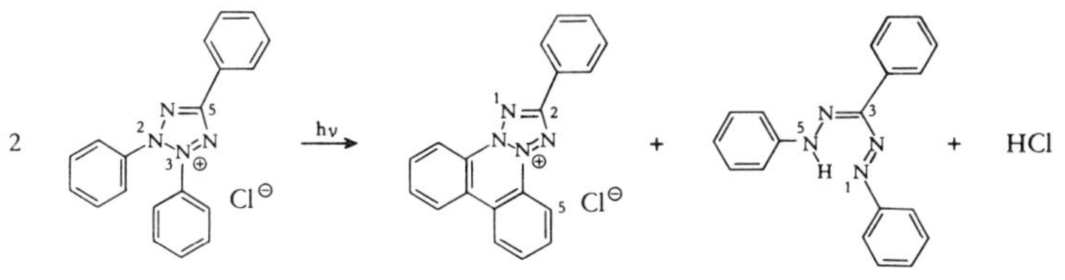

$$2 \quad \mathbf{41} \xrightarrow{h\nu} \mathbf{42} + \mathbf{43} + HCl$$

The nature of the disproportionation process, which proceeds in the absence of oxygen, was studied in more detail by JAMBOR [*16, 17*]. Substituents on the phenyl rings may alter the ease of ring closure considerably [*18*]. Besides tetrazolium salts with unsubstituted phenyl rings (cf. **41**) those in which the phenyls of the 2- or 3-position carry halogen, a carboxyl group or a methoxy residue are also suited to dehydrogenation. Also replacement of the C-5 phenyl group by methyl or hydrogen does not affect the ring closure tendency.

Tetrazolium salts which do not undergo ring formation [*18*] are listed below:

   2-(4-biphenylyl)-3,5-diphenyl-2H-tetrazolium chloride
   2-(1- or 2-naphthyl)-3,5-diphenyl-2H-tetrazolium bromide
   2-(4-nitrophenyl)-3,5-diphenyl-2H-tetrazolium chloride
   2,3-bis-(4-nitrophenyl)-5-phenyl-2H-tetrazolium chloride
   2,3,5-tris-(4-methoxyphenyl)-2H-tetrazolium chloride

Irradiation of bis-tetrazolium salts induced a smooth reaction with formation of bis-benzotetrazolocinnolinium salts [*19*] as illustrated by the photochemical transformation of 3,3'-(methylenedi-p-phenylene)-bis-[2,5-diphenyl-2H-tetrazolium nitrate] (**44**) or of **46** to the respective dehydrocyclization products, 7,7'-methylenebis-(benzo[c]tetrazolo[2,3-a]cinnolin-4-ium nitrate) (**45**) or **47**.

2-Phenylbenzo[c]tetrazolo[2,3-a]cinnolin-4-ium chloride (42) [*15*]. 2,3,5-Triphenyl-2H-tetrazolium chloride (41) (1.5 g) was dissolved in 1500 ml absolute alcohol and irradiated in 20 mm quartz cells for 14 hrs. with unfiltered light from a Quarzlampengesellschaft Hanau S 500 mercury high pressure burner at a distance of 60 cm. The solvent was evaporated to 10 ml. On addition of about 20 ml dry ether 1.27 g of product precipitated which were recrystallized from alcohol-ether to give colorless glistening needles of 42, m.p. 360—361°.

2,2′-p-Phenylenebis-(benzo[c]tetrazolo[2,3-a]cinnolin-4-ium nitrate) (47) [*19*]. Two g of 5,5′-p-phenylenebis-(2,3-diphenyl-2H-tetrazolium chloride) (46) were dissolved in 200 ml water and 10 ml 2 N HNO₃ and the solution irradiated for 20 hrs. with the aid of a water-cooled immersion lamp Heraeus PL 313 with an 80 W burner S 81. Workup gave 1.4 g (60%) of product which, after solution in glacial acetic acid-methanol, was reprecipitated with ethyl acetate to give pure 47, m.p. 420°.

References

[*1*] K.-H. GRELLMANN, G. M. SHERMAN and H. LINSCHITZ: J. Amer. chem. Soc. **85**, 1881 (1963).
[*2*] H. LINSCHITZ and K.-H. GRELLMANN: J. Amer. chem. Soc. **86**, 303 (1964).
[*3*] G. E. LEWIS: Tetrahedron Letters **9**, 12 (1960).
[*4*] P. HUGELSHOFER, J. KALVODA and K. SCHAFFNER: Helv. chim. Acta **43**, 1322 (1960).
[*5*] G. M. BADGER, R. J. DREWER and G. E. LEWIS: Austral. J. Chem. **16**, 1042 (1963).
[*6*] G. M. BADGER, R. J. DREWER and G. E. LEWIS: Austral. J. Chem. **17**, 1036 (1964).
[*7*] G. M. BADGER, N. C. JAMIESON and G. E. LEWIS: Austral. J. Chem. **18**, 190 (1965).

[8] G. M. Badger, C. P. Joshua and G. E. Lewis: Austral. J. Chem. **18**, 1639 (1965).
[9] F. B. Mallory and C. S. Wood: Tetrahedron Letters **1965**, 2643.
[10] G. M. Badger, C. P. Joshua and G. E. Lewis: Tetrahedron Letters **1964**, 3711.
[11] S. Searles jr. and R. A. Clasen: Tetrahedron Letters **1965**, 1627.
[12] M. P. Cava and R. H. Schlessinger: Tetrahedron Letters **1964**, 2109.
[13] R. E. Doolittle and C. K. Bradsher: Chem. and Ind. **1965**, 1631.
[14] F. Weygand and I. Frank: Z. Naturforsch. **3b**, 377 (1948).
[15] I. Hausser, D. Jerchel and R. Kuhn: Chem. Ber. **82**, 195 (1949).
[16] B. Jambor: Pharmazie **13**, 277 (1958); C. A. **53**, 4264 (1959) (comp. C. A. **51**, 1150 (1957)).
[17] B. Jambor: Pharmazie **13**, 282 (1958); C. A. **53**, 4264 (1959) (comp. C. A. **51**, 1151 (1957)).
[18] D. Jerchel and H. Fischer: Liebigs Ann. Chem. **590**, 216 (1954).
[19] D. Jerchel and H. Fischer: Chem. Ber. **88**, 1595 (1955).

Chapter 15

Photochemical dehydrodimerization

Photochemical dehydro-dimerization describes the following change:

$$2\,RH \xrightarrow{h\nu} R-R + 2\,H$$

This conversion takes place in the presence of oxygen, carbonyl compounds (ketones, quinones) or dyes, which act as hydrogen acceptors. Dyes are converted to leuco compounds and quinones give hydroquinones.

1. Dehydrogenation by oxygen

The formation of tetraarylethanes from diarylmethanes will be considered first. This concerns dehydrodimerizations of the following kind, where R is a monovalent group

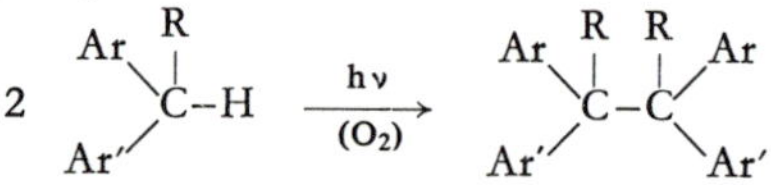

$$2\;\underset{Ar'}{\overset{Ar}{>}}\!\!\underset{}{\overset{R}{C}}\!-\!H \xrightarrow[(O_2)]{h\nu} \underset{Ar'}{\overset{Ar}{>}}\!\!\underset{}{\overset{R\;\;R}{C-C}}\!\!\underset{Ar'}{\overset{Ar}{<}}$$

SCHÖNBERG and MUSTAFA [1] found that it is possible to convert 3-phenyl-benzofuran-2(3H)-one (1), 2-phenylbenzo[b]thiophen-3(2H)-one (2), thioxanthene and anthrone, among others, to the corresponding dehydrodimers by illumination in the presence of oxygen.

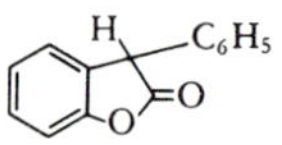

        1 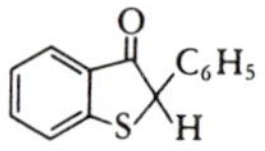   2

3,3′-Diphenyl-(3,3′-bibenzofuran)-2,2′(3H,3′H)-dione (3), 2,2′-diphenyl-(2,2′-bibenzo[b]thiophene)-3,3′(2H,2′H)-dione (4), 9,9′-bithioxanthene (5) and 10,10′-bianthrone (6) were obtained in this way. Experiments carried out in the dark, but under otherwise identical conditions, did not produce the dehydrodimers 3—6.

The formation of compounds 3—6 possibly involves hydroperoxide intermediates [1]. This theory is substantiated by the observation that certain monoarylmethanes and diarylmethanes form hydroperoxides with oxygen in the presence of light (cf. chapter 39).

References, p. 150

HOOKER [2] has developed a method for the photochemical conversion of 2-hydroxy-1,4-naphthoquinone (7) to the dehydrodimer 8 in the presence of oxygen. No reaction mechanism has been decided upon but the hydroxynaphthoquinone probably reacts in its keto-form (9).

9,9′-Bithioxanthene (5) [*1*]. Thioxanthene dissolved in dry, thiophene-free benzene in a Pyrex tube was exposed to the sun (1 day, Cairo) under conditions allowing access of dry air. 5 separated out and was recrystallized from xylene; m.p. 325°.

1,1′,4,4′-Tetrahydro-3,3′-dihydroxy-(2,2′-binaphthalene)-1,1′,4,4′-tetrone (8) [*2*]. Powdered hydroxynaphthoquinone (7) (4 g) was dissolved in boiling water (3 l) and the solution was transferred to an evaporating basin where it was maintained at a temperature of 70°. The reaction solution was irradiated for two hours with a UV lamp located above the dish. During this period 7 (0.5 g) dissolved in boiling water (400 ml) was added at intervals of 30, 60 and 90 minutes from the beginning of irradiation. Shortly after the onset of irradiation a product (yellow-brown, crystalline powder) separated out on the surface of the solution. This was filtered off immediately after the end of the experiment and was washed with water (yield 2.4 g). The substance (8) dissolves slowly in boiling acetic acid (1 g in 100 ml) and small orange-yellow crystals separate out on cooling. When heated in a melting point tube the substance begins to sublime at ca. 250° and melts at ca. 270—275° (decomposition).

2. Dehydrogenation by carbonyl compounds

Dehydrogenation can also be brought about by excited carbonyl compounds. Thus, when the diarylmethanes listed in table 10 were insolated in the presence of quinones and with exclusion of air, the corresponding 1,1,2,2-tetraarylethanes were obtained [*1,3*]. Diphenylmethane, fluorene, xanthene, thioxanthene and anthrone give rise to the following compounds in the presence of light (but not when light is absent): 1,1,2,2,-tetraphenylethane, 9,9-bifluorene (**10**), 9,9'-bixanthene (**11**), 9,9'-bithioxanthene (**5**) and bianthrone (**6**). In some cases the rate of conversion is remarkable (cf. the photochemical formation of **5** from xanthen-9-one and thioxanthene). Table 10 shows (in brackets) the carbonyl compounds found to promote the dehydrodimerization of each substance.

Table 10

diphenylmethane (p-benzoquinone, phenanthrenequinone, anthraquinone).
fluorene (p-benzoquinone).
xanthene (p-benzoquinone).
thioxanthene (p-benzoquinone, phenanthrenequinone, xanthen-9-one).
anthrone (p-benzoquinone, benzophenone).

MOORE and WATERS [*4*] were able to show that p-xylene with phenanthrenequinone forms not only an addition product under the action of UV light (cf. p. 183) but also gives rise to a small quantity (2 %) of 4,4'-dimethylbibenzyl by dehydrodimerization of xylene.

10   11

1,1,2,2-Tetraphenylethane [*3*]. Diphenylmethane (4 g) and p-benzoquinone (1.2 g) in thiophene-free, sodium dried benzene (20 ml) were exposed to sunlight for one month in Cairo. The reaction was carried out in a sealed glass tube (Monax glass) filled with carbon dioxide. Quinhydrone began to separate out after only an hour; after a month the complete precipitate was filtered off. The benzene was evaporated under vacuum and the oily residue steam-distilled, in order to remove diphenylmethane and p-benzoquinone, and then ether-extracted. The extract was dried (sodium sulfate) and the ether removed under vacuum. The residue was extracted with petroleum ether. A semi-solid mass was obtained which was recrystallized from petroleum (100—110°) to give colorless crystals of tetraphenylethane (m.p. 210°, yield 0.7 g).

9,9'-Bithioxanthene (5) [*1*]. Thioxanthene (1 g) and xanthone in equimolar proportions in benzene solution were exposed to sunlight (6 hours, July, Cairo) under the same experimental conditions and in the same type of reaction vessel as above. Afterwards, the bithioxanthene was filtered off, washed with benzene and recrystallized from xylene, m.p. 325°, yield 80%.

3. Dehydrogenations by dyes

Although ergosterol and related compounds form transannular peroxides (cf. chapter 39) in the presence of oxygen and sensitizers, an essentially different kind of reaction occurs in the presence of sensitizers but in the

References, p. 150

absence of oxygen. One such reaction was first discovered by WINDAUS and BORGEAUD [5] when they illuminated ergosterol in the presence of eosin and erythrosin. The reaction involves dehydrogenation of the sterol, with the sensitizing dye serving as hydrogen acceptor:

$$2\ C_{28}H_{44}O + \text{dye} \xrightarrow{h\nu} C_{56}H_{86}O_2 + \text{leuco-dye}$$

Since the product dissolves only with difficulty in most organic solvents it can be isolated with ease. The product was first described as a pinacol; INHOFFEN [6] has suggested formulation as (7,7'-biergosta-5,8,22-triene)-3β,3β'-diol (12) and this has found general approval (cf. WINDAUS and ROOSEN-RUNGE [7]).

12

(7,7'-Biergosta-5,8,22-triene)-3β,3β'-diol (12) [5]. Ergosterol (0.5 g) and erythrosin (0.5 g) were dissolved in 300 ml air-free alcohol and allowed to stand for two sunny days (June, Göttingen) with the exclusion of air. The reaction took place much more slowly than photooxidation. After this period, the dye had almost completely bleached out, the ergosterol had been used up (as shown by the digitonin test), and **12** (0.4 g) had separated out. On exposure to air, the color of the erythrosin was gradually restored.

12 is almost insoluble in alcohol, ether and acetone. It can be dissolved in boiling benzene, chloroform or pyridine and can be precipitated from these solutions by the addition of alcohol, m.p. 202—203°, with decomposition. On rapid heating it may melt at a few degrees higher.

4. Dianthracene from dihydroanthracene

The formation of dianthracene (cf. formula p. 98) from 9,10-dihydroanthracene (13) takes place with the elimination of hydrogen [8].

$$2\ \text{(13)} \xrightarrow{h\nu} \text{dianthracene} + 4\ H$$

Dianthracene [8]. Anthracene-free dihydroanthracene (13) in ethanolic solution is sealed off in a glass tube from which the air has been expelled by heating the alcohol just to its boiling point. On illumination, the dihydroanthracene rapidly converts to dianthracene, which separates from the solution as fine crystals. In sunlight, an abundant yield of dianthracene is formed even within 10—12 hours. Anthracene cannot be detected, even as traces, in the contents of the tube.

References

[1] A. SCHÖNBERG and A. MUSTAFA: J. Chem. Soc. **1945**, 657.
[2] S. C. HOOKER: J. Amer. chem. Soc. **58**, 1212 (1936).
[3] A. SCHÖNBERG and A. MUSTAFA: J. Chem. Soc. **1944**, 67.
[4] R. F. MOORE and W. A. WATERS: J. Chem. Soc. **1953**, 3405.
[5] A. WINDAUS and P. BORGEAUD: Liebigs Ann. Chem. **460**, 235 (1928).
[6] H. H. INHOFFEN: Naturwissenschaften **25**, 125 (1937).
[7] A. WINDAUS and C. ROOSEN-RUNGE: Ber. dtsch. chem. Ges. **73**, 321 (1940).
[8] H. MEYER and A. ECKERT: Monatsh. Chemie **39**, 241 (1918).

Chapter 16

Photochemical dehydrogenation

Relatively little work has been carried out on photochemical dehydrogenation; possibly this is due to the number of good methods which are available to the organic chemist for dehydrogenation.

1. Quinones as dehydrogenating agents

MOORE and WATERS [1] have dehydrogenated tetralin (1) photochemically by reaction with quinones, e.g. phenanthrenequinone, chloranil (2), or 1,4-naphthoquinone. Products were 1,2-dihydronaphthalene (3), detected as 1,2-dibromo-1,2,3,4-tetrahydronaphthalene, and the respective reduction products of the quinones, e.g., 9,10-phenanthrenediol, tetrachlorohydroquinone (4) or 1,4-naphthalenediol. The highest yields of 3 were obtained with phenanthrenequinone (50 %).

Irradiation of tetralin with chloranil in UV-light gave a 25 % yield of a 1 : 1 addition product, viz. 1,2,3,4-tetrahydro-1-(tetrachloro-4-hydroxyphenoxy)-naphthalene (5) [2] (comp. p. 182). The differing results obtained by MOORE and SCHENCK may be explained on the basis of the assumption that the photochemical dehydrogenation of tetralin proceeds via the ether 5 which decomposes thermally or under the influence of HCl into 3 and 4.

References, p. 154

Photodehydrogenation of 1,2,3,4,5-pentaphenyl-1,3-cyclohexadiene (**6**) with chloranil leads to the formation of pentaphenylbenzene (**7**) [*3*], the yield being 99 %.

$$\underset{6}{\text{H}_5\text{C}_6\text{-substituted cyclohexadiene}} \xrightarrow[\text{chloranil}]{h\nu} \underset{7}{\text{H}_5\text{C}_6\text{-substituted benzene}}$$

1,2-Dihydronaphthalene (3) [*1*]. Phenanthrenequinone (2 g) was added to 50 ml carefully purified **1** and the mixture irradiated under N_2 with a 500 W Hanovia mercury quartz lamp situated 25 cm from the reaction vessel. During irradiation (18 days) the thin-walled soft glass vessel was shaken. Afterwards, the phenanthrenediol (0.35 g) which had formed was filtered off, and the tetralin removed by distillation. The distillate was treated with a solution of bromine (10 % in chloroform) at 0°, until rapid disappearance of the color of the bromine ceased (about 0.4 ml bromine). After evaporation of the solvent, 1,2-dibromo-1,2,3,4-tetrahydronaphthalene (0.7 g) was obtained, m.p. 71°. Yield of **3**: 50%.

Pentaphenylbenzene (7) [*3*]. A solution of 2.03 g of **6** and 6.12 g of **2** in 1.1 l of distilled benzene, which was purged with nitrogen for 6 hrs., was irradiated by a Hanovia mercury vapor lamp equipped with a Vycor filter, under a nitrogen atmosphere for 140 min. After removal of the solvent, the red solid was chromatographed on an alumina column. The column was eluted with hexane-benzene (20 : 1) and the eluate was analyzed spectroscopically. In this manner, 2.01 g of a colorless solid were obtained along with 15 mg of a yellow solid as the only hydrocarbon fractions from the column. The white solid was recrystallized from benzene-hexane solution to give pure **7** in colorless prisms, m.p. 250.3—250.6°.

2. 1,2-Disulfides as dehydrogenating agents

The dehydrogenation of tetralin with the aid of diphenyl disulfide requires high temperatures and leads to the formation of thiophenol and naphthalene [*4*]. Under similar conditions anthracene has been obtained from dihydroanthracene.

Photochemical dehydrogenation, however, may be effected at room temperature using diphenyl disulfide and isoamyl disulfide. According to

References, p. 154

NAKAZAKI [5], benzhydrol and diphenyl disulfide, when irradiated in petroleum ether, produced benzopinacol. Tetralin was dehydrogenated photochemically in the presence of 2-benzothiazolyl disulfide (**8**).

MORIZUR [6] treated some sesquiterpene hydrocarbons (e.g. guaiene (**9**), aromadendrene (**10**), and α-gurjunene (**11**)) with thiyl radicals generated either thermally or photochemically from diaryl disulfides. In each case guaiazulene (**12**) was formed, yields from the photolytic procedure being generally slightly lower.

The dehydrogenating action of **8** and other diaryl disulfides in light is readily explainable by disulfide homolysis into free radicals which have been termed thiyl radicals [7]. These resonance stabilized radicals are the hydrogen abstracting species. The investigations of, amongst others, KHARASCH et al. [8] have rendered this assumption most plausible since thiyl radicals resulting from photoinitiated homolysis of disulfides were found to cause polymerization of vinyl compounds and dienes and to accelerate radical type addition of thiols to styrene. Reactions between photochemically produced thiyl radicals and 2,2-diphenyl-1-picrylhydrazyl were reported as well [9]. Finally, phenylthio radicals, prepared by photolysis of diphenyl disulfide, were found [10] to be stable at 77 °K.

3. Photosensitized dehydrogenation using dyes

In the absence of oxygen, irradiation of ergosterol in the presence of certain dyes leads to dehydrodimerization (comp. p. 149). A reaction of a different type has been observed by SCHULLER and LAWRENCE [11] in the case of levopimaric acid (**13**) and palustric acid (**14**).

Irradiation of deaerated ethanol solutions of **13** or **14** respectively with erythrosin B as sensitizer in varying ratios indicated that about one mole of dye was required for reaction with two moles of resin acid. Under these conditions, the product of the photodehydrogenation was in each case found to be dehydroabietic acid (**15**; 20 % isolable yield), indicating that dehydrogenation to an aromatic system had occurred.

Interestingly enough the greater part of the sensitizer could be replaced by an easily reducible compound, viz. nitromethane. Irradiation of **13** in the presence of a catalytic amount of erythrosin B and a molar amount of nitromethane resulted in a 17 % yield of **15** [11].

Dehydroabietic acid (15) [*11*]. A solution of 11.9 g of erythrosin B in 2700 ml of 95% ethanol was filtered and 8.17 g of levopimaric acid (**13**) dissolved in the filtrate (0.005 M in dye and 0.01 M in resin acid). The solution was placed in a 40 W reactor, purged with prepurified nitrogen, the reactor sealed, and irradiation commenced. Two external air blasts were directed on the reactor to keep the temperature around 30°. After 30 hrs. irradiation the solvent was removed under reduced pressure and the dry residue extracted with ether. The ether was filtered, washed with water, and the ether removed. The residue (8.0 g) exhibited no absorption maximum in the 272 mμ region. It was converted to a cyclohexylamine salt in acetone solution; yield 3.59 g (33%). The free acid was liberated from the salt using an aqueous phosphoric acid-ether mixture. The crude acid was placed on silica and eluted with 1200 ml of benzene. The effluent was collected in 75-ml aliquots and the solvent blown off with nitrogen. The residue from fractions 3—9 were combined and crystallized from 95% ethanol to give 0.98 g of dehydroabietic acid; m.p. 169—171°. Two further crops of 0.48 g and 0.14 g were obtained for a total of 1.60 g or 20% conversion from **13**.

References

[*1*] R. F. MOORE and W. A. WATERS: J. Chem. Soc. **1953**, 3405.
[*2*] G. O. SCHENCK: Dechema-Monographien **24**, 105 (1955).
[*3*] G. R. EVANEGA, W. BERGMANN and J. ENGLISH JR.: J. Org. Chem. **27**, 13 (1962).
[*4*] M. NAKAZAKI: Nippon kagaku zasshi **74**, 403 (1953); C. A. **48**, 12017 (1954).
[*5*] M. NAKAZAKI: Nippon kagaku zasshi **74**, 405 (1953); C. A. **48**, 12018 (1954).
[*6*] J.-P. MORIZUR: Bull. Soc. Chim. France **1964**, 1338.
[*7*] A. SCHÖNBERG, E. RUPP and W. GUMLICH: Ber. dtsch. chem. Ges. **66**, 1932 (1933).
[*8*] M. S. KHARASCH, W. NUDENBERG and T. H. MELTZER: J. Org. Chem. **18**, 1233 (1953).
[*9*] Y. SCHAAFSMA, A. F. BICKEL and E. C. KOOYMAN: Tetrahedron **10**, 76 (1960).
[*10*] U. SCHMIDT, A. MÜLLER and K. MARKAU: Chem. Ber. **97**, 405 (1964).
[*11*] W. H. SCHULLER and R. V. LAWRENCE: J. Org. Chem. **28**, 1386 (1963).

Chapter 17

Photochemical additions to carbon-carbon multiple bonds not resulting in ring formation

Reactions dealt with in this chapter usually involve free radicals. The importance of such reactions has recently given origin to a comprehensive survey on the use of radical reactions in synthetic organic chemistry by SOSNOVSKY [1].

The most commonly used radical sources in the dark are peroxides and azonitriles, the number of radical initiators increasing with higher temperatures applied. Compounds with covalent bonds frequently dissociate in light at room temperature. If the compounds themselves do not absorb light photosensitizers may be added which absorb the incident light.

1. Water

The photochemical addition reactions of water are worthy of detailed consideration because of their possible significance in biological processes.

The photochemical addition of water to olefins has been known for some time but this reaction has had little significance to date as far as preparative chemistry is concerned. As an example of such a reaction one may quote the conversion of crotonic acid to 3-hydroxybutyric acid by addition of water when irradiated with ultraviolet light for several weeks [2].

WANG et al. [3] subjected an aqueous solution of 1,3-dimethyluracil (1) to ultraviolet irradiation until a decrease of 80—90% in the optical density (260 mμ) had been observed. 6-Hydroxy-1,3-dimethylhydrouracil (2), m.p. 105—106°, was obtained in 60—75% yield. In a similar way, 5-fluorouracil (3) gave 5-fluoro-6-hydroxyhydrouracil (4) [4].

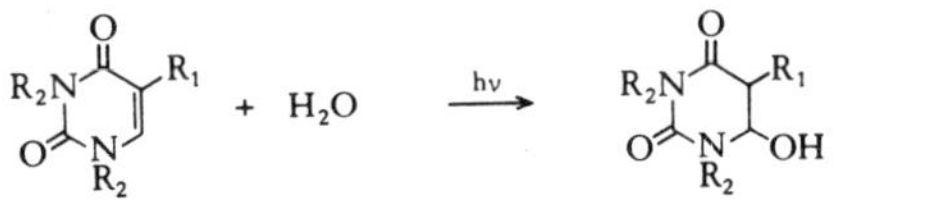

1: $R_1 = H$, $R_2 = CH_3$
3: $R_1 = F$, $R_2 = H$

2: $R_1 = H$, $R_2 = CH_3$
4: $R_1 = F$, $R_2 = H$

References, pp. 180—181

1,3-Dimethyluracil is regenerated from **2** by acid or heat. The scope and limitations of photochemical additions of water to pyrimidines have recently been reviewed by McLaren and Shugar [*5*] and by Smith [*6*].

Stoll and Schlientz [*7*] have studied the photochemical behavior of some lysergic and isolysergic acid derivatives of general structure **5** (only the former series shown). On irradiation of their solutions in dilute acid the authors obtained in each case two products which they named by attaching the prefix "lumi" to the names of the starting reagents. Lumi-products were only formed in the presence of light at temperatures below 30°, oxygen being absent.

5: $R=C_{17}H_{20}N_3O_4$
9: $R=OH$

6: $R=C_{17}H_{20}N_3O_4$

7: $R=C_{17}H_{20}N_3O_4$

By adding water across the Δ^9-double bond a new center of asymmetry is established at C-10. Thus both ergotamine (8α-COR) (**5**) and ergotaminine (8β-COR) each yield two stereoisomeric photoderivatives. The hydroxyl group of the lumi- compound in the two series can be in either the cis- or trans-position with respect to the hydrogen atom at C-5. The steric relationships are represented by the above formulae, the products originating from ergotaminine differing only in the configuration at C-8. 6-Methyl-Δ^8-ergolene-8-carboxylic acid (**8**), which is isomeric with lysergic acid (**9**), was shown to be completely inert towards water on irradiation [*8*]. The photoaddition of water to other unsaturated members of ergot alkaloids has systematically been studied by Hellberg [*9, 10*].

Lumiergotamines 6 and 7 [*7*]. 4 g of ergotamine (**5**) were dissolved in 50 ml glacial acetic acid in the cold. The solution was diluted with 450 ml of boiled distilled water which had been saturated with carbon dioxide and was then introduced into a thin-walled glass bottle. After the air had been displaced by carbon dioxide the vessel was illuminated by exposing it to direct sunlight for 30 hrs. (middle of August, Basle). The temperature of the solution did not rise above 30°.

The liquid became dark brown and no longer showed fluorescence under UV-light. It was then made alkaline with concentrated ammonia solution and thoroughly shaken with chloroform. The chloroform extracts were pooled, dried over sodium sulfate

References, pp. 180—181

and evaporated to dryness. The products were then isolated by chromatography for details of which the original literature should be consulted. Recrystallization finally yielded 2.6 g of lumiergotamine-(I) (6) of m.p. 247° and 205 mg of lumiergotamine-(II) (7) m.p. 192°.

2. Hydrogen peroxide (MILAS reaction)

Aqueous solutions of hydrogen peroxide react slowly, if at all, with allyl alcohol, crotonic or maleic acids, and similar unsaturated compounds in the dark. MILAS [11] has, however, established that hydrogen peroxide will add on to these compounds under the influence of ultraviolet light. It is assumed that the hydrogen peroxide is split photochemically to yield an activated free hydroxyl radical [12].

$$HO-OH \xrightarrow{h\nu} 2 \cdot OH$$

$$H_2C=CH-CH_2OH + H_2O_2 \xrightarrow{h\nu} HOH_2C-CHOH-CH_2OH$$

The photoaddition of two hydroxyl radicals to fumaric acid has been reported to be non-stereospecific [13]. Hydration of fumaric acid to malic acid occurred only with FENTONS reagent in a dark process, but not on photolysis of aqueous hydrogen peroxide.

Glycerol [11]. 10 g (1 mole) allyl alcohol were mixed in a quartz vessel with a 10% solution of hydrogen peroxide (1.05 mole). The mixture was exposed to the light of a Cooper-Hewitt Hg-vapor lamp and shaken vigorously (100 shakes per minute). After irradiation for 168 hrs. the mixture had lost 89% of its peroxide content. The products of the reaction were fractionally distilled and 6.8 g of a fraction boiling at 287—289° were obtained.

The reaction vessel was surrounded by an aluminium reflector to provide maximal utilization of rays from the lamp, which was located 50 cm from the vessel. Ventilation prevented the temperature of the reaction mixture from rising above room temperature.

3. Hydrogen bromide

One can bring about the photochemical addition of hydrogen bromide to olefinic double bonds in both acyclic and cyclic compounds. HASZELDINE and STEELE [14] state that hydrogen bromide will not add on to trifluoroethylene in the dark, even at 100°.

In the light, addition takes place rapidly at room temperature, and an almost quantitative yield of a mixture of 1-bromo-1,2,2-trifluoroethane and 1-bromo-1,1,2-trifluoroethane is obtained.

$$HFC=CF_2 + HBr \xrightarrow{h\nu} BrFHC-CHF_2 + BrF_2C-CFH_2$$

HASZELDINE [15] carried out the photochemical reaction of hexafluoropropene (10) with hydrogen bromide and obtained 1-bromo-1,1,2,3,3,3-

hexafluoropropane (**11**). This compound may also be synthesized by the dark reaction of HBr with **10** in the presence of $AlCl_3$. The photoaddition was rationalized as proceeding in the following manner:

$$HBr \xrightarrow{h\nu} H + \cdot Br$$
$$F_3C-CF=CF_2 \text{ (10)} + \cdot Br \longrightarrow F_3C-\overset{\cdot}{C}F-CF_2Br$$
$$F_3C-\overset{\cdot}{C}F-CF_2Br + HBr \longrightarrow F_3C-CFH-CF_2Br \text{ (11)} + \cdot Br$$

GOERING et al. [*16*] showed that 1-bromocyclohexene rapidly added on hydrogen bromide under UV irradiation yielding cis-1,2-dibromocyclohexane (m.p. 9.7—10.5°).

The stereochemistry of hydrogen bromide addition to olefins has been investigated by several authors. Under illumination (G. E. RS reflector sunlamp) propyne and HBr react rapidly in the liquid phase (—78 to —60°) in a stereospecific trans radical process producing cis-1-bromopropene [*17*].

$$H_3C-C\equiv CH + HBr \xrightarrow{h\nu} \underset{H_3C}{\overset{H}{\diagdown}}C=C\underset{Br}{\overset{H}{\diagup}}$$

The ultraviolet light initiated addition of hydrogen bromide to 2-bromo-2-norbornene (**12**) produced a mixture of trans-2,3-dibromonorbornane (**13**) and exo-cis-2,3-dibromonorbornane (**14**) [*18*]. These products were not interconverted under the reaction conditions. Exo-2-bromonorbornane (**15**) was shown to be a secondary product arising from photochemical debromination of **13** and/or **14**.

1-Bromo-1,1,2,3,3,3-hexafluoropropane (11) [*15*]. Hexafluoropropene (**10**) (2 g) and hydrogen bromide (5% excess) were sealed in a previously evacuated quartz tube (50 ml) and irradiated with a Hanovia UV lamp without filter. After irradiation for seven days the pale brown reaction product was washed with water and distilled. Distillation afforded 10% of unreacted **10**, 88% of 1-bromo-1,1,2,3,3,3-hexafluoropropane (**11**), b.p. 35.5° and 2% of 1,2-dibromohexafluoropropane, b.p. 70—72°.

2,3-Dibromonorbornanes (13 and 14) [*18*]. A solution of 3 g (17 mmole) of **12** in 200 ml of purified pentane was placed in a reaction flask and cooled to 0°, after which nitrogen was passed through for 15 minutes. The solution was irradiated with a Hanau

References, pp. 180—181

S 81 quartz immersion lamp during which time dry gaseous hydrogen bromide was bubbled through at a gentle rate. The reaction was allowed to proceed for 15 minutes. The excess hydrogen bromide was removed by washing with water and 10% sodium carbonate solution. After being dried over anhydrous potassium carbonate, the pentane was removed and the residue distilled to furnish 3.91 g (89%) of product, b.p.$_{0.7}$ 61—73°.

Careful fractionation of the mixture afforded **13** in the first fractions as shown by infrared spectrum and gas chromatography. Crystalline exo-cis material **14** separated from the last fraction. Recrystallization from methanol gave pure **14**, m.p. 58.5—60°.

4. Nitrosyl chloride

PARK et al. [*19*] succeeded with the photochemical addition of nitrosyl chloride to hexafluoropropene or to 1,1-dichloro-2,2-difluoroethylene (**16**).

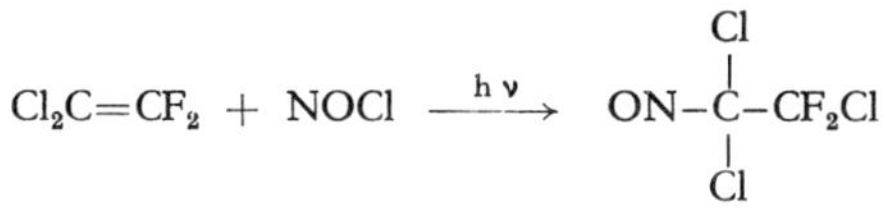

$$Cl_2C=CF_2 + NOCl \xrightarrow{h\nu} ON-\underset{\underset{Cl}{|}}{\overset{\overset{Cl}{|}}{C}}-CF_2Cl$$

       16   **17**

1,1,2-Trichloro-2,2-difluoro-1-nitrosoethane (17) [*19*]. 132 g (1 mole) of **16** and 65 g (1 mole) of nitrosyl chloride were placed in an evacuated heavy wall Pyrex tube of approximately 300 ml capacity, which was provided with a pressure gauge. The tube was irradiated with an ultraviolet lamp for about 1 hr. The reaction started in 15—20 min. and was extremely exothermic. The color of the contents changed progressively from red-brown to blue. At the end of 1 hr. the tube was cooled with Dry Ice and upon opening a small quantity of gas escaped. The crude blue product was washed with water and dried over calcium chloride. The product (162.5 g) was distilled at atmospheric pressure and gave 14.5 g (9.3%) of **17** together with a variety of other products.

In another similar experiment in which distillation was carried out at reduced pressures, 63 g of **17**, b. p.$_{200}$ 14—16° was obtained. This compound is thermally unstable and decomposes when distilled at normal pressure.

5. Alcohols, ethers, and tert. butyl hypochlorite

a) Alcohols

The addition of alcohols to olefins may proceed by either of the two routes depicted below. Products of the addition reaction will be ethers and/or alcohols.

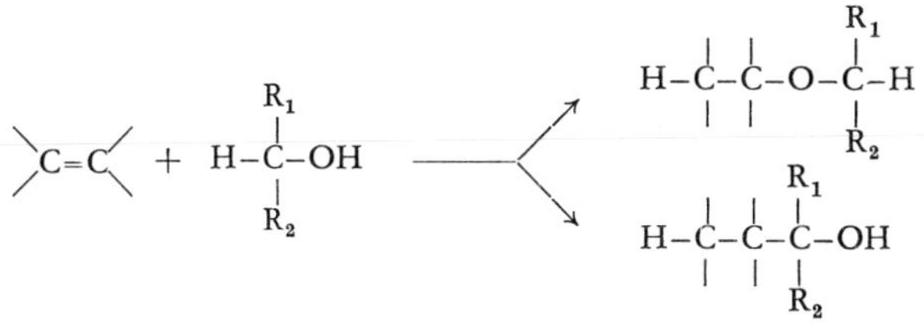

$$\underset{}{\overset{}{>}}C=C\overset{}{\underset{}{<}} + H-\underset{\underset{R_2}{|}}{\overset{\overset{R_1}{|}}{C}}-OH \longrightarrow \begin{array}{c} H-\underset{|}{\overset{|}{C}}-\underset{|}{\overset{|}{C}}-O-\underset{\underset{R_2}{|}}{\overset{\overset{R_1}{|}}{C}}-H \\ \\ H-\underset{|}{\overset{|}{C}}-\underset{\underset{R_2}{|}}{\overset{\overset{R_1}{|}}{C}}-\underset{}{\overset{}{C}}-OH \end{array}$$

The photochemical addition of alcohols across the olefinic double bond was already known 50 years ago but the reactions which were investigated at that time were found to proceed very slowly. An early example, investigated by STOERMER and STOCKMANN [2], is the addition of methanol to crotonic acid which led to the formation of 3-methoxybutyric acid (18). Ultraviolet irradiation for three weeks yielded only 8 g of this acid from 90 g of crotonic acid. Under similar conditions, 3-ethoxybutyric acid was formed in only 6.8% yield.

$$H_3C-CH=CH-COOH + CH_3OH \xrightarrow{h\nu} H_3C-CH-CH_2-COOH$$
$$\underset{\underset{\mathbf{18}}{OCH_3}}{|}$$

Our knowledge of the photochemical addition of alcohols across the olefinic double bond has been considerably augmented by the work of URRY and collaborators [20, 21]. For example, 2-methyl-2-decanol has been obtained from 1-octene and 2-propanol, and 2-octanol from ethanol and 1-hexene, addition being brought about either by UV-light or by tert. butyl peroxide. It is assumed that both the dark reaction and the photoreaction proceed via a free radical chain mechanism. This mechanism will be discussed in more detail when the addition of thiols to olefins is considered (cf. p. 162).

3-Methoxybutyric acid (18) [2]. 90 g crotonic acid were dissolved in methanol and the solution irradiated by an ultraviolet lamp for three weeks. Afterwards the solvent was carefully evaporated and the acid exactly neutralized with sodium bicarbonate. The residue from evaporating to dryness was repeatedly shaken with absolute alcohol, taking 12 g of a sodium salt into solution. This salt, which was deliquescent in air, was treated with dilute sulfuric acid and after extraction with ether, 8 g 3-methoxybutyric acid (18) were obtained, b. p.$_{20}$ 117—118°.

2-Octanol [21]. A solution of 1-hexene (22 g) in ethyl alcohol (360 g) was irradiated with a quartz mercury resonance lamp for 168 hours. Low-boiling substances and unconsumed reactants were distilled off yielding 1.5 g of octanol, b. p.$_{38}$ 97—98°.

b) Ethers

Tert. butyl peroxide induced free radical addition reactions of four, five, and six membered cyclic ethers to 1-octene have been reported by WALLACE and GRITTER [22]. These dark reactions require higher temperatures to occur thereby inducing rearrangement to open chain ketones which constitute the main products.

Exclusive formation of alkyl ethers was, however, brought about by irradiation at room temperature. Yields of alkylated products such as **19** could be further increased by use of dehydrogenating sensitizers such as acetone. Analogous reactions were carried out with dioxane [23].

$$\text{(tetrahydrofuran)} + H_2C=CH-(CH_2)_5-CH_3 \xrightarrow[\text{acetone}]{h\nu} \text{(tetrahydrofuranyl)}-(CH_2)_7-CH_3$$
$$\mathbf{19}$$

A reaction of similar type is the addition of tetrahydrofuran to maleic anhydride [24], which may be initiated by peroxides or by UV-light.

20

Diekmann and Pedersen [25] showed that tetrahydrofuran can add to α,α,α',α'-tetracyanoquinodimethane (21) yielding 22, the quinodimethane undergoing a 1,6-addition reaction. 22 may in inferior yield also be obtained in the dark by allowing tetrahydrofuran to act on 21 at 150° for 4 hrs. in the presence of tert. butyl peroxide.

21 22 23

Irradiation of tetracyanoethylene with tetrahydrofuran produced a 2 : 1 adduct (24) which is believed to derive from a primary addition product in a secondary reaction.

24

2-Octyltetrahydrofuran (19) [23]. A mixture of 1-octene, tetrahydrofuran (90 ml), and acetone (5 ml) was irradiated with a Hanau Q 81 high pressure mercury lamp in a quartz immersion apparatus for 1 hour. 1-Octene (5.1 g) and acetone (5 ml) in ten equal portions were then added at 1-hour intervals and the irradiation continued for another 12 hrs. Distillation in vacuo provided a fraction (5.4 g; b.p.$_{1.5}$ 80—140°), which was chromatographed on alumina. Elution with pentane gave 19, b.p.$_{0.4}$ 69—71°; yield 2.3 g = 25%.

2-Tetrahydrofuransuccinic anhydride (20) [24]. A solution of tetrahydrofuran (5 moles) and maleic anhydride (0.5 mole) was heated under reflux in a Pyrex flask and illuminated by a sunlamp (General Electric, CG 401—E 6) for 6 hours. 52 g (61%) of 20, b.p.$_{0.1}$ 105—110° and 28.6 g of residue was obtained.

α,α,α',α'-Tetracyano-α-(2-tetrahydrofuryl)-p-xylene (22). A solution of 3.5 g of 21 in 500 ml of tetrahydrofuran was irradiated with a GE sunlamp for 8 hours. The solvent was then removed under reduced pressure. The residual brown oil solidified on rubbing with benzene-petroleum ether mixture. On extraction with hot toluene and subsequent cooling 2.5 g (55%) of round, off-white crystals, m.p. 136—139°, were obtained. Recrystallization from a large volume of a benzene-cyclohexane mixture gave 22 as white rods, m.p. 138—140°. The insoluble residue, 0.3 g (8.5%) consisted of α,α,α',α'-tetracyano-p-xylene (23).

c) Tert. butyl hypochlorite

By the action of UV-light on chlorotrifluoroethylene or perfluoropropene and tert. butyl hypochlorite at lower temperatures (—20 to —40°) the corresponding perhaloalkyl tert. butyl ethers have been obtained [26].

$$F_2C=CFCl + (H_3C)_3COCl \xrightarrow{h\nu} (H_3C)_3CO-CF_2-CFCl_2$$
$$25$$

$$F_2C=CF-CF_3 \; (10) + (H_3C)_3COCl \xrightarrow{h\nu} (H_3C)_3CO-CF_2-CFCl-CF_3$$
$$26$$

1-tert. Butoxy-2,2-dichloro-1,1,2-trifluoroethane (25) [26]. In the course of 3 hours irradiation with UV light at —30° 120 g of chlorotrifluoroethylene were added to 100 g of tert. butyl hypochlorite. On completion of the reaction the products were washed with sodium carbonate solution and water, dried over sodium sulfate and distilled in vacuo. The ether **25** distills over as an almost colorless liquid at 47.5—53°/38.5 mm. Yield 171 g (88%).

1-tert. Butoxy-2-chloro-1,1,2,3,3,3-hexafluoropropane (26) [26]. Hexafluoropropene (**10**) (150 g) were treated dropwise with 70 g of tert. butyl hypochlorite at —30 to —20° under UV irradiation. At the end of the reaction, the products were washed with sodium thiosulfate ans sodium carbonate solution, dried over sodium sulfate and distilled in vacuo. Yield of crude **26** 130 g. Pure **26** has b.p.$_{104}$ 60.5°.

6. Sulfur compounds

a) Hydrogen sulfide, thiols and thiocarboxylic acids

It is assumed that the addition of hydrogen sulfide and of thiols to olefinic bonds proceeds according to the following scheme [27] (cf. also STACEY and HARRIS JR. [28]).

$$R-SH \xrightarrow{h\nu} R-S\cdot + H$$
$$H + H_2C=CH-R' \longrightarrow H_3C-\overset{\cdot}{C}H-R'$$
$$H_3C-\overset{\cdot}{C}H-R' + R-SH \longrightarrow H_3C-CH_2-R' + R-S\cdot \quad \text{or}$$
$$H + R-SH \longrightarrow H_2 + R-S\cdot$$
$$R-S\cdot + H_2C=CH-R' \longrightarrow R-S-CH_2-\overset{\cdot}{C}H-R'$$
$$R-S-CH_2-\overset{\cdot}{C}H-R' + R-SH \longrightarrow R-S-CH_2-CH_2-R' + R-S\cdot$$

In the case of hydrogen sulfide-olefin mixtures 1:2 addition products are observed as well.

Light of wavelengths shorter than ca. 280 mμ is required to cause the dissociation of the hydrogen-sulfur bond. Nevertheless light of longer wavelengths may be used if, in addition to the olefin and the sulfur compounds, the reaction mixture contains a sensitizer of dehydrogenating properties such as acetone [27]. If one works with light of wavelengths less than 280 mμ vessels must be of quartz or incorporate a quartz or calcium fluoride window [29, 30].

References, pp. 180—181

The addition of the sulfur group to the olefin takes place contrary to the MARKOWNIKOFF rule, that is the sulfur adds to the carbon atom which carries the greater number of hydrogen atoms.

The patent literature demonstrates that unsaturated compounds of many different types will combine photochemically with hydrogen sulfide [30] and thiols [31]. Among others mentioned are, ethylene, propene, 1-butene, styrene, cyclohexene, acetylene, and phenylacetylene. Oxygen containing olefins with which the reaction has been carried out are, among others, acrylates, divinyl ether, diallyl ether, and allyl alcohol. Aliphatic thiols which have been photochemically added to olefins comprise methyl to octyl thiols. Dithiols of the general formula $HS-(CH_2)_n-SH$ have been found to be reactive as well.

HARRIS JR. and STACEY [32] have investigated the light and X-ray initiated addition of hydrogen sulfide to methyl trifluorovinyl ether (27):

$$H_2S + F_2C=CF-O-CH_3 \xrightarrow{h\nu} HS-CF_2-CHF-O-CH_3 +$$
$$27 \qquad\qquad\qquad\qquad 28$$
$$S(F_2C-CHF-OCH_3)_2$$
$$29$$

The addition of hydrogen sulfide to tetrafluoroethylene and to chlorotrifluoroethylene on irradiation in the presence of acetone (as sensitizer) was reported by FOKIN and co-workers [33]; e.g. chlorotrifluoroethylene gave 43% of 2-chloro-1,1,2-trifluoroethanethiol (30), 10% bis-(2-chloro-1,1,2-trifluoroethyl) sulfide (31) and 31% of bis-(2-chloro-1,1,2-trifluoroethyl) disulfide (32).

$$HS-CF_2-CHFCl \qquad\qquad ClFHC-CF_2-S-CF_2-CHFCl$$
$$30 \qquad\qquad\qquad\qquad\qquad 31$$
$$ClFHC-CF_2-S-S-CF_2-CHFCl$$
$$32$$

Thioacetic acid combines photochemically with olefins, e.g. with 2-methyl-2-pentene, 1-methylcyclohexene and camphene, forming thioacetates [34, 35]. In the case of 1-methylcyclohexene approximately 85% cis- and 15% trans-2-methylcyclohexyl thioacetate were obtained as judged from the composition of the thiol mixture after hydrolysis. BORDWELL and HEWETT [34, 35] note that the photoaddition of thioacetic acid to olefins provides an excellent synthetic route to pure thiols since yields are usually high and orientation is exclusively anti-MARKOWNIKOFF. In a similar manner [36] thioacetic acid adds to 1-chlorocyclohexene to yield 66—73% of cis-2-chlorocyclohexyl thioacetate.

Bis-(2-chloroethyl) sulfide, mustard gas [27]. Hydrogen sulfide (9 ml) and vinyl chloride (10 ml) in a quartz tube were frozen in liquid air and the evacuated tubed sealed off. The mixture was irradiated with a quartz mercury lamp for about 10 mins. at 20°.

The tube was then opened when unreacted hydrogen sulfide and vinyl chloride escaped. The reaction product (70—80% calculated on the basis of vinyl chloride used) was composed of a mixture of 2-chloroethanethiol and bis-(2-chloroethyl) sulfide.

1-Butanethiol and dibutyl sulfide [27]. 1-Butene (44 mmole) and pure hydrogen sulfide (88 mmole) were brought into a quartz tube of internal diameter 10 mm. The tube was sealed and irradiated with a quartz mercury arc. During the reaction (4 mins.) the temperature of the contents of the tube was about 0°. The tube was then cooled in solid carbon dioxide and opened. Unconsumed reactants evaporated leaving a residue (3.8 ml; 80%) composed of a 85 : 15 mixture of 1-butanethiol and dibutyl sulfide. This was shaken with a large excess of aqueous sodium hydroxide solution (10%) and the supernatant liquid removed (0.5 ml). This was practically pure dibutyl sulfide. The alkaline solution was acidified and yielded 1-butanethiol, b. p. 98°.

1,1,2-Trifluoro-2-methoxyethanethiol (28) and bis-(1,1,2-trifluoro-2-methoxyethyl) sulfide (29) [32]. A mixture of 30 ml of **27** and an equal volume of liquid hydrogen sulfide was irradiated in a quartz reactor fitted with a Dry Ice condenser. Just a few minutes of irradiation was required to complete the reaction as evidenced by the slowing down of the reflux rate. After evaporation of the excess volatiles, there remained 37.82 g of residue. Distillation of a portion (26.7 g) of this through a helix-packed, low-temperature still gave 6 g of **28**, b.p.$_8$ 8—10°. The remainder of the reaction mixture was treated similarly, and the residues from these two fractionations were combined and distilled in vacuo to yield 15.02 g of **29**, b.p.$_{1.5}$ 65—66°.

2-Methyl-3-pentyl thioacetate [35]. Freshly distilled thioacetic acid (152.2 g) is added slowly with stirring to 336.6 g of 2-methyl-2-pentene. During the addition and 1 hr. afterwards the reaction mixture is irradiated with a 100 Watt bulb. Distillation of the mixture yields 308.5 g (96%) of 2-methyl-3-pentyl thioacetate, b.p.$_{13}$ 70°.

b) Sulfenyl chlorides

Trifluoromethanesulfenyl chloride adds to partially or fully halogenated olefins with formation of 1 : 1 adducts and varying amounts of bis-(trifluoromethyl) disulfide (**35**). Since these reactions can be accomplished not only by ultraviolet but also by X-ray irradiation and by catalytic amounts of azo initiators, it seems certain that they are free radical chain reactions. A striking feature of this series of experiments is the formation of both possible 1 : 1 adducts in each case [37] as shown below for 1,1-difluoroethylene:

$$F_3C-SCl + F_2C=CH_2 \xrightarrow{h\nu} \begin{cases} F_3C-S-CF_2-CH_2Cl & (33) \\ F_3C-S-CH_2-CF_2Cl & (34) \\ F_3C-S-S-CF_3 & (35) \end{cases}$$

2-Chloro-1,1-difluoroethyl trifluoromethyl sulfide (33) and 2-chloro-2,2-difluoroethyl trifluoromethyl sulfide (34) [37]. A mixture of 24 g of trifluoromethanesulfenyl chloride and 11 g of 1,1-difluoroethylene was sealed in a thick-walled Pyrex Carius tube and irradiated for 9.75 hours with a GE H-87-C 3 lamp placed 8 cm from the liquid portion of the reaction mixture. Upon distillation of the reaction mixture there was obtained 25.65 g of a fraction boiling at 23—82°. A study of this fraction by gas chromatography showed the presence of 54% **33**, 15% **34**, 17% **35** and 14% unidentified substance. The two 1 : 1 adducts were separated by preparative scale gas chromatography to yield **33** and **34**, b.p. 79° and 71° respectively.

References, pp. 180—181

c) Sulfonyl chlorides

The light induced formation of sulfones may be illustrated by the photochemical synthesis of **36** which was carried out by heating norbornene with p-toluenesulfonyl chloride under ultraviolet irradiation [*38*].

endo-3-Chloro-exo-2-p-toluenesulfonylbicyclo[2.2.1]heptane (36) [*38*]. 3 g (31.8 mmoles) of norbornene and 6.06 g (31.8 mmoles) of p-toluenesulfonyl chloride were mixed in a 125 ml Vycor flask and irradiated for 1 hr. with ultraviolet light from a Mazda AH-4 lamp placed about 2 cm away from the flask. Heat from the lamp was sufficient to maintain the temperature inside the flask at 120—130°. Vacuum distillation at room temperature and 0.2 mm pressure to remove unchanged starting material left a residue of m. p. 85—97°; yield 6.12 g = 92.5% based on norbornene consumed. Repeated recrystallization from methanol gave pure **36**, m.p. 114—115°.

7. Ammonia, amines and formamide

a) Ammonia and amines

According to STOERMER and ROBERT [*39*] ammonia, aniline, and p-toluidine add across the double bond of α,β-unsaturated acids with great ease under the influence of ultraviolet light. In the reaction between aniline and crotonic acid besides a little crotonic and isocrotonic anilide the main product is 3-anilinobutyric acid (**37**) together with some 3-anilinobutyric anilide (**38**). On reaction of crotonic acid with ammonia the principal products are 3-aminobutyric acid and 3,3′-iminodibutyric acid (**39**).

37: R = OH
38: R = NH−C₆H₅

URRY and co-workers [*40, 41*] have shown that piperidine undergoes a photochemical reaction with 1-octene with formation of 2-octylpiperidine (**40**). This observation is significant with respect to the photochemical synthesis of alkaloids.

3-Anilinobutyric anilide (38) [*39*]. On dissolving 10 g crotonic acid and 20 g distilled aniline in benzene and irradiating for three days with an external quartz lamp the solution became deep brown in color. The benzene was evaporated off and the residue from this process was extracted with ether, yielding 11.2 g of a mixture of aniline and a substance which later on crystallized (2 g). On treatment with hydrochloric acid this gave a salt which, after recrystallization from aqueous acetone, proved to be 3-anilinobutyric anilide hydrochloride, m.p. 212—213°.

b) Formamides

According to ELAD [*42*] the addition of formamide to terminal olefins may smoothly be carried out by irradiation, preferably in the presence of acetone as sensitizer [*43*]. The same type of process can also be effected by peroxides [*44*].

$$R_1-CH=CH_2 + H-C\overset{O}{\underset{N(H)(R_2)}{\diagup}} \xrightarrow{h\nu} R_1-CH_2-CH_2-C\overset{O}{\underset{N(H)(R_2)}{\diagup}}$$

41: $R_1 = C_5H_{11}$
43: $R_1 = H_2NCO-CH_2-CH_2-$

42: $R_1 = C_5H_{11}$; $R_2 = C(CH_3)_3$
44: $R_1 = H_2NCO-CH_2-CH_2-$; $R_2 = H$

Thus, irradiation of a solution of 1-octene and formamide in dry tert. butanol, containing acetone, at room temperature and with exclusion of

Table 11 [*43*]

olefin [a]	product, 1 : 1 adduct	[%][b]	source of light
1-hexene	heptanamide	50	sun
1-heptene	octanamide	57	sun
		61	ultraviolet [c]
1-octene	nonanamide	62	sun
		51	ultraviolet [c]
1-decene	undecanamide	67	ultraviolet [c]
methyl 10-undecenoate	methyl 11-carbamoyl-undecanoate	53	ultraviolet [c]
10-undecenamide	dodecanediamide	90	sun
methyl 4-pentenoate	methyl 5-carbamoyl-pentanoate	61	sun
		58	ultraviolet [c]
4-pentenamide (**43**)	adipamide (**44**)	77	sun

[a] The mole ratio of formamide-olefin in the experiments mentioned was 18 : 1. [b] Yields are based on the olefins employed. The conversions are nearly quantitative in most cases.
[c] Hanau Q 81 high pressure mercury vapor lamps fitted into Pyrex tubes were used as the radiation source for these acetone-initiated reactions.

oxygen gave nonanamide [*43*]. Similar treatment of a solution of 1-heptene (**41**) in N-tert. butylformamide gave N-tert. butyloctanamide (**42**) [*42*]. The feasibility of the photoamidation reaction becomes evident from table 11 [*43*].

The light-induced addition of formamide to alkyl maleates, fumarates and acetylenedicarboxylates was explored by ELAD [*45*] as well. Photoamidation of diethyl maleate or fumarate resulted in formation of diethyl carbamoylsuccinate (**45**). In the case of diethyl acetylenedicarboxylate two molecules of formamide were added and diethyl 1,2-dicarbamoylsuccinate (**46**) was obtained which on hydrolysis yielded 1,1,2,2-ethanetetracarboxylic acid.

$$H_5C_2OOC-CH_2-CH(CONH_2)(COOC_2H_5)$$
45

$$(H_5C_2OOC)(H_2NOC)CH-CH(CONH_2)(COOC_2H_5)$$
46

Acetone-initiated photoaddition of formamide to nonterminal olefins gave rise to mixtures of the isomeric amides in yields of up to 76% [*46*]. Thus, 3-heptene was converted to 32% of 2-ethylhexanamide and 22% of 2-propylpentanamide by UV irradiation in the presence of acetone. Oleamide, when similarly treated, gave 74% of 9-carbamoyloctadecanamide together with 10-carbamoyloctadecanamide. From cyclohexene, cyclohexanecarboxamide was obtained in 65% yield. Usually yields were higher with sunlight than with UV irradiation [*46*].

Octanamide [*43*]. A mixture of 1-heptene (**41**) (0.5 g), formamide (40 g) t.-butyl alcohol (35 ml) and acetone (5 ml) was irradiated for 45 min. A solution of 1-heptene (4.4 g), t.-butyl alcohol (10 ml), and acetone (7 ml) was then added in ten equal portions at 45-min. intervals, and irradiation was continued for another 6 hr. After removal of the solvents, formamide was distilled from the mixture at 0.2 mm. Treatment of the residue with acetone and filtration (to remove traces of oxamide), followed by the removal of the solvent and addition of water, led to an oily mixture which was crystallized from acetone-petroleum ether to give 3.2 g of octanamide, m.p. 98—103°. A pure sample showed m.p. 105—106°.

The aqueous layer was extracted with chloroform. Treatment of the residue left after removal of the solvent with a small volume of acetone caused the separation of 200 mg of *n*-pentylsuccinamide, which after crystallization from ethanol exhibited m.p. 218—219°.

The residue (2.4 g) from the combined mother liquors was chromatographed on alumina (120 g). Elution with acetone-petroleum ether (1:9) led to an oil (200 mg) which is believed to contain a mixture of telomers. Further elution with the same solvent mixture gave a 2:1 telomer (400 mg), m.p. 63—65° (*n*-pentane).

Acetone-petroleum ether (1:9) finally eluted 220 mg of 2-methylheptanamide. Crystallized from *n*-pentane it showed m.p. 76—77°. Elution with acetone-petroleum ether (3:17) yielded octanamide (850 mg). Ethanol-acetone (3:7) eluted a glassy oil (640 mg).

The recovered formamide distillate was diluted with saturated aqueous sodium chloride solution and extracted with chloroform. Removal of the solvent gave an oil (1.6 g) which furnished 320 mg of octanamide, m.p. 90—94° upon treatment with *n*-pentane. The residue was chromatographed on alumina (70 g). Petroleum ether eluted 2-decanone (810 mg).

Elution with benzene-petroleum ether (1 : 9) gave 2-methyl-2-nonanol (320 mg).

Adipamide (44) [*43*]. A mixture of 1 g of **43** and 40 g of formamide was dissolved in 20 ml of tert. butanol and 5 ml of acetone. This solution, contained in a Pyrex conical flask stoppered under nitrogen, was exposed to direct sunlight, precipitation of product starting at the end of the second day. After 2 days, a solution of 4 g of **43** in 50 ml of tert. butanol and 5 ml acetone was added in five equal portions at 2-day intervals, and the vessel was left in sunlight for another 2 days. The precipitate was washed with hot acetone to give 4.4 g of **44**, m.p. 224—226° (from ethanol). Another crop of 1.17 g of adipamide was obtained from the filtrate.

The residue from the mother liquors was treated with water and extracted with chloroform. Evaporation of the solvent afforded 620 mg of 7-oxooctanamide, m.p. 90—91° (from acetone-petroleum ether).

8. Dimethylmaleic anhydride

While the photosensitized addition of philodienes to olefins normally leads to cyclobutanes (comp. chapter 11), dimethylmaleic anhydride (**48**) shows a different behavior towards α-cedrene (**47**) [*47*]. On irradiation in acetone with or without benzophenone as sensitizer, the product was an open chain adduct, viz. **49**. The adduct, which on pyrolysis (250°) yielded **48** again, was imagined as arising through a photochemical ene-synthesis (for analogous ene-syntheses with oxygen see chapter 39).

Irradiation of α-cedrene with methylmaleic anhydride (**50**) gave the adduct **51** [*47*].

47

48: R = CH₃
50: R = H

49: R = CH₃
51: R = H

α,β-Dimethyl-β-cedrene-5-succinic anhydride (49) [*47*]. A solution of 0.1 mole of **47** and 0.04 mole of **48** in 100 ml of acetone was irradiated at 10° for 24 hours. The irradiation was conducted under argon in water-cooled glass apparatus, using a mercury high pressure lamp Philips HPK 125 W. Tetramethylcyclobutanetetracarboxylic dianhydride precipitated during the irradiation (9%). The filtrate was evaporated and the residue chromatographed over silica gel. Petroleum ether eluted unreacted **47**, and cyclohexane/benzene eluted **49**, m.p. 120—123°; yield 29.3% (calculated on the basis of **48** consumed). In some cases also dimeric α-cedrene could be isolated.

References, pp. 180—181

9. Aldehydes and ketones

a) Aldehydes

KHARASCH and co-workers [48] found that aliphatic aldehydes such as acetaldehyde, butyraldehyde and heptanal add to the olefinic double bond under the influence of light. The reaction proceeds according to

$$R-C{\overset{H}{\underset{O}{\diagdown}}} + H_2C=CHR' \xrightarrow{h\nu} R-C-CH_2-CH_2-R'$$
$$\phantom{R-C{\overset{H}{\underset{O}{\diagdown}}} + H_2C=CHR' \xrightarrow{h\nu} R-}\underset{O}{\|}$$

Similar addition reactions may be carried out in the dark if the olefin and the aldehyde are heated in the presence of acyl peroxides (e.g. acetyl peroxide). KHARASCH and co-workers suggested that free radicals are involved in both cases.

Polymeric products resulted from the irradiation of butyraldehyde with styrene and of crotonaldehyde with 1-octene [48].

The formation of a 1,4-diketone was observed by SCHLUBACH et al. [49] on allowing propionaldehyde to react with 1-hexyne in UV-light. They obtained a compound which was formed by the addition of 2 moles of the aldehyde to one mole of the acetylene, and which was formulated as the 1,4-diketone 4-butyl-3,6-octanedione (52) rather than as the isomeric 1,3-diketone.

$$2\ H_3C-CH_2-C{\overset{O}{\underset{H}{\diagdown}}} + HC\equiv C-C_4H_9 \xrightarrow{h\nu}$$

52

2-Decanone [48]. A solution of acetaldehyde (192.4 g) and 1-octene (123.6 g) were subjected to the action of UV-light (quartz mercury lamp) for 72 hrs. under a N_2-atmosphere; during the irradiation the reaction tube was cooled with ice-water. Gas evolved (2000 ml) consisted of 47% methane and 51% CO. Distillation at normal pressure yielded acetaldehyde (185 g) and 1-octene (115 g) as well as a small quantity of biacetyl. The distillation was continued at reduced pressure and a fraction obtained (7.5 g) which proved to be 2-decanone, b.p.$_{37}$ 117°.

b) Ketones

REUSCH [50] observed the stereospecific photochemical addition of acetone to bicyclo[2.2.1]heptene. Reflux of acetone solutions of norbornene in the dark produced no observable change in 50 hrs; poor yields of exo-2-acetonylbicyclo[2.2.1]heptane (53) may be obtained if azobisisobutyronitrile is added. The reaction leading to 53 is believed to proceed by chain addition of acetonyl radicals to bicyclo[2.2.1]heptene.

Processes similar to the formation of **53** are the photochemical addition of cyclohexanone to 1-octene yielding 2-octylcyclohexanone (**54**) [*51*] and the formation of 2-cyclohexylcyclohexanone (**55**) during illumination of cyclohexanone with cyclohexene [*52*].

exo-2-Acetonylbicyclo[2.2.1]heptane (53) [*50*]. A solution of 85% pure norbornene (10 g) in 120 ml dry acetone was irradiated for 48 hours, the solution being kept at reflux by a Hanovia SH mercury lamp placed 5 cm below the Pyrex vessel. The crude organic products were fractionated at reduced pressure to give 5.6 g (40%) of **53**, b.p.$_{12}$ 94—96°. The forerun (2.1 g) contained 50% volatile impurities in addition to **53**.

2-Cyclohexylcyclohexanone (55) [*52*]. A solution of cyclohexanone (25 g) in cyclohexene (100 ml) was irradiated at reflux temperature by a 500 W Hanovia lamp for 40 hrs. Distillation gave 15 g of a colorless oil, a portion of which (5 g) was chromatographed on alumina. Fractions eluted with light petroleum contained bi-2-cyclohexen-1-yl (**56**) (1 g) and cyclohexyl caproate (1.3 g). Benzene-light petroleum (1 : 4) eluted **55**, b.p.$_{20}$ 145—146°; yield 1.5 g.

10. Aliphatic polyhalides

For our knowledge of these addition reactions we are indebted above all to KHARASCH and collaborators who opened research in this field in 1945. Polyhalides mainly used were carbon tetrachloride and tetrabromide as well as bromotrichloromethane. HASZELDINE has worked successfully with trifluoroiodomethane.

It is generally accepted that the photochemical addition of polyhalides follows a radical chain mechanism as exemplified by the following case [*53*]:

$$BrCCl_3 \xrightarrow{h\nu} Br\cdot + \cdot CCl_3$$
$$R-CH=CH_2 + \cdot CCl_3 \longrightarrow R-\overset{\cdot}{C}H-CH_2-CCl_3$$
$$R-\overset{\cdot}{C}H-CH_2-CCl_3 + BrCCl_3 \longrightarrow R-CHBr-CH_2-CCl_3 + \cdot CCl_3$$

The photoaddition of trifluoroiodomethane across the carbon-carbon double bond is assumed to involve free trifluoromethyl radicals [*54*].

$$ICF_3 \xrightarrow{h\nu} \cdot CF_3 + I\cdot$$

References, pp. 180—181

The assumption that the photochemical addition of aliphatic polyhalides to the carbon double bond is coupled with the formation of free radicals is supported by the fact that similar additions may be carried out in the dark, provided acetyl peroxide be present. Thus with this initiator the addition of bromotrichloromethane to 1-octene, leading to the formation of 3-bromo-1,1,1-trichlorononane, may be carried out in the dark [53].

a) Polyhalides not containing fluorine

The photochemical addition of bromotrichloromethane to olefins takes place even under the action of visible light; under these conditions carbon tetrachloride will not react, since in this case ultraviolet light is necessary for the fission of the halide.

Photochemical addition of aliphatic polyhalides such as bromotrichloromethane can also take place with cyclic olefins. In general the addition is easier with compounds containing 5-membered rather than 6-membered rings, thus cyclopentadiene is more reactive than 1,3-cyclohexadiene, and cyclopentene more than cyclohexene. The scope of the reaction may be illustrated by the following examples (cf. table 12).

Table 12

olefin	halide	light source	product	ref.
1-octene	CCl_4	visible	no reaction	[55]
1-octene	CCl_4	UV	1,1,1,3-tetrachlorononane	[55]
1-octene	CBr_4	visible	1,1,1,3-tetrabromononane	[55]
styrene	CBr_4	visible	(1,3,3,3-tetrabromopropyl)-benzene (57)	[55]
propene	$BrCCl_3$	UV	3-bromo-1,1,1-trichlorobutane	[53]
β-methylstyrene	$BrCCl_3$	UV	58	[56]
ethyl cinnamate	$BrCCl_3$	UV	59	[56]
4-penten-2-ol acetate (60)	$BrCCl_3$	UV	4-bromo-6,6,6-trichloro-2-hexanol acetate (61)	[57]
1-phenyl-3-buten-1-ol acetate (62)	$BrCCl_3$	UV	3-bromo-5,5,5-trichloro-1-phenyl-1-pentanol acetate (63)	[57]
norbornene	$BrCCl_3$	UV	64	[58]
1,3-cyclohexadiene	$BrCCl_3$	UV	65 and 66	[58]
indene	$BrCCl_3$	UV	1-bromo-2-(trichloromethyl)-indan (67)	[58]
2,5-dihydrothiophene 1,1-dioxide	$BrCCl_3$	UV	68	[58]

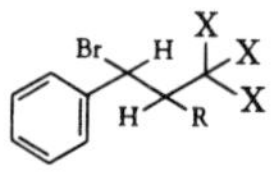

57: R=H, X=Br
58: R=CH$_3$, X=Cl
59: R=COOC$_2$H$_5$, X=Cl

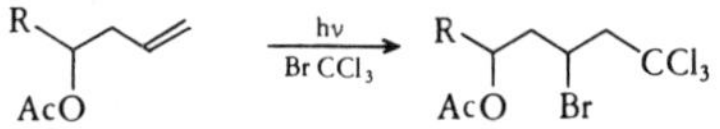

60: R=CH$_3$
62: R=C$_6$H$_5$

61: R=CH$_3$
63: R=C$_6$H$_5$

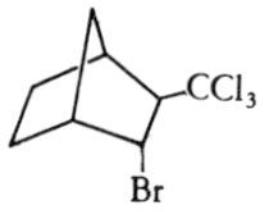

64 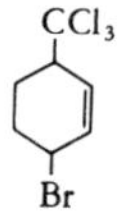   65 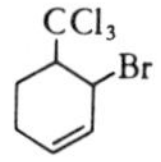   66 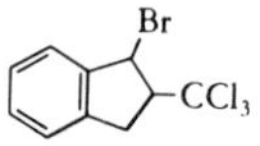   67

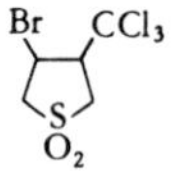

68 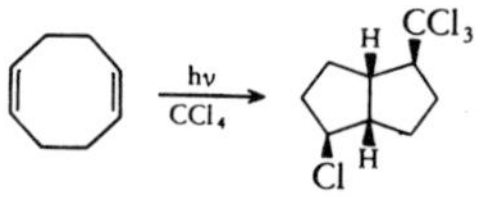   69   70

Irradiation of bromotrichloromethane with β-methylstyrene or with ethyl cinnamate affords only one of the possible isomers thus illustrating the selectivity of this addition. The respective products are, (1-bromo-3,3,3-trichloro-2-methylpropyl)-benzene (58) and ethyl β-bromo-α-(trichloromethyl)-hydrocinnamate (59) [56].

The photochemical addition product of bromotrichloromethane to norbornene [58] has recently been shown [59] to be 3-endo-bromo-2-exo-trichloromethylbicyclo[2.2.1]heptane (64).

Photoaddition of carbon tetrachloride to cis,cis-1,5-cyclooctadiene (69) resulted in formation of exo-2-trichloromethyl-exo-6-chloro-cis-bicyclo-[3.3.0]octane (70) via a transannular 1,5-cycloaddition [60].

Alkynes are also susceptible to photochemical addition reactions with polyhalides [61]. Thus bromotrichloromethane combines in light with 1-octyne with formation of 3-bromo-1,1,1-trichloro-2-nonene (71)

$$H_3C-(CH_2)_5-C\equiv CH + BrCCl_3 \xrightarrow{h\nu} H_3C-(CH_2)_5-CBr=CH-CCl_3 \quad (71)$$

With phenylacetylene or with 2-octyne the reaction is more complicated. Apart from the 1 : 1 addition products other compounds are formed which are derived from combination of one molecule of the halide and two molecules of the alkyne. With carbon tetrachloride and 1-octyne or 2-nonyne yields are considerably lower. Reaction products are high boiling liquids which decompose on distillation.

(1,3,3,3-Tetrabromopropyl)-benzene (57) [55]. A mixture of styrene (10 g; 0.1 mole), carbon tetrabromide (203 g, 0.61 mole) and carbon tetrachloride (173 g, 1.12 mole) were irradiated with visible light for 4 hrs. at 90°. The carbon tetrachloride and then most of the unreacted carbon tetrabromide were distilled off at normal and reduced pressure respectively. The residual brown oil (46 g) was distilled in a small Claisen flask. After a short forerun the main fraction was obtained (b.p.$_{0.1}$: 112—124°). The distillate crystallized after a short time; m.p. 57—59°, yield 41.8 g (96%).

1,1,1,3-Tetrabromononane [55]. A mixture of 1-octene (56 g, 0.5 mole) and carbon tetrabromide (600 g, 1.8 mol) was irradiated with visible light at 75° (7 hrs. stirring by bubbling nitrogen through the solution). The excess carbon tetrabromide was distilled off under reduced pressure; only a very small quantity of unchanged octene was found. The residue (205 g) was distilled under reduced pressure and yielded 196 g (88%) 1,1,1,3-tetrabromononane, b.p.$_{0.2}$: 125—130°.

exo-2-Trichloromethyl-exo-6-chloro-cis-bicyclo[3.3.0]octane (70) [60]. 27.1 g of cis,cis-1,5-cyclooctadiene (69), dissolved in 250 ml carbon tetrachloride, were irradiated in a Vycor flask with an UV lamp at 60° for 5 days. Distillation gave 22.0 g (33.5%) of **70**, b.p.$_{0.09}$ 83—84°. When the lamp was substituted by a GE sunlamp and the reaction vessel made of Pyrex yields of **70** dropped to 5.4%.

b) Polyfluoroalkyl iodides

According to HASZELDINE [54] the main product of the photochemical reaction of trifluoroiodomethane on ethylene is 1,1,1-trifluoro-3-iodopropane besides some 1,1,1-trifluoro-5-iodopentane.

$$H_2C=CH_2 + ICF_3 \xrightarrow{h\nu} IH_2C-CH_2-CF_3 + IH_2C-CH_2-CH_2-CH_2-CF_3$$

This reaction mode is common to terminal olefins which always add the CF_3 radical at the terminal methylene group [62]. Products of general formula

$$F_3C-CH_2-CHI-R \quad \text{(where R = } CH_3, Cl, F, COOCH_3, CN, CF_3)$$

have thus become available. None of the products form in the dark at room temperature. The significance of the addition reactions of the trifluoromethyl radical to olefinic systems may be illustrated by the synthesis of 4,4,4-trifluorocrotonic acid (**72**) [63]. The preparation is based on the photochemical addition of trifluoroiodomethane to acrylonitrile; the primary product is not isolated as such but directly converted to the acid

$$F_3CI + H_2C=CH-CN \xrightarrow{h\nu} F_3C-CH_2-CHI-CN$$

$$F_3C-CH_2-CHI-CN \xrightarrow{OH^\ominus} F_3C-CH=CH-COOH \quad (72)$$

Attack by a trifluoromethyl radical on perfluoro olefins ($R-CF=CF_2$) takes place at the terminal difluoromethylene group [15]. The reaction between trifluoroiodomethane and hexafluoropropene (**10**) on irradiation with UV-light of wavelength $\lambda < 220$ mμ leads to the formation of nonafluoro-2-iodobutane (**73**) as well as dodecafluoro-2-iodo-4-(trifluoromethyl)-hexane (**74**). If light of wavelength $\lambda > 300$ mμ is used, the reaction is very slow and only **73** is obtained. 1-Iodoheptafluoropropane results from light induced addition of trifluoroiodomethane to tetrafluoroethylene [54], oligomers accompanying the 1 : 1 adduct.

$$F_3C-CF=CF_2 \xrightarrow[F_3CI]{h\nu} F_3C-CFI-CF_2-CF_3 + F_3C-CFI-CF_2-CF-CF_2-CF_3$$
$$\hspace{7cm} | \hspace{0.5cm} CF_3$$
$$\hspace{0.5cm} 10 \hspace{4cm} 73 \hspace{4cm} 74$$

The photochemical action of F_3CI on acetylene yields 3,3,3-trifluoro-1-iodo-1-propene [64]. Pentafluoroiodoethane is likewise suitable to be added on to acetylene in UV-light [65], the main product is pentafluoro-1-iodo-1-butene (75) besides small quantities of 5,5,6,6,6-pentafluoro-1-iodo-1,3-hexadiene

$$HC\equiv CH + C_2F_5I \xrightarrow{h\nu} IHC=CH-CF_2-CF_3 + IHC=CH-HC=CH-CF_2-CF_3$$
$$\hspace{6cm} 75$$

Neither of the polyhalides adds on to acetylene in the dark unless heated to temperatures above 200°; nevertheless a smooth reaction is effected at room temperature in UV-light.

4,4,4-Trifluorocrotonic acid (72) [63]. Acrylonitrile (0.53 g) was sealed in a Pyrex tube (50 ml) with trifluoroiodomethane (8.5 g) and exposed to radiation from a UV lamp for 48 hrs. After removal of the excess trifluoroiodomethane unchanged acrylonitrile (0.01 g) was distilled off in vacuo. 4,4,4-Trifluoro-2-iodobutyronitrile was not isolated in a pure state but — after freeing from traces of polymeric acrylonitrile — immediately treated with 10% alcoholic potassium hydroxide (5% excess) first at room temperature, then at 50° (1 hour). The mixture was acidified with dilute sulfuric acid and extracted with ether, yielding 1.02 g (72%) of **72**, m.p. 50.5—51°.

3-Chloro-1,1,1-trifluoro-3-iodopropane [62]. Vinyl chloride (2.3 g) and trifluoroiodomethane (7.0 g) were irradiated for 4 days in a sealed Pyrex tube. The irradiation was conducted in such a way that only the liquid phase was exposed to radiation. A Hanovia UV lamp without filter was employed as the light source. At the end of the experiment the deep red solution (iodine) was distilled to give 4.4 g of 3-chloro-1,1,1-trifluoro-3-iodopropane, b.p. 120°.

Nonafluoro-2-iodobutane (73) [15]. Trifluoroiodomethane (10 g) and hexafluoropropene (10) (3 g) were sealed in a quartz tube (50 ml) and the vessel placed 10 cm. from a Hanovia UV lamp (without filter). The irradiation (14 days) was carried out in such a way that the liquid phase was protected from the light. During the irradiation the contents of the tube assumed a red color (due to iodine liberated). Distillation yielded 15% **10**, trifluoroiodomethane and nonafluoro-2-iodobutane (**73**), corresponding to 94% conversion, b.p. 65.5°.

Dodecafluoro-2-iodo-4-(trifluoromethyl)-hexane (74) [15]. 1.9 g **73** and 0.5 g **10** were irradiated as above in a 10 ml quartz tube. On working up the contents of the tube a 51% yield of **74** was obtained, b.p. 135—139°.

3,3,4,4,4-Pentafluoro-1-iodo-1-butene (75) [65]. A number of Pyrex Carius tubes of 50 ml capacity each were prepared containing acetylene (1.55 g, 15% excess) and pentafluoroiodoethane (12.8 g). These were sealed and exposed to the light of a Hanovia lamp (without filter) at a distance of 30 cm. After 9 days the contents of the tubes were collectively fractionated yielding unchanged acetylene (6%), pentafluoroiodoethane (3%), and higher boiling material (13 g) of which 57% was 3,3,4,4,4-pentafluoro-1-iodo-1-butene, b.p. 84.4°.

References, pp. 180—181

c) Photoaddition of polyhalides to conjugated systems

KHARASCH and FRIEDLANDER [58] have reported that conjugated dienes were also susceptible to photochemical addition reactions with polyhalides, but the reaction has received little attention. 1,4-Addition and 1,2-addition are observed with 1,3-dienes; thus, 1,3-cyclohexadiene produces 6-bromo-3-(trichloromethyl)-cyclohexene (65) and 3-bromo-4-(trichloromethyl)-cyclohexene (66) on irradiation with bromotrichloromethane [58].

More recently it was found that not only cyclic 1,3-dienes underwent photoinduced addition reactions with polyhalides but open chain 1,3-dienes and vinylcyclopropanes as well. Thus, while ethyl tribromoacetate (76) and 1,3-butadiene do not react in the dark a smooth reaction takes place in diffuse daylight [66]. Quantitative formation of ethyl 2,2,6-tribromo-trans-4-hexenoate (77), resulting from 1,4-addition, is observed. Quantum yields indicate a chain reaction of fairly high efficiency.

$$H_5C_2OOC\underset{Br}{\overset{Br}{C}}Br + H_2C=CH-CH=CH_2 \xrightarrow{h\nu} H_5C_2OOC\underset{Br}{\overset{Br}{C}}CH_2CH=CHCH_2Br$$
$$76\phantom{H_2C=CH-CH=CH_2\xrightarrow{h\nu}H_5C_2OOCCCH_2CH=CHCH_2Br}77$$

The light induced addition of bromotrichloromethane to 2-cyclopropylpropene (78) yielded 6-bromo-1,1,1-trichloro-3-methyl-3-hexene (79). HUYSER and TALIAFERRO [67] advanced the following mechanism to account for the ring opening of the vinylcyclopropane:

$$BrCCl_3 \xrightarrow{h\nu} \cdot Br + \cdot CCl_3$$

$$\underset{78}{\overset{H_3C}{\underset{H_2C}{>}}C-C\overset{CH_2}{\underset{HCH_2}{<|}}} + \cdot CCl_3 \longrightarrow \left[Cl_3C-CH_2-\overset{CH_3}{\underset{\cdot}{C}}-C\overset{CH_2}{\underset{HCH_2}{<|}}\right]$$

$$\left[Cl_3C-CH_2-\overset{CH_3}{\underset{\cdot}{C}}-C\overset{CH_2}{\underset{HCH_2}{<|}}\right] \longrightarrow \left[Cl_3C-CH_2-\overset{CH_3}{\underset{|}{C}}=CH-CH_2-\overset{\cdot}{C}H_2\right]$$

$$\left[Cl_3C-CH_2-\overset{CH_3}{\underset{|}{C}}=CH-CH_2-\overset{\cdot}{C}H_2\right] + BrCCl_3 \longrightarrow$$

$$Cl_3C-CH_2-\overset{CH_3}{\underset{|}{C}}=CH-CH_2-CH_2Br + \cdot CCl_3$$
$$79$$

Reactions involving free radicals need not always cause rearrangement of the vinylcyclopropane system. Thus light-induced reaction between **78** and methanethiol yields a mixture of unrearranged and rearranged addition products, (2-cyclopropyl)-1-propyl methyl sulfide and methyl 2-methyl-2-penten-1-yl sulfide [*67*].

While 2,3-di-tert. butyl-1,3-butadiene (**80**) was nearly irrespversive to bromotrichloromethane under irradiation [*68*], the monosubstituted diene afforded a total 36% of the 1,4-addition products expected within 2 hours' irradiation. The very low reactivity of **80** towards $BrCCl_3$ was ascribed to steric effects [*68*].

80

Ethyl 2,2,6-tribromo-trans-4-hexenoate (77) [*66*]. When butadiene was mixed with **76** (molar ratio 1 : 2 or less) at room temperature in daylight, quantitative conversion of butadiene to **77** took place within a few hours. After distillation through a Vigreux column whereby **76** was removed, the ester **77** was obtained as a pale yellow liquid, b.p.$_{0.04}$ 90—93°.

6-Bromo-1,1,1-trichloro-3-methyl-3-hexene (79) [*67*]. Bromotrichloromethane (7.5 g) and 2-cyclopropylpropene (**78**) (3 g) were sealed in a tube and illuminated with a 275 W Sylvania sun lamp at 5° in a water bath for 460 minutes. The reaction mixture was then distilled and 7.0 g (68% based on **78** consumed) of the 1 : 1 adduct (**79**) was collected, b.p.$_{0.09}$ 88°.

11. Organophosphorus compounds

a) Phosphines

Phosphine reacts with olefins in the dark if peroxides be present. The reaction may, however, also be initiated by light. STILES, RUST and VAUGHAN [*69*] used UV-light of $\lambda < 230$ mμ to effect addition, longer wavelengths being effective only in the presence of sensitizers such as acetone. Compounds of general formulae RPH_2, R_2PH, and R_3P were formed, the olefinic reactants being inter alia 1-butene, cyclohexene, and allyl alcohol. It was assumed that the radical chain is initiated by the photochemical decomposition of phosphine:

$$PH_3 \xrightarrow{h\nu} \cdot PH_2 + H$$

Butylphosphine and dibutylphosphine [*69*]. Phosphine and 1-butene (mole ratio 1 : 1) were enclosed in a sealed tube and irradiated with a general Electric Uviarc Lamp (360 W) from a distance of 25 cm. The tube was cooled with water to maintain room

References, pp. 180—181

temperature. The progress of the reaction could be followed by observing the volume of the liquid phase within the tube, which decreased in the course of the reaction. The experiment was discontinued when the change in volume became very small; the tube was then cooled in liquid N_2 and opened. The products were transferred to a distilling flask (under N_2 to prevent oxidation) and then fractionated. Butylphosphine (b.p. 82.2—87.8°; 38%), dibutylphosphine (b. p. 181—185°; 10%) and tributylphosphine (b.p. 240.4 to 242.2°; 2%) were obtained.

b) Phosphonates

Photochemical addition reactions of dialkyl phosphites or dialkyl phosphonates (cf. **81**) to unsaturated compounds have been investigated by several research groups [70—74]. The addition reactions may usually also be initiated with peroxides. The photochemistry of organophosphorus compounds will also be dealt with in chapter 44.

1-Octene and dibutyl phosphonate (**81**) if irradiated with UV-light in the presence of acetone combine with formation of dibutyl octylphosphonate (**82**) [70].

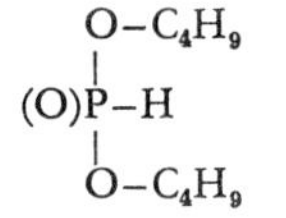

$$\begin{array}{cc} O-C_4H_9 & O-C_4H_9 \\ | & | \\ (O)P-H & (O)P-C_8H_{17} \\ | & | \\ O-C_4H_9 & O-C_4H_9 \\ \mathbf{81} & \mathbf{82} \end{array}$$

Trialkyl phosphonoundecanoates were prepared in 53—66% yield by adding dialkyl phosphonates to alkyl 10-undecenoates under ultraviolet radiation as illustrated below for the case of the butyl esters **81** → **83** [71]:

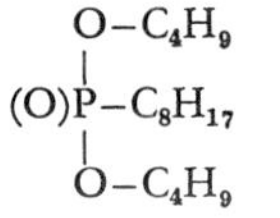

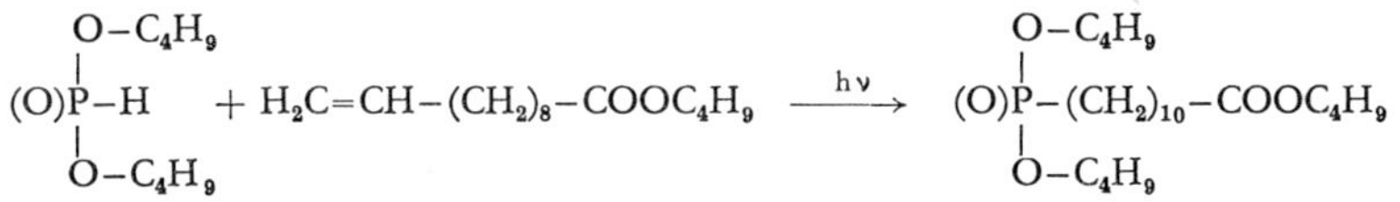

$$\begin{array}{c} O-C_4H_9 \\ | \\ (O)P-H \\ | \\ O-C_4H_9 \\ \mathbf{81} \end{array} + H_2C=CH-(CH_2)_8-COOC_4H_9 \xrightarrow{h\nu} \begin{array}{c} O-C_4H_9 \\ | \\ (O)P-(CH_2)_{10}-COOC_4H_9 \\ | \\ O-C_4H_9 \\ \mathbf{83} \end{array}$$

The addition of dialkyl phosphonothioates (cf. **84**) to olefins has been reported by PUDOVIK et al. [73, 74] to result in analogous formation of dialkyl alkylphosphonothioates (cf. **85**).

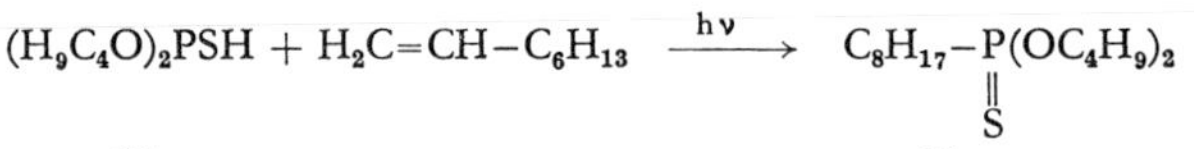

$$(H_9C_4O)_2PSH + H_2C=CH-C_6H_{13} \xrightarrow{h\nu} \underset{\underset{\mathbf{85}}{\overset{\|}{S}}}{C_8H_{17}-P(OC_4H_9)_2}$$

$$\mathbf{84}$$

Tributyl 11-phosphonoundecanoate (83) [71]. A mixture of butyl 10-undecenoate (0.2 mole) and dibutyl phosphonate (**81**) (0.6 mole) was placed in a quartz flask. The solution, flushed with nitrogen, was then irradiated from a distance of 3 cm by a 140 Watt high pressure quartz mercury arc for 6 hrs. at 100—110°. The entire reaction mixture was then fractionated under reduced pressure to afford pure **83**, b.p.$_{0.001}$ 150°; yield 66%.

Dibutyl octylphosphonothioate (85) [*73*]. An equimolecular mixture of dibutyl phosphonothioate (**84**) and 1-octene was placed in a quartz flask and irradiated with a PRK-2 mercury quartz lamp at a distance of 5—6 cm for 5—8 hrs. at 85—90°. Vacuum distillation gave a 66% yield of **85**, b.p.$_{2.5}$ 165—166°.

12. Organosilicon compounds

a) Trichlorosilane

HASZELDINE and MARKLOW [*75*] have shown that trichlorosilane (**86**) reacts photochemically with tetrafluoroethylene with formation of compounds of the general formula H-(CF$_2$—CF$_2$)$_n$—SiCl$_3$ where n = 1, 2, 3 etc. If light of λ > 220 mμ is used an almost quantitative yield is obtained. By a suitable choice of the relative quantities of reactants, the compound with n = 1 (**87**) may be synthesized in about 60% yield.

$$HSiCl_3 \ (86) \xrightarrow{h\nu} H + \cdot SiCl_3$$
$$F_2C=CF_2 + \cdot SiCl_3 \longrightarrow \cdot CF_2-CF_2-SiCl_3$$
$$\cdot CF_2-CF_2-SiCl_3 + HSiCl_3 \longrightarrow HF_2C-CF_2-SiCl_3 \ (87) + \cdot SiCl_3$$

In the photochemical reaction between trichlorosilane and tetrafluoroethylene molecular hydrogen is not evolved; thus the combination 2 H → H$_2$ does not take place [*75*].

Trichlorosilane and tetrafluoroethylene will also undergo thermal reaction with formation of the same compounds, but the photoinduced reaction is to be preferred since the thermal process causes a part of the tetrafluoroethylene to dimerize to perfluorocyclobutane.

A great variety of alkenes and alkynes is open to photochemical addition reactions with trichlorosilane. An interesting example of this UV-light induced addition was reported by CALAS et al. [*76*] when (+)-limonene (**88**) was converted to 1:1 and 1:2 adducts 9-(trichlorosilyl)-p-menth-1-ene (**89**) and 2,9-bis-(trichlorosilyl)-p-menthane (**90**), respectively. Compounds **89** and **90** were reported to have retained their optical activity.

Trichloro-(1,1,2,2-tetrafluoroethyl)-silane (87) [*75*]. 15.2 g of **86** and 6.85 g tetrafluoroethylene were sealed in a quartz tube under exclusion of air and moisture and irradiated with UV-light (24 hours). The liquid reaction product was fractionated and gave (besides unchanged **86**) 7.1 g (44%) of **87**, b.p. 84.5—85° as well as products with higher boiling points (4.0 g). The products were exclusively compounds of the general formula H—$(CF_2-CF_2)_n$—$SiCl_3$. The yields of **87** may be raised to 60% or higher by increasing the molar ratio of **86** : olefin to 4 : 1.

b) Organosilicon compounds

According to GEYER and HASZELDINE [77], dialkylsilanes may be added photochemically to tetrafluoroethylene; e.g., dimethylsilane combines with tetrafluoroethylene under irradiation to give a 83% yield of dimethyl-(1,1,2,2-tetrafluoroethyl)-silane (**91**) together with 7% of dimethyl-(1,1,2,2,3,3,4,4-octafluorobutyl)-silane (**92**) and 2% of dimethylbis-(1,1,2,2-tetrafluoroethyl)-silane (**93**). No reaction takes place in the dark.

$$\begin{array}{cc}
\text{CH}_3 & \text{CH}_3 \\
| & | \\
\text{H}_3\text{C}-\text{Si}-\text{CF}_2-\text{CF}_2\text{H} & \text{H}_3\text{C}-\text{Si}-\text{CF}_2-\text{CF}_2-\text{CF}_2-\text{CF}_2\text{H} \\
| & | \\
\text{H} & \text{H} \\
\textbf{91} & \textbf{92}
\end{array}$$

$$\begin{array}{c}
\text{CH}_3 \\
| \\
\text{HF}_2\text{C}-\text{CF}_2-\text{Si}-\text{CF}_2-\text{CF}_2\text{H} \\
| \\
\text{CH}_3 \\
\textbf{93}
\end{array}$$

Dimethyl-(1,1,2,2-tetrafluoroethyl)-silane (91) [77]. Tetrafluoroethylene (1.4 g, 14 mmole) and dimethylsilane (4.2 g, 70 mmole) were irradiated with a Hanovia S 250 U type arc without filter in a sealed quartz tube for 24 hours. Distillation yielded unchanged dimethylsilane and liquid products (2.71 g) from which 1.852 g (83%) of **91**, b. p. 62—64°, was obtained on fractionation. There remained a residue (160 mg) which contained 125 mg of **92** and 35 mg of **93**.

13. Organogermanium compounds

Photochemical addition to unsaturated compounds is not confined to the carbon and silicon derivatives among the group IV elements. Thus, FUCHS and GILMAN [78] found that triphenylgermane (**94**) adds to 1-octene if irradiated with UV-light. The addition reaction could also be initiated by benzoyl peroxide in a dark reaction.

$$(C_6H_5)_3\text{GeH} + H_2C=CH-C_6H_{13} \xrightarrow{h\nu} (C_6H_5)_3\text{Ge}-C_8H_{17}$$
$$\textbf{94} \qquad\qquad\qquad\qquad\qquad\qquad \textbf{95}$$

Octyltriphenylgermane (95) [78]. A mixture of triphenylgermane (**94**) (45 mmoles) and 1-octene (6 mmoles) were irradiated in 25 ml of heptane with a 125 Watt UV lamp for 48 hours (quartz vessel). The solid reaction product was distilled at 170—195°/0.15 mm. Hg. Recrystallization from ethanol yielded 2 g (80%) of **95**, m.p. 69—70.5°. From the mother-liquor 0.4 g of tetraphenylgermane was isolated, m.p. 232—234°.

References

[1] G. SOSNOVSKY: Free Radical Reactions in Preparative Organic Chemistry. New York: Macmillan 1964.
[2] R. STOERMER and H. STOCKMANN: Ber. dtsch. chem. Ges. 47, 1786 (1914).
[3] S. Y. WANG, M. APICELLA and B. R. STONE: J. Amer. chem. Soc. 78, 4180 (1956).
[4] H. A. LOZERON, M. P. GORDON, T. GABRIEL, W. TAUTZ and R. DUSCHINSKY: Biochemistry 3, 1844 (1964).
[5] A. D. MCLAREN and D. SHUGAR: Photochemistry of proteins and nucleic acids. Oxford: Pergamon Press 1964.
[6] K. C. SMITH, in: Photophysiology. Ed. by A. C. GIESE, Vol. 2, pp. 329—388. New York: Academic Press 1964.
[7] A. STOLL and W. SCHLIENTZ: Helv. Chim. Acta 38, 585 (1955).
[8] H. KOBEL, E. SCHREIER and J. RUTSCHMANN: Helv. Chim. Acta 47, 1052 (1964).
[9] H. HELLBERG: Acta Chem. Scand. 11, 219 (1957).
[10] H. HELLBERG: Acta Chem. Scand. 16, 1363 (1962).
[11] N. A. MILAS, P. F. KURZ and W. P. ANSLOW JR.: J. Amer. chem. Soc. 59, 543 (1937).
[12] W. A. WATERS, in: Organic Chemistry. An Advanced Treatise, ed. by H. GILMAN, Vol. 4, pp. 1120—1245; esp. p. 1154. New York: Wiley 1953.
[13] C. NOFRE, Y. LE ROUX, L. GONDOT and A. CIER: Bull. Soc. Chim. France 1964, 2451.
[14] R. N. HASZELDINE and B. R. STEELE: J. Chem. Soc. 1957, 2800.
[15] R. N. HASZELDINE: J. Chem. Soc. 1953, 3559.
[16] H. L. GOERING, P. I. ABELL and B. F. AYCOCK: J. Amer. chem. Soc. 74, 3588 (1952).
[17] P. S. SKELL and R. G. ALLEN: J. Amer. chem. Soc. 80, 5997 (1958).
[18] N. A. LE BEL: J. Amer. chem. Soc. 82, 623 (1960).
[19] J. D. PARK, A. P. STEFANI and J. R. LACHER: J. Org. Chem. 26, 4017 (1961).
[20] W. H. URRY, F. W. STACEY, O. O. JUVELAND and C. H. MCDONNELL: J. Amer. chem. Soc. 75, 250 (1953).
[21] W. H. URRY, F. W. STACEY, E. S. HUYSER and O. O. JUVELAND: J. Amer. chem. Soc. 76, 450 (1954).
[22] T. J. WALLACE and R. J. GRITTER: J. Org. Chem. 27, 3067 (1962).
[23] D. ELAD and R. D. YOUSSEFYEH: J. Org. Chem. 29, 2031 (1964).
[24] R. L. JACOBS and G. G. ECKE: J. Org. Chem. 28, 3036 (1963).
[25] J. DIEKMANN and C. J. PEDERSEN: J. Org. Chem. 28, 2879 (1963).
[26] K. WEISSERMEL and M. LEDERER: Chem. Ber. 96, 77 (1963).
[27] W. E. VAUGHAN and F. F. RUST: J. Org. Chem. 7, 472 (1942).
[28] F. W. STACEY and J. F. HARRIS, JR., in: Org. Reactions, ed. by A. C. COPE, Vol. 13, 150—376; esp. p. 164. New York: Wiley 1963.
[29] W. E. VAUGHAN and F. F. RUST: USP 2.398.479 of 16. 4. 1946.
[30] W. E. VAUGHAN and F. F. RUST: USP 2.398.480 of 16. 4. 1946.
[31] F. F. RUST and W. E. VAUGHAN: USP 2.392.294 of 1. 1. 1946.
[32] J. F. HARRIS, JR. and F. W. STACEY: J. Amer. chem. Soc. 85, 749 (1963).
[33] A. V. FOKIN, A. A. SKLADNEV and I. L. KNUNYANTS: Dokl. Akad. Nauk SSSR 138, 1132 (1961); Proc. Acad. Sci. USSR, Chem. Sect. 138, 579 (1961).
[34] F. G. BORDWELL and W. A. HEWETT: J. Amer. chem. Soc. 79, 3493 (1957).
[35] F. G. BORDWELL and W. A. HEWETT: J. Org. Chem. 22, 980 (1957).
[36] H. L. GOERING, D. I. RELYEA and D. W. LARSEN: J. Amer. chem. Soc. 78, 348 (1956).
[37] J. F. HARRIS, JR.: J. Amer. chem. Soc. 84, 3148 (1962).
[38] S. J. CRISTOL and J. A. REEDER: J. Org. Chem. 26, 2182 (1961).
[39] R. STOERMER and E. ROBERT: Ber. dtsch. chem. Ges. 55, 1030 (1922).
[40] W. H. URRY, O. O. JUVELAND and F. W. STACEY: J. Amer. chem. Soc. 74, 6155 (1952).
[41] W. H. URRY and O. O. JUVELAND: J. Amer. chem. Soc. 80, 3322 (1958).

References

[42] D. Elad: Chem. and Ind. **1962**, 362.
[43] D. Elad and J. Rokach: J. Org. Chem. **29**, 1855 (1964).
[44] A. Rieche, E. Schmitz and E. Gründemann: Angew. Chem. **73**, 621 (1961).
[45] D. Elad: Proc. Chem. Soc. **1962**, 225.
[46] D. Elad and J. Rokach: J. Org. Chem. **30**, 3361 (1965).
[47] C. H. Krauch and H. Küster: Chem. Ber. **97**, 2085 (1964).
[48] M. S. Kharasch, W. H. Urry and B. M. Kuderna: J. Org. Chem. **14**, 248 (1949).
[49] H. H. Schlubach, V. Franzen and E. Dahl: Liebigs Ann. Chem. **587**, 124 (1954).
[50] W. Reusch: J. Org. Chem. **27**, 1882 (1962).
[51] M. S. Kharasch, J. Kuderna and W. Nudenberg: J. Org. Chem. **18**, 1225 (1953).
[52] P. de Mayo, J. B. Stothers and W. Templeton: Can. J. Chem. **39**, 488 (1961).
[53] M. S. Kharasch, O. Reinmuth and W. H. Urry: J. Amer. chem. Soc. **69**, 1105 (1947).
[54] R. N. Haszeldine: J. Chem. Soc. **1949**, 2856.
[55] M. S. Kharasch, E. V. Jensen and W. H. Urry: J. Amer. chem. Soc. **69**, 1100 (1947).
[56] M. S. Kharasch and M. Sage: J. Org. Chem. **14**, 537 (1949).
[57] S. Doležal: Coll. Czech. Chem. Comm. **30**, 2638 (1965).
[58] M. S. Kharasch and H. N. Friedlander: J. Org. Chem. **14**, 239 (1949).
[59] E. Tobler and D. J. Foster: J. Org. Chem. **29**, 2839 (1964).
[60] R. Dowbenko: Tetrahedron **20**, 1843 (1964).
[61] M. S. Kharasch, J. J. Jerome and W. H. Urry: J. Org. Chem. **15**, 966 (1950).
[62] R. N. Haszeldine and B. R. Steele: J. Chem. Soc. **1953**, 1199.
[63] R. N. Haszeldine: J. Chem. Soc. **1952**, 3490.
[64] R. N. Haszeldine: J. Chem. Soc. **1950**, 3037.
[65] R. N. Haszeldine and K. Leedham: J. Chem. Soc. **1952**, 3483
[66] M. F. Leto and C. S. H. Chen: J. Org. Chem. **27**, 3708 (1962).
[67] E. S. Huyser and J. D. Taliaferro: J. Org. Chem. **28**, 3442 (1963).
[68] E. S. Huyser, F. W. Siegert and H. Wynberg: Tetrahedron Letters **1965**, 2569.
[69] A. R. Stiles, F. F. Rust and W. E. Vaughan: J. Amer. chem. Soc. **74**, 3282 (1952).
[70] A. R. Stiles, W. E. Vaughan and F. F. Rust: J. Amer. chem. Soc. **80**, 714 (1958).
[71] R. Sasin, W. F. Olszewski, J. R. Russell and D. Swern: J. Amer. chem. Soc. **81**, 6275 (1959).
[72] A. N. Pudovik and I. V. Konovalova: Zhur. obshchei Khim. **29**, 3342 (1959); J. Gen. Chem. USSR **29**, 3305 (1959).
[73] A. N. Pudovik and I. V. Konovalova: Zhur. obshchei Khim. **30**, 2348 (1960); J. Gen. Chem. USSR **30**, 2328 (1960).
[74] A. N. Pudovik, I. V. Konovalova and A. A. Guryleva: Zhur. obshchei Khim. **33**, 2924 (1963); J. Gen. Chem. USSR **33**, 2850 (1963).
[75] R. N. Haszeldine and R. J. Marklow: J. Chem. Soc. **1956**, 962.
[76] R. Calas, E. Frainnet and J. Valade: Bull. Soc. Chim. France **1953**, 793.
[77] A. M. Geyer and R. N. Haszeldine: J. Chem. Soc. **1957**, 1038.
[78] R. Fuchs and H. Gilman: J. Org. Chem. **22**, 1009 (1957).

Chapter 18

Photochemical addition reactions of 1,4- and 1,2-quinones with alkylbenzenes or with ethers

1. Addition of chloranil to hydrocarbons

According to MOORE and WATERS [1], chloranil (1) reacts photochemically with p-xylene (2) in a free radical manner to form 2,3,5,6-tetrachloro-4-(4-methylbenzyloxy)-phenol (3). Similar compounds, including 1:2-adducts of 1 with xylene, were obtained in dark reactions when 1 was employed as dehydrogenating agent in the steroid field [2], the formation of the ethers being explained in terms of an ionic reaction mechanism.

The photoaddition of chloranil to tetralin [3] (cf. also chapter 16) proceeds in a manner analogous to the formation of 3, giving 2,3,5,6-tetrachloro-4-(1,2,3,4-tetrahydro-1-naphthyloxy)-phenol (4) which is otherwise attainable in a slow dark process the reaction being initiated by nitrosoacetanilide [4].

2,3,5,6-Tetrachloro-4-(4-methylbenzyloxy)-phenol (3) [1]. 2 g 1 and 25 ml xylene, contained in a thin-walled glass bulb under nitrogen, were exposed to the radiation from a 500 W mercury Hanovia lamp placed about 25 cm away. After 40 days irradiation the bulb was opened. The precipitate (0.12 g) consisted of tetrachlorohydroquinone. Evaporation of the filtrate and digestion with ligroin gave 3, m.p. 125° (from acetic acid); yield 1.8 g.

2,3,5,6-Tetrachloro-4-(1,2,3,4-tetrahydro-1-naphthyloxy)-phenol (4) [3]. 7 g 1 and 9 g tetralin in 130 ml dry, thiophene-free benzene were irradiated with a water-cooled mercury immersion lamp Philips HQA 500 under nitrogen. Working up yielded 3.7 g (25%) of 4, m.p. 142.5°.

2. Addition of phenanthrenequinone to hydrocarbons

In the case of the photoaddition of 9,10-phenanthrenequinone (5) to o-xylene, p-xylene and 1,2,4-trimethylbenzene the products first observed by BENRATH and V. MEYER [5] are not alkylaryl ethers of 9,10-phenanthrenediol (cf. **6, 8**) as was assumed earlier [*1, 5*]. Instead the alternative formulation of the photoadducts being 10-alkylaryl-10-hydroxy-9(10H)-phenanthrones (cf. **7, 9**), originally mentioned by BENRATH and v. MEYER [*5*] could be substantiated. Thus it was eventually demonstrated [*6*] that the photoadduct of **5** with o-xylene was 10-hydroxy-10-(2-methylbenzyl)-9-(10H)-phenanthrone (**7**). In a similar fashion diphenylmethane, fluorene, and xanthene underwent 1,2-addition to phenanthrenequinone. These results could be confirmed and extended [*7*]. On reinvestigation [*8, 9*] these photoaddition reactions showed to be more complex in nature, the two isomeric adducts (cf. **8** and **9**) being in fact formed simultaneously. The relationship is visualized as shown below.

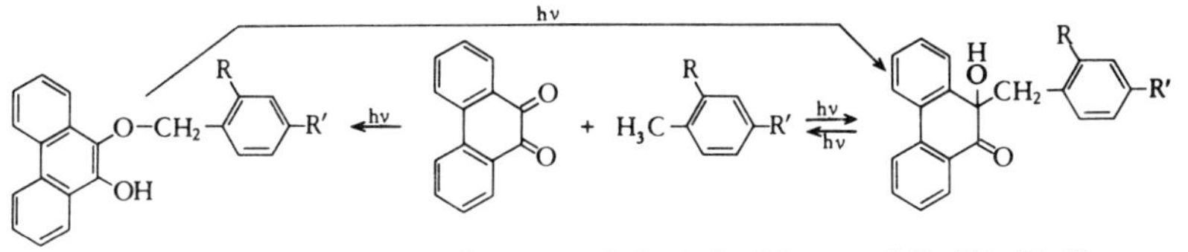

6: R=CH₃, R'=H
8: R=H, R'=CH₃

5

2: R=H, R'=CH₃

7: R=CH₃, R'=H
9: R=H, R'=CH₃

To effect the 1,2-addition of **5** to methyl or methylene groups the presence of oxygen as a catalyst appears to be a prerequisite [*6*]. Otherwise 1,4-addition and dehydrodimerization take place preponderantly (cf. chapter 15). The 1,2-addition products with diphenylmethane, fluorene, and xanthene (**10**) on heating yielded 1,1,2,2-tetraphenylethane, 9,9'-bifluorene, and 9,9'-bixanthene, respectively [*6*].

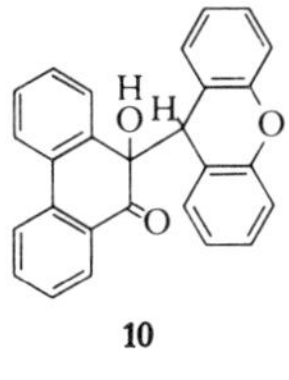

10

When 10-(4-methylbenzyloxy)-9-phenanthrol (**8**), the anticipated [*1, 5*] 1,4-addition product, was synthesized independently and irradiated [*7*], a rearrangement took place and the 1,2-adduct **9** was obtained as the sole product. On further irradiation this photolyzed to the starting materials **5** and p-xylene.

10-Hydroxy-10-(4-methylbenzyl)-9(10H)-phenanthrone(9)[*7*]. A suspension of 2 g of phenanthrenequinone in 20 ml of p-xylene was irradiated at 30° in a Pyrex vessel in an atmosphere of nitrogen; the light source was a 1000 W General Electric water-cooled, high pressure mercury vapor lamp (AH-6) with a Pyrex jacket. After 67 hrs. unreacted **5** (0.32 g) was removed by filtration and the filtrate adsorbed on 50 g. of Florisil. Elution with 250 ml of 10% benzene in petroleum ether afforded 60 mg of white solid, m.p. 82—83°, identical with authentic 1,2-di-p-tolylethane. Elution with 500 ml each of

90% benzene-petroleum ether and pure benzene gave 2 g (67%, 91% conversion) of light yellow **9**, m.p. 129—129.5° (from methylene chloride-petroleum ether). Elution with ethyl acetate gave an additional 0.28 g of **5**.

10-Hydroxy-10-(9-xanthenyl)-9(10H)-phenanthrone (10) [6]. 1 g **5** and 1.25 g xanthene were dissolved in 250 ml of benzene and the solution saturated with oxygen. After 20 mins. irradiation with a mercury lamp Philips HPK 125 W (immersion apparatus with a filter tube WG 1 (Schott)) the solvent was removed by freeze-drying. Brief digestion of the residue with warm methanol and filtration left yellow crystals (0.4 g) of **10**, m.p. about 150° (dec.).

When the solvent from the above irradiation was not removed by freeze-drying but was distilled off at 40° under reduced pressure major portions of 9,9'-bixanthene could be isolated.

3. Addition of phenanthrenequinone or tetrachloro-o-quinone to ethers

Phenanthrenequinone does not only undergo 1,2- but also 1,4-addition reactions with methyl or methylene groups on irradiation [6, 8—10] (for reactions with aldehydes see chapter 19).

Thus, ethers such as dioxane, tetrahydrofuran, and anisole add to phenanthrenequinone [10] in a 1,4-manner to give good yields of the adducts **11**—**13**. The addition to dioxane was reported to be photochemically reversible. PFUNDT and SCHENCK [6, 9] found that also in this addition reaction the presence of oxygen could change the reaction path: if oxygen was not excluded 1,2-addition products of **5** with ethers (in analogy with **7**, **9** and **10**) were prevailing.

3,4,5,6-Tetrachloro-1,2-benzoquinone on irradiation in dioxane gave the acetal **14** [10].

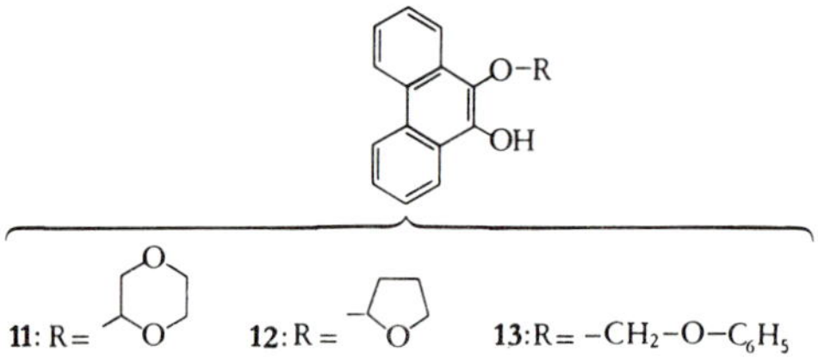

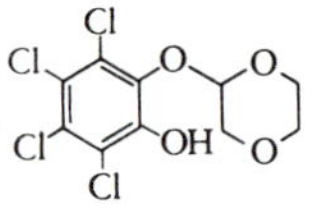

10-(2-Tetrahydrofuryloxy)-9-phenanthrol (12) [10]. A suspension of 1 g **5** in 20 ml tetrahydrofuran was irradiated under nitrogen for 20 hrs. by means of a General Electric 1000 W, water-cooled, high pressure mercury lamp (AH-6) with Corning 7—51 glass filter. Removal of excess solvent and chromatography of the residue on Florisil gave on elution with benzene 1.1 g (82%, 97% conversion) of nearly colorless **12**, m.p. 94—96° and 0.16 g of **5** (eluted with ethyl acetate).

2-(2,3,4,5-Tetrachloro-6-hydroxyphenoxy)-1,4-dioxane (14) [10]. A solution of 3 g of the quinone in 90 ml of dioxane was irradiated as above. After 14 hrs. the deep red color had faded to light orange. The excess dioxane was removed under reduced pressure without heating to give a tan solid which after one recrystallization from ethyl acetate gave 946 mg (20%) of **14**, m.p. 167—168°.

References, p. 185

References

[1] R. F. MOORE and W. A. WATERS: J. Chem. Soc. **1953**, 3405.
[2] H. DANNENBERG, H.-G. NEUMANN and D. DANNENBERG-VON DRESSLER: Liebigs Ann. Chem. **674**, 152 (1964).
[3] G. O. SCHENCK: Dechema Monographien **24**, 105 (1955).
[4] D. KOIKE: Nippon Kagaku Zasshi **77**, 1051 (1956); C. A. **53**, 5208 (1959).
[5] A. BENRATH and A. VON MEYER: J. prakt. Chem. [2] **89**, 258 (1914).
[6] G. PFUNDT: Diss. Göttingen **1962**.
[7] M. B. RUBIN and P. ZWITKOWITS: J. Org. Chem. **29**, 2362 (1964).
[8] M. B. RUBIN and P. ZWITKOWITS: Tetrahedron Letters **1965**, 2453.
[9] G. PFUNDT and G. O. SCHENCK: Unpublished results.
[10] M. B. RUBIN: J. Org. Chem. **28**, 1949 (1963).

Chapter 19

Photochemical additions of aldehydes to quinones, quinone imines and quinone oximes

1. Addition of aldehydes to 1,2-quinones

KLINGER [1, 2] found that 9,10-phenanthrenequinone reacted with aliphatic and aromatic saturated and unsaturated aldehydes e.g. acetaldehyde, benzaldehyde, and cinnamaldehyde under the influence of light. The photoadducts **1** may be cleaved hydrolytically to give 9,10-phenanthrenediol and RCOOH; it is possible thus to convert aldehydes to the corresponding acids.

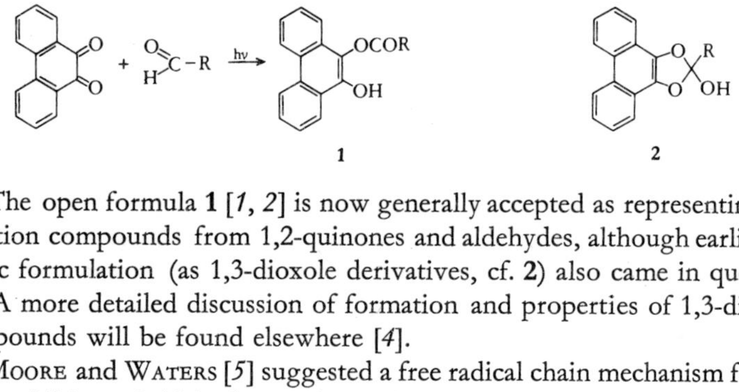

The open formula **1** [1, 2] is now generally accepted as representing the addition compounds from 1,2-quinones and aldehydes, although earlier the cyclic formulation (as 1,3-dioxole derivatives, cf. **2**) also came in question [3]. A more detailed discussion of formation and properties of 1,3-dioxole compounds will be found elsewhere [4].

MOORE and WATERS [5] suggested a free radical chain mechanism for the photoaddition of aldehydes to phenanthrenequinone:

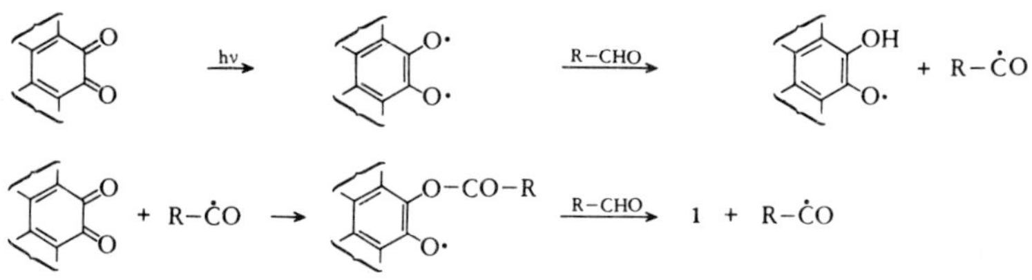

The number of aldehydes which add photochemically to 1,2-quinones is rather large so that the formation of such photoadducts may be taken as a general reaction of both aldehydes and 1,2-quinones. Table 13 contains some representative examples, the list being by no means comprehensive. It

References, p. 192

Addition of aldehydes to 1,2-quinones

Table 13 (numbers refer to citations)

Aldehydes \ 1,2-Quinones	Tetrachloro-o-quinone	4-Cyano-1,2-naphthoquinone (3)	4-Aryloxy-1,2-naphthoquinones	Acenaphthenequinone	Phenanthrenequinone	5,6-Chrysenedione (4)	Benzo[h]quinoline-5,6-dione (5)	3-Phenylbenzo[f]quinoxaline-5,6-dione (6)	2-Phenyl-2H-naphtho[1,2-d]-triazole-4,5-dione (7)
Acetaldehyde	3	6	7		1				
Benzaldehyde	3			8	1,9	10	11	12	
p-Chlorobenzaldehyde					9				
Salicylaldehyde				8	2				
p-Anisaldehyde	3	6	7	8	2,9	10	11	12	13
Phthalaldehyde					12				
Cinnamaldehyde	3	6		8	2				
2-Methoxy-1-naphthaldehyde					14				
9-Anthraldehyde					15	15			
2-Furaldehyde					2				
Quinaldehyde					15				

should be borne in mind that in some cases the structures given in the original literature should be revised in favor of the open form (cf. 1).

Usually photoproducts are of the o-hydroxy ester type. Exceptions are rare, a few of which will be discussed in the sequel.

In the case of the photochemical action of benzo[a]phenazine-5,6-dione (8) on benzaldehyde or p-anisaldehyde respectively [16] it was found that the components reacted in a molar ratio of 1:1. Nevertheless, the photoproducts were not yellow as expected (cf. the yellow color of benzo[a]-phenazine-5,6-diol diacetate (13) [17]), but were obtained in violet crystals. It is assumed [16] that the photoproducts have the constitutions 6-benzoyloxybenzo[a]phenazin-5(7H)-one (12) and 14 respectively produced from, for example, 11 by migration of hydrogen. Benzo[a]phenazine-

5,6-diol (**9**) itself is stable only as the blue violet tautomeric lactam 6-hydroxybenzo[a]phenazin-5(7H)-one (**10**) [*17*].

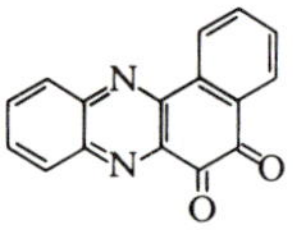

8 (yellow)

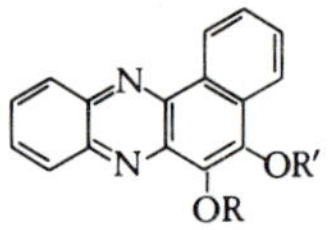

9: R=R'=H (yellow)
11: R'=H, R=COC$_6$H$_5$ (yellow)
13: R=R'=COCH$_3$ (yellow)

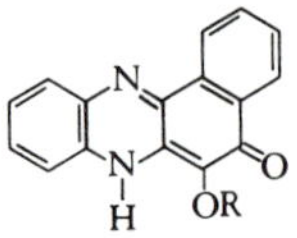

10: R=H (violet)
12: R=COC$_6$H$_5$ (violet)
14: R=CO−C$_6$H$_4$−OCH$_3$ (violet)

Colorless products of the ester type (cf. **1**) were obtained upon irradiation of p-anisaldehyde and cinnamaldehyde with 4-cyano-1,2-naphthoquinone (**3**) [*6*]. However, the action of acetaldehyde or propionaldehyde on **3** yielded orange colored products which were formulated as 3-acetyl-4-cyanonaphthalene-1,2-diol (**15**) and 4-cyano-3-propionylnaphthalene-1,2-diol respectively, spectroscopic evidence supporting this proposal [*7*]. The formation of **15** was imagined [*6*] to occur in a KHARASCH type addition of the aldehyde across the C=C-double bond:

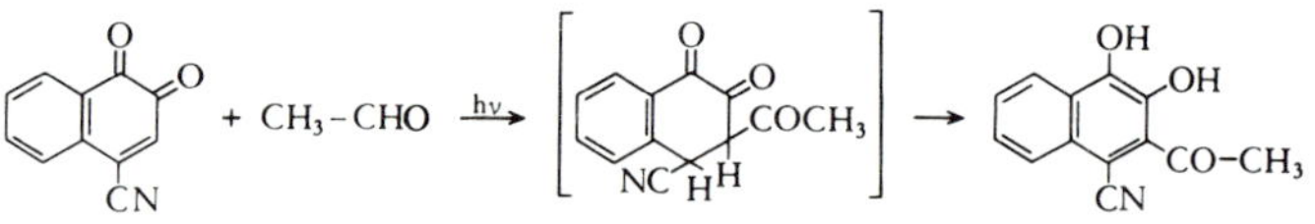

3 + CH$_3$−CHO $\xrightarrow{h\nu}$ [intermediate] → **15**

10-(2-Hydroxybenzoyloxy)-9-phenanthrol (**1**, R = salicyloyl) [*2*]. Phenanthrenequinone (5 g) and salicylaldehyde (25 ml), dissolved in 50 ml dry benzene were exposed to sunlight for 5 weeks (Königsberg). The crystals (4.5 g) which had separated were then sucked off and washed with benzene. A further 2.6 g was obtained from the benzene solution, the yield totalling 100%. By repeated recrystallization from hot benzene or alcohol the product was obtained as white needles, m.p. 188°.

6-(4-Methoxybenzoyloxy)-benzo[a]phenazin-5(7H)-one (**14**) [*16*]. 0.7 g of quinone **8**, 1 g anisaldehyde and 15 ml benzene (thiophene-free) were mixed in a Pyrex SCHLENK tube and the tube sealed in an atmosphere of dry carbon dioxide. The vessel was placed in sunlight for 10 days (Cairo). The precipitate which formed was filtered off, washed several times with acetone and then recrystallized several times from acetic acid to give ca. 70% of red-violet crystals of **14**, m.p. 249°.

3-Acetyl-4-cyanonaphthalene-1,2-diol (**15**) [*6*]. 0.6 g of **3** and 0.44 g acetaldehyde were dissolved in 20 ml dry, thiophene-free benzene and the solution sealed in a Pyrex tube under nitrogen. After 4 days insolation (Cairo) the yellow-brown precipitate was filtered off and recrystallized from benzene to afford 0.3 g of orange **15**, m.p. 228° (dec.).

2. Addition of aldehydes to 1,4-quinones

The studies on the photoaddition of aldehydes to p-quinones have been inaugurated by KLINGER as well. However, the mode of addition was found to be different in these cases since usually C-acylation took place

References, p. 192

instead of the O-acylation expected. Thus, from p-benzoquinone and benzaldehyde, 2,5-dihydroxybenzophenone (16) was obtained in the form of a quinhydrone addition compound [18]. In an analogous manner isovaleraldehyde and acetaldehyde were added to benzoquinone [18, 19], and acetaldehyde to 1,4-naphthoquinone [20]. The mechanism of these addition reactions has recently been elucidated [21] to involve addition of acyl radicals to the quinone with subsequent enolization of the primary adduct. In the irradiation solution of benzoquinone with acetaldehyde the reactive intermediate could be scavenged by 1,1-diphenylethylene. Thus, an enedione compound 17 accumulated which in ethanolic solution enolized to the corresponding quinol. In the absence of diphenylethylene a 70 % yield of 18 was obtained.

16: R=C_6H_5
18: R=CH_3

Acylation at the carbon atoms is, however, not the unique mode of addition with p-quinones. Thus, benzaldehyde adds in a 1,6-manner to chloranil [5, 22] and to khellinquinone (20) [23] respectively, products being 4-benzoyloxy-2,3,5,6-tetrachlorophenol (19) and 4-benzoyloxy-9-hydroxy-7-methyl-5H-furo[3,2-g]-1-benzopyran-5-one (khellinhydroquinone monobenzoate, 21) respectively.

2′,5′-Dihydroxyacetophenone (18) [19]. Finely powdered benzoquinone (5 g) was sealed in a tube with 30 ml freshly distilled acetaldehyde and exposed to sunlight. A thick glass rod was included within the tube so that the crust of quinhydrone which forms on the inner walls could be easily removed by shaking. After three months irradiation (Königsberg) the contents of the tube was poured into a basin, the quinhydrone adhering to the walls washed out with hot alcohol and the whole taken to dryness on a water bath. The brownish-green somewhat sticky residue was triturated, first with ether, then with aqueous sulfurous acid, in order to remove the glutinous matter, quinone and quinhydrone. Finally a brownish-yellow sandy powder remained, from which 18 was obtained in fair yield by sublimation; green-yellow needles, m.p. 202°.

Tetrachlorohydroquinone monobenzoate (19) [22]. Chloranil (5 g) and benzaldehyde (70 ml) were irradiated under nitrogen with a water-cooled mercury immersion lamp Osram HQA 500 (12°, 3 hrs.). Working up yielded 2.5 g (35%) of 19, m.p. [5] 180° (from acetic acid).

3. Addition of aldehydes to quinone imines and quinone oximes

a) Quinone imines

According to SCHÖNBERG and AWAD [24] aldehydes react with phenanthrenequinone imine (**22**) in light but not in the dark. Working with acetaldehyde, a product is obtained which PSCHORR [25] had previously synthesized from 10-amino-9-phenanthrol and for which he had suggested the formula of 10-acetamido-9-phenanthrol (**23**). The 2-hydroxy-2-methylphenanthro[9, 10-d]-4-oxazoline (**24**) structure also taken into consideration [24] cannot further be supported (cf. also [4]). Hence, the photoaddition reactions of quinone imines parallel those of the corresponding 1,2-quinones which lead to esters (comp. p. 186). Addition to the carbonyl group in quinone imines which would give rise to e.g. 10-acetoxy-9-phenanthrylamine was, however, not yet observed.

Apart from acetaldehyde, 9,10-phenanthrenequinone imine forms photoadducts [24] with benzaldehyde, p-anisaldehyde, p-chlorobenzaldehyde, piperonal, and 2-methoxy-1-naphthaldehyde. The reaction with the latter proceeds much more slowly than with benzaldehyde, which possibly may be explained in terms of steric hindrance. As investigated so far, the photoadducts are thermally unstable: thus, 10-benzamido-9-phenanthrol (**26**) goes to 2-phenylphenanthro[9,10-d]-oxazole (**25**), the cyclization possibly involving the intermediate **27**.

22 + R–CHO $\xrightarrow{h\nu}$

23: R = CH$_3$
26: R = C$_6$H$_5$

24: R = CH$_3$
27: R = C$_6$H$_5$

25: R = C$_6$H$_5$
28: R = C$_6$H$_4$–OCH$_3$

Photoadducts for which the structures originally proposed have to be revised in favor of the open chain constitution (cf. **23**, **26**) were also obtained on irradiating 7-isopropyl-1-methyl-9,10-phenanthrenequinone imine with acetaldehyde, benzaldehyde and p-anisaldehyde [26]. Analogously benzo[h]quinoline-5,6-dione imine (**29a** or **29b**) reacted with e.g. benzaldehyde, p-anisaldehyde or o-chlorobenzaldehyde to the corresponding N-acyl derivatives, e.g., 6-benzamidobenzo[h]quinolin-5-ol (**30**) or its isomer [11].

29a **29b** **30**: R = C$_6$H$_5$ **31** **32**

MUSTAFA and KAMEL [27] found that in light, but not in the dark, N-benzoyl-2,1-naphthoquinone imine (31) and the aromatic aldehydes benzaldehyde, p-anisaldehyde or p-tolualdehyde reacted to give adducts which were identified as e. g. 1-amino-2-naphthol dibenzoate (32).

10-Acetamido-9-phenanthrol (23) [24]. A solution of phenanthrenequinone imine (22) (1 g) and acetaldehyde (10 g) in benzene (thiophene-free and distilled over sodium) was sealed in a Monax glass SCHLENK tube under nitrogen and exposed to sunlight for 11 days (Cairo). Afterwards, the benzene was evaporated under reduced pressure and the residue washed with a little benzene. 23 was obtained from benzene in colorless crystals of m.p. 217° (red-brown melt, decomp.).

1-Amino-2-naphthol dibenzoate (32) [27]. N-Benzoyl-2,1-naphthoquinone imine (31) (1 g), benzaldehyde (1 ml) and dry, thiophene-free benzene (30 ml) were insolated (Cairo) for 10 days in a carbon dioxide atmosphere. The benzene solution was concentrated and the precipitate recrystallized from ethanol to afford 0.61 g of colorless 32, m.p. 234°.

b) Quinone oximes

The photochemical action of benzaldehyde or p-anisaldehyde on 9,10-phenanthrenequinone oxime led to products which on heating yielded 2-phenylphenanthro[9,10-d]oxazole (25) or 2-(4-methoxyphenyl)-phenanthro[9,10-d]oxazole (28) respectively [28]. Benzo[h]quinoline-5,6-dione oxime (33a or 33b) [11] or 2-phenyl-2H-naphtho[1,2-d]triazole-4,5-dione oxime (34) [13] have been added photochemically to aldehydes. The adducts thus formed on heating turned over to the corresponding oxazole derivatives, e.g. 35, with loss of 2 H and 2 O. No structure proposals have been made in the case of the photoaddition of aldehydes to quinone oximes, the reaction mode being still open to discussion. It may, however, be noted that the same oxazoles may be prepared, usually in higher yield, by heating the reactants in the presence of piperidine. Furthermore, as in the case of phenanthrenequinone oxime, it has been suggested [29] that some quinone oximes react as the corresponding nitroso phenols.

5-(4-Methoxyphenyl)-2-phenyl-2H-oxazolo[4′,5′:3,4]naphtho[1,2-d]triazole (35) [13]. A mixture of the oxime 34 (0.8 g) and p-anisaldehyde (0.9 g) in 80 ml benzene was exposed to sunlight for 28 days (Egypt). A colorless photoproduct separated during irradiation. It was recrystallized from xylene as almost colorless cyrstals, m.p. 250—251°; yield ca. 48%. This photoadduct was heated to 255° (bath temperature) under reduced pressure (oil-pump) for 30 minutes. The containing vessel was then cooled, the contents rubbed with benzene, filtered and washed with cold benzene. Recrystallization from xylene gave colorless crystals of 35, m.p. 275°.

References

[1] H. Klinger: Liebigs Ann. Chem. **249**, 137 (1888).
[2] H. Klinger: Liebigs Ann. Chem. **382**, 211 (1911).
[3] A. Schönberg, N. Latif, R. Moubasher and A. Sina: J. Chem. Soc. **1951**, 1364.
[4] G. Pfundt and G. O. Schenck, in: 1,4-Cycloaddition Reactions, Ed. by J. Hamer, pp. 345—417. New York: Academic Press 1967.
[5] R. F. Moore and W. A. Waters: J. Chem. Soc. **1953**, 238.
[6] A. Schönberg, W. I. Awad and G. A. Mousa: J. Chem. Soc. **1955**, 3850.
[7] W. I. Awad and M. S. Hafez: J. Amer. chem. Soc. **80**, 6057 (1958).
[8] A. C. Sircar and S. C. Sen: J. Indian Chem. Soc. **8**, 605 (1931).
[9] A. Schönberg and R. Moubacher: J. Chem. Soc. **1939**, 1430.
[10] A. Mustafa: J. Chem. Soc. **1949**, Suppl. 83.
[11] A. Mustafa, A. K. Mansour and A. F. A. M. Shalaby: J. Amer. chem. Soc. **81**, 3409 (1959).
[12] A. Mustafa, A. H. E. Harhash, A. K. E. Mansour and S. M. A. E. Omran: J. Amer. chem. Soc. **78**, 4306 (1956).
[13] A. Mustafa, A. K. Mansour and H. A. A. Zaher: J. Org. Chem. **25**, 949 (1960).
[14] A. Schönberg and A. Mustafa: J. Chem. Soc. **1947**, 997.
[15] A. Mustafa: J. Chem. Soc. **1951**, 1034.
[16] A. Schönberg, A. Mustafa and S. M. A. D. Zayed: J. Amer. chem. Soc. **75**, 4302 (1953).
[17] G. M. Badger, R. S. Pearce and R. Pettit: J. Chem. Soc. **1951**, 3204.
[18] H. Klinger and O. Standke: Ber. dtsch. chem. Ges. **24**, 1340 (1891).
[19] H. Klinger and W. Kolvenbach: Ber. dtsch. chem. Ges. **31**, 1214 (1898).
[20] G. O. Schenck and G. Koltzenburg: Naturwiss. **41**, 452 (1954).
[21] J. M. Bruce and E. Cutts: Chem. Comm. **1965**, 2.
[22] G. O. Schenck and G. Koltzenburg: Angew. Chem. **66**, 475 (1954).
[23] A. Schönberg and M. M. Sidky: J. Org. Chem. **22**, 1698 (1957).
[24] A. Schönberg and W. I. Awad: J. Chem. Soc. **1945**, 197.
[25] R. Pschorr: Ber. dtsch. chem. Ges. **35**, 2729 (1902).
[26] A. Schönberg and W. I. Awad: J. Chem. Soc. **1947**, 651.
[27] A. Mustafa and M. Kamel: J. Amer. chem. Soc. **77**, 5630 (1955).
[28] A. Schönberg, N. Latif, R. Moubasher and W. I. Awad: J. Chem. Soc. **1950**, 374.
[29] B. Eistert, R. Müller, H. Selzer and E.-A. Hackmann: Chem. Ber. **97**, 2469 (1964).

Chapter 20

Photoreductions with the aid of alcohols, ethers and other hydrogen donors

1. Photoreductions of C=C bonds

Cis-1,4-diphenyl-1,4-butenedione (cis-dibenzoylethylene, **1**) in degassed 2-propanol solution is photoreduced to 1,4-diphenyl-1,4-butanedione (dibenzoylethane, **2**) in the presence of benzophenone as sensitizer [*1*]. In the absence of sensitizers the irradiation takes a different course (comp. chapter 25).

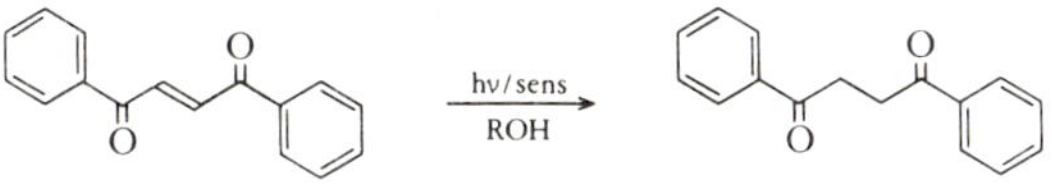

A similar photoreduction, though carried out without additional sensitizers, was reported by KÖLLER et al. [*2*] who found that irradiation of 1-phenalenone (**3**) in the presence of hydrogen donors, e.g., 2-propanol or diphenylmethane, produced 2,3-dihydro-1-phenalenone (**4**).

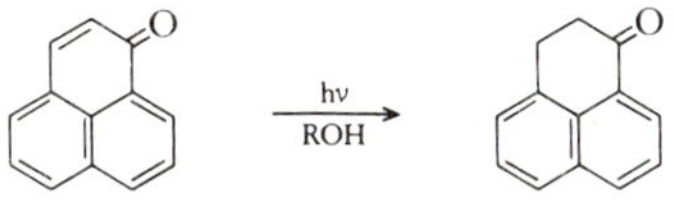

When 3β-acetoxypregna-5,16-dien-20-one (**5**) in ethanol solution was irradiated with UV light, a rapid reaction occurred [*3*] and two crystalline compounds were isolated. The first was the reduction product, 3β-acetoxy-pregn-5-en-20-one (**6**) while the second compound represented a solvent addition product, 3β-acetoxy-16α-(1-hydroxyethyl)-pregn-5-en-20-one (**7**). In the case of 3β-acetoxy-16-methylpregna-5,16-dien-20-one (**8**) analogous irradiation in ethanol or 2-propanol solution did not result in addition of solvent, but led to an almost quantitative yield of 3β-acetoxy-16β-methyl-pregn-5-en-20-one (**9**).

References, p. 197

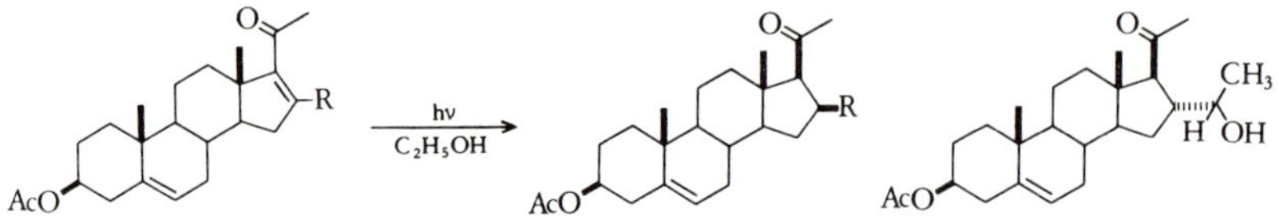

5: R = H
8: R = CH$_3$

6: R = H
9: R = CH$_3$

7

A similar reduction was observed by NANN et al. [4] when it was found that 15 hours' irradiation of a 0.006 molar ethanol solution of 17β-acetoxy-androst-4-en-3-one (testosterone acetate, 10) gave, among other products, 17β-acetoxy-5α-androstan-3-one (11) in 20% yield.

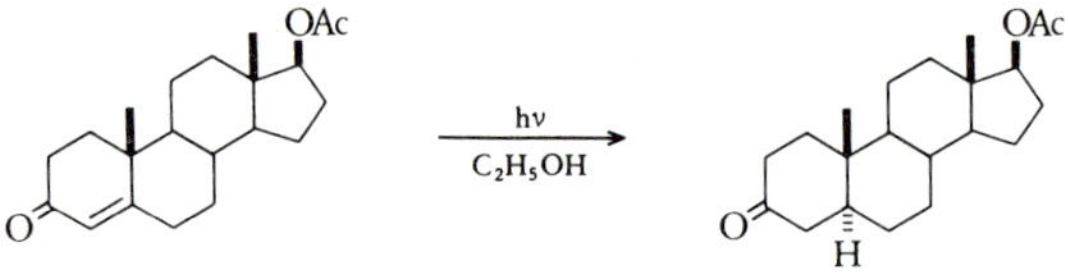

10

11

2. Photoreductions of C=O bonds. Formation of benzhydrols

BACHMANN [5] found that in 2-propanol containing a small amount of sodium 2-propanolate, benzophenone (12) is photochemically converted to benzhydrol (13) in excellent yield. The mechanisms proposed by COHEN [6] and BACHMANN [5] to account for the formation of benzhydrol from benzophenone in the presence of alkali are questioned [7].

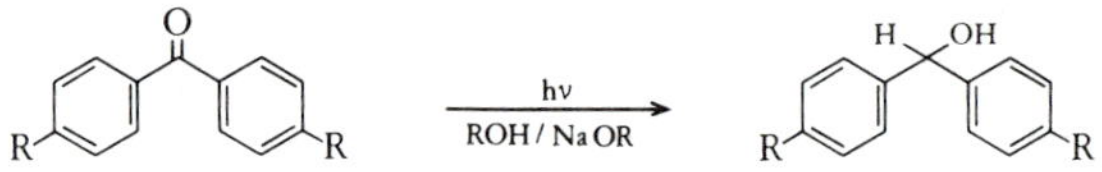

12: R = H

13: R = H

Similar reductions were carried out with the ketones listed in table 14. However, the formation of a carbinol compound was not observed [5] when 4,4'-bis-(dimethylamino)-benzophenone (MICHLERS ketone) or 1-naphthyl phenyl ketone were treated in a corresponding manner.

Table 14 [5]

ketone	secondary alcohol	yield
4-methylbenzophenone	4-methylbenzhydrol	95%
4,4'-dimethylbenzophenone	4,4'-dimethylbenzhydrol	90%
4-methoxybenzophenone	4-methoxybenzhydrol	90%
4-chloro-4'-methylbenzophenone	4-chloro-4'-methylbenzhydrol	80%
4-phenylbenzophenone	4-phenylbenzhydrol	95%
4-chlorobenzophenone	4-chlorobenzhydrol	80%

References, p. 197

In the absence of alkali but under otherwise identical conditions the ketones listed above will undergo photopinacolization. For this reaction, which may with more efficient hydrogen donors be accompanied by photoreduction, the reader is referred to chapter 22.

Phenyl 4-pyridyl ketone (14) on irradiation in 2-propanol [8] or in ethanol [9] gave α-phenyl-4-pyridinemethanol (15), alkali being not necessary to induce reduction in this case.

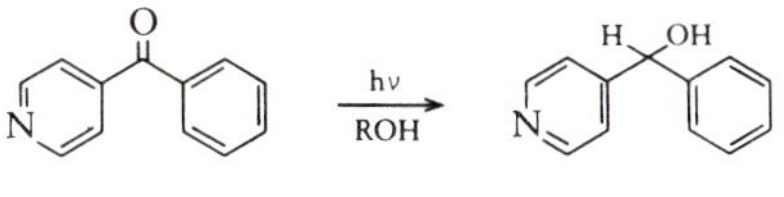

Benzhydrol (13) [5]. 25 g of benzophenone were placed in a 150 ml Pyrex bottle and 125 ml of 2-propanol containing a little sodium 2-propanolate (from 0.25 g sodium) added. After 7 days insolation the reaction mixture yielded 20 g of benzhydrol (13).

α-Phenyl-4-pyridinemethanol (15) [9]. A solution of 5 g of 14 in 60 ml of 95% ethanol was irradiated for 168 hrs. with a Pen-Ray quartz low-pressure mercury lamp. On removal of the solvent a reddish brown viscous oil was obtained, which after addition of ether and hexane crystallized (2.35 g). Distillation of this material at 140°/0.1 mm and two recrystallizations from a mixture of benzene and pentane gave pure 15, m.p. 125—126.5°.

3. Photoreductions of C=N bonds

Hydrogenation of a nitrogen-carbon double bond was observed by CERUTTI and SCHMID [10] on irradiation of 1,2,3,4-tetrahydro-4a-methyl-4aH-carbazole (16) in 2-propanol, 1,2,3,4,4a,9a-hexahydro-4a-methylcarbazole (17) being the product isolated in 45 % yield.

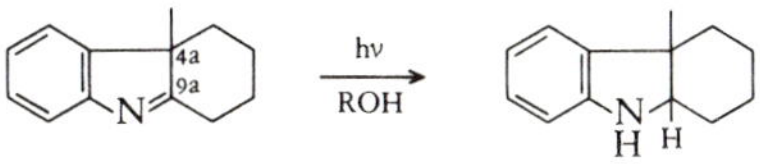

Carbazolium salts derived from 16 were found to be photoreducible as well [10] in even higher yields. The same relationship was encountered when, for example, acridine (18) and 10-methylacridinium chloride (22)

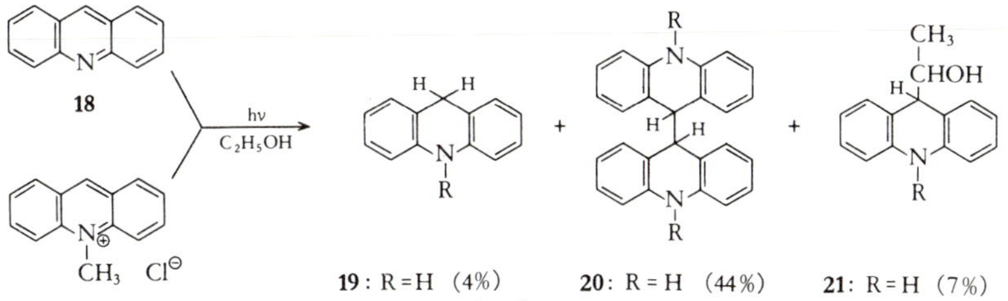

19: R=H (4%)
23: R=CH₃ (13%)
20: R=H (44%)
24: R=CH₃ (17%)
21: R=H (7%)
25: R=CH₃ (13%)

were each irradiated in ethanol solution [*11*]: the yield of acridan (**19**) (ca. 4 %) is contrasted with that of 10-methylacridan (**23**) (13 %). Respective yields of the other products isolated in the two irradiations [*11*] are given (p. 195) in parentheses.

A ca. 85 % yield of 9,9'-biacridan (**20**) was reported [*12*] to result from irradiation of **18** in ethanol. However, in this case neither **19** nor **21** could be isolated.

When 2-propanol solutions of phenazine (**26**) are exposed to sunlight in the absence of oxygen, deep colored crystals begin soon to separate which are regarded as molecular compounds of **26** and **27** [*13*]. The violet compound, m.p. 216—217°, is formed on mixing the components in the ratio 3:1, and the second, blue compound of m.p. 255—256° contains **26** and **27** in the ratio 1:1. On further irradiation complete conversion to colorless 5,10-dihydrophenazine (**27**) occurs while acetone may be detected. Since **27** is very sensitive to oxygen, the irradiation must be carried out under exclusion of oxygen.

Cyclohexane, ether or toluene may equally serve as hydrogen donors in the photoreduction **26** → **27** as was found by TOROMANOFF [*14*] who also observed the same sequence of stepwise reductions in the case of phenazines substituted in the 1- or 2-positions.

1,2,3,4,4a,9a-Hexahydro-4a-methylcarbazole (17) [*10*]. Compound **16** (672 mg) in 170 ml of isopropanol was irradiated with a Philips 93110 E mercury high-pressure lamp under nitrogen at about 23° for 18 hours. The crude product was chromatographed in benzene on silica gel and the rapidly migrating fraction rechromatographed in pentane on Al_2O_3. On twofold high-vacuum distillation at 70—80° (air bath) 305 mg (45%) of **17** were obtained as a colorless oil, which was characterized as the hydrochloride, m.p. 174—185° (from ethanol-ether).

10-Methylacridan (23) [*11*]. 10-Methylacridinium chloride (**22**) (1.21 g) was dissolved in 200 ml of 99.5% ethanol and irradiated for 3.5 hrs. with a Philips HPK 125 W mercury high-pressure lamp, the solution being agitated by argon. After 30 mins. irradiation silky crystals began to separate while the dark green fluorescence decreased. Concentration of the volume of the yellow solution to one third and chilling for one hour caused the precipitation of crystals, which were filtered off and washed with cold methanol. Recrystallization gave 151.7 mg (17%) of 10,10'-dimethyl-9,9'-biacridan (**24**), m.p. 272—274° (dec.). The filtrate was evaporated and chromatographed over silica gel. Elution with methylene chloride gave material which was purified by high-vacuum sublimation and recrystallization from aqueous methanol to give 112.8 mg (12.7%) of **23**, m.p. 92—93°. On further elution and rechromatography of the eluate 111.3 mg (12.6%) of 10-methyl-9-(1-hydroxyethyl)-acridan (**25**) were obtained, m.p. 89—90° (from ethyl acetate-pentane).

References, p. 197

4. Photoreductions of gem. chloronitroso compounds. Formation of oximes

Geminal chloronitroso compounds (comp. also chapter 29) may be reduced to the corresponding oximes not only by catalytic hydrogenation or by the action of $LiAlH_4$ and $NaBH_4$ but also photochemically [15]. It should, however, be noted that with the photochemical procedure yields are poor, thus rendering the method of little preparative value.

The photoreductions may be accomplished by exposing solutions of the chloronitroso compounds to sunlight, with ether and cyclohexane being used as solvents. In this way, 1-chloro-1-nitrosocyclohexane (**28**), 1-chloro-2-methyl-1-nitrosocyclohexane (**29**) and 4-chloro-4-nitrosoheptane (**30**) were reduced to the corresponding oximes.

2-Methylcyclohexanone oxime [15]. A solution of **29** (5.20 g) in 80 ml of ether is irradiated by direct sunlight in the absence of oxygen until the blue color has completely disappeared and the ether is of faint brown color. During the irradiation a dark brown resin precipitates clouding the solution, and hydrogen chloride is liberated. After pouring off the ether, the resin is dissolved in dilute HCl and neutralized. The resin which is thereby precipitated is taken up in chloroform and the aqueous phase shaken several times with chloroform. The purified chloroform solutions are dried with sodium sulfate. After removal of the solvent, 0.57 g (14%) 2-methylcyclohexanone oxime, m.p. 42−43°, is obtained by distillation of the black, viscous residue.

References

[1] G. W. Griffin and E. J. O'Connell: J. Amer. chem. Soc. **84**, 4148 (1962).
[2] H. Köller, G. P. Rabold, K. Weiss and T. K. Mukherjee: Proc. Chem. Soc. **1964**, 332.
[3] I. A. Williams and P. Bladon: Tetrahedron Letters **1964**, 257.
[4] B. Nann, D. Gravel, R. Schorta, H. Wehrli, K. Schaffner and O. Jeger: Helv. Chim. Acta **46**, 2473 (1963).
[5] W. E. Bachmann: J. Amer. chem. Soc. **55**, 391 (1933).
[6] W. D. Cohen: Rec. Trav. Chim. **39**, 243 (1920).
[7] G. O. Schenck, M. Pape, G. Matthias and G. von Bünau: unpublished results.
[8] M. R. Kegelman and E. V. Brown: J. Amer. chem. Soc. **75**, 4649 (1953).
[9] W. L. Bencze, C. A. Burckhardt and W. L. Yost: J. Org. Chem. **27**, 2865 (1962).
[10] P. Cerutti and H. Schmid: Helv. Chim. Acta **45**, 1992 (1962).
[11] H. Göth, P. Cerutti and H. Schmid: Helv. Chim. Acta **48**, 1395 (1965).
[12] F. Mader and V. Zanker: Chem. Ber. **97**, 2418 (1964).
[13] C. Dufraisse, A. Étienne and E. Toromanoff: Comptes rendus **235**, 759 (1952).
[14] E. Toromanoff: Ann. chim. [13], **1**, 115 (1956).
[15] E. Müller, H. Metzger and D. Fries: Chem. Ber. **87**, 1449 (1954).

Chapter 21

Formation of carbinols by photochemical addition of ketones and aldehydes to methylene groups

The photochemical reactions of carbonyl compounds dealt with in this chapter bear some resemblance to aldol condensation reactions and may be represented as follows:

$$\text{\textbackslash C=O} + \text{H-C-H} \xrightarrow{h\nu} \text{HO-C-C-H}$$

Though cases are known where ketones add photochemically to saturated hydrocarbons these reactions are synthetically of little value. A great variety of products arises thus and yields of carbinols are usually low. Thus when acetone was irradiated in cyclohexane or methylcyclohexane, the corresponding carbinols were obtained in yields of 12% or 21% respectively [1, 2].

Hydrocarbons with methylene groups adjacent to double bonds or aromatic systems are more prone to addition reactions leading to carbinols. Only these will be dealt with in the sequel. The interrelationships between reactions mentioned in this chapter and photopinacolization (comp. chapter 22) or photoaddition of hydrocarbons to quinone carbonyl groups (comp. chapter 18) should be noted. Formation of tertiary alcohols due to intramolecular cycloaddition reactions of carbonyl to methylene groups will be discussed in chapter 4.

1. Addition of ketones

DE MAYO [3] observed the formation of cyclohexenyldialkylcarbinols (cf. 1) on irradiation of dialkyl ketones, e.g. acetone, 3-pentanone and 4-heptanone, in cyclohexene. The addition of acetone was formulated as shown on page 199. Further products were bi-2-cyclohexen-1-yl (2) and, if the reaction was carried out at reflux temperature, trans-1,2-bis-(2-hydroxy-2-propyl)-cyclohexane (3).

References, p. 202

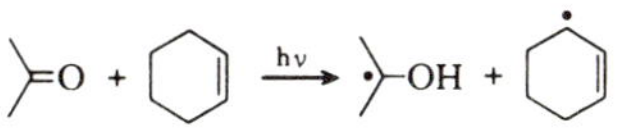

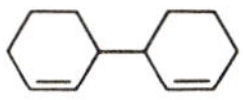

The scope of this light-induced addition reaction of ketones to hydrocarbons bearing an activated methylene group has been more fully explored by DE MAYO [*3—5, 7*]. In a manner analogous to the acetone case biacetyl was added to cyclohexene, the main product being 3-(2-cyclohexen-1-yl)-3-hydroxy-2-butanone (**4**) [*4*]. The same type of addition was observed when acenaphthene was irradiated with benzil. According to DE MAYO [*5*] the irradiation product first obtained by OLIVERI-MANDALÀ [*6*] has the structure **6** and not 6b,7,8,8a-tetrahydro-7,8-dihydroxy-7,8-diphenylcyclobut[a]acenaphthylene (**5**) as proposed by the Italian group. Further irradiation of **6** results in cleavage of the ketol to give 1-benzoylacenaphthene [*7*].

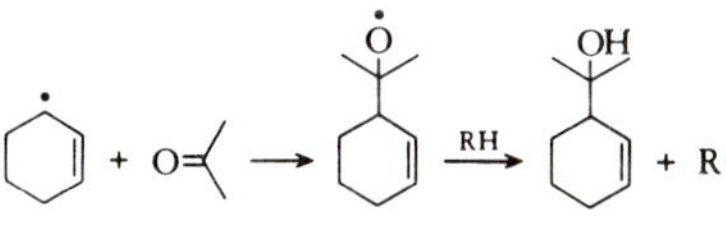

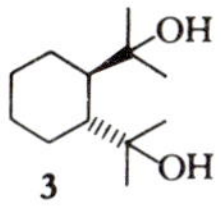

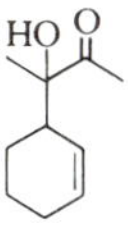

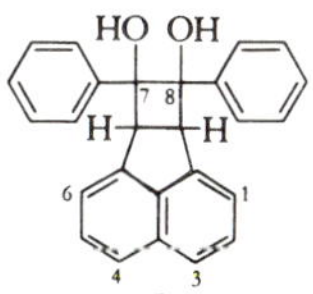

This behavior closely parallels the photolysis of 9-(xanthen-9-yl)-thioxanthen-9-ol (**9**) which was studied by SCHÖNBERG and MUSTAFA [*8*]. On placing a benzene solution of yellow thioxanthen-9-one (**7**) and xanthene (**8**) in sunlight, only a poor yield of 9-(xanthen-9-yl)-thioxanthen-9-ol (**9**) was obtained, the yield not being improved by extended irradiation. Further, irradiating in solution or heating to 270° caused **9** to decompose to the starting materials **7** and **8**. These observations were explained in terms of the establishment of a photochemical equilibrium in solution. It was assumed that the photochemical back-reaction consists of a fission of **9** into two dissimilar radicals, which then become stabilized by disproportionation [*8*].

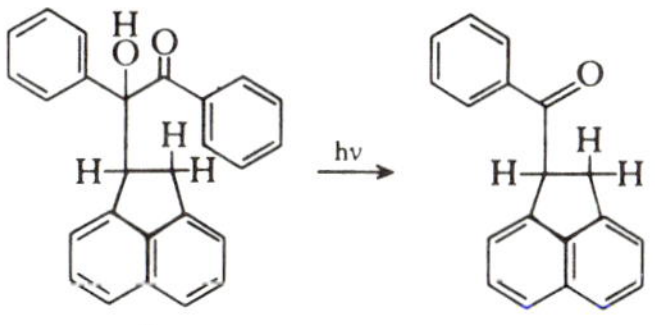

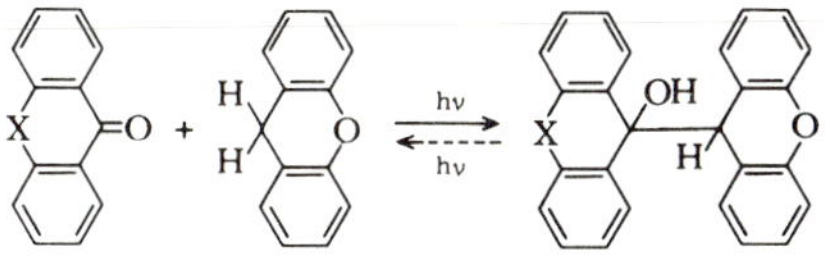

7: X = S
10: X = O

9: X = S
11: X = O

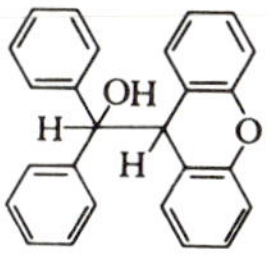

No such equilibrium was encountered when xanthen-9-one (**10**) was added to xanthene (**8**) in benzene solution: crystallization of the addition product 9-(xanthen-9-yl)-xanthen-9-ol (**11**) began after only a few hours irradiation [*9*]. The stability of **11** towards further irradiation is contrasted by the behavior of the corresponding pinacol, which was easily photolyzed in the presence of carbonyl compounds [*9*].

Benzophenone has frequently been added photochemically to activated methylene groups. Thus with xanthene the carbinol α,α-diphenyl-9-xanthenemethanol (**12**) was obtained [*9*]. Early in this century PATERNÒ and CHIEFFI [*10*] had already found that benzophenone (**13**) reacted with diphenylmethane (**14**) in sunlight to form 1,1,2,2-tetraphenylethanol (**15**). Irradiation of **13** with bis-(4-methoxyphenyl)-methane (**16**) [*11*] gave rise not only to 2,2-bis-(4-methoxyphenyl)-1,1-diphenylethanol (**17**), but also to the symmetrical dimers of the radicals involved in the photoreaction, benzopinacol and 1,1,2,2-tetrakis-(4-methoxyphenyl)-ethane. Only benzopinacol was, however, found on similar irradiation of di-p-tolylmethane (**18**).

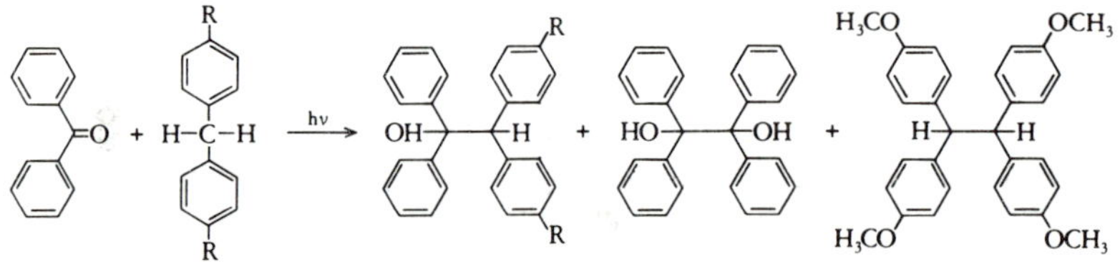

13
14: R = H
16: R = OCH$_3$
18: R = CH$_3$
15: R = H
17: R = OCH$_3$

The photochemical addition of 1,8-diazafluoren-9-one (9-oxo-9H-cyclopenta[1,2-b:4,3-b']dipyridine, **19**) to 1,8-diazafluorene (9H-cyclopenta[1,2-b:4,3-b']dipyridine, **20**) leads to the carbinol **21** [*12*].

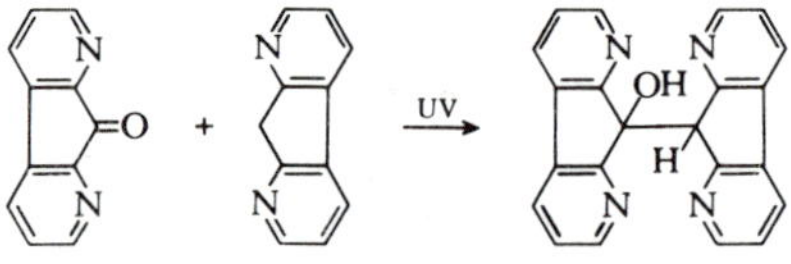

19   **20**   **21**

Photoaddition of benzophenone (**13**) to phenylacetic acid [*13*] yields 3-hydroxy-2,3,3-triphenylpropionic acid (**22**). This reaction was carried out by irradiating a benzene solution of the reactants with sunlight for 5 months. Apart from **22**, benzopinacol was also observed.

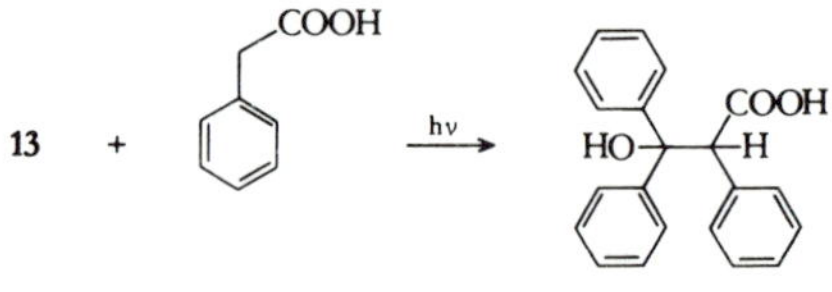

22

References, p. 202

α,α-Dimethyl-2-cyclohexene-1-methanol (1) [*3*]. A solution of acetone (50 g) in cyclohexene (275 ml) was irradiated under nitrogen in a quartz immersion apparatus at 20° for 100 hours. After evaporation of the solvents the residue was distilled in vacuo to give a colorless oil (70 g). Gas-liquid chromatography of a portion (0.65 g) gave two fractions, the first of which consisted of 1, b.p.$_{30}$ 82°; yield 0.34 g, 30%. The second fraction was constituted of bi-2-cyclohexen-1-yl (2) (0.28 g).

2-(1-Acenaphthenyl)-2-hydroxy-2-phenylacetophenone (6) [*5*]. Benzil (15 g) and acenaphtene (11 g) in 50 ml benzene were irradiated in a Pyrex flask under nitrogen with an 85 W HC3 (Hanovia) lamp. After 2 days the precipitate was collected (1.28 g) and crystallized from ethyl acetate to give 6, m.p. 237—239°.

9-(Xanthen-9-yl)-xanthen-9-ol (11) [*9*]. Equimolecular quantities of 8 (1 g) and 10 were dissolved in dry benzene (15 ml, thiophene-free) and placed in a Pyrex tube to sunlight. The tube was completely filled with the solution, closed with a cork and inverted in mercury to prevent entry of air. After one day of insolation (Cairo) the colorless crystals were filtered and recrystallized from ligroin. 11 had m.p. 194° (dec.); yield 80%.

2,2-Bis-(4-methoxyphenyl)-1,1-diphenylethanol (17) [*11*]. 10 g of bis-(4-methoxyphenyl)-methane (16) and 8 g 13 were dissolved in a small quantity of benzene and exposed to sunlight during 9 weeks. The slurry was triturated with methanol and filtered, the filtrate containing benzopinacol. The insoluble part (8 g) was extracted with boiling methanol and propanol, 1,1,2,2-tetra-(4-methoxyphenyl)-ethane remaining undissolved. The alcoholic solution was evaporated and the residue repeatedly recrystallized from petroleum ether to afford 17, m.p. 182—183°.

9-(1,8-Diazafluoren-9-yl)-1,8-diazafluoren-9-ol (21) [*12*]. The solution of 0.9 g 19 and 0.8 g 20 in 70 ml of benzene was irradiated with a mercury immersion lamp Q 81 (Hanau) at reflux temperature under nitrogen. After 3 hrs. the solvent was evaporated and the residue repeatedly recrystallized from acetone to yield 1.5 g (85%) of 21, m.p. 203—206° (dec.).

2. Addition of aldehydes

The photochemical addition of aldehydes to methylene groups activated by allylic double bonds or by aromatic systems has not found much interest in preparative photochemistry, since a variety of products is formed and carbinol yields usually are low. Thus when propionaldehyde is irradiated with cyclohexene [*1*] a 5% yield of α-ethyl-2-cyclohexene-1-methanol (23) was obtained together with 2 and the KHARASCH type addition product 1-cyclohexyl-1-propanone.

Unlike the formation of 23 the synthesis of 2-(benzyloxymethyl)-3-(4-nitrophenyl)-serine ethyl ester (25) by irradiation of 4-nitrobenzaldehyde with O-benzylserine ethyl ester (24) proceeds very smoothly [*14*].

$$\underset{24}{O_2N-\underset{H}{\overset{O}{\underset{|}{\overset{\|}{C}}}}- + \underset{NH_2}{\overset{\overset{\displaystyle C_6H_5}{\underset{|}{\overset{|}{CH_2}}}}{\underset{|}{\overset{|}{\underset{|}{\overset{O}{\underset{|}{CH_2}}}}}}-CH-COOC_2H_5}} \xrightarrow{h\nu} \underset{25}{O_2N-\underset{}{\overset{}{}}-\underset{}{\overset{HO}{\underset{|}{CH}}}-\underset{NH_2}{\overset{\overset{\displaystyle C_6H_5}{\underset{|}{\overset{|}{CH_2}}}}{\underset{|}{\overset{|}{\underset{|}{\overset{O}{\underset{|}{CH_2}}}}}}-C-COOC_2H_5}}$$

2-(Benzyloxymethyl)-3-(4-nitrophenyl)-serine ethyl ester (25) [*14*]. The hydrochloride of **24** (4 g) was stirred at 0° with 10 ml chloroform which contained 2% of gaseous ammonia. The filtered solution was evaporated to dryness at 20° in vacuo and the residue dissolved in absolute alcohol. 4-Nitrobenzaldehyde was added and the solution placed in the sun for 2 days (Jerusalem). 2 g of solid precipitated out, which were washed with ether and recrystallized from isopropanol to give **25**, m. p. 136°.

References

[*1*] N. C. Yang and D.-D. H. Yang: J. Amer. chem. Soc. **80**, 2913 [1958].
[*2*] K. Shima and S. Tsutsumi: Kogyo kagaku zasshi **64**, 460 [1961].
[*3*] P. De Mayo, J. B. Stothers and W. Templeton: Can. J. Chem. **39**, 488 [1961].
[*4*] P. W. Jolly and P. De Mayo: Can. J. Chem. **42**, 170 [1964].
[*5*] P. De Mayo and A. Stoessl: Can. J. Chem. **40**, 57 [1962].
[*6*] E. Oliveri-Mandalà, A. Giacalone and E. Deleo: Gazz. chim. Ital. **69**, 104 (1939).
[*7*] G. Kornis and P. De Mayo: Can. J. Chem. **42**, 2822 (1964).
[*8*] A. Schönberg and A. Mustafa: J. Chem. Soc. **1945**, 551.
[*9*] A. Schönberg and A. Mustafa: J. Chem. Soc. **1944**, 67.
[*10*] E. Paternò and G. Chieffi: Gazz. chim. Ital. **39**, II, 415 (1909).
[*11*] E. Bergmann and S.-I. Fujise: Liebigs Ann. Chem. **483**, 65 (1930).
[*12*] A. Schönberg and K. Junghans: Chem. Ber. **95**, 2137 (1962).
[*13*] E. Paternò and G. Chieffi: Gazz. chim. Ital. **40**, II, 321 (1910).
[*14*] E. D. Bergmann, H. Bendas and C. Resnick: J. Chem. Soc. **1953**, 2564.

Chapter 22

Photochemical formation and photolysis of 1,2-ethanediols

1. Formation of 1,2-ethanediols by the addition of alcohols to ketones

The simplest example is the photosynthesis of pinacol (2,3-dimethyl-2,3-butanediol, **1**) itself from acetone and 2-propanol; this was initially investigated by CIAMICIAN and SILBER [1] and later by SCHENCK [2]. Further examples are the formation of benzopinacol (tetraphenyl-1,2-ethanediol, **2**) from benzhydrol and benzophenone [3] and of triphenyl-1,2-ethanediol (**3**) from benzyl alcohol and benzophenone [4]. The formation of **2** may, however, proceed by a more complicated way [5] than by simple addition.

$$\begin{array}{c} R_1 \\ HO{-}H \\ R_2 \end{array} + \begin{array}{c} R_1 \\ O= \\ R_1 \end{array} \xrightarrow{h\nu} \begin{array}{cc} R_1 & R_1 \\ HO{-}{-}OH \\ R_2 & R_1 \end{array}$$

1: $R_1 = R_2 = CH_3$
2: $R_1 = R_2 = C_6H_5$
3: $R_1 = C_6H_5$, $R_2 = H$

Pinacol (1) [2]. 650 ml of an equimolecular mixture of acetone and isopropanol were irradiated with an air-cooled mercury vapor burner S 300 (Quarzlampengesellschaft). Additional external cooling maintained the reaction mixture at 48°. After the irradiation was complete, unreacted starting material was distilled off on the water-bath and the residue treated with the amount of water calculated to be necessary for the formation of pinacol hexahydrate. After standing for 12 hrs., the hydrate which had crystallized out was filtered off, dried in air and weighed. Yields after 24, 48 and 72 hrs. irradiation time were, 39.0, 74.4, and 119.0 g pinacol hexahydrate corresponding to 19.9, 38.0, and 60.8 g pinacol, respectively.

Benzopinacol (2) [3]. Benzhydrol (1 g) and benzophenone (ca. 1 g) in 20 ml dry, thiophene-free benzene were sealed in a Pyrex tube with exclusion of air and placed in sunlight for 2 days (December, Cairo). The solution was taken to dryness in vacuum and residue recrystallized from absolute ethanol. Benzopinacol was obtained with a yield of 80%, m.p. 187°.

2. Formation of 1,2-ethanediols via reductive dimerization

a) Aldehydes

The conversion of aldehydes to 1,2-ethanediols has been little investigated. CIAMICIAN and SILBER [6] observed the formation of the meso and

References, pp. 213—214

the racemic form of 1,2-diphenyl-1,2-ethanediol (hydrobenzoin, **4**) in the photochemical reaction of ethanol with benzaldehyde. The reaction with anisaldehyde proceeds analogously, leading to the formation of 1,2-bis-(4-methoxyphenyl)-1,2-ethanediol.

1-Naphthaldehyde was found not to undergo photoreductions with benzhydrol and other secondary alcohols [7]. The glycol 1,2-di-1-naphthyl-1,2-ethanediol (**5**) was however obtained when 1-naphthaldehyde was irradiated with 313 mµ light in the presence of tributylstannane, a more efficient hydrogen donor. In addition to **5**, 1-naphthalenemethanol (**6**) was formed, the molar ratio **5** : **6** being about 1 : 2.

1,2-Di-1-naphthyl-1,2-ethanediol (5) and 1-naphthalenemethanol (6) [7]. A solution was prepared by mixing 5 to 15 ml of a 1 molar solution of 1-naphthaldehyde in benzene with 5 to 10 ml of tributylstannane. The reaction mixture was diluted to 100 ml with benzene, degassed and irradiated, the light being filtered by a Corning 7-54 glass filter in series with an aqueous solution filter containing 145 g $NiSO_4 \cdot 6H_2O$ and 41.5 g $CoSO_4 \cdot 7H_2O$ per liter. Most of the benzene was stripped off, and the products were separated by chromatography on alumina. Two products were eluted; the first was **6**, a white solid of m.p. 60—61° (from petroleum ether), and the second was **5**, m.p. 185—186°.

b) Monoketones

The photopinacolization of alkyl aryl and diaryl ketones with the help of alcohols is a well investigated subject since CIAMICIAN and SILBER [8] opened research in this field. For preparative purposes it is customary to work with ethanol or isopropanol, sunlight or UV-light serving as the light sources. Table 15 lists ketones with which the pinacolization was success-

References, pp. 213—214

fully carried out on a preparative scale, alcohols used being given in brackets. It may be emphasized that in certain cases (e.g. the action of isopropanol on benzophenone) the reaction goes extremely smoothly. Use of methanol as solvent may lead to deviating results (cf. p. 207). Usually meso and racemic forms of the symmetrical pinacols are obtained.

Table 15

ketones (alcohols)	ref.
benzophenone (ethanol)	[6]
benzophenone (2-propanol)	[19]
4-phenylbenzophenone (2-propanol)	[20]
3-phenylbenzophenone (2-propanol)	[21]
4,4'-dimethoxybenzophenone (ethanol)	[22]
4,4'-dimethoxybenzophenone (2-propanol)	[23]
4,4'-dichlorobenzophenone (2-propanol)	[23]
3-benzoylpyridine (2-propanol)	[24]
acetophenone (ethanol)	[6]
acetophenone (butanol)	[25]
2-acetylpyridine (2-propanol)	[26]
3-acetylpyridine (2-propanol)	[26]
4-acetylpyridine (2-propanol)	[26]
deoxybenzoin (2-propanol)	[27]
3,4-dihydro-1(2H)-naphthalenone, α-tetralone (2-propanol)	[27]

The mechanism of the photoreduction of benzophenones to 1,2-ethanediols by isopropanol and related alcohols, the nature of the excited states involved and the feasible quantum yields have been the subject of numerous investigation during the past years; a discussion of these investigations lies outside the scope of this book. For reading references see, inter alia, [5,9—17]; the earlier literature has been summarized by SCHÖNBERG and MUSTAFA[18].

The photopinacolization of ketones by alcohols is, however, no general process. The ketones listed in table 16 are not converted to the corresponding pinacols by light in isopropanol solution.

Table 16

ketones	ref.
2-phenylbenzophenone	[27]
4-(methylthio)-benzophenone (7)	[28]
4,4'-bis-(methylthio)-benzophenone (8)	[28]
2-benzoylpyridine	[24]
4-benzoylpyridine	[24]
1-naphthyl phenyl ketone	[27]
di-1-naphthyl ketone	[27]
fluorenone (9)	[20]
xanthen-9-one (10)	[23]
o-dibenzoylbenzene	[29, 53]

The above statements on the non-reducibility of the said ketones should be taken with a certain reserve, since quite frequently the use of other solvents may facilitate the synthesis of pinacols also in cases when isopropanol fails. Thus, 4-aminobenzophenone and 4-hydroxybenzophenone remain unchanged on irradiation (356 mµ) in isopropanol. Yet the corresponding pinacols may be obtained on irradiation in pure cyclohexane with quantum yields of 0.2 and 1, respectively [*30*].

Photoreduction of 0.1 molar solutions of 4-dimethylaminobenzophenone in 2-propanol proceeds very slowly, about 0.003 the rate of benzophenone [*31*]. With addition of HCl the rate increases, 1,2-bis-(4-dimethylaminophenyl)-1,2-diphenyl-1,2-ethanediol dihydrochloride, m.p. 150—154° (dec.) being formed quantitatively.

4- and 2-aminobenzophenone may in isopropanol solution be photochemically converted to the pinacols if the solution contains acids. A 0.01 M solution of the para compound was reduced less than 7 % in the absence of hydrogen chloride after irradiation for 24 hrs. and was completely reduced in less than this time in the presence of 0.5 N HCl [*31*].

The failure of pinacol formation on irradiation of certain ketones has been extensively treated from the standpoint of theoretical photochemistry (cf. [*14*]). In this and related discussions one point does not seem to have been given sufficient attention: tetraaryl-1,2-ethanediols can only be isolated following the photochemical reduction of aryl ketones in 2-propanol if the pinacols themselves are stable in irradiated acetone solutions. This is by no means the case with all tetraarylglycols.

As stated earlier, the ketones **7—10** listed in table 16 are stable towards isopropanol in sunlight under CO_2 atmosphere. Their pinacols, being synthesized by dark reactions, were investigated [*23, 28*] in their behavior towards carbonyl compounds (for reactions with quinones see p. 211) under irradiation.

1,2-Bis-(4-methylthiophenyl)-1,2-diphenyl-1,2-ethanediol (**11**) [*28*], tetrakis-(4-methylthiophenyl)-1,2-ethanediol (**12**) [*28*], 9,9'-dihydroxy-9,9'-bifluorene (**13**) [*23*] and 9,9'-dihydroxy-9,9'-bixanthene (**14**) [*23*] are stable towards acetone in the dark, but are converted into the corresponding ketones **7—10** in sunlight. In each case the formation of 2-propanol in the irradiation experiments has been detected [*23, 28*]. All reactions were carried out under CO_2.

In order to explain the differing behavior of the ketones listed in tables 15 and 16, it will be assumed that the reaction is possible only because benzopinacol is stable towards acetone in light and does not cleave photochemically to the diphenylhydroxymethyl radical (**15**). On the other hand, the

References, pp. 213—214

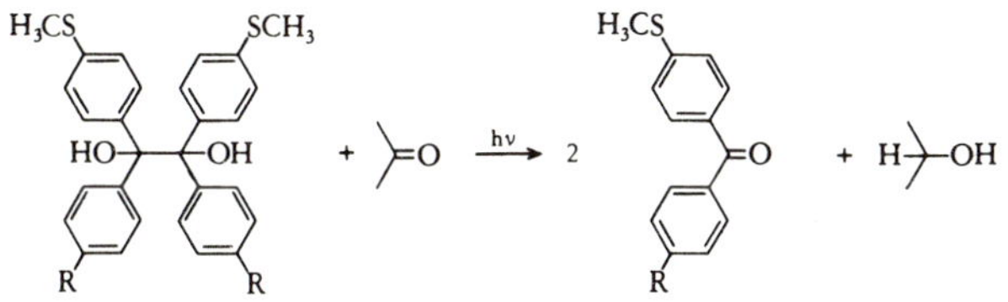

Formation of 1,2-ethanediols via reductive dimerization

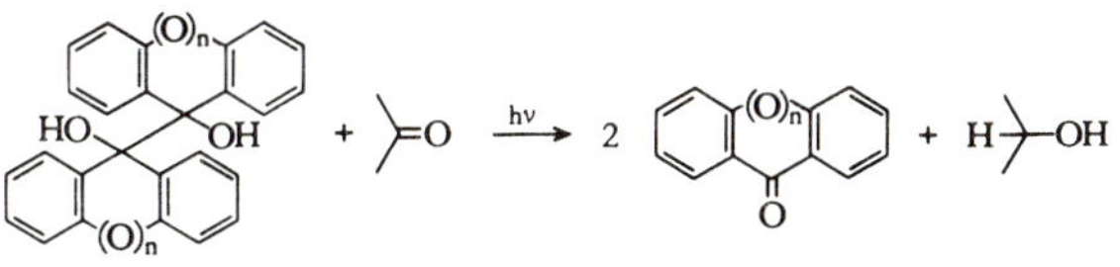

[Scheme showing pinacols 11 (R=H), 12 (R=SCH₃) reacting with acetone under hv to give diaryl ketones 7 (R=H), 8 (R=SCH₃) plus 2-propanol]

[Scheme showing pinacols 13 (n=0), 14 (n=1) reacting with acetone under hv to give xanthone-type ketones 9 (n=0), 10 (n=1) plus 2-propanol]

pinacols **11—14** ought to decompose into free radicals in light and in the presence of acetone to yield diaryl ketones and 2-propanol, as is illustrated below. These considerations may be applied also to other ketones which seem photochemically stable in isopropanol, taking their different optical absorptions into account.

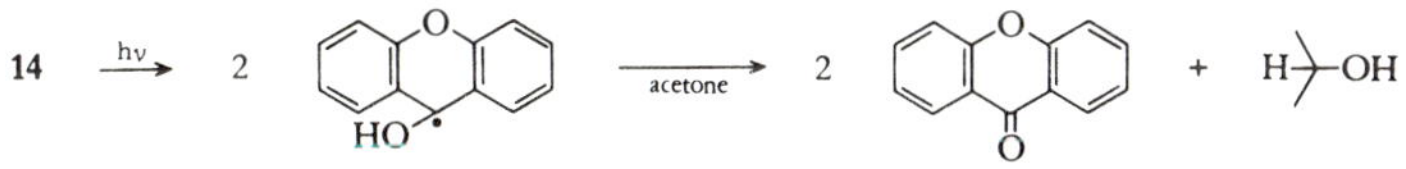

[Scheme: 14 → hv → 2 xanthenyl radicals → acetone → 2 × **10** + H-OH]

The formation of a radical analogous to **15**, viz. 1-hydroxyphenalen-1-yl (**17**), on irradiation of isopropanol solutions of 1-phenalenone (**16**) has recently been described [*32*]. Dimerization of two **17** moieties was, however, not observed.

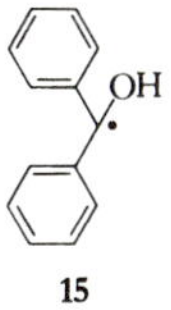

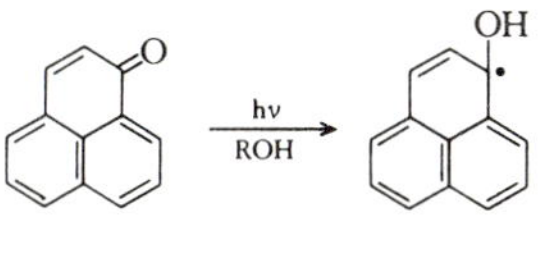

[Structures: **15** diphenylhydroxymethyl radical; **16** 1-phenalenone → hv/ROH → **17** 1-hydroxyphenalen-1-yl radical]

An instructive case of selectivity in pinacol formation has recently been published by GÖTH and collaborators [*33*]. While benzophenone or 4,4′-dichlorobenzophenone may in methanol be converted to the corresponding pinacols by irradiation with 254 mμ light (though in poor yield), 4,4′-dimethoxybenzophenone (**18**) yields only traces of the corresponding pinacol (**19**), the main product being 1,1-bis-(4-methoxyphenyl)-1,2-ethanediol (**20**), the 1:1 addition product. This result is especially surprising since **19** is produced from **18** by sunlight in both ethanol [*22*] and isopropanol [*23*].

3,4-Dihydro-1(2H)-naphthalenone (α-tetralone) gave, on irradiation in methanol [*33*], cis- and trans-1,1',2,2',3,3',4,4'-octahydro-1,1'-bi-1-naphthol (**21** and **22**) and 1,2,3,4-tetrahydro-1-hydroxy-1-naphthalenemethanol (**23**) in yields of 11, 8, and 20 %, respectively.

2-Methylbenzophenone (**24**) seemed stable towards irradiation in isopropanol since no pinacol could be detected. According to YANG and RIVAS [*34*] this result is caused by intramolecular photoenolization, **24** yielding 5-(α-hydroxybenzylidene)-6-methylene-1,3-cyclohexadiene (**25**). 2-Benzylbenzophenone (**26**) enolizes analogously to **27**. These reactions will be discussed further in chapter 3.

Photoenolization may also explain the non-formation [*35*] of a pinacol when 2,2'-bis-(4-methylphenylacetyl)-biphenyl (**28**) was irradiated in 2-propanol.

Methyl 2-naphthyl ketone may not be photochemically reduced by isopropanol [*7, 27*]. Nevertheless, according to HAMMOND and LEERMAKERS [*7*] the ketone is reduced by tributylstannane to α-methyl-2-naphthalenemethanol (**29**). For further photoreductions see chapter 20.

Benzopinacol (2) [*36*]. A warm solution of benzophenone (150 g) in 665 g 2-propanol, which had been treated with a drop of glacial acetic acid in order to neutralize traces of alkali, was set in the sunlight in a closed round bottomed flask the neck of which pointed downwards during the insolation. After 3—5 hrs. irradiation in strong sunlight crystals of benzopinacol began to separate, and after 8—10 days the reaction was complete. The vessel was cooled in ice and the crystals filtered off, washed with a little isopropanol and allowed to dry in air to give pure **2**, m.p. 188—195° (depending on the rate of heating; decomposition). The yield was 141—142 g (93—94%). The filtrate can be used for the reduction of further quantities of benzophenone.

c) 1,2,3-Triketones

Alloxan (**30**) is photochemically converted to alloxantin (**31**) by ethanol [*37*] or isopropanol [*38*]. Similarly 1,2,3-indantrione (**32**) is transformed to 2,2′-dihydroxy-2,2′-biindan-1,1′,3,3′-tetrone (**33**) [*39*].

The pinacol formula **33** [*39*] for hydrindantin has been proven beyond doubt [*40*] although occasionally hemiacetal structures have been proposed [*41*]. Formula **33** also explains well the fact that hydrindantin is obtained from **32** [*39*] by the action of the GOMBERG-BACHMANN reagent (Mg + MgI$_2$) [*42*] since many ketones which will not undergo enolization, are converted to the corresponding pinacols by this reagent. Also the ability of alloxantin to crystallize from water as a dihydrate supports formula **31**, since crystallization with several molecules of water is known in the case of pinacols.

Formula **35** has been taken into consideration [*43*] for the product formed, when 1,2,3-phenalenetrione was irradiated in isopropanol. **34**, 2,2′-dihydroxy-2,2′-biphenalene-1,1′,3,3′-tetrone, should however be regarded as the correct structure.

Alloxantin (31) [*37*]. A solution of alloxan (5 g) in 25 ml absolute ethanol was exposed to sunlight for 7 weeks (Winter, Bologna). After two weeks crystals of **31** began to separate (1.7 g). The alcohol distilled off contained aldehyde.

d) α-Ketocarboxylic acids and o-acylbenzoic acids

The photoreduction of phenylglyoxylic acid (**36**) in 2-propanol [*44*] yielded diphenyltartaric acid (**37**) which decarboxylated readily forming d,l-1,2-diphenyl-1,2-ethanediol (**4**). **37**, the racemic form, was obtained together with the meso form on cathodic reduction of **36** [*45*].

36: R=C$_6$H$_5$
38: R=CH$_3$

37: R=C$_6$H$_5$
39: R=CH$_3$

Irradiation of pyruvic acid (**38**) in the presence of benzhydrol as hydrogen donor yielded dimethyltartaric acid (**39**) [*46*]. Again irradiation in methanol resulted in formation of an addition product, 2-methylglyceric acid, a fact which emphasizes the particular role of methanol in these photopinacolization reactions.

According to HUYSER and NECKERS [*47*] alkyl phenylglyoxylates can be photochemically reduced by alcohols at ambient temperatures to the corresponding dialkyl diphenyltartrates (cf. **40→41**). However it was found [*47*] that the photochemical reductions of primary and secondary alkyl phenylglyoxylates in alcohols follow a different course at higher temperatures. Under these conditions mandelate esters of the alcohol in which the reaction is performed are obtained. Thus cyclohexyl phenylglyoxylate (**42**) in ethanol yields on illumination ethyl mandelate (**44**) and cyclohexanone. A possible interpretation of the formation of **44** is the assumption of the intermediacy of the hydroxyketene **43** [*47*] (comp. [*48*]).

40: R = C$_2$H$_5$
42: R = C$_6$H$_{11}$

Alkyl pyruvates were found [*49*] to undergo decarbonylation and subsequent formation of aldehydes on irradiation in benzene solution.

The photochemical conversion of 2-benzoylbenzoic acid (**45**) to 3,3'-diphenyl-3,3'-biphthalide (**47**), which was observed by LIMAYE [*50*], clearly proceeds via the hydroxy acid **46**. A similar reaction was reported by MUSTAFA [*51*] who obtained 3,3'-biphthalide (**49**) on photochemical reduction of 2-formylbenzoic acid (**48**) in 2-propanol.

45: R=C$_6$H$_5$
48: R=H

46: R=C$_6$H$_5$

47: R=C$_6$H$_5$
49: R=H

Diphenyltartaric acid (37) [*44*]. A suspension of 2 g of phenylglyoxylic acid (**36**) in 20 g isopropanol was insolated during 4 weeks (Summer, Cairo). The acid dissolved gradually to give a colorless solution. The reaction mixture was concentrated in vacuo and the residue recrystallized, first from benzene-light petroleum and then from ligroin, to give 1.2 g of **37**, m.p. 155°.

Diethyl diphenyltartrate (41) [*47*]. Ethyl phenylglyoxylate (**40**) (2.5 g) was dissolved in 2-butanol (8.0 g). The solution was sealed in a Pyrex tube and exposed to the sun for 2 days at about 30°. After the illumination the tube was kept at 0° for 6 hrs. whereupon 1.03 g crystallized from the solution. Recrystallization from petroleum ether gave pure **41**, m.p. 118—120°. 2-Butanone was detected by gaschromatography.

3,3'-Biphthalide (49) [*51*]. A solution of 1 g of **48** in 10 ml 2-propanol was sealed in a Pyrex tube (CO$_2$-atmosphere) and exposed to sunlight for 6 weeks (Autumn, Cairo). The crystals which separated were washed with petroleum ether and recrystallized from benzene to give colorless **49**, m.p. 257°.

3. Photochemical cleavage of 1,2-ethanediols by carbonyl compounds

SCHÖNBERG and MUSTAFA [*23*] found pinacol (**1**) to be stable to sunlight in benzene solution and in the presence of p-benzoquinone. In contrast, the pinacols listed in table 17, although stable in the dark under these conditions, are converted in sunlight to the corresponding ketones,

Table 17 [*23*]

2,3-diphenyl-2,3-butanediol
benzopinacol (**2**)
tetrakis-(4-methoxyphenyl)-1,2-ethanediol (**19**)
tetra-p-tolyl-1,2-ethanediol
tetrakis-(4-chlorophenyl)-1,2-ethanediol
9,10-diphenyl-9,10-phenanthrenediol (**50**)
1,2-diphenyl-1,2-acenaphthenediol (**53**)
9,9'-dihydroxy-9,9'-bixanthene (**14**)

while p-benzoquinone is reduced simultaneously to quinhydrone. Acetone may induce the cleavage of the C—C-bond as well, as was discussed earlier (comp. p. 206) for the case of the pinacols from xanthen-9-one and fluorenone. A similar degradation of glycols may be accomplished with lead tetraacetate as a dark reaction [52].

Recently RUBIN and ZWITKOWITS [35] carried out the photochemical cleavage of trans-9,10-bis-(4-methylbenzyl)-9,10-phenanthrenediol (52) which yielded 2,2'-bis-(4-methylphenylacetyl)-biphenyl (28) and quinhydrone.

50: R=C$_6$H$_5$
52: R=CH$_2$-C$_6$H$_4$-CH$_3$

51: R=C$_6$H$_5$
28: R=CH$_2$-C$_6$H$_4$-CH$_3$

53

Xanthen-9-one (10) [23]. 5g of **14** in 25 ml acetone were sealed in a glass tube under carbon dioxide and exposed to sunlight for 31 days (Cairo). Subsequently the solvent was evaporated in vacuo and the residue recrystallized from ligroin to give **10**. The acetone contained isopropanol which was detected with the aid of p-nitrobenzoyl chloride.

2,2'-Dibenzoylbiphenyl (51) [23]. 9,10-Diphenyl-9,10-phenanthrenediol (50) (4 g) and p-benzoquinone (2.4 g) were dissolved in ca. 20 ml dry, thiophene-free benzene and the solution exposed to sunlight for 14 days; during this time the reaction mixture was contained in a sealed Pyrex tube under CO$_2$. At the end of the illumination the quinhydrone was filtered off, the filtrate was evaporated in vacuo and the residue washed with petroleum ether or with cold alcohol. Recrystallization from alcohol afforded **51** in almost quantitative yield.

2,2'-Bis-(4-methylphenylacetyl)-biphenyl (28) [35]. A solution of 330 mg of trans-9,10-bis-(4-methylbenzyl)-9,10-phenanthrenediol (52) and 259 mg of sublimed p-benzoquinone in 14 ml benzene was irradiated with a General Electric 1000 W water-cooled high-pressure mercury lamp AH-6 (Pyrex vessel, nitrogen atmosphere, 30°). After 3.5 hrs. the irradiation was discontinued and 178 mg of quinhydrone were filtered off. The filtrate was concentrated on the steam bath under aspirator pressure until no further sublimation of p-benzoquinone occurred. Chromatography of the residue on alkaline alumina and elution with 50 and 90% benzene-petroleum ether yielded 84 mg of a colorless oil which was crystallized from isopropyl ether to give 54 mg of **28**, m.p. 83—84°.

Elution with benzene afforded 174 mg of starting material **52**.

References, pp. 213—214

References

[1] G. CIAMICIAN and P. SILBER: Ber. dtsch. chem. Ges. **44**, 1280 (1911).
[2] G. O. SCHENCK: Dechema-Monographien **24**, 105 (1955).
[3] A. SCHÖNBERG and A. MUSTAFA: J. Chem. Soc. **1943**, 276.
[4] G. CIAMICIAN and P. SILBER: Ber. dtsch. chem. Ges. **48**, 190 (1915).
[5] W. M. MOORE, G. S. HAMMOND and R. P. FOSS: J. Amer. chem. Soc. **83**, 2789 (1961).
[6] G. CIAMICIAN and P. SILBER: Ber. dtsch. chem. Ges. **34**, 1530 (1901).
[7] G. S. HAMMOND and P. A. LEERMAKERS: J. Amer. chem. Soc. **84**, 207 (1962).
[8] G. CIAMICIAN and P. SILBER: Ber. dtsch. chem. Ges. **33**, 2911 (1900).
[9] V. ERMOLAEV and A. TERENIN: J. Chim. phys. **55**, 698 (1958).
[10] G. O. SCHENCK, W. MEDER and M. PAPE: Proc. 2. UN Int. Conf. Peaceful Uses Atomic Energy **29**, 352 (1958).
[11] J. N. PITTS JR., R. L. LETSINGER, R. P. TAYLOR, J. M. PATTERSON, G. RECKTENWALD and R. B. MARTIN: J. Amer. chem. Soc. **81**, 1068 (1959).
[12] H. L. J. BÄCKSTRÖM and K. SANDROS: Acta Chem. Scand. **14**, 48 (1960).
[13] G. S. HAMMOND, W. P. BAKER and W. M. MOORE: J. Amer. chem. Soc. **83**, 2795 (1961).
[14] G. PORTER and P. SUPPAN: Pure appl. Chem. **9**, 499 (1964).
[15] G. PORTER and P. SUPPAN: Trans. Faraday Soc. **61**, 1664 (1965).
[16] C. WALLING and M. J. GIBIAN: J. Amer. chem. Soc. **87**, 3361 (1965).
[17] H. MAUSER, U. SPROESSER and H. HEITZER: Chem. Ber. **98**, 1639 (1965).
[18] A. SCHÖNBERG and A. MUSTAFA: Chem. Rev. **40**, 181 (1947).
[19] W. D. COHEN: Rec. Trav. chim. Pays-Bas **39**, 243 (1920).
[20] W. E. BACHMANN: J. Amer. chem. Soc. **55**, 391 (1933).
[21] H. H. HATT, A. PILGRIM and E. F. M. STEPHENSON: J. Chem. Soc. **1941**, 478.
[22] M. MIGATA: Bull. chem. Soc. Japan **7**, 334 (1932); C. A. **27**, 716 (1933).
[23] A. SCHÖNBERG and A. MUSTAFA: J. Chem. Soc. **1944**, 67.
[24] M. R. KEGELMAN and E. V. BROWN: J. Amer. chem. Soc. **75**, 4649 (1953).
[25] C. WEIZMANN, E. BERGMANN and Y. HIRSHBERG: J. Amer. chem. Soc. **60**, 1530 (1938).
[26] W. L. BENCZE, C. A. BURCKHARDT and W. L. YOST: J. Org. Chem. **27**, 2865 (1962).
[27] F. BERGMANN and Y. HIRSHBERG: J. Amer. chem. Soc. **65**, 1429 (1943).
[28] A. MUSTAFA: J. Chem. Soc. **1949**, 352.
[29] P. COURTOT and D. H. SACHS: Bull. Soc. Chim. France **1965**, 2259.
[30] G. PORTER and P. SUPPAN: Proc. Chem. Soc. **1964**, 191.
[31] S. G. COHEN and M. N. SIDDIQUI: J. Amer. chem. Soc. **86**, 5047 (1964).
[32] H. KÖLLER, G. P. RABOLD, K. WEISS and T. K. MUKHERJEE: Proc. Chem. Soc. **1964**, 332.
[33] H. GÖTH, P. CERUTTI and H. SCHMID: Helv. chim. acta **48**, 1395 (1965).
[34] N. C. YANG and C. RIVAS: J. Amer. chem. Soc. **83**, 2213 (1961).
[35] M. B. RUBIN and P. ZWITKOWITS: J. Org. Chem. **29**, 2362 (1964).
[36] W. E. BACHMANN: Org. Synth., Coll. Vol. **2**, 71 (1943).
[37] G. CIAMICIAN and P. SILBER: Ber. dtsch. chem. Ges. **36**, 1573 (1903).
[38] R. MOUBASHER and A. M. OTHMAN: J. Amer. chem. Soc. **72**, 2667 (1950).
[39] A. SCHÖNBERG and R. MOUBASHER: J. Chem. Soc. **1949**, 212.
[40] M. REGITZ, H. SCHWALL, G. HECK, B. EISTERT and G. BOCK: Liebigs Ann. Chem. **690**, 125 (1965).
[41] A. SCHÖNBERG and R. MOUBASHER: J. Chem. Soc. **1944**, 366.
[42] M. GOMBERG and W. E. BACHMANN: J. Amer. chem. Soc. **49**, 236 (1927).
[43] R. MOUBASHER and A. MUSTAFA: J. Chem. Soc. **1947**, 130.
[44] A. SCHÖNBERG, N. LATIF, R. MOUBASHER and A. SINA: J. Chem. Soc. **1951**, 1364.

[45] R. E. JUDAY: J. Org. Chem. **23**, 1010 (1958).
[46] P. A. LEERMAKERS and G. F. VESLEY: J. Amer. chem. Soc. **85**, 3776 (1963).
[47] E. S. HUYSER and D. C. NECKERS: J. Org. Chem. **29**, 276 (1964).
[48] N. C. YANG and A. MORDUCHOWITZ: J. Org. Chem. **29**, 1654 (1964).
[49] P. A. LEERMAKERS, P. C. WARREN and G. F. VESLEY: J. Amer. chem. Soc. **86**, 1768 (1964).
[50] D. B. LIMAYE: J. Univ. Bombay **1**, II, 52 (1932); C. A. **27**, 2097 (1933).
[51] A. MUSTAFA: J. Chem. Soc. **1949**, Suppl. 83.
[52] R. CRIEGEE: Ber. dtsch. chem. Ges. **64**, 260 (1931).
[53] A. SCHÖNBERG, N. LATIF, R. MOUBASHER and W. I. AWAD: J. Chem. Soc. **1950**, 374.

Chapter 23

Photochemistry of deoxybenzoin derivatives

Deoxybenzoin derivatives may, depending on their substituents, undergo a variety of photochemical reactions (cf. [1]). Reductive dimerization of pinacol radicals initially formed is predominant with solvents which may easily be dehydrogenated. This reaction, which is already well known [2, 3], will be dealt with in the chapter on pinacolization (chapter 22). In inert solvents such as benzene desyl compounds may photolyze to benzaldehyde and dimers of the fragments of the photoreaction [4, 5]. If substituents at the methylene group are prone to radical fission irradiation may then lead to desyl dimers and to dimers of the leaving group [4]. An additional mode of desyl decomposition was discovered recently [1], photolysis of the substituent being accompanied by cyclization and formation of a benzofuran structure. Some of the above reactions which are illustrated below will be dealt with in the sequel.

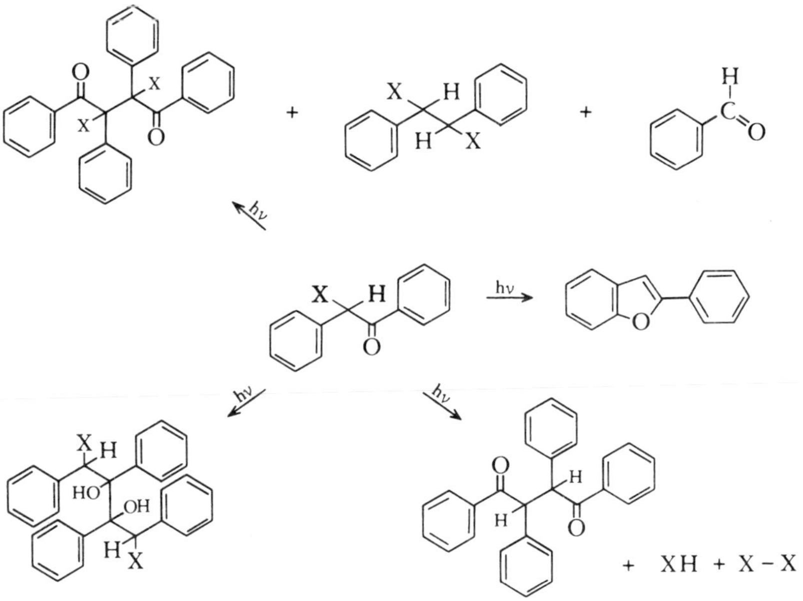

References, p. 217

In the absence of air, deoxybenzoin (1) and some of its aryl and methylene substituted derivatives undergo photolysis to 1,2-diphenylethane- (bibenzyl, 2), benzaldehyde and 1,2,3,4-tetraphenyl-1,4-butanedione (bidesyl, 3) [5]. Analogously, 2,2-diphenylacetophenone (4) under the action of sunlight yielded 1,1,2,2-tetraphenylethane (5) and benzaldehyde [4].

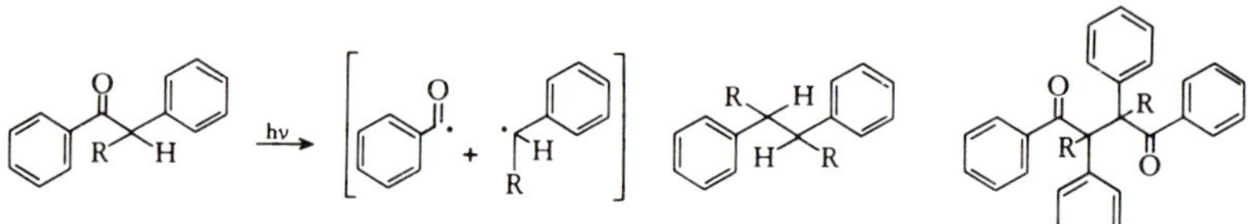

1: R = H
4: R = C$_6$H$_5$

2: R = H
5: R = C$_6$H$_5$

3: R = H

2 was envisaged [4] to arise either through direct dimerization of initially formed diphenylmethyl radicals or by the action of these radicals on starting material (4). Likewise the formation of benzaldehyde was thought to involve hydrogen abstraction from one of the compounds present in the system, e.g. 4.

According to the investigations of SCHÖNBERG et al. [4] certain aryl desyl sulfides undergo photolysis in sunlight. Thus, on irradiation of 2-phenyl-2-(phenylthio)-acetophenone (6) the stereoisomeric 1,2,3,4-tetraphenyl-1,4-butandiones (3) and benzenethiol were produced. The reaction was imagined to involve free radicals. This view is supported by the observation [6] that aryl desyl sulfides may serve as photoinitiators for the polymerization of olefins.

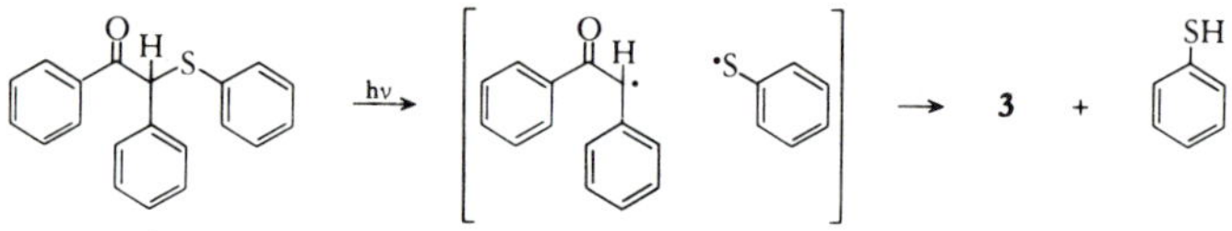

Quite a different behavior of some desyl derivatives was found by SHEEHAN and WILSON [1]. When appropriately substituted desyl esters,

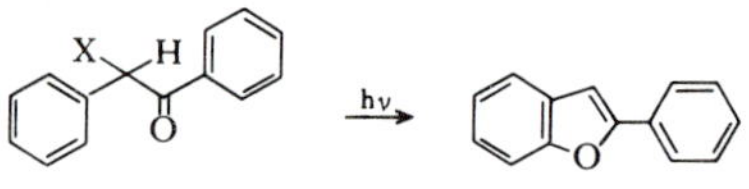

7: X = N(CH$_3$)$_2$ · HCl
9: X = OAc
10: X = Cl

8

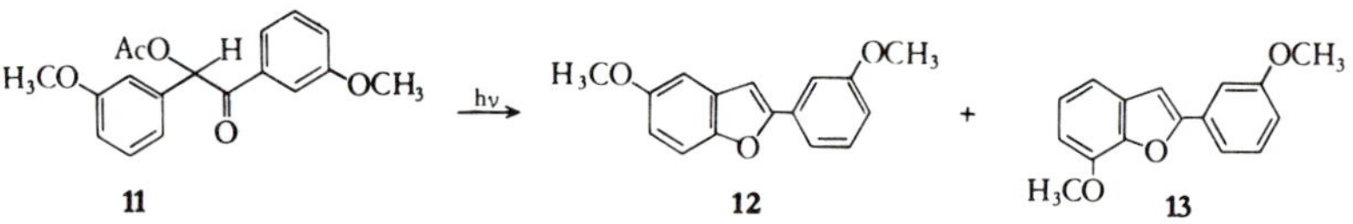

11

12

13

References, p. 217

e.g., 2-dimethylamino-2-phenylacetophenone hydrochloride (7) or 3,3'-dimethoxybenzoin acetate (11), were irradiated the products were 2-phenylbenzofuran (8) or the respective substitution compounds, e.g. 12. With 7 and benzoin acetate (9) the cyclization product 8 was always accompanied by over-irradiation products. In the irradiation of 2-chloro-2-phenylacetophenone (desyl chloride, 10) the main product (13%) was 3 as might be expected; the cyclization path to 8 was followed to 1%.

1,1,2,2-Tetraphenylethane (5) [4]. A solution of 4 (0.4 g) in 10 ml dry benzene (free from toluene and thiophene) was sealed in a Pyrex tube under carbon dioxide and placed in the sunlight for 2 weeks (Cairo). The solution, which had turned yellow, was concentrated to 2 ml and cooled, whereupon crystals of 5 precipitated (0.08 g.; m.p. 209°).

1,2,3,4-Tetraphenyl-1,4-butanedione (3) [4]. A solution of 2-phenyl-2-(phenylthio)-acetophenone (6) (2 g) in 15 ml dry, thiophene-free benzene was sealed as above and insolated (Cairo) for 30 days. 0.2 g of colorless 3 (bidesyl) separated as crystals, m.p. 255° (from benzene). The filtrate contained the 3-stereoisomer isobidesyl.

5-Methoxy-2-(3-methoxyphenyl)-benzofuran (12) [1]. A solution of 5.01 g of 11 in 800 ml of benzene was irradiated with a Hanovia S 200 W mercury immersion lamp for 17 hours (for details of the irradiation apparatus the original literature should be consulted). Upon removal of the solvent under reduced pressure, the resulting red oil deposited crystals after standing several days. These were recrystallized from hot methanol to give 1.062 g of colorless 12, m.p. 96—97°. Chromatography of the above filtrate on Al_2O_3 yielded further 0.665 g of a furan mixture and 0.530 g starting material. Careful fractional recrystallization afforded 7-methoxy-2-(3-methoxyphenyl)-benzofuran (13) of m.p. 74.5—76°. The over-all percentage yield, based on 89% conversion, was 46% for 12 and 2% for 13.

References

[1] J. C. Sheehan and R. M. Wilson: J. Amer. chem. Soc. **86**, 5277 (1964).
[2] E. Paternò, G. Chieffi and G. Perret: Gazz. chim. Ital. **44**, I, 151 (1914).
[3] W. D. Cohen: Chem. Weekblad **13**, 902 (1916).
[4] A. Schönberg, A. K. Fateen and S. M. A. R. Omran: J. Amer. chem. Soc. **78**, 1224 (1956).
[5] J. Kenyon, A. R. A. A. Rassoul and G. Soliman: J. Chem. Soc. **1956**, 1774.
[6] C. C. Petropoulos: J. Polymer Sci. A **2**, 69 (1964).

Chapter 24

Photochemical decarbonylation

Photodecarbonylation processes have not received much attention from organic chemists until recently. In the following chapter only a few outstanding examples of condensed phase decarbonylation processes will be discussed; for a survey on photolytic decarbonylation in the gas phase the reader is referred to SRINIVASAN [1].

1. Decarbonylation of ketones

a) Saturated cyclic ketones

Thujone (**1**) is converted smoothly and with unusual rapidity into carbon monoxide and 2-isopropyl-1,4-hexadiene (**3**) on exposure to ultraviolet radiation in the wavelength region 250—300 mµ [2]. The photolyses were carried out in quartz cells using either 150 W high-pressure Hanovia (510B1) xenon-mercury arcs or General Electric (H400A33-1) 400 watt "Dark Light" arcs with the external envelope removed.

Under given photochemical conditions the rates of carbon monoxide evolution (during first 10 % of reaction) from thujone, cyclopentanone, and cyclohexanone as pure liquids were respectively 0.79, 0.047 and 0.025 ml per minute. These results are taken to indicate that any intermediate radical such as **2** is capable of at best only very brief existence, and that the transition of thujone from its photoexcited state to the diene **3** and carbon monoxide is made unusually efficient by participation of the neighboring cyclopropane ring system.

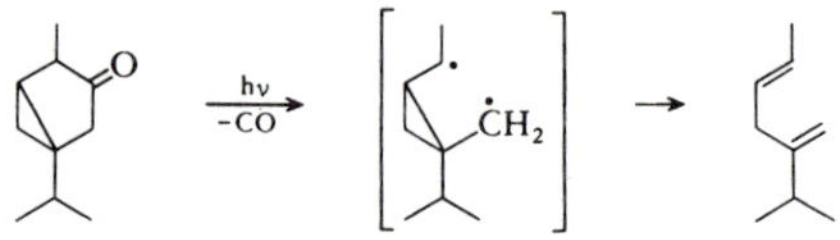

Irradiation of tetramethyl-1,3-cyclobutanedione (**4**) in benzene through Pyrex gives mainly carbon monoxide and 2,3-dimethyl-2-butene (tetramethylethylene, **8**) [*3*—*5*]. It is believed that the reaction proceeds via intermediates **5** and tetramethylcyclopropanone (**6**).

References, p. 224

When methanol is chosen as solvent, little **8** is obtained, the major products being methyl isobutyrate and methyl 2,2,3-trimethylbutyrate which arise through dimethylketene and 1-methoxy-1-hydroxy-2,2,3,3-tetramethylcyclopropane (**7**) [*3*, *5*]. When photolysis is carried out in furan [*4*] the liquid adduct **9** is formed in 15 % yield.

The photochemical reactions of **4** as studied by LEERMAKERS et al. [*5*] are illustrated in the scheme below:

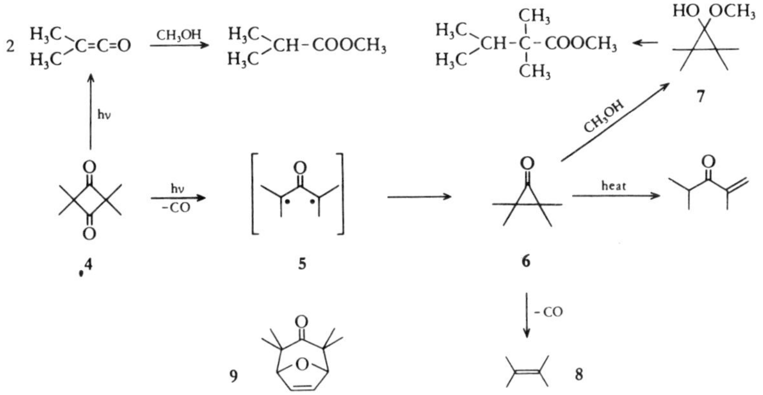

When dispiro[5.1.5.1]tetradecane-7,14-dione (**10**) was similarly photolyzed in benzene and methylene chloride under degassed conditions [*6*] a 61 % yield of bicyclohexylidene (**11**) was obtained.

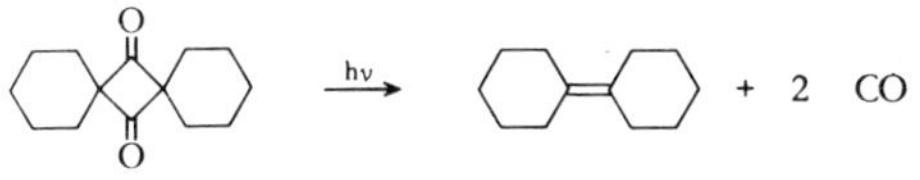

Irradiation of ethereal solutions of hexamethyl-1,3,5-cyclohexanetrione (**12**) [*7*] resulted in nearly quantitative decarbonylation. A variety of compounds was thus obtained, the formation of which may be envisaged as proceeding via biradicals which in themselves arise in several consecutive photoreactions (cf. scheme). The concentration of hexamethyl-1,3-cyclopentanedione (**13**) reached a maximum after 4 hrs. irradiation, when 83 % of **12** had reacted. Secondary products were 4-hydroxy-2,2,3,3,5-pentamethyl-4-hexenoic acid γ-lactone (**14**) and hexamethylcyclobutanone (**15**).

If the irradiation was extended beyond complete consumption of **12** (16 hours), the solution consisted mainly of **14** (about 65 %) and small quantities of 2-isopropylidene-3,3,4,4-tetramethyloxetane (**16**), hexamethylcyclopropane (**17**) and 2,3-dimethyl-2-butene (**8**).

b) Unsaturated cyclic ketones

Irradiation [8] of 2-methyl-3,5-cycloheptadienone (18) in ether gave carbon monoxide (95 %) and a mixture of geometrical isomers of 1,3,5-heptatrienes (19). In similar fashion irradiation of 3,5-cycloheptadienone (20) afforded isomeric 1,3,5-hexatrienes (21).

$$\xrightarrow{h\nu} CO + H_2C=CH-CH=CH-CH=CHR$$

18: R=CH$_3$ 19: R=CH$_3$
20: R=H   21: R=H

1,3,5-Heptatrienes (19) [8]. A solution of 0.73 g of 18 in 100 ml of ether in a quartz vessel equipped with an internal cooling coil and an outlet for gas collection was irradiated with a General Electric UA-3 mercury arc lamp for 6.5 hours. Carbon monoxide (95% of theory) was collected and identified by infrared comparison. The ethereal solution was concentrated by careful distillation, and the product was separated by preparative scale vapor phase chromatography.

c) Aromatic ketones

Photodecarbonylations of aromatic ketones have been studied mainly by QUINKERT and co-workers (comp. [9]). Thus, dibenzyl ketone (22) is rapidly and almost quantitatively converted to bibenzyl (23) on irradiation [10], and 1,1,3,3-tetraphenyl-2-propanone (24) yields 1,1,2,2-tetraphenylethane (25). Diphenylcyclopropenone, when irradiated, furnishes a 68 % yield of diphenylacetylene.

While, on irradiation of 2-indanone (26) dissolved in benzene, dibenzo[a,e]cyclooctadiene (27) is obtained only in poor yield [10], irradiation of

References, p. 224

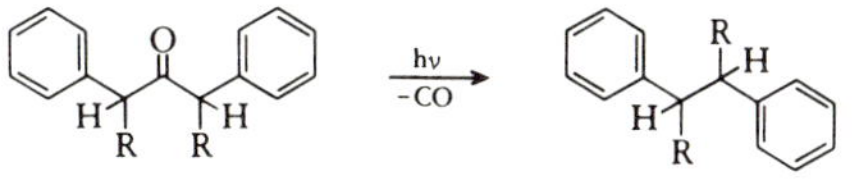

[Reaction scheme: 22/24 → 23/25]

22: R=H
24: R=C₆H₅

23: R=H
25: R=C₆H₅

[Reaction scheme: diphenylcyclopropenone → diphenylacetylene]

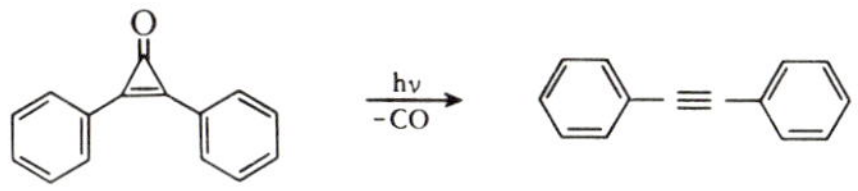

1-phenyl-2-indanone (**28**) affords a mixture of the rac. and meso forms of 3,4-diphenyldibenzo[a,e]cyclooctadiene (**29**) in a yield greater than 50%.

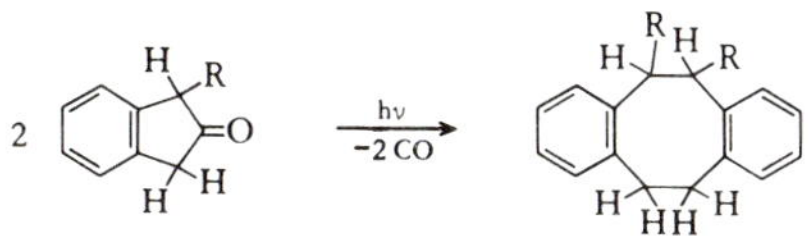

[Reaction scheme: 2 × 26/28 → 27/29]

26: R=H
28: R=C₆H₅

27: R=H
29: R=C₆H₅

While photolysis of **26** and **28** leads to dimers of the biradicals involved, similar irradiation of 1,3-diphenyl-2-indanone (**30**) results in quantitative formation of monomeric decarbonylation products [9], viz. cis-1,2-diphenylbenzocyclobutene (**31**) and trans-1,2-diphenylbenzocyclobutene (**32**). The ratio of the two compounds may be influenced by application of appropriate filter conditions.

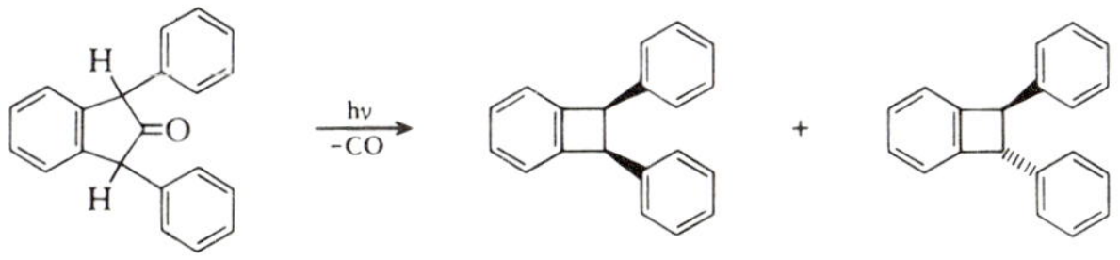

[Reaction scheme: 30 → 31 + 32]

30   31   32

A similar case of ring contraction was reported by CAVA and MANGOLD [11] who obtained a 82% yield of trans-1,2-dihydro-1,2-diphenylcyclobuta[l]-phenanthrene (**34**) on irradiation of benzene solutions of 1,3-dihydro-1,3-diphenyl-2H-cyclopenta[l]phenanthren-2-one (**33**).

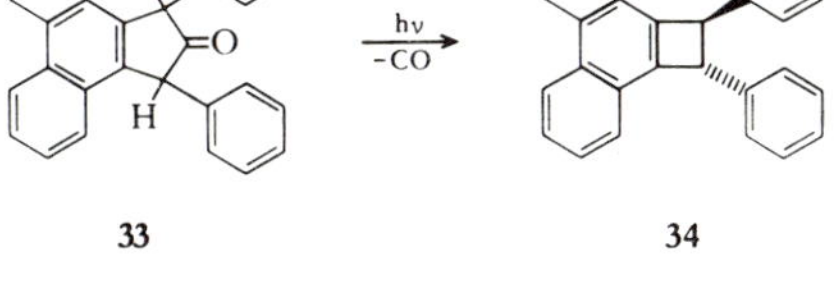

[Reaction scheme: 33 → 34]

33   34

2. Decarbonylation of a ketene

Irradiation of dimesitylketene (**35**) afforded tetramesitylethylene (**36**) in 19% yield as was found by ZIMMERMAN and PASKOVICH [*12*].

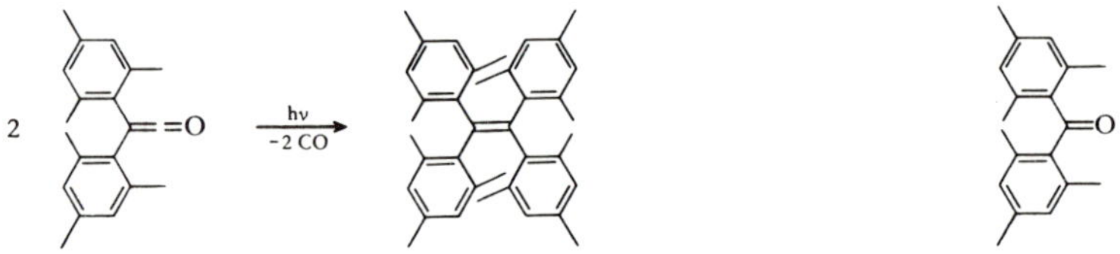

Tetramesitylethylene (36) [*12*]. A 300 ml solution of 2 g of **35** in dry cyclohexane was irradiated with a low-pressure mercury arc Hanovia immersion lamp under N_2 for 168 hours. Concentration of the solution in vacuo gave 2.173 g of tarry brown material which was dissolved in a minimum of chloroform and chromatographed on a silica gel column slurry packed with 5% ether in hexane. Elution with hexane yielded 0.351 g (19.4%) of **36**, m.p. 299—300°. Besides 0.127 g of 2,2′,4,4′,6,6′-hexamethylbenzophenone (**37**) were obtained with 5% ether in hexane.

3. Decarbonylation of aldehydes

Our knowledge of photodecarbonylation reactions of aldehydes is mainly based on the work of JEGER, SCHAFFNER and their collaborators (for a recent summary see SCHAFFNER [*13*]). Irradiation [*14*] of 3,17-bis-(ethylenedioxy)-androst-5-en-19-al (**38**) in ethanol at room temperature, using a high-pressure mercury lamp with Pyrex filter, caused evolution of carbon monoxide and conversion of the aldehyde into the 19-nor compound 3,17-bis-(ethylenedioxy)-estr-5-ene (**39**), m. p. 131—132°, in more than 90 % yield. The reaction was not sensitive to the presence of oxygen, but when a quartz vessel and a low-pressure mercury lamp were used, the rate of decarbonylation was significantly lower.

In order to trace the origin of the hydrogen atom at C-10 of compound **39**, 3,17-bis-(ethylenedioxy)-androst-5-en-19-al-19-d (**40**) was photolyzed. 3,17-Bis-(ethylenedioxy)-estr-5-ene-10-d (**41**) was formed in equally excellent yield, m.p. 133°. Mass-spectrometric examination revealed no loss of deuterium during the photochemical decarbonylation.

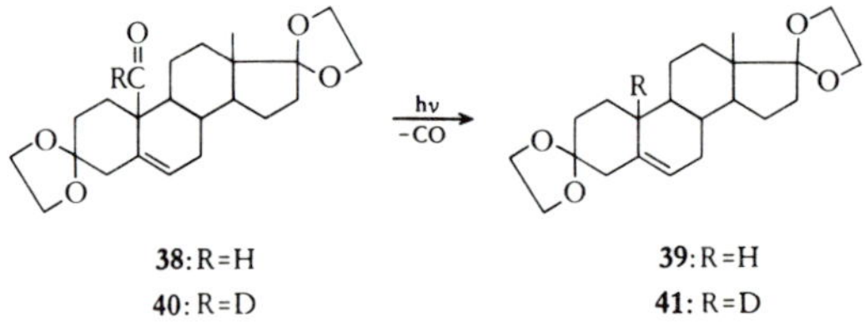

38: R=H
40: R=D

39: R=H
41: R=D

References, p. 224

Irradiation of the corresponding saturated aldehyde led to a complex mixture of products [14] (comp. also [13]), decarbonylation occurring only to a minor extent. Thus, the important role of the double bond in the **38 → 39** decarbonylation process was clearly established.

4. Decarbonylation of S-acyl xanthates

Acyl and arylacyl xanthates (comp. **42**) are smoothly decomposed by light from either a tungsten lamp or a mercury arc lamp [15]. The acyl xanthates have typical UV and IR absorption bands, the disappearance of which may be easily followed during photolysis.

Irradiation of O-ethyl S-phenylacetyl xanthate (**42**) in benzene solution under reflux gave a high yield of O-ethyl S-benzyl xanthate (**43**). The reaction proceeded faster in high boiling liquids; thus in refluxing toluene only two hours were required, whereas in boiling ether seven hours were needed to complete the reaction.

According to BARTON [15] the reactions are best explained by assuming the primary fission of **42** into acyl and xanthate radicals. The formation of **43** is illustrated below:

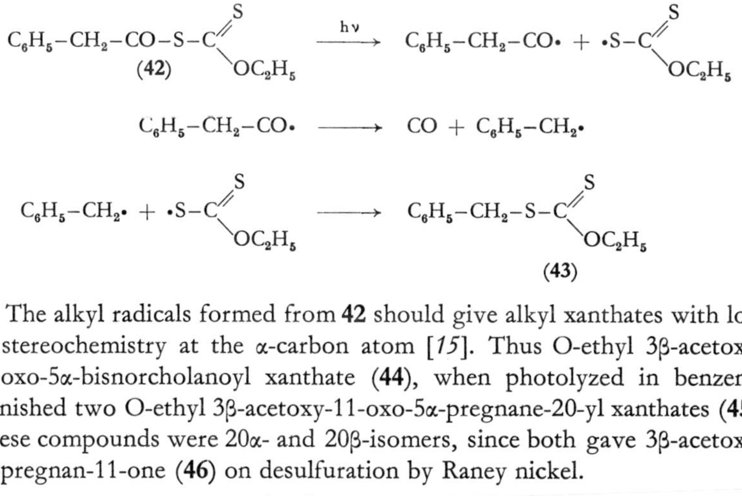

$$C_6H_5-CH_2-CO-S-C(\!\!=\!\!S)\!-\!OC_2H_5 \quad \xrightarrow{h\nu} \quad C_6H_5-CH_2-CO\cdot + \cdot S-C(\!\!=\!\!S)\!-\!OC_2H_5$$
$$(42)$$

$$C_6H_5-CH_2-CO\cdot \quad \longrightarrow \quad CO + C_6H_5-CH_2\cdot$$

$$C_6H_5-CH_2\cdot + \cdot S-C(\!\!=\!\!S)\!-\!OC_2H_5 \quad \longrightarrow \quad C_6H_5-CH_2-S-C(\!\!=\!\!S)\!-\!OC_2H_5$$
$$(43)$$

The alkyl radicals formed from **42** should give alkyl xanthates with loss of stereochemistry at the α-carbon atom [15]. Thus O-ethyl 3β-acetoxy-11-oxo-5α-bisnorcholanoyl xanthate (**44**), when photolyzed in benzene, furnished two O-ethyl 3β-acetoxy-11-oxo-5α-pregnane-20-yl xanthates (**45**). These compounds were 20α- and 20β-isomers, since both gave 3β-acetoxy-5α-pregnan-11-one (**46**) on desulfuration by Raney nickel.

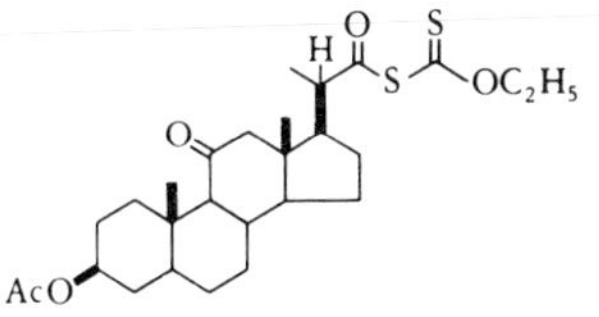

44

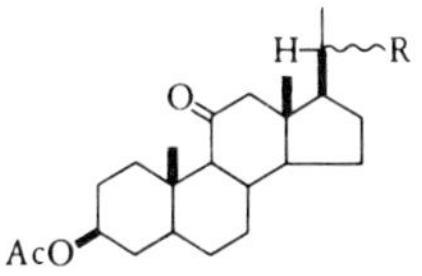

45: R = -S-CS-OC$_2$H$_5$
46: R = H

O-Ethyl S-benzyl xanthate (43) [*15*]. A solution of 2.403 g of **42** in 250 ml benzene was irradiated for 24 hrs. whilst under reflux. During the irradiation the yellow color faded. Removal of the solvent in vacuo gave 1.85 g of **43**, b. p.$_{.0.1}$ 95—97°.

O-Ethyl 3β-acetoxy-11-oxo-5α-pregnan-20-yl xanthates (45) [*15*]. 600 mg of **44** were dissolved in 36 ml of benzene and the solution irradiated with a tungsten lamp at 40° until the yellow color had disappeared (30 mins.).

The process was repeated four times. The combined product (2.89 g) was chromatographed over alumina. Elution with light petroleum-benzene gave two isomeric xanthates (**45**) of m. p. 193—195° and 138—140°, respectively.

References

[*1*] R. Srinivasan, in: Adv. Photochem., ed. by W. A. Noyes, jr., G. S. Hammond and J. N. Pitts, jr., Vol. **1**, pp. 83—113. New York: Interscience 1963.

[*2*] R. H. Eastman, J. E. Starr, R. S. Martin and M. K. Sakata: J. Org. Chem. **28**, 2162 (1963).

[*3*] N. J. Turro, G. W. Byers and P. A. Leermakers: J. Amer. chem. Soc. **86**, 955 (1964).

[*4*] R. C. Cookson, M. J. Nye and G. Subrahmanyam: Proc. Chem. Soc. **1964**, 144.

[*5*] N. J. Turro, P. A. Leermakers, H. R. Wilson, D. C. Neckers, G. W. Byers and G. F. Vesley: J. Amer. chem. Soc. **87**, 2613 (1965).

[*6*] P. A. Leermakers, G. F. Vesley, N. J. Turro and D. C. Neckers: J. Amer. chem. Soc. **86**, 4213 (1964).

[*7*] H. U. Hostettler: Tetrahedron Letters **1965**, 1941.

[*8*] O. L. Chapman, D. J. Pasto, G. W. Borden and A. A. Griswold: J. Amer. chem. Soc. **84**, 1220 (1962).

[*9*] G. Quinkert: Pure appl. Chem. **9**, 607 (1964).

[*10*] G. Quinkert, K. Opitz, W. W. Wiersdorff and J. Weinlich: Tetrahedron Letters **1963**, 1863.

[*11*] M. P. Cava and D. Mangold: Tetrahedron Letters **1964**, 1751.

[*12*] H. E. Zimmerman and D. H. Paskovich: J. Amer. chem. Soc. **86**, 2149 (1964).

[*13*] K. Schaffner: Chimia **19**, 575 (1965).

[*14*] J. Iriarte, J. Hill, K. Schaffner and O. Jeger: Proc. Chem. Soc. **1963**, 114.

[*15*] D. H. R. Barton, M. V. George and M. Tomoeda: J. Chem. Soc. **1962**, 1967.

Chapter 25

Photochemical formation and reactions of carboxylic acids and their derivatives

1. Formation of aliphatic carboxylic acids by the action of oxygen and water on alkyl halides

Photochemical oxidation of compounds of the general formula $F_3C-(CF_2)_n-CFXY$ (X or Y = H, F, Cl, Br or I) is surprisingly easy. In the propane series C—C cleavage readily occurs, for example, carbonyl fluoride (OCF_2) is formed from heptafluoro-1-iodopropane [1]. Nevertheless, the C—C cleavage is avoided if the oxidation is accomplished in the presence of water. In this way HASZELDINE [2] was able to obtain 3,4-dichloropentafluorobutyric acid (2) from 1,3,4-trichlorohexafluoro-1-iodobutane (1) by photochemical action in the presence of sodium hydroxide solution.

$$ClF_2C-CFCl-CF_2-CFClI \xrightarrow[O_2/H_2O]{h\nu} ClF_2C-CFCl-CF_2-COOH$$
$$\quad\quad\quad 1 \quad\quad\quad\quad\quad\quad\quad\quad\quad\quad\quad\quad\quad 2$$

3,4-Dichloropentafluorobutyric acid (2) [2]. The iodobutane 1 (4.3 g) was sealed in a quartz tube (200 ml) with oxygen (7 atmospheres) and aqueous sodium hydroxide (10%, 20 ml). The tube was shaken vigorously in a horizontal position and irradiated (12 hrs.) by a UV lamp 5 cm off the tube. The tube was then opened and more oxygen added; after a second irradiation period (12 hrs.) the excess oxygen was pumped off. The solution was acidified with hydrochloric acid; an emulsion was seen to form. The aqueous solution was extracted with ether (10 × 20 ml), the ethereal extracts dried and subjected to distillation giving 63% of 2, b.p.$_{25}$ 105—107°.

2. Formation of carboxylic acids by photolysis of cyclic ketones

a) Saturated ketones

CIAMICIAN and SILBER [3, 4] have converted a series of cyclic ketones to the corresponding open chain carboxylic acids by the action of sunlight in aqueous media. Thus caproic acid (4) is obtained from cyclohexanone (3) [4]. Similar reactions were carried out with 2-methylcyclohexanone [4] and menthone (5) [3].

References, pp. 240—241

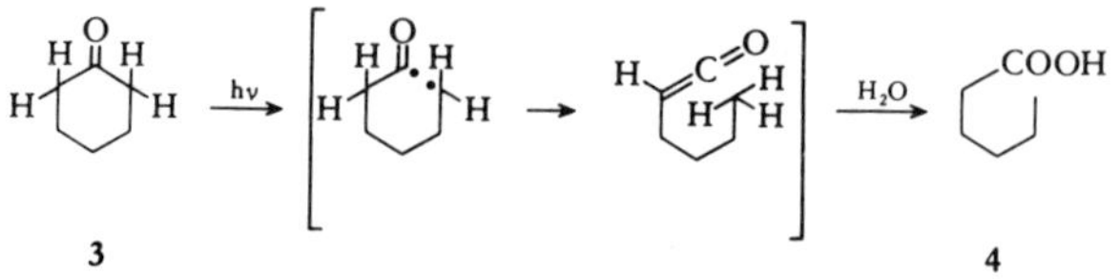

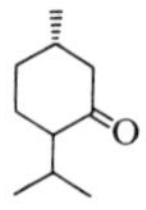

According to QUINKERT [5] the photochemical hydrolysis is envisaged as proceeding via disproportionation of the alkyl-acyl radical pair originating from the photochemical fission of the bond adjacent to the carbonyl group. In the case of cyclohexanone irradiation in deuterium oxide [6] led to caproic acid-2-d which supports the QUINKERT mechanism.

The photolysis reaction is restricted to ketones bearing at least one hydrogen atom at carbon α to the carbonyl group, as was found by QUINKERT [5, 7, 8] who systematically studied this hydrolysis reaction (for a recent review see [9]). Thus, 3β-methoxy-16,16-dimethylandrost-5-en-17-one (6) is stable towards irradiation in hydroxylic solvents [8], whereas 3β-methoxyandrost-5-en-17-one (7) readily photolyzes to the corresponding acid (8) (besides other products) [5, 7], inversion of the configuration at C-13 accompanying the ring opening reaction.

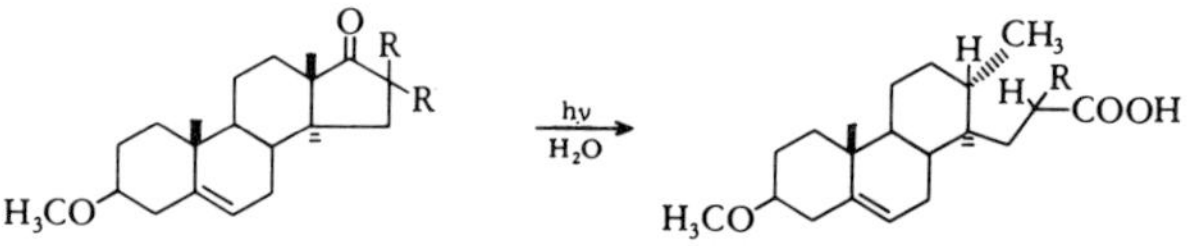

6: R=CH$_3$
7: R=H

8: R=H

CIAMICIAN's method of transforming cyclic ketones to the corresponding acid has found application also in triterpenoid chemistry. According to ARIGONI and collaborators [10] 5α-lanostan-3-one (9) was cleaved photochemically in aqueous acetic acid giving 3,4-seco-5α-lanostan-3-oic acid (10). By a similar procedure β-amyrone (11) was transformed into dihydronyctanthic acid (12).

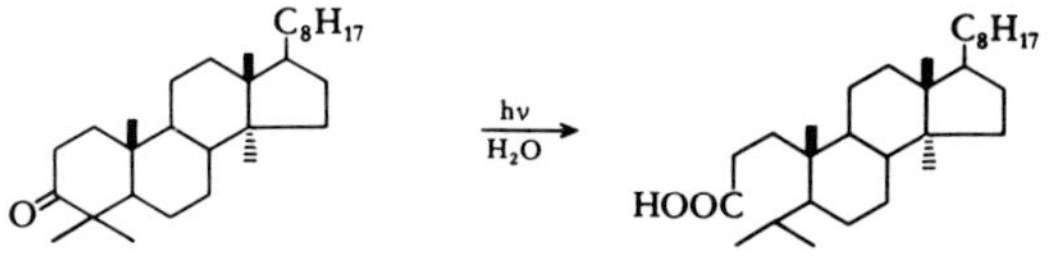

References, pp. 240—241

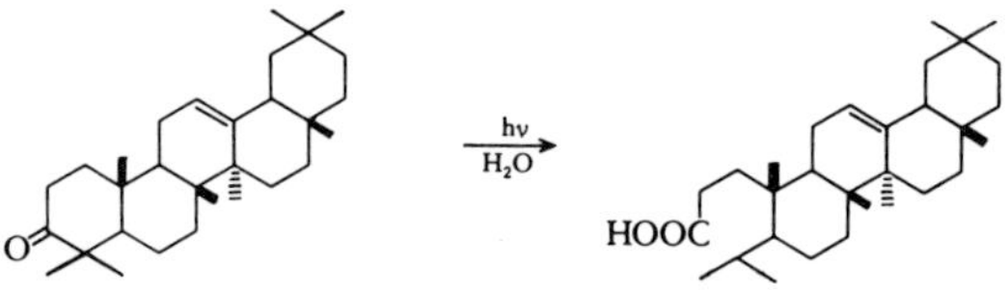

3β-Methoxy-13,17-seco-13α-androst-5-en-17-oic acid (8) [7]. 25 g 7 were dissolved in 450 ml dioxane and 150 ml water and this solution was irradiated with a xenon high pressure lamp (Osram XBF 6000) through quartz. After 14 days' irradiation working up afforded 6.5 g of acid product which was chromatographed on silica gel. Chloroform/3% acetone eluted an oil which crystallized (3.98 g). Recrystallization from petroleum ether and isopropyl ether yielded pure **8**, m.p. 113—115°.

3,4-Seco-5α-lanostan-3-oic acid (10) [10]. The ketone 9 (1.0 g) in acetic acid-water (9 : 1) (250 ml) was irradiated in a Pyrex flask at reflux temperature under nitrogen with a bare mercury-arc lamp (125 W) for 12 hours. The course of the reaction was followed by the decrease in intensity of the ZIMMERMANN test characteristic of triterpenoid 3-ketones. The solvent was removed in vacuo and the residue chromatographed over acid-washed alumina. Elution with chloroform-methanol afforded the seco-acid **10** (350 mg), m.p. 186—188° (from light petroleum).

b) Unsaturated ketones

It has been found [11, 12] that when α-tropolone (13) or its methyl ether (15) is irradiated with ultraviolet light in aqueous solution 4-oxo-2-cyclopentene-1-acetic acid (14), or its methyl derivative (16), can be obtained. For the mechanism of these remarkable reactions see CHAPMAN [13].

13: R=H
15: R=CH$_3$

14: R=H
16: R=CH$_3$

Purpurogallin tetramethyl ether (17) in aqueous alcoholic solution is converted by sunlight into a number of products including methyl 6,7,8-trimethoxy-1-naphthoate (18) [14]. This rearrangement was carried out by irradiation with a mercury vapor lamp or with sunlight, the latter being more efficacious. Since the methyl ester **18** is unaccompanied by the analogous ethyl ester, although ethanol is present in great excess, the photochemical transformation must be intramolecular.

On irradiation with ultraviolet light in the presence of water 2,4-cyclohexadienones may undergo fission to acids in high yield, as was found by BARTON [15, 16]. These reactions are most easily interpreted on the basis of an initial ring fission giving the cis-diene ketene followed by isomerization to the trans-diene ketene which can react with water giving the acid.

Depending on the nature of the substituents the products thus arising are either β,γ:δ,ε-conjugated or α,β:δ,ε-conjugated. Some examples of either type are listed in the table 18, the stereochemistry at the terminal double bond not being unambiguously settled in every case.

Table 18 [16]

starting material	product	yield
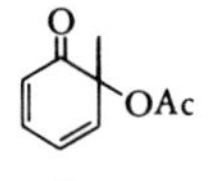 19	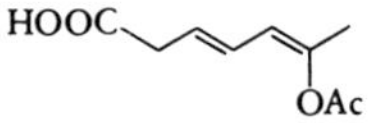 20	79%
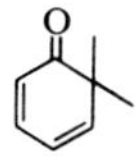 (structure)	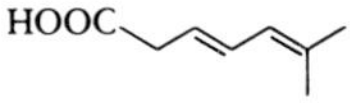 (structure)	64%
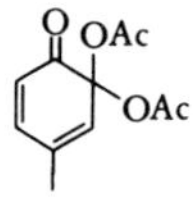 21	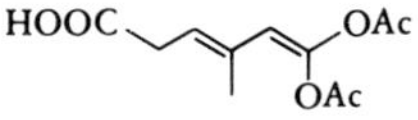 22	58%
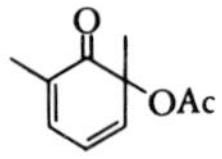 (structure)	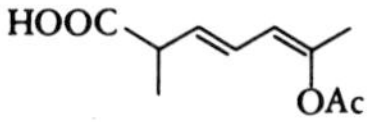 (structure)	50%
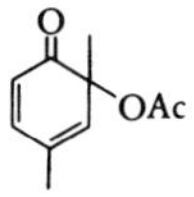 23	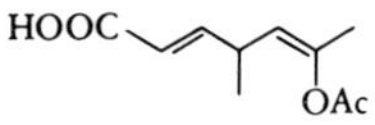 24	
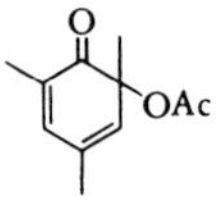 25	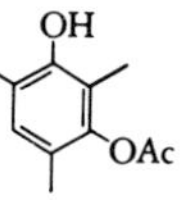 26	41%

When **19** was irradiated in the presence of aniline or cyclohexylamine the corresponding crystalline amides were obtained, e. g. **27** in the case of cyclohexylamine. However, the analog **23** furnished the **24**-cyclohexylamide (**28**) in 81 % yield. It is assumed by BARTON and QUINKERT [16] that the excess of amine used is sufficiently basic to isomerize the initial β,γ:δ,ε-conjugated diene into the isomer.

References, pp. 240—241

If **25** is irradiated in the presence of cyclohexylamine ring fission occurred smoothly affording 84 % of 6-acetoxy-2,4-dimethyl-2,5-heptadienoic acid cyclohexylamide (**29**) whereas aromatization was preferred with aniline or water, the main product being 3-acetoxymesitol (**26**) (comp. p. 234).

The photochemical transformation of lumisantonin (**30**) [*17*] to photosantonic acid (**32**) [*18, 19*] is a complicated process the details of which have recently been elucidated [*20, 21*]. Thus, **30** is on irradiation primarily transformed into the cyclohexadienone **31**, which in a subsequent photochemical reaction breaks down to photosantonic acid via a ketene.

Irradiation of 1-oxo-2,4-santadien-8α,12-olide (**33**) in moist ether yielded an acid mixture from which the solid acid **34** could be obtained in 47 % yield [*22*]. The remaining acidic oil could not be induced to crystallize and is believed to be a mixture of cis and trans isomers. Accepting the β,γ:δ,ε-dienic acid structure **34** for the "photoacid", the absence of any absorption characteristic of a conjugated diene must be due to the almost complete inhibition of coplanarity of the two olefinic groupings, as is also the case with photosantonic acid (**32**).

4-Oxo-2-cyclopentene-1-acetic acid (14) [*12*]. A solution of 5 g of tropolone in 500 ml of distilled water was irradiated with a G. E. A-H6 mercury arc for 2 hours. The aqueous solution was continuously extracted with ether for 24 hrs. and the ether extract was washed with 5% sodium bicarbonate. The bicarbonate wash was back-extracted with ether and then acidified. The acidic aqueous layer was continuously extracted with ether. The semi-solid (2.7 g) isolated from the ether extract was crystallized from chloroform to give 0.91 g (16%) of **14**, m.p. 102—103°.

Methyl 6,7,8-trimethoxy-1-naphthoate (18) [*14*]. Tetra-O-methylpurpurogallin (**17**) (3.0 g) was dissolved in a mixture of oxygen-free water (700 ml) and ethanol (300 ml). The warmed solution (ca. 50°) was sealed up in a long glass tube under nitrogen. After 28 days' exposure to moderate sunlight (England) the clear solution, which had darkened somewhat, was extracted with methylene chloride (3 × 200 ml). The combined extracts were dried rigorously (MgSO$_4$) and evaporated under reduced pressure at ca. 25°. The residual gum (2.9 g), which did not crystallize, was chromatographed in dry benzene on neutral alumina. Eluting the column with benzene gave 7 fractions (total 1.42 g). Further elution with benzene containing 2% of methanol gave only a non-crystalline solid.

When boiled with light petroleum, fraction 1 gave a solid which on recrystallization from light petroleum (charcoal) afforded colorless prisms of **18**, m.p. 81—82°. Total yield 0.45 g.

General procedure for photochemical cleavage of 2,4-cyclohexadienones [*16*]. The irradiations were carried out under oxygen-free nitrogen in a Pyrex flask by a mercury lamp (250 W), the heat of the lamp keeping the ethereal solutions at reflux. The flask was placed 10 cm from the light source. The solutions contained 0.1—1% of cyclohexadienone. Aliquot parts were removed at 30 min. intervals for ultraviolet absorption measurements. When there was no further change the reaction mixture was irradiated for a further 30 mins. and then worked up.

6-Acetoxy-3,5-heptadienoic acid (20) [*16*]. The cyclohexadienone **19** (1.24 g) in ether (1240 ml), previously saturated with water, was irradiated for 3 hours. The dried (MgSO$_4$) ethereal solution was evaporated and the residue crystallized twice from cyclohexane to give 1.09 g of **20**, m.p. 86—87°.

6,6-Diacetoxy-4-methyl-3,5-heptadienoic acid (22) [*16*]. 6,6-Diacetoxy-4-methyl-2,4-cyclohexadienone (**21**) (510 mg) in ether, previously saturated with water (510 ml), was irradiated for 2.5 hours. Crystallization from light petroleum gave 319 mg of **22**, m.p. 70—71°. The acid is sensitive to heat, especially when dissolved in polar solvents and should be crystallized with care.

6-Acetoxy-3,5-heptadienoic acid cyclohexylamide (27) [*16*]. The dienone **19** (1.22 g) and cyclohexylamine (1.51 g) in dry ether (800 ml) were irradiated for 2 hours. Removal of the excess of amine by shaking with 4N hydrochloric acid and working up in the usual way gave the amide. Purification by filtration in benzene solution over silicia gel and crystallization from ether gave 1.55 g of **27**, m.p. 86.5—88°.

Photosantonic acid (32) [*17*]. Lumisantonin (**30**) (500 mg) was dissolved in acetic acid (12 ml), and the mixture then diluted with water (14 ml). Irradiation of the mixture at —5° to +5° for 1.5 hrs. with a 125 W Crompton bare-arc mercury lamp, followed by isolation of the acidic portion and crystallization from chloroform-light petroleum, afforded 350 mg of photosantonic acid.

1-Carboxy-1,10-seco-3,5(10)-santadien-8α,10-olide (34) [*22*]. A solution of 2.0 g of **33** in 700 ml of ether previously saturated with water was irradiated with a G. E. AH-6 mercury lamp for 30 mins. The solution was concentrated to a small volume and extracted with 5% sodium bicarbonate. The alkaline extract was acidified, extracted with ether, and the ethereal solution dried. The solvent was evaporated and the oily residue crystallized from ether-petroleum ether to yield 1.0 g (47%) of **34**, m.p. 97—98°.

3. Formation of acid derivatives

a) Amides

ELAD, to whom we owe the discovery of the light induced addition of formamide to ethylenic and acetylenic bonds (cf. chapter 17), has also investigated the photosensitized amidation of aromatic hydrocarbons [*23*]. UV-irradiation of the hydrocarbons with formamide and acetone as sensitizer at room temperature gave the results summarized in table 19. Yields of the products isolated are based on the hydrocarbons consumed. Besides, considerable amounts of oxamide were obtained.

References, pp. 240—241

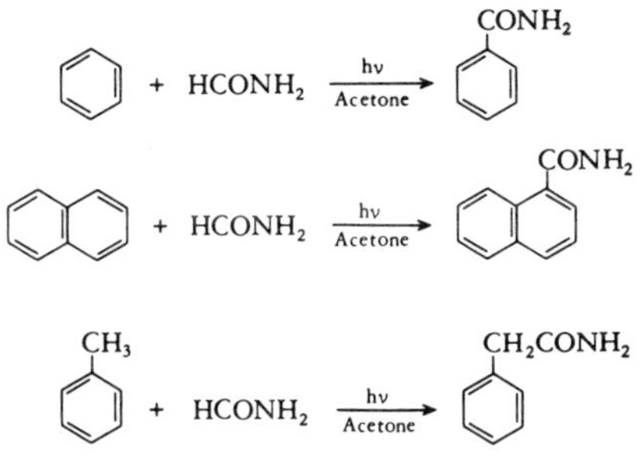

$$\text{C}_6\text{H}_6 + \text{HCONH}_2 \xrightarrow[\text{Acetone}]{h\nu} \text{C}_6\text{H}_5\text{CONH}_2$$

$$\text{C}_{10}\text{H}_8 + \text{HCONH}_2 \xrightarrow[\text{Acetone}]{h\nu} \text{C}_{10}\text{H}_7\text{CONH}_2$$

$$\text{C}_6\text{H}_5\text{CH}_3 + \text{HCONH}_2 \xrightarrow[\text{Acetone}]{h\nu} \text{C}_6\text{H}_5\text{CH}_2\text{CONH}_2$$

Table 19 [23]

hydrocarbon	product	yield
benzene	benzamide	15%
naphthalene	1-naphthamide	20%
toluene	phenylacetamide	23%
o-xylene	o-tolylacetamide	28%
m-xylene	m-tolylacetamide	26%
p-xylene	p-tolylacetamide	32%

The photolyses of 1,2-diketones in the presence of tetrafluorohydrazine at room temperature in Pyrex produced N,N-difluoroamides in good yield [24]. The reaction has been successfully applied to biacetyl, glyoxal and benzil.

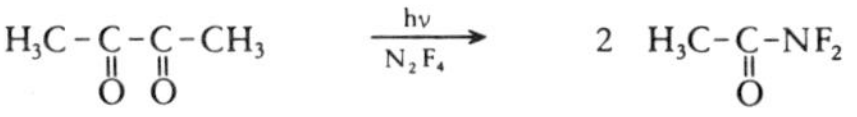

$$\text{H}_3\text{C-CO-CO-CH}_3 \xrightarrow[\text{N}_2\text{F}_4]{h\nu} 2\ \text{H}_3\text{C-CO-NF}_2$$

35

N,N-Difluoroacetamide (35) [24]. 60 mmole of 2,3-butanedione and 60 mmole of N_2F_4 were irradiated for 16 hrs. with a Hanovia EH-4 lamp to give 77 mmole of **35**, b.p. 45—47°; yield 80% based on N_2F_4 consumed. Tetrafluorohydrazine should be handled with great care (comp. [24]).

b) Lactones

α,β-Unsaturated acids, when irradiated in the presence of photosensitizers, may undergo a KHARASCH type of addition with alcohols. The γ-hydroxy acids thereby formed readily lose water with formation of γ-lactones. Primary as well as secondary alcohols have been added to maleic acid, fumaric acid and crotonic acid.

This reaction, first investigated by SCHENCK and coworkers [25], has opened a synthetic route to products otherwise only with difficulty accessible. Table 20 lists some compounds which have been thus synthesized. A more complete account will be found with PFAU [26].

Table 20

alcohol	acid	product	ref.
36, R = H	37, R'= COOH	2-methylparaconic acid (38)	[27]
36, R = CH$_3$	37, R'= COOH	terebic acid (39)	[25]
36, R = C$_2$H$_5$	37, R'= COOH	2-ethyl-2-methylparaconic acid (40)	[25]
36, R = C$_6$H$_{13}$	37, R'= COOH	2-hexyl-2-methylparaconic acid (41)	[27]
36, R = CH$_3$	37, R'= CH$_3$	4-hydroxy-3,4-dimethylvaleric acid γ-lactone (42)	[27]

In the case of methacrylic acid (**43**) a different course is observed [*28*]. Irradiation of **43** in 2-propanol solution with benzophenone as sensitizer gave rise to a 1 : 1 : 1 addition product of the lactone type, viz. 4-hydroxy-2,4-dimethyl-2-(diphenylhydroxymethyl)-valeric acid γ-lactone (**44**). On heating this turned over to 4-hydroxy-2,4-dimethylvaleric acid γ-lactone (**45**) and benzophenone.

α,β-Acetylenic acids are equally prone to photochemical addition reactions with alcohols. In this case two molecules of alcohol are added stepwise. Thus, acetylenedicarboxylic acid affords 24% of 2,3-bis-(1-hydroxy-1-methylethyl)-succinic acid dilactone (**46**) and 37% of 2,3-bis-(1-hydroxy-1-methylethyl)-succinic acid monolactone (**47**) on irradiation in 2-propanol in the presence of benzophenone [*29*]. With **47** lactonization is impossible for steric reasons since the side chains are positioned trans to each other; the alternative formulation of the monolactone as **48** [*30*] is inconsistent with the chemical properties of the compound [*29*].

PFAU [30] has also reported on the photosensitized addition of 2-propanol to propiolic acid to give 4-hydroxy-4-methyl-2-pentenoic acid γ-lactone (**49**); benzophenone serves as the sensitizer. **49** may then add a further molecule of alcohol to yield a compound which has been ascribed the constitution of 4-hydroxy-3-(1-hydroxy-1-methylethyl)-4-methylvaleric acid γ-lactone (**50**).

$$\begin{array}{c} H \\ C \\ \parallel\parallel \\ C \\ | \\ COOH \end{array} + \begin{array}{c} CH_3 \\ | \\ HC-OH \\ | \\ CH_3 \end{array} \xrightarrow[-H_2O]{h\nu/Sens.} \underset{\mathbf{49}}{\text{[lactone]}} \xrightarrow{\underset{CH_3CHOHCH_3}{h\nu/Sens.}} \underset{\mathbf{50}}{\text{[lactone-OH]}}$$

Terebic acid (39) [25]. A solution of 5 g of maleic acid and 1 g of benzophenone in 2-propanol (100 ml) was irradiated with a 125 W mercury high pressure burner at 16° for 18 hours. Working up (for details see the 1st. edition of this book, p. 243) afforded 4.56 g of **39** (96% calculated on the basis of maleic acid consumed), m.p. 176°.

4-Hydroxy-4-methyl-2-pentenoic acid γ-lactone (49) [30]. 12.1 g propiolic acid and 4 g. of benzophenone in 400 ml isopropanol were irradiated for 24 hrs. at reflux temperature. The solvent was evaporated, the residue dissolved in ether and washed with sodium carbonate solution. Distillation yielded 6.7 g of **49**, b.p.$_{0.5}$ 30—31°.

c) Acyl chlorides

KHARASCH and BROWN [31] were able to show that oxalyl chloride reacts photochemically with certain hydrocarbons, e.g., cyclohexane, methylcyclohexane and cyclopentane. In this reaction a hydrogen atom is replaced by the chloro-formyl group. Thus, cyclohexanecarboxylic acid chloride was obtained from cyclohexane. The yields with n-pentane, n-heptane and isooctane were poor [31], and negative results were obtained with toluene, m-xylene, p-chlorotoluene and 2-methylnaphthalene [32].

It is assumed [31] that product formation follows a free radical chain mechanism:

$$\left.\begin{array}{rl} (COCl)_2 & \xrightarrow{h\nu} 2 \cdot COCl \\ \text{or} \quad (COCl)_2 & \xrightarrow{h\nu} \cdot CO-COCl + \cdot Cl \end{array}\right\} \longrightarrow 2\,CO + 2\cdot Cl$$

$$\cdot Cl + RH \longrightarrow R\cdot + HCl$$
$$R\cdot + (COCl)_2 \longrightarrow RCOCl + \cdot COCl$$

The wavelength of the incident light appears to have an effect on the manner of photochemical decomposition of oxalyl chloride; light of wavelength 254 mμ favors the formation of ·COCl, light of wavelength 365 mμ that of ·CO—COCl.

Substitution of hydrogen by the chloro-formyl group takes place also in the dark at elevated temperatures, if peroxides such as benzoyl peroxide are present; in the absence of either light or peroxides no reaction will occur with oxalyl chloride.

With olefins the reaction proceeds by simple heating with oxalyl chloride, neither light nor peroxides having any apparent effect [33].

HASZELDINE and NYMAN have reported that photochemical oxidation of 1,1-dichloro-2,2,2-trifluoroethane by oxygen in the presence of chlorine gives trifluoroacetyl chloride in 90% yield, the chlorine behaving as a sensitizer [34, 35].

$$F_3C-CHCl_2 \xrightarrow[Cl_2/O_2]{h\nu} F_3C-C-Cl \atop \|\atop O$$

Cyclohexanecarboxylic acid chloride [31]. Cyclohexane (16.8 g) and oxalyl chloride (12.7 g) were irradiated for 20 hrs. in a Pyrex vessel at reflux temperature; the light source was a low-pressure mercury lamp. The weight of the reaction mixture decreased by 4.2 g corresponding to a 60% conversion. The reaction mixture was fractionally distilled and yielded 5.7 g of oxalyl chloride and 8.0 g of cyclohexanecarboxylic acid chloride, b.p. 180—181° (after repeated distillation).

4. Photochemical reactions of esters involving the acyloxy groups

a) Rearrangements of esters

As mentioned earlier (comp. p. 228), irradiation of 6-acetoxy-2,4,6-trimethyl-2,4-cyclohexadienone (**25**) in moist ether gives predominantly **26** as a result of acetoxy group migration [16]. The following scheme has been advanced to explain this transformation:

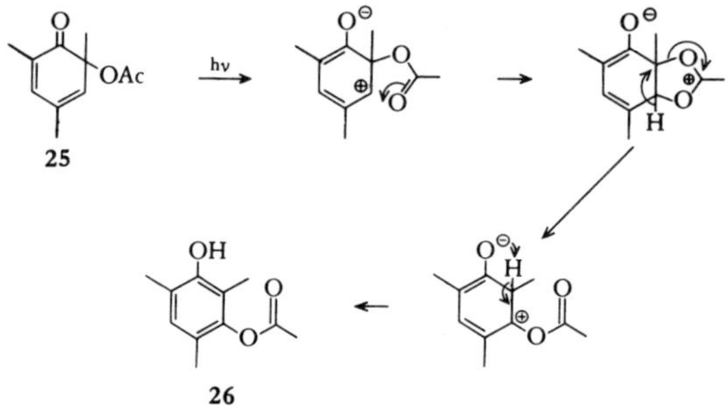

The irradiation [36] of 3β-acetoxycholest-5-en-7-one (**51**) yielded two photoproducts **52** and **53** which were easily separated from each other and from starting material on chromatography.

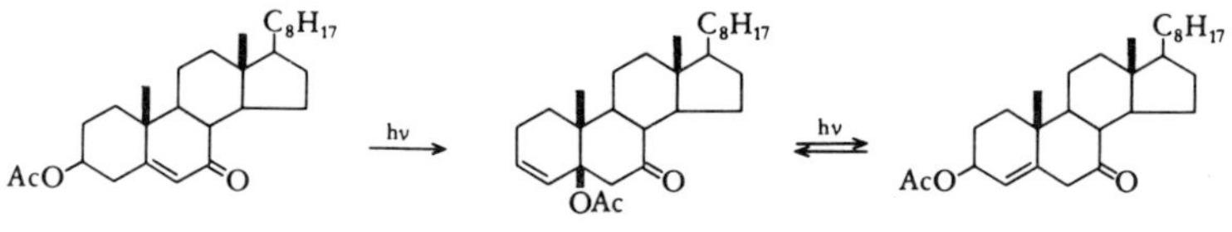

It has been established, that **52** is the primary product in the illumination of **51**, and that **53** is formed from **52** in a consecutive photochemical process. The irradiation of pure samples of either **52** or **53** gave an equilibrium mixture of the two.

3-Acetoxymesitol (26) [*16*]. The dienone **25** (787 mg) in ether (790 ml) saturated with water was irradiated for 23.5 hours. The product, on crystallization from light petroleum, gave 324 mg (41%) of **26** as needles, m.p. 90—90.5°.

5-Acetoxy-5β-cholest-3-en-7-one (52) and 3β-acetoxycholest-4-en-7-one (53) [*36*]. Solutions of **51** in tert.-butanol were irradiated through 2 mm Pyrex filter glass with a 200 W high-pressure mercury vapor lamp immersed into the liquid. Chromatography gave **52**, m.p. 114—114.5°, and **53**, m.p. 151—153°. At 20% conversion of **51** the product ratio **52**: **53** was about 80 : 20, and at 70% conversion the ratio was about 55 : 45.

b) Reductive elimination of an acetoxy group

Barton and Quinkert [*16*] have referred to a relatively simple example of this hydrogenolysis reaction. It concerns the formation of o-cresol (**54**) on irradiation of 6-acetoxy-6-methyl-2,4-cyclohexadienone (**19**) in anhydrous ether.

Warszawski et al. [*37*] were able to show that a similar reaction occurred on irradiation of 10β,17β-diacetoxyestra-1,4-dien-3-one (**55**) in dioxane with light of wavelength 254 mμ. Chromatographic separation of the crude irradiation mixture yielded 17β-acetoxy-3-hydroxyestra-1,3,5(10)-triene (**56**).

55: R=OAc
57: R=CH₃

This finding is in remarkable contrast to the results of irradiation of all other cross-conjugated dienones of type **57** which carry a β-orientated methyl group at C-10. Under the same reaction conditions, these compounds, without exception, yield ketonic and phenolic isomers of the starting materials (comp. p. 31). In particular, it has not been possible to isolate **56** following irradiation [*38*] of 17β-acetoxyandrosta-1,4-dien-3-one (**57**).

17β-Acetoxy-3-hydroxyestra-1,3,5(10)-triene (56) [*37*]. 500 mg of **55** in 12.5 ml purest dioxane were irradiated at room temperature, a Hanau NN 15/44 mercury low-pressure burner serving as the light source. The lamp was placed outside and 10 cm from the reaction vessel, which was of quartz. After 24 hrs. irradiation the solution was evaporated in vacuo and the residue chromatographed on silica gel. A total of 162 mg (38%)

crystals were eluted with benzene and benzene-ether (9:1) mixture; after three-fold precipitation from aqueous methanol **56** had m.p. 207—208°.

c) Light-induced FRIES rearrangement

ANDERSON and REESE [*39, 40*] reported that irradiation of pyrocatechol monoacetate (**58**) in alcoholic solution with UV-light yields a mixture of pyrocatechol and dihydroxyacetophenones **59** and **60**. The normal acid-catalyzed FRIES rearrangement of **58** yields predominantly 3,4-dihydroxyacetophenone (**60**), whereas the photochemical procedure preponderantly yields 2,3-dihydroxyacetophenone (**59**). Under the experimental conditions both isomers are stable, thus showing that **59** is not an intermediate in the formation of **60**.

In the same way, phenyl benzoate (**61**) gave 2-hydroxybenzophenone (**62**), 4-hydroxybenzophenone (**63**) and phenol.

KOBSA [*41*] investigated the photochemical FRIES rearrangement of a series of substituted 4-tert.-butylphenyl benzoates which, on UV-irradiation in benzene solution, gave good yields of substituted 5-tert.-butyl-2-hydroxybenzophenones. Table 21 lists some examples [*41*].

The reaction is of interest as a preparative method, as the aluminium chloride catalyzed rearrangement of **64** resulted in complex reaction mixtures in which 4-hydroxybenzophenone (**63**), obtained by elimination of the tert.-butyl group, predominated [*41*]. Similarly, the AlCl$_3$ catalyzed reaction of 4-tert.-butyl-2-chlorophenyl benzoate (**78**) produced 28 % of 3-chloro-4-hydroxybenzophenone (**79**) as the only well-defined product. The light-induced FRIES rearrangement of **78**, however, proceeded without elimination of the tert.-butyl group, giving rise to 5-tert.-butyl-3-chloro-2-hydroxybenzophenone (**80**, 7 %) and the chlorine-free product (**65**, 21 %).

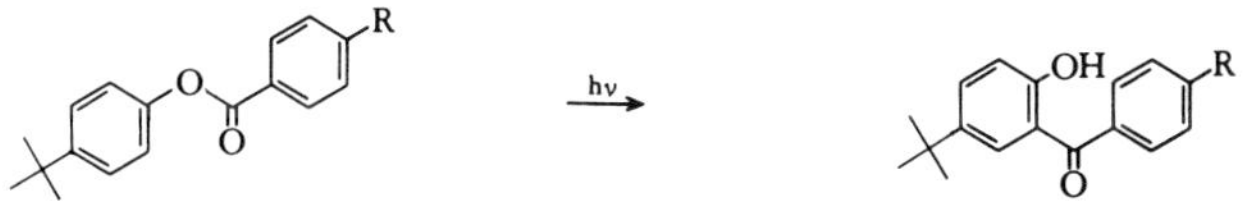

64: R=H
66: R=tert.-butyl
68: R=Cl
70: R=CN
72: R=NH$_2$
74: R=NO$_2$

65: R=H
67: R=tert.-butyl
69: R=Cl
71: R=CN
73: R=NH$_2$
75: R=NO$_2$

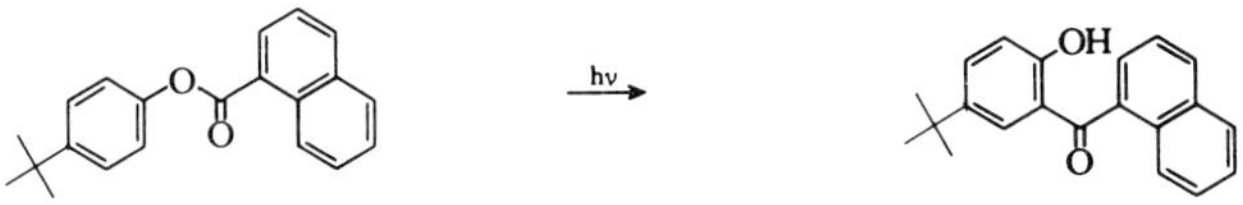

76

77

Table 21 [41]

starting material	product	yield
4-tert.-butylphenyl benzoate (64)	5-tert.-butyl-2-hydroxybenzophenone (65)	45%
4-tert.-butylphenyl 4-tert.-butylbenzoate (66)	4',5-di-tert.-butyl-2-hydroxybenzophenone (67)	48%
4-tert.-butylphenyl 4-chlorobenzoate (68)	5-tert.-butyl-4'-chloro-2-hydroxybenzophenone (69)	55%
4-tert.-butylphenyl 4-cyanobenzoate (70)	5-tert.-butyl-4'-cyano-2-hydroxybenzophenone (71)	48%
4-tert.-butylphenyl 4-aminobenzoate (72)	4'-amino-5-tert.-butyl-2-hydroxybenzophenone (73)	12%
4-tert.-butylphenyl 4-nitrobenzoate (74)	5-tert.-butyl-2-hydroxy-4'-nitrobenzophenone (75)	10%
4-tert.-butylphenyl 1-naphthoate (76)	1-naphthyl 5-tert.-butyl-2-hydroxyphenyl ketone (77)	44%

2- and 4-hydroxybenzophenone (62 and 63) [40]. Phenyl benzoate (61) (0.40 g) was dissolved in ethanol (50 ml) and the solution, contained in a stoppered quartz tube, irradiated from a distance of ca. 6 cm by a 500 W Hanovia mercury arc lamp. After 3 days' irradiation at ca. 30° the products were steam distilled and the yield of **62** in the distillate was estimated, spectrophotometrically, to be 20%. After extraction with chloroform the product was isolated as 2-hydroxybenzophenone 2,4-dinitrophenylhydrazone, m.p. 249—251°.

The steam-involatile material was also extracted with chloroform. The residue (0.213 g), after evaporation of the solvent, was chromatographed on silica gel and eluted with chloroform. Evaporation of the eluate and extraction of the solid residue with hot water gave pale yellow crystals of **63**, m.p. 130—132° (yield 28%, spectrophotometrically).

5. Further formation modes of esters

a) Formation of α-ketocarboxylates from α-keto acetals

The bromination of pyruvaldehyde diethyl acetal (**81**) with bromine has been reported to afford bromopyruvaldehyde diethyl acetal (**82**) [*42*]. When the bromination is carried out with N-bromosuccinimide in the presence of light [*43*] ethyl pyruvate (**83**) is formed.

$$BrCH_2-\underset{\underset{82}{}}{\overset{O}{\overset{\|}{C}}}-\underset{OC_2H_5}{\overset{OC_2H_5}{C}}-H \xleftarrow{Br_2} CH_3-\underset{\underset{81}{}}{\overset{O}{\overset{\|}{C}}}-\underset{OC_2H_5}{\overset{OC_2H_5}{C}}-H \xrightarrow[NBS]{h\nu} CH_3-\underset{\underset{83}{}}{\overset{O}{\overset{\|}{C}}}-\underset{OC_2H_5}{\overset{O}{C}}$$

3,3-Dimethyl-2-oxobutyraldehyde diethyl acetal yielded, by a similar photoprocess, ethyl 3,3-dimethyl-2-oxobutyrate and phenylglyoxal diethyl acetal gave ethyl phenylglyoxylate [*43*].

Ethyl pyruvate (83) [*43*]. 57.5 g of freshly distilled **81** and 70.2 g of N-bromosuccinimide were dissolved in 288 ml dry carbon tetrachloride in a flask fitted with a reflux condenser equipped with a calcium chloride tube. The mixture was heated by a 250 W drying lamp placed about 30 cm below the flask. As soon as the mixture began to reflux the light was turned off and switched on and off as required so that the mixture refluxed gently. As soon as the initial reaction was over, the light was brought closer to the reaction vessel and the mixture was refluxed for 3 hrs. and allowed to stand overnight.

After removal of the succinimide by filtration, the solvent was distilled through a column packed with glass helices. The residue was distilled in vacuo through the same column to give 36.3 g (78%) of ethyl pyruvate, b.p.$_{14}$ 48—52°.

b) Formation of esters from 2-butene-1,4-diones

If cis-1,4-diphenyl-2-butene-1,4-dione (cis-dibenzoylethylene, **84**) is irradiated in ethanol solution, ethyl 4-phenoxy-4-phenyl-3-butenoate (**86**) is formed in 36% yield [*44*]. With methanol, 2-propanol or benzhydrol the corresponding esters may be prepared [*45*].

84: R=H **85** **86: R=H**
87: R=C$_6$H$_5$ **88: R=C$_6$H$_5$**

1,2,4-Triphenyl-2-butene-1,4-dione (**87**) on analogous irradiation [*44*] in ethanol furnished ethyl 2,4-diphenyl-4-phenoxy-3-butenoate (**88**) in yields of up to 64% together with the free acid. The formation of **86** and **88** is visualized [*44, 45*] as proceeding via the ketene **85** which with the alcohols breaks down to the esters actually observed.

References, pp. 240—241

The reactions 84 → 86 and 87 → 88 bear a resemblance to the formation of 3-benzoyl-4-hydroxy-2-(α-phenoxybenzylidene)-4-phenyl-3-butenoic acid γ-lactone (90) from irradiated 2,3-dibenzoyl-1,4-diphenyl-2-butene-1,4-dione (tetrabenzoylethylene, 89) [46] which also proceeds with migration of a phenyl group from carbon to oxygen.

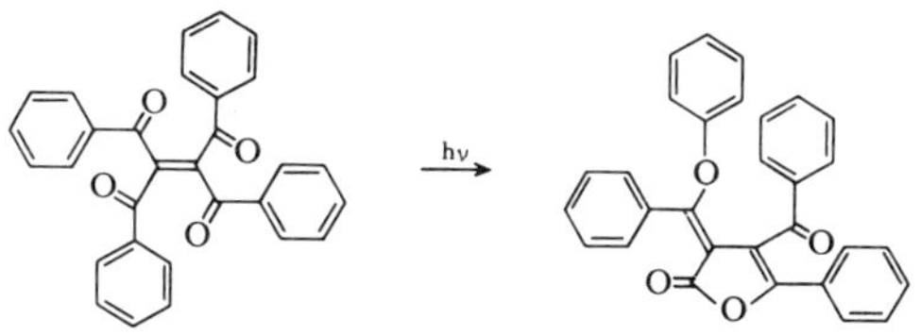

89   90

c) Formation of an ester from a diflavylene compound

When a benzene solution of 3,3'-epoxy-Δ⁴,⁴'-biflavylene (91) [47] was irradiated with a mercury vapor lamp in the presence of oxygen [48], α-(2-benzoyloxyphenyl)-3-hydroxy-2-phenyl-4H-1-benzopyran-Δ⁴,α-acetic acid γ-lactone (92) (yellow crystals of m.p. 224° from benzene-chloroform) was obtained. For a related oxidative fission of a double bond see chapter 39.

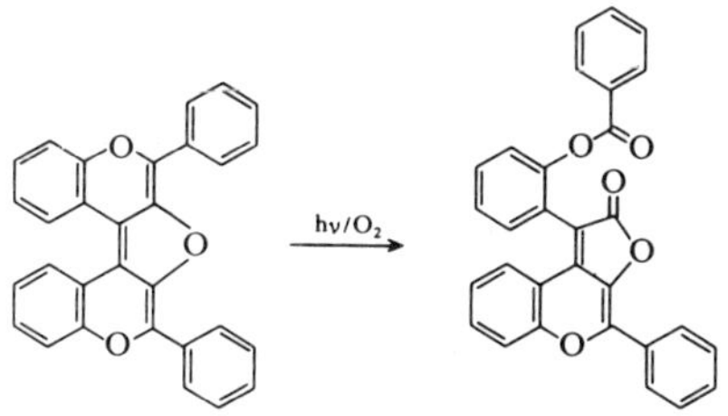

91   92

6. Formation of fluorenecarboxylic acids by photolysis of fluoranthenols

Several 3-fluoranthenols, when exposed to daylight in alkaline solutions, were converted to fluorenecarboxylic acids [49]. Thus, 3-fluoranthenol (93) itself gave 1-carboxyfluorene-9-acetic acid (94), and 3-hydroxy-1-fluoranthenecarboxylic acid (95) gave 1-carboxyfluorene-9-malonic acid (96).

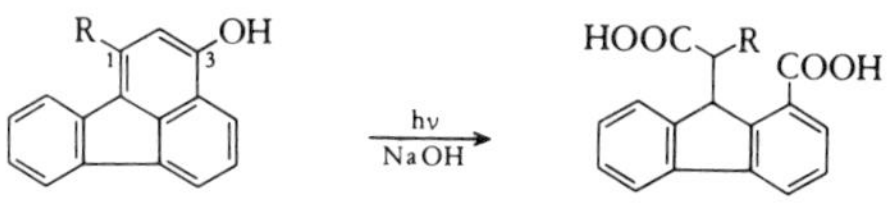

93: R = H
95: R = COOH

94: R = H
96: R = COOH

1-Carboxyfluorene-9-acetic acid (94) [*49*]. A solution of 1 g of compound **93** in 800 ml N NaOH was irradiated with a 200 W tungsten lamp for 14 days. The pale yellow solution, which would no longer couple with diazotized p-nitroaniline, was filtered from a few brown flocks and acidified with concentrated hydrochloric acid. The voluminous, almost colorless precipitate was filtered off, washed free from acid with water and dried. The raw product (1 g, 81.7%) was recrystallized from 50% acetic acid to give colorless needles of **94**, m. p. 231—232°.

References

[*1*] W. C. Francis and R. N. Haszeldine: J. Chem. Soc. **1955**, 2151.
[*2*] R. N. Haszeldine: J. Chem. Soc. **1955**, 4291.
[*3*] G. Ciamician and P. Silber: Ber. dtsch. chem. Ges. **40**, 2415 (1907).
[*4*] G. Ciamician and P. Silber: Ber. dtsch. chem. Ges. **41**, 1071 (1908).
[*5*] G. Quinkert, B. Wegemund and E. Blanke: Tetrahedron Letters **1962**, 221.
[*6*] G. O. Schenck and F. Schaller: Chem. Ber. **98**, 2056 (1965).
[*7*] G. Quinkert, B. Wegemund, F. Homburg and G. Cimbollek: Chem. Ber. **97**, 958 (1964).
[*8*] G. Quinkert, E. Blanke and F. Homburg: Chem. Ber. **97**, 1799 (1964).
[*9*] G. Quinkert: Angew. Chemie **77**, 229 (1965); internat. ed. **4**, 211 (1965).
[*10*] D. Arigoni, D. H. R. Barton, R. Bernasconi, C. Djerassi, J. S. Mills and R. E. Wolff: J. Chem. Soc. **1960**, 1900.
[*11*] W. G. Dauben, K. Koch and W. E. Thiessen: J. Amer. chem. Soc. **81**, 6087 (1959).
[*12*] W. G. Dauben, K. Koch, S. L. Smith and O. L. Chapman: J. Amer. chem. Soc. **85**, 2616 (1963).
[*13*] O. L. Chapman, in: Adv. Photochem., ed. by W. A. Noyes jr., G. S. Hammond and J. N. Pitts jr., Vol. 1, pp. 323—420. New York: Interscience 1963.
[*14*] E. J. Forbes and R. A. Ripley: J. Chem. Soc. **1959**, 2770.
[*15*] D. H. R. Barton: Helv. Chim. Acta **42**, 2604 (1959).
[*16*] D. H. R. Barton and G. Quinkert: J. Chem. Soc. **1960**, 1.
[*17*] D. H. R. Barton, P. de Mayo and M. Shafiq: J. Chem. Soc. **1958**, 140.
[*18*] E. E. van Tamelen, S. H. Levin, G. Brenner, J. Wolinsky and P. Aldrich: J. Amer. chem. Soc. **80**, 501 (1958).
[*19*] E. E. van Tamelen, S. H. Levin, G. Brenner, J. Wolinsky and P. Aldrich: J. Amer. chem. Soc. **81**, 1666 (1959).
[*20*] O. L. Chapman and L. F. Englert: J. Amer. chem. Soc. **85**, 3028 (1963).
[*21*] M. H. Fisch and J. H. Richards: J. Amer. chem. Soc. **85**, 3029 (1963).
[*22*] W. G. Dauben, D. A. Lightner and W. K. Hayes: J. Org. Chem. **27**, 1897 (1962).
[*23*] D. Elad: Tetrahedron Letters **1963**, 77.
[*24*] R. C. Petry and J. P. Freeman: J. Amer. chem. Soc. **83**, 3912 (1961).
[*25*] G. O. Schenck, G. Koltzenburg and H. Grossmann: Angew. Chemie **69**, 177 (1957).
[*26*] M. Pfau: Publ. Sci. Tech. Min. Air (France), Notes Tech. **143**. Paris: Serv. Doc. Sci. Tech. Armement 1965.
[*27*] R. Dulou, M. Vilkas and M. Pfau: Comptes rendus **249**, 429 (1959).
[*28*] M. Pfau: Comptes rendus **254**, 2017 (1962).
[*29*] G. O. Schenck and R. Steinmetz: Naturwissenschaften **47**, 514 (1960).
[*30*] M. Pfau, R. Dulou and M. Vilkas: Comptes rendus **251**, 2188 (1960).
[*31*] M. S. Kharasch and H. C. Brown: J. Amer. chem. Soc. **64**, 329 (1942).
[*32*] M. S. Kharasch, S. S. Kane and H. C. Brown: J. Amer. chem. Soc. **64**, 1621 (1942).

References

[33] M. S. Kharasch, S. S. Kane and H. C. Brown: J. Amer. chem. Soc. **64**, 333 (1942).
[34] R. N. Haszeldine and F. Nyman: Proc. Chem. Soc. **1957**, 146.
[35] R. N. Haszeldine and F. Nyman: J. Chem. Soc. **1959**, 387.
[36] P. D. Gardner and H. F. Hamil: J. Amer. chem. Soc. **83**, 3531 (1961).
[37] R. Warszawski, K. Schaffner and O. Jeger: Helv. Chim. Acta **43**, 500 (1960).
[38] H. Dutler, H. Bosshard and O. Jeger: Helv. Chim. Acta **40**, 494 (1957).
[39] J. C. Anderson and C. B. Reese: Proc. Chem. Soc. **1960**, 217.
[40] J. C. Anderson and C. B. Reese: J. Chem. Soc. **1963**, 1781.
[41] H. Kobsa: J. Org. Chem. **27**, 2293 (1962).
[42] J. H. Mowat: US pat. 2 436 073 of 17. 2. 1948.
[43] J. B. Wright: J. Amer. chem. Soc. **77**, 4883 (1955).
[44] H. E. Zimmerman, H. G. C. Dürr, R. G. Lewis and S. Bram: J. Amer. chem. Soc. **84**, 4149 (1962).
[45] G. W. Griffin and E. J. O'Connell: J. Amer. chem. Soc. **84**, 4148 (1962).
[46] H. Schmid, M. Hochweber and H. von Halban: Helv. Chim. Acta **30**, 1135 (1947).
[47] W. Dilthey and W. Höschen: J. Prakt. Chem. [2] **138**, 145 (1933).
[48] A. Schönberg and U. Friese: unpublished results.
[49] A. Sieglitz, H. Tröster and P. Böhme: Chem. Ber. **95**, 3013 (1962).

Chapter 26

Photochemical reactions with N-halogenated amines

1. Photochemical replacement of chlorine in N-chloroamines by hydrogen

WAWZONEK [1] has investigated the effect of UV-light on N-chlorodibutylamine (1) in carbon tetrachloride; irradiation caused the precipitation of dibutylamine hydrochloride. The free radicals (2) undergo mainly a bimolecular disproportionation reaction:

$$(C_4H_9)_2NCl \xrightarrow{h\nu} (C_4H_9)_2N\cdot + Cl\cdot$$
$$12$$

$$2\,(C_4H_9)_2N\cdot \longrightarrow (C_4H_9)_2NH + C_4H_9N=CHCH_2C_2H_5$$
$$2$$

Dibutylamine hydrochloride [1]. Dibutylamine (11.6 g) was chlorinated in pentane by the method of COLEMAN [2]. The yield of N-chlorodibutylamine (1) based on active halogen titration was 85%. This yield varied in other runs from 85—95%. The pentane was removed under reduced pressure at 0° and carbon tetrachloride (100 ml) was added to the residue. The resulting solution was irradiated with an ultraviolet lamp for 47 days, after which time no more active halogen could be detected. The white solid (5.93 g) obtained proved to be dibutylamine hydrochloride.

2. The light-induced HOFMANN-LÖFFLER reaction

a) Synthesis of pyrrolidines

The cyclization of N-halogenated amines is known as the HOFMANN-LÖFFLER reaction (cf. [3]). In 1883 HOFMANN [4] heated 1-bromo-2-propyl-piperidine (3) with acid and obtained a new base which later was shown to be octahydroindolizine (4).

References, p. 247

Ring closure according to this classical procedure was found to be a general route to pyrrolidines [5]. Thus N-chlorodibutylamine (1) yields 1-butylpyrrolidine (5) [2].

$$\underset{1}{\underset{\underset{CH_2CH_2CH_2CH_3}{|}}{\underset{NCl}{\overset{CH_2-CH_2}{\underset{|}{CH_2\quad CH_3}}}}} \xrightarrow[70-80\%]{H_2SO_4;\ 95°} \underset{\underset{CH_2CH_2CH_2CH_3}{|}}{\underset{5}{\boxed{N}}}$$

According to WAWZONEK [6] the formation of 5 may proceed in 85% sulfuric or trifluoroacetic acid at room temperature by irradiation of 1 with ultraviolet light, a fact which points to a free radical reaction [3]: N-chloroamine first forms a salt (6) with the acid which undergoes homolytic cleavage under the influence of light to afford amminium and chlorine free radicals. The amminium radical 7 then abstracts hydrogen intramolecularly from the δ-position in the chain to form a new radical 8 which by reaction with (6) gives a δ-chloro derivative (9). By terminal treatment with alkali, ring closure yields N-butylpyrrolidine (5).

$$C_4H_9-\overset{H}{\underset{Cl}{\overset{|\oplus}{N}}}-CH_2CH_2CH_2CH_3 \xrightarrow[-Cl\cdot]{h\nu} C_4H_9-\overset{H}{\underset{\bullet}{\overset{|\oplus}{N}}}-CH_2CH_2CH_2CH_3 \longrightarrow$$
$$\qquad\qquad 6 \qquad\qquad\qquad\qquad\qquad 7$$

$$C_4H_9-\overset{H}{\underset{H}{\overset{|\oplus}{N}}}-CH_2CH_2CH_2CH_2\cdot \xrightarrow{6} C_4H_9-\overset{H}{\underset{H}{\overset{|\oplus}{N}}}-CH_2CH_2CH_2CH_2Cl \xrightarrow{OH^-} \underset{\underset{C_4H_9}{|}}{\underset{5}{\overset{H_2C\quad CH_2}{\underset{H_2C\diagdown_N\diagup CH_2}{|\qquad\quad |}}}}$$
$$\qquad 8 \qquad\qquad\qquad\qquad\qquad 9$$

N-Butylpyrrolidine (5) [6]. *(a) 85% Sulfuric Acid.* — Dibutylamine (16.1 g) was converted into the N-chloro derivative by the method of COLEMAN [2]. Extraction from the petroleum-ether solution with 85% sulfuric acid (50 ml) was followed by irradiation in a quartz flask with an ultraviolet lamp at 20° for 48 hours. The irradiated sulfuric acid solution was poured onto an equal weight of crushed ice. Concentrated sodium hydroxide solution was added with vigorous stirring until the solution was strongly alkaline while keeping the temperature at 20°. The amines were steam distilled into excess hydrochloric acid and the resulting distillate, after concentration to 100 ml, was made alkaline. A HINSBERG separation gave N,N-dibutylbenzenesulfonamide (1.44 g, 4%) and N-butylpyrrolidine (5) which was isolated as the picrate (17.2 g, 78.5%). The irradiation of the N-chloro derivative from 6.2 g of amine in 100 ml of 85% sulfuric acid for 24 hours gave a 59% yield of N-butylpyrrolidine and 11.7% yield of dibutylamine.

(b) F_3CCOOH. — Dibutylamine (6.3 g) was converted into the N-bromo-derivative [7]. Removal of the pentane was followed by addition of the bromoamine dropwise to trifluoroacetic acid (40 ml) at 0—5°. The resulting red solution was irradiated for 24 hours at 25°, poured onto ice, and treated in a similar manner to that used for the sulfuric acid solution. In this manner 4.1 g of N,N-dibutylbenzenesulfonamide and 7.5 g of N-butylpyrrolidine picrate were obtained.

b) Synthesis of bridged nitrogen compounds

N-Chloro-N-methylcyclohexylamine (**10**) under irradiation in sulfuric acid solution yields 7-methyl-7-azabicyclo[2.2.1]heptane (**11**) [*8*]. The yield is only 11 % after a 30 hr. reaction period since for an efficient hydrogen abstraction to occur **10** is required to be in the thermodynamically unfavorable boat conformation.

HERTLER and COREY [*9*] have applied the HOFMANN-LÖFFLER reaction in the camphor series to functionalize one of the "unactivated" π-methyl groups. Irradiation of N-chlorocamphidine (**12**) in sulfuric acid solution gave after basification a tertiary amine which was isolated as the crystalline hydrobromide in 67 % yield. The product turned out to be cyclocamphidine (1,5-dimethyl-3-azatricyclo[3.3.1^{3,6}.0]nonane, **13**).

Cyclocamphidine (13) hydrobromide [*9*]. Camphidine hydrobromide (968.3 mg. 4.14 mmole) was chlorinated in pentane solution by the procedure of COLEMAN [*2*], Removal of the solvent left an oil to which was added 30 ml of 90 % sulfuric acid cooled to 0 °. The solution was placed in a quartz flask and irradiated with a mercury arc-lamp at 0 °. After 16 hrs. the solution was poured onto ice and made alkaline with sodium hydroxide. The resulting suspension was heated to boiling, allowed to cool, and extracted twice with ether. Dry hydrogen bromide was passed into the ether solution, and the oily precipitate was stirred with 3 N sodium hydroxide solution and 3 ml of benzenesulfonyl chloride overnight. The solution was acidified with hydrochloric acid, and the benzenesulfonamide of the secondary amine was removed by washing with ether. The aqueous solution was made alkaline with sodium hydroxide and extracted with ether. The ether was dried over magnesium sulfate, and dry hydrogen bromide was passed in. The amine hydrobromide was filtered and dried over phosphorus pentoxide at 0.1 mm. The product weighed 0.6444 g (67%) and crystallized from ethanol-ether as microcrystals, m.p. 353—357 ° (dec.).

c) Synthesis of conanines

Conanines (cf. **14** for conanine) are of biochemical interest as the conanine skeleton is found in many of the Holarrhena alkaloids. The synthesis of dihydroconessine (**17**) is outlined below; the irradiation of 3β-dimethylamino-20α-(N-chloro-N-methylamino)-5α-pregnane (**16**) produced a good yield of **17** [*10*].

References, p. 247

Dihydroconessine (17) [*10*]. 3β-Dimethylamino-20α-(N-chloro-N-methylamino)-5α-pregnane (**16**) (90 mg, prepared from 3β-dimethylamino-20α-methylamino-5α-pregnane (**15**) with N-chlorosuccinimide) was dissolved in 10 ml of 90% sulfuric acid at 0°. The resulting solution was irradiated in a quartz test-tube at 0° with a mercury arc lamp under a stream of nitrogen. After 70 minutes the solution was poured over ice, made alkaline with sodium hydroxide, and extracted with ether. The ether was evaporated, and the residue was refluxed with 10 ml of ethanol containing 1 g of potassium hydroxide for 0.5 hour. The solution was diluted with water and extracted with ether. The ether solution was dried over sodium sulfate and concentrated in vacuo to a slightly yellow semi-solid (80.3 mg) which was chromatographed on 7.5 g of Woelm neutral alumina. Elution with 2:1 benzene-ether gave 64.4 mg of oil (79%) which crystallized from aqueous acetone as flat needles, m.p. 101.5—102.5°.

d) Synthesis of pyrrolizidines

The HOFMANN-LÖFFLER ring closure seemed to be restricted to N-monohalogen compounds until 1960 when SCHMITZ [*11*] found that N,N-dibromo-4-heptanamine (**18**) underwent cyclization to pyrrolizidine (**19**) on warming in conc. sulfuric acid. The yield was only 2% but this was increased to 35% by conducting the cyclization at room temperature under UV irradiation.

Pyrrolizidine (19) [*11*]. A mixture of 11.5 g (0.1 mole) 4-aminoheptane and 100 ml 2N KOH was cooled in ice-salt freezing mixture and stirred vigorously while 32 g bromine were added over 3—4 minutes from a dropping funnel the stem of which dipped beneath the liquid. The oil which separated was taken up in ice-cold ether, and shaken successively with water, 6N H_2SO_4, water and sodium carbonate solution. Since the explosive properties of N-bromoamines have often found mention in the literature, the wash-liquids and the vessels used were precooled and all operations were conducted behind glass shields. Violent decomposition of **18** was never observed. After drying quickly with potassium carbonate the ether was removed in vacuo. The residue was covered with 100 ml ice-cold conc. sulfuric acid and carefully shaken. The yield of N,N-dibromo-4-heptanamine (**18**), determined iodometrically, was 83% of theory.

The sulfuric acid solution was placed in a 15 cm diameter crystallizing basin and while being cooled with ice-water irradiated by means of a 275 Watt mercury high pressure burner. The distance of the burner from the basin was 20 cm and the temperature of the liquid did not exceed 25 °. After $1^1/_2$ hours 90% of **18** had disappeared. Ice was added and the mixture made alkaline by cautious addition of 30% caustic soda solution. The resulting solution was continually extracted with ether and dried with potassium carbonate. The residue obtained by stripping off the ether yielded 9.35 g pyrrolizidine-picrate on the addition of 8 g picric acid in 300 ml alcohol and a further 0.4 g were obtained from the mother liquor. The yield was 34.6% (calculated on dibromoamine used) or 28.6% (calculated on 4-aminoheptane used), m.p. 257° (ethanol).

3. Formation of chloroalkylamines from N-chloroamines

a) N-Butyl-4-chlorobutylamine

By partial neutralization of a previously irradiated sulfuric acid solution of N-chlorodibutylamine (**1**) and subsequent treatment with barium chloride N-butyl-4-chlorobutylamine hydrochloride (**20**) was isolated in 37% yield [6].

$$(C_4H_9)_2N-Cl \xrightarrow[H_2SO_4]{h\nu} Cl-CH_2-CH_2-CH_2-CH_2-NH-C_4H_9 \cdot HCl$$
$$\mathbf{1} \qquad\qquad\qquad\qquad\qquad \mathbf{20}$$

N-Butyl-4-chlorobutylamine hydrochloride (20) [6]. N-Chlorodibutylamine (**1**) was prepared from dibutylamine (64.5 g) and extracted into pentane. The pentane solution was shaken with cold 85% sulfuric acid (200 ml) and the pentane removed under reduced pressure. The sulfuric acid was irradiated for 48 hours at 25° and then diluted to 500 ml with ice and water. After removal of unreacted chloroamine with hexane a 150-ml portion was treated slowly with dry sodium bicarbonate (110 g). The resulting solution was filtered from the sodium sulfate which crystallized and then was treated with an aqueous solution containing barium chloride (37 g). The barium sulfate was removed by filtration and the resultant solution was evaporated to dryness at 100° under reduced pressure using a water aspirator. The last traces of water were removed by azeotroping with benzene. The resulting hydrochloride was recrystallized from acetone three times and gave waxy plates melting at 211—212° with slight decomposition; yield 11.22 g (37.4%).

b) Molecular rearrangements of steroidal N-chloroamines

According to KERWIN [*12*] some steroidal N-chloroamines are extremely soluble in trifluoroacetic acid and the solutions prepared with thorough cooling are stable for several hours in the dark. Irradiation of the stirred reaction mixtures at 20—25° causes a rapid disappearance of positive halogen with only slight darkening of the solution. Evaporation of excess acid under reduced pressure circumvented the need for neutralization of large volumes of acid as required in the classical HOFMANN-LÖFFLER procedure. The formation of **22** is illustrated on the following page.

References, p. 247

18-Chloro-20α-methylamino-3β-trifluoroacetoxy-5α-pregnan-11-one trifluoroacetate (22) [*12*]. 3β-Hydroxy-20α-(N-chloro-N-methylamino)-5α-pregnan-11-one (**21**) (87 g) was irradiated (three General Electric 15 W Germicidal lamps) in four portions in trifluoroacetic acid and then the solutions were combined. Trifluoroacetic anhydride (25 ml) was added and the solution allowed to stand at room temperature for 1 hr. Excess acid and anhydride were removed under reduced pressure (Dry Ice trap) and the residual oil was dissolved in acetone. Addition of ether and petroleum-ether caused crystallization of product in nearly quantitative yield. Recrystallization from acetone yielded three crops of crystals, totalling 109.8 g (81%). The analytical sample was recrystallized from acetone, m.p. 156—160°.

When the trifluoroacetic anhydride treatment was omitted, the product was isolated in only 15—26% yield; the remainder of the reaction mixture persisted as an oil.

References

[*1*] S. WAWZONEK and J. D. NORDSTROM: J. Org. Chem. **27**, 3726 (1962).
[*2*] G. H. COLEMAN, G. NICHOLS and T. F. MARTENS: Org. Syntheses, Coll. Vol. **3**, 159 (1955).
[*3*] M. E. WOLFF: Chem. Rev. **63**, 55 (1963).
[*4*] A. W. HOFMANN: Ber. dtsch. chem. Ges. **16**, 558 (1883).
[*5*] K. LÖFFLER and C. FREYTAG: Ber. dtsch. Chem. Ges. **42**, 3427 (1909).
[*6*] S. WAWZONEK and T. P. CULBERTSON: J. Amer. chem. Soc. **81**, 3367 (1959).
[*7*] S. WAWZONEK, M. F. NELSON JR. and P. J. THELEN: J. Amer. chem. Soc. **73**, 2806 (1951).
[*8*] E. J. COREY and W. R. HERTLER: J. Amer. chem. Soc. **82**, 1657 (1960).
[*9*] W. R. HERTLER and E. J. COREY: J. Org. Chem. **24**, 572 (1959).
[*10*] E. J. COREY and W. R. HERTLER: J. Amer. chem. Soc. **81**, 5209 (1959).
[*11*] E. SCHMITZ and D. MURAWSKI: Chem. Ber. **93**, 754 (1960).
[*12*] J. F. KERWIN, M. E. WOLFF, F. F. OWINGS, B. B. LEWIS, B. BLANK, A. MAGNANI, C. KARASH and V. GEORGIAN: J. Org. Chem. **27**, 3628 (1962).

Chapter 27

Photochemical transformations of organic nitrites

1. Photochemical reactions of nitrites involving fission of oxygen-nitrogen bonds. The BARTON reaction

Though experiments on the photolysis of organic nitrites have been performed as early as 1936 [1] the importance of photochemical reactions in this field has been recognized only recently. A comprehensive review on earlier work in this field and on the scope of the reaction has been given some years ago by NUSSBAUM [2].

One of the photochemical reactions of organic nitrites is an isomerization process which has become known as the BARTON reaction. In this reaction a hydroxyl group forms and the nitroso group replaces hydrogen in the γ-position. Usually products are formed which are to be regarded as the result of consecutive reactions, e.g. dimeric nitroso compounds or oximes from isomerization of nitroso compounds. This reaction [3] has since been widely applied in organic syntheses; a recent account has been given by AKHTAR [4].

a) Simple aliphatic and alicyclic nitrites

Typical examples of the BARTON reaction carried through with relatively simple compounds are the conversion of n-octyl nitrite (1) to 4-nitroso-1-octanol dimer (2) [5], of 4-phenyl-1-butyl nitrite to 4-nitroso-4-phenyl-1-butanol dimer [6] and of d,l-menthyl nitrite (3) to d,l-10-nitrosomenthol dimer (4) [7].

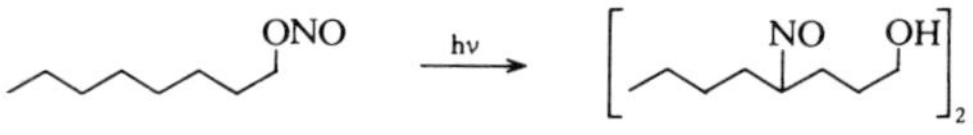

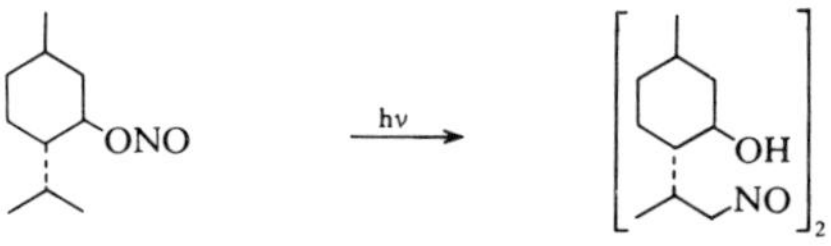

References, p. 254

The Barton reaction involves photolysis of the N—O bond in the nitrite ester giving an alkoxy radical and nitric oxide. The γ-hydrogen is abstracted by the alkoxy radical, and the resulting alkyl radical and NO combine according to the following scheme exemplifying photolysis of octyl nitrite (**1**):

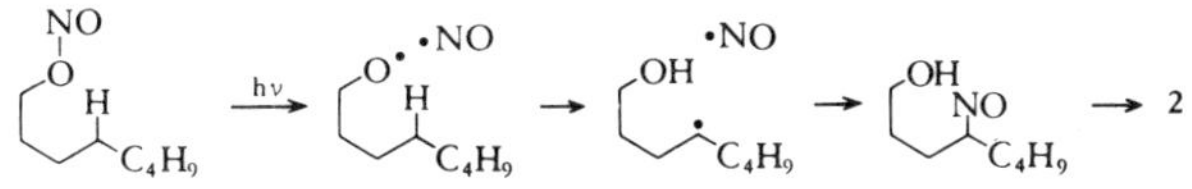

Cyclobutyl and cyclopentyl nitrites did not undergo the Barton reaction as formation of the prerequisite six membered transition state was physically impossible. Instead open chain nitroso aldehyde dimers were formed, viz. 4-nitrosobutanal and 5-nitrosopentanal dimers [*8*].

4-Nitroso-1-octanol dimer (2) [*5*]. Octyl nitrite (**1**) (22.5 g) was photolyzed in 1800 ml of heptane using an appropriate photolysis cell. The 4-nitroso-1-octanol dimer crystallized directly out of the reaction solution and was isolated by filtration, yield 7.0 g (25.9% by spectroscopic analysis). Recrystallization from heptane yielded pure **2** of m.p. 95—96°.

A 200 W Hanovia 654 A-36 ultraviolet lamp was utilized as the radiation source in this and the following experiment. The Pyrex jacket on the immersion well limited the light entering the photolysis cell to wavelengths greater than 300 mμ.

d,l-10-Nitrosomenthol dimer (4) [*7*]. d,l-Menthyl nitrite (**3**) (29.6 g) was photolyzed in 200 ml of heptane at 25° for 2.5 hrs. using the aforementioned light source. The crude nitroso dimer **4** precipitated as a gum, yield 16.0 g (39% by spectroscopic evidence). Crystallization from ether afforded 1.29 g (4.3%) of **4**, m.p. 127—128°.

b) Epimerization in nitrite photolysis

Nickon [*9*] discovered that the carbon-oxygen bond can change configuration during the Barton reaction. Nickon's findings should be taken into consideration, when this reaction is used as a degradative step in structure elucidation work.

When the nitrite ester of α-caryophyllene alcohol (**5**) is UV-irradiated in benzene solution rearrangement occurs according to the above established scheme. The hydroxyimino alcohol resulting, however, has its hydroxyl group in the epi configuration (**7**). The same hydroxyimino alcohol may be arrived at if the epi-α-caryophyllene alcohol nitrite (**6**) be irradiated. No mention was made as to which of the reaction steps entails epimerization.

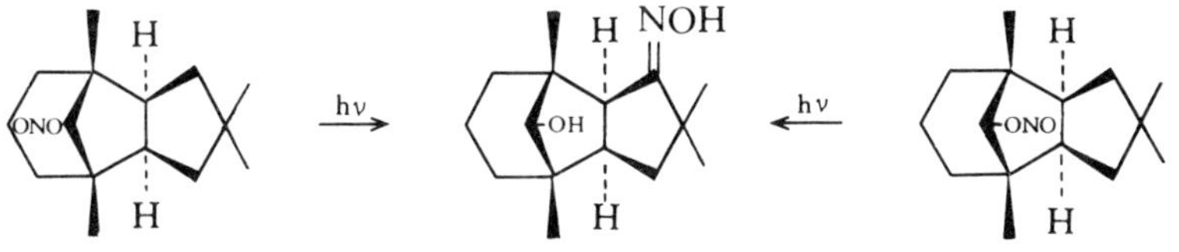

The stereochemistry shown in **5—7** was derived by NICKON et al. [*10*] from further degradative work, again making use of the BARTON reaction for the introduction of a third functional group into **7** or its respective hydrolysis product.

c) Steroidal nitrites

The BARTON reaction has found particularly wide application to synthesis in the steroid field; excellent reviews on this subject have appeared [*2,4,11*].

Research in this field was opened with the spectacular synthesis of aldosterone-21-acetate by BARTON [*12*]. Corticosterone 21-acetate 11-nitrite (**8**) when irradiated in toluene solution underwent rearrangement to aldosterone 21-acetate oxime (**9**) which was hydrolyzed with nitrous acid to aldosterone 21-acetate (**10**).

The above transformation as well as those mentioned below require the hydrogen to be replaced by the NO group to be "conformationally adjacent" to the nitrite group 4 carbon atoms away, so that the compulsory six membered transition state may form.

Further examples of the BARTON reaction in the steroid field are the synthesis of the hydroxyimino alcohol **12** from its progenitor **11**, and of the 19-hydroxyimino alcohol **14** from nitrite ester **13** [*13*].

Nitrites **15** and **17** which are epimeric at C-20 have by irradiation been converted to the corresponding 18-hydroxyimino alcohols **16** and **18** respectively

References, p. 254

[14]. The irradiation products are reported to have retained their original configuration at C-20. This would mean that photochemical epimerization as observed by NICKON (cf. preceding section) may be restricted to some peculiar cases.

A plausible interpretation has been offered by NUSSBAUM et al. [14] to account for the more complex reaction encountered in the irradiation of 20β-nitrites as compared to photolysis of 20α-nitrites. In the latter case a single product is isolated in high yield whereas with 20β-nitrites a variety of products are obtained, demonstrating that fragmentation reactions are competing with the BARTON reaction.

3α-Acetoxy-20β-hydroxy-18-hydroxyimino-5β-pregnan-11-one (12) [13]. A solution of 3α-acetoxy-20β-hydroxy-5β-pregnan-11-one 20-nitrite (11) (4.81 g) in 200 ml benzene was irradiated in a Pyrex vessel under nitrogen using a 200 W Hanovia high pressure mercury arc lamp at 20° for 2.5 hrs. Chromatography on Florisil (115 g) and elution with methylene chloride containing increasing amounts of methanol gave two distinct crystalline products. The less polar solid (1.36 g) was recrystallized from acetone-hexane to give 3α-acetoxy-20β-hydroxy-5β-pregnan-11-one. The more polar solid (2.16 g) was recrystallized from methylene chloride-hexane to give 12 as fine needles (0.86 g), m.p. 120—129°.

20β-Hydroxy-18-hydroxyiminopregn-4-en-3-one (16) [14]. 20β-Hydroxypregn-4-en-3-one nitrite (15) (18.0 g) was dissolved in 500 ml dry benzene and irradiated for $5^1/_2$ hrs. at 0° in a nitrogen atmosphere with a 200 W mercury lamp. The solution was concentrated to a thick oil which proved to contain at least 5 substances. Chromatography on 600 g of Florisil and elution with slightly polar solvents gave starting material. 30% ether-benzene eluted the oxime which after crystallization from ethyl acetate melted at 228—238° (3.7 g). Recrystallization from acetone afforded pure 16, m.p. 242—243°.

2. Photochemical reactions of nitrites involving fission of carbon-carbon bonds

a) Syntheses of nitrosoalkanes

Among the photochemical decomposition reactions of alkyl nitrites those of tert.-butyl nitrite have been the most extensively investigated. According to COE and DOUMANI [15] tert.-butyl nitrite disintegrates to acetone and nitrosomethane under the action of UV-light. The authors obtained nitrosomethane as a dimeric product from their experiments.

$$2 \ (CH_3)_3C-ONO \xrightarrow{h\nu} 2 \ H_3C-CO-CH_3 + (CH_3NO)_2$$

GOWENLOCK and TROTMAN [16] by the same procedure have obtained two different nitrosomethane dimers, which are thought to be cis-trans isomers. The following photochemical and thermal interconversions have been discussed:

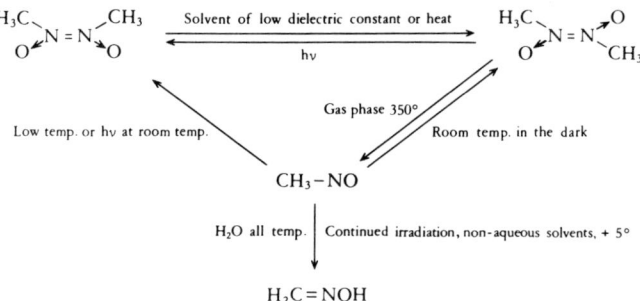

Mechanistic implications of nitrite photolysis have recently been reviewed by NUSSBAUM and ROBINSON [2] (comp. also GRAY and WILLIAMS [17]).

Nitrosomethane dimer [15]. 1 ml of liquid tert.-butyl nitrite was condensed into the bottom of a 350 ml quartz flask. The reaction vessel was then irradiated with a 360 W General Electric "Uviarc" lamp, a quartz flask filled with distilled water serving as a heat filter. Colorless crystals of nitrosomethane dimer (m.p. 122°) settled on the walls of the reaction vessel.

b) Hydroxamic acids

According to KABASAKALIAN [18] photolysis of the nitrites of either d,l-borneol (**19a**) or d,l-isoborneol (**19b**) in trichlorotrifluoroethane yields a cyclic hydroxamic acid, d,l-2-hydroxy-1,8,8-trimethyl-2-azabicyclo-[3.2.1] octan-3-one (**21**) together with α-campholenaldehydes **22** and **23**. d,l-3-Nitroso-α-campholanaldehyde (**20**) was shown to be the precursor of **21**, the cyclization step requiring activation by light. The two α-campholenaldehydes are believed to originate from **19** not via **20** but through a disproportionation reaction of alkoxy radicals initially formed.

References, p. 254

ROBINSON et al. [*19*] have reported the formation of steroidal hydroxamic acids (cf. **25**) during the photolysis of 17β-nitrites. The reactions are imagined to involve fission of a carbon-carbon bond following formation of the alkoxy radical during photolysis. 13α-compounds are also formed in some cases, which supports the view that C-13 radicals are intermediates in this rearrangement. The scheme below illustrates the photolysis of 3α-acetoxy-17β-hydroxy-5α-androstane nitrite (**24**) leading to the cyclic hydroxamic acid 3α-acetoxy-17a-hydroxy-17a-aza-D-homo-5α-androstan-17-one (**25**).

d,l-2-Hydroxy-1,8,8-trimethyl-2-azabicyclo[3.2.1]octan-3-one (21) [*18*]. A solution of 22.4 g of d,l-bornyl nitrite (**19a**) in 200 ml of trichlorotrifluoroethane was photolyzed using a Pyrex light filter. The solution was flash evaporated to an oil (20.5 g) and triturated with petroleum ether. The hydroxamic acid crystallized immediately; yield 4.2 g (19%). An analytical sample was obtained by slurrying with petroleum ether, m.p. 214—217°.

The petroleum ether extract contained d,l-α-1(6)-campholenaldehyde (d,l-2,3-trimethyl-3-cyclopentene-1-acetaldehyde) (**22**) and d,l-α-1(10)-campholenaldehyde (d,l-3-methylene-2,2-dimethylcyclopentane-1-acetaldehyde) (**23**).

c) Fragmentation reactions in the steroid series

Oxidative fragmentation reactions have been observed in the irradiation of steroidal nitrites bearing an oxygen function at the α-carbon [*20*].

Thus 3,3:21,21-bis-ethylenedioxy-20β-hydroxypregn-5-ene was converted to its nitrite (**26**) and photolyzed in benzene. The product was found to be 3,3-ethylenedioxy-17β-formylandrost-5-ene (**27**). By a similar procedure androst-4-ene-3,17-dione (**31**) was obtained when the 20α- (**28**), 20β- (**29**), and 17α- (**30**) nitrites were irradiated [*21*].

Androst-4-ene-3,17-dione (31) [*21*].

(a) From 17α,20α-dihydroxypregn-4-en-3-one 20-nitrite (**28**). The 20α-ol (300 mg) was dissolved in 8 ml of pyridine and treated with nitrosyl chloride-pyridine at —25° until a color change persisted. The reaction mixture was quenched in water, filtered, and dried in vacuo at room temperature. The 247 mg thus obtained were irradiated in benzene solution with a 200 W mercury lamp through Pyrex, under nitrogen, at 20°. Chromatography gave 148 mg of **31**.

(b) From 17α,20β-dihydroxypregn-4-en-3-one 20-nitrite (**29**). The 20β-ol (78 mg) was similarly nitrosated and irradiated. Chromatography again gave 27 mg of **31**.

(c) From 17α-hydroxypregn-4-ene-3,20-dione 17-nitrite (**30**). 17α-Hydroxyprogesterone (1 g) was nitrosated at —20° in an analogous prodecure. Irradiation of the benzene solution of **30** during 1 hr. gave 430 mg **31**.

References

[*1*] C. H. PURKIS and H. W. THOMPSON: Trans. Faraday Soc. **32**, 1466 (1936).
[*2*] A. L. NUSSBAUM and C. H. ROBINSON: Tetrahedron **17**, 35 (1962).
[*3*] D. H. R. BARTON, J. M. BEATON, L. E. GELLER and M. M. PECHET: J. Amer. chem. Soc. **82**, 2640 (1960).
[*4*] M. AKHTAR, in: Adv. Photochem. Vol. **2**, ed. by W. A. NOYES, JR., G. S. HAMMOND and J. N. PITTS, JR. New York: Interscience 1964, pp. 263—303.
[*5*] P. KABASAKALIAN and E. R. TOWNLEY: J. Amer. chem. Soc. **84**, 2711 (1962).
[*6*] P. KABASAKALIAN, E. R. TOWNLEY and M. D. YUDIS, J. Amer. chem. Soc. **84**, 2716 (1962).
[*7*] P. KABASAKALIAN and E. R. TOWNLEY: Amer. Perf. Cosmetics **78**, (2), 22 (1963).
[*8*] P. KABASAKALIAN and E. R. TOWNLEY: J. Org. Chem. **27**, 2918 (1962).
[*9*] A. NICKON, J. R. MAHAJAN and F. J. MCGUIRE: J. Org. Chem. **26**, 3617 (1961).
[*10*] A. NICKON, F. J. MCGUIRE, J. R. MAHAJAN, B. UMEZAWA and S. A. NARANG: J. Amer. chem. Soc. **86**, 1437 (1964).
[*11*] O. L. CHAPMAN, in: Adv. Photochem. Vol. **1**, ed. by W. A. NOYES, JR., G. S. HAMMOND and J. N. PITTS, JR., pp. 323—420. New York: Interscience 1963.
[*12*] D. H. R. BARTON and J. M. BEATON: J. Amer. chem. Soc. **82**, 2641 (1960).
[*13*] D. H. R. BARTON, J. M. BEATON, L. E. GELLER and M. M. PECHET: J. Amer. chem. Soc. **83**, 4076 (1961).
[*14*] A. L. NUSSBAUM, F. E. CARLON, E. P. OLIVETO, E. TOWNLEY, P. KABASAKALIAN and D. H. R. BARTON, Tetrahedron **18**, 373 (1962).
[*15*] C. S. COE and T. F. DOUMANI: J. Amer. chem. Soc. **70**, 1516 (1948).
[*16*] B. G. GOWENLOCK and J. TROTMAN: J. Chem. Soc. **1955**, 4190.
[*17*] P. GRAY and A. WILLIAMS: Chem. Rev. **59**, 239 (1959).
[*18*] P. KABASAKALIAN and E. R. TOWNLEY: J. Org. Chem. **27**, 3562 (1962).
[*19*] C. H. ROBINSON, O. GNOJ, A. MITCHELL, R. WAYNE, E. TOWNLEY, P. KABASAKALIAN, E. P. OLIVETO and D. H. R. BARTON: J. Amer. chem. Soc. **83**, 1771 (1961).
[*20*] A. L. NUSSBAUM, C. H. ROBINSON, E. P. OLIVETO and D. H. R. BARTON: J. Amer. chem. Soc. **83**, 2400 (1961).
[*21*] A. L. NUSSBAUM, E. P. YUAN, C. H. ROBINSON, A. MITCHELL, E. P. OLIVETO, J. M. BEATON and D. H. R. BARTON: J. Org. Chem. **27**, 20 (1962).

Chapter 28

Photochemical dealkylation of nitrogen compounds

1. Photolysis of N-alkyl and N-aralkyl amines

VLADIMIRTSEV et al. [1] found that exposure of alcoholic solutions of 3-chloro-2-N-ethylanilino-1,4-naphthoquinone (**1**) or 3-bromo-2-N-ethylanilino-1,4-naphthoquinone (**3**) to sunlight is accompanied by dealkylation and formation of 2-anilino-3-chloro-1,4-naphthoquinone (**2**) and 2-anilino-3-bromo-1,4-naphthoquinone (**4**), respectively. Loss of the acyl group on

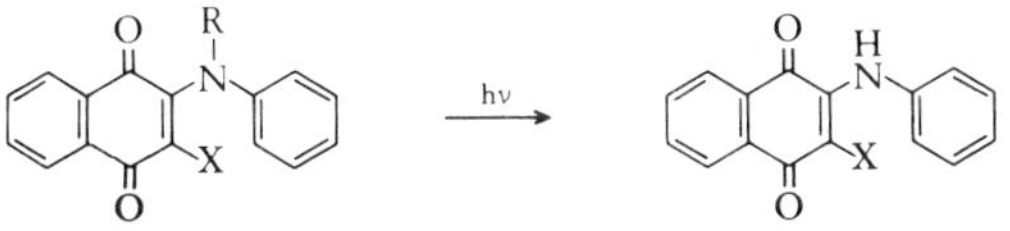

1: X = Cl, R = C_2H_5
3: X = Br, R = C_2H_5
5: X = Cl, R = CO-CH_3

2: X = Cl
4: X = Br

irradiation of 3-chloro-2-(N-phenylacetamido)-1,4-naphthoquinone (**5**) and formation of **2** was observed as well [1]. These reactions may also be accomplished in a carbon dioxide atmosphere.

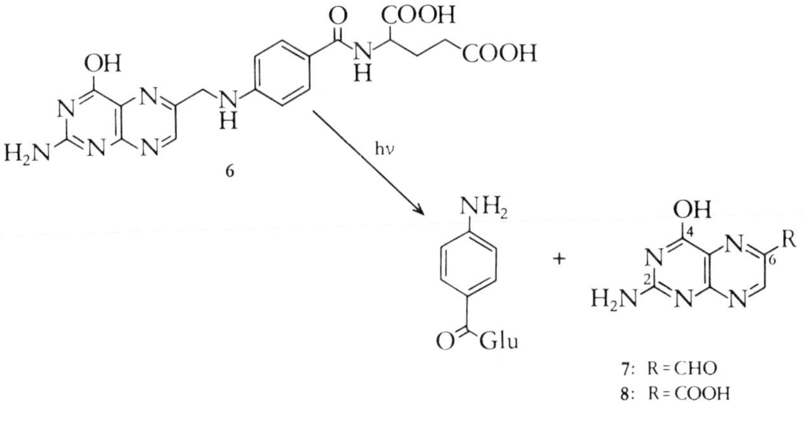

7: R = CHO
8: R = COOH

References, pp. 258–259

Photolysis of nitrogen-carbon bonds is of little significance from a preparative point of view but may be important with biological material, e.g., with amino acids and peptides. One such example will be briefly mentioned here. When pteroylglutamic acid (folic acid, **6**) is irradiated with UV-light in slightly acid solution [*2, 3*] oxidative cleavage occurs with formation of N-(p-aminobenzoyl)-glutamic acid and of 2-amino-4-hydroxy-6-pteridinecarboxaldehyde (**7**). This reaction was said to be photosensitized by riboflavine [*4*]. The aldehyde **7** undergoes a further oxidation to 2-amino-4-hydroxy-6-pteridinecarboxylic acid (**8**) which is known as one of the end products of the photolysis of naturally occurring pteridines (comp., among others, [*5*]).

BROCKMANN and co-workers [*6*] observed, that certain N-alkylated actinomycins were light sensitive in contrast to the actinomycins themselves. Exposure of a methanol solution of N-methylactinomycin C_2 (**9**) to sunlight resulted in demethylation and formation of actinomycin C_2 (**10**) (for the detailed peptide structures see [*7*]).

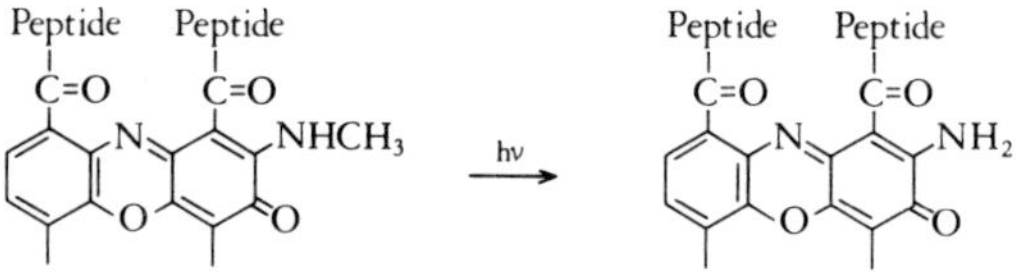

9 **10**

Investigations on this photodegradation have been resumed only recently when LEVINE and WANI [*8*] studied the photochemistry of simpler

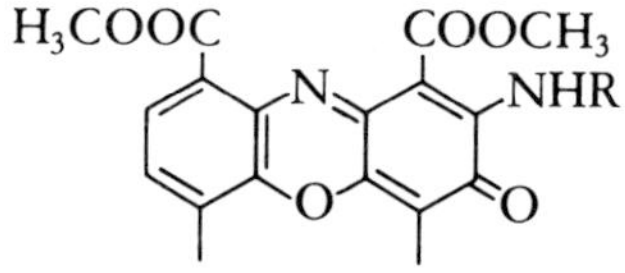

11: R = CH_3
13: R = CH $(CH_3)_2$
15: R = H

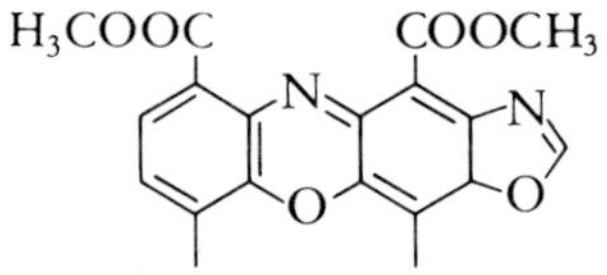

12

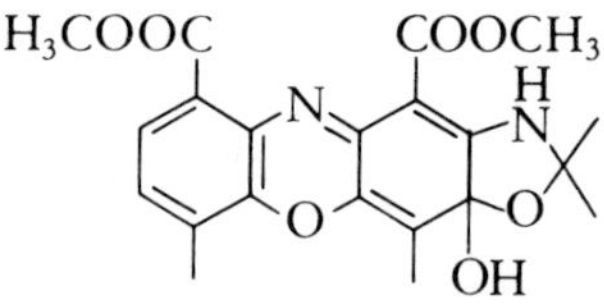

14

References, pp. 258–259

analogs of **9**. With dimethyl 4,6-dimethyl-2-methylamino-3-oxo-3H-phenoxazine-1,9-dicarboxylate (**11**) no dealkylation was observed but instead ring formation to dimethyl 9,11-dimethyl-5H-oxazolo[4,5-b]phenoxazine-4,6-dicarboxylate (**12**).

Scission of alkyl groups attached to the amino group was, however, found with higher alkyl groups; e.g., the isopropyl analog **13** gave, on irradiation in benzene-methanol (3 : 1) solution, 22 % of dimethyl 2-amino-4,6-dimethyl-3-oxo-3H-phenoxazine-1,9-dicarboxylate (**15**) besides a ring closure product, for which structure **14** was proposed [*8*]. For another photochemical oxazole synthesis the reader is referred to chapter 19.

2-Anilino-3-chloro-1,4-naphthoquinone (2) [*1*]. A solution of **1** (50 mg) in 100 ml of ethanol in a quartz flask was exposed to sunlight for 60 hours; a color change from violet to orange was observed, and at the end of the experiment a smell of isonitrile (but no acetaldehyde) could be perceived. The alcoholic solution was concentrated to 5 ml, and the red crystals (16 mg), which separated on cooling, were washed with a little alcohol to give **2**, m.p. 205—207°.

Actinomycin C$_2$ (10) [*6*]. A solution of 200 mg of **9** in 20 ml of methanol was exposed to sunlight for 3 days. Evaporation of the solvent in vacuo and chromatography, the details of which should be taken from the original paper, yielded 60 mg of red **10**, which after recrystallization from ethyl acetate had m.p. 235°.

2. Photolysis of N-alkyl nitrogen heterocycles

In the early 1930's the studies on the photochemistry of riboflavine or of simpler analogs reached their first climax; a summary of the work done by KARRER, KOSCHARA, KUHN, THEORELL, WARBURG and others prior to 1939 will be found elsewhere [*9*]. Since this book was first published a great many publications have been devoted to the more mechanistic aspects of riboflavine photochemistry; reviews have been given by OSTER [*10*], HOLMSTRÖM [*11*] and HEMMERICH [*12*] and their collaborators.

Flavine photochemistry is complicated due to the fact that (a) riboflavine may exert a photosensitizing influence and (b) that excited riboflavine may be photoreduced. These photoreactions leaving the riboflavine skeleton intact may be accompanied by irreversible reactions whereby the whole or part of the side-chain is split off the riboflavine molecule. It is these photolyses that will be dealt with here. The photolysis of riboflavine (**16**) and of similar flavines is complicated further by the fact that not only does it proceed differently in neutral or acid and in alkaline media, but takes a different course under aerobic or under anaerobic conditions (cf., among others, [*13—16*]).

The photochemical degradation of riboflavine (16) leads to lumiflavine (7,8,10-trimethylisoalloxazine, 17 [*13*]) in alkaline and to lumichrome (7,8-dimethylalloxazine, 18 [*17*]) in neutral or acid solution.

Investigations on flavine photochemistry were, in part, carried out with suitably substituted model compounds. Thus, KARRER [*14*, *18*] irradiated 10-(2-hydroxyethyl)-isoalloxazine (19) and 10-(2,3-dihydroxypropyl)-isoalloxazine (21) and obtained, in neutral solution, alloxazine (20) (comp. also HALWER [*15*]), and in alkaline solution, 10-methylisoalloxazine (22).

19: R = CH$_2$-CH$_2$OH
21: R = CH$_2$-CHOH-CH$_2$OH
22: R = CH$_3$

20

23

From this and related observations it was surmised [*12*, *14*, *18*] that the presence of a secondary hydroxy group in the 2-position of the side-chain was a prerequisite for the photochemical cleavage, but more recent investigations of YANG and McCORMICK [*19*] (comp. also [*15*]) showed that irradiation of, for example, 7,8-dimethyl-10-(5-hydroxypentyl)-isoalloxazine (23) produced lumichrome (18) as a major product.

Alloxazine (20) [*18*]. 10-(2,3-Dihydroxypropyl)-isoalloxazine (21) was dissolved in 75% methanol and the solution exposed to sunlight in a flat glass vessel. Within 1—2 hrs. irradiation the solution was bleached. On concentration of the solution alloxazine (20) precipitated and was filtered off.

10-Methylisoalloxazine (22) [*18*]. A solution of 21 in dilute aqueous NaOH was insolated for ca. 3 hours. Then the solution was acidified and extracted with chloroform; during this procedure some alloxazine (20) separated out and was filtered off. The chloroform extract, which showed a yellow-green fluorescence, was evaporated and the resulting residue was again taken up in chloroform. On evaporation of this second chloroform extract the remaining material was recrystallized from water to give bright yellow 10-methylisoalloxazine (22).

Lumichrome (18) [*14*]. Repeatedly recrystallized riboflavine (100 mg) is dissolved in 40 ml hot water and the solution while still warm, treated with 160 ml methanol. This solution is exposed to sunlight in a glass reaction vessel (72 × 6 × 1.2 cm), and decoloration occurs within 3 hrs. or 3 days, depending on weather conditions. The now faintly yellow solution is freed from methanol under reduced pressure and allowed to cool. Crude 18 precipitates from this solution and is filtered off (yield between 30—45%). Purification is effected by repeated extraction and recrystallization from chloroform. On cooling, lumichrome separates as almost colorless needles, m.p. above 300° [*17*].

References

[*1*] I. F. VLADIMIRTSEV, I. YA. POSTOVSKII and L. F. TREFILOVA: Zhur. Obshchei Khim. **24**, 181 (1954); J. Gen. Chem. USSR **24**, 183 (1954).
[*2*] O. H. LOWRY, O. A. BESSEY and E. J. CRAWFORD: J. Biol. Chem. **180**, 389 (1949).
[*3*] H. M. RAUEN and H. WALDMANN: Z. Physiol. Chem. **286**, 180 (1950).

References

[4] S. Scheindlin, A. Lee and I. Griffith: J. Am. Pharm. Assoc., Sci. Ed. **41**, 420 (1952).
[5] I. Ziegler, in: Pteridine Chemistry, Ed. by W. Pfleiderer and E. C. Taylor, p. 295. Oxford: Pergamon 1964.
[6] H. Brockmann, G. Pampus and R. Mecke: Chem. Ber. **92**, 3082 (1959).
[7] H. Brockmann: Pure Appl. Chem. **2**, 405 (1961).
[8] S. G. Levine and M. C. Wani: J. Org. Chem. **30**, 3185 (1965).
[9] H. Rudy, in: Fortschr. Chem. Org. Naturstoffe, Ed. by L. Zechmeister, Vol **2**, pp. 61—102. Wien: Springer 1939.
[10] G. Oster, J. S. Bellin and B. Holmström: Experientia **18**, 249 (1962).
[11] B. Holmström: Arkiv Kemi **22**, 329 (1964).
[12] P. Hemmerich, C. Veeger and H. C. S. Wood: Angew. Chem. **77**, 699 (1965); internat. ed. **4**, 671 (1965).
[13] R. Kuhn, H. Rudy and T. Wagner-Jauregg: Ber. dtsch. chem. Ges. **66**, 1950 (1933).
[14] P. Karrer, T. Köbner, H. Salomon and F. Zehender: Helv. Chim. Acta **18**, 266 (1935).
[15] M. Halwer: J. Amer. chem. Soc. **73**, 4870 (1951).
[16] B. Holmström: Arkiv Kemi **22**, 281 (1964).
[17] P. Karrer, H. Salomon, K. Schöpp, E. Schlittler and H. Fritzsche: Helv. Chim. Acta **17**, 1010 (1934).
[18] P. Karrer, H. Salomon, K. Schöpp and E. Schlittler: Helv. Chim. Acta **17**, 1165 (1934).
[19] C. S. Yang and D. B. McCormick: J. Amer. chem. Soc. **87**, 5763 (1965).

Chapter 29

Photochemical introduction of cyano and nitroso groups
1. Formation of nitriles

The action of cyanogen chloride on saturated open chain or cyclic hydrocarbons in the presence of UV light leads to the formation of nitriles [1].

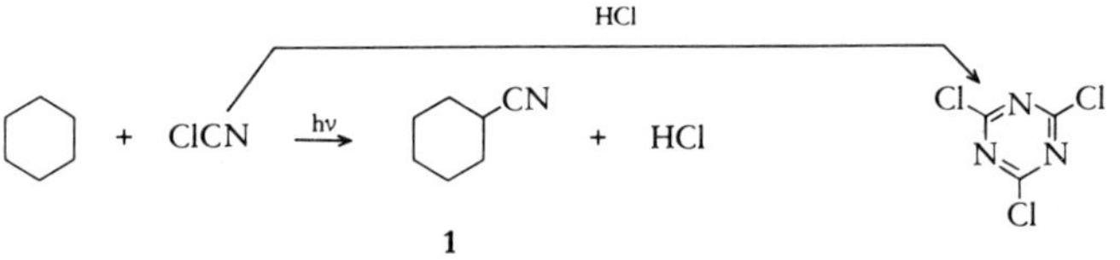

It has been repeatedly observed that the addition of small quantities of carbonyl compounds increases the conversion rate. This effect was observed, for example, in the synthesis of cyanocyclohexane (**1**) by the irradiation of cyclohexane with ClCN in the presence of acetyl chloride (0.05 mole % acetyl chloride/ClCN) [1].

In a similar way, saturated open chain or cyclic ethers afford the corresponding α-cyanoethers (cf. **2**) when irradiated with ClCN [2]. The introduction of the cyano group is conveniently carried out in the presence of sodium bicarbonate which serves to bind the hydrogen chloride produced in the reaction. Otherwise HCl will catalyze the trimerization of cyanogen chloride to cyanuric chloride, thus depriving the solution of the cyanation reactant.

2-Ethoxypropionitrile (**2**) is thus obtained from diethyl ether, and 2-cyano-1,4-dioxane from dioxane [2]. Tetrahydrofuran yields 2-cyanotetrahydrofuran as main product together with 3-cyanotetrahydrofuran. Analogously, 1,2-dimethoxyethane yields 91 % of a mixture containing a 53 : 47 ratio of 2,3-dimethoxypropionitrile (**3**) and (2-methoxyethoxy)-acetonitrile (**4**) [2].

$$H_5C_2-O-C_2H_5 + ClCN \xrightarrow{h\nu} H_5C_2-O-\underset{\underset{\textbf{2}}{CN}}{CH}-CH_3 + HCl$$

References, p. 265

$$H_3CO-CH_2-CH_2-OCH_3 \xrightarrow[ClCN]{h\nu} H_3CO-\underset{\underset{CN}{|}}{CH}-CH_2-OCH_3$$
$$\mathbf{3}$$
$$+ H_3CO-CH_2-CH_2-OCH_2-CN$$
$$\mathbf{4}$$

Yields range from 5 % to about 10 %, but conversions are high (ca. 90 %) since the unconsumed reactants may easily be recovered.

2-Ethoxypropionitrile (2) [*2*]. Finely powdered $NaHCO_3$ (18 g) was suspended in diethyl ether (74 g) in a cyclindrical reaction vessel, and 61 g of gaseous cyanogen chloride was introduced with magnetic stirring and cooling. The irradiation was carried out at 15—20° under a slow stream of nitrogen with a mercury high pressure lamp S 81 (details of the apparatus should be taken from the original paper). After 1 hour, excess ClCN and ether were distilled off. After removal of $NaHCO_3$ by filtration the working up yielded a residue which was distilled at normal pressure to give 9.2 g (95% conversion) of **2**, b.p. 129—131°. Yields may be considerably improved by addition of 3.9 g of acetone to the above reaction mixture before irradiation is started.

2. Formation of gem. chloronitroso and dimeric nitroso compounds

Our knowledge on the photonitrosation of hydrocarbons is mainly due to investigations of MÜLLER and co-workers. By the joint action of chlorine, nitrogen oxide (1 : 2 v/v) and UV light on hydrocarbons, MÜLLER and METZGER [*3*] obtained from cyclohexane, 1-chloro-1-nitrosocyclohexane (**5**) and from heptane, a mixture of isomeric gem. chloronitrosoheptanes. Toluene behaved differently as will be seen below. None of the geminal chloronitroso compounds mentioned were isolated in a pure state; however, from the reaction mixture obtained by treating cyclohexane with Cl_2 and NO, pure 1-chloro-1-nitrocyclohexane (**6**) was obtained by oxidation with concentrated nitric acid.

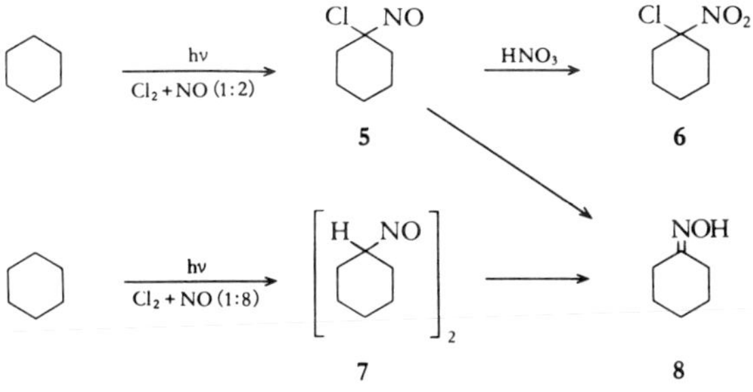

When the Cl_2/NO ratio was changed to 1 : 8 (v/v) another product prevailed, viz. nitrosocyclohexane dimer (**7**) which forms colorless crystals in contrast to **5** which is blue [*4*]. From heptane a mixture of isomeric nitrosoheptane dimers was obtained, while toluene gave only a very low yield of the corresponding compound, α-nitrosotoluene dimer.

1-Chloro-1-nitrosocyclohexane (5) as well as nitrosocyclohexane dimer (7) may easily be converted to cyclohexanone oxime (8) [*3—5*]. The isomerization of 7 to the oxime 8 may be catalyzed either by acids or, more conveniently, by amines [*5, 6*]. The dimerization kinetics of nitrosocyclohexane was studied by BURRELL, JR. [*7*]. A general survey on structure and properties of nitroso compounds has been given by GOWENLOCK and LÜTTKE [*8*] and by SMITH [*9*].

Nitrosocyclohexane dimer (7) [*4*]. Cyclohexane (300 ml) was gassed with 0.17 l/hr. chlorine and 1.40 l/hr. nitrogen oxide at 15—20° and simultaneously irradiated with a S 81 Quarzlampengesellschaft mercury high-pressure lamp; details of the irradiation apparatus are given elsewhere [*3*]. After 5 hours the reaction was interrupted, nitrogen passed through for 30 minutes, the solution washed with 2 N NaOH and dried with sodium sulfate. Working up yielded 4.6 g (60% based on Cl_2 consumed) of colorless 7, m.p. 116.5—117° (green melt). Other products were, 1-chloro-1-nitrosocyclohexane (5), cyclohexyl nitrate and chlorocyclohexane.

3. Formation of diphenylfuroxan via α-chloro-α-nitrosotoluene

Irradiation of toluene with chlorine and nitrogen oxide in the ratio of 1 : 2 yields diphenylfuroxan (11) [*3*]. Benzyl chloride and benzoic acid are obtained after hydrolysis of the reaction mixture.

MÜLLER and METZGER [*3*] assume the formation of 11 to proceed in the manner indicated; on treatment with alkali the chloride 9 yields benzonitrile oxide (10) from which 11 is formed by dimerization.

Diphenylfuroxan (11) [*3*]. Toluene (300 ml) was gassed with a mixture of 0.70 l/ hour chlorine and 1.40 l/hour nitrogen oxide at 20° and simultaneously irradiated as in the preceding experiment. After 5 hrs. the reaction was interrupted and nitrogen introduced for 30 minutes. The blood-red mixture was then washed with 2 N NaOH and water until the aqueous phase was colorless. The light blue colored organic layer turned yellow in the dark after some time; it was dried and evaporated under reduced pressure. The dark brown residue solidified on standing and was pressed on clay until the crystals were faintly yellow in color. Recrystallization from cyclohexane gave pure 11, m.p. 114—115°; yield 1.35 g (16%, based on Cl_2).

4. Formation of oximes from hydrocarbons through photolysis of NOCl or NO/Cl$_2$ mixtures

The direct photooximation of saturated hydrocarbons, especially that of cyclohexane, is of great technical importance, since ε-caprolactam (12) is easily obtainable from cyclohexanone oxime (8) via BECKMANN rearrange-

References, p. 265

ment. ε-Caprolactam (**12**) is the starting material for the German synthetic fibre "Perlon". For further literature in this field the reader is referred to SOSNOVSKY [*10*].

NOH   H
 ‖   N═O
(cyclohexane) → (7-membered ring)

   8   **12**

a) Photolysis of NOCl

According to NAYLOR and ANDERSON [*11*], cyclohexanone oxime (**8**) may be obtained (as the hydrochloride) in high yield by UV photolysis of nitrosyl chloride in cyclohexane provided the process is conducted at sufficiently low temperatures ($-25°$) and the NOCl is added very slowly. A nitroso compound was assumed to be formed in the first instance which then rearranges to **8** under the influence of acid. This assumption was substantiated by DONARUMA [*5*], who was able to produce 30—50 % yields of nitrosocyclohexane dimer (**7**) by expelling the hydrogen chloride evolved with the aid of nitrogen.

b) Photolysis of NOCl in the presence of HCl or NO

Photooximation may also be accomplished [*12, 13*] by photolysis of nitrosyl chloride in the presence of either hydrogen chloride or nitrogen oxide without the necessity of cooling. Whereas photolyses are usually carried out in the neat liquids to be oximated, for solid hydrocarbons a different procedure is required; e.g., cyclododecane is oximated in carbon tetrachloride, yielding 89.5 % of cyclododecanone oxime hydrochloride [*14*].

c) Photolysis of NO, Cl$_2$ and HCl

Nitrosyl chloride used in the aforementioned experiments may conveniently be replaced by mixtures of NO and Cl$_2$. As mentioned earlier, geminal chloronitrosoalkanes are thus accessible or, if another NO:Cl$_2$ ratio is applied, dimeric nitrosoalkanes. Oximes are not easily obtainable in this way. If, however, hydrogen chloride is present in the irradiation mixtures, oximes may be isolated directly. MÜLLER et al. [*15—17*] have developed a one-step synthesis of oximes, subjecting a number of substrates to the combined action of nitrogen oxide, chlorine and hydrogen chloride whilst irradiating.

A further improvement was brought about by passing the mixture of gases over a catalyst ("Kali-Perl-Katalysator-Alt" of Kalichemie; 10 % Al$_2$O$_3$ + 90 % SiO$_2$ + 0.15 % Cr$_2$O$_3$) shortly before introduction into the irradiation cell thus practically completely converting the NO/Cl$_2$ mixture to NOCl [18—21]. Cycloalkanone oximes prepared in this way comprise cyclobutanone through cyclododecanone oximes as well as oximes derived from bridged hydrocarbons. Some examples are given below (cf. table 22), yields being calculated on the basis of chlorine consumed. Irradiation was usually done with a mercury high-pressure lamp TQ 81 either neat or 30 % in CCl_4 solution. Other light sources such as fluorescent tubes or sodium vapor lamps may also be used.

Table 22

hydrocarbons	oximes	yields	ref.
cyclobutane	cyclobutanone oxime	65%	[18]
cyclopentane	cyclopentanone oxime	83%	[18]
cyclohexane	cyclohexanone oxime (8)	93%	[18]
cycloheptane	cycloheptanone oxime	87%	[18]
cyclooctane	cyclooctanone oxime	94%	[18]
cyclononane	cyclononanone oxime	90%	[18]
cyclodecane	cyclodecanone oxime	96%	[18]
cycloundecane	cycloundecanone oxime	90%	[18]
cyclododecane	cyclododecanone oxime	90%	[18]
bicyclo[2.2.1]heptane	bicyclo[2.2.1]heptan-2-one oxime	80%	[19]
pinane	3-pinanone oxime (13) and 4-pinanone oxime (14) (1:1 mixture)	37%	[19]
bicyclo[2.2.2]octane	bicyclo[2.2.2]octan-2-one oxime	70%	[19]
cis-bicyclo[3.3.0]octane	cis-bicyclo[3.3.0]octan-2-one oxime (15) and cis-bicyclo[3.3.0]octan-3-one oxime (16) (2:1 mixture)	60%	[19]
adamantane	2-adamantanone oxime (17)	63%	[19]
1,4-dimethylcyclohexane	2,5-dimethylcyclohexanone oxime	61%	[20]
1,3,5-trimethylcyclohexane	2,4,6-trimethylcyclohexanone oxime	20%	[20]
1,4-di-tert. butylcyclohexane	2,5-di-tert. butylcyclohexanone oxime	5%	[20]
spiro[5.5]undecane	spiro[5.5]undecanone oximes (18)	43%	[20]

Cyclooctanone oxime [13]. A solution of 15 g of NOCl and 6 g of HCl in 1 liter of cyclooctane was irradiated for 3 hrs. under cooling to 15° and stirring. The reaction was carried out in a cylindrical mixing vessel, metal vapor lamps, incandescent or fluorescent tubes, or sunlight serving as light sources. This treatment produced a viscous oil which was run off in portions. The collected oil was dissolved in a little water and carefully neutralized with a NaOH solution under cooling and stirring. Crystals precipitated were washed and dried to give 27 g (84% calculated on NOCl used) of cyclooctanone oxime, m.p. 42°.

Cyclohexanone oxime (8) [*15*]. A saturated solution of dry hydrogen chloride in 300 ml cyclohexane is prepared, and a mixture of chlorine gas and nitrogen oxide (750 ml/hour and 1500 ml/hour, respectively) is led in through a filter disc at about 15°; meanwhile the reaction mixture is irradiated with a mercury immersion lamp. It is advantageous to continuously add dry HCl to the gas mixture in order to maintain permanent saturation of the solution with hydrogen chloride. The oil which forms is run off from time to time, dissolved in water and neutralized with NaOH solution when **8** is precipitated in a crystalline form. Further quantities of the oxime are obtainable by treating the cyclohexane solution with water and neutralizing the aqueous extract. A 150 mins. irradiation period gives 5.35 g of cyclohexanone oxime (**8**), m.p. 86−88° together with 2.25 g of 1-chloro-1-nitrosocyclohexane (**5**). By variation of the $NOCl/Cl_2$ ratio yields of up to 6.5 g of **8** may be obtained [*17*].

References

[*1*] E. MÜLLER and H. HUBER: Chem. Ber. **96**, 670 (1963).
[*2*] E. MÜLLER and H. HUBER: Chem. Ber. **96**, 2319 (1963).
[*3*] E. MÜLLER and H. METZGER: Chem. Ber. **87**, 1282 (1954).
[*4*] E. MÜLLER and H. METZGER: Chem. Ber. **88**, 165 (1955).
[*5*] L. G. DONARUMA: J. Org. Chem. **23**, 1338 (1958).
[*6*] A. DI GIACOMO: J. Org. Chem. **30**, 2614 (1965).
[*7*] E. J. BURRELL, JR.: J. Phys. Chem. **66**, 401 (1962).
[*8*] B. G. GOWENLOCK and W. LÜTTKE: Quart. Rev. **12**, 321 (1958).
[*9*] P. A. S. SMITH: The Chemistry of Open-Chain Organic Nitrogen Compounds, Vol. **2**, pp. 355−389. New York: Benjamin 1966.
[*10*] G. SOSNOVSKY: Free Radical Reactions in Preparative Organic Chemistry, pp. 213−281; esp. p. 234, 248. New York: Macmillan 1964.
[*11*] M. A. NAYLOR and A. W. ANDERSON: J. Org. Chem. **18**, 115 (1953).
[*12*] W. REPPE, H.-J. RIEDL and O. VON SCHICKH (Badische Anilin- & Soda-Fabrik AG): Ger. Pat. 973 677 of 6. 8. 1955; C. **1961**, 7402.
[*13*] BASF. Brit. Pat. 788 436 of 26. 7. 1956; C. A. **52**, 16252 (1958).
[*14*] O. VON SCHICKH and H. METZGER (Badische Anilin- & Soda-Fabrik AG): Ger. Pat. 1 079 036 of 30. 5. 1958; C. A. **55**, 17545 (1961).
[*15*] E. MÜLLER, D. FRIES and H. METZGER: Chem. Ber. **90**, 1188 (1957).
[*16*] E. MÜLLER, H. METZGER and D. FRIES: Ger. Pat. 1 001 983 of 9. 8. 1955; C. A. **53**, 17926 (1959).
[*17*] E. MÜLLER, H. METZGER and D. FRIES: Brit. Pat. 789 732 of 8. 8. 1956; C. A. **52**, 13788 (1958).
[*18*] E. MÜLLER, H. G. PADEKEN, M. SALAMON and G. FIEDLER: Chem. Ber. **98**, 1893 (1965).
[*19*] E. MÜLLER and G. FIEDLER: Chem. Ber. **98**, 3493 (1965).
[*20*] E. MÜLLER and M. SALAMON: Chem. Ber. **98**, 3501 (1965).
[*21*] BASF. Fr. Pat. 1 337 512 of 21. 8. 1962; C. A. **60**, 2799 (1964).

Chapter 30

Photochemical transformations of unsaturated nitro compounds

1. Photolysis reactions of unsaturated nitro compounds

Irradiation of 9-nitroanthracene (**1**) with light of wavelengths 420 to 530 mµ afforded the dimer as expected (comp. chapter 9), but exposure to 370—410 mµ radiation gave 10,10'-bianthrone (**2**) and nitrogen oxide [*1*].

A reinvestigation [*2*] revealed this photolysis reaction to be more complex in nature. Illumination of degassed solutions of **1** with a mercury lamp through Pyrex produced anthraquinone (**3**) and **2**. When nitrogen was bubbled through the solution to sweep out gaseous products, a different product ratio was observed (see below) as well as in the case of NO being introduced into the solution under irradiation.

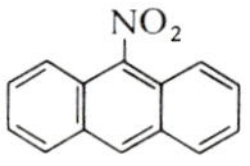

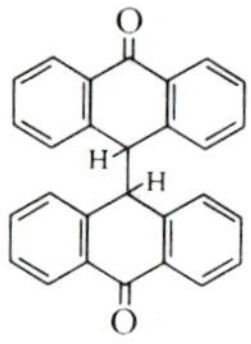

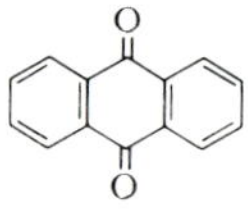

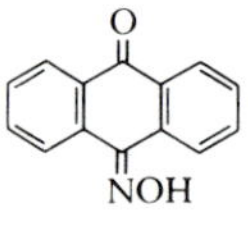

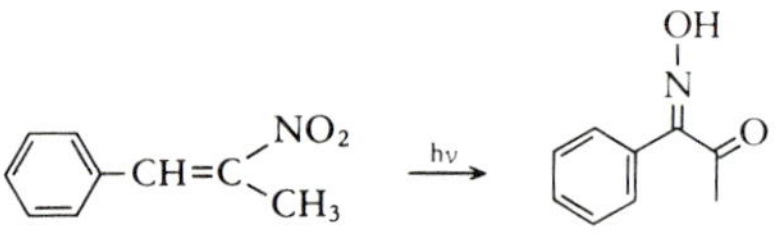

Irrad. [hrs.]	gas	yields [%]			
		2	3	4	NO
3.5	—	55	21	—	—
2	N$_2$	48	13	30	+
3.5	NO	9	77	—	—

References, p. 273

A reaction analogous to the **1 → 4** isomerization was observed [2] in the irradiation of β-methyl-β-nitrostyrene (**5**) in styrene which resulted in formation of a high yield of 1-phenyl-1,2-propanedione 1-oxime (**6**). For the interpretation of this reaction see CHAPMAN et al. [2].

2. Photochemical conversion of aromatic nitro compounds to nitroso compounds

The first photoisomerization of an aromatic nitro compound to a nitroso compound was discovered in 1901 by CIAMICIAN and SILBER [3]. These authors observed that 2-nitrobenzaldehyde (**7**) was converted to 2-nitrosobenzoic acid (**8**) when exposed to sunlight, and that this transformation took place both in solution (e.g. in benzene) and on irradiation of the crystals.

    7   **8**: R = H   **10**
              9: R = C_2H_5

No nitroso compounds were produced on analogous exposure of 3- and 4-nitrobenzaldehydes to light [3].

Since these initial studies many similar reactions have been carried out with a number of substances possessing the group **10**. A few examples are given below:

nitroterephthalaldehyde $\xrightarrow{h\nu}$ 4-formyl-2-nitroso-benzoic acid [4]

N-(2-nitrobenzylidene)-aniline (**11**) $\xrightarrow{h\nu}$ 2-nitrosobenz-anilide (**12**) [5]

(2-nitrophenyl)-diphenylmethane $\xrightarrow{h\nu}$ (2-nitrosophenyl)-diphenylmethanol [6]

1-arsenoso-2-nitro-
benzene (16)

2-nitrosobenzene-
arsonic acid (17) [7]

N-(2,4-Dinitrophenyl)-d,l-leucine (13), when irradiated in neutral or alkaline aqueous solution, undergoes degradation to 4-nitro-2-nitrosoaniline (14), 3-methylbutyraldehyde (15), and carbon dioxide [8]. This reaction seems to be a general one for N-(2,4-dinitrophenyl)-amino acids with the carboxy group adjacent to the light-sensitive moiety.

Bis-(2-nitrobenzylidene)-pentaerythritol (18), a dinitro compound, isomerizes under the influence of sunlight to give a mononitroso compound, (α-hydroxy-2-nitrosobenzylidene)-(2'-nitrobenzylidene)-pentaerythritol (19) [9]. The rate of isomerization is remarkable (15 mins. in benzene solution).

When alcoholic solutions of 2-nitrobenzaldehyde (7) are irradiated, the corresponding esters of 2-nitrosobenzoic acid are obtained with methanol and ethanol [3]. In this case, both isomerization and esterification occur, and both processes are photochemical in nature [10]. Acetal formation occurs first, and the resulting 2-nitrobenzaldehyde acetal (20) undergoes isomerization and loss of alcohol, yielding the alkyl 2-nitrosobenzoate (cf. 9). For another mechanistic interpretation see DE MAYO and REID [11].

BERSON and BROWN [12] investigated the photochemical reactions of 3,5-diacetyl-1,4-dihydro-2,6-dimethyl-4-(2-nitrophenyl)-pyridine (21) and analogous substances. In alcoholic solution, 21 was converted to two compounds by the action of light of wavelength 366 mμ: green 3,5-diacetyl-

2,6-dimethyl-4-(2-nitrosophenyl)-pyridine (**23**) and a yellow-brown substance considered to be the dimer of **23**. Compound **22** was taken to be the intermediate. The dimeric compound and **23** have identical UV-spectra in alcoholic solution.

21: $R_1 = R_2 = CH_3$
25: $R_1 = R_2 = OC_2H_5$
26: $R_1 = CH_3$, $R_2 = OC_2H_5$

22

23: $R_1 = R_2 = CH_3$

24

While the 4-nitrophenyl compound **24** was not affected by light (sun or mercury lamp), diethyl 1,4-dihydro-2,6-dimethyl-4-(2-nitrophenyl)-pyridine-3,5-dicarboxylate (**25**) and ethyl 5-acetyl-1,4-dihydro-2,6-dimethyl-4-(2-nitrophenyl)-pyridine-3-carboxylate (**26**) behaved photochemically as **21** [*12*]. No isomerization of **21** to **23** was observed in the dark. The sensitivity of **25** had previously been observed by HINKEL et al. [*13*].

Not all nitro compounds of type **10** isomerize to nitroso compounds under the action of light. To this latter group belongs 2-nitrocinnamaldehyde [*14*] (comp. p. 271). According to KRÖHNKE and VOGT [*15*] aromatic nitro compounds with ortho-carbonyl substituents do also not react in the sense indicated under the influence of light.

A general survey on the chemistry of o-nitrobenzene derivatives has recently been given by LOUDON and TENNANT [*16*].

2-Nitrosobenzoic acid (8) [*3*]. (*a*) A benzene solution of **7** is so light-sensitive that within 30 mins. exposure the entire vessel containing the solution becomes filled with a precipitate of fine white crystals. The product (generally completely pure as it stands) is recrystallized from alcohol to give **8**, m.p. 205—210° (dec.). The hot alcoholic solution has an emerald green color.

(*b*) Isomerization in the solid state. The walls of a flask are moistened with a saturated benzene solution of **7** in such a way that they become covered with a uniform layer of crystals following the complete evaporation of the solvent. If the flask is sealed and exposed to light, it can be observed that the crystals gradually lose their yellow color, finally becoming white. If cold benzene is then added, the white mass remains almost completely undissolved, since **8** is only very slightly soluble in benzene.

Ethyl 2-nitrosobenzoate (9) [*10*]. When exposed to diffuse light, the oily 2-nitrobenzaldehyde diethyl acetal (**20**) gradually becomes dark green in color in the course of 2—3 months. Finally shiny, colorless crystals separate out. In direct sunlight, **20** becomes green within 15 mins. and precipitation occurs in 1.5—2 hrs. After suction-filtration and removal from oil over clay, the crystals of **9** have m.p. 120—121°.

The colorless 2-nitrobenzaldehyde dimethyl acetal apparently reacts to direct sunlight even more rapidly than its homolog, while, after storing in darkness for 6—7 years, the acetals are not noticeably altered.

2-Nitrosobenzanilide (12) [5]. N-(2-Nitrobenzylidene)-aniline (11) (2.5 g) is dissolved in benzene (50 ml) and placed in direct sunlight in a sealed tube. After about 8 days' illumination, the pale yellow-brown precipitate is filtered off, washed with benzene and pressed on clay. Yield of **12**: 1.1 g (44%), m. p. 171°.

4-Nitro-2-nitrosoaniline (14) [8]. A mixture of 5.95 g of **13** and sodium hydrogen carbonate (20 g) in water (2 l) was illuminated for 3 days with two Osram MA mercury lamps in a flow apparatus for the construction of which the original article should be consulted. After 3 days' irradiation the crystals (2.7 g) were collected and recrystallized from 70% ethanol to afford green needles (2.4 g) of **14**, m. p. 185—186° (decomp.).

2-Nitrosobenzenearsonic acid (17) [7]. 1-Arsenoso-2-nitrobenzene (16) (3 g) is suspended in ordinary (wet) ether, and alcoholic HCl is added dropwise until solution is complete. The pale yellow liquid is poured into a bomb or similar vessel, sealed with a rubber stopper and paraffin cap, and exposed to sunlight for a few weeks. The yellow-brown precipitate is filtered off under suction and washed with ether.

3,5-Diacetyl-2,6-dimethyl-4-(2-nitrosophenyl)-pyridine (23) [12]. A solution of **21** (0.15 g) in 750 ml of 95% ethanol was illuminated with UV light (General Electric AH-4 mercury lamp) for 16 hours. Subsequently, the solvent was evaporated under reduced pressure and the residue dissolved in dilute HCl. Using sodium carbonate, the aqueous solution was made alkaline and a yellow-brown precipitate separated out. When the solution was concentrated, a mixture of buff needles and a few long aquamarine staves was obtained. A sample of the green form was isolated by hand-picking the crystals under the microscope. This material (**23**) melted at 129.5—130.5° (dec.).

3. Photoreduction of an aromatic nitro compound to an aniline derivative

A solution of 2,4-dinitrobenzenesulfenyl chloride (**27**) in glacial acetic acid (containing 0.14% of water) underwent no change when kept in the dark at 20—30° for 60 days [17]. When, however, solutions of **27** in acetic acid containing 0.14—2.5% of water were exposed to full sunlight, or to ultraviolet radiation from a quartz lamp, the color soon darkened, and a crystalline precipitate began to form after a few hours. The product obtained after exposure for 70 hrs. was shown to be 2-amino-4-nitrobenzenesulfonic acid (**30**). When the acetic acid solution of **27** was irradiated for relatively short periods, 2-acetoxyamino-4-nitrobenzenesulfonic acid (**29**) was obtained [17] in yields of up to 64%.

According to KALUZA and PEROLD [17] the overall reaction involves an intramolecular rearrangement of intermediately formed 2,4-dinitrobenzenesulfenic acid (28). This light-induced rearrangement requires water as a reactant and leads to 29 which is finally reduced photochemically to the aminosulfonic acid 30. This reduction again involves water as a reactant, as no smooth reaction is obtained in anhydrous medium.

2-Amino-4-nitrobenzenesulfonic acid (30) [*17*]. The sulfenyl chloride 27 (10 g) in acetic acid containing 2.5% water (200 ml) in a quartz flask, was exposed to full sunshine for 70 hrs. at temperatures up to 30°; the precipitate formed was filtered off daily. This material was extracted with water to give an insoluble residue (presumably bis-2,4-dinitrophenyl disulfide) (0.29 g) and a water-soluble acid (5.39 g; 58% yield). This was recrystallized from water as greyish small flat prisms decomposing towards 300°.

4. Photocyclization reactions of aromatic nitro compounds

The photochemical formation of indigotin (32) from 2′-nitrochalcone (31) was discovered by ENGLER and DORANT [*18*]. The reaction was reported to take the following course:

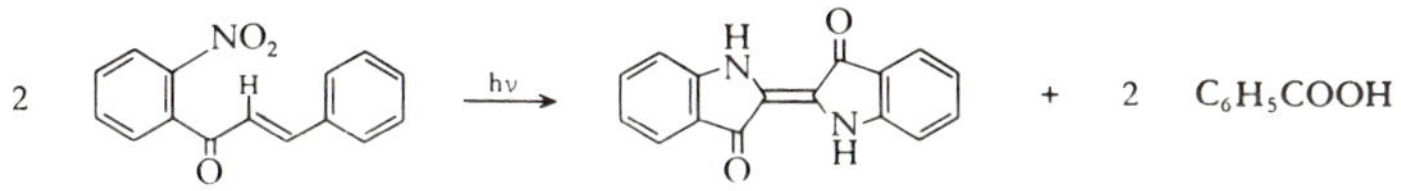

According to SACHS and HILPERT [*19*] an insolated solution of 4-hydroxy-4-(2-nitrophenyl)-2-butanone (33) in benzene becomes turbid with the separation of water. On removing the solvent a substance may be isolated which forms indigotin (32) on exposure to ammonia vapor. According to RIED and WILK [*20*] the said substance forms only after illumination in the solid state or in nonpolar solvents. In this connection, the production of indigotin photographs was mentioned [*20*].

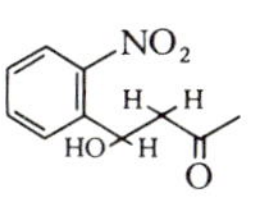

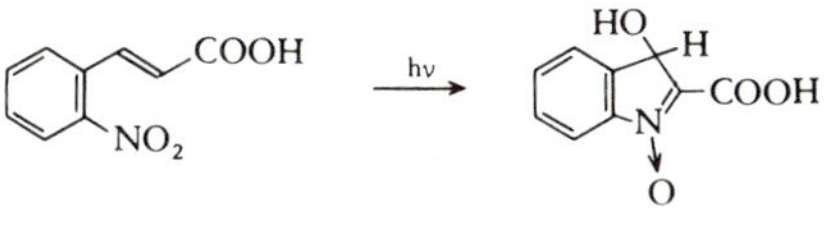

The photocyclization of 2-nitrocinnamic acid (34) was reported by TANASESCU [*21*] to yield 3-hydroxy-3H-indole-2-carboxylic acid 1-oxide (35). A 3-keto analogue of 35, viz. 2-phenylisatogen (37), was obtained on irradiation of pyridine solutions of 2-nitrotolan (36) [*22*], and 6-nitro-2-phenylisatogen (39) was obtained in similar treatment of 2,4-dinitrotolan (38) [*23*]. HUISGEN [*24*] has advanced the following scheme to account for the photochemical formation of 37 from 2-nitrotolan in pyridine. According to this scheme only the addition of pyridine to **36** requires the action of light. Attention is drawn [24] to the ease of formation of an isatogen from the pyridinium salt **40** [25].

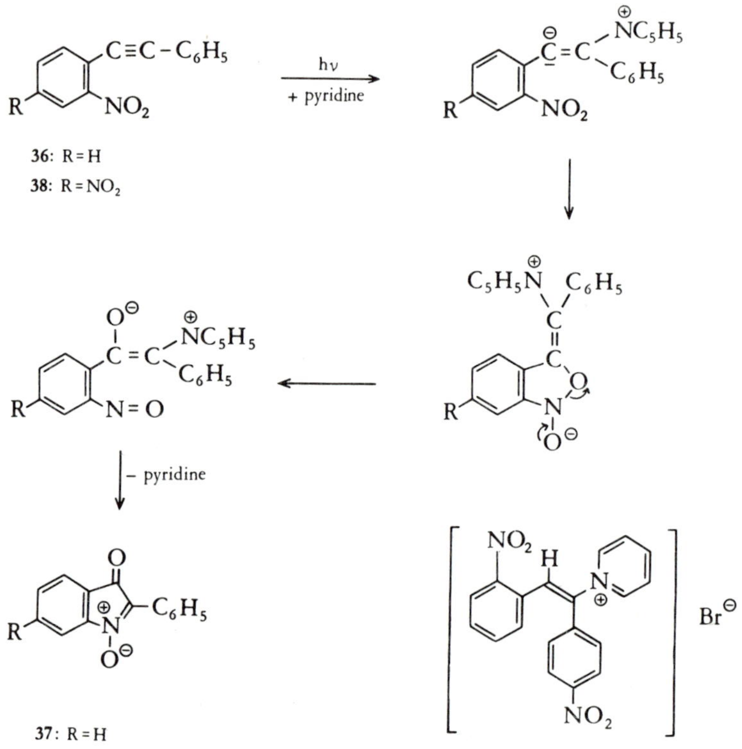

36: R = H
38: R = NO$_2$

37: R = H
39: R = NO$_2$

Indigotin (32) [18]. If a solution of **31** is allowed to evaporate in a shallow dish, a thin layer of the substance is obtained in the form of colorless needle-shaped crystals. The substance remains colorless in the dark, but if placed in direct sunlight the crystals adopt a green color in the course of one hour, whilst retaining their original form. With further illumination the color changes from green to blue-green and finally to blue-black with a characteristic coppery sheen. If the mass is then washed with alcohol and ether the indigotin remains undissolved in granular form and can be identified by its solubility in chloroform and the formation of a violet vapor.

When the same experiment is carried out using a sealed glass capsule filled with carbon dioxide, the same transformation may be observed. On opening the capsule, the smell of benzaldehyde can be perceived and benzoic acid can be detected in the reaction product.

6-Nitro-2-phenylisatogen (39) [23]. A yellow pyridine solution of **38** was exposed to sunlight whereupon the color changed to red-orange. After slow evaporation of the solvent a quantity of the ruby-red platelets of **39** remained, m.p. 205—206°; yield ca. 62%.

References, p. 273

References

[1] F. D. GREENE: Bull. Soc. Chim. France **1960**, 1356.
[2] O. L. CHAPMAN, A. A. GRISWOLD, E. HOGANSON, G. LENZ and J. REASONER: Pure Appl. Chem. **9**, 585 (1964).
[3] G. CIAMICIAN and P. SILBER: Ber. dtsch. chem. Ges. **34**, 2040 (1901).
[4] H. SUIDA: J. prakt. Chem. [2] **84**, 827 (1911).
[5] F. SACHS and R. KEMPF: Ber. dtsch. chem. Ges. **35**, 2704 (1902).
[6] I. TANASESCU: Bull. Soc. Chim. France [4] **39**, 1443 (1926).
[7] P. KARRER: Ber. dtsch. chem. Ges. **47**, 1783 (1914).
[8] D. W. RUSSELL: J. Chem. Soc. **1963**, 894.
[9] I. TANASESCU: Bul. Soc. Stiinte Cluj. **2**, 111 (1924/1925); C. A. **19**, 2932 (1915).
[10] E. BAMBERGER and F. ELGER: Liebigs Ann. Chem. **371**, 319 (1910).
[11] P. DE MAYO and S. T. REID: Quart. Rev. **15**, 393 (1961); esp. p. 415.
[12] J. A. BERSON and E. BROWN: J. Amer. chem. Soc. **77**, 447 (1955).
[13] L. E. HINKEL, E. E. AYLING and W. H. MORGAN: J. Chem. Soc. **1931**, 1835.
[14] G. CIAMICIAN and P. SILBER: Atti R. Accad. Lincei Roma [5] **11**, I, 277 (1902); C. **1902**, I, 1190.
[15] F. KRÖHNKE and I. VOGT: Chem. Ber. **85**, 376 (1952).
[16] J. D. LOUDON and G. TENNANT: Quart. Rev. **18**, 389 (1964).
[17] F. KALUZA and G. W. PEROLD: J. S. African Chem. Inst. **13**, 89 (1960); C. A. **55**, 11346 (1961).
[18] C. ENGLER and K. DORANT: Ber. dtsch. chem. Ges. **28**, 2497 (1895).
[19] F. SACHS and S. HILPERT: Ber. dtsch. chem. Ges. **37**, 3425 (1904).
[20] W. RIED and M. WILK: Liebigs Ann. Chem. **590**, 111 (1954).
[21] I. TANASESCU: Bull. Soc. Chim. France [4] **41**, 1074 (1927).
[22] P. PFEIFFER: Liebigs Ann. Chem. **411**, 72 (1916).
[23] P. PFEIFFER and E. KRAMER: Ber. dtsch. chem. Ges. **46**, 3655 (1913).
[24] R. HUISGEN: Angew. Chem. **75**, 604 (1963); internat. ed. **2**, 589 (1963).
[25] F. KRÖHNKE and M. MEYER-DELIUS: Chem. Ber. **84**, 932 (1951).

Chapter 31

Light-induced reactions of diazoalkanes, diazirines and related compounds

The photochemical decomposition of diazoalkanes is an important route to the formation of a reactive species R—C—R which has in the literature been referred to as carbene (**1a**) and as methylene (**1b**) (in the case of R = H).

$$\begin{array}{c} R \\ {\diagdown} \\ C=N_2 \\ {\diagup} \\ R \end{array} \xrightarrow[-N_2]{h\nu} \begin{array}{c} R \\ {\diagdown} \\ \ddot{C}\mathstrut \\ {\diagup} \\ R \end{array} \qquad\qquad R\text{-}\overset{\cdot}{\underset{\cdot}{C}}\text{-}R$$

$$\qquad\qquad\qquad\qquad\textbf{1a}\!:\!\mathrm{R}\!=\!\mathrm{H} \qquad\qquad\qquad \textbf{1b}\!:\!\mathrm{R}\!=\!\mathrm{H}$$

Since neither expression has found unequivocal acceptance no distinction will be made in this book. Hence in the following **1a** and **1b** will be referred to as carbene (**1**) for the sake of simplicity. No mechanistic implications should be drawn from this decision, however; a wealth of pertinent literature on the nature of the reactive species is at hand (cf. among others [1—5]). It has been questioned [6] whether reactions of R—C—R might not be simulated by so-called "pre-carbenes" since for reasons of energetics the existence of isolated carbene species seems improbable.

A variety of reaction modes of carbenes may be observed [1, 2, 7] depending upon the structure of the R in R—C—R. Though sporadic isomerization reactions are known, addition reactions seem to be preferred. Ring formation results from addition to unsaturated systems while with saturated addends insertion reactions are predominate. Examples for insertion into C—C, C—O, C-halogen, C—H, O—H, and N—H bonds are known and will be discussed in due course. Only the gross reaction will be formally classified in this context though closer inspection of reaction mechanisms will perhaps show the process to follow a different course. Occasionally different reaction modes of carbenes may be observed to proceed simultaneously.

References, pp. 292—293

1. Photoaddition reactions of carbenes to unsaturated systems resulting in ring formation

a) Formation of cyclopropane compounds

The light-induced formation of cyclopropanes from olefins and diazoalkanes — a reaction widely investigated since SKELL and DOERING opened the field — is believed to proceed via carbenes. The addition was first described to be stereospecific, cis-2-butene (2) thus on irradiation with diazomethane giving rise to cis-1,2-dimethylcyclopropane, and trans-2-butene (3) yielding trans-1,2-dimethylcyclopropane [8, 9].

More recent investigations using phenylcarbene (4) [10, 11] made clear that these results must not be generalized. Thus, phenylcarbene adds to 2 or 3 in a manner not completely stereospecific, perhaps indicating that 4 is intermediate between 1 (stereospecific addition) and diphenylcarbene (5) (non-stereospecific addition [12]) as concerns control of stereochemistry in the cyclopropane compounds resulting from addition to the 2-butenes.

Much work has been devoted to elucidating the degree of stereospecificity involved in carbene addition to unsaturated systems. As far as carbene 1 itself is concerned the pertinent literature has been summarized by DeMore and Benson [3], Frey [4] and Bell [5].

Irradiation of methyl diazoacetate results in formation of carbomethoxycarbene (6). This adds to cis-2-butene to yield a 5 : 2 mixture of methyl cis-2,3-dimethylcyclopropane-trans-carboxylate (7) and methyl cis-2,3-dimethylcyclopropane-cis-carboxylate (8), while with trans-2-butene methyl trans-2,3-dimethylcyclopropanecarboxylate (9) was formed exclusively [13].

The addition of **1** or of **4**, respectively, to cyclohexene has been reported to give rise to, inter alia, bicyclo[4.1.0]heptane (norcarane, **10**) and 7-phenylbicyclo[4.1.0]heptane (**11**), respectively [*14, 15*]. Sensitization of diazomethane photolysis by benzophenone [*16*] resulted in shift of product ratios in the sense indicated in table 23 beneath:

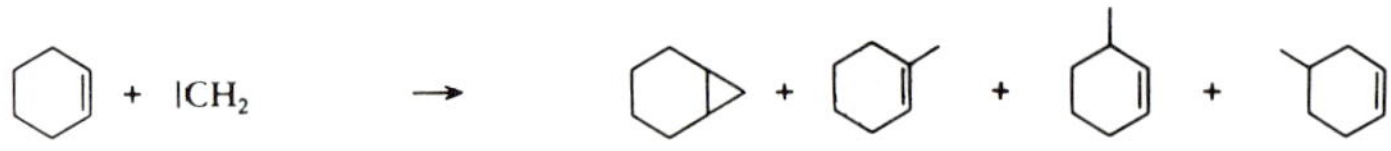

◯ + |CH₂ → ▷◯ + ◯ + ◯ + ◯

10

Table 23 [*16*]

method of CH_2N_2 decomposition	10	1-methyl-	3-methyl-cyclohexenes	4-methyl-
photolysis	1	0.24		1.3
sensitized photolysis	1	trace		0.42
thermal decomposition	1	0.43		1.7
thermal dec. with Cu-powder	1	0		0

An interesting example of spirocyclopropane ring formation was reported by MORICONI and MURRAY [*17*]. When 3-diazo-1-methyloxindole (**12**) was irradiated in very dilute cyclohexene solution two colorless geometric isomers, tentatively assigned 1′-methylspiro[bicyclo[4.1.0]heptane-7,3′-indolin]-2′-one structures **13** and **14** were obtained in a total yield of 27%. For further ring formation reactions of **12** comp. chapter 32.

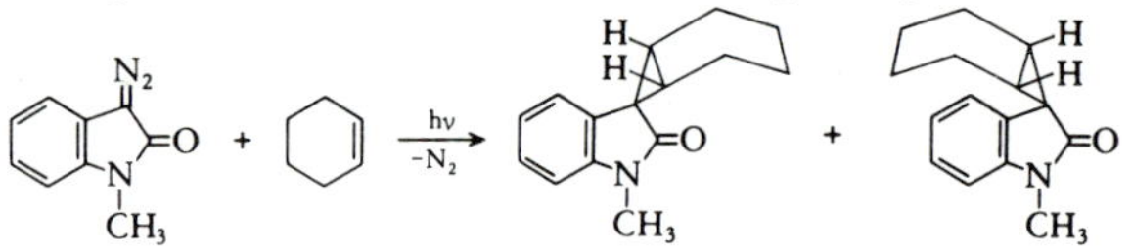

12   **13**   **14**

Aromatic compounds also react with carbenes under ring formation. Usually copper catalysis is preferred in effecting decomposition of the diazo component. The photochemical reaction of diazomethane with benzene, leading to the formation of a hydrocarbon, was first carried out by DOERING [*18*]. Formulae then discussed were cycloheptatriene (**15**) and bicyclo-[4.1.0]hepta-2,4-diene (norcaradiene, **16**).

15    **16** 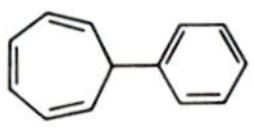   **17**

Besides, reasonable quantities of toluene were formed in the above reaction [*19*]. With phenylcarbene (**4**) phenylcycloheptatriene (**17**) is formed [*10, 15*]. The same compound is accessible also from biphenyl with diazomethane [*19*].

References, pp. 292—293

MEERWEIN [20] postulated that a mixture of cycloheptatriene (15) and norcaradiene (16) be produced in the photochemical reaction of diazomethane with benzene, and that 16 is converted to 15 on heating. However, the most widely accepted view would seem to be that all attempts to date to demonstrate the coexistence of the valence tautomers 15 and 16 have failed (comp. RHOADS [21]).

Even more complex is the case with the addition of carboalkoxycarbenes to benzene leading to esters of the so-called BUCHNER acids. Their

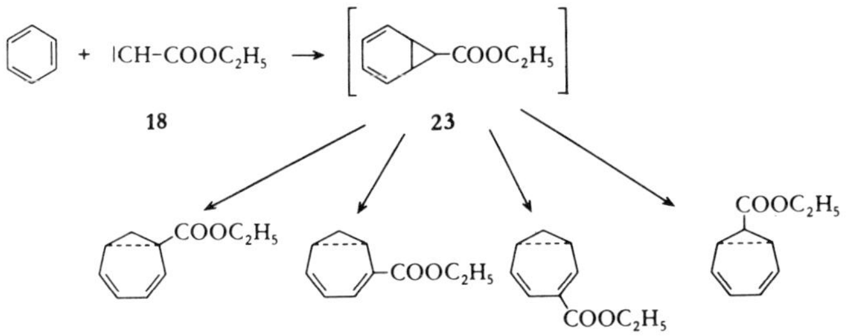

structures were imagined [22] to be as 19—22. Compounds 19, 20, 21 and 22 are usually (cf. [7]) taken as ethyl esters of cycloheptatriene-1-, 2-, 3-, or 7-carboxylic acids.

The existence of a discrete norcaradiene derivative (viz. ethyl bicyclo-[4.1.0]hepta-2,4-diene-7-carboxylate, 23) was questioned by DOERING [22]. In this conclusion it has obviously been overlooked that the primary addition product of carboethoxycarbene (18) to benzene [23] is easily photooxidized to an epidioxide in the presence of a sensitizer, which establishes the presence of a 1,3-diene system in the molecule (comp. chapter 39). Hence structure 23 is by no means invalidated by arguments pertaining to the valence tautomerism problem 15/16.

The photolysis of diazomethane with indan (24) has been investigated by ALDER [24]. This photoreaction was observed to yield two hydrocarbons, both of which produced azulene (25) on dehydrogenation with Pd/C. A similar procedure for the synthesis of 25 was reported by DOERING [25].

24: R=H
26: R=CH₃

25: R=H
27: R=CH₃

The analogous reaction utilizing 4,7-dimethylindan (26) as starting material afforded 4,8-dimethylazulene (27) [24].

GUTSCHE and JOHNSON [26] found that photolysis of 2-(diazomethyl)-bibenzyl (28) takes place with the liberation of nitrogen. Two hydrocarbons, 2-phenylindan (29) and 5,6-dihydro-6aH-cyclohepta[a]naphthalene (30) are isolated, arising from the intermediate carbene through insertion into a C—H bond or through addition to the phenyl ring as already described above for simpler systems.

The length of the chain separating the phenyl moiety from the reactive carbene seems to be critical. If the diazo derivative with a propane chain connecting the aromatic rings is photolyzed the ring enlargement reaction is almost completely suppressed [10].

Several unsaturated heterocyclic compounds have been subjected to photoaddition reactions with carboalkoxycarbenes [27]. The cyclopropane derivatives methyl 2-oxabicyclo[3.1.0]hexane-6-carboxylate (31), ethyl 2-oxabicyclo[3.1.0]hex-3-ene-6-carboxylate (32) and its sulfur analogue 33 were thus synthesized from 2,3-dihydrofuran, furan and thiophene, respectively.

Methyl cis-2,3-dimethylcyclopropane-trans- and -cis-carboxylate (7 and 8) [13]. After being irradiated for 40 hrs. with three 275 W General Electric sunlamps, a solution of 8 g methyl diazoacetate in 40 ml cis-2-butene (2) had become colorless. Distillation afforded 4.0 g (39% of theory) material, b.p.$_{50}$ 50—60°. Gas chromatography showed the product to consist of a mixture mainly of two substances, the ratio of which was about 5:2.

Methyl trans-2,3-dimethylcyclopropanecarboxylate (9) [13]. In the same manner as above, 8 g methyl diazoacetate was irradiated in 50 ml trans-2-butene. Distillation afforded 3.5 g (34%) material of b.p.$_{20}$ 40—50°. Pure 9 was obtained on gas chromatography.

Cycloheptatriene (15) [20]. Diazomethane (132 g) in benzene (7.5 liters) is irradiated with an Osram HQA 500 immersion lamp in three separate batches (10—14 hrs. per batch). After the addition of 1% hydroquinone, the greater part of the benzene is distilled off under reduced pressure (75 mm) and below 25° (bath) using a fractionating column. The residue (300 ml) is shaken for 12 hrs. with finely powdered mercuric chloride (30 g) in order to destroy nitrogen-containing compounds produced during irradiation. After filtration and washing with water, the solution is dried over calcium chloride and repeatedly fractionated at 35 mm, using a 25 cm WIDMER column to afford 210 g (72.5%) of a mixture of cycloheptatriene (15) and norcaradiene (16), b.p.$_{35}$ 29—31°.

References, pp. 292—293

100 g of the hydrocarbon mixture is heated at 210° in a sealed tube for 24 hrs. and finally fractionated twice using a 25 cm WIDMER column. Yield 93 g of **15**, b.p. 114.5—116°.

Ethyl 2,4-norcaradiene-7-carboxylate (23) [*23*]. A solution of 20 g ethyl diazoacetate in 200 ml benzene was irradiated with a mercury high pressure lamp Osram HQA 500 at 60°. After evolution of nitrogen had ceased (26 hrs.) the unreacted material was removed in vacuo at 30°. Distillation of the yellow-red residue (18.65 g) provided pure **23**, b.p.$_{0.1}$ 50—52°; yield 8.63 g = 30%.

5,6-Dihydro-6aH-cyclohepta[a]naphthalene (30) [*26*]. A solution of **28**, obtained by the oxidation of 51 g of 2-(2-phenylethyl)-benzaldehyde hydrazone with 71 g red mercuric oxide in 500 ml petroleum ether was used in the preparation.

The filtered red solution was diluted with further petroleum ether to 8 liters and the reaction mixture refluxed and illuminated by a General Electric RS sunlamp. After 1—3 days the liberation of nitrogen (60%) had ceased and the solution had turned pale yellow. After evaporation of the solvent, the residue was distilled from a Claisen flask to give 20 g. (45%) of a yellow liquid. Careful fractionation of this material yielded 13.1 g. of 2-phenylindan (**29**), b.p.$_{0.6}$ 94—96° and 4 g. of **30**, b.p.$_{0.65}$ 101—102°.

Ethyl 2-thiabicyclo[3.1.0]hex-3-ene-6-carboxylate (33) [*27*]. A solution of ethyl diazoacetate (20 g) in thiophene (100 ml) was irradiated under nitrogen at 200° in vessels with a water-cooled Solidex immersion lamp fitting, using a mercury high-pressure lamp Philips HPK 125 W. 2.75 liters of nitrogen was developed after 50 hrs. The dark-brown oil remaining after removal of excess thiophene was distilled to give 4.8 g (23% based on nitrogen evolved) of **33**, m.p. 36.5° (from aqueous ethanol).

b) Formation of oxide rings

According to MEERWEIN [*20*] ketones do not show enhanced reactivity towards diazomethane in light as compared to the dark reaction. Acetone thus on irradiation with diazomethane furnishes 1,2-epoxy-2-methylpropane (**34**), 2-butanone, 2-pentanone and 3-pentanone in essentially the same manner as in the dark reaction.

$$\underset{34}{\overset{O}{\underset{}{\lambda}}} + \mathrm{ICH_2} \rightarrow \overset{O}{\lambda} + \overset{O}{\lambda}\diagup + \overset{O}{\lambda}\diagdown + \overset{O}{\underset{}{\smile}}$$

The carbonyl group in carboxylic acid esters is less reactive with diazomethane than the carbonyl groups of aldehydes or ketones. This is demonstrated by the fact that esters react with diazomethane only under the influence of light. The conversions have a very complex mechanism as is exemplified below for the action of diazomethane on methyl formate. Different reaction modes are operative, leading either to oxide rings or to insertion reactions with C—H and C—O bonds. The following products were obtained [*20*]: ethyl formate, 1,2-epoxymethoxyethane and methoxyacetaldehyde, which could not, however, be isolated since it immediately underwent further reaction either with a second molecule of diazomethane to give methoxyacetone or with starting material to give 2,4-dimethoxy-1,3-dioxolane.

The large number of products formed is a disadvantage for preparative work, especially since not all of the compounds could be isolated in a pure state. Reference should be made to the original publication for experimental details.

c) Formation of aziridine rings

SHEEHAN and LENGYEL [28] found that phenyl isocyanate and diphenyldiazomethane react under the influence of ultraviolet light to yield 2,2-diphenylpseudoindoxyl (36). In addition, each reactant is converted individually into a characteristic irradiation product; that is, diphenyldiazomethane produces benzophenone azine (37) and phenyl isocyanate gives the cyclic dimer 1,3-diphenyluretidinedione. According to SHEEHAN and LENGYEL 1,3-diphenyloxindole (38) was also present in minor quantities.

The formation of 36 can be explained by assuming that the carbene (5) from diphenyldiazomethane adds to phenyl isocyanate to produce 1,3,3-triphenylaziridinone (35), which then collapses to the pseudoindoxyl 36 [28].

2,2-Diphenylpseudoindoxyl (36) [28]. An occasionally cooled solution of diphenyldiazomethane (1.2 g) in phenyl isocyanate (3.5 g), contained in a quartz tube, was irradiated with a 140 W Hanovia Utility Model high-pressure quartz mercury lamp from a distance of 12—15 cm. After 6 hrs., 115 ml of nitrogen had been evolved and the deep violet color of the diazo component had disappeared. The excess isocyanate was evaporated at room temperature and the residue was chromatographed on silica. Elution with n-pentane/benzene yielded benzophenone azine (37) (356 mg, m.p. 162—163°). Benzene eluted 1,3-diphenyluretidinedione (53 mg, m.p. 174—175°). Finally with benzene/ether (97:3) 36 was eluted (312 mg), which after recrystallization from ether/petroleum ether had m.p. 212—213°.

References, pp. 292—293

d) Formation of cyclopropene compounds

The addition of carbene (**1**) and carbomethoxycarbene (**6**) (generated photolytically from diazomethane and methyl diazoacetate respectively) to 2-butyne gives 1,2-dimethylcyclopropene (**39**) and methyl 1,2-dimethyl-cyclopropenecarboxylate (**40**) respectively [*13*].

Closs and co-workers [*29*] obtained 1,3,3-trimethylcyclopropene (**42**) in a smoothly proceeding photolysis of 1-diazo-2,3-dimethyl-2-butene (**41**). **42** had been synthesized previously [*30*] by the photolysis of 3,3,5-trimethyl-3H-pyrazole (**43**) (comp. chapter 35).

Methyl 1,2-dimethylcyclopropenecarboxylate (40) [*13*]. A mixture of 23 g 2-butyne and 7 g methyl diazoacetate was dissolved in 25 g isobutane (used as solvent to prevent the 2-butyne from crystallizing on the cold finger). This solution was placed in a flask fitted with a condenser of the cold finger type filled with Dry Ice. At the end of 30 hrs. irradiation with three 275 Watt General Electric sunlamps, excess diazoacetate was decomposed. Distillation afforded 3.5 g of material b.p.$_2$ 45—60° of which 80% (32% theoretical yield) consisted of pure **40**.

1,3,3-Trimethylcyclopropene (42) [*29*]. A solution of **41** (25 mmole) in n-heptane (50 ml) was irradiated in a Pyrex vessel at —20° with a 500 W Hanovia high pressure arc. After approximately 3 hrs. the solution was colorless. The reaction mixture was distilled and all volatile components boiling up to 95° were collected. Analysis of this distillate (4.6 g) by gas chromatography and infrared spectroscopy showed, besides n-heptane, 1,3,3-trimethylcyclopropene (**42**) (18 mmole = 70%, b.p. 42.8—43.3°) and 2,3-dimethyl-1,3-butadiene (1%).

2. Photoaddition reactions of carbenes to saturated compounds resulting in insertion into sigma bonds

a) Insertion into C—C, C—O and C-halogen bonds

Müller [*31*] has repeated Schlenk's experiment [*32*], in which the formation of 1,1,1,3,3,3-hexaphenylpropane (**44**) was observed in the reaction of diazomethane with hexaphenylethane (triphenylmethyl). When the experiment is carried out under irradiation with a quartz lamp, the same yield of **44** is obtained within 1 hr. as within 24 hrs. in the dark reaction. The experiment is included here since it formally represents an insertion reaction into a C—C bond. No other examples of this type of reaction seem to be known.

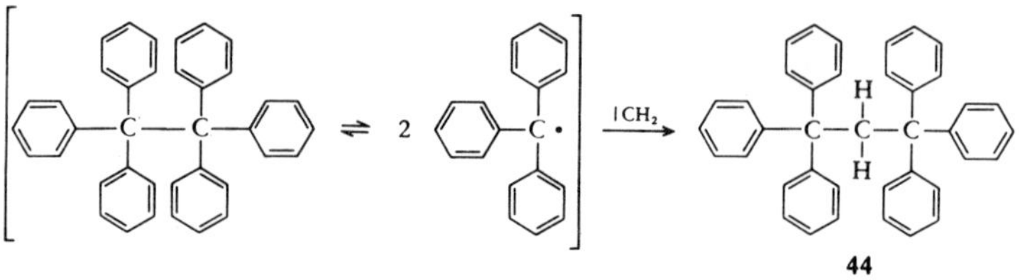

Photochemical interposition of carbenes into C—O bonds, however, is observed more frequently (comp. p. 279 for products arising in this way).

URRY and EISZNER [33] deserve credit for the discovery of a novel photoreaction involving diazomethane, in which carbene (1) reacts with carbon tetrachloride, chloroform, methyl trichloroacetate and similar compounds under over-all insertion into the carbon-halogen bonds:

$$-\overset{|}{\underset{|}{C}}-Hal + |CH_2 \longrightarrow -\overset{|}{\underset{|}{C}}-CH_2-Hal$$

Conversions of this type are illustrated below:

$$CCl_4 + 4\ CH_2N_2 \longrightarrow C(CH_2Cl)_4\ \textbf{(45)} \qquad (60\%)$$
$$BrCCl_3 + 4\ CH_2N_2 \longrightarrow BrCH_2C(CH_2Cl)_3\ \textbf{(49)} \qquad (38\%)$$
$$HCCl_3 + 4\ CH_2N_2 \longrightarrow H_3CC(CH_2Cl)_3 \qquad (45\%)$$
$$Cl_3CCOOCH_3 + 3\ CH_2N_2 \longrightarrow (ClCH_2)_3CCOOCH_3 \qquad (60\%)$$
$$H_3CCHBrCOOCH_3 + CH_2N_2 \longrightarrow H_3CCH(CH_2Br)COOCH_3\ (20\%)$$

When the concentration of diazomethane was lower, the product distribution changed significantly. Thus, bromotrichloromethane with a rather high diazomethane concentration gave a 38% yield of 2-(bromomethyl)-2-(chloromethyl)-1,3-dichloropropane (49). Medium CH_2N_2 concentration and irradiation at room temperature gave rise also to products 47 and 48, while at higher temperatures and with low concentrations of diazo compound only 46—48 were isolated [34].

	Cl BrĊCl Ċl **46**	Cl BrĊCH$_2$Cl Ċl **47**	Cl BrĊCH$_2$Cl ĊH$_2$Cl **48**	CH$_2$Cl BrĊCH$_2$Cl ĊH$_2$Cl **49**	CH$_2$Cl BrCH$_2$ĊCH$_2$Cl ĊH$_2$Cl
5—15°:	1%	16%	15%	14%	
40—70°:	13%	26%	18%	0%	

The photochemical reaction of methyl diazoacetate with chloroform or bromotrichloromethane produced compounds considered to be methyl 2,3,3-trichloropropionate (50) and methyl 3-bromo-2,3,3-trichloropropionate respectively [34,35].

$$\text{HCCl}_2\text{Cl} + |\text{CH}-\text{COOCH}_3 \rightarrow \text{Cl}_2\text{HC}-\text{CHCl}-\text{COOCH}_3 + \text{ClH}_2\text{C}-\text{COOCH}_3$$
$$(50)$$
$$\text{BrCCl}_2\text{Cl} + |\text{CH}-\text{COOCH}_3 \rightarrow \text{BrCl}_2\text{C}-\text{CHCl}-\text{COOCH}_3 + \text{Br}_2\text{HC}-\text{COOCH}_3 + \text{C}_2\text{Cl}_6$$

These photoreactions are of great theoretical interest; different reaction mechanisms involving free-radical chains and 1,2-shifts of chlorine [34, 36] (cf. GOULD [37] for a critical approach) have been discussed. Recent work [38] has shown that, at least in gas phase photolysis, chlorine abstraction by carbene is the predominating reaction.

2,2-Bis-(chloromethyl)-1,3-dichloropropane (45) [33]. Diazomethane (9.3 g) was swept in a stream of nitrogen (total 6 liters) over a period of 2 hrs. into carbon tetrachloride (185 g) illuminated in the reaction vessel equipped with the internal mercury discharge lamp. The yellow color of the diazomethane disappeared and nitrogen evolution ceased after an additional hour. The total nitrogen obtained from the diazomethane decomposition was 7.55 liters.

The reaction mixture was filtered to remove precipitated polymethylene (25 mg) and was distilled through a 12-plate fractionating column packed with single-turn glass helices. After unreacted carbon tetrachloride has distilled, a residue (7.2 g) remained, which solidified upon cooling. The solid was sublimed (10 mm) and pure **45**, m.p. 96.3—97° was obtained in a yield of 3.89 g = 60% based on crude product.

Methyl 2,3,3-trichloropropionate (50) [35]. A solution of methyl diazoacetate (12 g) in chloroform (248 g) was irradiated by a quartz mercury discharge tube immersed in it. Nitrogen (2.64 liters) was evolved over a period of 8 hours. Products identified after distillation were methyl chloroacetate (5.55 g) and methyl 2,3,3-trichloropropionate (**50**), b.p.$_{18}$ 85—90°; yield 2.35 g.

b) Insertion into C—H bonds

As early as 1942 chain elongation reactions with alkanes using carbenes were known [20, 39]. MEERWEIN [39] showed that diazomethane reacts with diethyl ether in the light, yielding ethyl propyl ether and ethyl isopropyl ether. Analogously tetrahydrofuran was converted to 2- and 3-methyltetrahydrofuran [39, 40]. With aliphatic and alicyclic hydrocarbons the methylene group was found to be randomly interposed between carbon and hydrogen [14, 41].

These reactions are currently of little preparative significance. Their mechanisms have been investigated by, inter alia, DOERING [14] and RICHARDSON [41].

Contrasting with the easy alkylation with carbene (**1**) diphenylcarbene (**5**) turned out to be rather unreactive in this respect as was found by KIRMSE et al. [42]. Irradiation of diphenyldiazomethane in cyclohexane gave only benzophenone azine (**37**) and 1,1,2,2-tetraphenylethane (**51**).

Exceptions are provided by fluorene, which produces 9-diphenylmethyl-fluorene (52) on irradiation with diphenyldiazomethane. Analogously 9-cyclohexylfluorene (53) and 5-cyclohexylcyclopentadiene (54) may be prepared from 9-diazofluorene and 5-diazocyclopentadiene, respectively, on irradiation in cyclohexane [*42*]. The alkylating power increases along the series from diphenyldiazomethane through 9-diazofluorene to 5-diazocyclopentadiene.

Phenylcarbene (4) appears to be intermediate between 1 and 5 in its reactivity towards hydrocarbons. When phenyldiazomethane reacted with n-pentane under irradiation [*10*] the ratio of the combined amount of (2-methylpentyl)-benzene and (2-ethylbutyl)-benzene to 1-phenylhexane was about 6, demonstrating that phenylcarbene is quite selective with respect to bond type, whereas carbene itself is not. Photolysis of phenyldiazomethane in cyclohexane produced cyclohexylphenylmethane. The intramolecular C—H insertion reaction with 2-(diazomethyl)-bibenzyl (28) which leads to 2-phenylindan (29) [*26*] has already been mentioned (comp. p. 278).

Sulfonylhydrazone salts may serve as suitable sources for the generation of carbenes. As compared to the thermal decomposition of such carbene progenitors the photochemical procedure is to be preferred since it allows one to work at room temperature where reaction sequences are more clearcut. Possibly photolysis produces diazo compounds first which then break down to carbenes. No reactions take place at room temperature in the dark.

The method was first introduced by DAUBEN and WILLEY [*43*] when camphor p-toluenesulfonylhydrazone was exposed to a quartz 500 W Hanovia high pressure mercury lamp, illumination leading to the formation of, inter alia, 55 and 56. The solvent composition was found to be critical as may be seen from table 24 [*43*].

References, pp. 292—293

Table 24

solvent	light source	hydrocarbon fraction	tricyclene (55)	camphene (56)
aqu. 0.1 N KOH	no filter	8%	18	82
meth. 0.1 N KOH	no filter	36%	36	64
meth. 0.1 N KOH	pyrex filter	44%	20	80
diglyme 0.1 N NaOCH$_3$	no filter	93%	99	1

The method has already been widely applied for the generation of carbenes, e.g. 4 [15]. Major differences between photolysis and pyrolysis of nortricyclanone p-toluenesulfonylhydrazone sodium salt (57) have been elaborated [44] which may shed some light on the kind of mechanism involved in each case.

Insertion of a carbene into aromatic C—H bonds has also been observed, though usually ring formation with subsequent valence isomerization is prevailing [10]. Proximity effects may play a significant role as becomes evident from the facile conversion of 2'-(diazomethyl)-biphenyl-2d (58) to fluorene-d$_1$ (59) on irradiation [45].

Both 3-benzoyl-4-diazo-5-phenylpyrazole (60) and 3-benzoyl-5-diazo-4-phenylpyrazole (61) yield 3-benzoyl-4,5-diphenylpyrazole (62) when photolyzed in benzene (yields being 100% and 78% respectively) [46].

No mechanism is discussed for these remarkable reactions by FARNUM and YATES. It seems possible — in view of the light-induced reactions between diazomethane and benzene (see above) — that **62** forms from **61** via **63**. A similar intermediate may be formulated for the formation of **62** from **60**.

5-Cyclohexylcyclopentadiene (54) [*42*]. 5-Diazocyclopentadiene (2.2 g) in 180 ml cyclohexane were irradiated with a mercury lamp S 81 (Quarzlampengesellschaft) at 15° under nitrogen. After 3.5 hrs. 73% nitrogen had been evolved and the solution had adopted a brown color. Brown flakes were removed by filtration and the solvent evaporated. Distillation of the residue in vacuo yielded 2.0 g (57%) of **54**, b.p.$_{17}$ 86.5—87°.

3-Benzoyl-4,5-diphenylpyrazole (62) [*46*]. A sample of **60** (270 mg) was dissolved in dry benzene and the resultant yellow solution (75 ml) was irradiated in a Pyrex flask stoppered with a soda-lime drying tube for 12 hrs. with a G. E. 275 W sunlamp. The pale yellow solution was then concentrated to a small volume on the hot plate and diluted with hexane to the cloud point. Crystallization was induced by scratching, and the mixture was cooled to 0°. Colorless crystalline **62** (320 mg, 100%) deposited, m.p. 177—177.5° (from 95% ethanol).

c) Insertion into O—H bonds

Photolysis of diazomethane with hydroxylic compounds [*39*] does not at present play a significant role in preparative organic chemistry. With substituted carbenes, however, addition to alcohols resulting in etherification may turn out to be of importance.

On illumination of diphenyldiazomethane in methanol, benzhydryl methyl ether (**64**) is obtained in a reaction which is formulated as a carbene reaction by KIRMSE [*42*]. Since carbenes have a strongly electrophilic nature because of their electron deficiency, the observed O-alkylation is imagined as an addition to the lone electron pair of the oxygen atom, with subsequent or simultaneous migration of the proton. The reaction is, however, formally classified here as an insertion reaction, whatever the detailed reaction mechanism may be.

3-Diazo-1-methyloxindole (**12**) on photochemical decomposition in very dilute ethanol solution gave a 21 % yield of 3-ethoxy-1-methyloxindole (**65**) [*17*] (cf. also chapter 32).

3-Benzoyl-4-diazo-5-phenylpyrazole (**60**) on illumination in acetic acid was converted to 4-acetoxy-3-benzoyl-5-phenylpyrazole (**66**) in 60 % yield [*46*].

A surprising difference in behavior towards water was encountered for **60** as compared to **61** [*46*]. The diazopyrazole **60** was found remarkably stable to acids in the dark: after treatment with boiling 50 % sulfuric acid for 30 mins. it was recovered to the extent of 90 %. Hydrolysis of **60** occurred readily in light, irradiation in aqueous acetone with a sunlamp converting it to 3-benzoyl-4-hydroxy-5-phenylpyrazole (**67**).

In contrast, the isomer **61** was not only completely destroyed after analogous treatment with hot acid in the dark but photolysis under comparable conditions gave the reduction product, 3-benzoyl-4-phenylpyrazole (**68**) in 88 % yield instead of the solvolysis product expected.

Benzhydryl methyl ether (64) [*42*]. Diphenyldiazomethane (3.9 g) in 100 ml methanol developed 440 ml. nitrogen within a 3-hr. period of irradiation with a quartz immersion lamp S 81 (Quarzlampengesellschaft). Following almost complete evaporation of the methanol, crystals separated out overnight. These were washed with a little methanol and proved to be benzophenone azine (**37**), m. p. 161°. The filtrate was distilled under vacuum, with the bulk (2.8 g) passing over at 150—154°/20 mm. Renewed distillation gave pure **64**, m. p. 18—19°.

3-Benzoyl-4-hydroxy-5-phenylpyrazole (67) [*46*]. The diazopyrazole **60** (270 mg) was dissolved in aqueous 75% acetone, and the solution was irradiated overnight in a Pyrex flask with a G. E. 275 W sunlamp. The pale yellow solution was poured into water and the pale yellow solid which separated was collected, dried and dissolved in hot chloroform/methanol. The solution was boiled until crystallization began, chilled to 0°, and the pale yellow needles were collected. The product thus obtained (170 mg = 65%) had m. p. 209—210°.

d) Insertion into N—H bonds

Primary and secondary amines can be photochemically alkylated using diphenylcarbene [*42*]. Diethylamine or isopropylamine when irradiated with diphenyldiazomethane produce N-(diphenylmethyl)-diethylamine (**69**) and N-(diphenylmethyl)-isopropylamine in 23 and 40 % yields, respectively.

N-(Diphenylmethyl)-diethylamine (69) [42]. The irradiation of diphenyldiazomethane (3.9 g) was carried out under nitrogen in cyclohexane containing 20% diethylamine. 440 ml nitrogen was developed during illumination (4 hrs.). The residue on evaporation was digested with a small quantity of ether and pure 1,1,2,2-tetraphenylethane separated out. Further quantities of ether were added and the mixture was repeatedly shaken with 5 N HCl. The acid extracts gave a crude base (1.1 g) — which did not at first crystallize — following addition of alkali, extraction with ether etc. Thereupon a small quantity was converted to the picrate, which on decomposition with ammonia yielded the crystalline base. Following seeding, the total material crystallized out and was pressed on clay to give 69, m. p. 59° (from methanol).

3. Photodimerization reactions of carbenes

If diazo compounds are photolyzed in the absence of other reaction partners dimerization and isomerization reactions of the intermediate carbenes may come into the foreground. Dimerizations will be treated in this section regardless of the reaction mechanism.

Photolysis of 2,2,2-trifluorodiazoethane (70) yields [47, 48], depending on reaction conditions, a variety of products, among which are isomers and dimers originating from the intermediate trifluoromethylcarbene (71). Products thus becoming available are, trifluoroethylene (72), cis- and trans-1,1,1,4,4,4-hexafluoro-2-butene (73 and 74), 1,1,2-trifluoro-3-(trifluoromethyl)-cyclopropane (75), 3,4,5-tris-(trifluoromethyl)-2-pyrazoline (76) and a polymer, polytrifluoromethylmethylene (77).

$$F_3C-CHN_2 \xrightarrow[-N_2]{h\nu} \left[F_3C-\underset{H}{C}|\right] \xrightarrow{71} F_3C-\underset{H}{C}=\underset{H}{C}-CF_3 + F_3C-\underset{H}{\overset{H}{C}}=\underset{H}{C}-CF_3$$

70   71   73   74

76 F₃C–C(CF₃)=N–N(H)–CF₃

72 F₂C=CHF

75 cyclopropane with H, CF₃, F, H, F, F

77 (F₃C–CH)ₙ

The initial concentration of 70 was found to control the product distribution [48] as is evinced below.

initial pressure of 70	72	73	74	75	76	77
2.5 atm. (liquid)	10	6	13	1	40	7
0.47 atm.	22	20	41	—	0	0
0.11 atm.	32	22	26	—	0	0

Gas phase photolysis of 2,2,3,3,4,4,4-heptafluorodiazobutane [47, 48] yields 31% 2H-heptafluoro-1-butene (78) and 47% trans-4H,5H-tetradecafluoro-4-octene (79). The isomerization reaction is remarkable insofar as the carbene rearranges with migration of a pentafluoroethyl group rather than of fluorine. For the dimerization of difluorocarbene see p. 291.

$$C_3F_7-CHN_2 \xrightarrow[-N_2]{h\nu} [F_5C_2-CF_2-HCl] \longrightarrow$$

$$F_5C_2-CH=CF_2 + F_7C_3-\underset{H}{\overset{H}{\underset{|}{C}}}=\underset{}{\overset{|}{C}}-C_3F_7$$
$$78 79$$

ZIMMERMAN and PASKOVICH [49] found that dimesityldiazomethane (80) on heating yields neither the corresponding ethylene 83 nor the azine, but instead produces 4,6-dimethyl-1-mesitylbenzocyclobutene (81) and the stilbene derivative 82. However, on irradiation at —75° under nitrogen, 80 afforded tetramesitylethylene (83) in virtually quantitative yield.

Benzaldehyde p-toluenesulfonylhydrazone on irradiation in the presence of sodium methoxide breaks down to phenylcarbene (4) via initially formed phenyldiazomethane [15]. If the photoreaction is conducted in cyclohexane or toluene solutions at 60—90°, the major product (46 % and 57 % yield, respectively) is trans-stilbene.

Photolysis of 3-diazo-1-methyloxindole (12) in carbon tetrachloride (cf. also chapter 32) gave 11 % of 1,1'-dimethylisoindigo (84) [17].

Tetramesitylethylene (83) [49]. Dimesityldiazomethane (80, 1.72 g) was dissolved in 50 ml dry benzene and the solution sealed under N_2 in a Pyrex combustion tube (wall thickness 3 mm). This tube was rotated in a slightly inclined brass trough lined with aluminium foil around a G. E. 1000 W AH 6 lamp at a distance of 15 cm. Cooling water at 17° was passed over the tube for temperature control. After 3 hrs. irradiation the solution was concentrated to give 1.643 g of a semi-crystalline residue. This material was crystallized from hexane/chloroform. The filtrates were combined and chromatographed on alumina. Hexane eluted 0.281 g of 83 and no other product was isolated. The total yield of pure 83 was 0.962 g (61.4%), m.p. 299—300°.

4. Miscellaneous photochemical reactions of carbenes

a) Isomerization reactions with formation of olefins

The rearrangement of trifluoromethylcarbene (**71**) and its homologue to trifluoroethylene (**72**) and 2H-heptafluoro-1-butene (**78**) [*48*] has already been described. Only a few other examples have been investigated, mostly by FREY (cf. [*4*]), in photolysis of diazo compounds. Thus, methylcarbene isomerizes to ethylene [*50*] and dimethylcarbene to propene [*51*] in the absence of suitable addends.

$$H_3C-CHN_2 \xrightarrow[-N_2]{h\nu} \left[H_3C-\underset{H}{C} | \right] \longrightarrow H_2C=CH_2$$

$$\underset{H_3C}{\overset{H_3C}{>}}\!\!\underset{N}{\overset{N}{\underset{\|}{C}}} \xrightarrow[-N_2]{h\nu} \left[\underset{H_3C}{\overset{H_3C}{>}}\!\!Cl \right] \longrightarrow H_2C=CH-CH_3$$

b) Univalent hydrogenation with subsequent dimerization

According to KIRMSE [*42*] diphenyldiazomethane is converted to 1,1,2,2-tetraphenylethane (**51**) when photolyzed in cyclohexane. This reaction is thought to involve diphenylcarbene as intermediate which, by abstraction of hydrogen from the solvent, is converted to diphenylmethyl radicals. For energetical reasons a recombination of unlike radicals seems to be unfavorable; instead **51** is obtained besides the azine (**37**).

Illumination of diphenyldiazomethane in benzene produces almost exclusively benzophenone azine; in cyclohexane and toluene, this constituent is less prominent, and in cyclohexene only tetraphenylethane is produced. When 9-diazofluorene is exposed to light in cyclohexane solution, 9,9'-bifluorene (**85**) is isolated as the major product. The insertion product, 9-cyclohexylfluorene (**53**), may also be obtained in this case (cf. p. 284).

85

9,9'-Bifluorene (85) [*42*]. 9-Diazofluorene (1.8 g) was irradiated in cyclohexane (180 ml) using an immersion lamp S 81 as light source. 66% of the theoretical amount of nitrogen was evolved in the course of 4.5 hrs. After concentration of the orange colored solution to 30 ml it was left to stand at room temperature for several days, whereupon rosettes of crystals (0.3 g) separated out (m.p. of crude product 225—227°). Recrystallization from acetone/water and cyclohexane gave pure **85**, m.p. 239—241°.

References, pp. 292—293

c) Addition to oxygen

Carbenes may be expected to react with oxygen. Indeed one such case was reported by BARTLETT and TRAYLOR [52]. When diphenyldiazomethane was irradiated at low temperatures in an oxygen atmosphere a low yield of dimeric benzophenone peroxide (3,3,6,6-tetraphenyl-s-tetroxane, **87**) was produced. This was cleaved on heating to benzophenone and oxygen.

$$5 + O_2 \rightarrow \left[\underset{86}{Ph_2C-O-O} \right] \xrightarrow{86} \underset{87}{\text{tetroxane}}$$

87 has been visualized as being the result of a head-to-tail dimerization of an intermediary benzophenone O-oxide (**86**) [42], for which a variety of electronic structures have been depicted (cf. [42, 52, 53]).

3,3,6,6-Tetraphenyl-s-tetroxane (87) [52]. 2.24 g of diphenyldiazomethane in chlorobenzene was photooxidized with a Hanovia 0802 N lamp in a sealed round bottomed flask containing oxygen. The reactor was cooled in Dry Ice and shaken for 30 mins., which caused the solution to decolorize. 0.175 g of an ethanol-insoluble residue melting at 160—170° was obtained. Recrystallization from acetone raised the melting point to 213.5—214°.

5. Photolysis of diazirines

Diazirine (**88**) is to be regarded as the cyclic isomer of diazomethane. The photochemistry of its derivatives has — due to their scarcity — not yet been investigated on a greater scale. Diazirine (**88**) as such has been shown on irradiation to behave analogously to diazomethane [54]. Some higher diazirines have also been investigated by FREY [54].

$$\underset{88}{\overset{N}{\underset{N}{\bigtriangleup}}} \qquad \underset{89}{\overset{F}{\underset{F}{\bigtriangleup}}\overset{N}{\underset{N}{}}} + X-Y \xrightarrow{h\nu} \overset{F}{\underset{F}{}}C\overset{X}{\underset{Y}{}}$$

Difluorodiazirine (**89**) which chemically is much less reactive than diazomethane or diazirine, is on photolysis with a 125 W ultraviolet lamp [55] completely converted to tetrafluoroethylene and nitrogen. Pyrolysis, however, gave hexafluorocyclopropane as the major product.

Irradiation of **89** with chlorine, iodine, dinitrogen tetroxide, or nitrosyl chloride, respectively [56], produces dichlorodifluoromethane, difluorodiiodomethane, difluorodinitromethane, or chlorodifluoronitromethane, respectively.

Difluorodiiodomethane [56]. 7 ml. of CF_2Cl_2 (solvent) and 78 mg of **89** was condensed into a 10 ml glass ampoule which contained 762 mg of iodine. The ampoule was sealed and warmed to 25° and the liquid phase was irradiated for 16 hrs. with a General

Electric BH-6 ultraviolet lamp fitted with a Corning No. 5840 filter. The entire mixture was fractionated through —78° and —196° receivers after the irradiation period. The —78° trap contained about 3×10^{-4} moles of difluorodiiodomethane contaminated with a small amount of 1,2-diiodotetrafluoroethane. Final purification by preparative vapor phase chromatography afforded a 20.7% yield of pure difluorodiiodomethane.

6. Addition of diazomethane to olefins with formation of a pyrazoline

Fluorinated olefins may occasionally show unusual behavior towards diazomethane. When 3,3,3-trifluoropropene (**90**) or its 2-methyl derivative (**92**) was allowed to stand with diazomethane for several hours in the dark, no reaction occurred [*57*]. UV-irradiation, however, effected cycloaddition of the diazo compound across the double bond, 5-(trifluoromethyl)-2-pyrazoline (**91**) and 5-methyl-5-(trifluoromethyl)-2-pyrazoline (**93**) thus becoming available.

$$F_3C-\underset{R}{C}=CH_2 \;+\; CH_2N_2 \;\xrightarrow{h\nu}\; \underset{H}{\overset{F_3C}{\underset{R}{\bigtriangleup}}}_{N-N}$$

90: R=H   91: R=H
92: R=CH₃   93: R=CH₃

5-(Trifluoromethyl)-2-pyrazoline (91) [*57*]. Reaction of 50 g of **90** with an approximately equivalent amount of diazomethane in ether, in a flask provided with a Dry Ice-acetone cooled condenser, was induced by irradiation with a quartz mercury arc ultraviolet lamp. After the color of diazomethane had disappeared (2 hrs.), distillation yielded 58 g (89%) of 5-(trifluoromethyl)-2-pyrazoline, b.p. 146.5°.

References

[*1*] J. HINE: Divalent Carbon. New York: Ronald 1964.
[*2*] W. KIRMSE: Carbene Chemistry. New York: Academic Press 1964.
[*3*] W. B. DE MORE and S. W. BENSON, in: Adv. Photochem., ed. by W. A. NOYES, G. S. HAMMOND and J. N. PITTS JR., Vol. **2**, 219—261 (1964). New York: Interscience.
[*4*] H. M. FREY, in: Progress in Reaction Kinetics, ed. by G. PORTER, Vol. **2**, 131—164 (1964). Oxford etc.: Pergamon.
[*5*] J. A. BELL, in: Progress in Physical Organic Chemistry, ed. by S. G. COHEN, A. STREITWIESER JR. and R. W. TAFT, Vol. **2**, 1—61 (1964). New York: Interscience.
[*6*] G. VON BÜNAU, P. POTZINGER and G. O. SCHENCK: Tetrahedron **21**, 1293 (1965).
[*7*] A. LEDWITH: Royal Inst. Chem. Lecture Series **1964**, No. 5.
[*8*] W. VON E. DOERING and P. LAFLAMME: J. Amer. chem. Soc. **78**, 5447 (1956).
[*9*] P. S. SKELL and R. C. WOODWORTH: J. Amer. chem. Soc. **78**, 4496 (1956).
[*10*] C. D. GUTSCHE, G. L. BACHMAN and R. S. COFFEY: Tetrahedron **18**, 617 (1962).
[*11*] G. L. CLOSS and R. A. MOSS: J. Amer. chem. Soc. **86**, 4042 (1964).
[*12*] R. M. ETTER, H. S. SKOVRONEK and P. S. SKELL: J. Amer. chem. Soc. **81**, 1008 (1959).
[*13*] W. VON E. DOERING and T. MOLE: Tetrahedron **10**, 65 (1960).
[*14*] W. VON E. DOERING, R. G. BUTTERY, R. G. LAUGHLIN and N. CHAUDHURI: J. Amer. chem. Soc. **78**, 3224 (1956).

[15] H. Nozaki, R. Noyori and K. Sisido: Tetrahedron 20, 1125 (1964).
[16] K. R. Kopecky, G. S. Hammond and P. A. Leermakers: J. Amer. chem. Soc. 84, 1015 (1962).
[17] E. J. Moriconi and J. J. Murray: J. Org. Chem. 29, 3577 (1964).
[18] W. von E. Doering and L. H. Knox: J. Amer. chem. Soc. 72, 2505 (1950).
[19] W. von E. Doering and L. H. Knox: J. Amer. chem. Soc. 75, 297 (1953).
[20] H. Meerwein, H. Disselnkötter, F. Rappen, H. von Rintelen and H. van de Vloed: Liebigs Ann. Chem. 604, 151 (1957).
[21] S. J. Rhoads, in: Molecular Rearrangements, ed. by P. de Mayo, Vol. 1, 655—706, esp. p. 700—703 (1963). New York: Interscience.
[22] W. von E. Doering, G. Laber, R. Vonderwahl, N. F. Chamberlain and R. B. Williams: J. Amer. chem. Soc. 78, 5448 (1956).
[23] G. O. Schenck and H. Ziegler: Liebigs Ann. Chem. 584, 221 (1954).
[24] K. Alder and P. Schmitz: Chem. Ber. 86, 1539 (1953).
[25] W. von E. Doering, J. R. Mayer and C. H. DePuy: J. Amer. chem. Soc. 75, 2386 (1953).
[26] C. D. Gutsche and H. E. Johnson: J. Amer. chem. Soc. 77, 5933 (1955).
[27] G. O. Schenck and R. Steinmetz: Liebigs Ann. Chem. 668, 19 (1963).
[28] J. C. Sheehan and I. Lengyel: J. Org. Chem. 28, 3252 (1963).
[29] G. L. Closs, L. E. Closs and W. A. Böll: J. Amer. chem. Soc. 85, 3796 (1963).
[30] G. L. Closs and W. Böll: Angew. Chem. 75, 640 (1963); intern. ed. 2, 399 (1963).
[31] E. Müller, A. Moosmayer and A. Rieker: Z. Naturforsch. 18 b, 982 (1963).
[32] W. Schlenk: Liebigs Ann. Chem. 394, 178 (1912), esp. p. 183.
[33] W. H. Urry and J. R. Eiszner: J. Amer. chem. Soc. 74, 5822 (1952).
[34] W. H. Urry and N. Bilow: J. Amer. chem. Soc. 86, 1815 (1964).
[35] W. H. Urry and J. W. Wilt: J. Amer. chem. Soc. 76, 2594 (1954).
[36] A. N. Nesmeyanov, R. Kh. Freidlina, V. N. Kost and M. Ya. Khorlina: Tetrahedron 16, 94 (1961).
[37] E. S. Gould: Mechanismus und Struktur in der organischen Chemie. 2. Ed., esp. p. 923—924, ref. 193 b. Weinheim: Verlag Chemie 1964.
[38] D. W. Setser, R. Littrell and J. C. Hassler: J. Amer. chem. Soc. 87, 2062 (1965).
[39] H. Meerwein, H. Rathjen and H. Werner: Ber. dtsch. chem. Ges. 75, 1610 (1942).
[40] W. von E. Doering, L. H. Knox and M. Jones jr.: J. Org. Chem. 24, 136 (1959).
[41] D. B. Richardson, M. C. Simmons and I. Dvoretzky: J. Amer. chem. Soc. 83, 1934 (1961).
[42] W. Kirmse, L. Horner and H. Hoffmann: Liebigs Ann. Chem. 614, 19 (1958).
[43] W. G. Dauben and F. G. Willey: J. Amer. chem. Soc. 84, 1497 (1962).
[44] D. M. Lemal and A. J. Fry: J. Org. Chem. 29, 1673 (1964).
[45] D. B. Denney and P. P. Klemchuk: J. Amer. chem. Soc. 80, 3289 (1958).
[46] D. G. Farnum and P. Yates: J. Amer. chem. Soc. 84, 1399 (1962).
[47] R. Fields and R. N. Haszeldine: Proc. Chem. Soc. 1960, 22.
[48] R. Fields and R. N. Haszeldine: J. Chem. Soc. 1964, 1881.
[49] H. E. Zimmerman and D. H. Paskovich: J. Amer. chem. Soc. 86, 2149 (1964).
[50] H. M. Frey: J. Chem. Soc. 1962, 2293.
[51] H. M. Frey and I. D. R. Stevens: J. Chem. Soc. 1963, 3514.
[52] P. D. Bartlett and T. G. Traylor: J. Amer. chem. Soc. 84, 3408 (1962).
[53] R. Huisgen: Angew. Chem. 75, 604 (1963), esp. p. 631; intern. Ed. 2, 565 (1963).
[54] H. M. Frey: Pure Appl. Chem. 9, 527 (1964).
[55] R. A. Mitsch: J. Heterocycl. Chem. 1, 59 (1964).
[56] R. A. Mitsch: J. Heterocycl. Chem. 1, 233 (1964).
[57] F. Misani, L. Speers and A. M. Lyon: J. Amer. chem. Soc. 78, 2801 (1956).

Chapter 32

Photochemical syntheses with diazoketones, quinone diazides and iminoquinone diazides

The diazoketone substances, which incorporate the $-CO-CN_2-$ group, can be divided into three classes: acyclic mono-diazoketones, acyclic bis-diazoketones and cyclic diazoketones. Their behavior in UV-light is similar due to the fact that all three classes react with the liberation of nitrogen.

1. Acyclic mono-diazoketones

a) Conversion to ketenes and α,β-unsaturated ketones

Photolysis of monodiazoketones frequently occurs with the formation of ketenes e.g. azibenzil (1) yields diphenylketene (2) and methyl benzoyldiazoacetate (3) is converted to methyl phenylketenecarboxylate (4) [1].

$$C_6H_5-CN_2-CO-C_6H_5 \xrightarrow[-N_2]{h\nu} (C_6H_5)_2=C=O$$
$$\qquad\qquad 1 \qquad\qquad\qquad\qquad\qquad 2$$

$$C_6H_5CO-CN_2-COOCH_3 \xrightarrow[-N_2]{h\nu} [C_6H_5CO-\overline{C}-COOCH_3] \longrightarrow \underset{\underset{COOCH_3}{|}}{\overset{\overset{C_6H_5}{|}}{C}}=C=O$$
$$\qquad\qquad 3 \qquad\qquad\qquad\qquad\qquad\qquad\qquad\qquad\qquad 4$$

In the photolysis of acyclic diazoketones containing the group $-CO-CN_2-CO-$, two isomeric ketenes may be formed e.g. in the case of 2-diazo-1-phenyl-1,3-butanedione the formation of A and B can be anticipated on theoretical grounds.

The experiment shows that the methyl group migrates (formation of B) as the ketene formed gives rise to ethyl 2-benzoylpropionate in absolute alcohol.

Many diazoketones with the formula $R-COCN_2CH_2-R'$ are able to form α,β-unsaturated ketones ($R-COCH=CHR'$) by photochemical decomposition. Such decomposition can also be catalyzed by silver oxide [3].

References, p. 312

$$C_6H_5-CO-CN_2-COCH_3 \xrightarrow[-N_2]{h\nu} \left[C_6H_5-CO-\overline{C}-COCH_3\right]$$

$$O=C=C\begin{subarray}{l}C_6H_5\\COCH_3\end{subarray} \quad A$$

$$O=C=C\begin{subarray}{l}CH_3\\COC_6H_5\end{subarray} \quad B$$

Diphenylketene (2) [2]. Azibenzil (1) (4.4 g) is dissolved in absolute ether (170 ml) and illuminated. A laboratory immersion lamp (S 81, Quarzlampengesellschaft, Hanau) is used as a light source, incorporated into a quartz socket with running water. By cooling in ice/water it is possible to work at 0° without impairing the life time of the lamp. The lamp, encased in this manner, is inserted into the irradiation vessel which is fitted with lateral gas inlets. The quantity of liquid required for irradiation amounts to 170 ml when the full capacity of this arrangement is utilized. Usually, the reaction solution is flushed with pure nitrogen during irradiation; this also guarantees sufficient mixing of the liquid. The irradiation vessel is itself held in a Dewar flask containing an appropriate cooling mixture to further enable a working temperature of 0° to be maintained. When nitrogen evolution ceases, (ca. 3 hrs.), the ether is removed under nitrogen and the diphenylketene distilled under reduced pressure, yield 3.5 g.

Benzylideneacetone [3]. 3-Diazo-4-phenyl-2-butanone (10 g) in 80% aq. dioxane (100 ml), containing a small quantity of triethylamine, was irradiated with an Osram bulb (200 Watt). The solution was cooled from the outside. After one hour, the diazoketone has entirely decomposed and the dioxane is evaporated under vacuum until two layers separate. 2 N sodium carbonate solution is added and the neutral fraction extracted with ether. After drying the ethereal solution is distilled and at 140°/16 mm an almost colorless liquid distils over. After standing for some time over ice this crystallizes (m.p.: 41°).

b) Photolysis in the presence of water or alcohol

When photolysis of acyclic diazoketones occurs in the presence of water or alcohol, the corresponding acids or esters are formed instead of the ketenes. The formation occurs by addition of water or alcohol to the ketene initially produced [2].

$$R-CN_2-\underset{\underset{O}{\|}}{C}-R \xrightarrow[-N_2]{h\nu} R_2C=C=O \begin{subarray}{l}\xrightarrow{H_2O} R_2\overset{H}{\underset{|}{C}}-COOH\\ \xrightarrow{R'OH} R_2\overset{H}{\underset{|}{C}}-COOR'\end{subarray}$$

The photolysis of diazoketones in the presence of water (alcohol) can be regarded as complementary to the ARNDT-EISTERT reaction. ARNDT and EISTERT [4] have demonstrated that rearrangement of diazoketones occurs with silver oxide giving homologous acids or their derivatives, depending on the reaction medium (water or alcohol). Both reactions proceed through ketene formation:

$$R-CO-CHN_2 \xrightarrow[-N_2]{h\nu \text{ or } Ag_2O} [R-CH=C=O] \xrightarrow{HX} R-CH_2-C\begin{smallmatrix}X\\\\O\end{smallmatrix}$$

The preparative importance of the photolysis of diazoketones in the presence of water (alcohol) lies in the very good yields that are usually obtained, as shown by HORNER, SPIETSCHKA and GROSS [2]. In the case of certain diazoketones, it has been found that they can undergo photolytic rearrangement whereas such rearrangement does not occur when the ARNDT-EISTERT method is used.

ROEDIG and LUNK [5] have carried out experiments with 3,3,4,4,4-pentachloro-1-diazo-2-butanone (5) and found that it remained unaltered after 48 hours boiling in methanol with a suspension of silver oxide. When it is exposed to filtered UV-light in aqueous dioxane, however, 3,4,4,4-tetrachlorocrotonic acid (7) is formed (yield 57%) probably via elimination of hydrogen chloride from 3,3,4,4,4-pentachlorobutyric acid (6) initially formed, during work-up. The methyl ester of 7 is obtained by photolysis of 5 in absolute methanol (yield 71%).

$$Cl_3C-CCl_2-CO-CHN_2 \xrightarrow[-N_2; +H_2O]{UV} Cl_3C-CCl_2-CH_2-COOH \xrightarrow{-HCl}$$
$$5 6$$

$$Cl_3C-CCl=CH-COOH$$
$$7$$

According to ROEDIG and LUNK [5] 3,4,4-trichloro-1-diazo-3-buten-2-one (8) cannot undergo the ARNDT-EISTERT reaction since the substance is sensitive to alkali. However, by photochemical treatment this can be easily converted to an acid which is either 3,4,4-trichloro-3-butenoic acid (9) or 3,4,4-trichlorocrotonic acid (10).

$$Cl_2C=CCl-COCHN_2 \xrightarrow[-N_2; +H_2O]{UV} Cl_2C=CCl-CH_2COOH \quad \text{or}$$
$$8 9$$

$$Cl_2CH-CCl=CHCOOH$$
$$10$$

Diphenylacetic acid [2]. When azibenzil (1) (4.4 g) in solution in dioxane (165 ml) and water (5 ml) is irradiated, working-up of the product gives diphenylacetic acid in excellent yield, m.p. 145°.

Ethyl diphenylacetate [2]. When the above irradiation is carried out in ethanol, the ethyl ester is obtained in good yield, m.p. 57°.

References, p. 312

3,4,4,4-Tetrachlorocrotonic acid (7) [*5*]. The light source is provided by a laboratory immersion lamp (S81, Quarzlampengesellschaft, Hanau) in the experimental arrangement described by HORNER [*2*] except that a filter solution is circulated through the jacket instead of the cooling water. The filter solution is of copper sulfate (2.5%) with a sixfold excess of ammonia. The solution forms a layer ca. 0.5 cm thick over the irradiation unit.

4 g of **5** is dissolved in a mixture of dioxane (140 ml) and water (10 ml) and irradiated with UV-light until the evolution of nitrogen ceases. In this case, cooling of the solution is not necessary. The solvent is distilled off under reduced pressure, the residue is dissolved in ether and the acid is extracted with sodium bicarbonate solution. On acidification, a dark oil separates out. This is extracted with ether and dried over calcium chloride. At 94—96°/0.1 mm, 1.8—2.0 g of a colorless liquid distils and solidifies in a cold bath. M.p. 83° (from petroleum ether).

Methyl 3,4,4,4-tetrachlorocrotonate [*5*]. Diazo ketone **5** (12 g) is dissolved in absolute methanol (150 ml) and exposed to filtered UV-light for eight hours at room temperature (as described above). The methanol is then distilled off at atmospheric pressure and the product is further heated to 100° under vacuum (water-pump) to ensure complete elimination of hydrogen chloride. Distillation yields a colorless liquid (7—8 g) with an agreeable smell, b.p.$_{0.1}$ 53—54°.

Acid 9 or 10 [*5*]. Diazo ketone **8** (3.5 g) is dissolved in a mixture of dioxane (140 ml) and water (10 ml) and irradiated with filtered UV-light until nitrogen evolution ceases (ca. 9 hrs.). This mixture is then worked up using the same procedure as that given for **7**. Distillation yields a colorless rapidly solidifying oil (2.6 g). This is pressed on clay and repeatedly recrystallized from petroleum ether. The acid **9** (or **10**) forms colorless tablets of m.p. 55—56°.

c) Photolysis in presence of N-methylaniline and ethanethiol

Diazoketones (**11**) can be photochemically converted to the N-methylanilides of homologous acids (cf. **12**) in the presence of N-methylaniline according to WEYGAND and BESTMANN [*6*].

Ethanethiol reacts similarly with **11**, forming the ethyl ester of the homologous thio-acids (cf. **13**). This reaction is of preparative importance because of the ease with which both **12** and **13** can be converted to the corresponding aldehydes **14**. This occurs with **12** by reaction with LiAlH$_4$ [*7*]; with **13** Raney nickel is used [*8*].

$$R-CO_2H \to \underset{\underset{O}{\|}}{\underset{11}{R-C-CHN_2}} \xrightarrow[C_6H_5NHCH_3]{h\nu} \underset{12}{RCH_2-\underset{\|}{\overset{O}{C}}-\underset{|}{\overset{CH_3}{N}}-C_6H_5} \xrightarrow{LiAlH_4}$$

$$\xrightarrow[C_2H_5SH]{h\nu} \underset{13}{RCH_2-\underset{\|}{\overset{O}{C}}-S-C_2H_5} \xrightarrow{Raney\ nickel} \underset{14}{RCH_2-\overset{H}{\underset{}{C}}=O}$$

N-Methyl-2-phenylacetanilide (12, R = C$_6$H$_5$) [*6*]. A stirred solution of diazoacetophenone (5 g) and N-methylaniline (15 g) in absolute benzene (160 ml) is irradiated for three days with a water cooled laboratory immersion lamp, which is cleared of a brown deposit every 12 hours. Finally benzene and N-methylaniline are removed from the solution under vacuum. Further distillation yields **12** (R = C$_6$H$_5$) as a yellow oil (b.p.$_{0.001}$ 121°; yield 77%).

Ethyl thiopropionate (13, R = CH$_3$) [6]. Diazoacetone (5 g) with ethanethiol (9 ml) in absolute ether (350 ml) is irradiated until the evolution of nitrogen ceases (ca. 8 hrs.). The ether is distilled off and the residue fractionated at 60 mm (b. p.$_{60}$: 160—165 °; yield 4.7 g = 67% theoretical).

d) Photolysis in the presence of azo compounds

HORNER et al. [2, 9] obtained compound **15** from azibenzil and azobenzene, compound **16** from benzoyldiazomethane and azobenzene, adduct **17** from diphenylketene and azodibenzoyl and compound **18** from azibenzil and α,α'-azotoluene. In all of these conversions, 1 mol. of the α-diazocarbonyl compound (or of the diphenylketene) reacts with 1 mol. of the azo compound.

There is no doubt that the α-diazocarbonyl compounds used are first converted to the corresponding ketenes, with liberation of nitrogen, and that the ketenes then react with the azo compounds to form the corresponding derivatives of 1,2-diazetidine. The adduct **15** had been previously prepared from diphenylketene and cis-azobenzene [10].

$$\begin{array}{c} R' \\ \diagdown \\ R'' \end{array} C=C=O + R'''-N=N-R''' \longrightarrow \begin{array}{c} R' \\ \diagdown \\ R''' \end{array} \begin{array}{c} C-C=O \\ | \quad | \\ R'''-N-N-R''' \end{array}$$

15: R' = R'' = R''' = C$_6$H$_5$
16: R' = H; R'' = R''' = C$_6$H$_5$
17: R' = R'' = C$_6$H$_5$; R''' = C$_6$H$_5$-CO
18: R' = R'' = C$_6$H$_5$; R''' = C$_6$H$_5$-CH$_2$

If azodibenzoyl and diphenylketene are irradiated together in 1:2 proportions, an adduct with the probable formula **19** is formed [9]. **20** is obtained when azibenzil (2 mol.) and diethyl azodicarboxylate are irradiated together. The same compound can also be obtained in a dark process from diethyl azodicarboxylate and diphenylketene in 1:2 proportions [11].

19: R = C$_6$H$_5$-CO
20: R = C$_2$H$_5$-O-CO

1,2,4-Triphenyl-1,2-diazetidin-3-one (16) [2]. Benzoyldiazomethane (3 g) is irradiated with azobenzene (3.6 g) in 170 ml benzene (for irradiation conditions cf. p. 295). When evolution of nitrogen has ceased, the solvent is distilled off and the red oily residue dissolved in hot methanol. After standing for some time, dense, yellow crystals (16) separate out and these can be easily purified by further recrystallization from methanol, m.p. 92°, yield 2 g.

1,2,4,4-Tetraphenyl-1,2-diazetidin-3-one (15) [*2*]. Azibenzil (4.4 g) is irradiated with azobenzene (3.6 g) in absolute ether (170 ml). The yellowish residue remaining after distillation of the ether is washed with a little alcohol and recrystallized from methanol. The adduct 15 (4.5 g) forms dense colorless crystals, m.p. 173°.

1,2-Dibenzyl-4,4-diphenyl-1,2-diazetidin-3-one (18) [*9*]. α,α'-Azotoluene (2.1 g; 0.01 mole) and azibenzil (2.2 g; 0.01 mole) are irradiated in benzene (170 ml). When evolution of nitrogen ceases, the solvent is distilled off and the residue dissolved in hot methanol. After a short time, yellow crystals separate out. Repeated recrystallization from methanol/ethyl acetate yields colorless crystals which melt at 154° (yield 2 g).

Diethyl hexahydro-3,5-dioxo-4,4,6,6-tetraphenylpyridazine-1,2-dicarboxylate (20) [*9*]. A solution of diethyl azodicarboxylate (1.7 g; 0.01 mole) and azibenzil (4.4 g; 0.02 mole) in benzene is irradiated in the usual way. After evaporation of the solvent, the residue is recrystallized from ethanol. Crystals soon separate out and, when recrystallized from methanol, give a melting point of 130° (yield 3 g).

e) Photolysis in the presence of azomethines

Ketenes formed as intermediate products in the photolysis of diazoketones can add on to azomethines (SCHIFF's bases) to give β-lactams (2-azetidinones (21)) [*12*].:

$$-CO-CN_2- \xrightarrow[-N_2]{h\nu} \quad \diagdown C=C=O$$

$$\diagdown C=N- \; + \; \diagdown C=C=O \quad \longrightarrow \quad \begin{array}{c} \diagdown C\!\!-\!\!N\!-\! \\ {}^{4}| \quad {}^{1}| \\ {}^{3}\diagup C\!\!-\!\!C=O \\ {}^{2} \end{array}$$

21

The ability of keto-ketenes to add on to SCHIFF's bases in a dark reaction had been observed previously [*13*] but the formation of β-lactams from aldoketenes and SCHIFF's bases can only take place photochemically.

KIRMSE and HORNER [*12*] point out that, in this connection, partially aliphatic-substituted azomethines show less reactivity than their completely aromatic substituted counterparts. Experiments with diazopyruvate ($N_2CH-CO-COOR$) did not result in addition to SCHIFF's bases.

1,3,3,4,4-Pentaphenyl-2-azetidinone (comp. 21) [*12*]. The photolysis experiments were carried out using a laboratory immersion lamp (S 81; Quarzlampengesellschaft, Hanau). Either a closed system was used (with stirring) or the diazo compound was irradiated whilst a solution of azomethine was added dropwise. With diazo compounds having a relatively short wavelength absorption, the azomethine added acts as a powerful light-filter and the reaction progresses very slowly. For this reason, the second procedure is to be recommended. Azibenzil (2.22 g) and benzophenone anil (2.57 g; i.e. 10 mMol. of each) were irradiated together in dry benzene (100 ml) until evolution of nitrogen had ceased (ca. 5 hrs.). The benzene was evaporated off under vacuum on a water bath and the residue was dissolved in chloroform and chromatographed on Al_2O_3 (Woelm neutral). The eluted solution was evaporated on a water bath, the residue digested in hot methanol (20 ml) and suction-filtered. The pale yellow β-lactam (3.25 g, 72%) was obtained, (m.p. 188—190°; after recrystallization from ethanol containing a little acetone, m.p. 190—191°).

3-Methyl-1,4-diphenyl-2-azetidinone (comp. 21) [*12*]. 10 mMol. of diazoacetone and benzalaniline in benzene (100 ml); drip-feed. The benzene is evaporated under vacuum and the residue chromatographed in benzene on an Al_2O_3 column. The pale yellow eluate was again evaporated and the residue digested with a small quantity of methanol. The crude product (0.75 g) had m.p. 111—113°. By mixing the mother-liquor with a little water and cooling, a further 0.5 g of a low melting point (60°) product was obtained. From this a further quantity of β-lactam (0.2 g) was obtained by recrystallization from methanol. (Total yield 47%). By recrystallization from dilute methanol with a little charcoal, white needles were obtained, m.p. 113°.

f) Replacement of the diazo group by hydrogen

If ethyl 2-diazo trifluoroacetoacetate in 200—350 ml of an alcohol is irradiated with UV, after nitrogen has been passed through the solution, dehydrogenation of the alcohol occurs producing an equivalent quantity of ethyl 4,4,4-trifluoroacetoacetate [*14*]. Ethanol forms acetaldehyde (41%) and isopropanol forms acetone (42%). According to WEYGAND, the reaction proceeds via the carbene **22**:

$$F_3C-CO-CN_2-COOC_2H_5 \xrightarrow[-N_2]{h\nu} \left[F_3C-CO-\overline{C}-COOC_2H_5\right] \xrightarrow{(CH_3)_2CHOH}$$
$$\textbf{22}$$
$$F_3C-CO-CH_2-COOC_2H_5 + CH_3-CO-CH_3$$

g) Intramolecular addition of a carbene

The photolysis [*15*] of 3-(diazoacetyl)-1,2-diphenylcyclopropene (**23**) has led to the formation of a new highly strained ring system, tricyclo-[2.1.0.0^{2,5}]pentane, probably by intramolecular addition of the ketocarbene generated photochemically.

1,5-Diphenyltricyclo[2.1.0.0^{2,5}]pentan-3-one (24) [15]. A 0.5% solution of **23** in tetrahydrofuran was irradiated at room temperature with a Hanovia 450 W mercury lamp using a Pyrex filter. Evolution of nitrogen subsided within 2 hrs., and chromatography on silica gel afforded pure **24**, m.p. 139—140°.

2. Acyclic bis-diazoketones

The photolysis of this group of compounds has been investigated by HORNER and SPIETSCHKA [1]. They found that the decomposition of bis-diazoketones takes place in a similar manner to that of mono-diazoketones which they had investigated. E.g. dimethyl 2,5-bis-(diazo)-ketipate (**25**), which gives no definite product on thermal decomposition, forms tetramethyl ethanetetracarboxylate (**26**) on irradiation in methanol.

$$H_3COOC-CN_2-CO-CO-CN_2-COOCH_3 \xrightarrow[-2\,N_2]{UV}$$
25

$$\left[\begin{array}{c} O \quad\quad O \\ \| \quad\quad \| \\ C \quad\quad C \\ / \quad\quad \backslash \\ H_3COOC \quad\quad COOCH_3 \end{array} \right] \xrightarrow{2\,CH_3OH} \begin{array}{c} H_3COOC \quad\quad COOCH_3 \\ H \quad\quad H \\ H_3COOC \quad\quad COOCH_3 \end{array}$$
26

1,8-Bis-(diazo)-2,7-octanedione (**27**) gives suberic acid (**28**) in the presence of water and forms the corresponding ester in the presence of ethanol.

$$N_2HC-CO-(CH_2)_4-CO-CHN_2 \xrightarrow[-2\,N_2]{h\nu} \left[O=C=CH-(CH_2)_4-CH=C=O \right] \xrightarrow{2\,H_2O}$$
27

$$HOOC-CH_2-(CH_2)_4-CH_2-COOH$$
28

Tetramethyl ethanetetracarboxylate (26) [1]. Dimethyl 2,5-bis-(diazo)-ketipate (**25**) (1.5 g) is irradiated with UV light (cf. the synthesis of **2** [2], p. 295) in absolute methanol (80 ml). On evaporation of the methanol a crystalline residue is left and this is repeatedly recrystallized from methanol and ether to give 0.8 g of **26**, m. p. 135°.

Suberic acid (28) [1]. 1,8-Bis-(diazo)-2,7-octanedione (**27**) (2 g) is dissolved in a mixture of dioxane and water (160 : 10 ml) and irradiated in the usual manner. A yellow residue is left after evaporation which almost completely dissolves in caustic soda. Addition of dilute acid yields suberic acid (1.5 g), m.p. 140° (from water).

3. Cyclic α-diazoketones

Photolysis of cyclic α-diazoketones usually involves ring contraction with concomitant formation of ketenes. In an aqueous medium the reaction may proceed further to the formation of acids as shown in the scheme on page 302.

Photolysis of diazocarbonyl compounds

$$(C_xH_y)_n \underset{C=O}{\overset{C=N_2}{\diagdown\diagup}} \xrightarrow[-N_2]{h\nu} (C_xH_y)_n\ \diagup\!\!\!\diagdown C=C=O \xrightarrow{+H_2O} (C_xH_y)_n \underset{COOH}{\overset{H}{\diagdown C \diagup}}$$

Photolyses not involving ring contraction will be discussed on p. 306

a) Ring contraction of five-membered rings

CAVA [16] has studied the photolysis of 2-diazo-1-indanone (**29**) which leads to the formation of benzocyclobutene-1-carboxylic acid (**30**). Similar results were reported by HORNER [17].

CAVA carried out the photolysis in aqueous tetrahydrofuran; he recommends the presence of sodium bicarbonate in the photolysis mixture.

$$\mathbf{29} \xrightarrow[-N_2;\,+H_2O]{h\nu} \mathbf{30}$$

In 1961 it was shown by MATEOS [18] that the irradiation of 16-diazo-3β-hydroxyandrostan-17-one (**31**) leads to a D-ring contraction (**32**). In the following year reports from several research groups on similar ring contractions were published [19—21].

$$\mathbf{31} \xrightarrow[-N_2;\,+H_2O]{h\nu} \mathbf{32}$$

MULLER [22] reported on the synthesis of **34** and **35** which was carried out by illumination of **33** in moist ether.

$$\mathbf{33} \xrightarrow[-N_2;\,+H_2O]{h\nu} \mathbf{34} + \mathbf{35}$$

A-, B- and C-norsteroids (five-membered instead of six-membered rings) had been prepared previously; the successful synthesis of the D-norsteroids which have a cyclobutane instead of a cyclopentane D-ring completes the preparation of modified steroids in which any of the four rings of the steroid nucleus is contracted. Besides, the formation of **32** constituted the first synthesis of a trans-bicyclo[4.2.0]octane system.

Benzocyclobutene-1-carboxylic acid (30) [16]. A solution of 2-diazo-1-indanone (**29**) (2.0 g, 0.0126 mole) in tetrahydrofuran (200 ml) and water (100 ml) containing

References, p. 312

sodium bicarbonate (2.0 g) was irradiated at the boiling point of the solution for 10 hours. The ultraviolet light source was a low-pressure argon-filled mercury discharge tube. Removal of the tetrahydrofuran by distillation left a tarry aqueous residue which, after extraction with methylene chloride, was acidified and extracted with ether. The ether layer was washed with water, dried over sodium sulfate and evaporated. The tan solid residue was sublimed at 90° (2 mm) to yield small white prisms of benzocyclobutene-1-carboxylic acid (0.400 g, 21%), m.p. 71—74°. Recrystallization from petroleum ether gave white needles, m.p. 74—75°.

3β-Hydroxy-D-norandrostane-16ξ-carboxylic acid (32) [23]. The diazoketone (31) (4 g) dissolved in 160 ml distilled tetrahydrofuran and 40 ml water was irradiated by immersing in the solution a Hanau S 700 mercury lamp in a jacketed flask and maintaining the temperature of the solution at $10° \pm 1°$. After one hour, a sample showed complete removal of the 2075 cm^{-1} band from the I. R. spectrum and that the reaction was, therefore, complete. The mixture was poured into 600 ml water and extracted with ether; the organic layer was extracted with a saturated solution of sodium bicarbonate and on acidification with hydrochloric acid the D-nor acid 32 precipitated as a foamy solid. After being kept for six hours at 5° the product was filtered off, washed with water, decolorized with Norit charcoal and recrystallized from methanol, m.p. 200—205° (3 g, 75% yield). By recrystallization from methanol the m.p. was raised to 205—206°.

b) Ring contraction of six-membered ring systems

HORNER [24] drew attention to the difference between the photolysis and pyrolysis of diazocamphor (36). Ring contraction only occurred in the photolytic process yielding 1,6,6-trimethylbicyclo[2.1.1]hexane-5-carboxylic acid (37a) when photolysis was carried out in aqueous medium. This reaction provided the first ready entry into the bicyclo[2.1.1]hexane series. The stereochemistry of the photoproduct was established by MEINWALD [25]. The product is the exo acid 37a. This may be epimerized into the less crowded 5-endo-configuration (37b).

In addition to the acid 37a, 1,7,7-trimethyltricyclo[2.2.1.0^{3,5}]heptan-2-one (38) and camphoric anhydride (39) were formed in the photolysis of 36. 38 is the main product in the pyrolysis of diazocamphor.

WIBERG [26] irradiated diazonorcamphor (40) in methanol solution and obtained methyl bicyclo[2.1.1]hexane-5-carboxylate (41).

Methyl bicyclo[2.1.1]hexane-5-carboxylate (41) [*26*]. A solution of 73.6 g (0.54 mole) of crude diazonorcamphor (**40**) in 4500 ml of anhydrous methanol was irradiated with a water-cooled Hanovia 500 W. immersion quartz lamp using a Corex filter. After 24 hours 95% of the diazo compound had reacted. Most of the solvent (4 l) was removed by distillation through a 30 inch column packed with Heli-Pak. To the residue was added 1 litre of water and chopped ice, and the mixture was extracted with four 1 litre portions of pentane. The pentane solution was dried over anhydrous sodium sulfate and the solvent was removed by distillation through the above-mentioned column. Distillation of the residue gave 41 g (54%) of methyl bicyclo[2.1.1]hexane-5-carboxylate, b.p. 67—71° at 17—18 mm. The material thus obtained contained a small amount of ketonic impurity. A pure sample was obtained by vapor phase chromatography.

c) Ring contraction of o-quinone diazides (SÜS reaction)

The main members of the cyclic diazoketone group are the quinone diazides of the benzene and naphthalene series. The course of the photoreactions of these compounds, which have achieved importance in diazo-type manufacture, has been explained by the fundamental research of Süs [*27, 28*]. Decomposition leads to the formation of five-membered rings (e.g. **44**), as is shown in the example of 2,1-naphthoquinone diazide (**42**) [*27*]. The intermediate product is taken to be the carbene **43**.

The versatility of the Süs reaction may also be illustrated by the conversion of 10-diazo-9(10H)-phenanthrone (**45**) and 6-diazo-5(6H)-chrysenone (**46**), respectively, to compounds of the fluorene series [*29*].

According to HORNER [*30*], 3-diazo-2,3,6,7,8,9-hexahydro-5H-benzo-cyclohepten-2-one (**47**) and related compounds undergo ring contraction upon illumination. **47** forms bicyclo[5.3.0]deca-1(10),7-diene-9-carboxylic acid (**48**).

On photolysis of 6-diazo-2,4-cyclohexadien-1-one (**49**) in an aqueous medium, one does not obtain the expected monomeric cyclopentadiene carboxylic acid (**50**), but the dimer of this compound. It can be assumed that the monomeric acid is initially formed on photolysis and that this then dimerizes [*28*]. The acid is also produced in the dimeric form when prepared according to THIELE's method [*31*].

The same effect is observed in the photolysis of the diazo compound **51** in water. In this case, both the monomeric acid (**51**) and the dimeric acid are formed [*32*].

The photolysis of quinone diazides of heterocyclic ring systems is another field which has been investigated mainly by Süs and his school. In the case of 3-diazo-2(3H)-pyridone (**53**) the intermediate product (**54**) undergoes rearrangement through migration of a nitrogen-carbon bond [*32*]. In the presence of water, the pyrrole-2-carboxylic acid is formed (**55**).

The photosynthesis of indole carboxylic acid (**57**) from 3-diazo-4(3H)-quinolone (**56**) is worthy of note, since the photoreaction proceeds very smoothly in this case. The reaction product, indole-3-carboxylic acid (**57**) separates out from an aqueous acidic solution during illumination [*33*].

Süs and MÖLLER [*34*] have obtained 6-azaindole-3-carboxylic acid (**59**) by irradiation of aqueous solutions of 3-diazo-1,7-naphthyridin-4(3H)-one

(58). This synthesis is important because harmyrin(60) can be readily produced on this basis, by heating of 59 in a paraffin-bath previously heated to 200°. Harmyrin (6-azaindole) has been obtained as a degradation product of certain harmala-alkaloids by PERKIN JR. et al. [*35, 36*].

$$58 \xrightarrow[-N_2; +H_2O]{h\nu} 59 \xrightarrow{200°} 60$$

Conversions analogous to **58** → **59** were carried out by MÖLLER [*37*] with 3-diazo-1,6-naphthyridin-4(3H)-one (**61**) and 3-diazo-3,4-dihydro-4-oxo-1,8-naphthyridine-7-carboxylic acid (**62**).

61   **62**

Indene-1-carboxylic acid (44) [*27*]. The quinonediazide **42** (0.5 g) (prepared from 1-amino-2-naphthol) was finely ground in order to dissolve it in a mixture of water (200 ml) and hydrochloric acid (20 ml). Ethyl alcohol (60 ml) was added and the mixture illuminated with an arc lamp (12 amp) in a photo-dish whilst shaken and cooled in ice. The diazo-compound decomposes within about an hour of illumination (no further coupling with phloroglucin) and indene-1-carboxylic acid separates as flakes from the solution, which smells of aniseed, after standing for a short time (yield 0.3 g). After filtration, the crude product is recrystallized from benzene. The carboxylic acid **44** crystallizes as long needles to rod-shaped prisms (m.p. 161°).

Indole-3-carboxylic acid (57) [*33*]. 3-Diazo-4(3H)-quinolone (**56**) (2 g) is dissolved in glacial acetic acid (40 ml) and exposed to sunlight or irradiated with a sealed arc lamp, after the addition of water (160 ml). The vessel is externally cooled with ice during the experiment and the pale yellow solution gradually becomes brown in color as the product separates out as a pale brown solid. When no more diazo-compound can be detected by alkaline coupling of the solution with phloroglucin, the reaction mixture is suction-filtered. The residue is dissolved in bicarbonate solution and this is treated with animal charcoal. On addition of hydrochloric acid, the product separates out and indole-3-carboxylic acid is obtained as white needles by recrystallization from aqueous acetone (m.p. 218°, yield 0.9 g).

d) Photolysis of cyclic α-diazoketones not leading to ring contraction

By analogy, photolysis of 1-diazo-3,3-diphenyl-2-indanone (**63**) in an aqueous medium should lead to the formation of **64**. However, according to CAVA [*38*] ultraviolet irradiation of **63** in a two-phase solvent system of ether and aqueous sodium bicarbonate yielded the lactone **65**; similarly photolysis of 2-diazo-3,3-diphenyl-1-indanone (**66**) yielded only **65**. For the mechanism of these reactions see CAVA.

References, p. 312

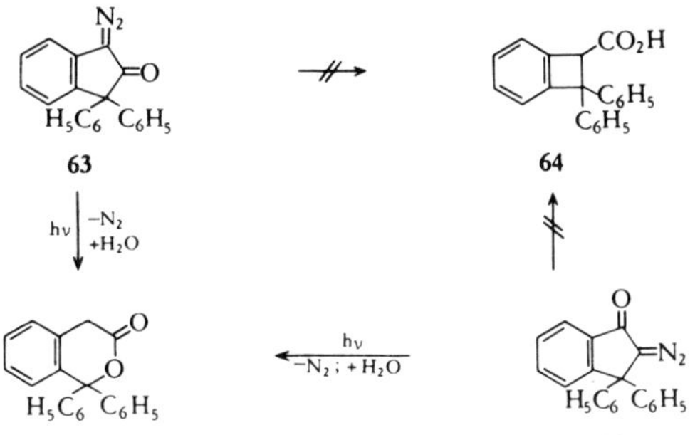

The photolysis of 3-diazo-1-methyl-oxindole (**67**) shows that, as in the preceding example, reaction products are formed without WOLFF rearrangement. Instead the intermediary carbene adds to solvents or to olefins present in the reaction mixture. The following scheme depicts the formation of, e.g. 3-ethoxy-1-methyloxindole (**68**), 3-chloro-3-trichloromethyl-1-methyloxindole (**69**), and 1′-methyl-2,2-diphenylspiro[cyclopropane-1,3′-indolin]-2′-one (**70**) [*39*] (comp. also chapter 31).

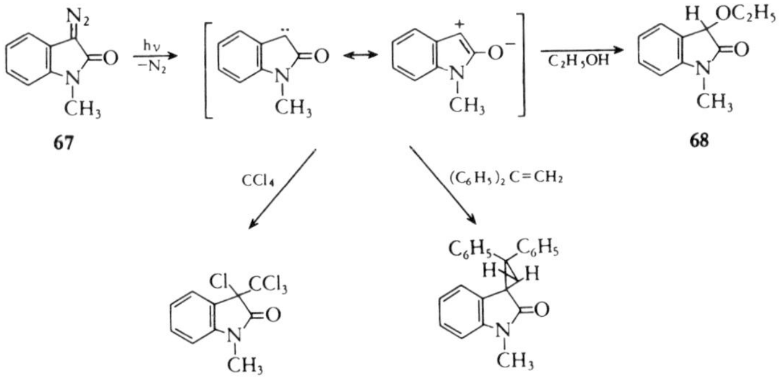

The aromatic ketocarbene which results from 3,4,5,6-tetrachloro-o-benzoquinone diazide (**71**) by thermal or photochemical loss of nitrogen, is no longer capable of the WOLFF rearrangement [*40*]. This explains the course of the photolysis of **71** in methyl alcohol, which yields 3,4,5,6-tetrachloro-2-methoxyphenol (**72**) as the only definite product. The photolysis of **71** in phenyl mustard oil yields 4,5,6,7-tetrachloro-2-phenylimino-1,3-benzoxathiole (**73**).

The photolysis of **71** in cyclohexanone or acetone yielded 4,5,6,7-tetrachlorospiro[1,3-benzodioxole-2,1′-cyclohexane] (**74**) and 4,5,6,7-tetrachloro-2,2-dimethyl-1,3-benzodioxole (**75**) respectively. 4,5,6,7-Tetrachloro-2-methylbenzoxazole (**76**) is formed by the photolysis of **71** in acetonitrile.

1,1-Diphenyl-3-isochromanone (65) [*38*]. A solution of 500 mg of **63** in 200 ml of ether was mixed with 200 ml of 1% aqueous sodium bicarbonate solution in a 1 litre Pyrex flask. The resulting suspension was stirred under nitrogen and irradiated for 12 hrs. with a Westinghouse 100 W. mercury spotlight, without filter. The two phases of the reaction mixture were separated and subjected to several counter-extractions with ether and aqueous sodium bicarbonate solution. The aqueous extracts were combined and acidified with dilute HCl; resulting suspension was extracted several times with ether, the ether extracts were evaporated to dryness, and the residue was dissolved in methylene chloride and chromatographed with the same solvent on Woelm Grade II acid alumina to give 234 mg (48%) of lactone **65**, m.p. 163—164°.

3-Ethoxy-1-methyloxindole (68) [*39*]. A stirred solution of 1.00 g (5.77 mole) of **67** in 250 ml of absolute ethanol was irradiated at room temperature for 18 hrs. under nitrogen. The solution was evaporated to dryness on a steam bath, and the resulting residue was dissolved in 15 ml of chloroform. Chromatography on aluminium oxide and elution with carbon tetrachloride, followed by evaporation of the eluate to dryness, afforded a pale yellow oil which ultimately crystallized to give 0.268 g (24%) of crude **68**. One recrystallization from petroleum ether gave pure 3-ethoxy-1-methyloxindole (**68**), m.p. 57—58°, as compact, white needles.

3,4,5,6-Tetrachloro-2-methoxyphenol (72) [*40*]. A solution of 2.54 g **71** (9.85 mmole) in 80 ml methanol was irradiated by a watercooled mercury quartz immersion lamp Q 81. Within 2 hrs., 91 mol % of N_2 had come off. After distilling off the methanol, 1.79 g pale yellow oil came over at 110—120°/0.01 mm Hg; on cooling this solidified with crystallization (70% crude product). After repeated extractions from petroleum ether (30—40°) the colorless needles of the readily soluble 3,4,5,6-tetrachloro-2-methoxyphenol (**72**) melted at 123—124°.

4,5,6,7-Tetrachlorospiro[1,3-benzodioxole-2,1′-cyclohexane] (74) [*40*]. 2.68 g **71** (10.4 mmole) were dissolved in 10 ml warm benzene, mixed with 60 ml cyclohexanone and irradiated by the quartz immersion lamp; after evolution of 75% nitrogen during 3.5 hrs. the reaction came to a halt. After distillation of the cyclohexanone in a water-pump vacuum, 0.79 g of a light yellow oil passed over at 135—150°/0.001 mm Hg. Crystallized from acetone, **74** had m.p. 150—152°, yield 0.34 g (10%).

References, p. 312

4,5,6,7-Tetrachloro-2-methylbenzoxazole (76) [*40*]. 2.02 g **71** (7.85 mmole) were photolyzed in 60 ml acetonitrile using the water-cooled immersion lamp; 1.04 mole equivalents of gas were liberated in 3 hours. From the dark oil which came over at 160-180°/ 0.005 mm Hg, by triturating with methanol, 0.35 g crystalline **76** (16.5%) was obtained. High vacuum sublimation at 120° and repeated crystallization from methanol gave colorless needles, m. p. 140.5—141.5°.

4. Azo dyes from quinone diazides

The normal course of photolysis of quinone diazides leads to the formation of products no longer containing the diazo nitrogen (cf. p. 304). In some cases, however, photolysis involves a considerably more complex reaction and the end-product contains an azo group with nitrogen derived from the diazo group of the starting material. These reactions will be described using the example of 3-diazo-4-oxo-1,5-cyclohexadiene-1-sulfonic acid (**77**) [*27*].

When **77** (obtained from 2-aminophenol-4-sulfonic acid) is irradiated in aqueous solution, red-yellow crystals of the azo compound **81** are formed. The formation of this compound takes place via the intermediate which undergoes ring contraction to form the cyclic ketene **79**. The latter adds on 1 mol. of water, in accordance with the familiar addition properties of ketenes, forming 5-carboxy-1,3-cyclopentadiene-2-sulfonic acid (**80**), which gives the azo dye **81** by addition to unconverted quinone diazide **77**.

The scarlet azo dye **84**, obtained from the photolysis of 4-diazo-3,4-dihydro-3-oxo-1-naphthalenesulfonic acid (**82**), is of interest because it is formed not only photochemically but also by direct action of **82** on the 1-carboxyindene-3-sulfonic acid (**83**) in a dark reaction [*27*]. This dark reaction supports the assumption that the photochemical formation of **84** also proceeds via **83**.

The photochemical properties of the diazo compound **85** from 3-amino-4-hydroxy-2,6-dimethylpyridine are of particular interest [*33*]. When this compound is subjected to photolysis, ring contraction gives rise to 2,5-dimethylpyrrole-3-carboxylic acid (**86**) but only in good yield when carried out at low temperature (ice-cooling) and with a strong light source.

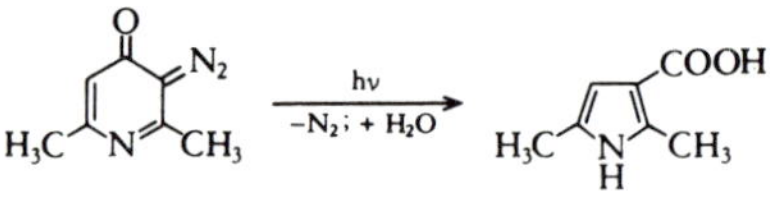

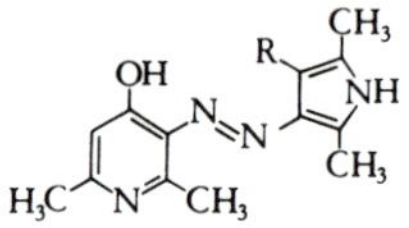

When this is done without cooling, **85** reacts with **86** to give the azo dye **87** which readily loses carbon dioxide leaving the azo compound **88**. It is not necessary to isolate the intermediate products in the photochemical preparation of **88** since 4-hydroxy-2,6-dimethylpyridine-3,3′-azo-2,′5′-dimethylpyrrole (**88**) can be obtained directly by solar irradiation of a solution prepared by diazotization of 3-amino-4-hydroxy-2,6-dimethylpyridine.

5. p-Quinone diazides and p-iminoquinone diazides

If a thin layer of p-benzoquinone diazide (**89**) is exposed to light in the absence of water and organic solvents, products are obtained which are completely insoluble in organic solvents. The following reaction is assumed to take place [*41*].

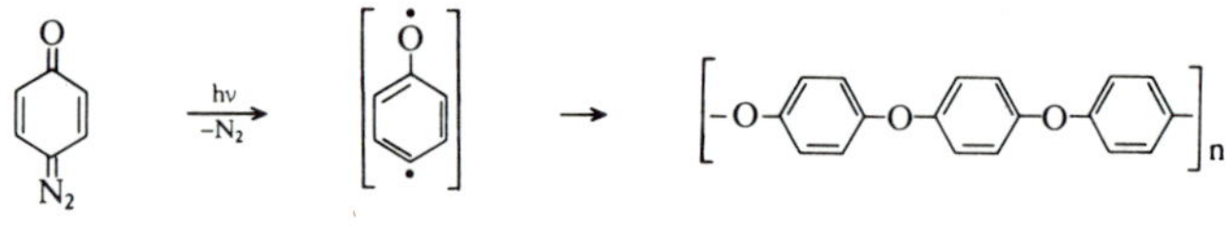

On the other hand, photochemical decomposition of p-benzoquinone diazide (**89**) in the presence of primary aliphatic alcohols results in the addition of the alcohol to the molecule **90** giving p-alkoxyphenols (e.g. **91**). p-Iminoquinone diazides behave in the same way (see **92** → **93**).

References, p. 312

If irradiation of p-quinone diazides is carried out in the presence of aromatic hydrocarbons, nuclear arylation occurs giving p-hydroxybiphenyls [41].

Pyridine reacts in the same manner as aromatic hydrocarbons. Bleaching of 4-diazo-2,6-dichloro-2,5-cyclohexadien-1-one in the presence of pyridine gives 2,6-dichloro-4-(x-pyridyl)-phenol (**94**) [41].

Nuclear arylation also occurs in the photolysis of p-iminoquinone diazide in the presence of aromatic hydrocarbons [41].

Hydroquinone monoethyl ether (91, R = C$_2$H$_5$) [41]. 1,4-Benzoquinone diazide (**89**) (0.8 g) was dissolved in absolute alcohol (80 ml) and the solution irradiated in a shallow dish with an arc lamp. As soon as the diazo-compound could no longer be detected, the solution was evaporated under reduced pressure. Hydroquinone monoethyl ether was obtained giving a melting point of 66° after two recrystallizations from water (yield 0.5 g).

4-Ethoxydiphenylamine (93, R = C$_2$H$_5$) [41]. 4-Diazo-N-phenyl-2,5-cyclohexadien-1-imine (**92**) (2 g) was illuminated, in absolute ethanol (800 ml), with an arc lamp, with efficient cooling. After evaporation of the solvent, the residue was digested four times with 50 ml portions of ether. The residue from the ether extractions was boiled with gasoline and the solution reduced to a small volume. On cooling, 4-ethoxydiphenylamine crystallized out in colorless flakes (m.p. 72—73°, yield 0.12 g).

4-Hydroxybiphenyl [41]. 1,4-Benzoquinone diazide (**89**) (0.4 g) was suspended in benzene (60 ml) and illuminated with the arc lamp. The diazide gradually dissolved. The residue from the mixture was dissolved in benzene (20 ml) and, after treatment with charcoal, mixed with hot gasoline until turbidity set in. On cooling, colorless crystals separated out, m.p. 164° (from benzene/gasoline) yield 0.16 g.

References

[1] L. Horner and E. Spietschka: Chem. Ber. **85**, 225 (1952).
[2] L. Horner, E. Spietschka and A. Gross: Liebigs Ann. Chem. **573**, 17 (1951).
[3] V. Franzen: Liebigs Ann. Chem. **602**, 199 (1957).
[4] F. Arndt and B. Eistert: Ber. dtsch. chem. Ges. **68**, 200 (1935).
[5] A. Roedig and H. Lunk: Chem. Ber. **87**, 971 (1954).
[6] F. Weygand and H. J. Bestmann: Chem. Ber. **92**, 528 (1959).
[7] F. Weygand, G. Eberhardt, H. Linden, F. Schäfer and I. Eigen: Angew. Chem. **65**, 525 (1953).
[8] M. L. Wolfrom and J. V. Karabinos: J. Amer. chem. Soc. **68**, 1455 (1946).
[9] L. Horner and E. Spietschka: Chem. Ber. **89**, 2765 (1956).
[10] A. H. Cook and D. G. Jones: J. Chem. Soc. **1941**, 184.
[11] C. K. Ingold and S. D. Weaver: J. Chem. Soc. **127**, 378 (1925).
[12] W. Kirmse and L. Horner: Chem. Ber. **89**, 2759 (1956).
[13] H. Staudinger: Die Ketene. Stuttgart: Enke 1912.
[14] F. Weygand, W. Schwenke and H. J. Bestmann: Angew. Chem. **70**, 506 (1958).
[15] S. Masamune: J. Amer. chem. Soc. **86**, 735 (1964).
[16] M. P. Cava, R. L. Little and D. R. Napier: J. Amer. chem. Soc. **80**, 2257 (1958).
[17] L. Horner, W. Kirmse and K. Muth: Chem. Ber. **91**, 430 (1958).
[18] J. L. Mateos and O. Chao: Bol. inst. quim. univ. nat. autón. Méx. **13**, 3 (1961).
[19] J. Meinwald, G. G. Curtis and P. G. Gassman: J. Amer. chem. Soc. **84**, 116 (1962).
[20] A. Hassner, A. W. Coulter and W. S. Seese: Tetrahedron Letters **1962**, 759.
[21] M. P. Cava and E. Moroz: J. Amer. chem. Soc. **84**, 115 (1962).
[22] G. Muller, C. Huynh and J. Mathieu: Bull. Soc. Chim. France **1962**, 296.
[23] J. L. Mateos, O. Chao and H. Flores R: Tetrahedron **19**, 1051 (1963).
[24] L. Horner and E. Spietschka: Chem. Ber. **88**, 934 (1955).
[25] J. Meinwald, A. Lewis and P. G. Gassman: J. Amer. chem. Soc. **84**, 977 (1962).
[26] K. B. Wiberg, B. R. Lowry and T. H. Colby: J. Amer. chem. Soc. **83**, 3998 (1961).
[27] O. Süs: Liebigs Ann. Chem. **556**, 65 (1944).
[28] O. Süs: Liebigs Ann. Chem. **556**, 85 (1944).
[29] O. Süs, H. Steppan and R. Dietrich: Liebigs Ann. Chem. **617**, 20 (1958).
[30] L. Horner and K.-H. Weber: Chem. Ber. **95**, 1227 (1962).
[31] J. Thiele: Ber. dtsch. chem. Ges. **34**, 68 (1901).
[32] O. Süs and K. Möller: Liebigs Ann. Chem. **593**, 91 (1955).
[33] O. Süs, M. Glos, K. Möller and H.-D. Eberhardt: Liebigs Ann. Chem. **583**, 150 (1953).
[34] O. Süs and K. Möller: Liebigs Ann. Chem. **599**, 233 (1956).
[35] W. H. Perkin, jr. and R. Robinson: J. Chem. Soc. **101**, 1775 (1912).
[36] W. Lawson, W. H. Perkin, jr. and R. Robinson: J. Chem. Soc. **125**, 626 (1924).
[37] K. Möller and O. Süs: Liebigs Ann. Chem. **612**, 153 (1958).
[38] M. P. Cava, D. G. McConnell, K. Muth and M. J. Mitchell: J. Org. Chem. **27**, 1908 (1962).
[39] E. J. Moriconi and J. J. Murray: J. Org. Chem. **29**, 3577 (1964).
[40] R. Huisgen, G. Binsch and H. König: Chem. Ber. **97**, 2868 (1964).
[41] O. Süs, K. Möller and H. Heiss: Liebigs Ann. Chem. **598**, 123 (1956).

Chapter 33

Photochemical syntheses with diazonium salts and diazosulfonates

1. Reductive deamination of diazonium salts

ORTON and COATES [1] investigated the action of sunlight on aqueous solutions of diazonium salts and found a noticeable increase in the rate of decomposition as compared to that of the dark reaction.

Alcohols react with diazonium salts according to (a) and (b), light favoring the reductive deamination (cf. b) [2].

$$[ArN_2]^{\oplus}X^{\ominus} + HO-CH_2R \begin{array}{c} \text{(a)} \nearrow Ar-O-CH_2R + N_2 + HX \\ \\ \searrow ArH + N_2 + HX + OCHR \\ \text{(b)} \end{array}$$

Light-induced reductive deamination proceeds via free radicals; this hypothesis is based on the observation that photodecomposition of diazonium salts in the presence of acrylonitrile causes polymerization of the latter with formation of characteristic end-groups [2].

The following table summarizes the yields of the UV-photolyses of the compounds $[RC_6H_4N_2]^{\oplus}Cl^{\ominus}$ in methanol solution [2]:

Table 25

R	reduction %	ether %
p-NO$_2$	60	0
p-Cl	60	0
p-CH$_3$	40—50	20—30
p-OCH$_3$	40—50	20—25

In some cases it has been found advantageous to conduct the photolysis in isopropanol; thus, anisole is obtained in 70 % yield by photolysis of p-methoxybenzenediazonium chloride in isopropanol.

References, p. 317

Two procedures are used for the photolysis of the diazonium salt. In the first, the diazonium salt is isolated, dissolved in alcohol and this solution irradiated. From the preparative point of view it is, however, better to avoid the isolation; the diazonium salt is prepared in situ from the arylamine salt and ethyl nitrite in the desired alcohol and then irradiated directly.

Chlorobenzene [2]. p-Chlorobenzenediazonium chloride (5—10 g) is dissolved in 150 ml methanol and irradiated with an immersion lamp S 81 (Quarzlampengesellschaft, Hanau), at 0° (ice-cooling). Under these conditions thermal decomposition is negligible. Photolysis is complete after 3—4 hours.

To isolate the product, the reaction solution is poured into a large volume of water, extracted with ether and finally fractionally distilled. A disadvantage of this method of working up is that some diazo-resin is present in the reaction product and this makes the fractional distillation of small quantities difficult. The diazo-resin may be separated, however, by steam-distilling the irradiated solution when the reaction products (reduction product and ether) distill over, leaving the resin as a residue. The steam-distillate is extracted with ether and finally fractionated; chlorobenzene is obtained in 60% yield.

2. Replacement of the diazonium group by halogen or the hydroxy group

A series of aromatic diazonium salts react with hydrochloric acid or with hydrobromic acid under the influence of sunlight yielding halogenated aromatic compounds [3]. In this way, **1** is converted to **2** or **3**, respectively, and 3-diazocarbazole chloride is transformed into 3-chlorocarbazole by irradiation in the presence of HCl.

4   1   2 : X = Cl
            3 : X = Br

The photoreactions of diazonium salts with very dilute sulfuric acid take another course. By analogy with the thermal decomposition of diazonium compounds in aqueous acid solutions whereby the diazonium group is replaced by hydroxyl, on irradiation the corresponding hydroxy compounds are formed. This photoreaction proceeds very smoothly in many cases, while the corresponding thermal decomposition often leads to resin formation. In this manner, 4-anilinobenzenediazonium sulfate (**1**) afforded **4**, and 4-dimethylaminobenzenediazonium chloride gave 4-dimethylaminophenol [3].

The behavior of 2-dimethylaminobenzenediazonium chloride deviates somewhat. On irradiation by sunlight in the presence of concentrated hydrochloric acid the expected 2-chloro-N,N-dimethylaniline is formed and, in the presence of dilute sulfuric acid, 2-dimethylaminophenol, again as expected. Nevertheless, in both cases N-methylaniline may be isolated (30 %) [*3*].

4-Chlorodiphenylamine (2) [*3*]. 10 g 4-anilinobenzenediazonium sulfate (**1**) (technical product) are dissolved in 500 ml water and 750 ml concentrated HCl and the solution exposed to sunlight in a flat dish. The temperature of the solution is maintained at about 12° by ice-cooling. The decomposition of the diazo compound may be followed by observing the rapid decoloration of the yellow solution and the incapacity of coupling with azo-components. After filtration from a small amount of a blue dye, the solution is diluted with water and extracted with ether. The residue from the ethereal solution is crystalline and consists of 4 g of nearly pure 4-chlorodiphenylamine. After recrystallizing from dilute methanol, the product melted at 74° and was identical to a comparison preparation obtained by a SANDMEYER reaction.

3-Chlorocarbazole [*3*]. 20 g 3-diazocarbazole chloride are dissolved in 2 litres concentrated hydrochloric acid and the solution placed in sunlight. 3-Chlorocarbazole, which precipitates, must be repeatedly recrystallized from ethanol in order to purify it. Glistening, mother-of-pearl-like plates, m.p. 199°.

4-Bromodiphenylamine (3) [*3*]. 5 g of **1** are dissolved in 250 ml hydrobromic acid and placed in sunlight. The 4-bromodiphenylamine (3.5 g) which forms, precipitates from the decolorized solution. The raw product contains traces of **4** and is recrystallized from aqueous ethanol to give prismatic needles of **3**, m.p. 88°.

4-Anilinophenol (4) [*3*]. A solution of 10 g of **1** in 1 l water and 10 ml 50 % sulfuric acid is exposed to sunlight. A small quantity of a black-blue precipitate is formed in the bleached solution. After filtration the filtrate is extracted with ether. The residue from the ethereal solution contains about 3 g of **4** which crystallizes in shining platelets resembling mother-of-pearl. It is purified by recrystallizing from a mixture of benzene and petroleum ether, m.p. 70°.

3. Photolysis of diazonium salts as a method of cyclization

HUISGEN and ZAHLER [*4*] have described the preparation of fluoren-9-one (**6**) by irradiation of 2-diazobenzophenone (**5**) in aqueous sulfuric acid.

Fluoren-9-one (6) [*4*]. 1.97 g 2-aminobenzophenone are boiled with 50 ml water and 5 ml concentrated sulfuric acid and diazotized at —5° with 1.38 g sodium nitrite in 15 ml water. After decomposing the excess nitrite with 1.2 g urea and diluting the solution to 180 ml. with water, the reaction mixture was photolyzed in a quartz irradiation apparatus UVM (Quarzlampengesellschaft Hanau), at 50°. The insoluble product separated as a film on the quartz walls and brought the photolysis to a standstill at 80—90 % nitrogen evolution. The combined product from three runs was dissolved in benzene, washed with 2 N NaOH and water and dried. The neutral part, distilled in high vacuum, yielded 2.17 g fluorenone (40 %).

4. Change in reactivity of aryl diazosulfonates

Aryl diazosulfonates, formed by the reaction of aryl diazonium salts with potassium sulfite in alkaline solution, exist in stable and labile forms. The trans azo structure (cf. 7) is assumed for the stable species. However, there is no general agreement as to the constitution of the labile form (cf. [5, 6]).

Unlike the labile form, the stable form of the aryl diazosulfonates does not show the characteristic properties of diazonium salts, e. g., the compounds in this form do not couple with naphthol in alkaline medium. It has been shown in a series of cases that certain diazosulfonates, which show no coupling behavior in the dark, exhibit coupling if the solutions are irradiated. For example, a 1 % solution of the stable form of sodium 3-bromo-p-toluenediazosulfonate (7), which will not couple with 2-naphthol in the dark, yields a coupling product (8) after the solution has been irradiated [7].

$$\text{7} \xrightarrow{h\nu} \left[\text{photo product} \right] \xrightarrow{\text{2-naphthol}} \text{8}$$

The coupling reaction of diazosulfonates with phenols under the influence of irradiation has already been reported upon by FEER [8] in 1889.

1-(2-Bromo-p-tolylazo)-2-naphthol (8) [7]. 3-Bromo-p-toluenediazosulfonate **7** (33 mMole) is dissolved in 100 ml of 0.04 N HCl. This solution, contained in a water-cooled stainless steel basin, was irradiated for 3 mins. by means of a UV lamp Philips SP 500 from a distance of 15 cm from the basin. The irradiated solution was added to a solution of 29 mMole of 2-naphthol in 20 ml 0.4 N NaOH. After addition of sodium chloride (1 g) the red precipitate was filtered off and washed, first with water, then with 50 ml of 0.5 N HCl and finally again with water. After drying the product (0.4 g) in vacuum it was recrystallized from ligroin containing a little benzene to afford pure **8**, m.p. 172°; yield 0.26 g.

5. Photolysis of a 1,2,3-thiadiazine S,S-dioxide

According to HOFFMANN and SIEBER [9] photolysis of naphtho[1,8-de]-[1,2,3]thiadiazine 1,1-dioxide (9) in benzene solution gave, besides "naphthothiam blue", naphtho[1,8-bc]thiete 1,1-dioxide (10) and dinaphtho-[1,8-bc:1',8'-fg][1,5]dithiocine 7,7,14,14-tetroxide (11). Separate irradiation of either **10** or **11** in ethanol showed the substances to be photochemically interconvertible.

References, p. 317

Naphtho[1,8-bc]thiete 1,1-dioxide (10) [9]. A benzene solution of **9** was irradiated with a Quarzlampengesellschaft Q 81 lamp at 15° for 36 hours. Concentration of the reaction mixture caused the precipitation of 3% of **11**, decomposing above 200°. Removal of the solvent and extraction of the residue with cyclohexane gave a 25% yield of **10**, m.p. 184°.

References

[1] K. J. P. ORTON, J. E. COATES and F. BURDETT: J. Chem. Soc. **91**, 35 (1907).
[2] L. HORNER and H. STÖHR: Chem. Ber. **85**, 993 (1952).
[3] O. SÜS: Liebigs Ann. Chem. **557**, 237 (1947).
[4] R. HUISGEN and W. D. ZAHLER: Chem. Ber. **96**, 736 (1963).
[5] H. H. HODGSON and E. MARSDEN: J. Chem. Soc. **1943**, 470.
[6] H. C. FREEMAN and R. J. W. LE FÈVRE: J. Chem. Soc. **1951**, 415.
[7] J. DE JONGE and R. DIJKSTRA: Rec. Trav. chim. Pays-Bas **75**, 290 (1956).
[8] A. FEER: DRP 53455 of 5. 12. 1889.
[9] R. W. HOFFMANN and W. SIEBER: Angew. Chem. **77**, 810 (1965); internat. ed. **4**, 786 (1965).

Chapter 34

Synthetic applications of light-induced reactions of azides

Under the action of light, azides are capable of losing nitrogen to give **1**, where R may be (among others) an alkyl group, an aryl group, an acyl group or a sulfonyl group.

$$RN_3 \xrightarrow{h\nu} [RN] + N_2$$
$$\quad\quad\quad\quad\quad \mathbf{1}$$

The intermediate may theoretically exist either in the triplet biradical state (**2**) or in the singlet state (**3**).

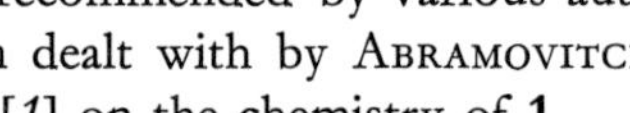

$$R-\dot{\ddot{N}}\cdot \quad\quad\quad R-\ddot{\ddot{N}}$$
$$\mathbf{2} \quad\quad\quad\quad \mathbf{3}$$

The terminology used to denote **1** is not uniform. The term "imenes" is used in analogy to "carbenes", and the terms "azenes", "nitrenes" and "imidogens" are recommended by various authors. The problem of nomenclature has been dealt with by ABRAMOVITCH and DAVIS in their comprehensive survey [1] on the chemistry of **1**.

The formation of **1** from azides can take place thermally as well as photochemically. It is consequently not surprising that photolysis and pyrolysis of azides frequently lead to the same products. HORNER et al. [2] observed the formation of 3-phenyl-2H-azirine (**5**) on thermal or photochemical decomposition of α-azidostyrene (**4**).

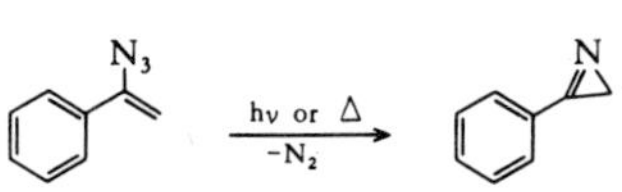

References, p. 327

1. Photolysis of alkyl azides

According to BARTON [*3*] alkyl azides yield "nitrenes" on photolysis and these can react in different ways. Isomerization via 1,2-hydrogen shift may lead to the imine (cf. **7**, route A) and hydrogen abstraction from the solvent would give the amine (cf. **8**, route B). A cyclization route (C) which would afford pyrrolidine (**9**) from **6** [*3*] was not verifiable [*4, 5*].

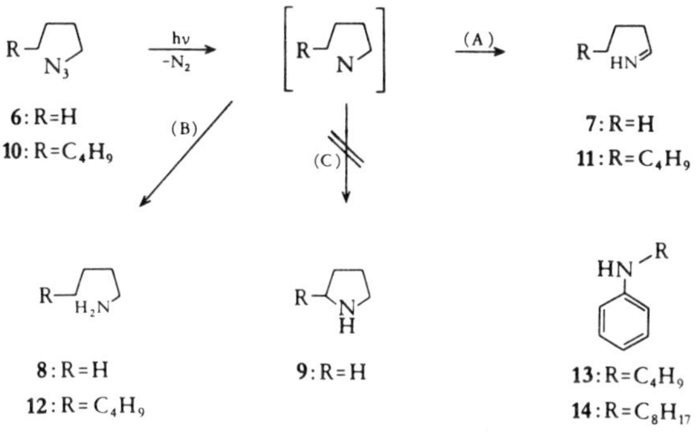

6: R=H
10: R=C$_4$H$_9$

7: R=H
11: R=C$_4$H$_9$

8: R=H
12: R=C$_4$H$_9$

9: R=H

13: R=C$_4$H$_9$
14: R=C$_8$H$_{17}$

Thus, 1-azidobutane (**6**), when photolyzed in a variety of solvents, gave mainly butylidenimine (**7**) besides some 1-butylamine (**8**), and 1-azido-octane (**10**) on similar treatment gave 1-octanimine (**11**) as the major product together with 1-octylamine (**12**) [*5*].

Irradiation of **6** or **10** in benzene resulted in addition of solvent, N-butylaniline (**13**) and N-phenyl-1-octylamine (**14**) being the respective products while azidocyclohexane, when irradiated in cyclohexane solution, was reduced to cyclohexylamine [*3*].

No products other than imines resulting from 1,2-shifts were isolated following photolyses of triarylmethyl azides [*6*]. Thus, irradiation of azidotriphenylmethane (**15**) in hexane with a low-pressure mercury lamp produced N-(diphenylmethylene)-aniline (**16**) in good yield. 1-Azido-1,1-diphenylethane (**17**) and 2-azido-2-phenylpropane (**18**) were photolyzed as well, but product mixtures were more complex in these cases.

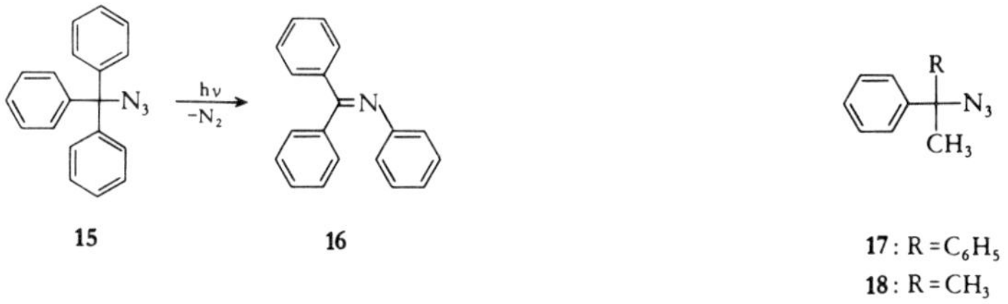

15

16

17: R=C$_6$H$_5$
18: R=CH$_3$

Cyclohexylamine [*3*]. A solution of azidocyclohexane (5 g) in 350 ml of cyclohexane, contained in a quartz flask under nitrogen, was irradiated for two hrs. with a 500 W lamp at room temperature. Removal of the solvent in vacuo, chromatography of the resulting oil over alumina and elution with light petroleum gave 2 g (51%) of cyclohexanone. Further elution with chloroform-benzene (2 : 3) afforded cyclohexylamine, isolated as the hydrochloride (1.5 g, 33%).

N-(Diphenylmethylene)-aniline (16) [*6*]. A solution of 202.9 mg of **15** in 12 ml hexane was placed in a cylindrical cell with quartz windows of ca. 12 ml capacity. The solution was irradiated for 2.5 hrs. with a Hanovia SC 2537 low-pressure mercury vapor lamp without filters. Chromatography over alumina and elution with hexane-benzene (19 : 1) gave 129.1 mg (64%) of unreacted azide. Elution with benzene gave 55.1 mg (83% based on **15** consumed) of the imine **16**.

2. Photolysis of aryl azides

a) Carbazoles and 4-phenylbenzofuroxan from 2-azidobiphenyl derivatives

When solutions of certain 2-azidobiphenyls (cf. **19**) are exposed to UV-light, carbazoles are formed [*7*]. The synthesis of carbazole (**20**), 2,7-dinitrocarbazole and related compounds takes the same path as the synthesis of 1,3-dibromocarbazole (**22**) from 2-azido-3,5-dibromobiphenyl (**21**), described below in more detail. Carbazole formation also occurs in the thermal decomposition of substances in the 2-azidobiphenyl series, e.g. **20** is obtained from **19** in 1,2,4-trichlorobenzene at about 180° [*7*]. Higher yields are usually obtained in the thermal processes.

2-Azido-3-nitrobiphenyl (**23**) reacts differently in forming 4-phenylbenzofuroxan (**24**) as the main product of photolysis.

19: R=H
21: R=Br

20: R=H
22: R=Br

23

24

1,3-Dibromocarbazole (22) [*7*]. A solution of 1 g of **21** in 20 ml tetralin was irradiated in a quartz vessel using a mercury vapor lamp (100 W). Evolution of nitrogen began after a few minutes. After 3 hrs., when no more nitrogen was evolved, the solution was concentrated to 5 ml using a hot-plate and a stream of dry air. The light tan oil resulting was crystallized from aqueous ethanol to afford pale yellow crystals of **22**, m.p. 106—107°; yield 0.53 g.

References, p. 327

b) Photochemical conversion of aryl azides to azo compounds

According to Horner [2] both 4-azidoanisole and 4-azidobiphenyl (25) can be converted to the corresponding azo compounds on irradiation.

$$2 \; C_6H_5{-}C_6H_4{-}N_3 \xrightarrow[-2\,N_2]{h\nu} C_6H_5{-}C_6H_4{-}N=N{-}C_6H_4{-}C_6H_5$$

25 → 26

4,4′-Diphenylazobenzene (26) [2]. When a solution of 4-azidobiphenyl (25) (1.95 g) in 100 ml benzene is irradiated with a Quarzlampengesellschaft high-pressure mercury lamp S 81, flakes with a golden sheen separate out within a short time. In the course of 4—5 hours, the expected quantity of 240 ml of nitrogen has been generated. The solvent is distilled off and the residue crystallized from chloroform, yielding 26 as orange flakes, m.p. 250°; yield 1.35 g (81%).

3. Photolysis of acyl azides

a) The light-induced Curtius rearrangement

Many acyl azides exhibit Curtius rearrangement on heating, yielding isocyanates. This rearrangement can also take place photochemically, e.g. photolysis of benzoyl azide (27) yields phenyl isocyanate (28) [8].

$$R-CON_3 \xrightarrow[-N_2]{h\nu} R-N=C=O$$

27: R = C_6H_5 28: R = C_6H_5

According to Horner and Gross [9], phenylacetyl azide (29) in methanol can be converted to methyl benzylcarbamate (30), and adipic acid diazide (31) in ethanol can be converted to diethyl tetramethylenedicarbamate (32).

$$C_6H_5-CH_2-CON_3 + CH_3OH \xrightarrow[-N_2]{h\nu} C_6H_5-CH_2-NH-COOCH_3$$

29 → 30

$$N_3\underset{\underset{O}{\|}}{C}-(CH_2)_4-\underset{\underset{O}{\|}}{C}N_3 + 2\,C_2H_5OH \xrightarrow[-2\,N_2]{h\nu} H_5C_2O-\underset{\underset{O}{\|}}{C}-\overset{H}{\underset{}{N}}-(CH_2)_4-\overset{H}{\underset{}{N}}-\underset{\underset{O}{\|}}{C}-OC_2H_5$$

31 → 32

A more systematic study [10] revealed, that the intermediate acyl imene 33 originating from 27 via loss of nitrogen may undergo a variety of reactions with alcohols. Thus, photolysis of benzoyl azide in ethanol gives rise to 42% of ethyl carbanilate (34), 32% of ethyl benzohydroxamate (35) and 24% of benzamide. Acetaldehyde, the oxidized product corresponding to benzamide, is isolated in 21% yield. With benzophenone as sensitizer a quantitative conversion to benzamide and acetaldehyde is observed.

$$27 \xrightarrow[-N_2]{h\nu} \left[H_5C_6-\overset{\overset{O}{\|}}{C}-N \atop 33 \right] \xrightarrow{C_2H_5OH} H_5C_6-CONH_2 + CH_3-CHO$$

$$\downarrow \qquad\qquad | \xrightarrow{C_2H_5OH} \qquad\qquad \downarrow$$

$$\underset{28}{H_5C_6-N=C=O} \xrightarrow{C_2H_5OH} \underset{34}{H_5C_6-NH-COOC_2H_5} \qquad \underset{35}{H_5C_6-CO-NH-OC_2H_5}$$

Salicyloyl azide (**36**), when irradiated in benzene solution, undergoes an interesting rearrangement to 2-benzoxazolinone (**37**) [*9*].

Phenyl isocyanate (28) [*8*]. Dry benzoyl azide (**27**) (14.7 g) is dissolved in absolute benzene (100 ml) and irradiated for 12 hours with a lamp S 81 (Quarzlampengesellschaft, Hanau). This is done at 6° and nitrogen is passed through the solution during the experiment. The solution, which has a pungent odor, is mixed with aniline (10 ml) and left to stand overnight; diphenylurea crystallizes out, m.p. 232.5°, yield 12 g (56% theoretical).

The mother liquor is evaporated to dryness and the semi-crystalline viscous residue is treated with petroleum ether. A crystalline product (6 g) is obtained and this is pressed on clay. Colorless rhomboids of benzanilide are obtained after two recrystallizations (m.p. 165°, yield 30.4%). There is, therefore, an observed transformation of 87% of the original acyl azide to the decomposition product. Phenyl isocyanate can be obtained with petroleum ether as solvent, and ethyl phenylcarbamate with alcohol.

Methyl benzylcarbamate (30) [*9*]. Phenylacetyl chloride (7.7 g) is dissolved in ether (70 ml) and vigorously shaken for $1/_2$ hour with a concentrated aqueous solution of sodium azide (4 g). The separated and dried ethereal solution of the acid azide is mixed with dry methanol (50 ml), diluted with absolute ether to 170 ml and irradiated for $7^1/_2$ hours (immersion lamp S 81) with external ice-cooling. After evaporating the solvent under reduced pressure, an oily mixture (9 g) of urethan and methyl phenylacetate is obtained, probably as a result of reaction of the acyl azide with methanol. The ester is steam distilled and the residue recrystallized from ether-petroleum ether to give 1.2 g of **30**, m.p. 61—62°.

2-Benzoxazolinone (37) [*9*]. A solution of 2 g salicyloyl azide (**36**) in 160 ml of dry benzene is irradiated as above for 4.5 hours under nitrogen. After evaporation of the solvent, a crystalline residue (1.5 g) remains. Two recrystallizations from benzene give pure **37**, m.p. 139°, identical with the compound prepared by STOERMER [*11*].

b) Formation of lactams

According to APSIMON and EDWARDS [*12*] photochemical reactions of acyl azides are useful in organic synthetic work, an example being the formation of lactams. Thus, the perhydrophenanthrene derivative **38**, when refluxed for one hour in dry hexane, produced the corresponding isocyanate **39**. When the same solution was irradiated with UV-light at room temperature, the δ-lactam **40** (25%) was obtained.

References, p. 327

Photolysis of O-methyl podocarpic acid azide (**41**) gave 4aβ-aminomethyl-1,2,3,4,4a,9,10,10a-octahydro-6-methoxy-1α-methyl-trans-phenanthrene-1β-carboxylic acid lactam (**42**) [*12*] which was related to a degradation product of the alkaloid atisine.

When a structurally related azide, viz. 1,1-dimethyl-trans-decahydro-4a-naphthalenecarboxylic acid azide (**43**) was photolyzed in hexane at 0°, a greater variety of compounds was obtained [*13—15*], possibly due to the enhanced flexibility of the decalin system as compared to that of the perhydrophenanthrene system. Thus photolysis produced 35% of 4a-isocyanato-1,1-dimethyl-trans-decahydronaphthalene (**44**), 9% of 1-aminomethyl-1-methyl-trans-decahydro-4a-naphthalenecarboxylic acid lactam (**45**), 14% of a γ-lactam which was regarded as 6-amino-1,1-dimethyl-trans-decahydro-4a-naphthalenecarboxylic acid lactam (**46**) and 8% of 1,1-dimethyl-trans-decahydro-4a-naphthalenecarboxamide (**47**) [*14*].

Application of the photolysis method to acyl azides of the triterpene series [16] again proved the usefulness of the method in functionalizing angular methyl groups.

4aβ-Aminomethyl-1α-methyl-trans,anti,cis-perhydrophenanthrene-1β-carboxylic acid lactam (40) [12]. When a hexane solution of 1α,4aβ-dimethyl-trans,anti,cis-perhydrophenanthrene-1β-carboxylic acid azide (38) was irradiated in a 1 cm quartz cell using a Hanovia ultraviolet lamp at room temperature, there was a steady evolution of nitrogen. After completion of the reaction the products were separated by chromatography over alumina. 1β-Isocyanato-1α.4aβ-dimethyl-trans,anti,cis-perhydrophenanthrene (39) which was the main product (65%) was eluted very readily with hexane. Benzene-chloroform eluted the δ-lactam 40 (25%). Recrystallization from ether-pentane and sublimation at 120°/0.001 mm yielded colorless needles of 40, m.p. 183°.

c) Formation of amides

In the experiments mentioned above the lactams were usually accompanied by amides which were formed from the intermediary acyl imenes, the respective solvents serving as hydrogen donors. Analogously hexanamide (49) was formed from hexanoyl azide (48) on irradiation in cyclohexane [12].

$$C_5H_{11}-C\underset{N_3}{\overset{O}{\diagup}} \xrightarrow[C_6H_{12}]{h\nu} N_2 + C_5H_{11}-C\underset{NH_2}{\overset{O}{\diagup}}$$

$$\qquad 48 \qquad\qquad\qquad\qquad\qquad 49$$

Hexanamide (49) [12]. Hexanoyl chloride (2 g) in dioxane (25 ml) and sodium azide (2.5 g) in water (10 ml) were shaken vigorously for 5 mins. and extracted with cyclohexane. The organic layer was well washed with water, 5% sodium bicarbonate, and water again, then dried with anhydrous sodium sulfate. This solution was irradiated at room temperature for 12 hours. A Hanovia ultraviolet lamp was used and the photolysis cell (thickness 1 cm, 150 ml capacity) was fitted with a circulating device. The only product that could be readily separated from the irradiation mixture was hexanamide (230 mg), which after sublimation at 70°/0.0001 mm had m.p. 99—100°.

4. Photochemical reaction of ethyl azidoformate

a) Reaction with cyclic hydrocarbons

When a dilute cyclohexane solution of ethyl azidoformate (50) was irradiated in the absence of oxygen an almost quantitative yield of nitrogen was evolved. A 51% yield of ethyl cyclohexanecarbamate (51) was obtained besides 12% of urethan [17, 18].

$$N_3-C\underset{OC_2H_5}{\overset{O}{\diagup}} + \bigcirc \xrightarrow[-N_2]{h\nu} \bigcirc\!\!-\!\!N(H)\!-\!C(=\!O)\!-\!OC_2H_5 + H_2N-C\underset{OC_2H_5}{\overset{O}{\diagup}}$$

$$\qquad 50 \qquad\qquad\qquad\qquad 51$$

With cyclohexene not only insertion reactions and reduction of the intermediate carboethoxy imene occurred but also addition, affording, as the main product, 7-carboethoxy-7-azabicyclo[4.1.0]heptane (52). Relative yields of the ethyl 1-, 2-, and 3-cyclohexenylcarbamates were, 1%, 3% and 9%, respectively, while **52** was obtained in a yield of 56%. Urethan and bi-2-cyclohexen-1-yl were present in trace and 4% amounts, respectively [*18*]. Lowering the temperature (−75°) raised the yield of **52** to 75%. On the other hand, addition of a sensitizer greatly reduced the amount of insertion and addition products in favor of urethan and bicyclohexenyl.

An interesting synthesis of azepine derivatives using photolysis of **50** was reported by HAFNER and KÖNIG [*19*]. When ethyl azidoformate was dissolved in benzene and irradiated with UV light, a 70% yield of ethyl 1H-azepine-1-carboxylate (**54**) was obtained. This ring enlargement reaction is formulated as proceeding via the aziridine **53** in close analogy to the reaction of benzene with diazomethane to yield cycloheptatriene (comp. p. 276) [*19*].

b) Reaction with alcohols

Photolysis of **50** in tert.-butanol gives 60% of ethyl N-tert.-butoxycarbamate (**55**) [*20*]. With lower aliphatic alcohols yields of the corresponding ethyl N-alkoxycarbamates drop significantly while high yields of ethyl carbamate are obtained (see below).

$H_5C_2O-\overset{O}{\underset{\|}{C}}-N_3 \xrightarrow[ROH;\ -N_2]{h\nu} H_5C_2O-\overset{O}{\underset{\|}{C}}-NH-OR\ +\ H_5C_2O-\overset{O}{\underset{\|}{C}}-NH_2$

50

R = CH$_3$	28%	52%
R = C$_2$H$_5$	3%	97%
R = (CH$_3$)$_2$CH	10%	90%
55: R = (CH$_3$)$_3$C	60%	

Irradiation of tert.-butyl azidoformate (**56**) in tert.-butanol does not only lead to addition of the alcohol to the acyl imene with formation of 29% of tert.-butyl N-tert.-butoxycarbamate (**57**) but also to 5,5-dimethyl-2-oxazolidinone (**58**) via an intramolecular insertion reaction of the imene [*20*].

$$(CH_3)_3CO-\overset{\overset{O}{\|}}{C}-N_3 \xrightarrow[(CH_3)_3COH]{h\nu} (CH_3)_3CO-\overset{\overset{O}{\|}}{C}-NH-OC(CH_3)_3 + \underset{N}{\overset{O}{\underset{|}{\bigtriangleup}}}=O$$
56   57   58

5. Photochemical syntheses with sulfonic acid azides

a) Photolysis in methanol

According to HORNER and CHRISTMANN [21], the photolysis of p-toluenesulfonyl azide (59) in methanol has the following mechanism:

$$p\text{-}CH_3\text{-}C_6H_4\text{-}SO_2\text{-}N_3 \xrightarrow[-N_2]{h\nu} \left[p\text{-}CH_3\text{-}C_6H_4\text{-}SO_2\text{-}\overline{N}\cdot \right]$$
59

$$\xrightarrow{CH_3OH} p\text{-}CH_3\text{-}C_6H_4\text{-}SO_2\text{-}NH\text{-}OCH_3$$
60

O-Methyl-N-(p-toluenesulfonyl)-hydroxylamine (60) is formed. Its constitution was verified by an independent synthesis via reaction of p-toluenesulfonyl chloride with O-methylhydroxylamine.

O-Methyl-N-(p-toluenesulfonyl)-hydroxylamine (60) [22]. p-Toluenesulfonyl azide (7.88 g) is dissolved in methanol (270 ml); 10 ml are withdrawn, diluted with 10 ml of water and a pH-value of 6.65 is found. After allowing to stand for 3 hours, the pH drops to 5.25. If the low pressure lamp is then switched on evolution of gas begins immediately. When ca. 90% of the nitrogen has been generated, the solution has a pH-value of 2.4. The irradiated solution is evaporated and the residue worked up with ether. 0.6 g (8.5% theoretical) of ammonium p-toluenesulfonate remains undissolved. The ethereal solution is then extracted by shaking twice with sodium hydroxide solution. On acidification of the alkaline solution with 2 N HCl a substance (still brown in color) separates out. This is dissolved in ethanol and treated with animal charcoal. After distilling off the ethanol, 60 remains as a colorless mass of crystals which can be recrystallized from ethanol/water or from ligroin, m.p. 113°; yield 3.3 g (44%).

b) Photolysis in sulfoxides

When p-toluenesulfonyl azide (59) is photolyzed in sulfoxides, N-sulfonyl sulfoximides (cf. 61) are formed [22].

$$\underset{H_3C}{\overset{H_3C}{>}}SO + N_3\text{-}SO_2\text{-}\underset{}{\bigcirc}\text{-}CH_3 \xrightarrow[-N_2]{h\nu} \underset{H_3C}{\overset{H_3C}{>}}\underset{\overset{\downarrow}{O}}{S}=N\text{-}SO_2\text{-}\underset{}{\bigcirc}\text{-}CH_3$$
                59   61

This reaction can take place thermally, at the decomposition temperature of the azide, as well as photochemically.

S,S-Dimethyl-N-(p-toluenesulfonyl)-sulfoximide (61) [22]. A solution of 59 (5.91 g) in 270 ml of dimethylsulfoxide is irradiated in a flow-apparatus with a Quarzlampengesellschaft Hanau 30/89 low-pressure lamp. The calculated quantity of nitrogen is evolved within a 4-hour period, after which the solution is distilled off and the residue crystallized out by scratching in methanol. The crystals are suction-filtered. A brown oil separates out from the deep red-brown mother liquor when mixed with water, and eventually crystallizes after long standing and treatment in a refrigerator. Repeated recrystallization from water gives colorless crystals of **61**, m. p. 168°; yield 2.4 g (32%).

c) Photolysis in sulfides

Sulfides undergo a similar reaction to that of sulfoxides with sulfonyl azides when exposed to light, yields being, however, higher in most cases [22]. Thus, sulfimides (cf. **62**) may be obtained in yields of ca. 50%.

$$H_3C-S-CH_3 + N_3-SO_2-\langle\rangle-CH_3 \xrightarrow[-N_2]{h\nu} \begin{array}{c}H_3C\\H_3C\end{array}\!\!>\!\!S\!=\!N\!-\!SO_2-\langle\rangle-CH_3$$

$$\qquad\qquad 59 \qquad\qquad\qquad\qquad\qquad\qquad\qquad 62$$

S,S-Dimethyl-N-(p-toluenesulfonyl)-sulfimide (62) [22]. A solution of p-toluenesulfonyl azide (**59**) (4.93 g) in 100 ml of dimethyl sulfide is irradiated with a S 81 high-pressure lamp. The irradiation apparatus is cooled in ice-water. Evolution of gas begins immediately and colorless crystals separate from the pale yellow solution after some time. When 400 ml (ca. 40%) of nitrogen have been evolved, irradiation is discontinued. The precipitate is suction-filtered, the solvent distilled off and the residue boiled with hot water. The residue from the aqueous extraction is combined with the precipitate and recrystallized from water in the presence of charcoal to give pure **62**, m. p. 158—159°; yield 2.1 g (54.5%).

References

[1] R. A. ABRAMOVITCH and B. A. DAVIS: Chem. Rev. **64**, 149 (1964).
[2] L. HORNER, A. CHRISTMANN and A. GROSS: Chem. Ber. **96**, 399 (1963).
[3] D. H. R. BARTON and L. R. MORGAN JR.: J. Chem. Soc. **1962**, 622.
[4] D. H. R. BARTON and A. N. STARRATT: J. Chem. Soc. **1965**, 2444.
[5] R. M. MORIARTY and M. RAHMAN: Tetrahedron **21**, 2877 (1965).
[6] W. H. SAUNDERS JR., and E. A. CARESS: J. Amer. chem. Soc. **86**, 861 (1964).
[7] P. A. S. SMITH and B. B. BROWN: J. Amer. chem. Soc. **73**, 2435 (1951).
[8] L. HORNER, E. SPIETSCHKA and A. GROSS: Liebigs Ann. Chem. **573**, 17 (1951).
[9] L. HORNER and A. GROSS: personal communication to A. S.
[10] L. HORNER, G. BAUER and J. DÖRGES: Chem. Ber. **98**, 2631 (1965).
[11] R. STOERMER: Ber. dtsch. chem. Ges. **42**, 3133 (1909).
[12] J. W. APSIMON and O. E. EDWARDS: Can. J. Chem. **40**, 896 (1962).
[13] W. L. MEYER and A. S. LEVINSON: Proc. Chem. Soc. **1963**, 15.
[14] W. L. MEYER and A. S. LEVINSON: J. Org. Chem. **28**, 2859 (1963).
[15] R. F. C. BROWN: Austral. J. Chem. **17**, 47 (1964).
[16] S. HUNECK: Chem. Ber. **98**, 2305 (1965).
[17] W. LWOWSKI and T. W. MATTINGLY: Tetrahedron Letters **1962**, 277.
[18] W. LWOWSKI and T. W. MATTINGLY JR.: J. Amer. chem. Soc. **87**, 1947 (1965).
[19] K. HAFNER and C. KÖNIG: Angew. Chem. **75**, 89 (1963); internat. edition **2**, 96 (1963).
[20] R. KREHER and G. H. BOCKHORN: Angew. Chem. **76**, 681 (1964); internat. edition **3**, 589 (1964).
[21] L. HORNER and A. CHRISTMANN: Angew. Chem. **75**, 707 (1963); internat. edition **2**, 599 (1963).
[22] L. HORNER and A. CHRISTMANN: Chem. Ber. **96**, 388 (1963).

Chapter 35

Photolysis of pyrazolines, pyrazoles, azo compounds, 1,2,3-thiadiazoles, and p-benzoquinone diimine N,N'-dioxides

1. Photolysis of pyrazolines

Though thermal decomposition of pyrazolines is a well-known route to cyclopropanes, the synthetic value of this reaction is reduced by the extensive formation of olefinic products and by the lack of stereospecificity.

RINEHART JR. and VAN AUKEN [1, 2] found, however, that when methyl cis-3,4-dimethyl-1-pyrazoline-3-carboxylate (1) was irradiated with a sunlamp at room temperature, methyl cis-1,2-dimethylcyclopropanecarboxylate (2) was obtained in 63—76% yield. The major side product was methyl tiglate (3). Analogously methyl trans-3,4-dimethyl-1-pyrazoline-3-carboxylate (4) gave methyl trans-1,2-dimethylcyclopropanecarboxylate (5) and methyl angelate (6).

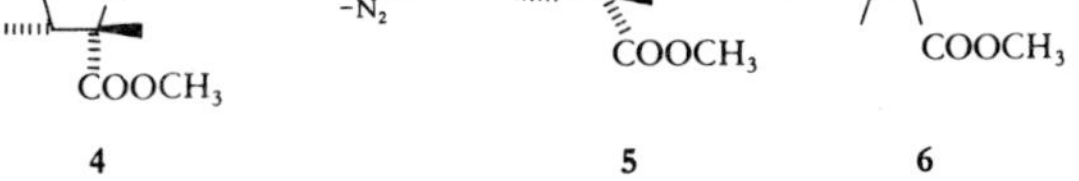

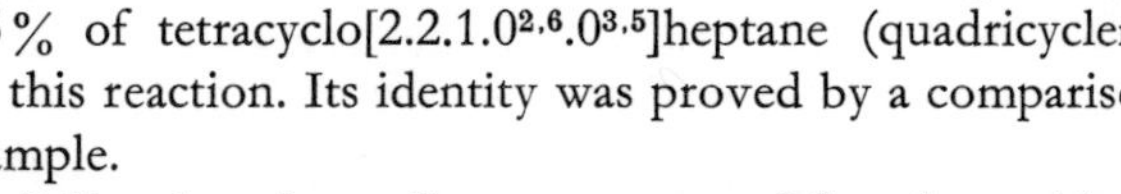

Irradiation of an ethereal solution of 8,9-diazatetracyclo[4.3.0.0^{2,4}.0^{3,7}]-non-8-ene (7) led to a loss of the azo ultraviolet absorption [3]. A total yield of 35% of tetracyclo[2.2.1.0^{2,6}.0^{3,5}]heptane (quadricyclene, 8) was obtained in this reaction. Its identity was proved by a comparison with an authentic sample.

The possibility that the cyclic azo compound first forms bicyclo[2.2.1]-heptadiene (norbornadiene, 9) which then isomerizes to 8, was ruled out by the fact that no conversion of 9 to quadricyclene took place under the reaktion conditions employed.

References, p. 333

7   8   9

Investigations on the photolytic decomposition of steroid pyrazolines were made by Kocsis and collaborators [4]. 5α-Cholestane-3,6-dione 4α,5-pyrazoline (10) and similar compounds were irradiated with UV light in dioxane solution. In the photolysis of 10 the authors succeeded in isolating 4-methylcholest-4-ene-3,6-dione (11) which is the known [5] pyrolysis product of 10. Irradiation also produced a substance isomeric with 11, in approximately the same yield. This compound was assigned the structure of 4α,5-methylene-5α-cholestane-3,6-dione (12) on the basis of UV and IR spectra [4].

10   11   12

Tetracyclo[2.2.1.0^{2,6}.0^{3,5}]heptane (8) [3]. A solution of 0.5 g of 7 in 100 ml of dry ether was contained in a Pyrex flask under nitrogen. The vessel was externally irradiated at reflux temperature for 8.5 hrs. with an uncooled Quarzlampengesellschaft Q 81 immersion lamp. At the end of this period, the ether solution was concentrated to about 15 ml, at a bath temperature of 50°, and washed with a saturated solution of silver nitrate and then with water. The solution was dried and distilled to give a yellow oil. This was distilled in a micro distillation apparatus at 95°. IR-spectroscopic examination of the middle portion of five fractions showed that it consisted of essentially pure 8.

4α,5-Methylene-5α-cholestane-3,6-dione (12) [4]. A solution of 10 (460 mg) in 100 ml of dioxane was irradiated for 1 hour. The residue from evaporation of the solvent (from 5 identical batches) was chromatographed on alumina. Petroleum ether/benzene (3 : 1) eluted a crystalline compound (739 mg), which after repeated recrystallization from methylene chloride/methanol had m.p. 171°.

2. Photolysis of pyrazoles

While some steroidal 1-pyrazolines furnish cyclopropane compounds on irradiation, the isomeric 2-pyrazolines are stable towards light [4]. The unsaturated congeners of 1-pyrazolines, viz. 3H-pyrazoles (cf. 13), give cyclopropenes on photolysis. For a review on cyclopropene chemistry see CARTER and FRAMPTON [6].

CLOSS and BÖLL [7] have reported briefly on the photochemical formation of 1,3,3-trimethylcyclopropene (14) from 3,3,5-trimethyl-3H-pyrazole (13), by irradiation with a Hanovia UV-lamp at 13° (yield 65 %).

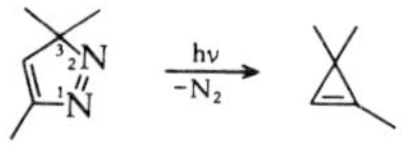

EGE [8] photolyzed 4',5'-dicarbomethoxyspiro[fluorene-9,3'-[3H]- pyrazole] (15) in benzene or tetrahydrofuran solution, using a Quarzlampengesellschaft Q 81 mercury high-pressure lamp, and obtained 2,3-dicarbomethoxyspiro[cyclopropene-1,9'-fluorene] (16) as an almost colorless product after repeated recrystallization from methanol or ether/petroleum ether (m.p. 146°; yield ca. 70%).

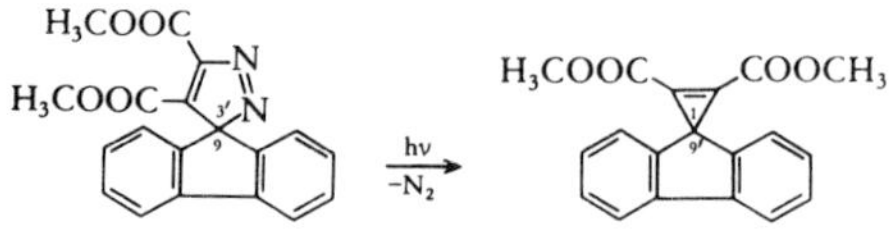

EGE [8] also carried out the photolysis of dimethyl 3,3-diphenyl-3H-pyrazole-4,5-dicarboxylate (17) in methanol, benzene, or tetrahydrofuran at 30°. The product was not the cyclopropene derivative anticipated but colorless crystals of dimethyl 3-phenylindene-1,2-dicarboxylate (18, m.p. 93—94°; yield 65%).

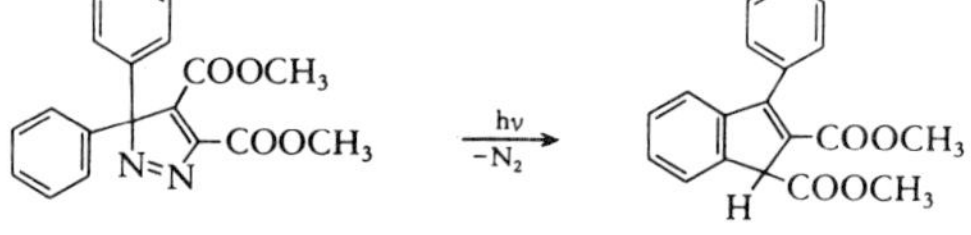

3. Photolysis of diaroyl azo compounds

No appreciable evolution of nitrogen occurred when a benzene solution of azodibenzoyl (dibenzoyldiimide, 19) was irradiated with UV light, but bis-(4-chlorobenzoyl)-diimide (20), under the same conditions, decomposed to form 4,4'-dichlorobenzil (21) [9]. The photolysis of bis-(2-chlorobenzoyl)-diimide (22) followed the same course, but no p-anisil (24) was obtained from di-p-anisoyldiimide (23) [9]. At higher temperatures, however, also 19 was found photolyzable [10].

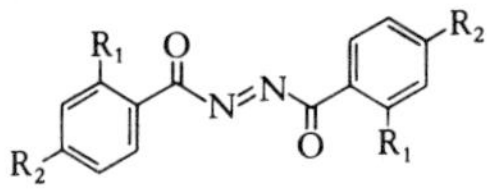

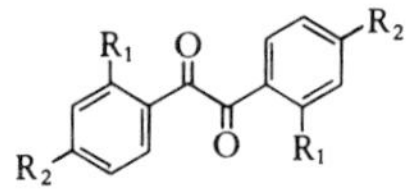

19: $R_1 = R_2 = H$
20: $R_1 = H, R_2 = Cl$
22: $R_1 = Cl, R_2 = H$
23: $R_1 = H, R_2 = OCH_3$

21: $R_1 = H, R_2 = Cl$

24: $R_1 = H, R_2 = OCH_3$

The photolysis of diaroyl azo compounds was investigated with regard to the possibility of free radicals being formed. The simplest proof of the existence of free radicals is provided by their ability to induce polymerization. During the photodecomposition of **20** in the presence of acrylonitrile and benzene, only a small quantity of polymer was formed; about as much as was formed from **21** alone [*9*]. It was, therefore, assumed that in the photolysis of **20** free radicals are only formed in a very small quantity, if at all.

More recent investigations [*10*] showed that indeed aroyl radicals may play a role in the photolytic decomposition of diaroyl azo compounds. A high number of fission and recombination products was obtained, among which were also 1,3,4-oxadiazole derivatives.

4,4′-Dichlorobenzil (21) [*9*]. Photolysis of **20** was carried out under nitrogen with the exclusion of moisture, using a Quarzlampengesellschaft S 81 mercury immersion lamp. Reference is made to the original paper for experimental details.

The solution of **20** (4.5 g) in absolute benzene (150 ml) changed color from red to yellow with considerable evolution of nitrogen within 4 hrs. of irradiation. On concentration of the solution under reduced pressure a small amount of 1,2-bis-(4-chlorobenzoyl)-hydrazine separated out first. On further concentration of the filtrate to about one third of the original volume, a yellow solid (1.8 g) precipitated. Recrystallization from benzene gave pure **21**, m. p. 193°.

4. Photolysis of 1,2,3-thiadiazoles

According to KIRMSE and HORNER [*11*] (comp. also [*12*]), 1,2,3-thiadiazoles (cf. **25**) are very stable thermally but decompose readily on photolysis with the liberation of nitrogen. Photolysis yields products which can be regarded as derived from **26**. Thus the formation of 1,4-dithiins (cf. **27**) can easily be explained; an example is the formation of tetraphenyl-1,4-dithiin (**27**) from 4,5-diphenyl-1,2,3-thiadiazole (**25**) [*11*]. For the photolysis of a 1,2,3-thiadiazine derivative see chapter 33.

25: $R_1 = R_2 = C_6H_5$
30: $R_1 = H, R_2 = C_6H_5$

26

27: $R_1 = R_2 = C_6H_5$

28

29: $R_1 = R_2 = C_6H_5$
31: $R_1 = H, R_2 = C_6H_5$

Apart from the 1,4-dithiins formed in the photolysis of 1,2,3-thiadiazoles, compounds of a novel type (cf. **29**) are also obtained. The name dithiafulvenes was suggested [*11*] for these substances, but the correct designation is as derivatives of 2-methylene-1,3-dithiole. Their formation can be explained as follows: the primary decomposition product **26** is supposed to rearrange to a thioketene (cf. **28**) which then reacts with a further molecule of the kind **26** to give the 1,3-dithiole derivative. In this way, 2-benzylidene-4-phenyl-1,3-dithiole (**31**) was obtained from 4-phenyl-1,2,3-thiadiazole (**30**) [*11*].

Tetraphenyl-1,4-dithiin (27) and 4,5-diphenyl-2-diphenylmethylene-1,3-dithiole (29) [*11*]. 4,5-Diphenyl-1,2,3-thiadiazole (**25**) (2.4 g) in boiling benzene (70 ml) was irradiated with an immersion lamp S 81 until 200 ml of nitrogen had been liberated. The benzene was distilled off and the residue digested with 100 ml of ethanol. After cooling the product was filtered off, dried and treated with 20 ml of concentrated sulfuric acid. It was then suction filtered through a sintered glass filter and washed with a little concentrated H_2SO_4. The residue on the filter was recrystallized from chloroform/ethanol after being washed with water and methanol. Thus, 0.1 g of **27** were obtained, m.p. 184° (dec.).

The acid filtrate was poured into 500 ml of water and the yellow precipitate so formed was suction filtered and thoroughly washed with methanol. Recrystallization from dimethylformamide afforded fine yellow needles (0.4 g) of **29**, m.p. 234°.

2-Benzylidene-4-phenyl-1,3-dithiole (31) [*11*]. A quartz flow-apparatus (UVM) with a glass centrifugal pump was filled with a solution of **30** (16.2 g) in 1.1 liters of benzene. Irradiation with a Quarzlampengesellschaft S 700 mercury lamp was continued until 1000 ml of N_2 had been evolved (1.5—2 hrs.). Evaporation of the solvent in vacuo and crystallization from 100 ml of ether gave a 55% yield of yellow **31**, which after recrystallization from toluene or glacial acetic acid had m.p. 207°. The above ethereal filtrate on evaporation and steam-distillation gave 8.3 g of starting material **30**.

5. Photolysis of p-benzoquinone diimine N,N'-dioxides

N,N'-Disubstituted p-benzoquinone diimine N,N'-dioxides (cf. **32**) are rapidly decomposed by light of wavelengths 300—450 mµ. p-Benzoquinone imine N-oxides (cf. **34**) are formed at first, and then these are converted to p-benzoquinone and azo compounds (cf. **33**) [*13*].

32: R = H
35: R = OCH₃   **33**: R = H   **34**: R = H

When solutions containing two different quinone diimine N,N'-dioxides are photolyzed, mixed azo compounds are formed in addition to the symmetrical azo derivatives, e.g., when N,N'-diphenyl-p-benzoquinone diimine N,N'-dioxide (**32**) and N,N'-bis-(4-methoxyphenyl)-p-benzoquinone diimine N,N'-dioxide (**35**) are irradiated together, azobenzene (**33**), 4-methoxyazobenzene and 4,4'-dimethoxyazobenzene may be obtained [*13*].

References, p. 333

N-Phenyl-p-benzoquinone imine N-oxide (34) *[13]*. Compound **32** (1 g) was dissolved in 200 ml of dry, thiophene-free benzene in an Erlenmeyer flask. This solution was exposed for 4 hrs. to the radiation from a Hanovia type A (385—580 W) high-pressure mercury quartz lamp through a Corning 3060 filter. Four such solutions were combined after irradiation and concentrated under reduced pressure at room temperature. The concentrate (40 ml) was heated to 50° in order to ensure complete solution, and 80 ml of petroleum ether were added. Brown crystals (**A**) separated out (2.415 g), and azobenzene (**33**) was obtained from the filtrate by evaporation of the solvent and recrystallization from aqueous ethanol (m.p. 68°; yield 1.3 g).

The crystals **A** were dissolved in benzene and 3 vols. of petroleum ether were added. A brown precipitate formed which was immediately filtered off. At room temperature, red-brown crystals of **34** (m.p. 141—142°) separated from the filtrate in the course of 24 hours.

References

[1] K. L. RINEHART JR. and T. V. VAN AUKEN: J. Amer. chem. Soc. **82**, 5251 (1960).
[2] T. V. VAN AUKEN and K. L. RINEHART JR.: J. Amer. chem. Soc. **84**, 3736 (1962).
[3] R. M. MORIARTY: J. Org. Chem. **28**, 2385 (1963).
[4] K. KOCSIS, P. G. FERRINI, D. ARIGONI and O. JEGER: Helv. Chim. Acta **43**, 2178 (1960).
[5] L. F. FIESER: J. Amer. chem. Soc. **75**, 4386 (1953).
[6] F. L. CARTER and V. L. FRAMPTON: Chem. Rev. **64**, 497 (1964).
[7] G. L. CLOSS and W. BÖLL: Angew. Chem. **75**, 640 (1963); internat. ed. **2**, 399 (1963).
[8] G. EGE: Tetrahedron Letters **1963**, 1667.
[9] L. HORNER and W. NAUMANN: Liebigs Ann. Chem. **587**, 93 (1954).
[10] D. MACKAY, U. F. MARX and W. A. WATERS: J. Chem. Soc. **1964**, 4793.
[11] W. KIRMSE and L. HORNER: Liebigs Ann Chem. **614**, 4 (1958).
[12] W. KIRMSE: Angew. Chem. **71**, 537 (1959).
[13] C. J. PEDERSEN: J. Amer. chem. Soc. **79**, 5014 (1957).

Chapter 36

Miscellaneous light-induced reactions of organic nitrogen compounds

1. Incorporation of C_1 or C_2 fragments by the photochemical reaction of various nitrogen compounds with alcohols

a) Benzo[f]quinolines from SCHIFF's bases

Ultraviolet irradiation [1] of the SCHIFF's base N-benzylidene-2-naphthylamine (1) in ethanol in the presence of air gave as the main product (30—40 % yield) a crystalline material of m.p. 188°. This was shown to be 3-phenylbenzo[f]quinoline (2). Analogous irradiation of N-(4-methoxybenzylidene)-2-naphthylamine (3) gave 3-(4-methoxyphenyl)-benzo[f]quinoline (4).

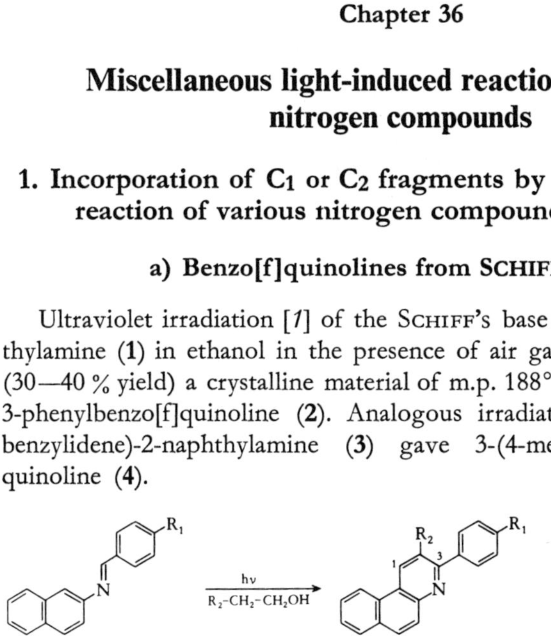

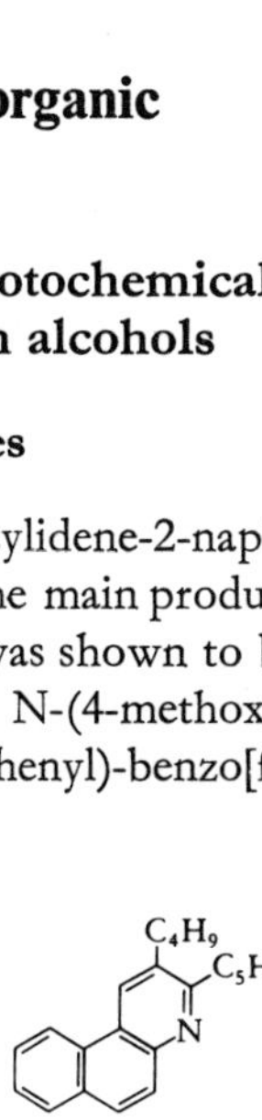

1: R_1=H
3: R_1=OCH$_3$

2: R_1 = R_2=H
4: R_1=OCH$_3$, R_2=H
5: R_1=H, R_2=CH(CH$_3$)$_2$
6: R_1=H, R_2=n-C$_4$H$_9$

7

SHANNON et al. [1] assumed that the C_2 fragment which had been incorporated in the course of these photoreactions was derived from acetaldehyde formed by the photooxidation of ethanol.

This hypothesis concerning a novel participation of the solvent could be substantiated [2] as follows. When higher primary alcohols were used as solvents with the above SCHIFF's bases, side chains of appropriate length were introduced into the resulting benzo[f]quinolines. In this way 2-isopropyl-3-phenylbenzo[f]quinoline (5) was produced from 1 on irradiation in isopentyl alcohol. Irradiation of 1 in n-hexanol gave a 25 % yield of 2-butyl-3-phenylbenzo[f]quinoline (6) besides 37 % of 2-butyl-3-pentylbenzo[f]quinoline (7) [2].

References, pp. 339—340

b) Imidazolidines from diamines or from SCHIFF's bases

According to CERUTTI and SCHMID [3] irradiation of the methanol solution of N,N'-diphenylethylenediamine (8) leads to the formation of 1,3-diphenylimidazolidine (9), the reaction not being sensitive towards oxygen. This addition of a C_1 fragment was shown to require either light of wavelength < 190 mµ or of wavelength 254 mµ in conjunction with a sensitizer. The formation of 9 could not be detected on heating 8 in methanol for 6 hrs. at 130° in a nitrogen or air atmosphere. The intermediacy of ·CH_2OH radicals was supposed [3].

Similarly the action of methanol on N-benzylidenemethylamine (10) in light [4] leads to the formation of meso- and d,l-1,3-dimethyl-4,5-diphenylimidazolidine (11 and 12).

When the benzylidene moiety was contained in a ring system an analogous incorporation reaction occurred. Thus 3,4-dihydro-1-methylisoquinoline (13) on irradiation in methanol solution in the presence of a sensitizer furnished meso- and d,l-5,6,10,11,15b,15c-hexahydro-15b,15c-dimethyl-8H-imidazo[5,1-a:4,3-a']diisoquinoline (cf. 14) [4].

The same compounds were also obtainable in a similar way from the 1,1',2,2',3,3',4,4'-octahydro-1,1'-dimethyl-1,1'-biisoquinolines (15) [4]. This photocondensation parallels that of 8 to 9.

c) Oxazolidines from SCHIFF's bases

The formation of oxazolidine derivatives was observed [3] on irradiation of certain 3H-indoles in methanol. In this way 9,9a-dihydro-9,9-dimethyl-9a-phenyl-1H,3H-oxazolo[3,4-a]indole (17) was prepared from 3,3-dimethyl-2-phenyl-3H-indole (16) in methanol, acetone or benzophenone serving as sensitizers. This photoreaction was found to be sensitive towards oxygen, while irradiation of 3,3-dimethyl-2-phenyl-2-indolinemethanol (**18**), affording **17** as well, was not. Compound **18** was, however, not regarded an intermediate in the **16** → **17** photoconversion.

A reaction similar to **16** → **17** takes place on irradiation of 2,3,4,4a-tetrahydro-4a-methyl-1H-carbazole (**19**) in pure methanol [*3*]: 5,6,7,7a-tetrahydro-7a-methyl-1H,3H,4H-oxazolo[4,3-k]carbazole (**20**) is formed in 50 % yield. Acid hydrolysis of the latter yielded 1,2,3,4,4a,9a-hexahydro-4a-methylcarbazole-9a-methanol (**21**) which on irradiation in methanol regenerated **20**.

The transformation **19**→**20** was inhibited by oxygen whereas that of **21**→**20** was not. In explanation of the conversion **19**→**20** it is assumed [*3*] that the ·CH_2OH residue attaches itself twice to the starting material followed by elimination of the elements of water. In the case of unsensitized reactions, the radical ·CH_2OH may be formed by direct photolysis of methanol; in a sensitized reaction the excited sensitizer abstracts a hydrogen atom with formation of this radical.

The photochemical reaction of 2,3,4,4a-tetrahydro-4a,9-dimethyl-1H-carbazolium iodide (**22**) in methanol-acetone yielded the oily 1,2,3,4,4a,9a-hexahydro-4a,9-dimethylcarbazole-9a-methanol (**23**) [*3*].

1,3-Diphenylimidazolidine (9) [*3*]. N,N'-Diphenylethylenediamine (**8**) (1.1g) in 700 ml methanol and 5 ml acetic acid was irradiated for 14 hrs. by a low pressure mercury burner Hanau NK 6/20 under nitrogen in a flow apparatus. Finally the reaction mixture was evaporated under vacuum and the residue in benzene solution chromatographed twice on silica gel. The fraction eluted first contained 220 mg (18%) of **9**, m. p. 125.5—126° (from pentane). Starting material was eluted next (20—30%) while resinous substances remained on the column.

Meso- and d,l-1,3-dimethyl-4,5-diphenylimidazolidine (11 and 12) [*4*]. 1.0 g of N-benzylidenemethylamine (**10**) in 500 ml methanol was irradiated under argon in a quartz flow apparatus by a low pressure mercury lamp Hanau NK 6/20. Finally the mixture was evaporated in vacuo and the oily residue chromatographed on silica gel. At first

References, pp. 339—340

by-products and later two main products were eluted with chloroform-methanol (100:1.5). The fast moving fraction was distilled at 80—90°/0.001 mm Hg, whereupon it crystallized. Recrystallization from aqueous methanol gave 96 mg (13%) of **11**, m. p. 48—49°. The slower moving main product was chromatographed again and purified by distillation at 70—80°/0.001 mm Hg to give 92 mg (12.4%) of **12** as a colorless oil.

5,6,7,7a-Tetrahydro-7a-methyl-1H,3H,4H-oxazolo[4,3-k]carbazole (20) [*3*]. A solution of 1.22 g of **19** in 500 ml. methanol was irradiated in four batches with a Philips 93109 E low pressure mercury lamp. After 48 hrs. the mixture was evaporated in vacuum and the residue chromatographed in benzene solution on silica gel. A small quantity of a by-product was followed by the main product which in turn was followed by some starting material. Distillation at 100—110°/0.001 mm Hg gave 575 mg (51%) of an oil which crystallized on cooling. Recrystallization from pentane and methanol-water and high-vacuum sublimation gave pure **20**, m.p. 83.5°.

2. Photolysis of oxadiazolinones

The photolysis of ethyl 5-oxo-4-phenyl-Δ^2-1,2,4-oxadiazoline-3-carboxylate (**24**) in dry dioxane yielded ethyl 2-benzimidazolecarboxylate (**25**) [*5*]. This reaction was also carried out with the corresponding isopropyl, butyl and cyclohexyl esters.

24: R=COOC$_2$H$_5$ 25: R=COOC$_2$H$_5$
26: R=C$_6$H$_5$ 27: R=C$_6$H$_5$

Pyrolysis of 3-carbalkoxy-Δ^2-1,2,4-oxadiazolin-5-ones (cf. **24**) proceeds with loss of two molecules of carbon dioxide and formation of disubstituted cyanamides (cf. [*5*]). In the case of phenyl substituted 1,2,4-oxadiazolinones both pyrolysis and photolysis produce benzimidazoles [*6*]. Thus, 3,4-diphenyl-Δ^2-1,2,4-oxadiazolin-5-one (**26**) may be pyrolyzed to 2-phenylbenzimidazole (**27**) in good yield, the decarboxylation being catalyzed by peroxides [*6*]. The same compound is also obtained at a very much lower temperature by UV-irradiation of **26** in dry dioxane.

Ethyl benzimidazole-2-carboxylate (25) [*5*]. The ester **24** (3 g) in 250 ml absolute dioxane was irradiated with a Quarzlampengesellschaft S 81 Hg-vapor lamp at 40—50° while being agitated by a stream of nitrogen. After 5—6 hrs. the evolution of carbon dioxide ceased. Evaporation of the solution yielded 60—70% of **25**, m.p. 221° (from aqueous ethanol).

3. Light-induced abnormal benzidine rearrangement

The photochemical conversion of 4-methylazobenzene (**28**) to 2-methylbenzo[c]cinnoline (**29**) in 22 N sulfuric acid [*7*] has already been noted (comp. chapter 14). In this reaction, 4-(4-aminophenyl)-4-methylcyclohexa-2,5-dienone (**33**) is produced as a second product.

The formation of **29** was formulated [7] as a photochemical disproportionation process, 4-methylhydrazobenzene (**30**) being the hydrogenated reaction partner. In the dark this rearranges to 2-amino-5-methyl-diphenylamine (**31**). In light, however, the benzidine rearrangement seems to take an abnormal course leading to the imine **32** which hydrolyzes to the dienone (**33**) actually observed.

4-(4-Aminophenyl)-4-methylcyclohexa-2,5-dienone (33) [7]. 4-Methylazobenzene (**28**) (3.01 g) in 135 ml 22 N sulfuric acid was illuminated for 105 hrs. with a 125 W Philips mercury lamp. The mixture was then partially neutralized with an aqueous solution of 80 g NaOH with cooling. Extraction with benzene gave 1.50 g (50%) of **29**, m.p. 137—138°. The acidic solution, after extraction with benzene, was made weakly alkaline with aqueous sodium hydroxide and steam distilled. The residue was extracted with ether and after removal of solvent from this extract 1.216 g of a red-brown solid was obtained. Recrystallization from ethanol afforded pale yellow prisms of **33**, m.p. 167—168°.

4. Light-induced condensations involving primary amines and aldehydes

The synthesis of lobelanine (**34**) from glutaraldehyde, methylamine hydrochloride and benzoylacetic acid at pH 4 and 25° [8] usually takes several days for completion. Irradiation was reported [9] to accelerate this condensation considerably, a short illumination being usually sufficient to induce reaction.

In a similar manner the condensation reactions of aniline and related bases, e.g. 1-naphthylamine and benzylamine, with aldehydes, viz. benzaldehyde, p-anisaldehyde and cinnamaldehyde in a buffered medium were reported to be accelerated by irradiation [10].

References, pp. 339—340

5. Aromatic nitriles by photochemical cleavage of aromatic aldazines

When benzaldehyde azine (35) in cyclohexane was photolyzed in quartz with a low-pressure mercury arc at room temperature, only about 1 % of benzonitrile (36) could be isolated [*11*]. Irradiation was carried out in quartz (3 days) since in Pyrex the reaction time was three times longer. However, if two equivalent weights of benzophenone were added, an 85 % yield of benzonitrile was obtained in the same time. Thus, with cyclohexane, benzene, or dioxane as solvents, a number of azines were converted to the corresponding nitriles and benzopinacol (37). The yields indicated in table 26 are based on 1—5 g experiments; the best procedure was to irradiate the solutions until the yellow color of the azines disappeared or became very faint.

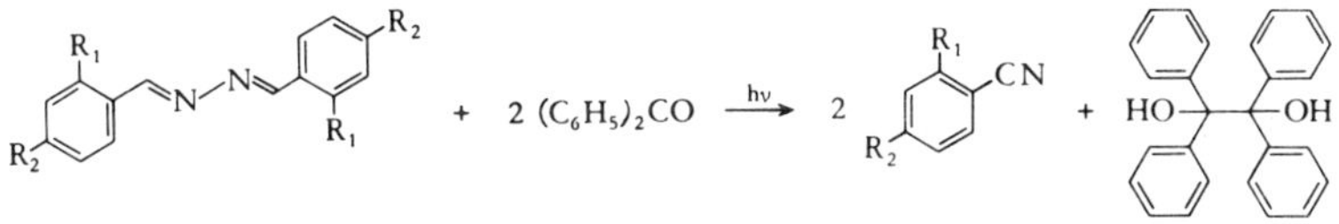

35: $R_1=R_2=H$
38: $R_1=Cl, R_2=H$
40: $R_1=H, R_2=Cl$
42: $R_1=H, R_2=OCH_3$
44: $R_1=H, R_2=NO_2$
46: $R_1=H, R_2=N(CH_3)_2$

36: $R_1=R_2=H$
39: $R_1=Cl, R_2=H$
41: $R_1=H, R_2=Cl$
43: $R_1=H, R_2=OCH_3$
45: $R_1=H, R_2=NO_2$
47: $R_1=H, R_2=N(CH_3)_2$

Table 26

azine	nitrile (yield)		37 (yield)
benzaldehyde azine (35)	36	(85%)	(82%)
2-chlorobenzaldehyde azine (38)	39	(88%)	(84%)
4-chlorobenzaldehyde azine (40)	41	(82%)	(80%)
anisaldehyde azine (42)	43	(82%)	(79%)
4-nitrobenzaldehyde azine (44)	45	(95%)	(93%)
4-dimethylaminobenzaldehyde azine (46)	47	(80%)	(78%)

References

[*1*] J. S. SHANNON, H. SILBERMAN and S. STERNHELL: Tetrahedron Letters **1964**, 659.
[*2*] P. J. COLLIN, H. SILBERMAN, S. STERNHELL and G. SUGOWDZ: Tetrahedron Letters **1965**, 2063.
[*3*] P. CERUTTI and H. SCHMID: Helv. chim. Acta **45**, 1992 (1962).
[*4*] P. CERUTTI and H. SCHMID: Helv. chim. Acta **47**, 203 (1964).

[5] T. BACCHETTI and A. ALEMAGNA: Gazz. chim. Ital. **91**, 1475 (1961); C. A. **57**, 7252 (1962).
[6] T. BACCHETTI and A. ALEMAGNA: Rend. Ist. Lombardo Sci. Lettere **94 A**, 242 (1960); C. A. **55**, 16527 (1961).
[7] G. M. BADGER, R. J. DREWER and G. E. LEWIS: Austral. J. Chem. **16**, 1042 (1963).
[8] C. SCHÖPF and G. LEHMANN: Liebigs Ann. Chem. **518**, 1 (1935).
[9] J. KLOSA: Pharmazie **8**, 1030 (1953); C. A. **50**, 15550 (1956).
[10] J. KLOSA: Arch. Pharm. **287**, 62 (1954); C. A. **51**, 14741 (1957).
[11] J. E. HODGKINS and J. A. KING: J. Amer. Chem. Soc. **85**, 2679 (1963).

Chapter 37

Photohalogenation

1. Photohalogenation. Scope of the reaction

The effect of light on the reactions of chlorine and bromine with organic compounds is a phenomenon ranking as one of the oldest and most frequently investigated in organic photochemistry. It is noteworthy that one of these well-known reactions, the photochlorination of benzene, has achieved great industrial importance since it was found that one of the stereoisomers from the group of hexachlorocyclohexanes formed is an excellent insecticide (Gammexane).

The photochemical action of chlorine takes place in principle in the same manner as for bromine, and the following discussion of the mechanism of photochlorination is also valid for that of photobromination. A more complete account on free-radical halogenations will be found elsewhere (cf. [1]).

The mechanism of photochlorination proceeds according to the following pathways:

$$Cl_2 \xrightarrow{h\nu} 2\ Cl\cdot$$
$$\cdot Cl + R_3CH \longrightarrow HCl + R_3C\cdot$$
$$R_3C\cdot + Cl_2 \longrightarrow R_3CCl + Cl$$

Termination of the chain reaction can be brought about either by the combination of two chlorine radicals or by the combination of two free alkyl groups. The length of the chain reaction is adversely affected by the reaction of chlorine radicals with any impurity, e.g. oxygen, phenols and amines.

The addition of chlorine to olefines also takes place via the formation of free chlorine radicals:

$$Cl_2 \xrightarrow{h\nu} 2\ Cl\cdot$$
$$R_2C=CR_2 + Cl\cdot \longrightarrow R_2\overset{}{C}-\overset{\cdot}{C}R_2$$
$$\!|$$
$$Cl$$

$$R_2\overset{}{C}-\overset{\cdot}{C}R_2 + Cl_2 \longrightarrow R_2C-CR_2 + Cl\cdot$$
$$\ \ ||\ \ |$$
$$\ ClCl\ Cl$$

It is therefore not surprising that in the chlorination of unsaturated compounds, e. g. aromatic hydrocarbons, chlorination under irradiation leads to the formation of products involving both addition and substitution reactions.

The photochlorination (photobromination) of alkyl aromatic compounds takes place primarily in the side-chain. The hydrogen atoms attached to the α-carbon atom are replaced with particular ease by halogens.

Of particular interest from the preparative standpoint are those photohalogenations which produce better results than the corresponding dark reactions. This is the case, for example, in the preparation of chlorocyclopropane from cyclopropane. Both thermal chlorination and photochlorination lead to the formation of chlorocyclopropane, but the photochlorination is to be preferred since the product of the thermal reaction contains considerable quantities of allyl chloride [2].

For preparative purposes it is also significant that photohalogenation frequently follows a very selective course. Hass et al. [3] observed that in the photochlorination of propane at —60° 2-chloropropane is obtained in a yield of 73 %. Russell and Brown [4] obtained 2-bromo-2,3,3-trimethylbutane in almost quantitative yield from the photochemical action of bromine on 2,2,3-trimethylbutane at 80°.

Examples of the photohalogenation of acetylene compounds have also become known. According to Haszeldine [5] the photochlorination of hexafluoro-2-butyne with UV-light produces 2,3-dichloro-1,1,1,4,4,4-hexafluoro-2-butene.

$$F_3C-C\equiv C-CF_3 + Cl_2 \xrightarrow{h\nu} F_3C-CCl=CCl-CF_3$$

Derivatives of butadiene react with chlorine photochemically in such a fashion that first one double bond and then the other is saturated with chlorine.

Faseeh [6] was able to show that 5-phenyl-2,4-pentadienoic acid (**1**) does not react with chlorine in the dark, but that in the light the reaction is such that at first 2,3-dichloro-5-phenyl-4-pentenoic acid (**2**) and then 2,3,4,5-tetrachloro-5-phenylvaleric acid (**3**) is formed.

$$C_6H_5CH=CH-CH=CH-COOH \xrightarrow[Cl_2]{h\nu} C_6H_5CH=CH-CHCl-CHCl-COOH$$
$$\mathbf{1} \qquad\qquad \mathbf{2}$$

$$\xrightarrow[Cl_2]{h\nu} C_6H_5CHCl-CHCl-CHCl-CHCl-COOH$$
$$\mathbf{3}$$

The photochlorination of benzene is, as was mentioned previously, a process of great technical significance since, as was found in 1943, one of the hexachlorocyclohexanes, known as the γ-isomer, is a powerful insecticide [8]. The γ-isomer was first obtained by van der Linden [9].

In order to obtain the best possible yield of the γ-isomer, extensive experiments regarding the operative wavelengths, the choice of solvent, the reaction temperature and the concentration of the reactants were undertaken.

Although there are many descriptions of the formation of 1,2,3,4,5,6-hexachlorocyclohexane in the patent literature (cf. [10]), examples suited to serve as a laboratory experiment do not appear to be readily available. Hence, the authors are indebted to Mr. J. H. BROWN (ICI, General Chemicals Division) [11] for details of the irradiation procedure.

1,2-Dichloro-1,1,3,3,3-pentafluoro-2-methylpropane [7]. A sealed Pyrex tube (30 ml) containing 1,1,3,3,3-pentafluoro-2-methylpropene (2.7 g) and a slight excess of chlorine was exposed to sunlight (20 min). The tube was filled to the exclusion of air and moisture by using a vacuum system. 1,2-Dichloro-1,1,3,3,3-pentafluoro-2-methylpropane was obtained in a yield of 100%, b.p.: 75—76°.

2,3-Dichloro-5-phenyl-4-pentenoic acid (2) [6]. Chlorine was led into a solution of **1** (43.5 g; 0.25 Mol) in carbon tetrachloride (435 g) and the solution was irradiated with sunlight at 37°. The absorption of the Cl_2 (0.25 Mol) took 1 hr., during which time the reaction mixture became clearer and the temperature rose to 57°. Chlorine (0.125 Mol) was then passed into the solution for a further hour and afterwards the vessel was placed in sunlight for a third hour. The yellow color disappeared and a small quantity of colorless crystals separated out. After filtration, the solvent was evaporated and the viscous residue was stirred with light petroleum/xylene. The colorless precipitate which formed was recrystallized from light petroleum/xylene and yielded 2,3-dichloro-5-phenyl-4-pentenoic acid (2) (yield 57.7 g; m.p. 126—127°).

1,2,3,4,5,6-Hexachlorocyclohexanes [11]. "500 g benzene (which should be of high purity — free of thiophene and cycloparaffins) are placed in a round bottomed glass flask supported in a cold water bath. The flask should be fitted with a stirrer, chlorine inlet pipe dipping below the benzene level, a thermometer, and reflux condenser.

The system is illuminated by a source of light containing light of wavelength less than 500 mμ. Medium to high pressure mercury arc lamps rich in the Hg blue (454 mμ) line are particularly suitable and are exemplified by the 400 W 'Mercra' lamp manufactured by British Thomson Houston Ltd. This is placed about 20 cm from the reactor. Air is removed from the system by a nitrogen purge and chlorine is then admitted to the flask at about 25 l/hour, the temperature of the system being maintained at about 25°.

At this temperature crystals of the less soluble (α, β) isomers will begin to separate after some 90 g chlorine has been absorbed into the system (corresponding to some 5% conversion of the benzene). At this point it is convenient to stop the chlorine feed and after allowing a short further period of illumination to consume chlorine dissolved in the system, the excess benzene along with traces of monochlorobenzene formed in the reaction, is removed by steam distillation. The solid product is ground and dried in an oven at 80°. The product obtained contains approximately 63% α, 7% β, 14% γ, and 8% δ-benzene hexachloride isomers with smaller quantities of the ε-isomer and higher polychlorocyclohexanes. Efficiency of chlorine conversion to benzene hexachloride exceeds 95%."

3-Chlorocoumarin [12]. Coumarin (219 g; 1.5 Mol) in carbon tetrachloride (450 g) is heated, whilst stirring, to 75° in a glass flask fitted with a reflux condenser. Cl_2 is passed into the solution, which is stirred and irradiated (Hg-lamp; the emitted light should include the wavelengths 280 mμ to 540 mμ). The temperature is kept at 75° and the influx of the chlorine is so regulated that 106 g (1.5 Mol) is passed in within one hour.

Following this, the solvent is rapidly evaporated and the residue heated (ca. 200°, 60 to 90 min.). After the recommended period of heating, no further generation of hydrogen chloride occurs. The pale raw product so obtained (m.p. ca. 118°) is recrystallized from isopropanol (800 ml) to give a 95.2% yield of 3-chlorocoumarin, m.p. 122°.

9-Bromofluorene [*13*]. Over the course of 4 hours, 8 g bromine in carbon tetrachloride (20 ml) is added dropwise to a gently boiling solution of pure fluorene (8.3 g; 0.05 Mol) in 45 ml carbon tetrachloride. (Bath temperature not above 85°). The red-brown solution is further heated until it becomes yellow in color. After distilling off the solvent under reduced pressure, the residue (12 g) is twice recrystallized from cyclohexane. Colorless needles are obtained (m.p. 103—104°; yield 70% of theoretical).

In direct sunlight, the same bromination is complete within 15 min. In boiling chloroform the 2-bromofluorene forms in the dark, and only in sunlight is the 9-bromofluorene formed.

p-Bromobenzyl bromide [*14*]. The reaction is carried out in a three-necked Pyrex flask, fitted with dropping funnel, stirrer and reflux condenser. p-Bromotoluene (102 g) (0.60 Mol) is heated to 120° and then exposed to the radiation from a 100 Watt lamp. Bromine (102 g; 0.64 Mol) is added dropwise over three hours, and the mixture is stirred throughout. When the generation of hydrobromic acid ceases, the mixture is cooled and after some time the solid reaction product is filtered off. The product is washed three times with 30 ml alcohol (m.p. 61°; yield 80 g). A further quantity of the product (18—20 g) can be obtained by cooling of the filtrate (ice/salt).

2. Photochlorination

a) Chlorination of benzene in the presence of iodine

The photochlorination of benzene in the presence of iodine produces, apart from substitution products (mono-, di- and trichlorobenzenes) and hexachlorocyclohexane, various stereoisomeric partial addition products (chlorocyclohexenes), e. g. pentachlorocyclohexene and 3,4,5,6-tetrachlorocyclohexenes. The isomers of this latter substance can be separated by chromatography to obtain a stereoisomerically homogeneous 3,4,5,6-tetrachlorocyclohexene [*15*].

The iodine catalyzed photochlorination of benzene was carried out with the aid of an infrared incandescent lamp. A solution of iodine (6.25 g) in dry, thiophene-free benzene (2200 g) was irradiated whilst chlorine (995 g) was passed through the solution (21—30°). Reference should be made to the original publication [*15*] for further details.

It was found [*16*] that the photochlorination of monochlorobenzene and o-dichlorobenzene (in the presence of iodine) could also be carried out so as to produce an additive partial chlorination. In this way, a mixture of isomers of 1,3,4,5,6-pentachlorocyclohexene was obtained from monochlorobenzene. Four isomers were separated using distribution chromatography.

References, pp. 360—361

b) Chlorination of benzene in the presence of maleic anhydride

Although the photochlorination of benzene and maleic anhydride separately leads to the formation of hexachlorocyclohexane and 2,3-dichlorosuccinic anhydride respectively, the photochlorination of benzene in the presence of maleic anhydride gives rise to the formation of 2-chloro-3-phenylsuccinic anhydride (4) and a second compound which is probably 2-chloro-3-(2,3,4,5,6-pentachlorocyclohexyl)-succinic anhydride (5) [17].

Since 4 can undergo loss of HCl it is an easily obtainable intermediate in the synthesis of phenylmaleic anhydride, especially since the starting material is readily available.

ECKE et al. [17] suggest the following free radical chain mechanism, pointing out that the addition of resonance stabilized radicals to maleic anhydride is a well known phenomenon:

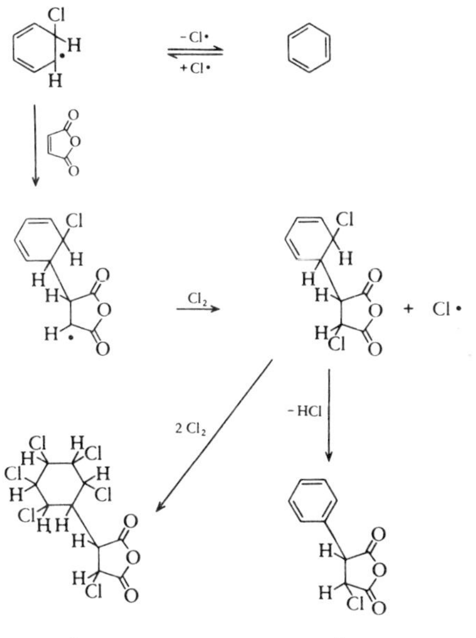

2-Chloro-3-phenylsuccinic anhydride (4) [17]. The Pyrex reaction flask (500 ml) was fitted with a stirrer, a gas inlet, a thermometer, and a reflux condenser which in turn exhausted to a water scrubber for evolved hydrogen chloride. A solution of 88 g of maleic anhydride in benzene (281 g) was placed in the flask, and the contents were heated to 70°. This temperature was maintained during irradiation (2 General Electric Sun lamps, CG 401 CX), and chlorine was passed in over 2.2 hrs. (64 g; 0.9 Mol). As soon as all the chlorine had reacted, N_2 was passed through the reaction vessel in order to drive off dissolved hydrogen chloride. Unconverted benzene was distilled off and the distillation was discontinued on attainment of a temperature of 120°; the remaining volatile material was then driven off under reduced pressure. A viscous residue remained, and this was dissolved in ether (135 ml). On cooling the solution to 0°, a colorless solid product was obtained (66 g; m.p. 90—98°). 4 was purified by recrystallizing a number of times from benzene and ether (34 g; m.p. 103—104.5°).

c) Replacement of the sulfonyl chloride group by chlorine

RONDESTVEDT et al. [*18*] used the photolysis of chlorine in carbon tetrachloride to convert styrene-β-sulfonyl chloride (6) to 1,1,2-trichloro-2-phenylethane (7). No reaction took place in the dark.

$$C_6H_5-CH=CH-SO_2Cl + Cl_2 \xrightarrow{h\nu} C_6H_5-CHCl-CHCl_2 + SO_2$$
$$6 7$$

MILLER and WALLING [*19*] have made similar observations. The authors obtained chlorobenzene by treatment of benzenesulfonyl chloride with chlorine under irradiation. In this process chlorine is regenerated, and 0.25 Mol chlorine was sufficient for the conversion of 1 Mol benzenesulfonyl chloride in carbon tetrachloride at 70°, the yield of chlorobenzene being 92 %.

$$C_6H_5-SO_2Cl + Cl_2 \xrightarrow{h\nu} C_6H_5Cl + SO_2 + Cl_2$$

When 4-bromobenzenesulfonyl chloride (8) was similarly treated [*19*], not only the sulfonyl group but also the bromine atom was replaced by chlorine, the product thus obtained being p-dichlorobenzene (9).

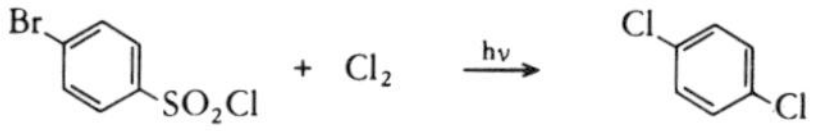

Reaction of chlorine with diphenyl sulfone in carbon tetrachloride produced chlorobenzene. Surprisingly, methyl benzenesulfonate proved to be stable against attack by chlorine.

$$C_6H_5-SO_2-C_6H_5 + Cl_2 \xrightarrow{h\nu} 2\,C_6H_5Cl + SO_2$$

MILLER and WALLING [*19*] suggest the following reaction mechanism:

$$C_6H_5SO_2Cl + Cl\cdot \longrightarrow C_6H_5Cl + \cdot SO_2Cl$$
$$\cdot SO_2Cl \longrightarrow SO_2 + Cl\cdot$$
$$C_6H_5SO_2C_6H_5 + Cl\cdot \longrightarrow C_6H_5Cl + C_6H_5SO_2\cdot$$
$$C_6H_5SO_2\cdot + Cl_2 \longrightarrow C_6H_5SO_2Cl + Cl\cdot$$

or

$$C_6H_5SO_2\cdot \longrightarrow C_6H_5\cdot + SO_2$$
$$C_6H_5\cdot + Cl_2 \longrightarrow C_6H_5Cl + Cl\cdot$$

References, pp. 360—361

1,1,2-Trichloro-2-phenylethane (7) [*18*]. A solution of **6** (10.1 g; 0.05 Mol) in carbon tetrachloride (125 ml) was treated with chlorine until a weight increase of 3.6 g (0.05 Mol) had been achieved. Irradiation with UV-light brought the solution to the boiling point, so the solution was cooled during the irradiation period of 2 hrs. During this time, the color of the chlorine disappeared. The solvent was evaporated and the residue distilled to give 9 g of **7**, b.p.$_{1.4}$ 84—85°.

p-Dichlorobenzene (9) [*19*]. A solution of **8** (32.4 g) in 250 ml carbon tetrachoride was irradiated at 70° with a 200 Watt lamp, while a stream of chlorine was passed through the solution. After 2 hrs., the solvent and the bromine were driven off, and p-dichlorobenzene (**9**), m.p. 47—50°, was obtained in 97% yield (18.1 g).

d) Replacement of the nitroso group by chlorine

Trifluoronitrosoethylene (**10**) reacts smoothly with chlorine in the dark to give 1,2-dichlorotrifluoro-1-nitrosoethane. Photochemical chlorination leads to the formation of chlorotrifluoroethylene (**12**) and 1,1,2-trichlorotrifluoroethane (**13**) by way of the trifluorovinyl radical (**11**) [*20*].

$$CF_2=CFNO \xrightarrow[-NO]{h\nu} [CF_2=CF\cdot] \xrightarrow{Cl_2} CF_2=CFCl \xrightarrow{Cl_2} CF_2Cl-CFCl_2$$
$$\quad\quad 10 \quad\quad\quad\quad\quad 11 \quad\quad\quad\quad\quad 12 \quad\quad\quad\quad\quad 13$$

Chlorotrifluoroethylene (12) and **1,1,2-trichlorotrifluoroethane (13)** [*20*]. Exposure of trifluoronitrosoethylene (**10**) (2 mMole) and chlorine (2 mMole) to ultraviolet light for 3 hrs. gave unchanged **10** (51%), **12** (21%) and **13** (15% based on **10**), but no 1,2-dichlorotrifluoro-1-nitrosoethane.

e) Replacement of alkyl groups by chlorine

Strong steric interactions take place between the trichloromethyl groups and the ortho chlorine atoms in octachlorotoluene (**14**) and decachloro-p-xylene (**16**). According to BALLESTER [*21*], **14** and **16**, on chlorination with chlorine in light readily undergo chlorinolysis yielding hexachlorobenzene (**15**). In the hexachlorobenzene molecule only comparatively small steric interactions take place, the chlorine atoms being displaced alternately above and below the mean plane of the benzene ring [*22*].

The product mixture obtained by free radical chlorination of spiropentane (**17**) has been shown to consist in part of the normal substitution product, chlorospiropentane (**18**), but mainly of chlorides which can be accounted for by initial ring opening attack of a chlorine atom on carbon. The latter products include 1,1-bis-(chloromethyl)-cyclopropane (**20**), 4-chloro-2-(chloromethyl)-1-butene (**21**) and 1,2,4-trichloro-2-(chloromethyl)-butane (**22**). A mechanistic scheme which accounts for the reactions [*23*] is shown below.

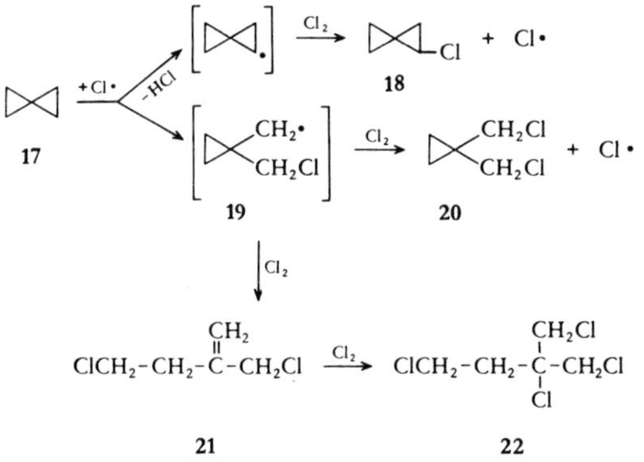

$$17 \xrightarrow[-HCl]{+Cl\cdot} \begin{cases} [\text{cyclopropane radical}] \xrightarrow{Cl_2} \text{18-Cl} + Cl\cdot \\ [\text{19: } \underset{CH_2Cl}{\overset{CH_2\cdot}{\diagup}}] \xrightarrow{Cl_2} \text{20: } \underset{CH_2Cl}{\overset{CH_2Cl}{\diagup}} + Cl\cdot \end{cases}$$

$$\downarrow Cl_2$$

$$\underset{21}{ClCH_2-CH_2-\overset{CH_2}{\underset{\|}{C}}-CH_2Cl} \xrightarrow{Cl_2} \underset{22}{ClCH_2-CH_2-\overset{CH_2Cl}{\underset{Cl}{C}}-CH_2Cl}$$

The ring opening of radical **19** to yield completely open chain products finds an analogy in the formation of 4-chloro-1-butene (**24**) by photochlorination of methylcyclopropane (**23**) [*24*].

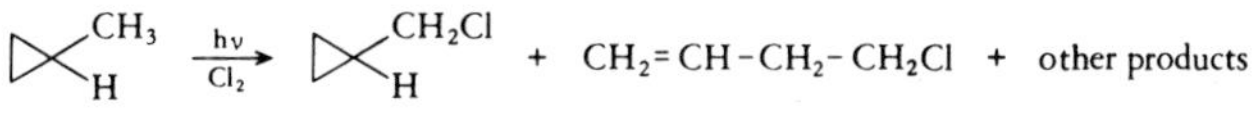

$$\underset{23}{\overset{CH_3}{\underset{H}{\diagup}}} \xrightarrow[Cl_2]{h\nu} \underset{24}{\overset{CH_2Cl}{\underset{H}{\diagup}}} + CH_2=CH-CH_2-CH_2Cl + \text{other products}$$

Hexachlorobenzene (15) [*21*]. A solution of **14** (0.2 g) in 7.5 ml carbon tetrachloride was placed in a 50 ml round-bottomed, Pyrex flask equipped with a reflux condenser and an inlet tube. The solution was heated on a steam-bath and illuminated by a 500 Watt incandescent lamp at a distance of 20 cm while a slow stream of dry chlorine was introduced for 6 hours. Upon evaporation of the solvent a quantitative yield of hexachlorobenzene, m.p. 228°, was obtained.

f) Solvent effects on the site of attack by chlorine

According to RUSSELL [*25, 26*] the selectivity in photochlorination of alkanes can be altered and controlled by the choice of the solvent. In the case of benzene and related compounds the solvent effect is thought to arise from association of the chlorine atom with the π-electrons of the aromatic nucleus. This association produces a complexed chlorine atom (a π-complex) which has a lower reactivity and hence a greater selectivity than a free chlorine atom.

Thus in the presence of a number of aliphatic solvents the photochlorination of 2,3-dimethylbutane produces 60 % of 1-chloro-2,3-dimethyl-

References, pp. 360—361

butane and 40 % of the 2-isomer. In 8 M benzene the 2-isomer is formed in 90 % yield while in 12 M carbon disulfide the yield of the 2-isomer is greater than 95 % of the total yield of chloro-2,3-dimethylbutanes [*25*].

These results have been confirmed and extended by WALLING [*27*]. Further studies on orientation effects in the photochlorination of aliphatic compounds were carried out by BRUYLANTS and co-workers (cf. [*28*]).

g) Chlorination of a cyclic trisulfone

EL-HEWEHI and HEMPEL [*29*] converted s-trithiane (**25**) to hexachloro-s-trithiane 1,1,3,3,5,5-hexaoxide (**27**) without the isolation of the intermediate trisulfone **26**.

$$\underset{25}{\overset{S}{\underset{S\smile S}{\bigcap}}} \xrightarrow{[O]} \left[\underset{26}{\overset{O_2}{\underset{O_2S\smile SO_2}{\bigcap}}}\right] \xrightarrow[Cl_2]{h\nu} \underset{27}{\overset{Cl\;O_2Cl}{\underset{Cl\;\;\;\;Cl}{\underset{O_2S\;\;\;SO_2}{\overset{S}{Cl\;\;Cl}}}}}$$

Hexachloro-s-trithiane 1,1,3,3,5,5-hexaoxide (27) [*29*]. s-Trithiane (**25**) (34 g, 0.25 mole) was rubbed to a paste with dilute sulfuric acid and then slowly mixed with a cold saturated solution of $KMnO_4$ by stirring at room temperature. When the reaction was almost complete, excess of $KMnO_4$ was added and the mixture was stirred over several days. Afterwards, the oxidation-products were filtered off under suction, the excess $KMnO_4$ was washed out with water and the brown filter cake suspended in water (1.5 l). After the addition of a few drops of silicone oil, in order to avoid excess foaming, a strong current of chlorine was passed into the suspension. During the chlorination (ca. 15 hrs.), the flask was irradiated with UV-light. Following the irradiation, the brown precipitate was filtered off under suction, washed with water, dried and then recrystallized from glacial acetic acid. White needles of **27** ((33.4 g; 31.8% based on **25**) were obtained (m.p. 253—254° dec.).

3. Photobromination

a) Migration of alkyl groups during bromination

In the photobromination of 2,2,4,4-tetramethylpentane (**28**), an interesting migration of a methyl group has been observed [*30*]. At 100° (gas phase), little or no bromine-containing products were obtained from photochemical bromination, but at 200° conversion occurred and, apart from the formation of a dibromide $C_9H_{18}Br_2$ (5 %), 2-bromo-2,3,4,4-tetramethylpentane (**29**) was produced in 72 % yield.

$$Br_2 \xrightarrow{h\nu} 2\;Br\cdot$$

$$(CH_3)_3C-CH_2-C(CH_3)_3 + Br\cdot \longrightarrow (CH_3)_3C-\overset{\cdot}{C}H-C(CH_3)_3 + HBr$$
28

$$(CH_3)_3C-\overset{\bullet}{C}H-C(CH_3)_3 \longrightarrow (CH_3)_2\overset{\bullet}{C}-\underset{\underset{CH_3}{|}}{\overset{\overset{H}{|}}{C}}-C(CH_3)_3$$

$$(CH_3)_2\overset{\bullet}{C}-\underset{\underset{CH_3}{|}}{\overset{\overset{H}{|}}{C}}-C(CH_3)_3 + Br_2 \longrightarrow (CH_3)_2\underset{\underset{Br}{|}}{C}-\underset{\underset{CH_3}{|}}{\overset{\overset{H}{|}}{C}}-C(CH_3)_3 + Br\bullet$$
$$\mathbf{29}$$

b) Bromination in the presence of oxygen

The action of oxygen in the photobromination of saturated hydrocarbons is very complex. According to KHARASCH et al. [*31*], small quantities of oxygen accelerate the photobromination of cyclohexane, methylcyclohexane and isobutane whilst with larger concentrations the effect diminishes. The combined effects of oxygen and light are much greater than the sum of the individual effects.

In the photobromination of tetrachloroethylene [*32*] oxygen at low concentrations was found to have an enhancing effect on the reaction. At higher concentrations, however, the formation of dibromotetrachloroethane is suppressed.

If the photobromination of styrene-β-sulfonyl chloride (**6**) in carbon tetrachloride is carried out in a nitrogen atmosphere, the sulfonyl-free products **30** and **31** are obtained (cf. p. 346). In the photobromination of **6** in the presence of small quantities of oxygen the formation of **32** could also be detected [*18*].

$$C_6H_5-CH=CH-SO_2Cl \xrightarrow[Br_2]{h\nu} C_6H_5-CHBr-CHBrCl + C_6H_5-CHBr-CHBr_2$$
$$\mathbf{6}\mathbf{30}\mathbf{31}$$

The formation of **32** in the presence of oxygen is explained by the following mechanism:

$$C_6H_5-CH=CH-SO_2Cl + Br\bullet \longrightarrow C_6H_5-\overset{\bullet}{C}H-CHBr-SO_2Cl$$
$$\mathbf{6}$$

$$C_6H_5-\overset{\bullet}{C}H-CHBr-SO_2Cl + O_2 \longrightarrow C_6H_5-\underset{\underset{O-O\bullet}{|}}{C}H-CHBr-SO_2Cl$$

$$C_6H_5-\underset{\underset{O-O\bullet}{|}}{C}H-CHBr-SO_2Cl + Br_2 \longrightarrow C_6H_5-CHBr-CHBr-SO_2Cl + O_2 + Br\bullet$$
$$\mathbf{32}$$

References, pp. 360—361

c) Bromination in the allylic position with the aid of bromine

Relatively little is known about the photobromination of substances in the allylic position. SCHALTEGGER [*33, 34*] reported on the conversion of the cholesteryl esters **33** (R = benzoyl or tosyl) to the corresponding 7β-bromo compounds **34** (yield 60 %) using a 200 Watt lamp as light source. The bromination was carried out in carbon tetrachloride solution. In the dark process, an addition reaction with bromine occurs at the double bond.

$$\text{33} \xrightarrow[\text{Br}_2]{h\nu} \text{34}$$

d) Bromination with chlorine-bromine mixtures

SPEIER [*35*] noted that chlorotrimethylsilane (**35**) failed to react with bromine in light at reflux temperature. However, when chlorine was passed into the solution containing bromine and **35**, (bromomethyl)chlorodimethylsilane (**36**) was obtained in 62 % yield together with polybrominated products.

$$2\ H_3C-\underset{\underset{CH_3}{|}}{\overset{\overset{CH_3}{|}}{Si}}-Cl\ +\ Br_2\ +\ Cl_2\ \xrightarrow{h\nu}\ 2\ BrCH_2-\underset{\underset{CH_3}{|}}{\overset{\overset{CH_3}{|}}{Si}}-Cl\ +\ 2\ HCl$$

           35   **36**

By a similar procedure toluene gave benzyl bromide (71 %), and cyclohexane gave bromocyclohexane (45 %) [*35*].

(Bromomethyl)chlorodimethylsilane (36) [*35*]. Chlorotrimethylsilane (**35**) (119 g; 1.1 moles) was illuminated with a 60 W incandescent lamp in a 500 ml flask equipped with a reflux condenser topped by a Dry Ice cooled knock down condenser. From a separatory funnel a small amount of bromine was added. Chlorine was then passed through the mixture until the dark red color of bromine disappeared. More bromine was then added and the process was continued until 46.5 g, 0.64 equivalent, was added. The mixture then weighed 168 g. The product was distilled to yield 61 g (51%) of recovered **35** and 75 g of (bromomethyl)chlorodimethylsilane (**36**), b.p.$_{740}$ 130°.

e) Orienting effects in the photobromination of alkyl bromides

THALER's investigations [*36*] of the halogenation of alkyl and cycloalkyl halides have revealed that the photobromination of alkyl bromides is quite different from the bromination of other alkyl halides, and from alkyl halide halogenations in general. Unlike other free radical halogenations of alkyl halides, which frequently give a multitude of products, the bromination of alkyl bromides was highly selective, giving 84—94 % of the vicinal dibromide isomer. The influence of the bromine substituent directing the attack of a halogen atom to the adjacent carbon is contrary to other radical halogenations of substituted alkanes, in which this position has been generally demonstrated to be one of the least reactive positions in the molecule. THALER's results of the bromination of bromocyclohexane (table 27) and of the halogenation of 1-halobutanes (table 28) are shown below.

Table 27 (tr.= trans).

$C_6H_{10}Br_2$	1,1-and/or cis-1,2-	tr.-1,2-	tr.-1,3-	tr.-1,4-	cis-1,3-	cis-1,4-
% of total dibromide	3.53	94.0	0.92	0.50	0.52	0.53

Table 28

reagent	substrate	distribution of dihalide isomers [%]			
		1,1	1,2	1,3	1,4
Br_2	1-bromobutane	0.9	84.5	14.6	—
Br_2	1-chlorobutane	22.8	25.3	51.9	—
Cl_2	1-bromobutane	5.0	21.8	50.3	22.9

SKELL and READIO [37] reported the photobromination of cis-1-bromo-4-tert. butylcyclohexane (37) to be a highly selective reaction, yielding cis-1-trans-2-dibromo-4-tert. butylcyclohexane (38). By contrast, trans-1-bromo-4-tert.- butylcyclohexane (39) is considerably less reactive and less selective towards attack by bromine.

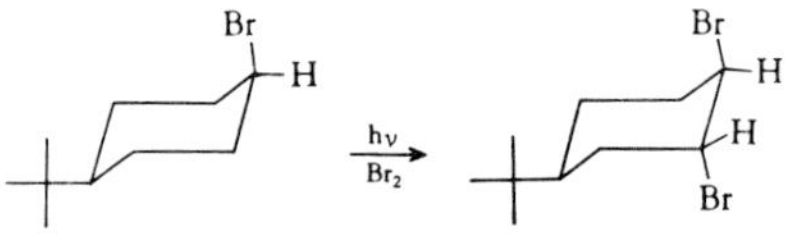

Trans-1,2-dibromocyclohexane [36]. A 1 : 5 molar bromine to bromocyclohexane mixture was transferred to a Pyrex tube and degassed by alternatively evacuating on a water pump and flushing with nitrogen three times while cooling in a trichloroethylene-Dry Ice bath. The sealed tube was placed in a thermostated bath at 60° and irradiated with a 150 Watt unfrosted incandescent bulb, which was also immersed in the bath about 10 cm from the reaction tube. The irradiation was continued until the color of bromine disappeared.

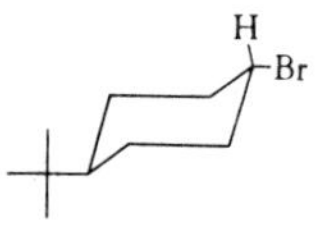

Distillation of a preparative scale run yielded a fraction, b.p. 116—118.5° (25 mm), which was demonstrated by G. L. C. to contain dibromocyclohexanes in 83 % yield based on bromine. Trans-1,2-dibromocyclohexane was isolated by fractionation.

f) Bromination of aryl selenocyanates

HÖLZLE and JENNY [*38*] have shown that 2,4-dinitro-1-selenocyanatobenzene (**40**), which does not react with bromine in diffuse daylight, can be converted to 2,4-dinitrobenzeneselenenyl bromide (**41**) (75 % yield) by irradiation with light of short wavelength. The selenobromide so formed is identical to that obtained from the bromination of bis-(2,4-dinitrophenyl) diselenide.

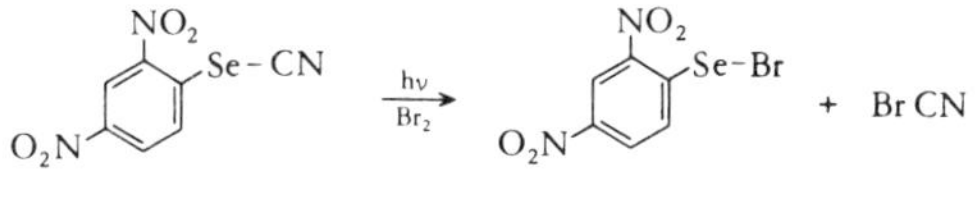

Anthraquinone-1-selenenyl bromide (**43**), which could not previously be obtained from 1-selenocyanatoanthraquinone (**42**), can now be prepared by means of a strictly analogous procedure [*38*]:

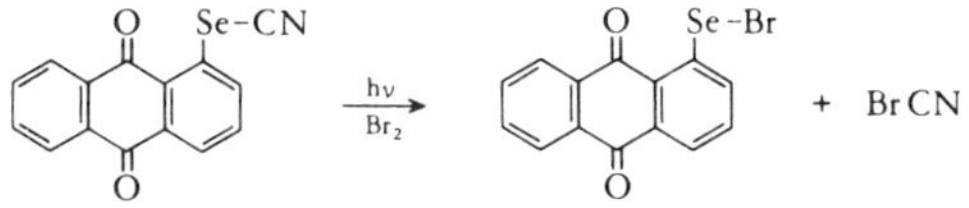

Anthraquinone-1-selenenyl bromide (43) [*38*]. 1-Selenocyanatoanthraquinone (**42**) (1 g) and bromine (1 ml) were suspended in 25 ml chloroform and irradiated for 3 hrs. with a UV-lamp, under reflux. The solution was filtered hot and the solvent, excess bromine and cyanogen bromide distilled off under vacuum. Red, glossy platelets crystallized out from glacial acetic acid or dioxane (yield: 0.5 g; m. p. 220°).

4. Photoiodination

Relatively little is known on the photochemical iodination of organic substrates. The earlier literature on the photochemical addition of iodine to unsaturated compounds has recently been summarized by ROEDIG [*39*].

Low temperature irradiation of iodine with cis- or trans-2-butene gives quantitative yields of the diastereomeric 2,3-diiodobutanes **44** or **45**, respectively [*40*]. The vicinal diiodides proved to be thermally and photochemically labile as was to be expected (cf. [*39*]).

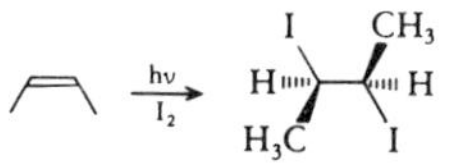

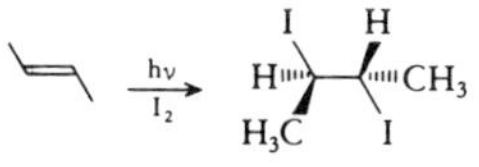

d,l-2,3-Diiodobutane (44) and **meso-2,3-diiodobutane (45)** [*40*]. Mixtures of 1 g of iodine and 10 g of cis- or trans-2-butene, respectively, in 50 ml of propane are illuminated at reflux temperature (—42°). Decolorization usually occurs within 30 mins. if the reactants are sufficiently pure. The colorless diiodides are isolated in quantitative yield by pumping off solvent and unreacted olefin. The crystalline diiodides have m.p. —24 to —23° (**44**) and —11° (**45**), respectively.

5. Photohalogenation with the aid of inorganic and organic halides

a) Experiments with iodine chloride

The transformation of iodobenzene or bromobenzene to chlorobenzene under irradiation in the presence of iodine chloride was carried out by MILLIGAN [*41*].

Chlorobenzene [*41*]. To 8.20 g of iodobenzene dissolved in 4 l of CCl_4 was added 10.7 ml of 0.373 M ICl (10.0 mole %) in CCl_4. The solution was stirred and irradiated with a 275 Watt mercury arc sunlamp 15 cm away for 12 hours. The violet colored solution was washed with dilute thiosulfate and water and then dried with calcium carbide. The solvent was removed by distillation through a 1.5 × 75 cm helix packed column until the residue was reduced to about 80 ml. The residue was transferred to a 100 ml volumetric flask and was diluted to volume with CCl_4. Analysis of the solution by gas chromatography showed that a quantitative yield, based on ICl, of chlorobenzene was obtained.

b) Chlorination with sulfuryl chloride

The photochemical reaction of organic compounds with SO_2Cl_2 is frequently used as a method of sulfochlorination (cf. chapter 43). On the other hand, VORONKOV and DAVYDOVA [*42*] found that ultraviolet radiation induces the chlorination of organosilicon compounds with sulfuryl chloride to a significantly greater extent than do organic peroxides. Thus, it has been possible to chlorinate trichloromethylsilane (**46**) and dichlorodimethylsilane, compounds which are difficult to chlorinate otherwise.

Intense ultraviolet irradiation of a boiling mixture of **46** and sulfuryl chloride leads to the formation of mono-, di-, and trichloro derivatives **47**—**49** [*42*]. The yield of these compounds varies with the mole ratio of the starting materials.

$$H_3C-SiCl_3 \xrightarrow[SO_2Cl_2]{h\nu} ClH_2C-SiCl_3 + Cl_2HC-SiCl_3 + Cl_3C-SiCl_3$$
$$\textbf{46} \qquad\qquad \textbf{47} \qquad\qquad \textbf{48} \qquad\qquad \textbf{49}$$

In contrast to the photochlorination of **46** with sulfuryl chloride, analogous treatment of trichlorophenylsilane (**50**) resulted in low yields of the expected trichloro(chlorophenyl)silane (**51**), the main product being hexachlorobenzene [*42*].

References, pp. 360—361

$$C_6H_5-SiCl_3 \xrightarrow[SO_2Cl_2]{h\nu} C_6H_4Cl-SiCl_3 + C_6Cl_6$$
$$\qquad\;\,\textbf{50} \qquad\qquad\qquad\quad\;\; \textbf{51} \qquad\quad \textbf{15}$$

Trichloro(chloromethyl)silane (47), trichloro(dichloromethyl)silane (48) and **trichloro(trichloromethyl)silane (49)** [*42*]. A mixture of 149.5 g (1 mole) of **46** and 270.0 g (2 moles) of SO_2Cl_2 was irradiated and heated to boiling with a PRK 2 mercury lamp. The reaction was conducted in a quartz flask fitted with a reflux condenser joined to a bottle containing concentrated sulfuric acid. The reaction was interrupted when the evolution of gaseous products ceased (38 hours). Fractional distillation of the reaction mixture gave 20 g of **47**, b.p. 118—123°, 22 g of **48**, b.p. 143—148°, and 18 g of **49**, b.p. 156—160°. Further purification was effected by a second fractionation.

Trichloro(chlorophenyl)silane (51) and **hexachlorobenzene (15)** [*42*]. A mixture of 105.8 g (0.5 mole) of **50** and 67.5 g (0.5 mole) of SO_2Cl_2 was irradiated and heated as above for 53 hours. Distillation of the reaction mixture in a column gave 18 g (14%) of **51** with a boiling range of 220—245°. The high-boiling residue crystallized. Recrystallization from a mixture of benzene and ether gave 70 g (50%) of hexachlorobenzene, m.p. 229°.

c) Chlorination with trichloromethanesulfonyl chloride

Trichloromethanesulfonyl chloride (**52**) can be used to chlorinate alkanes and alkyl aromatics, sulfur dioxide and chloroform being produced as by-products.

$$RH + \underset{\textbf{52}}{Cl_3CSO_2Cl} \xrightarrow{h\nu} RCl + HCCl_3 + SO_2$$

This light-induced reaction shows a very high degree of selectivity as to the site of chlorination on certain compounds compared to the more random chlorination observed when chlorine or sulfuryl chloride are used as the chlorinating agents [*43*]. Thus chlorination of ethylbenzene with **52** gave a monochlorinated product, the infrared analysis of which showed only (1-chloroethyl)-benzene. The alkyl side chain of p-bromotoluene was chlorinated with **52** without any substitution of bromine by chlorine. This substitution is reported, however, to occur with Cl_2 or SO_2Cl_2 as chlorinating agents.

In the photoinitiated reactions, the free radicals that initiate the chain reaction very likely result from the photochemical cleavage of **52** into free radicals which react with the hydrocarbon [*43*]:

$$\textbf{52} \xrightarrow{h\nu} Cl_3C\cdot + SO_2 + Cl\cdot$$
$$Cl_3C\cdot + RH \longrightarrow HCCl_3 + R\cdot$$
$$R\cdot + \textbf{52} \longrightarrow RCl + Cl_3CSO_2\cdot$$
$$Cl_3CSO_2\cdot + RH \longrightarrow Cl_3CSO_2H + R\cdot$$
$$Cl_3CSO_2H \longrightarrow Cl_3CH + SO_2$$

(1-Chloroethyl)-benzene [43]. Trichloromethanesulfonyl chloride (52) (0.115 mole) was dissolved in 1 mole ethylbenzene in a round-bottomed Pyrex flask fitted with a spiral condenser. A 275 W General Electric sunlamp was placed about 10—15 cm from the bottom of the flask to furnish both the necessary radiation and heat. After 6 hrs. irradiation at 130—133° the chloroethylbenzene (0.10 mole, 87% conversion) was isolated by distillation. The product was free from (2-chloroethyl)-benzene.

p-Bromobenzyl chloride [43]. p-Bromotoluene (0.50 mole) and 52 (0.115 mole) were irradiated as above (10 hrs. at 110—115°) to yield 0.11 mole (97% conversion) of p-bromobenzyl chloride.

d) Chlorination with trichloromethanesulfenyl chloride

When solutions of trichloromethanesulfenyl chloride (53) in alkanes are exposed to light, alkyl chlorides, hydrogen chloride, and bis-(trichloromethyl) disulfide (54) are formed [44].

$$RH + 2\ Cl_3C-SCl \xrightarrow{h\nu} RCl + HCl + Cl_3C-SS-CCl_3$$
$$53 \phantom{-SCl \xrightarrow{h\nu} RCl + HCl + Cl_3C-SS-}54$$

One advantage of this chlorination method is the high selectivity, which is even further increased in the presence of benzene. The disulfide may be reconverted to the sulfenyl chloride by molecular chlorine in a separate step.

The formation of the products is accounted for by the following scheme [44]:

$$Cl_3C-SCl \xrightarrow{h\nu} Cl_3C-S\cdot + Cl\cdot$$
$$53$$
$$Cl_3C-S\cdot + RH \longrightarrow Cl_3C-SH + R\cdot$$
$$R\cdot + Cl_3C-SCl \longrightarrow RCl + Cl_3C-S\cdot$$
$$Cl_3C-SH + Cl_3C-SCl \longrightarrow Cl_3C-SS-CCl_3 + HCl$$
$$54$$

2-Chloro-2,3-dimethylbutane and **1-chloro-2,3-dimethylbutane** [44]. A solution of 74.4 g (0.4 mole) of 53 in 260 g (3 mole) of 2,3-dimethylbutane was exposed to sunlight at 0° until the mixture had become colorless. Fractional distillation yielded 20.5 g of 2-chloro-2,3-dimethylbutane (b.p. 112—113°) and, using p-xylene as a booster in the fractionating column, a fraction of 0.9 g (b.p. 123—125°) consisting mainly of 1-chloro-2,3-dimethylbutane. GLC analysis of the original reaction mixture showed that the total yield of the chlorides is 98%, of which 94.9% is the tertiary and 5.1% the primary chloride.

e) Chlorination and bromination with the aid of N-chloro- and N-bromosuccinimide

MARTIN and his co-workers have carried out extensive experiments in analysis of the effect of light on the halogenation of aromatic compounds (e.g. toluene, ethylbenzene, 1- and 2-methylnaphthalene and 2,3-dimethyl-

References, pp. 360—361

naphthalene) using N-chlorosuccinimide and N-bromosuccinimide. In many cases, illumination was observed to produce a great increase in the rate of reaction. The following comparison illustrates the effect of light on the conversion of toluene to benzyl chloride using N-chlorosuccinimide. Toluene was present in excess, and the times indicated refer to the completion of the conversion [45].

Dark reaction	8 hrs. 47 min.
Daylight	2 hrs. 31 min.
Mercury lamp (Philips "Philera")	4 min.

It is interesting to note that hydroquinone has a retarding effect on the photoreaction between N-chlorosuccinimide and toluene. This observation indicates that the reaction involves a chain mechanism, in which free radicals play an important part [45].

The bromination of 2,3-dimethylnaphthalene in glacial acetic acid in daylight takes a different course to the dark reaction in carbon tetrachloride in the presence of benzoyl peroxide.

If the reaction is carried out in chlorobenzene under UV-light, a mixture of the two dibromo compounds, which are difficult to separate, is formed [45].

PRIJS and co-workers [46] have discovered that it is possible to obtain 8-(bromomethyl)-quinoline (56) from 8-methylquinoline (55) by bromination with N-bromosuccinimide, if the reaction is carried out in the presence of dibenzoyl peroxide or light. When large quantities are involved, the photochemical reaction is to be recommended. In this reaction, 8-(dibromomethyl)-quinoline is produced as the by-product (57).

55: R=CH$_3$ 56: R=CH$_2$Br   57: R=CHBr$_2$

In preparative organic chemistry, N-bromosuccinimide serves mainly to replace a hydrogen atom located in the allylic position with respect to a double bond (ZIEGLER Reaction). It has been repeatedly observed that light has a favorable effect on this reaction.

$$-CH_2-C=C- \xrightarrow[NBS]{h\nu} -CHBr-C=C-$$

The bromination of cholesteryl benzoate (**58**) in the allylic position with the formation of **59** using N-bromosuccinimide was accelerated by UV-light. In the synthesis of **60** the dehydrohalogenation can be undertaken without isolation of **59** [*47*]. For the photobromination of **58** with Br_2 see p. 351.

MEYSTRE et al. [*48*] have shown that the action of N-bromosuccinimide on C-22 of 24,24-diphenylchola-4,23-dien-3-one (**61**) under illumination leads very smoothly to the bromo compound **62** (partial formula). **62** was not isolated but converted to 24,24-diphenylchola-4,20,23-trien-3-one (**63**) using dimethylaniline for the elimination of HBr. **63** was used by the Swiss workers [*48*] as an intermediate in the synthesis of progesterone (pregn-4-ene-3,20-dione) (**64**).

The photochemical reaction of N-bromosuccinimide with dialkyl acetals (**65**) of ketoaldehydes has led to a simple synthesis of α-ketoesters (**67**) [*49*]. Although the first step in this synthesis is a bromination — the intermediate bromo compound (**66**) not being isolated, however — this reaction will be dealt with in more detail in chapter 25.

8-(Bromomethyl)-quinoline (56) and 8-(dibromomethyl)-quinoline (57) [*46*]. A solution of 8-methylquinoline (14.3 g; **55**) in carbon tetrachloride (200 ml) was mixed with bromosuccinimide (17.8 g) and heated to 70° by illumination from beneath with a 60 Watt bulb. The mixture was simultaneously irradiated from one side with an 80 Watt UV-lamp. The reaction was complete after one hour, and the solution was cooled to 0° and filtered. The residue was washed with a small quantity of carbon tetrachloride, while the filtrate was shaken with ice-cold 2 N-NaOH and then finally washed with water until neutral. After drying and distilling off the solvent, the residue was extracted with hot petroleum ether and 8-(bromomethyl)-quinoline (56) was obtained after cooling (yield 13.9 g; m.p. 83—84°).

If the petroleum ether extract is concentrated further, colorless crystals of 8-(dibromomethyl)-quinoline (57) (ca. 2.5 g) are obtained, (m.p. 107—108° after recrystallization from ligroin).

24,24-Diphenylchola-4,20,23-trien-3-one (63) [*48*]. Dienone 61 (2 g) and N-bromosuccinimide (725 mg) in 50 ml carbon tetrachloride were refluxed and irradiated strongly for 15 minutes. The suspension was then cooled, freed from succinimide formed by suction filtration and the filtrate evaporated under vacuum. The residue was dissolved in 10 ml dimethylaniline and the solution was boiled for 10 minutes. The cooled solution was diluted with ether, washed with hydrochloric acid and water, dried, filtered through alumina and then evaporated. The residue was dissolved in hot ethanol, treated with charcoal and the solution concentrated in vacuo. Compound 63 separated out and was recrystallized from ethanol until pure; m.p. 106—109°.

f) Bromination with the aid of dibromodimethylhydantoin

In the attempted side chain bromination of 2,4-dibenzamido-6-methylquinazoline (**68**), only unsatisfactory results were obtained with N-bromosuccinimide as brominating agent [*50*]. Irradiation of a refluxing solution of **68** in chloroform in the presence of bromine did also not lead to **70** but caused the elimination of a benzoyl group, 4-amino-2-benzamido-6-methylquinazoline (**69**) being formed quantitatively. The desired bromomethyl derivative **70** was finally obtained in good yield by treatment of **68** with 1,3-dibromo-5,5-dimethylhydantoin (**71**) in boiling carbon tetrachloride under irradiation [*50*].

2,4-Dibenzamido-6-(bromomethyl)-quinazoline (70) [*50*]. A rigorously dried suspension of 2 g of **68**, 1,3-dibromo-5,5-dimethylhydantoin (**71**) (0.8 g) and benzoyl peroxide (0.15 g) in 90 ml CCl$_4$ was heated under reflux whilst being irradiated by a 500 W tungsten lamp. After 1 hr., an almost clear, reddish-brown solution was obtained, which on further heating decolorized and deposited a yellow solid. After 105 mins., the solution being negative to starch-iodide, the solution was allowed to cool. The precipitate was collected, washed with ether and extracted with warm water (2 × 20 ml) to leave 1.91 g (79%) of **70** undissolved. Recrystallized from toluene, **70** had m.p. 213°.

References

[1] G. SOSNOVSKY: Free Radical Reactions in Preparative Organic Chemistry, pp. 282—413. New York: Macmillan 1964.
[2] J. D. ROBERTS and P. H. DIRSTINE: J. Amer. chem. Soc. 67, 1281 (1945).
[3] H. B. HASS, E. T. MCBEE and P. WEBER: Ind. Eng. Chem. 28, 333 (1936).
[4] G. A. RUSSELL and H. C. BROWN: J. Amer. chem. Soc. 77, 4025 (1955).
[5] R. N. HASZELDINE: J. Chem. Soc. 1952, 2504.
[6] S. A. FASEEH: J. Chem. Soc. 1953, 3708.
[7] R. N. HASZELDINE: J. Chem. Soc. 1953, 3565.
[8] R. E. SLADE: Chem. and Ind. 1945, 314.
[9] T. VAN DER LINDEN: Ber. dtsch. chem. Ges. 45, 231 (1912).
[10] L. A. MILLER, J. H. DUNN, C. M. NEHER and S. N. HALL: USP 2,622,105 of 16.12.1952.
[11] J. H. BROWN: Personal communication to A. SCHÖNBERG.
[12] J. T. RUCKER: USP 2,687,417 of 24.8.1954.
[13] G. WITTIG and F. VIDAL: Chem. Ber. 81, 368 (1948).
[14] M. WEIZMANN and S. PATAI: J. Amer. chem. Soc. 68, 150 (1946).
[15] G. CALINGAERT, M. E. GRIFFING, E. R. KERR, A. J. KOLKA and H. D. ORLOFF: J. Amer. chem. Soc. 73, 5224 (1951).
[16] A. J. KOLKA, H. D. ORLOFF and M. E. GRIFFING: J. Amer. chem. Soc. 76, 1244 (1954).
[17] G. G. ECKE, L. R. BUZBEE and A. J. KOLKA: J. Amer. chem. Soc. 78, 79 (1956).
[18] C. S. RONDESTVEDT JR., R. L. GRIMSLEY and C. D. VER NOOY: J. Org. Chem. 21, 206 (1956).
[19] B. MILLER and C. WALLING: J. Amer. chem. Soc. 79, 4187 (1957).
[20] C. E. GRIFFIN and R. N. HASZELDINE: J. Chem. Soc. 1960, 1398.
[21] M. BALLESTER, C. MOLINET and J. CASTAÑER: J. Amer. chem. Soc. 82, 4254 (1960).
[22] O. BASTIANSEN and O. HASSEL: Acta Chem. Scand. 1, 489 (1947).
[23] D. E. APPLEQUIST, G. F. FANTA and B. W. HENRIKSON: J. Amer. chem. Soc. 82, 2368 (1960).
[24] J. D. ROBERTS and R. H. MAZUR: J. Amer. chem. Soc. 73, 2509 (1951).
[25] G. A. RUSSELL: J. Amer. chem. Soc. 80, 4987 (1958).
[26] G. A. RUSSELL: J. Amer. chem. Soc. 80, 4997 (1958).
[27] C. WALLING and M. F. MAYAHI: J. Amer. chem. Soc. 81, 1485 (1959).
[28] H. MAGRITTE and A. BRUYLANTS: Ind. chim. belge 22, 547 (1957); C.A. 51, 13735 (1957).
[29] Z. EL-HEWEHI and D. HEMPEL: J. Prakt. Chem. [4] 22, 1 (1963).
[30] M. S. KHARASCH, Y. C. LIU and W. NUDENBERG: J. Org. Chem. 20, 680 (1955).
[31] M. S. KHARASCH, W. HERED and F. R. MAYO: J. Org. Chem. 6, 818 (1941).
[32] J. WILLARD and F. DANIELS: J. Amer. chem. Soc. 57, 2240 (1935).
[33] H. SCHALTEGGER: Experientia 5, 321 (1949).
[34] H. SCHALTEGGER: Helv. Chim. Acta 33, 2101 (1950).
[35] J. L. SPEIER: J. Amer. chem. Soc. 73, 826 (1951).
[36] W. THALER: J. Amer. chem. Soc. 85, 2607 (1963).
[37] P. S. SKELL and P. D. READIO: J. Amer. chem. Soc. 86, 3334 (1964).
[38] G. HÖLZLE and W. JENNY: Helv. Chim. Acta 41, 356 (1958).
[39] A. ROEDIG, in: Methoden der organischen Chemie, ed. by E. MÜLLER, 4th ed., Vol. 5, part 4, pp. 517—678; esp. p. 530. Stuttgart: Thieme 1960.
[40] P. S. SKELL and R. R. PAVLIS: J. Amer. chem. Soc. 86, 2956 (1964).

[41] B. MILLIGAN, R. L. BRADOW, J. E. ROSE, H. E. HUBBERT and A. ROE: J. Amer. chem. Soc. **84**, 158 (1962).
[42] M. G. VORONKOV and V. P. DAVYDOVA: Dokl. Akad. Nauk SSSR **125**, 553 (1959); Proc. Acad. Sci. USSR, Sect. Chem. **125**, 230 (1959).
[43] E. S. HUYSER and B. GIDDINGS: J. Org. Chem. **27**, 3391 (1962).
[44] H. KLOOSTERZIEL: Rec. Trav. chim. Pays-Bas **82**, 497 (1963).
[45] M. F. HEBBELYNCK and R. H. MARTIN: Bull. Soc. Chim. Belges **59**, 193 (1950).
[46] B. PRIJS, R. GALL, R. HINDERLING and H. ERLENMEYER: Helv. Chim. Acta **37**, 90 (1954).
[47] S. BERNSTEIN, L. J. BINOVI, L. DORFMAN, K. J. SAX and Y. SUBBAROW: J. Org. Chem. **14**, 433 (1949).
[48] C. MEYSTRE, A. WETTSTEIN and K. MIESCHER: Helv. Chim. Acta **30**, 1022 (1947).
[49] J. B. WRIGHT: J. Amer. chem. Soc. **77**, 4883 (1955).
[50] V. OAKES, H. N. RYDON and K. UNDHEIM: J. Chem. Soc. **1962**, 4678.

Chapter 38

Photochemical conversions of organic halides
1. Replacement of bromine by hydrogen, chlorine or ^{82}Br
a) Replacement of bromine by hydrogen

The transformation of 11a-bromo-6-demethyl-6-deoxytetracycline (**1**) to 6-demethyl-6-deoxytetracycline (**2**) has been described by HLAVKA and KRAZINSKI [*1*]. Irradiation of **1** was carried out in the presence of 1-naphthol, which acts as a scavenger for the bromine atom produced.

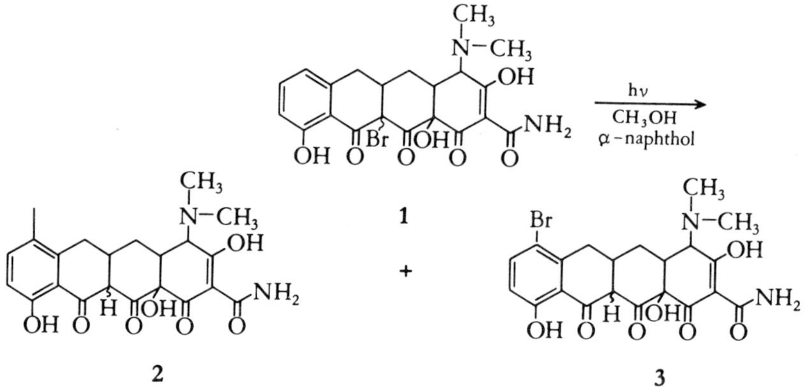

6-Demethyl-6-deoxytetracycline (2) [*1*]. A solution of 100 mg of the hydrochloride of **1** and of 90 mg of 1-naphthol in 15 ml methanol was irradiated with a Hanovia 30600 lamp in a Hanovia double-walled immersion well. After 4 hrs. the solution was evaporated to a volume of 2 ml and diluted with 100 ml of ether. The solid weighed 90 mg. This material was separated by fractional crystallization from methanol/ether. Paper strip chromatography showed the main fraction (90%) to be **2** and the second fraction (10%) to be 7-bromo-6-demethyl-6-deoxytetracycline (**3**).

b) Replacement of bromine by chlorine or ^{82}Br

3-Bromopentafluoropropene does not react with chlorine in the dark but in the presence of light the bromine-free compound 1,2,3-trichloropentafluoropropane is formed in 22 % yield, together with four other compounds [*2*].

References, pp. 371–372

VOEGTLI et al. [*3*] showed the substitution of bromine by chlorine in aromatic halides to be a frequently occurring phenomenon. Table 29 shows a few examples:

Table 29

starting material	product	yield
p-dibromobenzene	p-dichlorobenzene	90%
1-bromo-4-chlorobenzene	p-dichlorobenzene	90%
1-bromo-3-chlorobenzene	m-dichlorobenzene	85%
1-bromo-2-chlorobenzene	o-dichlorobenzene	82%
1,1-bis-(4-bromophenyl)-2,2,2-trichloroethane	1,1-bis-(4-chlorophenyl)-1,2,2,2-tetrachloroethane	85%
1,1-bis-(4-bromophenyl)-2,2-dichloroethylene	1,1-bis-(4-chlorophenyl)-1,2,2,2-tetrachloroethane	90%

MILLER and WALLING [*4*] have treated bromobenzene with excess chlorine both in darkness and in light, under otherwise identical conditions. Replacement of the bromine by chlorine took place in both cases, but the yields were very different: 39% (in the dark) vs. 81% (in the light). A similar substitution also took place in the case of bromobenzoic acids (m- and o-), but not in the case of 1-bromo-4-nitrobenzene. It is possible to replace the bromine atom in bromobenzene by ^{82}Br. Although no reaction takes place in the dark, under strong illumination a 30% exchange with ^{82}Br can be brought about in 17 hours [*4*].

p-Dichlorobenzene [*3*]. p-Dibromobenzene (25 g) in carbon tetrachloride (100 ml) is chlorinated under irradiation for 3—4 hours. For details of the irradiation procedure the reader is referred to VOEGTLI [*3*]. Immediately after the introduction of the chlorine, the contents of the flask become dark brown in color, and after a short time a part of the bromine can be seen in the reflux, together with the carbon tetrachloride. The rest of the bromine escapes. At the end of the experiment, air is sucked through the apparatus until both chlorine and bromine have been driven off. Following this, first the carbon tetrachloride and then the p-dichlorobenzene is distilled off using a distillation flask with a short column. Yields of p-dichlorobenzene are 14 g, b.p. 170°, m.p. 52°.

2. Replacement of iodine in iodides by hydrogen, nitric oxide or chlorine

a) Experiments with aliphatic iodides

Irradiation [*5*] of trifluoroiodomethane with UV-light in the presence of mercury and ethyl alcohol (or n-hexane, or water) at room temperature produces trifluoromethane in good yield. It is assumed that the trifluoroiodomethane is photochemically dissociated:

$$F_3CI \xrightarrow{h\nu} F_3C\cdot + I\cdot$$

The re-formation of trifluoroiodomethane is prevented by reaction between the mercury and the iodine atoms (formation of mercuric iodide).

Following the dissociation, reaction occurs between the trifluoromethyl free radicals and the ethyl alcohol (or n-hexane, or water):

$$F_3C\cdot + CH_3CH_2OH \longrightarrow HCF_3 + CH_3\dot{C}HOH$$
$$2\ CH_3\dot{C}HOH \longrightarrow CH_3CH_2OH + CH_3CHO$$

Experiments carried out in the presence of n-hexane produced hexene, and the following transformation took place with water:

$$F_3C\cdot + H_2O \longrightarrow HCF_3 + \cdot OH$$

HASZELDINE [6,7] reported that perfluoroalkyl iodides can be converted to the corresponding nitroso compounds by the photochemical action of nitrogen oxide. He carried out the experiments in the presence of mercury, which served to remove amounts of NO_2 formed.

Under the action of light at first perfluoroalkyl radicals were formed, as a result of the extrusion of iodine from the perfluoroalkyl iodides. This was demonstrated in the formation of nonafluoro-2-nitrosobutane (5) from nonafluoro-2-iodobutane (4) [7]:

$$CF_3-CF_2-CFI-CF_3 \xrightarrow{h\nu} CF_3-CF_2-\overset{\cdot}{C}F-CF_3 \xrightarrow{NO}$$
$$\qquad 4$$
$$\qquad\qquad CF_3-CF_2-CF(NO)-CF_3$$
$$\qquad\qquad\qquad 5$$

Using this method, a series of blue substances were prepared, some of which are listed (together with boiling-points) in table 30 [8].

Table 30

CF_3NO	−86°	CF_2ClCF_2NO	− 2°
C_2F_5NO	−42°	CF_2BrNO	−12°
C_3F_7NO	−12°	CF_2ClNO	−35°
C_4F_9NO	17°		

MASON and DUNDERDALE [9] obtained trifluoronitrosomethane using a similar method. In contrast to the experiments of HASZELDINE, however, the reaction of the nitrogen oxide with the trifluoroiodomethane was carried out without the addition of mercury.

HASZELDINE [10] observed the replacement of the iodine atoms in aliphatic iodides by chlorine when chlorine was irradiated with 1,1,1-trifluoro-3-iodopropane (6):

$$F_3C-CH_2-CH_2I \xrightarrow{h\nu} F_3C-CH_2-\dot{C}H_2 + I\cdot \xrightarrow{Cl_2} F_3C-CH_2-CH_2Cl + ICl$$
$$\qquad 6 \qquad\qquad\qquad\qquad\qquad\qquad\qquad\qquad\qquad 7$$

Trifluoromethane [5]. A sealed Pyrex glass tube containing mercury (1 ml), trifluoroiodomethane (2.0 g) and ethyl alcohol (20 ml) was irradiated with UV-light.

References, pp. 371−372

Throughout the irradiation period, the tube was vigorously shaken mechanically (3 days, room temperature). Mercuric iodide separated out and the liquid became brown in color. Trifluoromethane (93%) together with some acetaldehyde was produced.

Nonafluoro-2-nitrosobutane (5) [7]. A sealed glass tube (quartz; 50 ml) containing nonafluoro-2-iodobutane (4) (2.1 g), nitrogen oxide (80% excess) and mercury (3 g) was exposed to UV-light (Hanovia lamp) for 8 days. The light source was situated 10 cm from the tube, and the tube was vigorously shaken throughout the irradiation. Fractionation of the contents of the tube under reduced pressure gave, besides unconverted 4, nonafluoro-2-nitrosobutane (5). The latter substance is a blue liquid, b.p.: 23—25°; yield: 23%.

3-Chloro-1,1,1-trifluoropropane (7) [10]. The reaction of chlorine with 1,1,1-trifluoro-3-iodopropane (6) under photochemical conditions was carried out in a quartz tube (200 ml capacity), which was cooled with running water. The tube was combined with an efficient reflux-condenser cooled to 0°. Chlorine was conducted through 6 (11.2 g), and the flow was regulated so that the temperature in the reaction-vessel did not rise above 25°, whereby the greater part of the chlorine was absorbed. During the reaction, the tube was irradiated with a Hanovia UV-lamp (without filter), which was placed 60 cm from the tube. Gases passing out of the reflux-condenser were conducted through water. With careful procedure, only traces of organic material escaped from the reflux-condenser. The reaction was discontinued after 1.2 mol Cl_2 had been absorbed. Two such runs were combined and fractionated to give, 3-chloro-1,1,1-trifluoro-3-iodopropane (11.2 g; 43%; b.p.: 119°), 3-chloro-1,1,1-trifluoropropane (7) (2.1 g; 16%; b.p.: 45—46°), 3,3-dichloro-1,1,1-trifluoropropane (2 g; 12%; b. p.: 72—74°) and 3,3,3-trichloro-1,1,1-trifluoropropane (2.2 g; 11%; b.p.: 93.5—96.5°).

b) Chlorobenzene from iodobenzene by photochemical decomposition of iodine chloride

This displacement reaction was carried out by MILLIGAN [11], and a quantitative conversion was obtained.

$$\text{C}_6\text{H}_5\text{I} \xrightarrow[\text{ICl}]{h\nu} \text{C}_6\text{H}_5\text{Cl}$$

Chlorobenzene [11]. To 8.20 g of iodobenzene dissolved in 4 l of CCl_4 was added 10.7 ml of 0.373 M ICl (10.0 mole %) in CCl_4. The solution was stirred and irradiated by a 275 Watt mercury arc sunlamp, 15 cm away, for 12 hours. The violet colored solution was washed with dilute thiosulfate and water and then dried with calcium carbide. The solvent was removed by distillation through a 1.5 × 75 cm helix packed column until the residue was reduced to about 80 ml. The residue was transferred to a 100 ml volumetric flask and was diluted to volume with CCl_4. Analysis of the solution by gas chromatography showed that a quantitative yield, based on ICl, of chlorobenzene was obtained.

3. Deiodination of aliphatic iodides

Only a very small quantity of hexafluoroethane is formed on irradiation of trifluoroiodomethane vapor with UV-light [5]. However, if other substances are added which react with the dissociation products $\cdot CF_3$ and $I \cdot$ (e.g. mercury and cyanogen), the free radicals are prevented from recombining and hexafluoroethane may be obtained.

The use of cyanogen in the photochemical deiodination of aliphatic iodides — whereby the cyanogen is converted into ICN — is of less importance preparatively than the use of mercury. This latter has proved to be particularly useful in the deiodination of compounds with the end groups CClFI. Using this method, HASZELDINE [12] obtained 9 from 1,2,4-trichlorohexafluoro-4-iodobutane (8).

$$CClF_2-CClF-CF_2-CClFI \xrightarrow[Hg]{h\nu}$$
$$8$$

$$CClF_2-CClF-CF_2-CClF-CClF-CF_2-CClF-CClF_2$$
$$9$$

HASZELDINE points out in this connection that in the photochemical deiodination of **8** with mercury, a product is formed which is free from olefins. This is not the case with the customary deiodination method using zinc.

1,2,4,5,7,8-Hexachlorododecafluorooctane (9) [12]. This compound was obtained by vigorously shaking 5.1 g of **8**, diluted with 5 ml of 1,1,2-trichlorotrifluoroethane, and 20 ml of mercury for 8 days under UV-irradiation in a quartz tube. After extraction of the reaction product with ether and distillation, a 81% yield of **9** was obtained, b.p.$_{20}$ 142—144°.

4. Debromination of 1,1-diaryl-2-bromoethylenes

TADROS et al. [13] carried out the following dimerization in sunlight in acetic acid. This transformation did not take place in the dark.

1,1,4,4-Tetrakis-(4-methoxyphenyl)-1,3-butadiene (11) [13]. A solution of 1,1-bis-(4-methoxyphenyl)-2-bromoethylene (**10**) in acetic acid (10 ml) was exposed to direct sunlight under an atmosphere of nitrogen (3 months, Cairo). The reaction vessel was a sealed Pyrex tube. The contents of the tube assumed a brown color after 24 hrs., and this color deepened on further irradiation. A precipitate was formed (0.4 g) which was shown to be **11**.

5. Formation of organomercury compounds by the action of mercury on alkyl iodides

EMELÉUS and HASZELDINE [14] have shown that trifluoroiodomethane and pentafluoroiodoethane differ from iodomethane and iodoethane in

References, pp. 371—372

their behavior towards metals. Trifluoroiodomethane cannot be grignardized with the aid of the usual catalysts and no conversion to organometallic derivatives of Zn, Cd or Li was possible.

Trifluoroiodomethane and pentafluoroiodoethane will, on the other hand, form stable compounds with mercury, e.g. F_3CHgI, when the reaction is initiated by heat or illumination. If the reaction is carried out in the dark, relatively high temperatures must be used (260—290°). In the light reactions lower temperatures suffice. The conversion will take place in light with or without the presence of a solvent.

(Trifluoromethyl)-mercury iodide [*14*]. Trifluoroiodomethane (7.5 g) dissolved in 4 ml of perfluoromethylcyclohexane was heated to 110° in a Pyrex tube and shaken with mercury for 36 hours. Throughout this period, the tube was irradiated with a Hanovia lamp. Evaporation of the solvent and extraction with ether affords 7.5 g (80% conversion) of (trifluoromethyl)-mercury iodide, which sublimes at 80°.

6. Preparation of hexaarylethanes by the action of triarylmethyl halides on triarylmethanes

SCHLENK and HERZENSTEIN [*15*] have shown that 9,9'-diphenyl-9,9'-bifluorene (**14**) can be easily obtained photochemically from 9-phenylfluorene (**12**) and 9-chloro-9-phenylfluorene (**13**).

When 4-biphenylyldiphenylmethane (**15**) and 4-biphenylylchlorodiphenylmethane (**16**), contained in a sealed tube under CO_2 atmosphere, are exposed to intense sunlight for several hours, the following equilibrium is attained which is, however, heavily biased to the left side [*15*]:

9,9′-Diphenyl-9,9′-bifluorene (14) [*15*]. A cold, concentrated solution of **12** and **13** in molecular proportions in benzene is exposed to sunlight or the light of a mercury vapor lamp for a few days. The reaction is carried out in a thin-walled glass vessel and with exclusion of air. The colorless crystals of **14** which separate out in large quantity are filtered off and the filtrate is once more exposed until the separation of crystals is complete. The liquid starts fuming when the reaction vessel is opened to the air, on account of the HCl formed. Compound **14** so prepared is found to be completely pure.

7. Photolysis of aromatic iodo compounds

a) Iodobenzene and related substances

The problem in adapting the photolysis of iodobenzene and related substances for synthetic purposes [*16*] was to establish conditions which promote the scission of the carbon-iodine bonds without causing subsequent reactions of products, or undesired side reactions. By conducting the reaction at, or near, room temperature and using an ultraviolet source which provides energy at a wave length which essentially cleaves only the carbon-iodide bond, the side reactions are minimized.

In synthetically useful yields Wolf [*16*] obtained, among others, the following products:

2-hydroxybiphenyl (from 2-iodophenol in benzene)
4-hydroxybiphenyl (from 4-iodophenol in benzene)
4-nitrobiphenyl (from 1-iodo-4-nitrobenzene in benzene)
2-methoxybiphenyl (from iodobenzene in anisole).

In some cases yields were surprisingly good. Thus Kharasch [*17*] obtained a 90% yield of 4-phenylbenzaldehyde after irradiation of a benzene solution of 4-iodobenzaldehyde with a mercury low pressure lamp for 48 hours.

2-Hydroxybiphenyl [*16*]. 2-Iodophenyl (1 g) in 50 ml pure, dry benzene, was irradiated in a "Vycor"-7100 tube, by a helical cold-cathode, low pressure mercury lamp (manufactured by Dallons Laboratories, Los Angeles, California). After twenty hours of irradiation, the release of iodine, conveniently measured by titration with thiosulfate, was complete. Isolation of the product from the organic layer and purification by two passes through an alumina column gave 0.46 g (60%) yield) of pure 2-hydroxybiphenyl, fully characterized by its m. p. and infrared spectrum.

b) Formation of benzyne on photolysis of 1,2-diiodobenzene or (2-iodophenyl)-mercury iodide

The photolysis of 1,2-diiodobenzene (**17**) has been investigated by Kampmeier and Hoffmeister [*18*] who found evidence for the formation of benzyne (**18**) in this irradiation since, in the presence of tetracyclone (**19**), 1,2,3,4-tetraphenylnaphthalene (**20**) was obtained.

References, pp. 371–372

WITTIG [*19*] obtained a similar result by UV-irradiation of (2-iodophenyl)-mercury iodide (**21**) in benzene in the presence of **19**.

1,2,3,4-Tetraphenylnaphthalene (20) [*18*]. Equimolar amounts of **17** and **19** were dissolved in benzene in a cyclindrical quartz cell. This solution was irradiated for 41 hrs. with a Hanovia SC 2537 low pressure mercury vapor lamp at room temperature. 2-Iodobiphenyl was the major product besides a 10% yield of 1,2,3,4-tetraphenylnaphthalene (**20**), the expected DIELS-ALDER adduct of benzyne (**18**) and **19**. This must arise from **17** since the photolysis of tetracyclone (**19**) in benzene gave no **20**.

8. Photolysis of alkyl hypoiodites, acyl hypoiodites and N-iodoamides

a) Photolysis of alkyl hypoiodites. Preparation of ethers

The oxidation of non-activated methylene groups with lead tetraacetate and iodine in organic solvents has become known as the "hypoiodite reaction". Our knowledge of the process is mainly attributable to Swiss research workers. The term "hypoiodite reaction" refers to the probable formation of hypoiodites (cf. **22**) as primary products, but these have until now not been isolated in pure form. The formation of these hypoiodites and their secondary reactions have been reported by HEUSLER et al. [*20—23*]. A recent account has been given by AKHTAR [*24*].

It is sometimes profitable to employ irradiation with strong visible light in order to accelerate the reaction, e.g. in the case of 20-hydroxypregnane compounds. Under these conditions, substitution reactions at activated sites on the substrates are rare, or, at the most, occur only as side-reactions [*20*]. In this manner, 3β-acetoxy-5-chloro-6β-hydroxy-5α-androstan-17-one (**23**) was converted to **24** [*25*].

3β-Acetoxy-5-chloro-6β,19-oxido-5α-androstan-17-one (24) [*25*]. A suspension of 90 g lead tetraacetate (thoroughly freed from acetic acid under high vacuum) and 30 g CaCO$_3$ in 4 l cyclohexane, briefly heated to 80°, was mixed with 20 g iodine and 15 g of **23**. The mixture was then boiled under a reflux condenser, whilst stirring and irradiating with a 500 Watt lamp, for 60 min. After cooling, the still lightly colored mixture was filtered through celite and the residue was thoroughly washed with ether. The filtrate was extracted by shaking with 1 l 10% sodium thiosulfate solution and with water, dried over sodium sulfate and evaporated under vacuum (water-pump). The crystalline raw-product was recrystallized from ether-methanol to give 11.48 g of pure **24**, m.p. 181–182°. A further 1.35 g impure **24** was obtained from the mother-liquor.

b) Photolysis of acyl hypoiodites. Replacement of the carboxyl group by iodine

According to BARTON [*26*] the following photochemical decarboxylation may be effected, which probably involves formation of an intermediate acyl hypoiodite (RCOOI):

$$\text{RCOOH} \xrightarrow[\text{Pb(OAc)}_4\,;\,\text{I}_2]{h\nu} \text{RI}$$

To a 5% w/v suspension of lead tetraacetate (1 mole) in refluxing carbon tetrachloride (illuminated by a tungsten lamp) is added the carboxylic acid (1 mole) and then iodine (1 mole) in the same solvent until the iodine color persists. The reaction proceeds more slowly and less cleanly (lower yields) in the dark.

This reaction is a convenient procedure for the decarboxylation of primary and secondary carboxylic acids in high yield [*26*]. Illustrative examples of this are given in table 31.

Table 31

carboxylic acids	nor-iodides
primary acid: 12-acetoxystearic	11-acetoxy-1-iodoheptadecane (82%)
secondary acid: cyclohexanecarboxylic	iodocyclohexane (91%)
aromatic acid: benzoic	iodobenzene (56%) and 1,4-diiodobenzene (17%)
dicarboxylic acid: adipic	1,4-diiodobutane (33%)

c) Photolysis of N-iodoamides. Preparation of lactones

BARTON and BECKWITH [*27*] treated acid amides with lead tetraacetate and iodine whilst irradiating with UV-light, and obtained lactones following alkaline hydrolysis of the reaction product.

References, pp. 371–372

The mechanism of the formation of γ-lactones from amides involves photolysis of an N-iodoamide (**25**) followed by intramolecular hydrogen transfer (**26** → **27**) and coupling of the resultant radical with iodine. Hydrolysis of the γ-iodoamide (**28**) so produced would then give a γ-lactone via an intermediate γ-iminolactone [27, 28].

Thus, stearamide was transformed into γ-stearolactone (m. p. 49—50°) and 3β-acetoxy-11-oxo-5α-pregnane-20-carboxamide (**29**) yielded **30** [27].

3β-Acetoxy-16β-hydroxy-11-oxo-5α-pregnane-20-carboxylic acid lactone (30) [27]. 1 Mole of **29** and 3 mole of lead tetraacetate were dissolved with 4 mole iodine in chloroform and this solution irradiated at 15° for 5 hours. Alkaline hydrolysis and acetylation of the crude product afforded 0.55 mole of **30**, m. p. 265—267°.

References

[1] J. J. HLAVKA and H. M. KRAZINSKI: J. Org. Chem. **28**, 1422 (1963).
[2] A. H. FAINBERG and W. T. MILLER JR.: J. Amer. chem. Soc. **79**, 4170 (1957).
[3] W. VOEGTLI, H. MUHR and P. LÄUGER: Helv. Chim. Acta **37**, 1627 (1954).
[4] B. MILLER and C. WALLING: J. Amer. chem. Soc. **79**, 4187 (1957).
[5] J. BANUS, H. J. EMELÉUS and R. N. HASZELDINE: J. Chem. Soc. **1950**, 3041.
[6] R. N. HASZELDINE: J. Chem. Soc. **1953**, 2075.
[7] R. N. HASZELDINE: J. Chem. Soc. **1953**, 3559.
[8] R. N. HASZELDINE: Angew. Chem. **66**, 693 (1954).
[9] J. MASON and J. DUNDERDALE: J. Chem. Soc. **1956**, 754.
[10] R. N. HASZELDINE: J. Chem. Soc. **1951**, 2495.
[11] B. MILLIGAN, R. L. BRADOW, J. E. ROSE, H. E. HUBBERT and A. ROE: J. Amer. chem. Soc. **84**, 158 (1962).
[12] R. N. HASZELDINE: J. Chem. Soc. **1955**, 4302.
[13] W. TADROS, A. B. SAKLA and Y. AKHOOKH: J. Chem. Soc. **1956**, 2701.
[14] H. J. EMELÉUS and R. N. HASZELDINE: J. Chem. Soc. **1949**, 2948.
[15] W. SCHLENK and A. HERZENSTEIN: Ber. dtsch. chem. Ges. **43**, 3541 (1910).
[16] W. WOLF and N. KHARASCH: J. Org. Chem. **26**, 283 (1961).
[17] N. KHARASCH and L. GÖTHLICH: Angew. Chem. **74**, 651 (1962); internat. ed. **1**, 459 (1962).
[18] J. A. KAMPMEIER and E. HOFFMEISTER: J. Amer. chem. Soc. **84**, 3787 (1962).
[19] G. WITTIG and H. F. EBEL: Liebigs Ann. Chem. **650**, 20 (1961).

[20] C. Meystre, K. Heusler, J. Kalvoda, P. Wieland, G. Anner and A. Wettstein: Helv. Chim. Acta **45**, 1317 (1962).
[21] K. Heusler, J. Kalvoda, C. Meystre, G. Anner and A. Wettstein: Helv. Chim. Acta **45**, 2161 (1962).
[22] K. Heusler, J. Kalvoda, P. Wieland, G. Anner and A. Wettstein: Helv. Chim. Acta **45**, 2575 (1962).
[23] K. Heusler and J. Kalvoda: Angew. Chem. **76**, 518 (1964); internat. ed. **3**, 525 (1964).
[24] M. Akhtar, in: Adv. Photochem., Vol. **2**, ed. by W. A. Noyes jr., G. S. Hammond and J. N. Pitts jr., pp. 263—303. New York: Interscience 1964.
[25] H. Überwasser, K. Heusler, J. Kalvoda, C. Meystre, P. Wieland, G. Anner and A. Wettstein: Helv. Chim. Acta **46**, 344 (1963).
[26] D. H. R. Barton and E. P. Serebryakov: Proc. Chem. Soc. **1962**, 309.
[27] D. H. R. Barton and A. J. L. Beckwith: Proc. Chem. Soc. **1963**, 335.
[28] D. H. R. Barton, A. L. J. Beckwith and A. Goosen: J. Chem. Soc. **1965**, 181.

Chapter 39

Photochemical formation of hydroperoxides and peroxides

1. Replacement of hydrogen by the hydroperoxide group

a) Unsensitized photooxidation of unsaturated compounds

Unsensitized photochemical replacement of a carbon-bonded hydrogen by the hydroperoxide group by the direct action of oxygen is only of small preparative significance as far as open chain compounds are concerned. The experiments seldom lead to homogeneous hydroperoxide products. Thus FARMER and SUTTON [1] obtained a mixture of two hydroperoxides by the light-induced autoxidation of methyl oleate. Closer examination [2] revealed the methyl oleate autoxidation to be more complex in nature; for details of the mechanism involved the pertinent autoxidation literature (e.g. [3—5], see also [6]) should be consulted.

A series of homogeneous hydroperoxides may be obtained by the action of oxygen on cyclic hydrocarbons under irradiation. For these syntheses we are mainly indebted to HOCK and his school. Successful experiments were carried out with, among others, indan [7], p-xylene [8], cumene and diphenylmethane [9]. In every case the oxidation was carried out in UV-light at temperatures between 60° and 85°. The products obtained have the formulae **1—4**.

Isolation of the hydroperoxides **1—4** was effected either by distillation in high vacuum or by shaking with alkali, the latter method being applied only when the hydroperoxide showed markedly acid properties. Reaction of the hydroperoxides with aqueous sulfite was found a suitable reduction method.

FARMER and SUTTON [10] obtained a mixture of two hydroperoxides (**5** and **6**) by the photooxidation of 1,2-dimethyl-1-cyclohexene.

References, pp. 402—406

It must be emphasized that a series of hydrocarbons with saturated side chains or rings react with oxygen already in the dark to form hydroperoxides; tetralin may be cited as an example [*11*]. The question of whether irradiation with UV-light is necessary for the synthesis of compounds **1—4** (or some of them) does not appear to have been fully resolved.

On irradiation of aerated solutions of 1,2,3,4,4a,9a-hexahydrofluorene (**7**) in absolute methanol in the presence of sulfuric acid, TREIBS [*12*] obtained 1,2,3,4,4a,9a-hexahydro-4a-methoxyxanthene (**8**).

Tetralin moieties are photooxidizable also when incorporated in more complex ring systems. The photooxidation of (+)-3-methoxy-N-methylmorphinane (**9**) on allowing aqueous solutions of this substance to stand in daylight was recognized by the development of an intense yellow color [*13*]. The oxidation product, **10**, isolated in about 10 % yield, could also be obtained in a dark reaction by the action of chromium trioxide and sulfuric acid on **9**. Considering the constitutional similarities between **9** and tetralin it seems very probable that the joint action of oxygen and light on **9** leads to a hydroperoxide which by loss of water breaks down to the ketone **10**.

1-Hydroperoxyindan (1) [*7*]. 500 g freshly distilled indan were shaken with dry oxygen for 24 hrs. whilst simultaneously irradiated by a mercury vapor lamp and an incandescent lamp (240 W, acting as a heat supply). The contents of the reaction vessel were maintained at 60° throughout the experiment. After 4.7 l oxygen was taken up, the peroxide mixture was cooled to —5° and 40 g 25% sodium hydroxide solution added in a thin stream with stirring. The precipitated sodium salt was filtered off and washed a few times with ether, then suspended in a little ice-cold water and treated with cold 2N hydrochloric acid with cooling. The oil which formed was taken up in ether and the resulting yellow solution washed with sodium bicarbonate solution. In this way the ether solution became almost colorless and the bicarbonate solution, from which it was possible to isolate a little homophthalic acid (m.p. 175—177°), became brown. The ethereal solution was dried over Na_2SO_4, the ether driven off, and **1** distilled in high vacuum; b.p.$_{0.01}$ 64—65°, yield 28 g.

References, pp. 402—406

1,2,3,4,4a,9a-Hexahydro-4a-methoxyxanthene (8) [*12*]. Oxygen was passed through a solution of 250 g of hexahydrofluorene **7** in 1500 ml absolute methanol and 3 ml conc. sulfuric acid during irradiation with UV-light at 60° for 50 hours. Finally the solution was neutralized with ammonia and the methanol distilled off. After cooling, the product was separated from precipitated ammonium sulfate and distilled in high vacuum. After separation of unreacted starting material, 50 g (70% calculated on **7** consumed) distilled over, which after several distillations through a Vigreux column had b.p.$_{0.08}$ 82—83°.

(—)-3-Methoxy-N-methyl-10-oxomorphinane (10) [*13*]. 1000 ml. of a 1% aqueous solution of the hydrobromide of **9** was stirred and saturated with oxygen for 30 minutes; the solution was then divided equally among 8 Erlenmeyer flasks (each of 500 ml capacity) and placed on a white base in sunlight for 3 days (Basle). Finally, the contents of the flasks were combined, treated with 100 ml conc. HCl, washed with ether and the acid solution brought to pH of 9.3 by addition of ammonia solution. The precipitate which formed was allowed to stand for three hours, filtered, dried, precipitated three times from acetone and finally sublimed in high vacuum (0.1 mm, 180°). 1.13 g pale yellow prisms of **10** were obtained, m.p. 189—190°.

b) Photosensitized oxidation of olefins

SCHENCK and co-workers [*14*] found that the photosensitized oxidation of olefins proceeded according to the following scheme:

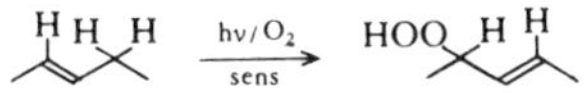

In the sensitized photooxidation the oxygen molecule adds to one C-atom of the double bond, then a hydrogen from the allylic position migrates to the oxygen with concomitant shift of the double bond. The addition according to the above scheme was termed [*14, 15*] "indirect substituting addition in the allylic position" in close analogy to the principles governing in ene-synthesis as elaborated by ALDER [*16*].

In this manner trans-pinocarveyl hydroperoxide (**12**) is formed from α-pinene (**11**) and myrtenyl hydroperoxide (**15**) from β-pinene (**14**) [*14*].

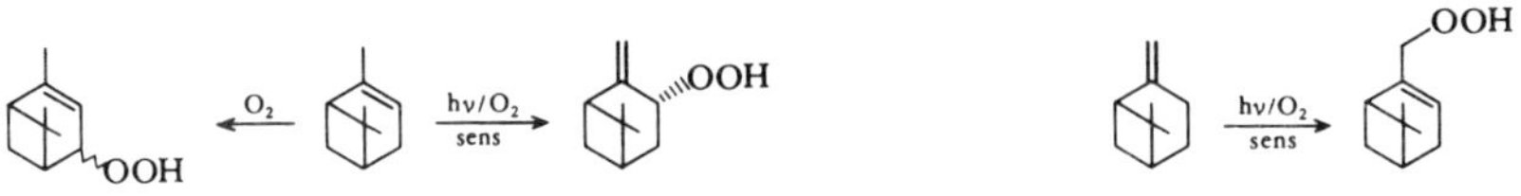

13   **11**   **12**   **14**   **15**

The products of photosensitized oxygenation and of autoxidation are identical only in exceptional cases [*14, 15, 17, 18*]. Thus, autoxidation of α-pinene leads to verbenyl hydroperoxide (**13**) instead of **12** [*14*]. Photosensitized oxygen transfer to (+)-limonene (**16**) furnishes, after reduction of the hydroperoxides primarily formed, six optically active alcohols [*17*]. Autoxidation in the dark, however, affords a different product mixture, the composition of which (in %) is given below as compared to that of the photooxidation procedure [*18*]. For another preparation of the same products involving singlet oxygen in a dark reaction see FOOTE and co-workers [*19*].

Structures **16**, **17**, **18**, **19**, **20**, **21**, **22**, **23**, **24**.

	(+)-17	(+)-18	(+)-19	(−)-20	(±)-20	(−)-21	(±)-21	(+)-22	(+)-23	(±)-24
photoox.	34	10	20	10	0	5	0	21	0	0
autox.	11	12	0	0	22	0	19	0	13	13

Differing views — which will not be commented on here — have been expressed as to the mechanisms and stereochemical requirements involved in the photosensitized oxygenation of olefins [15, 17, 20—27].

Substrates which readily undergo oxygenation on irradiation in the presence of a dye, comprise tri- and tetraalkyl substituted olefins of the aliphatic, alicyclic and aralkyl types. Semicyclic double bonds usually react only reluctantly, if at all [28]. No complete account on the preparative aspects of photosensitized oxidation of olefins can be given here, and only a few examples will be mentioned. For more information the reader is referred to the pertinent papers of, inter alia, NICKON, SCHENCK and their schools (see also [4, 5]).

The dye-sensitized oxidation [29] of 2,3-dimethyl-2-butene (tetramethylethylene, **25**) and 1,2,3,4,5,6,7,8-octahydronaphthalene (Δ^9-octalin, **27**) produces 1-hydroperoxy-2,3-dimethyl-3-butene (**26**) and 2,3,4,5,6,7-hexahydro-4a(1H)-naphthyl hydroperoxide (**28**) respectively.

$$\mathbf{25} \xrightarrow{h\nu/O_2,\ sens} \mathbf{26}$$

$$\mathbf{27} \xrightarrow{h\nu/O_2,\ sens} \mathbf{28}$$

When citronellol (**29**) is irradiated in the presence of rose bengal under oxygen two hydroperoxides are obtained [*30*], which on reduction yield 60% of 2,6-dimethyl-3-octene-2,8-diol (**30**) and 35% of 2,6-dimethyl-1-octene-3,8-diol (**31**). The alcohol **30** undergoes an allylic rearrangement very easily with the diol resulting spontaneously cyclizing to a mixture of stereoisomeric "rose oxides". These cis- and trans-4-methyl-2-(2-methyl-1-propenyl)-tetrahydropyrans (**32**), which have thus become readily available, constitute the active principle of the rose scent.

Olefinic steroids proved very suitable for studies on the stereochemical requirements of the photosensitized oxygen transfer reaction [*24, 31*].

With hematoporphyrin as sensitizer for the photochemical oxygen transfer reaction on cholesterol SCHENCK and co-workers [*32*] were able to obtain a cholesterol hydroperoxyde, namely 5-hydroperoxy-3β-hydroxy-5α-cholest-6-ene (**33**) which in chloroform solution underwent an allylic rearrangement with formation of 7α-hydroperoxy-3β-hydroxycholest-5-ene (**34**) [*33*]. If the photooxidation is carried out in chloroform solution, **34** may be obtained directly. Another synthesis has been elaborated by NICKON and BAGLI [*25*] by sensitized O_2-transfer to 3β-hydroxy-5α-cholest-6-ene (**35**).

Sensitized photooxygenation again evinced its usefulness when applied to steroidal allylic alcohols e.g. 3β-hydroxycholest-4-ene (**36**) [*34, 35*]. The intermediate hydroperoxide breaks down to 4α,5-epoxy-5α-cholestan-3-one (**37**) along with some cholest-4-en-3-one. Higher light intensities were, however, required in this experiment as compared with cholesterol photo-oxidation.

36   **37**

(—)-trans-Pinocarveyl hydroperoxide (12) [*14*]. 136 g (+)-α-pinene (**11**) and 1.5 g of methylene blue as sensitizer were dissolved in 2800 ml isopropanol. The reaction took place in an irradiation apparatus fitted with an internal water-cooling system at 18—21° employing two mercury vapor lamps Osram HgH 1000. Oxygen was circulated by a diaphragm pump entering the apparatus through a fritted glass gas inlet. After about 60 hrs., when 11.7 l O_2 (52%) had been consumed, the sensitizer began to fade. The illumination was discontinued and the solution worked up. Evaporation of the solvent and unreacted starting material under reduced pressure (bath below 50°) gave a residue which was distilled in high vacuum. Thus, 45 g of **12** were obtained, b.p.$_{0.01}$ about 64°.

(—)-Myrtenyl hydroperoxide (15) [*14*]. 217 g of (—)-β-pinene (**14**) and 2 g of methylene blue, dissolved in 2800 ml isopropanol were irradiated as above. After consumption of 30% of oxygen (46 hours) the reaction came to a standstill as a result of decolorization of the sensitizer. Working up as above gave 61.9 g of **15**, b.p.$_{0.01}$ about 62°.

1-Hydroperoxy-2,3-dimethyl-3-butene (26) [*29*]. Tetramethylethylene (**25**) (8.4 g) in 80 ml methanol was photooxidized in the presence of 50 mg of rose bengal. After 50 mins. the oxygen uptake was 2200 ml. After removal of the solvent at 12 mm Hg (bath below 25°), distillation yielded 9.5 g (82%) of **26**, b.p.$_{12}$ 54—55°.

10-Hydroperoxy-$\Delta^{1(9)}$-octalin (28) [*29*]. A solution of 13.6 g of **27** in 100 ml isopropanol to which 50 mg rose bengal had been added, took up 2080 ml O_2 in 90 mins. irradiation (Philipps HPK 125 W lamp). After removing the solvent at 12 mm Hg (bath 30°) the resulting crystals were sublimed at 0.001 mm Hg (bath 50—55°) to give colorless needles (14.8 g) of **28**, m.p. 59—60°.

5-Hydroperoxy-3β-hydroxy-5α-cholest-6-ene (33) [*36*]. A solution of 15 g cholesterol and 50 mg hematoporphyrin in 100 ml pyridine was irradiated under oxygen with an Osram HQA 500 mercury burner, contained in a glass water-cooled immersion shaft. When 1.1 mole O_2 had been consumed the solution was concentrated to 20 ml in vacuo, diluted with 20 ml of methanol and briefly boiled with charcoal. Water was then added dropwise to the hot filtered solution until turpidity was almost reached. Slow evaporation of the alcoholic solvent caused 11.8 g (73%) of **33** to precipitate; m.p. 148—149° (from aqueous methanol).

7α-Hydroperoxy-3β-hydroxycholest-5-ene (34) [*33*]. 3 g of cholesterol and 30 mg protoporphyrin methyl ester as sensitizer were dissolved in 100 ml. chloroform and the solution irradiated as above. After absorption of 195 ml oxygen (90 minutes) the solution was set aside for 36 hrs. and then evaporated in vacuo. Crystallization of the residue gave 1.37 g (42%) of **34**, m.p. 150—152°.

4α,5-Epoxy-5α-cholestan-3-one (37) [*35*]. 360 mg of **36** and 5 mg of hematoporphyrin, dissolved in 75 ml pyridine, were irradiated under oxygen with four 15 W fluorescent bulbs (72 hrs., 40°). Work-up gave an oil (340 mg) which was dissolved in petroleum ether and chromatographed on alumina. Benzene-petroleum ether (1:3)

References, pp. 402—406

eluted 180 mg (49%) of **37**, which after recrystallization from ethanol melted at 123°. Further elution with benzene-petroleum ether (2:1) gave 13% of cholest-4-en-3-one, and benzene-ether (5:1) yielded 26% of starting material **36**.

c) Photosensitized oxidation of secondary alcohols

When secondary alcohols are irradiated under oxygen in the presence of dehydrogenating sensitizers, e.g. benzophenone, a radical chain reaction is initiated leading to alkyl hydroperoxides carrying the OOH function at the same carbon atom as the hydroxy group [*37*]. For instance, 2-hydroperoxy-2-hydroxypropane (**38**), 2-hydroperoxy-2-hydroxybutane (**39**) and 2-hydroperoxy-2-hydroxypentane (**40**) are obtained by this method.

$$\begin{array}{c} CH_3 \\ | \\ HO-C-H \\ | \\ R \end{array} \xrightarrow{h\nu/O_2}_{\text{sens}} \begin{array}{c} CH_3 \\ | \\ HO-C-OOH \\ | \\ R \end{array} \qquad \begin{array}{cc} CH_3 & CH_3 \\ | & | \\ H_3C-C-OO-C-CH_3 \\ | & | \\ HOO & OOH \end{array}$$

38: R=CH₃
39: R=C₂H₅
40: R=C₃H₇

41

The hydroperoxides thus synthesized should be handled with care because of their explosiveness. Trace amounts of water convert them to the corresponding ketones and hydrogen peroxide — substances heretofore regarded as oxidation products proper.

2-Hydroperoxy-2-hydroxypropane (38) [*37*]. 78 g isopropanol, in which 1.5 g benzophenone were dissolved, were irradiated with a Philips HPK 125 W mercury burner in an apparatus incorporating a water-cooled glass shaft for immersing the lamp. Oxygen, which was passed through a calcium chloride drying tower, was circulated entering the solution through a fritted glass gas inlet. 3.9 l O₂ were absorbed in 17.7 hrs. (12°), the reactants adopting a light yellow color during the irradiation. An aliquot of the solution was analyzed: 82% of the oxygen consumed titrated for active oxygen, and acetone 2,4-dinitrophenylhydrazone precipitated in 92% yield. The irradiated solution was concentrated at room temperature/0.1 mm Hg and the residue distilled at 0.01 mm Hg (bath temperature 20—25°) when 0.6 g bis-(2-hydroperoxy-2-propyl) peroxide (**41**), m.p. 37°, crystallized in the condenser. 4 g **38** (25% based on O₂ consumed) was obtained as "distillate". The hydroperoxide is explosive, and the experiment should therefore be carried out with the utmost caution (explosion screen, use of water bath only!).

d) Photooxidation of ethers

Aliphatic and alicyclic ethers were converted to their α-hydroperoxides on irradiation under oxygen, benzophenone serving as sensitizer [*37*]. Thus tetrahydrofuran gave **42**, p-dioxane gave hydroperoxy-p-dioxane and isochromane (**43**) yielded 1-hydroperoxyisochromane (**44**). The latter had already earlier been obtained by RIECHE and SCHMITZ [*38*] on shaking isochromane with oxygen at room temperature under irradiation. Hydroperoxide formation could similarly be observed in the cases of 1-methylisochromane (**45**), yielding 1-hydroperoxy-1-methylisochromane (**46**) with

UV-light, and of 1-benzylisochromane (**47**) forming **48** in sunlight [*38*]. Nevertheless, the photosensitization procedure seems to be preferable for its wider applicability.

43: R=H
45: R=CH$_3$
47: R=CH$_2$–C$_6$H$_5$

44: R=H
46: R=CH$_3$
48: R=CH$_2$–C$_6$H$_5$

2-Hydroperoxytetrahydrofuran (42) [*37*]. In 28.5 hrs., 88 g tetrahydrofuran in which 1.5 g benzophenone had been dissolved, absorbed 13.38 l O$_2$ when irradiated. After distilling off the unconsumed tetrahydrofuran, high vacuum distillation of the residue gave 28.6 g of the hydroperoxide **42**, b.p.$_{0.01}$ 45—50° (yield 46% according to O$_2$-uptake).

1-Hydroperoxyisochromane (44) [*38*]. 10 g Isochromane (**43**) was shaken with oxygen at room temperature and irradiated by a UV-fluorescent tube. After 30 hrs. 850 ml O$_2$ had been absorbed (51%). The substance was dissolved in 50 ml. ether and shaken, first with 40 ml then with 20 ml 2N NaOH. The collected alkaline washings were extracted with a little ether and then carefully brought to pH 8—9 by the addition of 2N H$_2$SO$_4$ with stirring and cooling with ice. The oily precipitate was taken up in ether, the ether dried with potassium carbonate and removed through a drying tower under reduced pressure. The residue crystallized on triturating with petroleum ether and seeding (yield 4.95 g, 40%). For the purification, **44** was dissolved in 4 parts of ether, treated with 15 parts of petroleum ether, seeded and cooled to 0°; m.p. 68—70°.

1-Benzyl-1-hydroperoxyisochromane (48) [*38*]. In allowing 5 g **47** to stand in sunlight for 6 days in a crystallizing dish, a 12% theoretical conversion to **48** was obtained. It was diluted with ether-petroleum ether and shaken with 10 ml 30% sodium hydroxide solution. This resulted in the precipitation of a semi-solid mass which was separated from the alkaline solution, diluted with water and neutralized with dilute sulfuric acid. The oil which separated was taken up in ether, the ether dried and then removed under reduced pressure. The residue crystallized to give 0.11 g **48**, clustering needles from ether-petroleum ether, m.p. 70—72°.

e) Photooxidation of nitrogen compounds

The autoxidation of substituted hydrazones (cf. **49**) as observed by Busch and Dietz [*39*] led to the formation of colored substances which were at first regarded to be cyclic 1,2-peroxides (cf. **50**). Re-examination [*40, 41*] showed the oxidation products to be hydroperoxides derived from the appropriate azo compounds (cf. **51**).

References, pp. 402—406

The migration of the double bond from the hydrazone to the azo position [42] may easily be effected by subjecting the hydrazones to photosensitized oxygen transfer which, as stated earlier, follows the principles of indirect substituting addition in the allylic position [15]. According to SCHENCK and WIRTH [43] this reaction occurs if solutions of the substituted hydrazones are exposed to both light and oxygen in the presence of a sensitizer (e.g. methylene blue, rubrene). It could be shown that the photosensitized oxygenation reaction proceeded much more rapidly than the dark reaction.

Aliphatic and alicyclic amines, when irradiated in aerated solutions in the presence of dyes, are converted to hydroperoxides [20]. Absorption of oxygen is dependent on the number of methylene groups adjacent to the nitrogen atom. Thus, propylamine and isobutylamine absorbed 1 equivalent of O_2, pyrrolidine, piperidine and diethylamine consumed 2 O_2, and triethylamine about 3 O_2. Nicotine was found photooxidizable too [20, 44], and theophylline (54) yielded. on methylene blue sensitized photo-oxidation [45], 1,3-dimethylallantoin (55).

2-(4-Bromophenylazo)-2-hydroperoxypropane (53) [20]. Acetone p-bromophenylhydrazone (52) (2.3 g) was dissolved in benzene (80 ml) which had been previously saturated with chlorophyll (from Oleander) and filtered. Oxygen was circulated through the solution which was irradiated by means of a water-cooled glass immersion lamp Osram HQA 500. 201 ml oxygen were absorbed in 7 mins. (80.5%). Work-up yielded 1.9 g (73%) of yellow 53, m.p. 45—47°.

1,3-Dimethylallantoin (55) [45]. Theophylline (54) (4 g) was dissolved in 200 ml of water, and the pH was brought to 9.2 with NaOH. After addition of 10 mg. of methylene blue the solution was irradiated with two 150 W Sylvania floodlamps in a glass-wall temperature bath of 25°. Oxygen was bubbled through the solution and the pH was kept at 9.2 by automatic titration with 1N NaOH. After 22 hrs. 2 volumes of acetone were added and the mixture kept at —20° for 48 hours. 500 mg of precipitate were removed and the supernatant was evaporated to dryness. The resulting solid was crystallized from water and ethanol to give 1.3 g of **55**, m.p. 212—214°.

f) Photooxidation of a phenol

SCOTT and BEDFORD [46] discovered the photooxidative conversion of 7-chloro-anhydrotetracycline (**56**) to 7-chloro-6-deoxy-6-hydroperoxy-dehydrotetracycline (**57**).

The addition of small quantities of 3,4-benzopyrene as sensitizer greatly accelerated this process, especially when weak light sources were employed [47]. The oxidation of N-tert. butyl-7-chloro-anhydrotetracycline (**58**) and 9,N-di-tert. butyl-7-chloro-anhydrotetracycline (**60**) respectively proceeded analogously, to give **59** and **61**, respectively. Cyclohexane and benzene-cyclohexane mixtures proved best suited as solvents for the reaction.

No definite statement as to the mechanism involved was made [47]. Formation of an 6,11-epidioxide with subsequent rearrangement would account for product formation as would "indirect substituting addition" of oxygen at C-6 with synchronous migration of hydrogen from the C-11 hydroxy group. For a similar case of hydrogen migration see the photooxidation of tetraphenylpyrrole (comp. p. 396).

56: $R_1=R_2=H$
58: $R_1=$tert.butyl, $R_2=H$
60: $R_1=R_2=$tert.butyl

57: $R_1=R_2=H$
59: $R_1=$tert.butyl, $R_2=H$
61: $R_1=R_2=$tert.butyl

7-Chloro-6-deoxy-6-hydroperoxydehydrotetracycline (57) [47]. 5 g of **56** and 25 mg of 3,4-benzopyrene were dissolved in 1400 ml of benzene. The solution was irradiated in a Pyrex vessel with a 450 W Hanovia laboratory lamp while oxygen was blown through. The solution was seeded after the first hour, and after 5 hrs. the irradiation was discontinued. The crystalline product (4.2 g) was collected and recrystallized from dioxane-benzene to furnish pure **57**.

2. Transannular peroxides from cyclic 1,3-dienes

The autoxidation of acyclic 1,3-dienes in light has previously found little preparative significance except in the chemistry of the drying oils. The situation is quite different for cyclic 1,3-dienes which on irradiation readily add on one molecule of oxygen, extra sensitizers being generally necessary to effect the oxygenation. The reaction, which leads to transannular peroxides, may formally be regarded as a diene synthesis with oxygen as a dienophile [*15*]. Throughout this chapter the term "epidioxide" will be preferred for this class of compounds; this corresponds to the German usage of "Endoperoxyd".

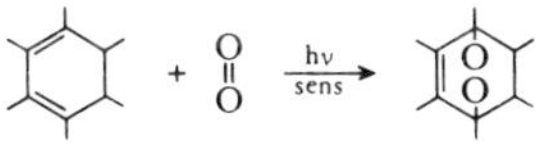

$$\text{diene} + \underset{O}{\overset{O}{\|}} \xrightarrow[\text{sens}]{h\nu} \text{endoperoxide}$$

ALDER and SCHUMACHER [*48*] have also pointed out the great similarity between the above reaction and adduct formation between cyclic 1,3-dienes and dienophiles. The analogy is stressed by the fact that certain epidioxides decompose into their components on heating. The classical example of this is rubrene peroxide (cf. p. 390). These parallels will not be discussed further here. Also mention will only be made on differing views regarding the mechanisms involved in photosensitized oxidation of cyclic 1,3-dienes [*21—23, 25—27, 49*].

The transannular peroxides as mentioned in this chapter will for convenience be classified in 4 groups; the first is comprised of those epidioxides which can be obtained from alicyclic 1,3-dienes. The second group contains peroxides of fused hydrocarbons, particularly out of the steroid series, the third peroxides of acenes and related compounds. The fourth group comprises epidioxides originating from cyclopentadienes. A rather complete account on formation and properties of this class of compounds will be found elsewhere [*50*]. Partial aspects have been covered by BERGMANN and McLEAN [*51*], SCHÖNBERG [*52*] and ARBUZOV [*53*].

Acene epidioxides are obtained directly by action of oxygen in light, the addition of extra photosensitizers being unnecessary since the acenes exhibit both sensitizer and acceptor properties [*21*]. However, sensitizers are indispensable in the photooxidation of the first and second group though formation of epidioxides on direct autoxidation of cyclic 1,3-dienes has been claimed [*54*].

a) Epidioxides from alicyclic 1,3-dienes

The investigation on the formation of these compounds was initiated by SCHENCK and ZIEGLER [*55*] who succeeded in synthesizing ascaridole (**63**) from α-terpinene (**62**) on illuminating solutions of the 1,3-diene in the presence of oxygen and chlorophyll as sensitizer. This method deserves mention since it was a technically feasible process in Germany during the second world war. The synthesis was necessitated by a shortage of chenopodium oil from which ascaridole is isolated [*56*]. Similarly cantharidine (**66**), a vesicatory compound obtained from the Spanish fly, was synthesized [*57*]. 1,2-Dimethyl-3,5-cyclohexadiene-1,2-dicarboxylic anhydride (**64**) was oxygenated to the epidioxide **65**, which was then converted in three steps to cantharidine (**66**).

62 → **63** (hv/O₂, sens)

64 → **65** (hv/O₂, sens) → **66** (3 steps)

67 → **68** + **69** (hv/O₂, sens)

70 → **71** (hv/O₂, sens)

72: R=H; **74**: R=C₆H₅ → **73**: R=H; **75**: R=C₆H₅ (hv/O₂, sens)

76 → **77** (hv/O₂, sens)

78: R=H; **80**: R=C₆H₅ → **79**: R=H; **81**: R=C₆H₅ (hv/O₂, sens)

Besides the already mentioned photooxidation of α-terpinene (**62**) a series of cyclic 1,3-dienes was investigated on their ability to form epidioxides. In this manner compounds **68, 69, 71, 73, 75, 77, 79** and **81** were synthesized from α-phellandrene (**67**) [58, 59], methyl norcaradiene-7-carboxylate (**70**) [60], cyclopentadiene (**72**) [61], 1,4-diphenylcyclopentadiene (**74**) [62], 1,3-cyclohexadiene (**76**) [63, 64], 1,3-cycloheptadiene (**78**) [65] and 1,4-diphenyl-1,3-cycloheptadiene (**80**) [66] respectively.

These epidioxides, as well as those originating from fused systems, may undergo a variety of interesting rearrangements on heat treatment, e.g. formation of epoxy-ketones and epoxy-1,4-oxides. Reduction gives unsaturated 1,4-diols or 1,4-oxides, depending upon the reducing agent. The reader is referred to the pertinent literature [50, 51] for details.

Phellandrene trans-epidioxide (69) [58]. 50 g (—)-α-phellandrene (**67**) in 1500 ml isopropanol was treated with oxygen in the presence of 0.5 g methylene blue during irradiation by a 300 W incandescent lamp; the reactants were shaken mechanically. After 6 days, the uptake of O_2 came to a halt (8100 ml). The solvent was removed under reduced pressure at 40° through a Vigreux column leaving a blue liquid with a characteristic odor. By double distillation in vacuo the reaction products were separated, **69** distilling between 46—59°/0.01 mm Hg; yield 52%.

2,3-Dioxabicyclo[2.2.1]hept-5-ene (73) [61]. Irradiation of pure cyclopentadiene containing rose bengal as sensitizer at —100° with the aid of a sodium lamp Philips SO 140 W and subsequent high vacuum distillation at temperatures between —30 and —20° yielded **73**, m.p. below —30°, which explodes violently above 0°.

2,3-Dioxabicyclo[2.2.2]oct-5-ene, norascaridole (77) [63]. 48.5 g 1,3-cyclohexadiene (**76**) and 1 g of methylene blue were dissolved in 2500 ml isopropanol and irradiated as above. After consumption of 6.96 l oxygen (10 days) the reaction came to a stillstand. The mixture was freed from solvent and unreacted **76** in vacuo and the residue finally distilled in high vacuum to give 9.4 g of **77**, b.p.$_{0.01}$ 36°. According to another preparation [*64*], which is fully described in the 1st edition of this book (p. 55), **77** has m.p. 88.5° (dec.).

6,7-Dioxabicyclo[3.2.2]non-8-ene (79) [*65*]. An aerated solution of 4.2 g of 1,3-cycloheptadiene (**78**) and 0.3 g of eosin sodium salt in 950 ml of absolute ethanol was irradiated by two tungsten filament lamps (totalling 350 W) below 39°. After 144 hrs. the reaction mixture was worked up in two parts. The ethanol was driven off under reduced pressure, the water-bath being maintained at 35—40°. The bright red crystalline residue was sublimed at 40°/0.3 mm onto a cold finger (Dry Ice). The total weight of the white, mealy sublimate was 1.65 g (29%), m. p. 119—122.4°.

1,5-Diphenyl-6,7-dioxabicyclo[3.2.2]non-8-ene (81) [*66*]. 300 mg **80** was treated with oxygen in the presence of 5 mg eosin-Na in absolute ethanol (120 ml) to which 1 ml chloroform had been added. The solution was illuminated with a 1000 W tungsten lamp for 75 minutes. After evaporation of the solvent under reduced pressure the residue is dissolved in ether. The ether extract after removal of traces of dye with water, furnishes 339 mg of crude **81**, which after being rinsed with methanol has m.p. 96—96.5°.

b) Epidioxides from fused 1,3-cyclohexadienes

Hoping to obtain a substance active against rickets, WINDAUS and BRUNKEN [*67*] placed alcoholic solutions of ergosterol (**82**) containing photosensitizers such as eosin, in the sunlight. Whether sunlight or light from an incandescent lamp was used the authors observed that oxygen was absorbed and a peroxide was formed. Initially a four-membered ring peroxide was formulated (cf. **83**); the correct formula (**84**) was later suggested by FIESER [*68*]. A summary of the early work on steroid epidioxides is to be found elsewhere [*51*].

Japanese workers were able to isolate further products in the sensitized photooxidation of ergosterol [*69*]. Depending on the solvent chosen for the oxygenation reaction varying amounts of the products **85**—**88** were obtained (following esterification).

Among the C-9,C-10-isomers of ergosterol, viz. lumisterol (**89**), pyrocalciferol (9α-lumisterol, **90**) and isopyrocalciferol (9β-ergosterol, **91**), the correct configurations of which have only lately been established [70], response towards photosensitized oxygen transfer is not uniform. In the anti series photooxidation is always accompanied by side reactions leading to tetraenes, while in the syn series (**90** and **91**) neat formation of the corresponding epidioxides occurs with a greater reaction velocity than in the anti series [71].

In the case of lumisterol derivatives photooxidation is particularly complicated as was found by BLADON [72]. For steric reasons addition of oxygen across the α-side of the molecule ("rear attack") is strongly hindered. Instead lumisteryl acetate β-epidioxide (**92**) is isolated. 3β-Acetoxylumista-5,7,9(11),22-tetraene (**93**), being a by-product of the sensitized photooxidation, gives rise in turn to α- as well as β-epidioxides (**94** and **95**). Thus the introduction of an additional double bond serves to flatten the molecule, the diene system becoming accessible from the front and the rear side [72].

Apart from ergosterol and its isomers, other steroid dienes are also converted to the corresponding transannular peroxides on sensitized photooxygenation. In this way, the epidioxides **96—98** have been synthesized from, 3β-hydroxyergosta-5,7,9(11),22-tetraene (dehydroergosterol)

[73], cholesta-2,4-diene [74] and 3β-acetoxyergosta-6,8(14),9(11),22-tetraene [75], respectively. **98** has turned out to be a useful intermediate in the synthesis of compounds related to cortisone. A rather complete coverage of this subject will be found with GOLLNICK and SCHENCK [50] (see also [51—53]).

Pine gum resin acids palustric acid (**99**) and levopimaric acid (**100**) undergo facile oxygenation on irradiation in the presence of photosensitizers (for a summary see [76]). An interesting variation was found by SCHULLER and LAWRENCE [77] when neoabietic acid (**101**) was subjected to the same reaction. Addition of oxygen to the exocyclic double bond proceeds with shift of the latter to the cis-dienoid position. The intermediate hydroperoxide absorbs a second molecule of oxygen to give the hydroperoxyepidioxide **102**.

Heteroannular cisoid dienes have also been examined on their tendency to form epidioxides [31, 78, 79] (cf. also [50]). When 3β-acetoxy-5α-ergosta-7,14,22-triene (ergosterol B$_3$ acetate, **103**) was photooxidized in the presence of erythrosin B as sensitizer [78] four compounds were obtained, one of which showed peroxidic properties. Its constitution remained, however, unresolved. The formation of the other substances is easily understood on the basis of the assumption that the hypothetical 3β-acetoxy-7,15-epidioxy-5α-ergosta-8(14),22-diene (**104**) breaks down to the epoxy- and hydroxyketones actually observed. The analog compound, 3β-acetoxy-5α-ergosta-

7,14-diene, was found [*31*] to consume two equivalents of O_2 when photo-oxidized.

When the model compound **105**, bi-1-cyclohexen-1-yl, was similarly treated [*79*], a stable epidioxide (**106**) could be isolated in 51% yield.

105   hv/O₂, sens →   **106**

5,8α-Epidioxy-3β-hydroxy-5α-ergosta-6,22-diene (84) [*67*]. A 200 W Osram nitralamp was placed in a water-filled beaker, fitted with inlet and outlet. This vessel was surrounded by a second beaker, which contained the solution to be irradiated. In order to make the most efficient use of the radiant energy, the apparatus was placed in a box, the walls of which were covered with mirror-glass.

A solution of 4 g of ergosterol in 1200 ml 95% alcohol, heated to 60°, was treated with 6 mg eosin and introduced into the apparatus. During the irradiation (3 hours) a slow stream of oxygen was led through the liquid. At the end of the experiment the solution was concentrated to small bulk under reduced pressure; the crystals which separated were filtered off and repeatedly crystallized from alcohol or acetone to give pure **84**, m.p. 178°; yield 70%.

3β-Acetoxy-5,8α-epidioxy-5α-ergosta-6,22-diene (85) [*69*]. 3 g of **82** and 35 mg of eosin blue as sensitizer were dissolved in 150 ml benzene and 15 ml methanol, and the solution was irradiated at 30° with two 500 W Iwasaki-Denki tungsten lamps while oxygen was bubbled through. After 2.5 hrs. the solution was brought to dryness in vacuo and the residue acetylated with pyridine-acetic anhydride. Chromatography on Al_2O_3 yielded 1.8 g of **85**, m.p. 200—201°, 0.82 g of **86**, m.p. 202—203°, 0.38 g of **87**, m.p. 144—146°, and 0.22 g of **88**, m.p. 169—171°.

2α,5-Epidioxy-5α-cholest-3-ene (97) [*74*]. Cholesta-2,4-diene (9 g) was dissolved in 1000 ml warm absolute ethanol and the solution treated with 14 mg eosin. The photooxidation was carried out at 25° in the apparatus described in the oxidation of ergosterol (see above); the light source was a 200 W Mazda lamp. After the irradiation (9 hours) the solvent was evaporated in vacuo at temperatures below 30°. The crystalline residue was repeatedly crystallized from dilute acetone and from 95% alcohol to give pure **97**, m.p. 113—114°; yield 60—70%.

6,14-Epidioxy-18-hydroperoxy-9,14-dihydroabietic acid (102) [*77*]. A solution of 8.16 g of neoabietic acid (**101**) and 0.135 g of erythrosin B in 2700 ml of 95% ethanol was irradiated under oxygen in a 40 W reactor for 2 hours. The solution was then concentrated under reduced pressure to about 100 ml, chilled in an ice-bath, and 3.35 ml of freshly distilled cyclohexylamine added slowly with stirring and cooling, final pH 9. On standing, the crystalline salt slowly appeared. It was collected by filtration, washed thoroughly with pentane and dried over Drierite; yield 8.64 g (69%). After recrystallization from 95% ethanol the **102**-cyclohexylamine salt had m.p. 181—181.5° (dec.).

2.55 g of this salt was suspended in ether and shaken with dilute aqueous phosphoric acid. The ether layer required only two water washings to be free of mineral acid. Evaporation of the ether gave 2.03 g (80%) of the free acid which was crystallized from aqueous methanol, yield 1.77 g of needles. Recrystallization from the same medium gave star clusters, m. p. 176° (dec.); yield 0.67 g.

1,2,3,4,4a,6a,7,8,9,10-Decahydrodibenzo-o-dioxin (106) [*79*]. A solution of 16.2 g of **105** and 100 mg rose bengal in 50 ml isopropanol and 100 ml methanol absorbed 2990 ml of oxygen (244 minutes) when irradiated with a Philips HPK 125 W mercury burner at 20° (immersion apparatus). Evaporation of the solvents in vacuo left 22.1 g residue which was dissolved in 10 ml benzene and chromatographed on alumina. Elution with petroleum ether gave 9.8 g (51%) of crystals which were recrystallized from ethanol at −30° to give pure **106**, m. p. 54°.

c) Epidioxides from acenes

While benzene and naphthalene do not react with oxygen in light to form peroxides, anthracene (**107**) and a number of its substitution and annelation derivatives do so. For a survey on the epidioxides thus available see again [*50—53*] and, more specifically, ETIENNE [*80*].

It is remarkable that 9,10-epidioxy-9,10-dihydroanthracene (**108**), the anthracene peroxide, was discovered only in 1935 [*81*], especially as anthracene had already long been the subject of a very large number of photochemical experiments. Peroxide formation in the acene series takes place only under illumination, while in the dark, anthracene, for example, is stable towards oxygen even when the experiments are continued for many years [*82*]. The production of **108** and **111** using electrodeless discharge generated singlet oxygen has recently been reported by COREY and TAYLOR [*27*].

107: $R_1=R_2=H$

108: $R_1=R_2=H$
109: $R_1=R_2=OCH_3$
110: $R_1=H, R_2=CH_3$
111: $R_1=R_2=C_6H_5$

The major part of our present day knowledge on this class of compounds is due to DUFRAISSE and his school. The photochemical formation of these acene epidioxides is so simple that the method will retain its significance even if preparations via dark reactions should become feasible in the future (comp. the preparation of 9,10-epidioxy-9,10-diphenylanthracene (**111**) from 9,10-dihydroperoxy-9,10-diphenylanthracene [*64*]).

Basically the formation of the acene epidioxides takes place in solution; it has been found expedient to work in sunlight and with very dilute solutions. The oxygenation being a photosensitized process, the mechanism of which will not be discussed here (cf. [*83*]), the addition of extra sensitizers is usually unnecessary since the acenes do not only exhibit acceptor but also sensitizer properties. In many cases irradiation of the acene solution in the presence of oxygen leads not only to peroxide formation but also to photodimerization of the acene; naturally high dilution favors the formation of the epidioxides.

The nature of the solvent has a pronounced effect on the velocity of peroxide formation. With rubrene (**113**), for example, the following relative reaction rates have been established [*84*]: carbon disulfide 9, chloroform 3, benzene 1, ether 0.5, and nitrobenzene 0.1. By far the greater number of acene peroxides have been synthesized in carbon disulfide and only occasional use has been made of other solvents. The significance of the solvent choice is again demonstrated by the rapid formation of an epidioxide from 1,4-dimethoxy-9,10-di-(2-pyridyl)-anthracene (**112**) in ether which fails however in acid media (pH < 5.9) [*85*].

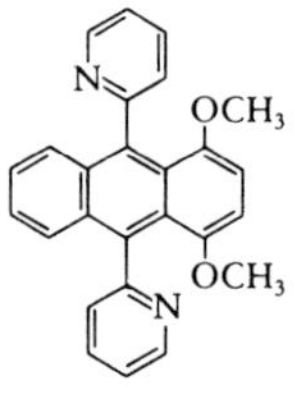

112

It has been questioned whether the light-induced action of oxygen on the anthracenes does not lead to compounds which should be regarded as molecular compounds of the particular anthracene and oxygen. This question is well justified since a number of the substances obtained in this way decompose on heating with the liberation of elementary oxygen. The classic example is the thermal decomposition of rubrene epidioxide (5,12-epidioxy-5,6,11,12-tetraphenylnaphthacene **114**) [*86*]. Nevertheless, chemical degradation studies have shown beyond doubt that the peroxides are not molecular compounds.

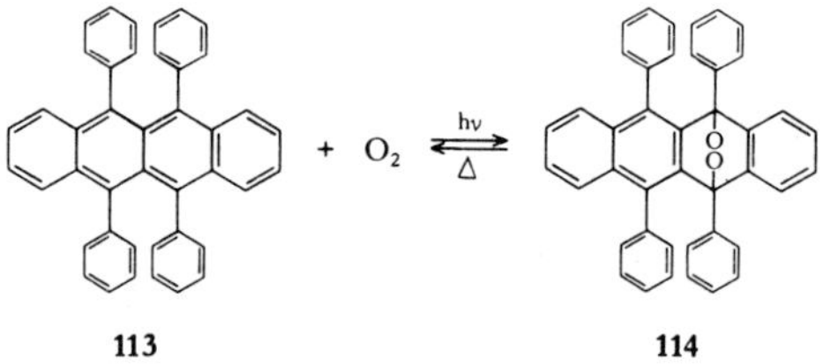

113 + O_2 ⇌ **114**

For preparative work it is important to remember that the acene peroxides show considerable variation in respect of their thermal stability. 9,10-Epidioxy-9,10-dimethoxyanthracene (**109**) is so stable that it may be sublimed without decomposition whereas the epidioxide **116** from 1,4-dimethoxy-9,10-diphenylanthracene (**115**) [*87*] — which was formerly regarded as the 9,10-epidioxide **117** [*88*] — is exceptionally thermolabile. This compound loses 25 % of its peroxidic oxygen in 10 days at 20°.

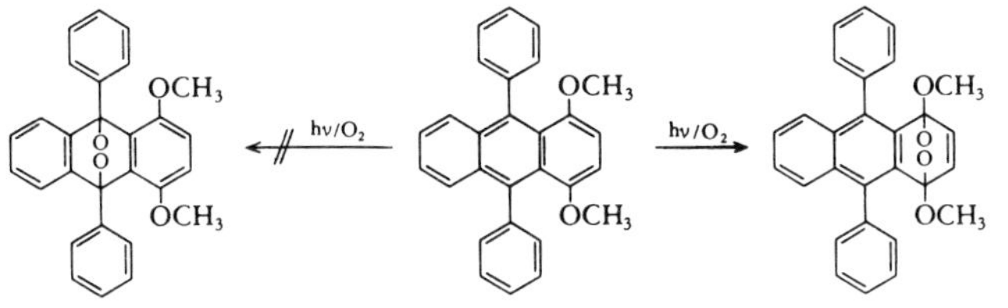

117   115   116

Thermal lability of these peroxides may also be the reason that in certain cases only p-quinones could be isolated following photooxidation of acenes [*89*]. The following table 32 shows the decomposition temperatures of some acene epidioxides.

Table 32

epidioxide	decomposition temperature	ref.
9,10-epidioxyanthracene (**108**)	120° (explos.)	[*81*]
9,10-epidioxy-9-methylanthracene (**110**)	ca. 80° (explos.)	[*90*]
9,10-epidioxy-2-methoxy-9,10-diphenyl-anthracene	ca. 160°	[*91*]
5,12-epidioxynaphthacene (**118**)	ca. 120° (explos.)	[*92*]
5,12-epidioxy-5,12-diphenyl-naphthacene (**119**)	ca. 160°	[*93*]

DUFRAISSE [*94*] found 5,12-dichloro-6,11-diphenylnaphthacene to yield two isomeric epidioxides, viz. 5,12-dichloro-5,12-epidioxy-6,11-diphenylnaphthacene (**120**) and 5,12-dichloro-6,11-epidioxy-6,11-diphenylnaphthacene (**121**), on treatment with oxygen in light. Also bis-epidioxides may be prepared: 9,9′,10,10′-tetraphenyl-1,1′-bianthryl (**122**) yielded 9,10;9′,10′-bis-epidioxy-9,9′,10,10′-tetraphenyl-1,1′-bianthryl (**123**) on exposure to both light and oxygen, the reaction rate being much higher than usually encountered with phenyl substituted anthracenes [*95, 96*].

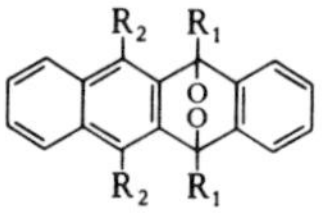

118: $R_1=R_2=H$

119: $R_1=C_6H_5$, $R_2=H$

120: $R_1=Cl$, $R_2=C_6H_5$

121: $R_1=C_6H_5$, $R_2=Cl$

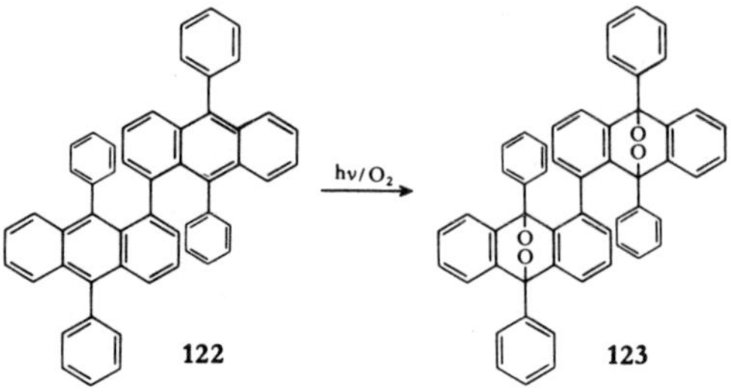

122 **123**

7,16-Diphenyldibenzo[a,o]perylene (meso-diphenylhelianthrene **124**) reacts extraordinarily quickly with atmospheric oxygen: the dilute violet solution in carbon disulfide is decolorized in less than 1 second when placed in direct sunlight [*96, 97*] and a yellow peroxide is obtained. The constitution of this compound is not yet established unambiguously but it is assumed that the epidioxide is to be regarded as a biradical (cf. **125** or **126**). Formulae **127** and **128** have also been suggested [*97, 98*].

It is remarkable that 10,10′-diphenyl-9,9′-bianthryl (**129**) and 7,14-diphenylphenanthro[1,10,9,8-opqra]perylene (meso-diphenylnaphthodianthrene **130**), which differ from **124** merely by two hydrogen atoms more or less, do not form peroxides in light [*96, 99, 100*].

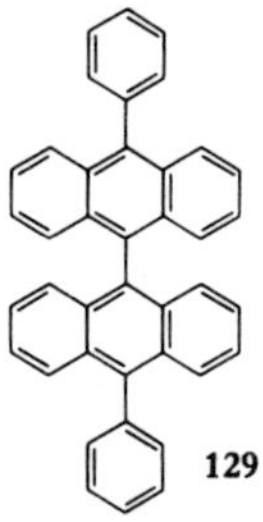

129

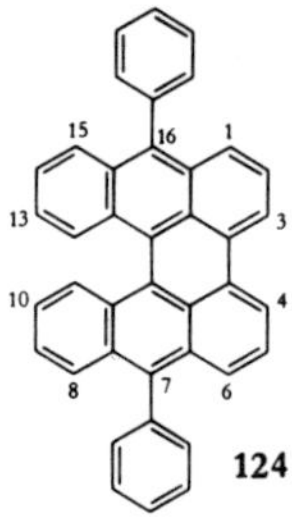

124

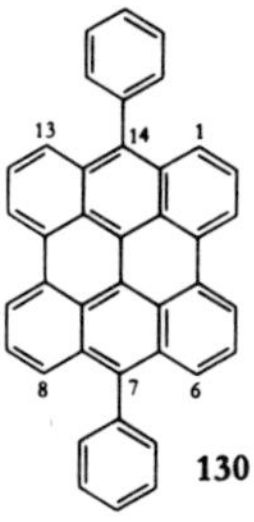

130

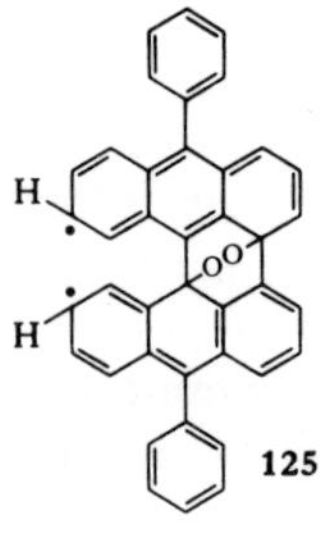

125

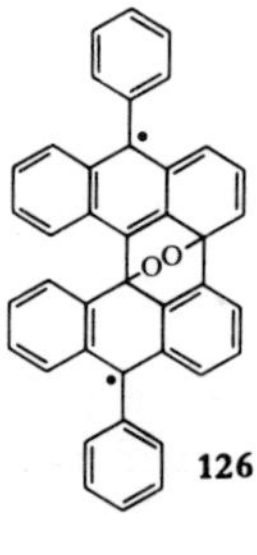

126

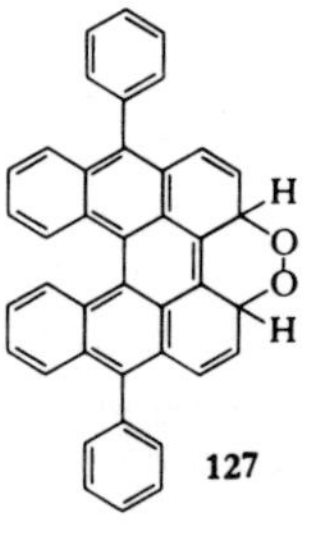

127

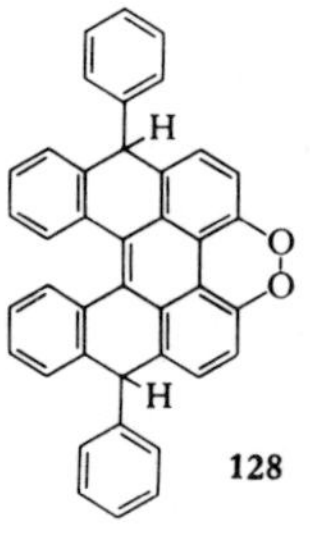

128

Presented below are epidioxides derived from ring systems which have so far received no mention; a characteristic example is given for each system.

7,12-Epidioxy-7,12-diphenylbenz[a]anthracene (**131**) [*101*]
5,10-Epidioxy-5,10-diphenylbenzo[g]quinoline (**132**) [*102*]
5,10-Epidioxy-5,10-diphenylbenzo[g]isoquinoline (**133**) [*103*]
5,10-Epidioxy-5,10-diphenylbenzo[g]quinazoline (**134**) [*104*]
5,14-Epidioxy-5,7,12,14-tetraphenylpentacene (**135**) [*105*]
8,13-Epidioxydibenz[b,h]acridine (**136**) [*106*]
4b,9-Epidioxy-9,10-diphenylindeno[1,2,3-fg]naphthacene (**137**) [*107*]
9,13b-Epidioxy-9-phenylnaphtho[3,2,1-kl]xanthene (**138**) [*108*]
9,13b-Epidioxy-9-phenylnaphtho[3,2,1-kl]thioxanthene (**139**) [*109*]
4b,12b-Epidioxy-8H,16H-dibenzo[a,j]perylene-8,16-dione (**140**) [*110, 111*]

9,10-Epidioxy-9,10-diphenylanthracene (111) [*112*] (as quoted in [*64*], p. 21). 100 mg of 9,10-diphenylanthracene in 60 ml pure carbon disulfide contained in a 500 ml Uviol glass flask was placed in the summer sun for 30 minutes. By concentration to small bulk and addition of petroleum ether the peroxide was obtained in small colorless prisms, decomposition temperature about 200°. Yield almost quantitative.

1,4-Epidioxy-1,4-dimethoxy-9,10-diphenylanthracene (116) [*87*]. A dilute (2 g/liter) solution of **115** in ether is irradiated at —50° with a Philips SP 500 mercury lamp while oxygen is bubbled through the solution. The peroxide which precipitates from the solution is filtered off (yield 94%) and washed with ether, m.p. 180—185° (dec.).

4b,12b-Epidioxy-8H,16H-dibenzo[a,j]perylene-8,16-dione (hetero-cerodianthrone epidioxide (140) [*111*]. If a few milliliters of a solution, containing 500 mg 8H,16H-dibenzo[a,j]perylene-8,16-dione per liter carbon disulfide, are exposed to bright sunshine, the originally red-violet solution becomes completely decolorized in less than three seconds. The solution is then concentrated under reduced pressure to yield colorless crystals of **140**, decomposition temperature about 150°.

These crystals rapidly turn reddish-violet when illuminated, due to regeneration of starting material; the isolation procedure of **140** has, therefore, to be carried out under exclusion of strong light.

d) Photooxidation of carbo- and heterocyclic cyclopentadienes

The photooxidation of cyclopentadiene and its 1,4-diphenyl derivative has already been mentioned (comp. p. 384). Arylated cyclopentadienes carrying functional groups have also been subjected to the same procedure, photosensitization being usually brought about by the respective substrates. Thus DUFRAISSE and coworkers [*113*] irradiated several pentaaryl cyclopentadienols (cf. **141**) under oxygen. The products resulting were claimed to be peroxides (cf. **142**) though no typical peroxide properties were exhibited by these compounds. In the first edition of this book (p. 56), therefore, another formulation (cf. **143**) was put forward to explain the facile transformation of the photoproducts into tetraarylfurans [*113*]. This proposal has recently been substantiated [*114*] on the basis of spectroscopic

evidence; formula **144**, 5-benzoyl-2,5-dihydro-2,3,4,5-tetraphenyl-2-furanol, is better in accord with the experimental data than is **142**. Nevertheless, such a peroxide is to be postulated as the primary product. Application of low temperature irradiation techniques may facilitate the synthesis of these epidioxides.

Substituted cyclopentadienones (cyclones), the chemistry of which has recently been reviewed upon [*115*] yield 1,4-diketones when exposed to both light and oxygen (see [*115*], p. 292). Tetraphenylcyclopentadienone (tetracyclone, **145**) was thus converted to cis- and trans-dibenzoylstilbenes **146** and **147** (1,2,3,4-tetraphenyl-2-butene-1,4-diones), respectively and tetraphenyl-2H-pyran-2-one (**148**). The intermediacy of an epidioxide was presumed [*116*] which by loss of carbon monoxide turns over to the products eventually observed.

References, pp. 402—406

ALLEN and VANALLAN [*117*] subjected a number of cyclopentadienones to photochemical oxidation and observed opening of the cyclopentadienone ring: 7,9-diisopropyl-8H-cyclopent[a]acenaphthylen-8-one (**149**) gave 1,2-diisobutyrylacenaphthylene (**150**) and 14,15-acenaphtheno[12](2,5)cyclopentadienophan-17-one (dodecahydro-7,20-methanocyclohexadec[a]acenaphthylen-21-one, **151**) yielded [14](1,2)acenaphthylenophane-1,14-dione (tetradecahydrocyclohexadec[a]acenaphthylene-7,20-dione, **152**). Again it is to be assumed that photooxygenation first leads to the formation of an epidioxide which spontaneously decarbonylates to the end product.

Investigations on photochemical oxidation in the heterocyclic series are mainly due to DUFRAISSE and SCHENCK and their schools (for summaries see [*50, 53*]). Photooxygenation in the furan series will, however, be dealt with in another context in chapter 42.

Photooxidation of cyclopentadienes containing nitrogen has as yet received little attention. One process of considerable preparative value has been elaborated by DE MAYO and REID [*118*]. Irradiation of aerated dilute solutions of pyrrole containing dyes gave a 32 % yield of **154**, probably via 7-aza-2,3-dioxabicyclo[2.2.1]hept-5-ene (**153**). Manganese dioxide oxidation of **154** gave maleimide. Thus sensitized photooxidation has opened a ready access to this valuable synthetic agent. N-Methylpyrrole behaved analogously [*118*].

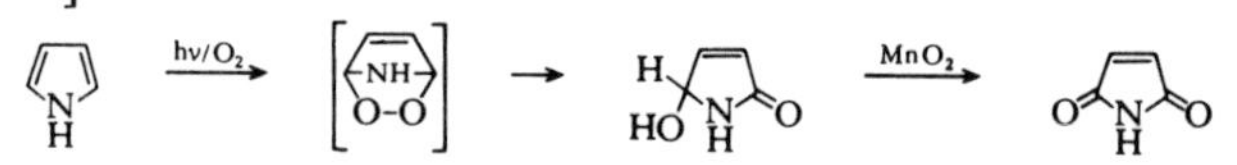

153   154

Photooxidation in the imidazole series [*119*] showed some peculiarities which have received an explanation only recently. 2,4,5-Triphenylimidazole (lophine, **155**) on irradiation in the presence of oxygen absorbs one equivalent of O_2 to yield, amongst others (cf. N,N'-dibenzoylbenzamidine **157**), a product which was formulated as the epidioxide 1,4,6-triphenyl-2,3-dioxa-5,7-diazabicyclo[2.2.1]hept-5-ene (**156**) [*119, 120*]. A reinvestigation [*121, 122*] revealed this product, however, to be a hydroperoxide, viz. 4-hydroperoxy-2,4,5-triphenyl-4H-imidazole (**158**). This unusual result was explained [*123*] on the basis of the assumption that **156** was in fact the primary oxygen adduct this undergoing a spontaneous rearrangement to the hydroperoxide actually isolated. It may be noted here that photosensitized oxygen transfer following the principles of "indirect substituting addition of O_2 in the allylic position" [*15*] would also account for the formation of **158**. Thus **156** need not necessarily be an intermediate.

The same, though tentative, explication holds also for the production of 2-hydroperoxy-2,3,4,5-tetraphenyl-2H-pyrrole (**160**) on photosensitized oxygenation of 2,3,4,5-tetraphenylpyrrole (**159**) [*123*]. In earlier investigations [*124*] epoxy compounds and ring fission products had been observed which now appear to have resulted from secondary reactions of the hydroperoxide **160**.

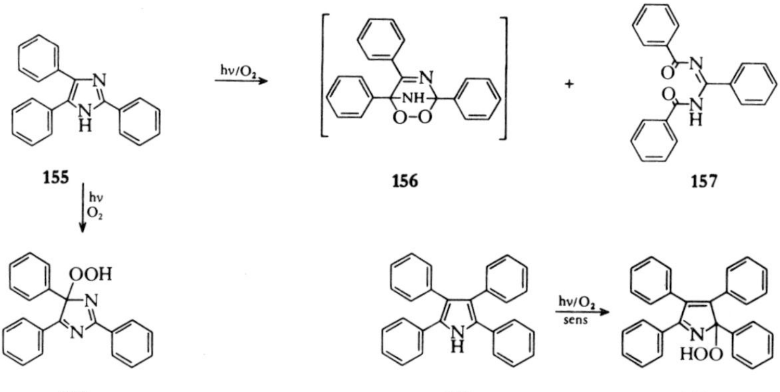

155   156   157

158   159   160

For N-substituted pyrroles a different behavior is to be expected. According to THEILACKER and SCHMIDT [*125*], if a dilute solution of 1,2,3-triphenylisoindole (**161**) in dry carbon disulfide be irradiated under simultaneous passage of carefully dried oxygen, 1,3-epidioxy-1,2,3-triphenyl-isoindole (**162**) is formed. Contrasting with the thermal lability of the peroxide derived from diphenylisobenzofuran (cf. chapter 42), **162** is much more stable and melts at 80—90° with blackening and evolution of gas. Reduction or reaction with dilute acid converts **162** into 1,2-dibenzoyl-benzene.

161   hv/O₂   162

[14] **(1,2)Acenaphthylenophane-1,14-dione (152)** [*117*]. A solution of **151** (4 g) in isooctane (200 ml) was irradiated with a sunlamp while air was bubbled through. The initially red solution became yellow at the end of the reaction and the diketone partially separated out. After concentrating the solution the yellow product was filtered off and recrystallized from butanol to give pure **152**, m.p. 137°; yield 3.9 g (97%).

5-Hydroxy-3-pyrrolin-2-one (154) [*118*]. A dilute (0.1%) aqueous solution of pyrrole, containing 2.5 mg % eosin, was shaken under oxygen while being irradiated with a 100 W tungsten lamp. One mole O_2 was rapidly consumed and the solution brought to dryness at room temperature. Extraction of the residue with acetone and crystallization from the same solvent gave 32% of **154**, m.p. 102—102.5°.

2-Hydroperoxy-2,3,4,5-tetraphenyl-2H-pyrrole (160) [*123*]. Solutions of **159** in chloroform or methanol, to which methylene blue or 8H,16H-dibenzo[a,j]perylene-8,16-dione as sensitizers were added, were irradiated with incandescent lamps (3 kW) for 5—10 minutes. Work-up yielded 80% of **160**, m.p. 170—172° (dec.).

3. Photochemical formation of bis-aralkyl and bis-acyl peroxides

a) Formation of six-membered cyclic peroxides

WITTIG and GAUSS [*126*] reported the autoxidation of 1,1-bis-(4-methoxyphenyl)-ethylene in benzaldehyde to result in formation of 3,3,6,6-tetrakis-(4-methoxyphenyl)-1,2-dioxane (**163**).

163

A similar process, which only took place in light, however, was observed by MUSTAFA and ISLAM [*127*] on irradiation of 10-methyleneanthrone in benzene solution in air. The peroxide **164** decomposed to anthraquinone on heating (300°).

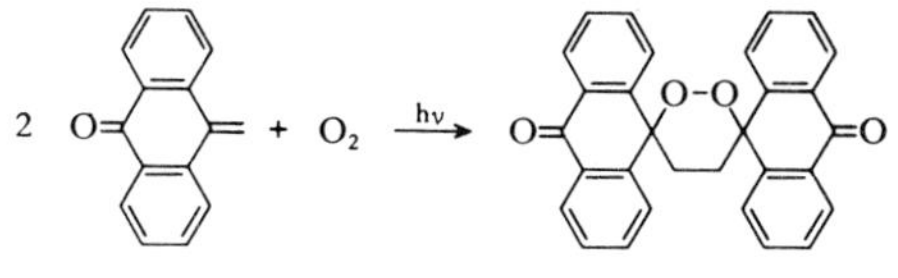

$$2 \text{ (10-methyleneanthrone)} + O_2 \xrightarrow{h\nu} \textbf{164}$$

164

10-Methyleneanthrone-oxygen 2:1-adduct (164) [*127*]. A Pyrex vessel containing a solution of 10-methyleneanthrone (1 g) in 25 ml dry, thiophene-free benzene was placed in the sunlight for 10 days (September, Cairo); the vessel was open to air. Almost colorless crystals separated which were recrystallized from benzene-petroleum ether to give **164**, m.p. 200° (dec.).

b) Formation of open chain peroxides

If bromotriphenylmethane in cyclohexane is exposed to the action of light (sun or mercury lamp) in the presence of oxygen, triphenylmethyl peroxide is formed. It is assumed that the reaction proceeds via the free triphenylmethyl radical [*128*].

$$(C_6H_5)_3C\text{—Br} \xrightarrow{h\nu} (C_6H_5)_3C\cdot + Br$$

$$2 (C_6H_5)_3C\cdot + O_2 \longrightarrow (C_6H_5)_3C\text{—OO—}C(C_6H_5)_3$$

SCHÖNBERG and MUSTAFA [*129, 130*] found that peroxide formation did not occur if a benzene solution of triphenylmethane, 9-phenylfluorene or 9-benzylxanthene was exposed to sunlight in the air. Nevertheless, similar experiments with 9-phenylxanthene (**165**), 9-phenylthioxanthene, 9-(1-naphthyl)-xanthene, 9-(1-naphthyl)-thioxanthene, 9-(4-methoxyphenyl)-xanthene, 9-p-tolylxanthene, 9-m-tolylxanthene, and 9-o-tolylxanthene gave rise to peroxide formation. 4-Chloro-9-phenylxanthene behaved analogously [*131*]. No peroxide formation could, however, be observed with 14-phenyl-14H-dibenzo[a,j]xanthene (**167**), a fact which was ascribed to steric hindrance [*129*].

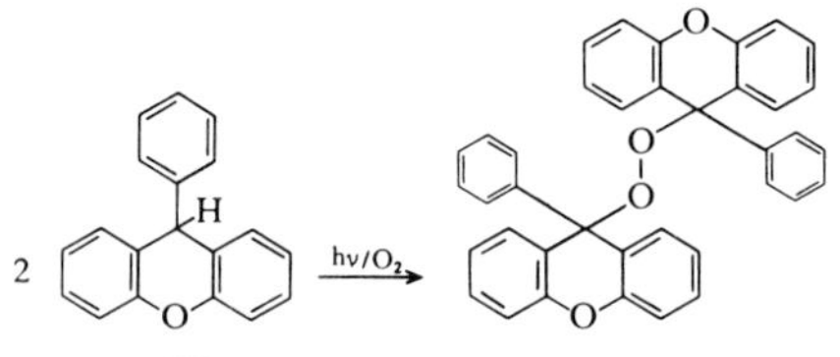

$$2 \text{ (165)} \xrightarrow{h\nu/O_2} \text{ (166)}$$

165 **166** 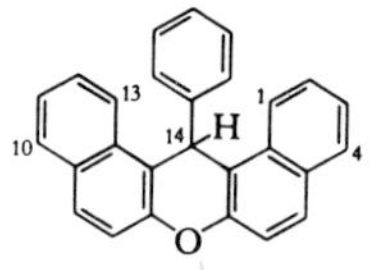 **167**

References, pp. 402—406

Two mechanisms a) and b) have been considered [129] to account for peroxide formation:

a) $2\ (Ar)_3C-H + 2\ O_2 \longrightarrow 2\ (Ar)_3C-OOH \longrightarrow$
$(Ar)_3C-OO-C(Ar)_3 + H_2O_2$

b) $2\ (Ar)_3C-H \longrightarrow [2\ (Ar)_3C\cdot + 2\ H] \xrightarrow{O_2} (Ar)_3C-OO-C(Ar)_3$

N-Aryl-1,2,3,4-tetrahydroisoquinolines (cf. **168**) are converted to peroxides on irradiation in the presence of oxygen as was found by RIECHE and collaborators [132].

168: R = CH$_3$ **169**: R = CH$_3$

Reactions according to the above equation have been performed with compounds R = H, CH$_3$ (**168**), NO$_2$, Cl, or with the 2-naphthyl derivative, the peroxides being colorless compounds reducible with triphenylphosphine.

Bis-(9-phenyl-9-xanthenyl) peroxide (**166**) [129]. 9-Phenylxanthene (**165**) was dissolved in dry thiophene-free benzene, the solution placed in an open Pyrex tube and insolated (November, Cairo) during 15 days. The precipitate was recrystallized from benzene to give pure **166**, m.p. 230° (dec.). More of the peroxide was obtainable from the benzene solution, the yield totalling 80%.

Bis-(N-p-tolyl-1,2,3,4-tetrahydro-1-isoquinolyl) peroxide (**169**) [132]. 1 g of N-p-tolyl-1,2,3,4-tetrahydroisoquinoline (**168**), dissolved in 4—5 ml benzene, was shaken with oxygen at room temperature under irradiation by a UV-fluorescent tube. 30 ml oxygen were taken up within 24 hours (60%, calculated on dimeric peroxide). The benzene solution was treated with the fourfold volume of isopropanol; after several days, the pale yellowish, crystalline product was filtered off (220 mg, 21%). Dissolved in 1 part chloroform and reprecipitated by the addition of 9 parts isopropanol, **169** had m.p. 103—105°.

c) Formation of acyl peroxides

Unsymmetrical acyl peroxides are formed by air oxidation of aromatic aldehydes in acetic anhydride under the influence of light [133]. An improved preparation was reported by OLDEKOP and co-workers [134].

If benzaldehyde (**170**), dissolved in acetone or benzene is treated with oxygen in sunlight, perbenzoic acid (**175**) is produced [135]. **175** is used in preparative chemistry to convert olefins to the corresponding epoxides. If the light-induced autoxidation of benzaldehyde is carried out in the presence

$$R\text{-}C_6H_4\text{-}CHO \xrightarrow{h\nu/O_2} R\text{-}C_6H_4\text{-}C(=O)\text{-}OOH \xrightarrow{(CH_3CO)_2O} R\text{-}C_6H_4\text{-}C(=O)\text{-}OO\text{-}C(=O)\text{-}CH_3 + CH_3COOH$$

170: R = H
171: R = Cl
173: R = CH$_3$

175: R = H

172: R = Cl
174: R = CH$_3$

of such olefins, epoxides may be synthesized directly. RAYMOND [*136*] and SWERN and coworkers [*137*] showed that oleic acid, methyl oleate and 9-octadecen-1-ol gave rise to the corresponding epoxides when subjected to UV-light induced co-oxidation with benzaldehyde in acetone.

Acetyl 4-chlorobenzoyl peroxide (172) [*134*]. 4-Chlorobenzaldehyde (**171**) and acetic anhydride were mixed in a molar ratio of 1 : 3 and aerated (air flow of 2.5—3 liter/min.) under illumination with a 75 W lamp for 3—6 hours. The reaction was carried out at 30—40° in the presence of 0.2—0.3% sodium acetate or 10—15% calcium carbonate. Then the mixture was quenched in water to hydrolyze excess acetic anhydride whereupon the peroxide separated as a white crystalline mass. This was washed with aqueous soda solutions and water and dried. Recrystallization from petroleum ether gave pure **172**, m.p. 49.5°; yield 83%.

Acetyl p-toluoyl peroxide (174) [*134*]. This was prepared from p-tolualdehyde (**173**) in an analogous manner. The peroxide had m. p. 65—65.5°; yield 53%.

9,10-Epoxyoctadecanoic acid [*137*]. A mixture of 426 g **170**, 120.3 g of oleic acid and 730 ml acetone was placed in a two-liter three necked Pyrex flask equipped with thermometer, Dry Ice condenser and two fritted glass gas inlet tubes. A constant stream of dry air was led through the solution while it was irradiated by means of two 125 W mercury vapor lamps. The temperature was kept between 25—35° by external air cooling with a fan. After 8 hrs. the reaction was discontinued since no more peroxide was consumed. Cooling of the mixture to —50° and filtering with the aid of a cold funnel gave 106 g of a white powder which smelled of benzaldehyde. Recrystallization from acetone at —20° (10 ml acetone/g of product) afforded 84 g (70%) of 9,10-epoxyoctadecanoic acid, m.p. 53.5—56°. The product is the higher melting isomer (m.p. 59.5°) but its melting point is lowered by impurities to a value below that of the low melting isomer (m.p. 55.5°).

4. Miscellaneous photochemical oxidation reactions

Among the great number of oxidation reactions reported to require both oxygen and light quite a few may on thorough investigation turn out to be merely autoxidative; mention has already been made on current literature in the autoxidation field. In the following some reactions will be dealt with which were reported to be definitely photochemical.

The reader is also referred to chapter 43 for the photooxidation of thioketones, to chapter 25 for the oxidation of halides and to chapter 42 for sensitized photooxidation of furan compounds.

References, pp. 402—406

a) Oxidative cleavage of C—C bonds

ISMAIL [*138*] found that 10-(fluoren-9-ylidene)-anthrone (**176**) was photochemically cleaved by oxygen into fluoren-9-one and anthraquinone. This reaction is mentioned here although it is of no interest for the preparation of either fluorenone or anthraquinone. It is however possible that analogous cleavages of ethylenes may be of preparative significance in the future (comp. p. 239).

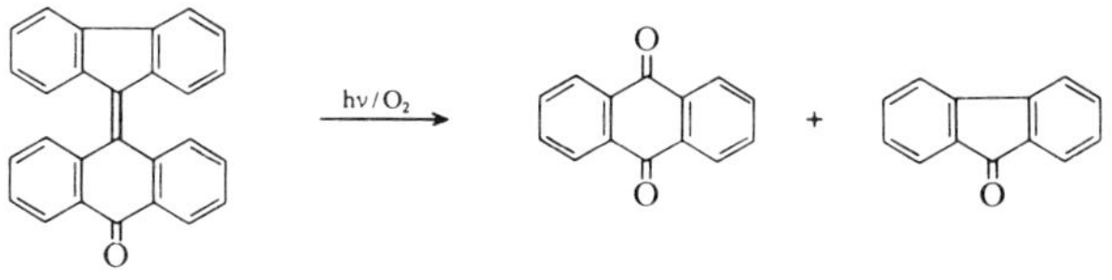

When 2-tert. butyl-2-hydroxy-1,3,3-trimethylindoline (**177**) was exposed to air and sunlight [*139*] it soon became pink and after 24 hrs. was completely converted into a red oil. The transformation was shown to require both oxygen and light. Work-up of the oil proved this to consist of 1,3,3-trimethyloxindole (**178**). Gaseous products liberated during the oxidation were, inter alia, methane, ethylene or ethane, isobutene, and formaldehyde.

Both 2-hydroxy-1,3,3-trimethyl-2-phenylindoline (**179**) and 2-hydroxy-1,3,3-trimethylindoline (**180**) were recovered unchanged after exposure to air and sunlight [*139*].

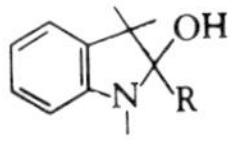

177: R = tert. butyl
179: R = C_6H_5
180: R = H

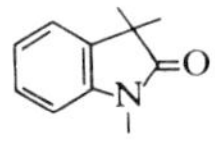

b) Photooxidation of Curare alkaloids

The photosensitized oxygen transfer reaction was applied to several unsaturated alkaloids of the Calabash Curare type by BERNAUER, SCHMID, KARRER and collaborators (for recent summaries on the chemistry of Curare alkaloids see [*140, 141*]).

Thus, C-dihydrotoxiferine-I (**181**) and toxiferine-I (**184**) when irradiated in aerated solutions (aqueous methanol) in the presence of eosin as sensitizer gave C-calebassine (**182**) and C-alkaloid-A (**185**), respectively [*142, 143*]. With methylene blue sensitizing the photoreaction no **182** was formed from **181** but instead a "lumi-dihydrotoxiferine-I" was obtained, the constitution of which was not resolved [*140*].

A different reaction was observed when **181** or **184** were exposed to light and air in the solid state, sensitizers being absent. In this case C-curarine-I (**183**) or C-alkaloid-E (**186**) could be isolated [*142, 143*].

In neither case could a definite reaction mechanism be established; to our knowledge transannular reactions of the type **181** → **182** have not been encountered before in photooxidation work.

References

[*1*] E. H. FARMER and D. A. SUTTON: J. Chem. Soc. **1943**, 119.
[*2*] J. ROSS, A. I. GEBHARDT and J. F. GERECHT: J. Amer. chem. Soc. **71**, 282 [1949].
[*3*] W. O. LUNDBERG: Autoxidation and Antioxidants. New York: Interscience 1961.
[*4*] A. G. DAVIES: Organic Peroxides. London: Butterworths 1961.
[*5*] E. G. E. HAWKINS: Organic Peroxides. Their Formation and Reactions. London: Spon 1961.

[6] H. WEXLER: Chem. Rev. 64, 591 (1964).
[7] H. HOCK and S. LANG: Ber. dtsch. chem. Ges. 75, 1051 (1942).
[8] H. HOCK and S. LANG: Ber. dtsch. chem. Ges. 76, 169 (1943).
[9] H. HOCK and S. LANG: Ber. dtsch. chem. Ges. 77, 257 (1944).
[10] E. H. FARMER and D. A. SUTTON: J. Chem. Soc. 1946, 10.
[11] H. HOCK and W. SUSEMIHL: Ber. dtsch. chem. Ges. 66, 61 (1933).
[12] W. TREIBS and R. SCHÖLLNER: Chem. Ber. 94, 42 (1961).
[13] O. HÄFLIGER, A. BROSSI, L. H. CHOPARD-DIT-JEAN, M. WALTER and O. SCHNIDER: Helv. Chim. Acta 39, 2053 (1956).
[14] G. O. SCHENCK, H. EGGERT and W. DENK: Liebigs Ann. Chem. 584, 177 (1953).
[15] G. O. SCHENCK: Angew. Chem. 64, 12 (1952).
[16] K. ALDER, H. SÖLL and H. SÖLL: Liebigs Ann. Chem. 565, 73 (1949).
[17] G. O. SCHENCK, K. GOLLNICK, G. BUCHWALD, S. SCHROETER and G. OHLOFF: Liebigs Ann. Chem. 674, 93 (1964).
[18] G. O. SCHENCK, O.-A. NEUMÜLLER, G. OHLOFF and S. SCHROETER: Liebigs Ann. Chem. 687, 26 (1965).
[19] C. S. FOOTE, S. WEXLER and W. ANDO: Tetrahedron Letters 1965, 4111.
[20] G. O. SCHENCK: Angew. Chem. 69, 579 (1957).
[21] G. O. SCHENCK: Z. Elektrochem. 57, 675 (1953).
[22] G. O. SCHENCK and K. GOLLNICK: Forschungsber. Land Nordrhein-Westfalen No. 1256. Köln: Westdeutsch. Verlag 1963.
[23] K. GOLLNICK and G. O. SCHENCK: Pure Appl. Chem. 9, 507 (1964).
[24] A. NICKON and J. F. BAGLI: J. Amer. chem. Soc. 83, 1498 (1961).
[25] C. S. FOOTE and S. WEXLER: J. Amer. chem. Soc. 86, 3879 (1964).
[26] C. S. FOOTE and S. WEXLER: J. Amer. chem. Soc. 86, 3880 (1964).
[27] E. J. COREY and W. C. TAYLOR: J. Amer. chem. Soc. 86, 3881 (1964).
[28] E. KLEIN and W. ROJAHN: Tetrahedron 21, 2173 (1965).
[29] G. O. SCHENCK and K.-H. SCHULTE-ELTE: Liebigs Ann. Chem. 618, 185 (1958).
[30] G. OHLOFF, E. KLEIN and G. O. SCHENCK: Angew. Chem. 73, 578 (1961).
[31] W. EISFELD: Diss. Universität Göttingen 1965.
[32] G. O. SCHENCK, K. GOLLNICK and O.-A. NEUMÜLLER: Liebigs Ann. Chem. 603, 46 (1957).
[33] G. O. SCHENCK, O.-A. NEUMÜLLER and W. EISFELD: Liebigs Ann. Chem. 618, 202 (1958).
[34] A. NICKON and W. L. MENDELSON: J. Amer. chem. Soc. 85, 1894 (1963).
[35] A. NICKON and W. L. MENDELSON: J. Amer. chem. Soc. 87, 3921 (1965).
[36] G. O. SCHENCK and O.-A. NEUMÜLLER: Liebigs Ann. Chem. 618, 194 (1958).
[37] G. O. SCHENCK, H.-D. BECKER, K.-H. SCHULTE-ELTE and C. H. KRAUCH: Chem. Ber. 96, 509 (1963).
[38] A. RIECHE and E. SCHMITZ: Chem. Ber. 90, 1082 (1957).
[39] M. BUSCH and W. DIETZ: Ber. dtsch. chem. Ges. 47, 3277 (1914).
[40] K. H. PAUSACKER: J. Chem. Soc. 1950, 3478.
[41] R. CRIEGEE and G. LOHAUS: Chem. Ber. 84, 219 (1951).
[42] H. C. YAO and P. RESNICK: J. Org. Chem. 30, 2832 (1965).
[43] G. O. SCHENCK and H. WIRTH: Naturwissenschaften 40, 141 (1953).
[44] L. WEIL and J. MAHER: Arch. Biochem. Biophys. 29, 241 (1950).
[45] M. I. SIMON and H. VAN VUNAKIS: Arch. Biochem. Biophys. 105, 197 (1964).
[46] A. I. SCOTT and C. T. BEDFORD: J. Amer. chem. Soc. 84, 2271 (1962).
[47] M. SCHACH VON WITTENAU: J. Org. Chem. 29, 2746 (1964).
[48] K. ALDER and M. SCHUMACHER, in: Fortschr. Chem. org. Naturstoffe, ed. by L. ZECHMEISTER, vol. 10, 1—118; esp. p. 94—96. Wien: Springer 1953.
[49] G. O. SCHENCK: Z. Elektrochem. 56, 855 (1952).

[50] K. GOLLNICK and G. O. SCHENCK, in: 1,4-Cycloaddition Reactions, ed. by J. HAMER, pp. 255—344. New York: Academic 1967.
[51] W. BERGMANN and M. J. McLEAN: Chem. Rev. **28**, 367 (1941).
[52] A. SCHÖNBERG, in: Le Mécanisme de l'Oxydation. **8**. Conseil de Chimie, ed. by l'Inst. Int. Chimie Solvay, pp. 217—259. Bruxelles: Stoops 1950.
[53] YU. A. ARBUZOV: Russ. Chem. Rev. **34**, 558 (1965); Uspekhi khimii **34**, 1332 (1965).
[54] H. HOCK and F. DEPKE: Chem. Ber. **84**, 349 (1951).
[55] G. O. SCHENCK and K. ZIEGLER: Naturwissenschaften **32**, 157 (1944).
[56] G. O. SCHENCK and H. SCHULZE-BUSCHOFF: Dtsch. med. Wschr. **73**, 29 (1948).
[57] G. O. SCHENCK and R. WIRTZ: Naturwissenschaften **40**, 581 (1953).
[58] G. O. SCHENCK, K. G. KINKEL and H.-J. MERTENS: Liebigs Ann. Chem. **584**, 125 (1953).
[59] G. O. SCHENCK and K. ZIEGLER, in: Festschrift Arthur Stoll. Basel: Birkhäuser 1957.
[60] G. O. SCHENCK and H. ZIEGLER: Naturwissenschaften **38**, 356 (1951).
[61] G. O. SCHENCK and D. E. DUNLAP: Angew. Chem. **68**, 248 (1956).
[62] G. O. SCHENCK, W. MÜLLER and H. PFENNIG: Naturwissenschaften **41**, 374 (1954).
[63] G. O. SCHENCK: DBP 913 892, 21. 6. 1954; Chem. Abstr. **52**, 14 704 (1958).
[64] R. CRIEGEE, in: Methoden der Organischen Chemie, ed. by E. MÜLLER, vol. **8**, part 3, pp. 1—74. Stuttgart: Thieme 1952.
[65] A. C. COPE, T. A. LISS and G. W. WOOD: J. Amer. chem. Soc. **79**, 6287 (1957).
[66] P. COURTOT: Ann. chim. [13] **8**, 197 (1963).
[67] A. WINDAUS and J. BRUNKEN: Liebigs Ann. Chem. **460**, 225 [1928].
[68] L. F. FIESER: The Chemistry of Natural Products Related to Phenanthrene, p. 147. New York: Reinhold 1936.
[69] S. IWASAKI and K. TSUDA: Chem. Pharm. Bull. **11**, 1034 (1963).
[70] J. CASTELLS, E. R. H. JONES, G. D. MEAKINS and R. W. J. WILLIAMS: J. Chem. Soc. **1959**, 1159.
[71] W. G. DAUBEN and G. J. FONKEN: J. Amer. chem. Soc. **81**, 4060 (1959).
[72] P. BLADON: J. Chem. Soc. **1955**, 2176.
[73] M. MÜLLER: Z. physiol. Chem. **231**, 75 (1935).
[74] E. L. SKAU and W. BERGMANN: J. Org. Chem. **3**, 166 (1938).
[75] G. D. LAUBACH, E. C. SCHREIBER, E. J. AGNELLO and K. J. BRUNINGS: J. Amer. chem. Soc. **78**, 4746 (1956).
[76] W. H. SCHULLER, J. C. MINOR and R. V. LAWRENCE: Ind. Eng. Chem., Prod. Res. Develop. **3**, 97 (1964).
[77] W. H. SCHULLER and R. V. LAWRENCE: J. Amer. chem. Soc. **83**, 2563 (1961).
[78] D. H. R. BARTON and G. F. LAWS: J. Chem. Soc. **1954**, 52.
[79] K.-H. SCHULTE-ELTE: Diss. Universität Göttingen 1961.
[80] A. ÉTIENNE, in: Traité de Chimie Organique, ed. by V. GRIGNARD, G. DUPONT and R. LOQUIN, vol. **17**, 1299—1332. Paris: Masson 1949.
[81] C. DUFRAISSE and M. GÉRARD: Comptes rendus **201**, 428 (1935).
[82] C. DUFRAISSE, J. LE BRAS and A. ALLAIS: Comptes rendus **216**, 383 (1943).
[83] E. J. BOWEN, in: Adv. Photochem., ed. by W. A. NOYES JR., G. S. HAMMOND and J. N. PITTS JR., vol. **1**, 23—42. New York: Interscience 1963.
[84] C. DUFRAISSE and M. BADOCHE: Comptes rendus **200**, 1103 (1935).
[85] A. ÉTIENNE and Y. LEPAGE: Comptes rendus **236**, 1498 (1953).
[86] C. MOUREU, C. DUFRAISSE and P. M. DEAN: Comptes rendus **182**, 1584 (1926).
[87] C. DUFRAISSE, J. RIGAUDY, J.-J. BASSELIER and N. K. CUONG: Comptes rendus **260**, 5031 (1965).

[88] C. Dufraisse, L. Velluz and L. Velluz: Comptes rendus **208**, 1822 (1939).
[89] A. Étienne and A. Staehelin: Comptes rendus **234**, 1453 (1952).
[90] M. A. Willemart: Bull. Soc. Chim. France [5] **5**, 556 (1938).
[91] C. Dufraisse, R. Demuynck and A. Allais: Comptes rendus **215**, 487 (1942).
[92] C. Dufraisse and R. Horclois: Bull. Soc. Chim. France [5] **3**, 1880 (1936).
[93] C. Dufraisse and R. Horclois: Bull. Soc. Chim. France [5] **3**, 1894 (1936).
[94] C. Dufraisse, A. Étienne and C. Winnick: Comptes rendus **236**, 2133 (1953).
[95] C. Dufraisse and G. Sauvage: Comptes rendus **221**, 665 (1945).
[96] G. Sauvage: Ann. Chim. [12] **2**, 844 (1947).
[97] C. Dufraisse and G. Sauvage: Comptes rendus **225**, 126 (1947).
[98] G. M. Badger: The structures and reactions of the aromatic compounds. Cambridge: University Press 1957.
[99] C. Dufraisse, L. Velluz and L. Velluz: Bull. Soc. Chim. France [5] **5**, 600 (1938).
[100] G. Sauvage: Comptes rendus **225**, 247 (1947).
[101] L. Velluz: Bull. Soc. Chim. France [5] **6**, 1541 (1939).
[102] A. Étienne: Comptes rendus **217**, 694 (1943).
[103] A. Étienne and J. Robert: Comptes rendus **223**, 331 (1946).
[104] M. Legrand: Comptes rendus **237**, 822 (1953).
[105] C. F. H. Allen and A. Bell: J. Amer. chem. Soc. **64**, 1253 (1942).
[106] A. Étienne and A. Staehelin: Bull. Soc. Chim. France **1954**, 748.
[107] C. Dufraisse and M.-T. Mellier: Comptes rendus **215**, 576 (1942).
[108] C. Dufraisse and J. Baget: Comptes rendus **220**, 47 (1945).
[109] M. R. Panico: Comptes rendus **234**, 852 (1952).
[110] C. Dufraisse and M.-T. Mellier: Comptes rendus **215**, 541 (1942).
[111] M.-T. Mellier: Ann. Chim. [12] **10**, 666 (1955).
[112] C. Dufraisse and A. Étienne: Comptes rendus **201**, 280 (1935):
[113] C. Dufraisse, A. Étienne and J. Aubry: Comptes rendus **239**, 1170 (1954).
[114] J.-J. Basselier and M.-J. Scholl: Comptes rendus **258**, 6463 (1964).
[115] M. A. Ogliaruso, M. G. Romanelli and E. I. Becker: Chem. Rev. **65**, 261 (1965); erratum 717.
[116] N. M. Bikales and E. I. Becker: J. Org. Chem. **21**, 1405 (1956).
[117] C. F. H. Allen and J. A. VanAllan: J. Org. Chem. **18**, 882 (1953).
[118] P. De Mayo and S. T. Reid: Chem. and Ind. **1962**, 1576.
[119] C. Dufraisse, A. Étienne and J. Martel: Comptes rendus **244**, 970 (1957).
[120] C. Dufraisse and J. Martel: Comptes rendus **244**, 3106 (1957).
[121] J. Sonnenberg and D. M. White: J. Amer. chem. Soc. **86**, 5685 (1964).
[122] E. H. White and M. J. C. Harding: J. Amer. chem. Soc. **86**, 5686 (1964).
[123] C. Dufraisse, G. Rio, A. Ranjon and O. Pouchot: Comptes rendus **261**, 3133 (1965).
[124] H. H. Wasserman and A. Liberles: J. Amer. chem. Soc. **82**, 2086 (1960).
[125] W. Theilacker and W. Schmidt: Liebigs Ann. Chem. **605**, 43 (1957).
[126] G. Wittig and W. Gauss: Chem. Ber. **80**, 363 (1947).
[127] A. Mustafa and A. M. Islam: J. Chem. Soc. **1949**, Suppl. 81.
[128] J. O. Halford and L. C. Anderson: Proc. Nat. Acad. Sci. U.S.A. **19**, 759 (1933).
[129] A. Schönberg and A. Mustafa: J. Chem. Soc. **1945**, 657.
[130] A. Schönberg and A. Mustafa: J. Chem. Soc. **1947**, 997.
[131] A. Mustafa, W. Asker and M. E. D. Sobhy: J. Org. Chem. **25**, 1519 (1960).
[132] A. Rieche, E. Höft and H. Schultze: Chem. Ber. **97**, 195 (1964).
[133] J. U. Nef: Liebigs Ann. Chem. **298**, 202 (1897).
[134] Yu. A. Oldekop, A. N. Sevchenko, I. P. Zyatkov, G. S. Bylina and A. P. Elnitskii: Dokl. Akad. Nauk SSSR **128**, 1201 (1959); Proc. Acad. Sci. USSR. Chem. Sect. **128**, 907 (1959).

[135] P. A. A. VAN DER BEEK: Rec. Trav. chim. Pays-Bas **47**, 286 (1928).
[136] E. RAYMOND: J. Chim. phys. **28**, 480 (1931).
[137] D. SWERN, T. W. FINDLEY and J. T. SCANLAN: J. Amer. chem. Soc. **66**, 1925 (1944).
[138] A. F. A. ISMAIL and Z. M. EL-SHAFEI: J. Chem. Soc. **1957**, 3393.
[139] B. ROBINSON: J. Chem. Soc. **1963**, 586.
[140] K. BERNAUER, in: Fortschr. Chem. Org. Naturstoffe, ed. by L. ZECHMEISTER, vol. 17, 183—247. Wien: Springer 1959.
[141] A. R. BATTERSBY and H. F. HODSON, in: The Alkaloids. Chemistry and Physiology, ed. by R. H. F. MANSKE, vol. **8**, 515—579. New York: Academic Press 1965.
[142] K. BERNAUER, H. SCHMID and P. KARRER: Helv. Chim. Acta **40**, 1999 (1957).
[143] K. BERNAUER, F. BERLAGE, H. SCHMID and P. KARRER: Helv. Chim. Acta **41**, 1202 (1958).

Chapter 40

Photochemical formation and transformations of epoxides

1. Photochemical formation of epoxides

Epoxides appearing among the photooxidation products of olefins are usually resulting from secondary reactions of the intermediate hydroperoxides or epidioxides. Epoxides may form from these intermediates either by rearrangement or via a PRILESHAJEW type reaction with unconsumed olefin. The formation of steroid epoxyketones [1] and of epoxides in the pyrrole [2] and furan [2, 3] series has been mentioned in the pertinent chapters (39 and 42). Hence, only a few further examples of photochemical epoxide formation will be discussed in the following text.

According to FRANKEL et al. [4], the photochemical action of oxygen on tetrachloroethylene in the presence of chlorine leads to the formation of trichloroacetyl chloride, phosgene, hexachloroethane, and epoxytetrachloroethane, which is obtained in 9 % yield. The solutions were insolated for 12 hrs. at 36—40°, oxygen and chlorine being present in excess. Details of the apparatus may be taken from the original paper.

$$Cl_2C=CCl_2 \xrightarrow[Cl_2, O_2]{h\nu} Cl_3C-C\overset{O}{\underset{Cl}{\diagdown}} + Cl_3C-CCl_3 + ClC\overset{O}{\underset{Cl}{\diagdown}} + Cl_2C\overset{O}{-\!\!-}CCl_2$$

DUFRAISSE et al. [5] were able to convert 1,2,3,4-tetraphenylfulvene (**1**) into a colorless compound, for which structure **3** was proposed, by irradiation of dilute aerated solutions of **1** in ether. Within 150 mins. of irradiation a 90 % yield of **3** was obtained. The primary product of the action of oxygen on **1** was suspected to be **2**, from which **3** might form by rearrangement.

1   2   3

Lately, BASSELIER [6] succeeded in fully elucidating the course of the formation of the bis-epoxide **3**. 7-Methylene-1,4,5,6-tetraphenyl-2,3-dioxabicyclo[2.2.1]hept-5-ene (**2**) was eventually isolated following illumination of dilute solutions of **1** containing sensitizers (preferably rubrene or 8H,16H-dibenzo[a,j]perylene-8,16-dione) and oxygen. Under these conditions a 3 min. irradiation period sufficed for the production of a 85 % yield of **2**. Exposure of the latter to the same radiation in the absence of sensitizer resulted then in isomerization to 5-methylene-1,2,4,6-tetraphenyl-3,7-dioxatricyclo[4.1.0.0^{2,4}]heptane (1,2:3,4-diepoxy-5-methylene-1,2,3,4-tetraphenylcyclopentane, **3**). This photoisomerization took place at room temperature as well as at $-50°$.

2. Photoisomerization of epoxyketones

The photoisomerization of chalcone oxides to 1,3-diketones (comp. the transformation of 2,3-epoxy-3-phenylpropiophenone (**4**) to 1,3-diphenyl-1,3-propanedione (**5**)), discovered as early as 1918 by BODFORSS [7], has

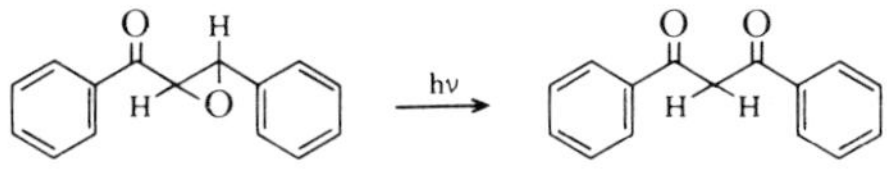

long been neglected. It is only since 1962, due especially to the work of JEGER and co-workers on steroid epoxides, that the significance of this isomerization has been established, i.e., that rearrangement of α,β-epoxyketones of unsaturated as well as saturated cyclic compounds is a feasible reaction. An excellent review of the work of the Zurich group and of related work has been given by JEGER, SCHAFFNER and WEHRLI [8]. Reaction mechanisms of the photochemical transformations of epoxyketones are extensively treated [8—10], so that discussion of this aspect will not be included here.

a) Acyclic α,β-epoxyketones

The migratory aptitudes of substituents at the epoxyketone moiety were explored by ZIMMERMAN and collaborators [10]. Thus, while irradiation of 3,4-epoxy-4-phenyl-2-pentanone (**6**) resulted in formation of 25 % of 2-methyl-1-phenyl-1,3-butanedione (**7**) in complete analogy to the findings

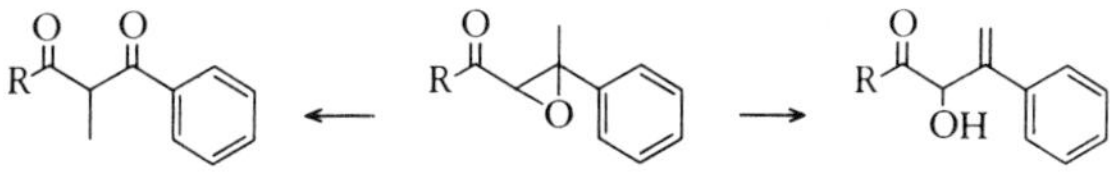

7: R= CH$_3$

6: R= CH$_3$
8: R= C$_6$H$_5$

9: R = C$_6$H$_5$

of BODFORSS, the closely related compound, 2,3-epoxy-3-methyl-3-phenyl-propiophenone, (trans-dypnone oxide, **8**) afforded 2-hydroxy-1,3-diphenyl-3-buten-1-one (**9**) in 25 % yield.

Another variation in the reactivity of α,β-epoxyketones was encountered when pulegone α-oxide (**10**) was subjected to photolysis in ethereal solution [*11*]. Rearrangement to 2-acetyl-cis- and 2-acetyl-trans-2,5-dimethylcyclohexanones (**11** and **12**, respectively) was observed, as well as a novel isomerization to pulegone β-oxide (**13**) (for the most likely configurations of **10** and **13** comp. [*12*]).

b) Acyclic β,γ-epoxyketones

While almost all photoisomerizations discussed in this chapter have been carried out with α,β-epoxyketones, one example of a β,γ-epoxyketone has been investigated by PADWA [*13*]. Trans-3,4-epoxy-1,4-diphenyl-1-butanone (**14**) on irradiation in benzene solution through Pyrex gave 18 % of 1,4-diphenyl-1,4-butanedione (**15**) and 23 % of an alcohol which was attributed the structure of 2,3-epoxy-cis-1,2-diphenyl-1-cyclobutanol (**16**).

c) Cyclic α,β-epoxyketones

The photochemistry of 4,5-epoxy-3,4-diphenyl-2-cyclopenten-1-one (**17**) was studied by PADWA [*14, 15*]. Irradiation in ether for 10 mins. gave a photolabile intermediate which in its turn underwent a further photoreaction. Choice of appropriate filter conditions facilitated the isolation [*14*] of this intermediate, viz. 4,5-diphenyl-2H-pyran-2-one (**18**). Irradiation of the latter [*15*] under identical conditions caused the production of three major components, 1,2,4,7-tetraphenylcyclooctatetraene (**19**), p-terphenyl (**20**), and diphenylacetylene (**21**). The ratio of these secondary photolysis products of **17** was shown [*15*] to depend on the length of the irradiation time. For the structures suggested for the connecting links between **18** and **19** see PADWA and HARTMAN [*15*].

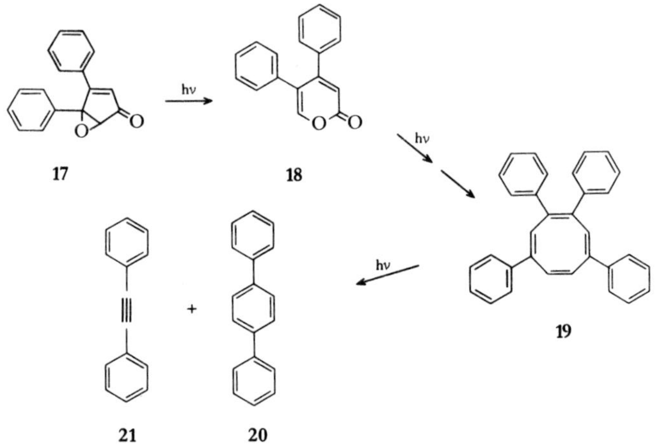

Surprisingly enough, two compounds closely related to **17** by the surplus of two or one phenyl groups (cf. **22** and **25**) differed considerably from **17** in behavior. 4,5-Epoxy-2,3,4,5-tetraphenyl-2-cyclopenten-1-one (**22**), for example, did still partly rearrange to tetraphenyl-2H-pyran-2-one (**23**) on irradiation in ethanol, benzene [*16*] or acetone [*17*] solutions, but concomitant formation of a valence tautomer of **22**, viz. 2,4,5,6-tetraphenyl-pyrylium-3-oxide (**24**) was considerable [*16*]. The red solution containing **24** bleached slowly on discontinuation of the irradiation; the photochemical equilibrium **22** ⇌ **24** could be established from either side [*16*].

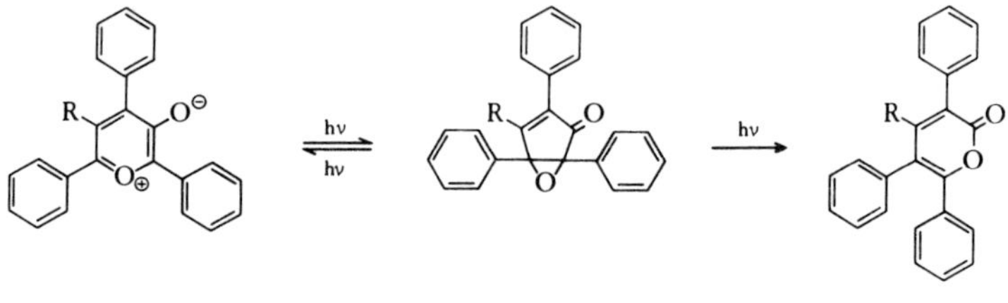

24: R = C$_6$H$_5$
27: R = H

22: R = C$_6$H$_5$
25: R = H

23: R = C$_6$H$_5$
26: R = H

In the case of the red 2,4,6-triphenylpyrylium-3-oxide (**27**), studied by ULLMAN [*18*], irradiation in aprotic solvents caused a decoloration of the solution. Working up yielded, in the absence of hydroxylic solvents, 4,5-epoxy-2,4,5-triphenyl-2-cyclopenten-1-one (**25**) and, in the presence of methanol, 3,5,6-triphenyl-2H-pyran-2-one (**26**).

More photochemical valence tautomerization reactions were investigated by ULLMAN and MILKS [*19, 20*] and by ZIMMERMAN and SIMKIN [*21*]. Appearance of a red coloration on illumination of solid or dissolved **28** had

References, p. 413

been noted by WEITZ and SCHEFFER [22] already in 1921. Irradiation of deoxygenated benzene solutions of 2,3-epoxy-2,3-diphenyl-1-indanone (**28**) [20] or of the closely related 2,3-epoxy-2-methyl-3-phenyl-1-indanone (**31**) [21] produced intensely colored pyrylium compounds **29** and **32**, respectively. At least **31** did not only undergo valence tautomerization but rearrangement as well [21], 3-methyl-4-phenylisocoumarin (**34**) being formed in 27 % yield.

28: R = C_6H_5
31: R = CH_3

29: R = C_6H_5
32: R = CH_3

30: R = C_6H_5
33: R = CH_3

The color intensity attained by irradiation was found [20] to be dependent on the wave length of the incident light: intensely colored solutions could be produced by long irradiation in oxygen-free benzene solutions with filtered (320—390 mµ) light. Unfiltered light was less effective while visible radiation (450 mµ) led to rapid and complete decoloration of the previously colored solutions. Heating to nearly 200° also gave rise to intense coloration.

The valence tautomers, 1,3-diphenyl-2-benzopyrylium-4-oxide (**29**) and 3-methyl-1-phenyl-2-benzopyrylium-4-oxide (**32**) could not be isolated directly. Reaction with dienophiles was found a suitable trapping method, however. On ultraviolet irradiation of solutions of **28** in dimethyl acetylenedicarboxylate no color was developed and a 1:1 adduct, dimethyl 8,9-dihydro-9-oxo-5,8-diphenyl-5,8-epoxy-5H-benzocycloheptene-6,7-dicarboxylate (**30**) was obtained [20]. On analogous treatment, **31** gave **33** [21]. A norbornadiene adduct (**35**) was formed from **28**; the isolation of the same adduct by heating the two components provided conclusive evidence that the colored species formed by heating and by irradiation were identical.

34: R = CH_3

35: R = C_6H_5

Tetraphenyl-2H-pyran-2-one (23) [*17*]. 4,5-Epoxy-2,3,4,5-tetraphenyl-2-cyclopenten-1-one (**22**, for a revision of the original structure see PÜTTER and DILTHEY [*23*]) (500 mg) was dissolved in acetone and irradiated with a daylight lamp for 10 hours. A red coloration developed which faded again in the course of the irradiation. Evaporation of the solvent and crystallization of the residue from methanol gave 400 mg of **23**, m.p. 165°.

1,4,4a,10,11,11a-Hexahydro-5,11-diphenyl-5,11-epoxy-1,4-methano-5H-dibenzo-[a,d]cyclohepten-10-one (35) [*20*]. A solution of 1.0 g of the indenone oxide **28** in 50 ml of freshly distilled bicyclo[2.2.1]heptadiene contained in a water-cooled quartz reaction vessel was irradiated with light from a 100 W medium-pressure mercury arc for 5 hours. Evaporation of the excess reagent in vacuo and trituration of the residue with 25 ml of hot ethanol gave 0.83 g (63%) of the adduct **35**. Recrystallization from ethanol gave pure material, m.p. 215—216°.

Yields could be increased to over 90% by irradiation with 320—390 mμ light from a B-H6 source equipped with suitable filters.

d) Steroid α,β-epoxyketones

Photoisomerization reactions of epoxyketones are less complex in the steroid series, 1,3-diketones being formed exclusively. Thus, 17β-acetoxy-4β,5-epoxy-5β-androst-1-en-3-one (**36**) and its isomer **37** are cleanly converted [*24*] to 17β-acetoxy-10(5 → 4)-abeo-androst-1-ene-3,5-dione (**38**) (only the enolized forms shown). Solvents suited for the isomerization were dioxane or ethanol, and mercury low-pressure lamps were used.

The 1,2-dihydro analogs of **36** or **37** (cf. **39**, **40**) underwent the same type of photolysis, though more slowly [*24*]. In the case of 17β-acetoxy-4β,5-epoxy-5β-androstan-3-one (**39**) the photolysis was found to be accelerated by using the full UV-spectrum of a mercury high-pressure burner instead of monochromatic light 254 mμ. On the other hand, secondary reactions of the diketones formed were more frequent with high-pressure lamps [*9*].

No enolization was encountered when 17β-acetoxy-4β,5-epoxy-4α-methyl-5β-androstan-3-one (**41**) or 17β-acetoxy-4α,5-epoxy-4β-methyl-5α-androstan-3-one (**43**) were irradiated [*9*]. The 17β-acetoxy-5-methyl-A-nor-B-homoandrostane-3,6-diones **42** and **44**, being formed in yields of 31% and 21%, respectively, were shown not to be identical.

39: R = H
41: R = CH$_3$

42: R = CH$_3$

44: R = CH$_3$

40: R = H
43: R = CH$_3$

17β-Acetoxy-10(5 → 4)-abeo-androst-1-ene-3,5-dione (38) [*24*]. A 0.029 molar solution of **36** in dioxane was irradiated with a Quarzlampengesellschaft NK 6/20 low-pressure mercury lamp for 40 hours. Working up gave 80% of pure crystalline **38**, m.p. 171–172°.

References

[*1*] A. Nickon and W. L. Mendelson: J. Amer. chem. Soc. **87**, 3921 (1965).
[*2*] H. H. Wasserman and A. Liberles: J. Amer. chem. Soc. **82**, 2086 (1960).
[*3*] R. E. Lutz, W. J. Welstead, jr., R. G. Bass and J. I. Dale: J. Org. Chem. **27**, 1111 (1962).
[*4*] D. M. Frankel, C. E. Johnson and H. M. Pitt: J. Org. Chem. **22**, 1119 (1957).
[*5*] C. Dufraisse, A. Étienne and J.-J. Basselier: Comptes rendus **244**, 2209 (1957).
[*6*] J.-J. Basselier: Comptes rendus **258**, 2851 (1964).
[*7*] S. Bodforss: Ber. dtsch. chem. Ges. **51**, 214 (1918).
[*8*] O. Jeger, K. Schaffner and H. Wehrli: Pure Appl. Chem. **9**, 555 (1964).
[*9*] H. Wehrli, C. Lehmann, K. Schaffner and O. Jeger: Helv. Chim. Acta **47**, 1336 (1964).
[*10*] H. E. Zimmerman, B. R. Cowley, C.-Y. Tseng and J. W. Wilson: J. Amer. chem. Soc. **86**, 947 (1964).
[*11*] C. K. Johnson, B. Dominy and W. Reusch: J. Amer. chem. Soc. **85**, 3894 (1963).
[*12*] C. Djerassi, W. Klyne, T. Norin, G. Ohloff and E. Klein: Tetrahedron **21**, 163 (1965).
[*13*] A. Padwa: J. Amer. chem. Soc. **87**, 4205 (1965).
[*14*] A. Padwa: Tetrahedron Letters **1964**, 813.
[*15*] A. Padwa and R. Hartman: J. Amer. chem. Soc. **86**, 4212 (1964).
[*16*] J. M. Dunston and P. Yates: Tetrahedron Letters **1964**, 505.
[*17*] R. Pütter and W. Dilthey: J. prakt. Chem. [2] **149**, 183 (1937).
[*18*] E. F. Ullman: J. Amer. chem. Soc. **85**, 3529 (1963).
[*19*] E. F. Ullman and J. E. Milks: J. Amer. chem. Soc. **84**, 1315 (1962).
[*20*] E. F. Ullman and J. E. Milks: J. Amer. chem. Soc. **86**, 3814 (1964).
[*21*] H. E. Zimmerman and R. D. Simkin: Tetrahedron Letters **1964**, 1847.
[*22*] E. Weitz and A. Scheffer: Ber. dtsch. chem. Ges. **54**, 2327 (1921).
[*23*] R. Pütter and W. Dilthey: J. prakt. Chem. [2] **150**, 40 (1937).
[*24*] C. Lehmann, K. Schaffner and O. Jeger: Helv. Chim. Acta **45**, 1031 (1962).

Chapter 41

Photochemical formation of four membered rings with one oxygen atom (PATERNÒ-BÜCHI reaction)

1. Formation of oxetanes

a) Cycloaddition of aldehydes or ketones to olefins

As early as 1909 PATERNÒ and CHIEFFI [1] had noticed that trimethylene oxides (oxetane, **1**) were formed by the photochemical addition of aldehydes or ketones to olefins. At that time, this work did not find the attention it

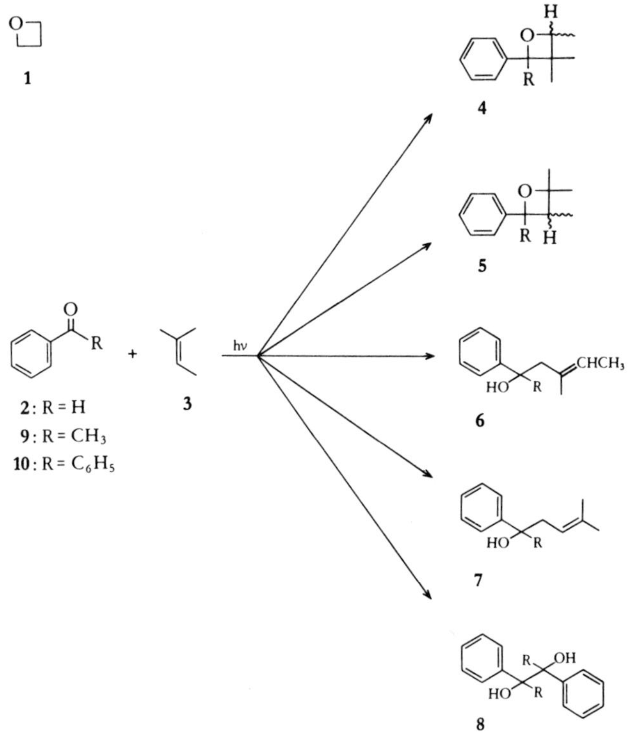

2: R = H
9: R = CH$_3$
10: R = C$_6$H$_5$

References, pp. 424—425

deserved. The present interest in the reaction discovered by PATERNÒ, and in the reaction mechanism stems from the work of BÜCHI et al. [2], and the reaction is therefore known (cf. YANG [3]) as the PATERNÒ-BÜCHI reaction. A general survey on preparation and properties of oxetanes will be found elsewhere [4].

The PATERNÒ-BÜCHI reaction will be illustrated by the reaction of 2-methyl-2-butene (3) with benzaldehyde (2) and with acetophenone (9) as investigated first by PATERNÒ [1] and BÜCHI [2] and later studied by YANG and his co-workers [5].

In the case of acetophenone or benzophenone (10) oxetane formation with 3 was found [5] to be more stereoselective than with benzaldehyde, since more than 90% of the oxetanes belonged to the 4 type while with 2 the ratio of 4 : 5 was 1.6 : 1, with all four possible isomeric oxetanes being

Table 33

carbonyl compound	olefin	oxetane (yield)	ref.
benzaldehyde (2)	2-methyl-2-butene (3)	4 + 5 (64%)	[1,2,5]
1-naphthaldehyde	3	isomers (70%)	[5]
2-naphthaldehyde	3	isomers (70%)	[5]
9-anthraldehyde	3	11	[5]
acetaldehyde	styrene	12 (49%)	[9]
propionaldehyde	furan	13 (80%)	[10]
benzaldehyde (2)	furan	14 (88%)	[10]
acetophenone (9)	2-methyl-2-butene (3)	4, R = CH_3	[1,2,5]
benzophenone (10)	3	4, R = C_6H_5 (58%)	[1,5,8]
10	propene	15 (5%)	[8]
10	cis- or trans-2-butene	16 (79%)	[8]
10	2-methylpropene	17 (93%)	[8]
10	2,3-dimethyl-2-butene	18 (70%)	[8]
10	bicyclo[2.2.1]heptene	19 (80%)	[11]
10	furan	20 (94%)	[12]
2-naphthyl phenyl ketone	3	4, R = $C_{10}H_7$ (62%)	[5]
fluoren-9-one	3	21 (unstable)	[5]
4,4'-dimethylbenzophenone	2-methylpropene	22 (74%)	[8]
4-chlorobenzophenone	2-methylpropene	23 (76%)	[8]
4-methylbenzophenone	2-methylpropene	24 (81%)	[8]
4-aminobenzophenone	2-methylpropene	no oxetane	[8]
1'-acetonaphthone	3	no oxetane	[5]
2'-acetonaphthone	3	no oxetane	[5]
1-naphthyl phenyl ketone	2-methylpropene	no oxetane	[8]
2-naphthyl phenyl ketone	2-methylpropene	no oxetane	[8]
xanthen-9-one	2-methylpropene	no oxetane	[8]
benzophenone (10)	1,2-dichloroethylene	no oxetane	[8]
10	1,1-diphenylethylene	no oxetane	[8]
10	stilbene	no oxetane	[8]

present. Studies on the photochemical formation of oxetanes are complicated by the fact that the oxetanes themselves are not stable towards irradiation (comp. [4, 6, 7]). A similar lability was observed with thietanes, the sulfur analogs of oxetanes (cf. chapter 43).

As will be seen from table 33, yields of oxetanes are usually good; the table, however, also lists examples where irradiation fails to produce oxetanes. A discussion of the mechanisms implicated in the C_3O-cycloaddition lies outside the scope of this book, and the reader is referred to YANG [4, 5] and ARNOLD [8] for pertinent references. Oxetanes bearing functional groups will be dealt with on page 420.

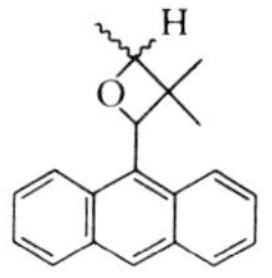

11

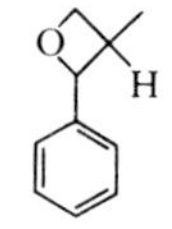

12

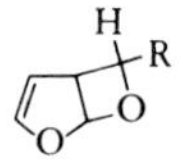

13: $R = C_2H_5$
14: $R = C_6H_5$

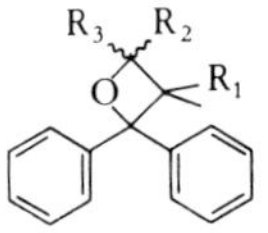

15: $R_1 = R_3 = H$
16: $R_1 = R_2 = H, R_3 = CH_3$
17: $R_1 = CH_3, R_2 = R_3 = H$
18: $R_1 = R_3 = CH_3$

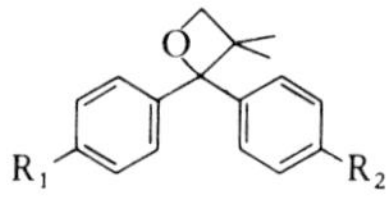

22: $R_1 = R_2 = CH_3$
23: $R_1 = Cl, R_2 = H$
24: $R_1 = CH_3, R_2 = H$

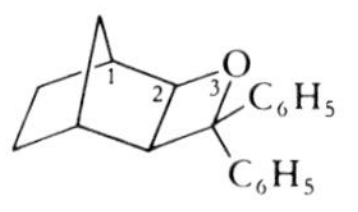

19

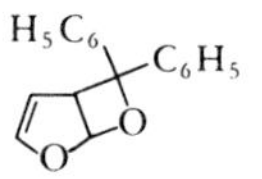

20

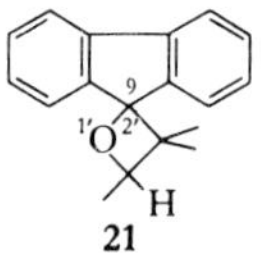

21

When 9-anthraldehyde was irradiated in 2-methyl-2-butene (**3**) with a mercury medium-pressure lamp [5], the main product was an oxetane (**11**, m.p. 160—163°), but by cutting off wavelengths greater than 410 mμ by using a 2,2'-dihydroxybenzophenone solution as filter, oxetane formation was completely suppressed.

Fluoren-9-one and **3** reacted photochemically to give 9-isopropylidenefluorene (**25**) [3]. 3',3',4'-Trimethylspiro[fluorene-9,2'-oxetane] (**21**) was assumed to be the intermediate in this reaction which finally breaks down to **25** and acetaldehyde.

References, pp. 424—425

2,3,3-Trimethyl-4-phenyloxetane (4) and 2,2,3-trimethyl-4-phenyloxetane (5) [*5*]. The solution of benzaldehyde (2) in 2-methyl-2-butene (3) was flushed with nitrogen for 15 minutes and then irradiated with a Hanovia 450 W medium-pressure mercury arc in a Vycor immersion well. The oxetanes 4 and 5 were obtained in 64% yield, the ratio 4:5 being 1.6:1. Other products isolated were 3-methyl-1-phenyl-3-penten-1-ol (6) or 4-methyl-1-phenyl-3-penten-1-ol (7) (15%) and hydrobenzoin (1,2-diphenyl-1,2-ethanediol, 8) (11%). Büchi et al. [*2*] give b.p.$_{0.2}$ 44° for 4.

2-Methyl-3-phenyloxetane (12) [*9*]. An equimolar mixture of acetaldehyde and styrene was irradiated under nitrogen at 10—15° with a 350 W mercury high-pressure lamp. After 8 hrs. irradiation the unconsumed materials were distilled off to give, besides 13 g of high boiling residue, 8 g of 12, b.p.$_2$ 69—70°. Thus, from 42 g of aldehyde and 104 g of styrene there was obtained a yield of 49.4% of 12 (based on aldehyde consumed).

6-Phenyl-2,7-dioxabicyclo[3.2.0]hept-3-ene (14) [*10*]. A mixture of 112 g of furan and 42.5 g of 2 was irradiated at 5—10° in a nitrogen atmosphere for 6 hrs. The unchanged starting materials were then distilled off and a fraction boiling at 100—115°/2 mm Hg (24.5 g, 35%) was collected. Redistillation gave pure 14, m. p. 49—50°; the yield, based on 2 consumed, being 88%.

3,3,4,4-Tetramethyl-2,2-diphenyloxetane (18) [*8*]. A 200 ml benzene solution, being 0.1 molar in benzophenone (10) and 2,3-dimethyl-2-butene was cooled to 5—10° and irradiated with a 450 W mercury high-pressure lamp through Pyrex until the ketone had disappeared (one to two days). Work-up and crystallization gave 70% of 18, m.p. 123—125°.

4,4-Diphenyl-3-oxatricyclo[4.2.1.0^{2,5}]nonane (19) [*11*]. Bicyclo[2.2.1]heptene (norbornene) and benzophenone are dissolved in benzene and irradiated with a Philips HPK 125 W mercury high-pressure lamp for 40—80 hours. The adduct 19, m. p. 121°, is obtained in 80% yield. In ether solution no cycloaddition occurred but only reduction with formation of benzopinacol.

6,6-Diphenyl-2,7-dioxabicyclo[3.2.0]hept-3-ene (20) [*12*]. A solution of 4 g of 10 in 180 ml of furan (freshly distilled from triphenylphosphine) was irradiated with a Philips HPK 125 W high-pressure mercury burner at 12°. During the irradiation the solution was stirred magnetically under nitrogen. After 44 hrs. illumination the furan was evaporated when 5.2 g (94%) of 20 remained, m. p. 105—106° (from aqueous methanol).

b) Intramolecular cycloaddition leading to oxetanes

According to Srinivasan [*13*] 5-hexen-2-one (26) on vapor phase photolysis at 313 mμ and up to 139° was converted to the isomeric 1-methyl-2-oxabicyclo[2.2.0]hexane (27), albeit in low yield. An analogous reaction was reported by Morrison [*14*] who found that trans-5-hepten-2-one (28) was photoisomerized to 30% each of 1,3-dimethyl-2-oxabicyclo-[2.2.0]hexane (29) and cis-5-hepten-2-one.

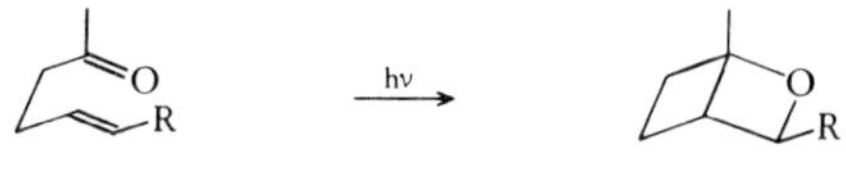

26: R = H
28: R = CH₃

27: R = H
29: R = CH₃

Closer investigations by YANG and co-workers [15] revealed that photolysis of γ,δ-unsaturated ketones eventually gave rise to both 2-oxabicyclo[2.2.0]hexanes and 5-oxabicyclo[2.1.1]hexanes (cf. the formation of **30** and **31** from 6-methyl-5-hepten-2-one illustrated below).

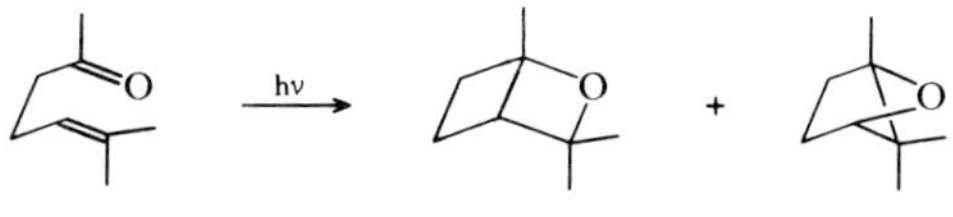

1,3,3-Trimethyl-2-oxabicyclo[2.2.0]hexane (30) and **1,6,6-trimethyl-5-oxabicyclo[2.1.1]hexane (31)** [15]. 6-Methyl-5-hepten-2-one (15 g) in 115 ml of pentane was irradiated with a Hanovia 450 W mercury lamp for 137 hours. Work-up yielded, besides 24% of non-volatile material and 7% of unreacted ketone, a 56% crop of oxetanes **30** and **31** in the ratio 2 : 5, respectively, which were separated by preparative vapor phase chromatography.

c) Cycloaddition of p-quinones to olefins leading to spirooxetanes

Irradiation of cyclooctene and p-benzoquinone gave a 1 : 1 adduct having structure **32** [16]. Similar spirooxetanes were obtained from p-benzoquinone and cyclohexene, 1,5-cyclooctadiene or bicyclo[2.2.1]heptadiene. A different reaction was, however, observed on irradiation of p-benzoquinone with 2,3-dimethyl-1,3-butadiene (cf. p. 424).

Irradiation of cyclooctene with chloranil [17] gave, depending on the proportions of the reactants, varying ratios of the adducts **33** and **34**. For some oxetanes arising through C₃O-cyclodimerization of dialkyl-p-benzoquinones see chapter 8.

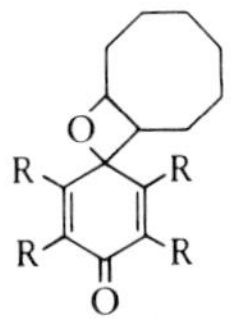

32: R = H
33: R = Cl

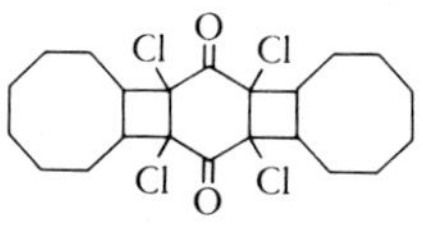

34

2,3,5,6-Tetrachlorospiro[2,5-cyclohexadiene-1,10′-[9′]-oxabicyclo[6.2.0]decan]-4-one (33) [17]. Irradiation of a solution of 300 ml of cyclooctene and 20 g of chloranil in 150 ml benzene in a borosilicate glass flask at 30° for 64 hours with a Hanovia S-500 medium-pressure mercury lamp gave 4.8 g of colorless crystalline cyclooctene chloranil 2:1 adduct (34) of m.p. 235°, and 4.4 g of 33 as an orange-yellow oil, b.p. $_{0.05}$ 189—191°. Unchanged chloranil (14 g) was recovered.

d) Cycloaddition of 1,2-dicarbonyl compounds to olefins leading to α-ketooxetanes

While the photochemical addition of olefins to 1,2-dicarbonyl compounds, leading to the formation of dihydro-1,4-dioxins (cf. A), has been known since 1944 (cf. chapter 12), the C_3O-cycloaddition which leads to the formation of α-ketooxetanes (cf. B) was discovered only recently [18] when furocoumarins and furochromones were irradiated with 9,10-phenanthrenequinone (35).

In this way, phenanthrenequinone on irradiation reacted with benzofuran or isocoumarin to give the adducts 36 and 37, respectively, and benzil gave, with visnagin (38) the oxetane 39 [18]. It has occasionally been observed that the addition of phenanthrenequinone to olefins may lead to the formation of both type A and B products; experiments describing the formation of 7,7a-dihydrospiro[indeno[2,1-b]oxete-2(2aH),9′(10′H)-phenanthren]-10′-one (40, m.p. 174°) from 35 and indene, and of 2a,8b-dihydro-2a-phenylspiro[1H,4H-oxeto[2,3-c][2]benzopyran-1,9′-(10′H)-phenanthrene]-4,10′-dione (41, m.p. 203—206°) from 35 and 3-phenylisocoumarin (42) will be found in chapter 12.

36: X = O
40: X = CH$_2$

37: R = H
41: R = C$_6$H$_5$

6-Benzoyl-5b,7a-dihydro-5-methoxy-2-methyl-6-phenyl-4H,6H-oxeto[3′,2′:4,5]-furo[3,2-g]-1-benzopyran-4-one (benzil visnagin 1:1 adduct, 39) [*18*]. Benzil (2.1 g) and 0.575 g of visnagin (**38**) were dissolved in benzene and irradiated under argon with a Philips HPK 125 W high-pressure mercury lamp, the water-cooled lamp immersion shaft consisting of Schott WG 1 filter glass. After 10 hrs. irradiation at 15—20° the solvent was removed and the residue triturated with 20 ml of ether to give 0.62 g (56%) of **39**, m.p. 227—231°. Chromatography of the ether phase over silica gel afforded 1.4 g of unconsumed benzil.

e) Formation of oxetanes bearing functional groups

When fumaronitrile (**43**) was irradiated in acetone an 1:1 adduct of the components was formed besides maleonitrile (**44**) [*19*]. The correct structure of the adduct as being 4,4-dimethyl-trans-2,3-oxetanedinitrile (**45**) was, however, established only recently [*20*]. Maleonitrile (**44**) being formed through stereoisomerization reacted in its turn with acetone to give 4,4-dimethyl-cis-2,3-oxetanedinitrile (**46**).

The formation of iminooxetanes, observed by SINGER and BARTLETT [*7*], results from the photocycloaddition of aldehydes and ketones to dimethyl-N-(2-cyano-2-propyl)-ketenimine (**47**).

References, pp. 424—425

Two types of adducts were observed; for example, 4-methoxybenzaldehyde (anisaldehyde) gave a 3-iminooxetane (cf. **49**) while benzophenone gave 2- as well as 3-iminooxetanes (cf. **48** and **49**). No adduct was obtained from **47** and 2'-acetonaphthone or from **47** and 1-naphthaldehyde [7]. The adduct **50** was photochemically and thermally (125°) cleaved to 2-isocyanato-2-methylpropionitrile and 9-isopropylidenefluorene (**25**). The latter reaction parallels the cleavage of **21** to yield **25** and acetaldehyde (comp. p. 416).

On irradiation of fluoroaldehydes, fluoroketones and fluoroacyl fluorides with fluoroolefines, HARRIS and COFFMAN [21] have been able to obtain good yields of polyfluorooxetanes. Similar fluorine containing oxetanes have since been synthesized through C_3O cycloaddition from acetaldehyde and tetrafluoroethylene [22] or from hexafluoroacetone and ethylene [23]. Table 34 gives a survey, the adducts isolated being frequently mixtures of cis and trans isomers in a 1 : 1 ratio.

4,4-Dimethyl-2,3-oxetanedinitrile (45 and 46) [20]. A solution of 90 g of **43** in 1350 ml of acetone was irradiated with a Hanovia 450 W mercury lamp with water cooling and nitrogen gas stream agitation. Removal of the solvent after 56 hrs. irradiation gave an oil consisting of 14% **43**, 12% **44**, 64% **45** and 22.5% **46**. Distillation through a spinning band column afforded, following 26.7 g of **43** + **44**, a fraction (b.p.₁ 88—91°) which crystallized from ether-hexane to give 45 g of the trans-oxetane **45**, m.p. 41—42.3°. The fraction boiling at 117—118°/1 mm Hg when crystallized from ether gave 19.7 g of cis-oxetane **46**, m.p. 59.5—60.2°.

4'-[N-(2-Cyano-2-propyl)-imino]-3',3'-dimethylspiro[fluoren-9,2'-oxetane] (50) [7]. A degassed solution of **47** and fluoren-9-one (ca. 0.3 M in each), contained in a sealed Pyrex tube, was irradiated at 7—12° with a 550 W Hanovia medium-pressure mercury lamp for 10—40 hours. By direct crystallization from the reaction mixture the adduct **50**, m. p. 134—137°, was isolated besides tetramethylsuccinonitrile, the ratio of the two products being 80 : 20.

Table 34

fluorocarbonyl cpd.	olefin	oxetane	yield	ref.
F_3C-CHO	$F_2C=CF-CF_3$	oxetane: O–C(H)(CF₃)–C(CF₃)(F)–C(F)(F) ring	(32%)	[21]
$HF_2C-(CF_2)_3-CHO$	$F_2C=CF-CF_3$	oxetane: O–C(H)((CF₂)₃CF₂H)–C(CF₃)(F)–C(F)(F) ring	(59%)	[21]
H_3C-CHO	$F_2C=CF_2$	oxetane: O–C(H)(CH₃)–C(F)(F)–C(F)(F) ring	(3%)	[22]
H_3C-CHO	$Cl_2C=CF_2$	oxetane: O–C(H)(CH₃)–C(Cl)(Cl)–C(F)(F) ring	(2%)	[22]
$ClF_2C-CO-CF_2Cl$	$F_2C=CF-CF_3$	oxetane: O–C(CF₂Cl)(CF₂Cl)–C(CF₃)(F)–C(F)(F) ring **51**	(56%)	[21]
$F_3C-CO-CF_3$	$F_2C=CF-CF_3$	oxetane: O–C(CF₃)(CF₃)–C(CF₃)(F)–C(F)(F) ring	(50%)	[21]
$F_3C-CO-CF_3$	$H_2C=CH_2$	oxetane: O–C(CF₃)(CF₃)–C(H)(H)–C(H)(H) ring		[23]
$F_3C-CO-CF_3$	$H_2C=CF_2$	two isomers: O–C(CF₃)(CF₃)–C(F)(F)–C(H)(H) + O–C(CF₃)(CF₃)–C(H)(H)–C(F)(F)		[23]
perfluorocyclobutanone (F,F / F,F / F,F / C=O)	$F_2C=CF-CF_3$	spiro-oxetane with cyclobutane ring and O–C–C(CF₃)(F)–C(F)(F)	(33%)	[21]
$F_7C_3-C(=O)-F$	$F_2C=CF-CF_3$	oxetane: O–C(F)(C₃F₇)–C(CF₃)(F)–C(F)(F) ring	(73%)	[21]

References, pp. 424–425

2,2-Bis-(chlorodifluoromethyl)-3,4,4-trifluoro-3-(trifluoromethyl)-oxetane (51) [21]. The ultraviolet radiation source consisted of a helix-shaped low-pressure mercury lamp constructed of 37 mm quartz tubing and powered by a 5000 V, 60 mA transformer. The lamp fitted around the quartz reaction tube so that its radiation impinged primarily on the liquid portion of the reaction mixture. For details of the low-temperature irradiation assembly the reader should consult the original paper. Hexafluoropropene (135 g) and 242 g of 1,3-dichlorotetrafluoroacetone were loaded in the tube. The mixture was irradiated for 12 days. Upon distillation of the reaction mixture through a spinning band still, there was obtained 176.4 g (56%) of crude 51, distilling at 105—109°. A second distillation through a small Podbielniak column gave pure 51, b.p. 107.5°.

2. Formation of oxetes as intermediates in the cycloaddition of carbonyl compounds to acetylenes

BÜCHI et al. [24] assumed the formation of oxetes (cf. 53) in the photoreaction between 5-decyne (52) and benzaldehyde (2) or acetophenone (9). The products isolated were 6-benzylidene-5-decanone (54) and 3-butyl-2-phenyl-2-octen-4-one (55), respectively, and both were regarded to be mixtures of cis and trans forms.

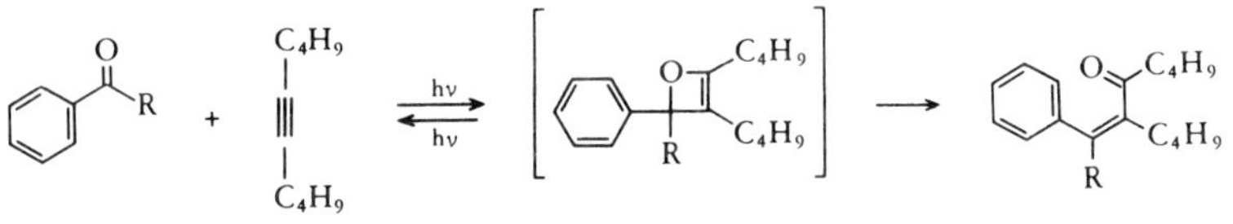

2: R = H
9: R = CH₃
52
53
54: R = H
55: R = CH₃

An oxete (56) has also been postulated as intermediate in the photoaddition of p-benzoquinone to diphenylacetylene in benzene [25] or acetonitrile [26] solution. In this case, 2-(4-oxo-2,5-cyclohexadienylidene)-2-phenylacetophenone (57) was formed in strict analogy to the 53 → 54 reaction.

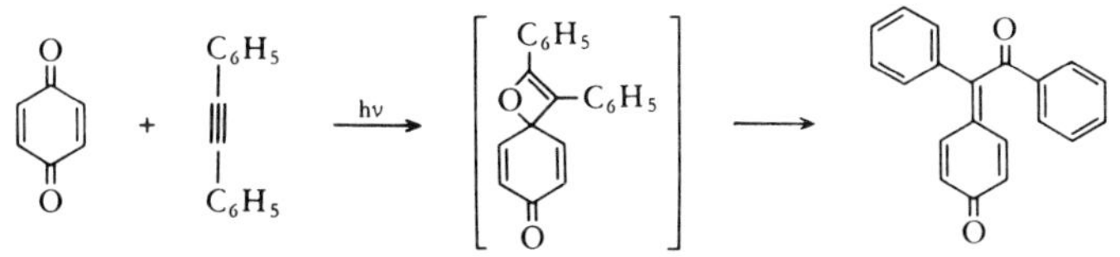

56
57

6-Benzylidene-5-decanone (54) [24]. A 3:1 mixture of 2 and 52 was irradiated with a mercury resonance arc at 40° for 96 hours. Starting materials were removed by distillation in vacuo, and chromatography of the residue gave 13% of 54, b. p.$_{0.1}$ 108—111°.

2-(4-Oxo-2,5-cyclohexadienylidene)-2-phenylacetophenone (57) [25]. A solution of 5 g of benzoquinone and 12 g of tolan in 180 ml benzene was irradiated under air in a borosilicate glass flask for 18 hrs. at 38—42° at a distance of 12 cm from a Hanovia S-500 medium-pressure mercury lamp. The product (**57**) was obtained in 60% yield (7.9 g) as yellow hexagonal plates, m.p. 104°.

3. Cycloaddition of p-benzoquinone to conjugated dienes not resulting in formation of oxetanes

On irradiation with p-benzoquinone, unconjugated dienes afforded C_3O-cycloadducts (cf. p. 418). 1,3-Dienes such as 1,3-butadiene, isoprene and 2,3-dimethyl-1,3-butadiene, on analogous treatment, showed a different behavior. Thus, illumination of cooled (0°) deoxygenated solutions of p-benzoquinone in 2,3-dimethyl-1,3-butadiene by a Hanovia medium-pressure mercury lamp in conjunction with a filter transmitting only light of wavelength > 300 mµ gave, after distillation and vapor phase chromatography, a pale yellow solid (m.p. 43—44°), which was shown to be the C_5O-cycloadduct **58** [*27*]. A cleaner product was obtained when the irradiation was carried out with light of wavelength > 410 mµ.

58   **59**

3,4-Dimethyl-1-oxaspiro[5.5]undeca-3,7,10-trien-9-one (58) [*27*]. p-Benzoquinone (1.2 g) and 2,3-dimethyl-1,3-butadiene (75 ml) were deoxygenated separately, mixed, and irradiated at 0° for 1.5 hrs. with light of wavelength above 410 mµ (for details of the apparatus and the filter solutions the original paper should be consulted). After removal of the excess of diene by distillation, and of quinone by sublimation, the dark brown product was distilled to give a yellow oil (1.1 g), b.p.$_{0.1}$ 90—140° (bath). The oil, when dissolved in 10 ml of petroleum and kept at 0° for 12 hrs., deposited 4a,5,8,8a-tetrahydro-6,7-dimethyl-1,4-naphthoquinone (**59**) (0.12 g) as yellow crystals, m.p. 114—115°. The filtrate was chromatographed on silica gel. Benzene-ether (9 : 1) eluted a fraction which on distillation gave 0.69 g (33%) of **58**.

References

[*1*] E. Paternò and G. Chieffi: Gazz. Chim. Ital. **39**, I, 341 (1909).
[*2*] G. Büchi, C. G. Inman and E. S. Lipinsky: J. Amer. chem. Soc. **76**, 4327 (1954).
[*3*] N. C. Yang: Pure Appl. Chem. **9**, 591 (1964).
[*4*] S. Searles, jr. in: Heterocyclic compounds with three- and four-membered rings, ed. by A. Weissberger, Part. 2, pp. 983—1068; esp. p. 990, 1045 (The Chemistry of Heterocyclic Compounds, Vol. **19**). New York: Interscience 1964.
[*5*] N. C. Yang, M. Nussim, M. J. Jorgenson and S. Murov: Tetrahedron Letters 1964, 3657.
[*6*] J. D. Margerum, J. N. Pitts, jr., J. G. Rutgers and S. Searles: J. Amer. chem. Soc. **81**, 1549 (1959).

[7] L. A. Singer and P. D. Bartlett: Tetrahedron Letters **1964**, 1887.
[8] D. R. Arnold, R. L. Hinman and A. H. Glick: Tetrahedron Letters **1964**, 1425.
[9] H. Sakurai, K. Shima and I. Aono: Bull. Chem. Soc. Japan **38**, 1227 (1965).
[10] S. Toki, K. Shima and H. Sakurai: Bull. Chem. Soc. Japan **38**, 760 (1965).
[11] D. Scharf and F. Korte: Tetrahedron Letters **1963**, 821.
[12] G. O. Schenck, W. Hartmann and R. Steinmetz: Chem. Ber. **96**, 498 (1963).
[13] R. Srinivasan: J. Amer. chem. Soc. **82**, 775 (1960).
[14] H. Morrison: J. Amer. chem. Soc. **87**, 932 (1965).
[15] N. C. Yang, M. Nussim and D. R. Coulson: Tetrahedron Letters **1965**, 1525.
[16] D. Bryce-Smith and A. Gilbert: Proc. Chem. Soc. **1964**, 87.
[17] D. Bryce-Smith and A. Gilbert: Tetrahedron Letters **1964**, 3471.
[18] C. H. Krauch, S. Farid and G. O. Schenck: Chem. Ber. **98**, 3102 (1965).
[19] J. Jennen: Bull. Soc. Chim. Belg. **46**, 258 (1937); C. A. **32**, 500 (1938).
[20] J. J. Beereboom and M. Schach von Wittenau: J. Org. Chem. **30**, 1231 (1965).
[21] J. F. Harris, jr. and D. D. Coffman: J. Amer. chem. Soc. **84**, 1553 (1962).
[22] E. R. Bissell and D. B. Fields: J. Org. Chem. **29**, 249 (1964).
[23] E. W. Cook and B. F. Landrum: J. Heterocycl. Chem. **2**, 327 (1965).
[24] G. Büchi, J. T. Kofron, E. Koller and D. Rosenthal: J. Amer. chem. Soc. **78**, 876 (1956).
[25] D. Bryce-Smith, G. I. Fray and A. Gilbert: Tetrahedron Letters **1964**, 2137.
[26] H. E. Zimmerman and L. Craft: Tetrahedron Letters **1964**, 2131.
[27] J. A. Barltrop and B. Hesp: J. Chem. Soc. **1965**, 5182.

Chapter 42

Photochemical formation and reactions of furans

1. Photoisomerization of quinoid compounds to furan derivatives

According to SCHULTE-FROHLINDE [1] 1,1',4,4'-tetrahydro-(2,2'-binaphthalene)-1,1',4,4'-tetrone (1) rearranges to 5-hydroxydinaphtho[1,2-b: 2',3'-d]furan-7,12-dione (2) on irradiation with light of wavelength 365 mµ. The rearrangement only takes place in hydroxylic solvents, e.g. methanol, ethanol, aqueous tetrahydrofuran, and acetic acid. In benzene, toluene, tetralin, carbon tetrachloride, and anhydrous tetrahydrofuran no photoreaction could be observed. The transformation 1 → 2 can also be effected thermally.

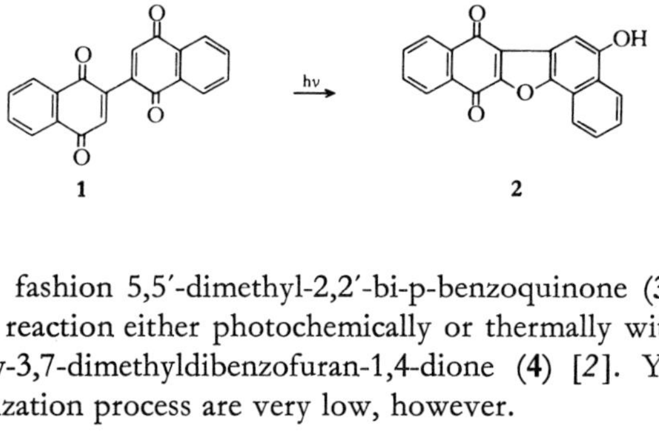

In a similar fashion 5,5'-dimethyl-2,2'-bi-p-benzoquinone (3) undergoes a cyclization reaction either photochemically or thermally with formation of 8-hydroxy-3,7-dimethyldibenzofuran-1,4-dione (4) [2]. Yields in the thermal cyclization process are very low, however.

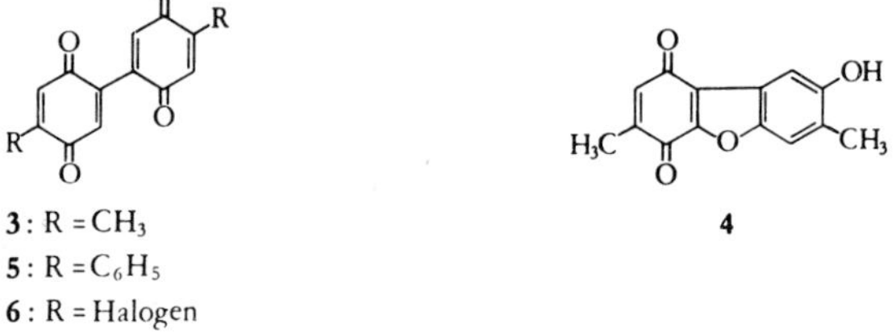

3: R = CH$_3$
5: R = C$_6$H$_5$
6: R = Halogen

The nature of the substituents appears to play a distinct role in the tendency of biquinones to undergo the 3 → 4 rearrangement. Thus the diphenyl

References, p. 433

derivative **5** is stable to irradiation, and compounds **6** (with R = Cl or Br) are stable to both heat and light.

On irradiation in hydroxyl-free solvents, the indigoid 4,4′-dimethyl-($\Delta^{2,2'(1H, 1'H)}$-binaphthalene)-1,1′-dione (**7**) rearranges to 7,12-dihydro-5,12-dimethyldinaphtho[1,2-b : 2′,3′-d]furan-7-one (**8**). This light-induced cyclization is immediately followed by a dark reaction in which 7-hydroxy-5,12-dimethyldinaphtho[1,2-b:2′,3′-d]furan (**9**) is formed by enolization [*3*].

5-Hydroxydinaphtho[1,2-b:2′,3′-d]furan-7,12-dione (2) [*1*]. 250 mg **1** are dissolved in 2.5 l methanol and irradiated for several hours by means of an immersion lamp (Heraeus, Hanau) or with a 500 W incandescent lamp. The solvent is evaporated and the residue recrystallized from aqueous tetrahydrofuran to afford red crystals of **2**, m.p. 360°; yield 90—100%.

7-Hydroxy-5,12-dimethyldinaphtho[1,2-b:2′,3′-d]furan (9) [*3*]. 50 to 100 mg **7** are dissolved in 400 ml benzene and irradiated by means of two 200 watt incandescent lamps while being cooled with water. The dark violet solution of **7**, which must be well stirred during the irradiation, slowly becomes decolorized and after about 30 minutes is light yellow. The benzene is evaporated off at 35° in vacuo. The residue is of a slightly greenish, sometimes somewhat reddish hue. Its spectrum shows that it consists of a mixture of keto and enol forms (**8** and **9**). To obtain **9** pure, the crude irradiation product is dissolved in 200 ml hot alcohol, filtered and precipitated with water, m.p. 238° (dec.).

The keto form (**8**) may be obtained in fairly pure state if dilute solutions of **7** in benzene are irradiated at 6°. When the color of the solution has faded the solvent is removed at low temperature to afford a reddish residue consisting of **8** contaminated with a little **9** (spectroscopic analysis).

2. Photochemical reactions of furans with oxygen

Systematic studies in this field were begun by SCHENCK in the early 1940's [*4*—*7*]. The scope of photosensitized reactions of dienes with oxygen will be dealt with in more detail in chapter 39.

a) Simple furan derivatives

The products of unsensitized photooxidation of furans are frequently polymeric substances which explode on heating. Thus 2-methylfuran forms a polymeric photoperoxide which detonates violently on heating.

Photosensitized oxidation of furans is of greater significance for preparative chemistry since in a series of cases compounds are obtainable which are otherwise accessible only with difficulty or not at all and which are in themselves valuable starting materials for further syntheses.

Recent work on the photosensitized oxidation of furan derivatives in solution has revealed that initially an ozonide type peroxide is formed in what may be regarded as a diene synthesis with oxygen [8—10]. The products isolated, the major part of which were hitherto unreported, were shown to have resulted from consecutive dark reactions as visualized in the scheme below, conversion to **14** resp. **15** not being possible with $R_2 = CH_3$:

a: $R_1 = R_2 = H$ b: $R_1 = CH_3$, $R_2 = H$ c: $R_1 = R_3 = CH_3$

The monomeric ozonide type peroxides (**11**) which are formed with quantum yields of about unity are stable at sufficiently low temperatures only ($< -90°$). Reduction of the primary ozonide (**11**) or of **13** leads to ketones or aldehydes (**16**) respectively.

If the photooxidation is conducted in benzene-petroleum ether mixtures at low temperatures, the solution mixed with alcohols and allowed to warm up, alkoxy-hydroperoxy-dihydrofurans (**13**) may in some cases be obtained. The same products are accessible if the photooxidation is carried out in alcoholic solution (comp. also [*11*]). A slow polymerization process was found to compete with alcoholysis. Thus, in the case of 2,5-dimethylfuran (**10c**) a dimeric product (**12c**, x = 2) could be isolated besides 2,5-dihydro-5-hydroperoxy-2-methoxy-2,5-dimethylfuran (**13c**).

A third process which can be observed with **11a, b** is a spontaneous isomerization with formation of pseudo acids (**15**). These 2-alkyl-2,5-dihydro-2-hydroxyfuran-5-ones are formed if solutions of **11** in inert solvents are allowed to warm up to $> -20°$.

References, p. 433

Final products of the photosensitized oxidation of furans as reported earlier [6] are usually pseudo esters **14** which arise from **11** either via **13** through elimination of water or via **15** through esterification. In the preparation of e.g. the pseudo ester of malealdehydic acid **14a** [7] furan may be replaced by furfural, pyromucic acid or suitable furfuryl compounds, the side chains being split off in secondary reactions. A variety of useful synthetic intermediates has thus become readily available [6, 7, 10].

2-Ethoxy-2,5-dihydrofuran-5-one (14a, $R_3 = C_2H_5$) [6]. 200 g Furfural and 2 g eosin were irradiated in 850 ml ethanol by a mercury high pressure burner HgH 1000 (Osram) in a glass immersion lamp arrangement [12] while air was bubbled through the mixture. After 24 hrs. the solvent was distilled off through a Vigreux column at a bath temperature up to 145°. The remaining dark product was fractionated using a small column. Besides 29.2 g furfural 174.4 g (76.6%) of **14a**, $R_3 = C_2H_5$, was obtained distilling as an almost colorless liquid, b.p.$_{13}$ 97—101°. The danger of explosions may be diminished by addition of V_2O_5 prior to irradiation.

2,5-Dihydro-5-hydroperoxy-2-methoxy-2-methylfuran (13b, $R_3 = CH_3$) [9]. A solution of 4.1 g **10b** in a 1:1 mixture of benzene-petroleum ether (120 ml) was irradiated with a Philips mercury lamp 125 W at —30° in the presence of dinaphtho[2,3-b:2′,3′-d]thiophene as sensitizer. 1.15 l O_2 were absorbed during 90 mins. The reaction mixture was added to 200 ml methanol and kept at room temperature for 3 hours. Evaporation of the solvent and digestion of the residue with an ether-petroleum ether mixture at —40° afforded 2.5 g. (35%) of **13b**, $R_3 = CH_3$, m.p. 79—80°. The same product is obtainable in considerably higher yield (80%) if the photooxidation is carried out in methanol solution.

4-Oxo-2-pentenal (16b) [9]. 8.2 g **10b** in 50 ml benzene and 100 ml petroleum ether were treated as above. The crude oxidation mixture was combined with a solution of 27 g triphenyl phosphine in 300 ml petroleum ether at —20°. After filtration and distillation 3.4 g of **16b** resulted, b.p.$_{13}$ 71—73°.

2,5-Dimethylfuran peroxide dimer (12c, x = 2) [9]. Irradiation of a solution (120 ml) of **10c** (4.8 g) in benzene-petroleum ether (2:1) with dinaphtho[2,3-b:2′,3′-d]thiophene as sensitizer while being flushed with oxygen resulted in absorption of 1148 ml O_2 during 55 mins. 40 ml of this solution were kept at room temperature. When after 5 hrs. the pungent odor had disappeared the solvent was removed and the residue treated with ether-petroleum ether to yield 0.6 g (32%) of colorless needles of **12c**, x = 2, m.p. 154°.

b) Fused furan systems

The transformations described above can also be expected to occur in the analogous treatment of fused furans. Thus photosensitized oxygen transfer on 1,2,3,4,6,7,8,9-octahydrodibenzofuran (**17**) in benzene-petroleum ether solution at $-20°$ and subsequent addition of methanol resulted in a 70% yield of 1,2,3,4,4a,5a,6,7,8,9-decahydro-4a-hydroperoxy-5a-methoxydibenzofuran (**18**) which must have arisen from an intermediate ozonide type peroxide [*9, 10*]. Menthofuran (**19**) when subjected to photosensitized oxidation undergoes similar reactions. Products which can with suitable methods be isolated are, 3,9-dihydro-9-hydroperoxy-3-methoxymenthofuran (**20**) [*9, 10*] and a pseudo acid, 3,9-dihydro-3-hydroxy-9-oxo-menthofuran (**21**) [*13*].

3,9-Dihydro-9-hydroperoxy-3-methoxymenthofuran (20) [*9*]. 7.5 g of **19** in 150 ml methanol absorbed 1165 ml of oxygen when illuminated with a Philips HPK 125 W mercury lamp under oxygen in the presence of 100 mg rose bengal (31 mins.). The solvent was distilled off and residual sensitizer removed by filtration of an ethereal solution through Al_2O_3. Treatment of the eluate with petroleum ether at $-30°$ afforded 7.8 g (70%) of **20**, m.p. 76°.

c) Aryl furans

The aryl substituted furans seem to differ in their behavior on photooxidation from simpler furans since the products eventually isolated do not parallel those in constitution which may be obtained with simpler members. More detailed investigations, however, reveal that in all cases examined ozonide type peroxides have to be postulated as primary reaction products. The different products actually observed arise then in consecutive reactions the course of which depends on the nature of the solvent.

DUFRAISSE and ECARY [*14*] succeeded in isolating such a primary peroxide after exposure of 1,3-diphenylisobenzofuran (**22**) in dilute carbon

disulfide solution to sunlight. The resultant 1,3-epidioxy-1,3-dihydro-1,3-diphenylisobenzofuran (**23**) which explodes at about 20° readily decomposes in solution yielding 1,2-dibenzoylbenzene (**24**) which had formerly been regarded as the actual reaction product of the photooxidation [15]. An autoxidation process was later shown to compete with the photooxidation leading to a polymeric peroxide of different properties [16].

The solvent dependence of the product ratio in photosensitized oxidations of furans follows also from results obtained by WASSERMAN and LIBERLES [17]. Thus when a methanolic solution of tetraphenylfuran (25) was irradiated and exposed to air in the presence of methylene blue main products were, cis-dibenzoylstilbene (26) and 2,5-dihydro-2,5-dimethoxy-2,3,4,5-tetraphenylfuran (27). In acetone as solvent, however, cis-2,3-epoxy-1,2,3,4-tetraphenyl-1,4-butanedione (28) and benzoyl-benzoyloxystilbene (29) were formed in yields of 45% and 20% respectively.

Similar results were reported by LUTZ et al. [18] who on methylene blue sensitized photooxidation of 25 in acetone succeeded in isolating not only 28 and 29 but also 2,3:4,5-diepoxytetrahydro-2,3,4,5-tetraphenylfuran (30). The formation of this and other products is accounted for on the basis of the intermediary formation of an ozonide type peroxide. The isolation of one representative of this class of compounds viz. 3,4-bis-(p-bromophenyl)-2,5-epidioxy-2,5-dihydro-2,5-diphenylfuran (31) was reported, and plausible mechanisms have been offered by LUTZ [18] to account for the formation of products not encountered in photosensitized oxidation of simple furan derivatives.

1,3-Epidioxy-1,3-dihydro-1,3-diphenylisobenzofuran (23) [14]. A solution of 22 (2 g) in 1 l carbon disulfide is under agitation exposed to bright sunshine during 70 seconds. On evaporation of the solvent at reduced pressure and below 0° crystals of 23 begin to separate. The pure peroxide explodes near 18°, but impure preparations have decomposition points up to 50°. Gases evolved on decomposition consist of carbon dioxide with traces of oxygen.

d) [2.2](2,5)Furanophane

In an attempt to synthesize a macrocyclic tetraketone, [2.2](2,5)furanophane (**32**) was exposed to air in an ethanolic solution containing methylene blue and irradiated with a 150 W floodlamp. The main product (42 % yield) was, however the tetracyclic diketone 13-oxatetracyclo[7.3.1^{1,4}.0.0^{4,8}]tridec-2-ene-7,10-dione (**34**). The reaction appears to take place through intermediate **33**, which incorporates diene and dienophil in a particularly favorable geometry for an intramolecular DIELS-ALDER reaction [*19*].

3. Photoaddition of methanol to a furan derivative

The diterpene furan derivatives cafestol and kahweol (**35**) are main constituents of the unsaponifiable matter in the coffee-bean oil. The latter substance, which is 1-dehydrocafestol, is very sensitive to light. KAUFMANN [*20*] found that **36** was converted to 3,19-dihydro-3,19-dimethoxy-

cafestol 17-acetate (**39**) by the action of light if aqueous methanol was used as a solvent. Irradiation in anhydrous methanol, however, left **36** unaltered.

References, p. 433

The above reaction scheme was suggested which explains why **39** can be obtained from **36** only by using aqueous methanol. Oxygen did not affect the reaction.

After exposure of a solution of **36** in aqueous methanol to diffuse daylight during 4 weeks minor quantities of **38** could be isolated by chromatography. The ketoaldehyde underwent a slow cyclization reaction in methanol with formation of **39**.

Irradiation of **36** in ethyl acetate solution followed by ozonolysis at —50° produced a ketoacid which is believed to derive from **37** [21].

3,19-Dihydro-3,19-dimethoxy-cafestol 17-monoacetate (39) [20]. A solution of 20 mg kahweol monoacetate (**36**) in 5 ml methanol (from which water had not been removed) was irradiated for 1 hour by a mercury vapor lamp at 20°. The solution was evaporated in vacuo, the resulting oily residue dissolved in 20 ml benzene and the solution chromatographed on 2 g silica gel. **39** (12 mg) was eluted with benzene/ether (3:1) and after twofold recrystallization from ether/petroleum ether, melted at 178—182°. The action of direct sunlight for 30 hours (3 days) on 1 g kahweol monoacetate in 100 ml methanol, with subsequent purification as above, yielded 800 mg crystalline **39**, m.p. 180° (from ether).

References

[1] D. SCHULTE-FROHLINDE and V. WERNER: Chem. Ber. **94**, 2726 (1961).
[2] A. J. SHAND and R. H. THOMSON: Tetrahedron **19**, 1919 (1963).
[3] D. SCHULTE-FROHLINDE and F. ERHARDT: Chem. Ber. **93**, 2880 (1960).
[4] G. O. SCHENCK, in: FIAT review of German Science 1939—1946. Preparative Organic Chemistry (Part II, senior author K. ZIEGLER), Vol. 37, 167—208 (1953).
[5] G. O. SCHENCK: Angew. Chem. **64**, 12 (1952).
[6] G. O. SCHENCK: Liebigs Ann. Chem. **584**, 156 (1953).
[7] S. H. SCHROETER, R. APPEL, R. BRAMMER and G. O. SCHENCK: Liebigs Ann. Chem. **697**, 42 (1966).
[8] E. KOCH and G. O. SCHENCK: Chem. Ber. **99**, 1984 (1966).
[9] K.-H. SCHULTE-ELTE: Diss. Universität Göttingen 1961.
[10] C. S. FOOTE, M. T. WUESTHOFF, S. WEXLER, I. G. BURSTAIN, R. DENNY, G. O. SCHENCK and K.-H. SCHULTE-ELTE: Tetrahedron **23**, 2583 (1967).
[11] C. S. FOOTE and S. WEXLER: J. Amer. chem. Soc. **86**, 3879 (1964).
[12] G. O. SCHENCK, H. EGGERT and W. DENK: Liebigs Ann. Chem. **584**, 177 (1953).
[13] G. O. SCHENCK and C. FOOTE: Angew. Chem. **70**, 505 (1958).
[14] C. DUFRAISSE and S. ECARY: Comptes rendus **223**, 735 (1946).
[15] A. GUYOT and J. CATEL: Bull. Soc. Chim. France [3] **35**, 1124 (1906).
[16] A. LE BERRE and R. RATSIMBAZAFY: Bull. Soc. Chim. France **1963**, 229.
[17] H. H. WASSERMAN and A. LIBERLES: J. Amer. chem. Soc. **82**, 2086 (1960).
[18] R. E. LUTZ, W. J. WELSTEAD, JR., R. G. BASS and J. I. DALE: J. Org. Chem. **27**, 1111 (1962).
[19] H. H. WASSERMAN and A. R. DOUMAUX: J. Amer. chem. Soc. **84**, 4611 (1962).
[20] H. P. KAUFMANN and A. K. SEN GUPTA: Chem. Ber. **96**, 2489 (1963).
[21] H. P. KAUFMANN and A. K. SEN GUPTA: Chem. Ber. **97**, 2652 (1964).

Chapter 43

Photochemical formation and transformations of organic sulfur compounds

Of the many photochemical reactions of organic sulfur compounds which have become known in the past years, only a limited number will be discussed in the following text. For a more complete survey of the literature on organic sulfur compounds, the reader is referred to the books and reviews of, inter alia, KHARASCH [1], MUSTAFA [2], SOSNOVSKY [3] and GILBERT [4].

1. Photochemical syntheses using SO_2, SO_2Cl_2 and SCl_2

a) Sulfochlorination

The conversion of aliphatic hydrocarbons to the corresponding alkyl sulfonyl chlorides is a process of considerable industrial importance. Sulfochlorination may be carried out using sulfur dioxide/chlorine mixtures or sulfuryl chloride.

Thus by the simultaneous reaction of chlorine and sulfur dioxide on paraffin hydrocarbons it is possible to obtain paraffin sulfonyl chlorides — a process which has become known as the REED reaction [5, 6]. The sulfochlorination may be carried out in diffuse daylight but it is only under the action of UV-light that the sulfochlorination excels the competing chlorination [7].

The sulfochlorination proceeds according to the overall equation [8]:

$$RH + SO_2 + Cl_2 \xrightarrow{h\nu} RSO_2Cl + HCl$$

SCHUMACHER and STAUFF [9] suggested the following free radical mechanism:

$$Cl_2 \xrightarrow{h\nu} 2\,Cl\cdot$$
$$Cl\cdot + RH \longrightarrow R\cdot + HCl$$
$$\begin{cases} R\cdot + SO_2 \longrightarrow RSO_2\cdot \\ R\cdot + Cl_2 \longrightarrow RCl + Cl\cdot \end{cases}$$
$$RSO_2\cdot + Cl_2 \longrightarrow RSO_2Cl + Cl\cdot$$

References, pp. 451—452

This scheme explains the formation of alkyl chlorides which are always found amongst the reaction products; further, it becomes understandable that the sulfochlorination may be initiated not only by short-wavelength light but also by certain chemical substances serving as radical sources, e.g. peroxides and tetraethyl lead.

This method which was reviewed some years ago by ECKOLDT [10] holds less interest for the preparative chemist since it is often difficult to obtain a homogeneous product. Thus ASINGER et al. [11] reported that by the UV-light induced action of chlorine and sulfur dioxide on propane in carbon tetrachloride solution, they obtained a mixture of the isomeric 1- and 2-propanesulfonyl chlorides, and these were not separable into the two pure components even by delicate fractionation. Working with n-butane, ASINGER [12] had a similar experience; he obtained a mixture of isomeric butanemonosulfonyl chlorides which could not be separated by rectification. His experiments also yielded a mixture of butanedisulfonyl chlorides from which butane-1,4-disulfonyl chloride could be obtained by fractional crystallisation.

As with all other substitution reactions of higher paraffin hydrocarbons, so in the sulfochlorination, with progressing reaction, di- and polysubstitution occurs with increasing significance [13].

The products of disulfochlorination of propane, n-butane, isobutane and n-dodecane contain neither geminal- nor vicinal-disubstituted compounds. The first sulfochloride group in the molecule must therefore hinder a second substitution on the same or the adjacent carbon atom.

Sulfochlorination of the hydrochlorides of n-propyl-, isobutyl-, and n-butylamines using chlorine and sulfur dioxide leads to the formation of the corresponding aminoalkanesulfonyl chloride hydrochlorides:

$$HCl \cdot H_2N - C_nH_{2n+1} \xrightarrow[SO_2 + Cl_2]{h\nu} HCl \cdot H_2N - C_nH_{2n} - SO_2Cl + HCl$$

Thus with ethylamine hydrochloride, substitution of hydrogen by the SO_2Cl group leads to 2-aminoethanesulfonyl chloride hydrochloride since 2-aminoethanesulfonic acid (taurine) is found on hydrolysis of the reaction mixture [14].

The photochemical sulfochlorination of saturated hydrocarbons may also be carried out with the aid of sulfuryl chloride, the reaction requiring the presence of catalytic amounts of bases. Thus reaction of n-heptane, tert.-butylbenzene, cyclohexane, and methylcyclohexane with SO_2Cl_2 in the presence of pyridine, quinoline or piperidine gives a mixture of the chlorinated derivative and the sulfonyl chloride [15]. For example, irradiation of cyclohexane/SO_2Cl_2 mixtures in benzene solution containing some 2-mercaptothiazole as catalyst yielded 28.6% of chlorocyclohexane and 34.0% of cyclohexanesulfonyl chloride; with pyridine as catalyst the ratio was 9.4% : 54.8% respectively.

KHARASCH et al. [*16*] suggested the following mechanism for the photochemical sulfochlorination in the presence of pyridine:

$$SO_2Cl_2 \xrightarrow{pyridine} SO_2 + Cl_2$$
$$Cl_2 \xrightarrow{h\nu} 2\ Cl\cdot$$
$$Cl\cdot + RH \longrightarrow R\cdot + HCl$$
$$R\cdot + SO_2 \longrightarrow RSO_2\cdot$$
$$RSO_2\cdot + Cl_2 \longrightarrow RSO_2Cl + Cl\cdot$$

SCHUMACHER and STAUFF [*9*] came to a somewhat different conclusion; the reader is referred to the original work.

Sulfuryl chloride also reacts with organosilicon compounds if irradiated with UV-light in the presence of trace amounts of pyridine. The corresponding sulfonyl chlorides are obtained, viz. (trimethylsilyl)methanesulfonyl chloride from tetramethylsilane [*17*].

KHARASCH and co-workers [*16*] have shown that the photochemical sulfochlorination of lower aliphatic acids leads to substitution in the β- or β- and γ-positions. In higher aliphatic acids substitution is random, but in no case were α-derivatives observed. On the other hand α-chlorinated acids were formed simultaneously.

Acetic acid cannot be sulfochlorinated, while propionic acid yields **1** which is the cyclic anhydride of 3-sulfopropionic acid. It is a crystalline compound which, like other compounds of analogous constitution, is very reactive towards water, alcohols, and amines forming the free sulfoacids, esters and amides:

$$H_3C-CH_2-COOH + SO_2Cl_2 \xrightarrow{h\nu} HCl + \left[\begin{array}{c} H_2C-CH_2-COOH \\ | \\ SO_2Cl \end{array}\right] \xrightarrow{-HCl} O_2S\overset{\frown}{\underset{O}{}}\!=\!O$$
$$\mathbf{1}$$

The technique of photochemical sulfochlorination [*10*]. It is customary to use a mercury vapor lamp, either an immersion lamp or one operating outside the reaction vessel; however, the latter arrangement is only possible if the reaction vessel is made of quartz or of UV-glass. Light of wavelengths between 300 and 360 mμ is especially favorable for the sulfochlorination process.

As a rule it has not been found practical to carry out the sulfochlorination in the gas phase; it is better to work in the liquid phase, whereby the reactants are led into an inert solvent, e. g. carbon tetrachloride, and the solution is stirred and irradiated at a temperature of 10—20°. In this way intimate mixing of the gases is ensured and the reaction products are separated without using lower temperatures.

Sulfochlorination of liquid hydrocarbons may conveniently be carried out in a vertical quartz tube containing a thermometer and a cooling coil. The sulfur dioxide and chlorine in a molar ratio of 1.1 : 1 are led in through a frit at the bottom of the tube while the exit gases pass through a reflux condenser to which the quartz tube is attached and thence into the fume chamber. Provided the diameter of the quartz tube is not greater than about 60 mm, mechanical stirring is not necessary since the bubbles of gas cause sufficient disturbance of the reaction mixture. A mercury vapor lamp 25 cm distant from the reaction vessel is used as the light source.

References, pp. 451—452

Propanesulfonyl chlorides [*11*]. 1000 g of a solution was prepared by passing 2.5 parts by volume of propane, 1.1 part by vol. sulfur dioxide and 1 part by volume of chlorine into carbon tetrachloride while irradiating with a quartz mercury lamp. The reaction was continued until an approximately 20% solution of the reaction products was obtained. The solvent was removed and the liquid reaction products separated from 15% residue (propane-1,3-disulfonyl chloride, m.p. 48°) by distillation. The fraction boiling between 70—150° at 15 mm was rectified using a 50 cm Raschig column to yield a mixture of the isomeric 1- and 2-propanesulfonyl chlorides, b.p.$_{15}$ 72—79°; yield 88%. The residue from this rectification consisted of sulfonyl chlorides of chloropropane and dichloropropane.

(Trimethylsilyl)methanesulfonyl chloride [*17*]. 88 g (1.0 mole) of tetramethylsilane, 67.5 g (0.5 mole) of sulfuryl chloride and 0.5 ml of pyridine were placed in a Pyrex flask equipped with a water-cooled reflux condenser surmounted by a reflux condenser cooled with Dry Ice. The flask was irradiated by means of a 1500 W tungsten filament lamp 25 cm away until evolution of hydrogen chloride ceased. The mixture was then fractionated at reduced pressure to yield 49.5 g (53% bsaed on SO$_2$Cl$_2$) of trimethylsilylmethanesulfonyl chloride, b.p.$_1$ 57°. In addition some monochlorinated material and approximately 0.5 mole of unreacted tetramethylsilane were obtained.

3-Sulfopropionic acid anhydride (1) [*16*]. A mixture of 74 g (1.0 mole) anhydrous propionic acid and 185.6 g (1.375 mole) sulfuryl chloride were irradiated with a 200—300 W tungsten lamp at 50—60° in a round-bottomed flask fitted with reflux condenser. A colorless precipitate was formed during the irradiation. Although the reaction appeared to be over after 5 hours, the irradiation was continued for a few hours more. Dissolved gases and unreacted sulfuryl chloride were removed in vacuo and 200 ml ligroin-benzene (80 : 20) was added to complete the precipitation and facilitate the filtration which was carried out under exclusion of moisture. The solid reaction product was washed with small quantities of dry ligroin and then dried over fused NaOH, P$_2$O$_5$ and paraffin wax in vacuo. The yield of **1** was 58 g, m.p. 76—77°. The ligroin-benzene filtrate contained 40 g of 2-chloropropionic acid.

b) Sulfenylchlorination

MÜLLER [*18*] has investigated the light-induced reaction of sulfur dichloride with cycloalkanes such as cyclopentane, cyclohexane and cycloheptane, and he has established that sulfenyl chlorides (cf. **2**) are formed. This reaction was termed sulfenylchlorination.

$$RH + SCl_2 \xrightarrow{h\nu} R-SCl + HCl$$
$$\mathbf{2}$$

The sulfenylchlorination cannot be brought to completion according to the above equation since the sulfenyl chlorides formed undergo further photochemical reactions. Thus it is possible to obtain dicyclohexyl disulfide directly from sulfur dichloride and cyclohexane by prolonged irradiation:

$$3\ C_6H_{12} + 2\ SCl_2 \xrightarrow{h\nu} C_6H_{11}-S-S-C_6H_{11} + C_6H_{11}Cl + 3\ HCl$$

Acyclic aliphatic hydrocarbons, e.g. n-heptane, if irradiated together with SCl$_2$ yield mixtures of isomeric sulfenyl chlorides.

Sulfenylchlorination procedure [*18*]. The light source used was a mercury high pressure lamp S 81 (Quarzlampengesellschaft Hanau) although the reaction may also be conducted using visible light. After irradiation the excess reactants were distilled from the reaction mixture at the lowest possible temperature in vacuo. The residues are yellow, evil smelling and easily decomposing liquids, essentially composed of sulfenyl chlorides (cf. 2). Since these compounds are difficult to identify as such they may be treated inter alia with potassium cyanide to give cycloalkyl thiocyanates. The yields (calculated on the basis of cyclopentane, cyclohexane and cycloheptane consumed) were 55, 58, and 56% respectively. For preparative details the reader is referred to the original literature.

c) Sulfoxidation

According to DAINTON and IVIN [*19*] sulfinic acids are formed by the photochemical action (UV-light) of sulfur dioxide on aliphatic hydrocarbons. Reaction occurs most readily with long chain paraffins, tertiary hydrogen being attacked more easily than secondary and secondary more easily than primary.

$$RH + SO_2 \xrightarrow{h\nu} R\text{-}SO_2H$$

This reaction does not yet appear to have been of preparative significance. More attention has been paid to photochemical sulfoxidation [*20*]:

$$RH + SO_2 + {}^1/_2 O_2 \xrightarrow{h\nu} R\text{-}SO_3H$$

ASINGER et al. [*21*] have studied the sulfoxidation of higher paraffin hydrocarbons in UV-light. A mixture is produced of all the theoretically possible isomeric secondary sulfonic acids. Primary sulfonic acids are obtained in lower yields since the reaction rate for hydrogen abstraction is lower for primary than for secondary hydrogen atoms.

It is noted in a patent [*22*] that pure cyclohexanesulfonic acid can be obtained photochemically from cyclohexane, sulfur dioxide and oxygen. The reaction involves the intermediary formation of cyclohexanesulfonic peracid which cannot be isolated. Working with acetic acid or acetic anhydride as a solvent for the sulfoxidation process, GRAF [*20*] obtained cyclohexanesulfonyl acetyl peroxide ($C_6H_{11}\text{-}SO_2\text{-}O\text{-}O\text{-}CO\text{-}CH_3$).

Cyclohexanesulfonic acid [*22*]. The apparatus consists of a tall narrow cyclinder of Jena glass, fitted at the bottom with a glass frit through which the gases are led, and in the upper part with an inbuilt cooling coil. The top is connected with a reflux condenser. In this apparatus, 600 parts by weight of cyclohexane are treated with a gaseous mixture consisting of 2 moles of sulfur dioxide and 1 mole of oxygen. The reaction vessel is irradiated by means of a mercury quartz lamp and the temperature within the vessel maintained between 15—25° by cooling. After a short time the contents of the apparatus become cloudy and soon afterwards an oil begins to collect in rapidly increasing amount on the bottom of the vessel. The precipitated oil may be continuously removed and replaced by fresh cyclohexane without interrupting the process. Passing in the mixed gases at a rate of 21 litres per hour, about 65 g oil is obtained in an hour, corresponding to an 80—90% consumption of the gas.

References, pp. 451—452

The precipitated oil is principally cyclohexanesulfonic acid with some sulfuric acid; in moist air a hydrate of cyclohexanesulfonic acid crystallizes out which may be obtained pure by filtration. The main product may be worked up by adding the oil to water, removing cyclohexane, sulfur dioxide and volatile decomposition products by passing steam through the mixture and neutralizing with caustic soda. On evaporating to dryness a powder is obtained which consists of the sodium salt of cyclohexanesulfonic acid (about 90%) and sodium sulfate (10%). The sodium cyclohexanesulfonate may be obtained in shining leaflets by recrystallization from water.

A similar result is obtained if the reaction is carried out in apparatus made of UV-glass or quartz. In this case the reaction may be brought about, albeit more slowly, by means of an incandescent lamp. The sulfoxidation of cyclohexane may also be performed as described above but using the irradiation merely to initiate the reaction and switching off the light source after the oil begins to precipitate. The reaction continues in the dark and affords practically the same yields as when the irradiation is continued throughout.

d) Cyclic sulfates from sulfur dioxide and o-quinones

Schenck and Schmidt-Thomée [23] found that a series of o-quinones (cf. table 35) reacted with SO_2 in sunlight or under the action of UV-light to form cyclic sulfates (cf. 3). No reactions take place in the dark.

3

Table 35

tetrachloro-o-benzoquinone	2-nitro-9,10-phenanthrenequinone
1,2-naphthoquinone	3-nitro-9,10-phenanthrenequinone
3-nitro-1,2-naphthoquinone	4-nitro-9,10-phenanthrenequinone
9,10-phenanthrenequinone	5,6-chrysenedione

9,10-Sulfonyldioxyphenanthrene (phenanthro[9,10-d]-1,3,2-dioxathiole 2,2-dioxide, 3)[23]. 25 g powdered phenanthrenequinone is suspended in 1.2 l absolute benzene saturated with SO_2 and irradiated with a mercury lamp Osram HgH 2000 for 80 hrs. with magnetic stirring. The bulk of the cyclosulfate precipitates during the reaction as almost white needles which are sucked dry and washed with ether (22—23 g). The filtrate is concentrated to 150 ml by distillation; on cooling, a further quantity (6—7 g) of yellowish cyclosulfate is precipitated. This product is filtered off, washed with benzene and ether, powdered and boiled with 100 ml 40% sodium bisulfite solution in order to remove phenanthrenequinone. After filtration the product is washed with water, methanol, and ether. Recrystallized from dioxane the sulfate 3 has m.p. 202—203° (decomp.); yield 28—30 g (86—92%).

2. Photochemical formation and transformations of sulfides

a) Thioethers from di- and trisulfides

Brandt, Eméléus and Haszeldine [24] irradiated bis-(trifluoromethyl) disulfide (4) with UV-light and observed the formation of bis-(trifluoromethyl) sulfide (**5**). They discussed whether the disulfide had the structure **6** which would make the loss of sulfur readily understandable. However X-ray photographic studies and studies of the infrared spectra proved that the disulfide is constituted as in structure **4** [*25*]. It is necessary to carry out the photochemical conversion of **4** → **5** in quartz vessels since **4** is not changed if irradiated in Pyrex apparatus.

$$F_3C-S-S-CF_3 \xrightarrow{UV} F_3C-S-CF_3 + S \qquad F_3C-S-CF_3$$
$$\downarrow$$
$$S$$

$$\textbf{4} \qquad\qquad\qquad \textbf{5} \qquad\qquad \textbf{6}$$

The formation of **5** must be explained in terms of a photolysis of the disulfide **4** with formation of free thiyl radicals (cf. **7**) [*24*].

$$F_3C-S-S-CF_3 \xrightarrow{h\nu} 2\ F_3C-S\cdot \quad (7)$$
$$F_3C-S\cdot + F_3C-S-S-CF_3 \longrightarrow F_3C-S-CF_3 + F_3C-S-S\cdot$$
$$F_3C-S-S\cdot \longrightarrow F_3C-S\cdot + S$$
$$2\ F_3C-S-S\cdot \longrightarrow F_3C-S-S-S-S-CF_3$$
$$F_3C-S\cdot + F_3C-S-S-S-S-CF_3 \longrightarrow F_3C-S-CF_3 + F_3C-S-S-S-S\cdot$$

HASZELDINE and KIDD [*26*] subjected a quartz tube containing bis-(trifluoromethyl) trisulfide (**8**) to UV-radiation and observed the following reaction:

$$2\ F_3C-S-S-S-CF_3 \xrightarrow{h\nu} F_3C-S-CF_3 + F_3C-S-S-CF_3$$
$$\textbf{8} \qquad\qquad\qquad \textbf{5} \qquad\qquad \textbf{4}$$

Bis-(trifluoromethyl) disulfide (4) and bis-(trifluoromethyl) sulfide (5) [*26*]. The trisulfide (**8**) (1.07 g) was irradiated with UV-light for 17 days in a sealed quartz tube; a viscous oil formed which slowly deposited crystals of sulfur (66%). Fractionation of the volatile reaction products yielded the monosulfide **5** (49%) and the disulfide **4** (34%). The unreacted disulfide was sealed in a quartz tube (capacity 10 ml) and irradiated again for 5 days to yield further **5**. Total yield of the monosulfide 70%.

Bis-(trifluoromethyl) sulfide (5) [*24*]. Bis-(trifluoromethyl) disulfide (**4**) (1.140 g) was sealed under dry N_2 in a quartz tube (100 ml) and irradiated for 13 days with the light from a Hanovia lamp, situated 10 cm from the reaction tube. During the course of the irradiation, an oil formed on the walls of the reaction vessel and on further irradiation crystals appeared (sulfur). The volatile products were subjected to vacuum distillation which yielded (besides 0.346 g **4**) bis-(trifluoromethyl) sulfide (**5**) of b.p. —22.2°; yield 0.449 g.

By repeating the irradiation process on the regained disulfide (0.364 g), further quantities of **5** were obtained, total yield 0.635 g (66%).

b) Photolysis of 9,9'-bis-(phenylthio)-9,9'-bifluorene

If 9,9'-bis-(phenylthio)-9,9'-bifluorene (**9**) is dissolved in benzene and irradiated with UV-light, decomposition occurs with formation of $\Delta^{9,9'}$-bifluorene (**10**) [*27*].

References, pp. 451—451

Photochemical formation and transformations of sulfides

$\Delta^{9,9'}$-**Bifluorene (10)** [27]. The colorless solution of **9** (1.1 g) in benzene (200 ml) was placed in a quartz vessel under exclusion of air and subjected to the radiation from an externally placed mercury quartz lamp. After only a few minutes the solution assumed an orange-red color and by the end of the irradiation (11 hrs.) the liquid was intensely red. During the irradiation the temperature did not rise above 50°. The benzene was removed under reduced pressure and the residue, a red oil, boiled with approx. 75% alcohol. The insoluble residue was pressed dry on a clay tile and dissolved in a little chloroform. The solution was filtered and treated with alcohol which caused red crystals to separate. By repeating this treatment **10** was obtained as red crystals (m.p. 186°).

The decomposition of **9** may also be accomplished thermally, e.g. in boiling ethyl benzoate ($2^1/_2$ hours).

c) Insertion reaction of mercury with disulfides

If bis-(trifluoromethyl) disulfide (**4**) in the presence of mercury is subjected to radiation from a UV-lamp, colorless bis-(trifluoromethylthio)-mercury (**11**) [24] is formed.

$$F_3C-S-S-CF_3 + Hg \xrightarrow{h\nu} F_3C-S-Hg-S-CF_3$$
$$\qquad 4 \qquad\qquad\qquad\qquad\qquad 11$$

A similar insertion reaction may also be carried out with bis-(heptafluoropropyl) disulfide (**12**) [28].

Bis-(trifluoromethylthio)-mercury (11) [24]. Bis-(trifluoromethyl) disulfide (**4**) (0.892 g) and mercury (30 g) were shaken and irradiated (4 days) in a sealed quartz tube (capacity 100 ml). A Hanovia lamp (without filter) acted as the radiant source and was placed 10 cm from the reaction tube. The walls of the tube became covered with a black substance (mercuric sulfide). The solid reaction product was taken up into ether, the ethereal solution dried (Na_2SO_4) and evaporated at room temperature. The solid residue was purified by sublimation at 65° (normal pressure) yielding 0.931 g (53%) bis-(trifluoromethylthio)-mercury, m.p. 37.5°.

d) Conversion of a disulfide to a sulfenyl chloride

KOBER [29] has carried out the photochlorination of bis-(heptafluoropropyl) disulfide (**12**) and obtained heptafluoro-1-propanesulfenyl chloride (**13**).

$$C_3F_7-S-S-C_3F_7 + Cl_2 \xrightarrow{h\nu} 2\ C_3F_7SCl$$
$$\qquad 12 \qquad\qquad\qquad\qquad 13$$

Heptafluoro-1-propanesulfenyl chloride (13) [29]. The apparatus consisted of a flask (2000 ml) connected with a chlorine cylinder and (through a condenser) with a Dry Ice cooled trap. Bis-(heptafluoropropyl) disulfide (**12**) was placed in the flask and irradiated with ultraviolet light, while chlorine was passed over the disulfide **12** in intervals or continuously. At the end of the reaction the contents of the flask and the Dry Ice trap

e) Action of mercaptoacetic acid on benzo[a]pyrene

Conway and Tarbell [30] photochemically reacted benzo[a]pyrene (14) with mercaptoacetic acid (15) and obtained benzo[a]pyrene-6-acetic acid (17), but only when the reagents were irradiated with UV-light in quartz vessels; in a Pyrex apparatus no reaction occurred.

It is very probable that 16 is formed initially. This assumption is supported by the observation that the methyl ester of 16 in hexane transforms to the methyl ester of 17 in UV-light. The combined action of thioacetic acid and oxygen on 14 (without irradiation) is reported to result in formation of fairly stable benzo[a]pyrene-6-thiol acetate via thiyl radicals [31].

Benzo[a]pyrene-6-acetic acid (17) [30]. Benzo[a]pyrene (14) (1 g) was dissolved in 20 ml of mercaptoacetic acid (15) by heating to 90°. The solution was irradiated with a Hanovia C 2055 quartz mercury arc in a quartz vessel (N_2 atmosphere). The irradiation (10 hours) was conducted so that the temperature of the solution was maintained at 90—95° by the heat of the lamp. When the irradiation was finished, 50 ml water were added and the mixture allowed to stand overnight. The yellow precipitate was filtered off and then dried in vacuo (1.08 g). Extraction with two 25 ml portions of boiling benzene left a yellow residue of 17 (0.23 g), which decomposed slowly above 205°. The benzene solution yielded 0.57 g of benzo[a]pyrene.

f) Photolysis of a dixanthate

Photolysis of S-acyl xanthates results in formation of acyl radicals which in turn may undergo a decarbonylation process (cf. p. 223). Since primary alkanecarbonyl radicals thus produced are stable at room temperature the synthesis of α-diketones from suitable precursors was possible. Thus 1,2-cyclopentanedione (19) was obtained on photolysis of O,O-diethyl S,S-glutaryldixanthate (18) [32].

1,2-Cyclopentanedione (19) [32]. 170 g of 18 in 750 ml benzene was irradiated at 40° with a mercury arc lamp for 24 hrs. during which time the yellow color faded. Treatment

References, pp. 451—452

of the product in ethanol with 2,4-dinitrophenylhydrazine reagent gave 1,2-cyclopentanedione bis-2,4-dinitrophenylhydrazone (55%).

3. Photochemical thiocyanation

Our knowledge of the photochemical thiocyanation of organic compounds by means of thiocyanogen or thiocyanogen chloride is relatively small (cf. BACON [33]).

BACON et al. [34] suggested that the photoinitiated reactions of thiocyanogen (20) and thiocyanogen chloride (21) involve the thiocyanate radical and that its relatively easy formation is associated with resonance stabilisation of SCN radicals through the cyano group:

$$N\equiv C-S-S-C\equiv N \ (20) \xrightarrow{h\nu} \left[N\equiv C-S\cdot \longleftrightarrow \cdot N=C=S\right]$$

$$Cl-S-C\equiv N \ (21) \xrightarrow{h\nu} Cl\cdot + \cdot S-C\equiv N$$

FREDERIKSEN and LIISBERG [35] have found that cholesterol (22) can be thiocyanated in the 7-position by reaction with free thiocyanogen under irradiation with UV-light. No reaction is observed in darkness or in diffuse daylight. This substitution reaction is reminiscent of the UV-light induced bromination of cholesteryl esters by free bromine in which case analogously the 7-bromo compound is formed [36].

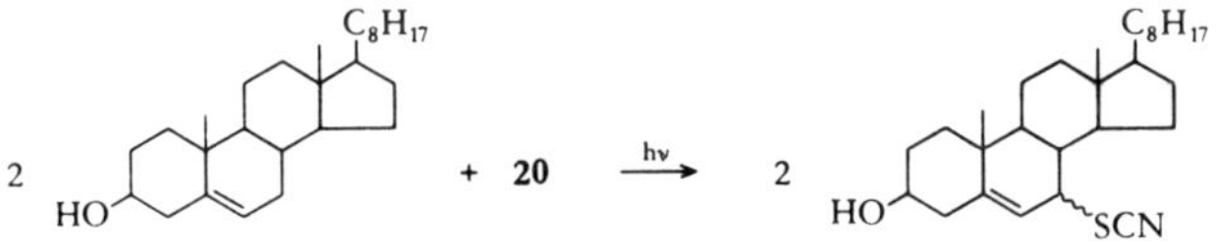

Thiocyanogen reacts with aryl alkyl hydrocarbons, e.g. toluene, ethylbenzene, m-xylene and isopropylbenzene, in a similar fashion as with cholesterol; in each case SCN radicals resulting from photolytic scission of the disulfide bond cause substitution of hydrogen on the α-carbon atom [37]. Normal products are thiocyanates but isomerization to isothiocyanates during isolation may occur. Thus isopropylbenzene is first converted to 2-phenyl-2-thiocyanato-propane (24) which on distillation isomerizes to yield 2-isothiocyanato-2-phenylpropane (25):

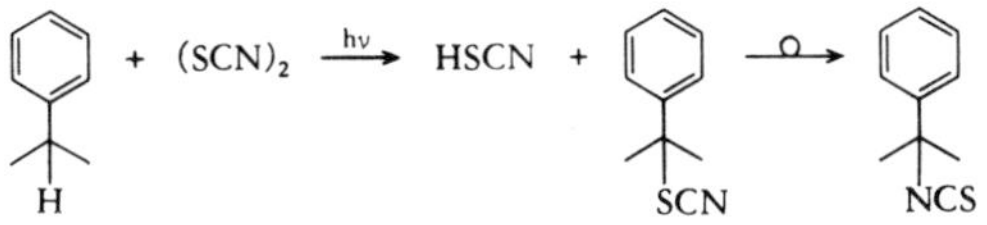

Thiocyanogen chloride (21) has been used as a source of SCN radicals by BACON and GUY [38]. Though in organic solvents and with no added catalyst, thiocyanogen chloride effects nuclear thiocyanation, with ultraviolet irradiation homolytic reaction occurs between thiocyanogen chloride and aryl alkyl hydrocarbons, the results parallelling those obtained with thiocyanogen (**20**). With irradiated solutions of **21** in acetic acid thiocyanation occurs partly in the nucleus and partly in the side chain as was shown with m-xylene [*38*].

7-Thiocyanatocholest-5-en-3β-ol (23) [*35*]. 250 ml of a solution of thiocyanogen in ether was mixed with a solution of 25 g cholesterol (**22**) in 250 ml ether. 2.5 ml glacial acetic acid was added and the solution was irradiated with mercury light for one hour at 10–15°. The crystalline precipitate which formed was filtered off and dissolved in 200 ml chloroform. After treatment with charcoal the solution was concentrated to 70 ml and petroleum ether added until crystallization began. Recrystallization from 50 ml ethyl acetate yielded 20 g **23**, m.p. 139–140°.

4. Photooxidation

a) Sulfoxides

SCHENCK and KRAUCH [*39*] irradiated dimethyl sulfoxide (**26**) and diethyl sulfoxide under oxygen in the presence of O_2-transferring photosensitizers (sens). Peroxides could not be detected, but the corresponding sulfones (cf. **27**) were formed according to the overall equation:

$$2 \text{ H}_3\text{C-SO-CH}_3 + \text{O}_2 \xrightarrow{\text{sens}/h\nu} 2 \text{ H}_3\text{C-SO}_2\text{-CH}_3$$
$$\quad\quad\quad\mathbf{26} \quad\quad\quad\quad\quad\quad\quad\quad\quad\quad \mathbf{27}$$

Irradiation of the solutions containing sensitizers in the absence of oxygen left the sulfoxides unchanged, as did saturation with O_2 in the dark.

26 behaved quite differently when irradiated under oxygen in the absence of sensitizers. 58 % of the oxygen consumed was found to be peroxidic. On working up by distillation the peroxide properties disappeared and **27** was obtained in 55 % yield. Thus the unsensitized oxygenation takes a reaction path different from that of the sensitized oxygen transferring reaction. A more detailed discussion of photooxidation will be found in chapter 39.

Dimethyl sulfone (27) [*39*]. Dimethyl sulfoxide (**26**) (164 g), in which 0.1 g Rose Bengal was dissolved, absorbed 5 l O_2 during irradiation for 13.8 hours (mercury immersion lamp Philips HPK 125 W, 20°). After vacuum distillation of the unconsumed **26** and recrystallization of the residue from acetone/petroleum ether, 41.3 g (97 % based on O_2 consumed) dimethyl sulfone (**27**) were obtained, m.p. 106—110°.

b) Thiourea

SCHENCK and WIRTH [*40*] obtained aminoiminomethanesulfinic acid (**28**) from thiourea by photochemical reaction with oxygen in the presence of sensitizers:

$$\underset{\underset{\text{S}}{\|}}{\text{H}_2\text{N-C-NH}_2} + \text{O}_2 \xrightarrow{\text{sens}/h\nu} \underset{\underset{\text{SO}_2\text{H}}{|}}{\text{H}_2\text{N-C=NH}}$$
$$\quad\quad\quad\quad\quad\quad\quad\quad\quad\quad\quad\quad\quad\quad \mathbf{28}$$

If the reaction is continued to the consumption of 2 moles O_2 the reaction products are cyanamide and sulfuric acid [41]:

$$H_2N-\underset{\underset{S}{\|}}{C}-NH_2 + 2\,O_2 \xrightarrow{sens/h\nu} N\equiv C-NH_2 + H_2SO_4$$

Aminoiminomethanesulfinic acid (28) [42]. If a solution of 0.76 g thiourea (0.01 mole) in 100 ml ethanol be irradiated by a 125 W mercury immersion lamp Osram HQA 500 at 20° under O_2 in the presence of 50 mg Rose Bengal, 0.01 mole O_2 is taken up in 20 mins. After only 14 minutes the solution turns cloudy owing to crystallisation of **28**. After filtration and washing with warm alcohol 0.56 g (60.2%) are obtained, m.p. 142—143° (dec.).

c) Conversion of thioketones into ketones

SCHÖNBERG and MOSTAFA [43] investigated the stability of several thioketones towards oxygen with or without irradiation. The thioketones listed on the left in table 36 are stable towards O_2 in the dark but by the action of sunlight are converted into the corresponding ketones with concomitant formation of SO_2. The ketones listed in the righthand column are not only stable in the dark under these experimental conditions but also in light. This stability has been ascribed to their tendency to form zwitter-ions [43].

Table 36

unstable to O_2 in light	stable to O_2 in light
4,4′-dimethoxythiobenzophenone	10-phenyl-9-acridanthione (**31**)
4,4′-bis-(dimethylamino)-thio-benzophenone	4-thioflavone (**32**)
xanthene-9-thione (**29**)	2,6-diphenyl-4H-thiopyran-4-thione (**33**)
thioxanthene-9-thione (**30**)	

Thiobenzophenone (**34**) differs from the thioketones of table 36 in being autoxidable [44—46] even in the dark and in the solid state. Besides benzophenone, sulfur and sulfur dioxide, products include 3,3,5,5-tetraphenyl-1,2,4-trithiolane (**35**). The wavelength dependence of the photochemical reaction of **34** with O_2 was studied by OSTER et al. [*47*]. Thus only visible light appeared to be responsible for the photooxidation since light of wavelength 365 mµ left **34** unaltered if oxygen was present. In deaerated alcoholic solutions, however, photoreduction took place to yield a variety of products.

4,4′-Dimethoxybenzophenone [*43*]. A deep blue solution of 4,4′-dimethoxythiobenzophenone in dry benzene was placed in the sun (Cairo) for a week, and during this period dry oxygen was blown through. During the irradiation the color of the solution changed to red and then to orange-yellow. The solvent was removed and the yellow residue found to contain elementary sulfur and 4,4′-dimethoxybenzophenone. SO_2 was detected in the exit gases from the benzene solution under irradiation.

d) Co-oxidation of thiols and olefins by oxygen

KHARASCH [*48*] was the first to discover the co-oxidation of thiols and olefins in the presence of oxygen at room temperature to yield substituted β-hydroxy sulfoxides (**37**). This co-oxidation has attracted the interest of the petroleum industry in connection with gum formation in cracked gasolines.

$$R'-SH + CHR=CR_2 + O_2 \longrightarrow R'-S-CHR-CR_2-OOH$$
$$36$$
$$\longrightarrow R'-SO-CHR-CR_2-OH$$
$$37$$

OSWALD [*49*] has shown that the 2-sulfinyl alcohols (**37**) are not primary products of the co-oxidation reaction but result from rearrangement of the β-hydroperoxy sulfides (**36**) initially formed. It was demonstrated that in some cases co-oxidation of thiols and olefins occured in light, but not in the dark. Thus with 1-dodecanethiol and styrene no reaction was evident even after 6 hrs. aeration at room temperature in the dark. However, when the reaction mixture was irradiated by an ultraviolet lamp a solid co-oxidation product (9 % yield) with a peroxide content of 13 % was obtained after 30 mins. reaction time.

OSWALD's investigations [*50*] also showed that by light-induced co-oxidation of 4-chlorobenzenethiol (**39**) and exo-dicyclopentadiene at −5°, the unstable compound **40** was formed.

References, pp. 451—452

The corresponding reaction with aldrin (41) proceeded similarly. In this case the hydroperoxide 42 could be isolated [50].

5,6,7,8,9,9-Hexachloro-2-(p-chlorophenylthio)-1,2,3,4,4a,5,8,8a-octahydro-3-hydroperoxy-1,4:5,8-dimethanonaphthalene (42) [50]. Into 162 ml of n-pentane solution of 3.62 g (25 mmole) of 4-chlorobenzenethiol (39) and 9.12 g (25 mmole) of aldrin (41) air was introduced while it was irradiated from 5-cm distance with an ultraviolet lamp and cooled by an icewater bath. After 5 min., the precipitation of a white fluffy crystalline product started. The ultraviolet initiation was removed after 5 more mins. After a total of 30 mins. of air introduction, the crystals were removed by filtration, washed with n-pentane and dried in a vacuum desiccator. In this manner, 2 g (15.4%) of 42 was obtained, which melted at 116—119°. Then it solidified again and melted again with decomposition between 240—242°.

5. Miscellaneous photochemical reactions of sulfur compounds

a) Photolysis of sulfones

The elimination of sulfur dioxide from cyclic sulfones on irradiation with light of 280—320 mµ was observed by CAVA et al. [51, 52]. Thus, 7,12-dihydro-7,12-epithiopleiadene 13,13-dioxide (43) was transformed by irradiation into pleiadene dimer (44), and 1,3-dihydro-1,3-diphenylnaphtho-[2,3-c]thiophene 2,2-dioxide (45) gave trans-1,2-dihydro-1,2-diphenylcyclobuta[b]naphthalene (46).

45: R=C$_6$H$_5$
47: R=H

1,3-Dihydronaphtho[2,3-c]thiophene 2,2-dioxide (47) remained unchanged by light in the absence or in the presence of sensitizers such as benzophenone or acetophenone [51].

Pleiadene dimer (44) [*52*]. The sulfone **43** (0.30 g) was dissolved in 200 ml of benzene and the solution irradiated with a medium pressure mercury arc at 10° for 75 minutes. After chromatography over alumina, white needles (0.114 g, 50%) of **44** were obtained, which after recrystallization from toluene melted at 350—370° (dec.).

b) Photolysis of unsaturated sultones

KING et al. [*53*] (comp. also [*54*]) have irradiated the cyclic unsaturated sultone 9,10-diphenylacenaphth[1,2-c] [1,2]oxathiin 7,7-dioxide (**48**) in methanol and have obtained the methyl sulfonate **49**. Likewise irradiation of **50** gave a mixture of the geometrical isomers **51** and **52** [*53*].

The esterification clearly is a secondary step following photochemical cleavage of the O—S-bond in the 1,2-oxathiin dioxide (cf. **53**) which results in formation of a sulfene (cf. **54**). The photolysis of **48** or of **50** thus parallels the photochemical cleavage of cyclohexa-2,4-dienones leading to unsaturated esters (comp. p. 227).

Methyl 1-(α-benzoyl-α-phenylmethyl)-2-acenaphthylenesulfonate (49) [*53*]. 474 mg of the sultone **48** were dissolved in 700 ml of methanol and the solution transferred to 2 Pyrex flasks, which were irradiated for 40 mins. with a 80 W Hanovia CH-3 lamp. The reaction mixture was worked up by pouring into excess water and extracting with chloroform; the extracts were washed successively with dilute sodium bicarbonate solution and water, and dried over sodium sulfate. Evaporation of the solvent left a brown semicrystalline residue from which most of the color was removed by washing with a small amount of ether. Chromatography of the crude ester (348 mg = 73%) on silica gel followed by four recrystallizations from chloroform-ether gave **49**, m. p. 162—179°.

c) Photolysis of thiobenzophenone in the presence of olefins

KAISER and WULFERS [*55*] have studied the irradiation of thiobenzophenone (**34**) in the presence of olefins e.g. cis-2-butene, trans-2-butene, and 1-hexene. With the butenes, 1,1-diphenyl-1-propene was obtained while with 1-hexene a 3:2 mixture of 1,1-diphenyl-1-hexene and 1,1-diphenylethylene was obtained. A possible mechanistic route involves the addition of thiobenzophenone to the olefin to give thietanes, followed by the decomposition of the cyclic sulfides to the products.

$$H_5C_6-\overset{C_6H_5}{\underset{S}{\|}} + \overset{R}{\underset{R'}{\|}} \xrightarrow{h\nu} \left[H_5C_6 \overset{C_6H_5}{\underset{S}{\square}} \overset{R}{\underset{R'}{}} \right] + \left[H_5C_6 \overset{C_6H_5}{\underset{S}{\square}} \overset{R'}{\underset{R}{}} \right]$$

34

$$\downarrow h\nu \qquad \qquad \downarrow h\nu$$

$$H_5C_6 \overset{C_6H_5}{=} R \qquad \qquad H_5C_6 \overset{C_6H_5}{=} R'$$
$$+ \qquad \qquad +$$
$$S=CH-R' \qquad \qquad S=CH-R$$

1,1-Diphenyl-1-propene [55]. cis-2-Butene and nitrogen were bubbled continuously through a solution of thiobenzophenone (2.2 g) in cyclohexane (170 ml) which was irradiated with a Hanovia No. 608A-36 lamp. After 4 hrs. the thioketone had completely reacted, as evidenced by the disappearance of its blue color. The solvent and excess cis-2-butene were removed under reduced pressure, and the residue was distilled at 0.5 mm giving a light blue-tinged liquid. Chromatography of the liquid on acid alumina with petroleum ether as the eluent gave a white crystalline solid (0.71 g), m.p. 47—49°. Recrystallization of the solid from 95% ethanol yielded pure 1,1-diphenyl-1-propene (0.5 g, 23% yield based on thiobenzophenone), m.p. 49—50°.

d) Photochemical formation of a dipyridyl sulfide from a 1,4-dihydropyridinethione

The photochemical conversion of several 1-substituted 4-pyridinethiones to di-4-pyridyl sulfide (**56**) was observed by COMRIE [*56*]. When ethanolic solutions of ethyl 1,4-dihydro-4-thioxopyridine-1-carboxylate (**55**) were exposed to ultraviolet light or sunlight, **56** was obtained in 90% yield. Irradiation of solid **55** gave, besides **56**, diethyl carbonate and O,S-diethyl thiocarbonate. This photolysis was visualized as involving photodimerization of **55** as primary step.

$$H_5C_2OOC-N\overset{}{\diagup}=S \xrightarrow{h\nu} N\overset{}{\diagup}-S-\overset{}{\diagdown}N + OC\overset{SC_2H_5}{\diagdown OC_2H_5} + OC\overset{OC_2H_5}{\diagdown OC_2H_5}$$

55   56

Di-4-pyridyl sulfide (56) [*56*]. The thione **55** (0.92 g) was exposed to ultraviolet light until the product failed to show an absorption peak at 370 mμ. Fractional distillation of the resulting brown oil under reduced pressure gave 0.12 g of diethyl carbonate and 90 mg of diethyl thiocarbonate, b.p.$_{20}$ 65°. Recrystallization of the residue from light petroleum (charcoal) gave pure **56**, m.p. 72° (0.2 g).

e) Photochemical reactions involving extrusion of sulfur from dithietanes and thiiranes

Irradiation of O-ethyl thioacetate (**57**) with 254 mμ radiation [*57*] gave 63% of 2,3-diethoxy-2-butene (**58**), sulfur and a small amount of 2,3-diethoxy-2,3-dimethylthiirane (**59**).

As in the foregoing experiment (cf. the irradiation of **55**) photodimerization at the carbon-sulfur double bond was envisaged as being the primary reaction. Breakdown of the 1,2- or 1,3-dithietanes would then give either **58** or **59** [*57*].

The photolysis of a thiirane was investigated by PADWA [*58*] when it was found that trans-2,3-dibenzoyl-2,3-diphenylthiirane (**60**) afforded a high yield of a mixture of cis- and trans-1,2,3,4-tetraphenyl-1,4-butanedione when irradiated in benzene solution through Pyrex. The desulfurization appeared to be highly stereoselective since **60** gave at least 90% of the trans isomer.

2,3-Diethoxy-2-butene (58) and 2,3-diethoxy-2,3-dimethylthiirane (59) [*57*]. The ester **57** (300 g, in 5 portions) was irradiated with a Quarzlampengesellschaft NN 15/44 mercury low-pressure lamp with exclusion of air. After 30 hrs. irradiation distillation through a VIGREUX column gave 170 g of starting material (b.p.$_{80}$ 50—55°), 71 g of a light yellow liquid (b.p.$_{80}$ 82—84°, containing the cis and trans isomers of **58** in a ratio of about 2:3) and 12 g of a yellow oil, b.p.$_{80}$ 90—120°. The distillation residue contained 29 g of sulfur. The high boiling oily fraction was diluted with 10 ml of petroleum ether and chilled in a Dry Ice bath. After one hour the mixture was filtered with suction and the solid material crystallized twice from petroleum ether to give 4.1 g of colorless **59**, m.p. 51—52°.

References, pp. 451—452

f) Photochemical aryl migration in arylthiophenes

When 2-phenylthiophene (**61**) was irradiated with UV-light for 38 hrs. at 80° in benzene solution, 3-phenylthiophene (**62**) was formed in good yield [*59*]. No re-isomerization of **62** to **61** could be observed.

Similarly, when 2,2'-bithiophene (**63**) was illuminated in benzene at 60° for 10 hrs., 2,3'-bithiophene (**64**) and 3,3'-bithiophene (**65**) were formed. It was established [*59*] that **64** was the precursor of **65**.

References

[*1*] Organic Sulfur Compounds. Ed. by N. KHARASCH, vol. **1** and **2**. Oxford: Pergamon 1961, 1966.
[*2*] A. MUSTAFA, in: Adv. Photochem., ed. by W. A. NOYES JR., G. S. HAMMOND and J. N. PITTS JR., vol. **2**, pp. 63—136. New York: Interscience 1964.
[*3*] G. SOSNOVSKY: Free Radical Reactions in Preparative Organic Chemistry. New York: Macmillan 1964; esp. pp. 62—119.
[*4*] E. E. GILBERT: Sulfonation and related reactions. New York: Interscience 1965.
[*5*] C. F. REED: USP 2,046,090 of 30. 6. 1936.
[*6*] C. F. REED: USP 2,174,492 of 26. 9. 1939.
[*7*] I. G. FARBENINDUSTRIE AG: Fr. Pat. 842,509 of 6. 3. 1939.
[*8*] J. H. HELBERGER: Die Chemie **55**, 172 (1942).
[*9*] H.-J. SCHUMACHER and J. STAUFF: Die Chemie **55**, 341 (1942).
[*10*] H. ECKOLDT, in: Methoden der Organischen Chemie, ed. by E. MÜLLER, Vol. **9**, pp. 407—428; esp. p. 414. Stuttgart: Thieme 1955.
[*11*] F. ASINGER, W. SCHMIDT and F. EBENEDER: Ber. dtsch. chem. Ges. **75**, 34 (1942).
[*12*] F. ASINGER, F. EBENEDER and E. BÖCK: Ber. dtsch. chem. Ges. **75**, 42 (1942).
[*13*] F. ASINGER, B. FELL and H. SCHERB: Chem. Ber. **96**, 2831 (1963).
[*14*] H. FEICHTINGER: Chem. Ber. **96**, 3068 (1963).
[*15*] M. S. KHARASCH and A. T. READ: J. Amer. chem. Soc. **61**, 3089 (1939).
[*16*] M. S. KHARASCH, T. H. CHAO and H. C. BROWN: J. Amer. chem. Soc. **62**, 2393 (1940).
[*17*] G. D. COOPER: J. Org. Chem. **21**, 1214 (1956).
[*18*] E. MÜLLER and E. W. SCHMIDT: Chem. Ber. **96**, 3050 (1963).
[*19*] F. S. DAINTON and K. J. IVIN: Trans. Faraday Soc. **46**, 374 (1950).
[*20*] R. GRAF: Liebigs Ann. Chem. **578**, 50 (1952).
[*21*] F. ASINGER, G. GEISELER and H. ECKOLDT: Chem. Ber. **89**, 1037 (1956).
[*22*] I. G. FARBENINDUSTRIE AG: Ger. Pat. 735096 of 1. 4. 1943; C. A. **38**, 1249 (1944).

[23] G. O. SCHENCK and G. A. SCHMIDT-THOMÉE: Liebigs Ann. Chem. **584**, 199 (1953).
[24] G. A. R. BRANDT, H. J. EMELÉUS and R. N. HASZELDINE: J. Chem. Soc. **1952**, 2198.
[25] G. R. A. BRANDT, H. J. EMELÉUS and R. N. HASZELDINE: J. Chem. Soc. **1952**, 2549.
[26] R. N. HASZELDINE and J. M. KIDD: J. Chem. Soc. **1953**, 3219.
[27] A. SCHÖNBERG and T. STOLPP: Liebigs Ann. Chem. **483**, 90 (1930).
[28] R. N. HASZELDINE and J. M. KIDD: J. Chem. Soc. **1955**, 3871.
[29] E. KOBER: J. Amer. chem. Soc. **81**, 4810 (1959).
[30] W. CONWAY and D. S. TARBELL: J. Amer. chem. Soc. **78**, 2228 (1956).
[31] A. L. J. BECKWITH and L. B. SEE: Austral. J. Chem. **17**, 109 (1964).
[32] D. H. R. BARTON, M. V. GEORGE and M. TOMOEDA: J. Chem. Soc. **1962**, 1967.
[33] R. G. R. BACON, in: Organic Sulfur Compounds, ed. by N. KHARASCH, Vol. **1**, pp. 306—325; esp. p. 318. Oxford: Pergamon 1961.
[34] R. G. R. BACON, R. G. GUY, R. S. IRWIN and T. A. ROBINSON: Proc. Chem. Soc. **1959**, 304.
[35] E. FREDERIKSEN and S. LIISBERG: Chem. Ber. **88**, 684 (1955).
[36] H. SCHALTEGGER: Helv. Chim. Acta **33**, 2101 (1950).
[37] R. G. R. BACON and R. S. IRWIN: J. Chem. Soc. **1961**, 2447.
[38] R. G. R. BACON and R. G. GUY: J. Chem. Soc. **1961**, 2428.
[39] G. O. SCHENCK and C. H. KRAUCH: Chem. Ber. **96**, 517 (1963).
[40] G. O. SCHENCK and H. WIRTH: Naturwissenschaften **40**, 141 (1953).
[41] O. WARBURG and V. SCHOCKEN: Arch. Biochem. Biophys. **21**, 363 (1949).
[42] G. O. SCHENCK and H. WIRTH: unpublished results.
[43] A. SCHÖNBERG and A. MOSTAFA: J. Chem. Soc. **1943**, 275.
[44] L. GATTERMANN and H. SCHULZE: Ber. dtsch. chem. Ges. **29**, 2944 (1896).
[45] H. STAUDINGER and H. FREUDENBERGER: Ber. dtsch. chem. Ges. **61**, 1836 (1928).
[46] A. SCHÖNBERG, O. SCHÜTZ and S. NICKEL: Ber. dtsch. chem. Ges. **61**, 2175 (1928).
[47] G. OSTER, L. CITAREL and M. GOODMAN: J. Amer. chem. Soc. **84**, 703 (1962).
[48] M. S. KHARASCH, W. NUDENBERG and G. J. MANTELL: J. Org. Chem. **16**, 524 (1951).
[49] A. A. OSWALD: J. Org. Chem. **26**, 842 (1961).
[50] A. A. OSWALD and F. NOEL: J. Org. Chem. **26**, 3948 (1961).
[51] M. P. CAVA, R. H. SCHLESSINGER and J. P. VAN METER: J. Amer. chem. Soc. **86**, 3173 (1964).
[52] M. P. CAVA and R. H. SCHLESSINGER: Tetrahedron **21**, 3073 (1965).
[53] J. F. KING, P. DE MAYO, E. MORKVED, A. B. M. A. SATTAR and A. STOESSL: Can. J. Chem. **41**, 100 (1963).
[54] E. HENMO, P. DE MAYO, A. B. M. A. SATTAR and A. STOESSL: Proc. Chem. Soc. **1961**, 238.
[55] E. T. KAISER and T. F. WULFERS: J. Amer. chem. Soc. **86**, 1897 (1964).
[56] A. M. COMRIE: J. Chem. Soc. **1963**, 688.
[57] U. SCHMIDT, K. KABITZKE, I. BOIE and C. OSTERROHT: Chem. Ber. **98**, 3819 (1965).
[58] A. PADWA and D. CRUMRINE: Chem. Comm. **1965**, 506.
[59] H. WYNBERG and H. VAN DRIEL: J. Amer. chem. Soc. **87**, 3998 (1965).

Chapter 44

Photochemical reactions of organophosphorus and organoarsenic compounds

1. Organophosphorus compounds

Photochemical reactions have not played a significant role in organophosphorus chemistry until recently. As far as radiation is used in the production of free radicals in organophosphorus chemistry, the reader is referred to SOSNOVSKY [1] for the pertinent literature. A complete account of organophosphorus chemistry will be found elsewhere (cf. [2]).

a) Light-induced addition of dialkyl phosphonates to quinones

The addition of alkyl esters of phosphonic acid (cf. 1) to quinones is accelerated by light, as was found by RAMIREZ and DERSHOWITZ [3]. Thus dimethyl phosphonate (1), when irradiated with chloranil under nitrogen in a quartz vessel (Hanovia 100 Watt Utility Model UV-lamp with filter) afforded a 64% yield of 4-hydroxy-2,3,5,6-tetrachlorophenyl dimethyl phosphate (**2**, m.p. 236—238°) within 4.5 hrs. and a quantitative yield within 15—20 hours; the dark reaction yielded 26% in 4.5 hrs. and 100% in 3 days.

b) Photochemical reactions of trialkyl phosphites

PLUMB and GRIFFIN [4] investigated the light-induced air oxidation of several trialkyl phosphites (cf. **3**) and found that quantitative uptake of oxygen occurred readily, the corresponding trialkyl phosphates being formed (R = methyl, ethyl, isopropyl, butyl).

$$\begin{array}{c} RO \\ RO-P \\ RO \end{array} \quad \xrightarrow{h\nu/O_2} \quad \begin{array}{c} RO \\ RO-PO \\ RO \end{array}$$

3: $R = C_4H_9$ **4**: $R = C_4H_9$

In the absence of solvent the conversion of tributyl phosphite (**3**) to tributyl phosphate (**4**) went to 45% during 20 hrs. oxidation by dry air in the dark at 50—55°, while in the presence of ultraviolet radiation, but under otherwise identical conditions, **4** was formed quantitatively in 5 hours [4].

When the same substrates were irradiated under nitrogen at 20° for 48 hrs. (450 Watt Hanovia lamp in a quartz immersion well), a fair number of photolysis products was formed [5]. Thus, **3** gave 39% of dibutyl butylphosphonate (**5**), 2% of dibutyl phosphonate (**6**), and 11% of **4**, while 46% of starting material was recovered (GLC analysis). Hence, the ARBUSOV rearrangement of **3** to **5** took place with a 72% conversion rate.

$$\begin{array}{c} RO \\ RO-P \\ RO \end{array} \xrightarrow{h\nu} \begin{array}{c} RO\;O \\ \diagdown\uparrow \\ P-R \\ \diagup \\ RO \end{array} + \begin{array}{c} RO\;O \\ \diagdown\uparrow \\ P-H \\ \diagup \\ RO \end{array} + \begin{array}{c} RO \\ RO-PO \\ RO \end{array}$$

3: $R = C_4H_9$ **5**: $R = C_4H_9$ **6**: $R = C_4H_9$ **4**: $R = C_4H_9$

A photochemical desulfuration of alkanethiols to the corresponding alkanes by the action of triethyl phosphite was observed by HOFFMANN et al. [6]:

$$(C_2H_5O)_3P + RSH \xrightarrow{h\nu} \underset{\textbf{7}}{(C_2H_5O)_3PS} + RH$$

In this unusual desulfuration triethyl phosphorothionate (**7**) is formed. The reaction proceeds practically to completion at reflux temperature during several hours, or at a lower rate at room temperature, and is effectively catalyzed by ultraviolet radiation. For the mechanism of the reaction see WALLING and RABINOWITZ [7] and BUNYAN and CADOGAN [8].

References, p. 458

Tributyl phosphate (4) [4]. Tributyl phosphite (3) (7.1 g) was placed in a 125 ml two-necked Vycor flask equipped with an air inlet-tube and a thermometer both of which dipped into the liquid, a magnetic stirrer and a reflux condenser. Air was passed successively through concentrated KOH solution and concentrated H_2SO_4 and dried over phosphorus pentoxide before passing into the reaction vessel. With a continuous stream of dry air at a flow rate of 20 ml/min., the phosphite was stirred and irradiated for 20 hrs. with an unfiltered 100 W mercury resonance lamp (Hanovia 8A-1 quartz lamp). The burner was placed 5 cm. from the flask keeping the liquid at a temperature of 50—60° during the reaction. The weight of product was 7.3 g (97%). Reduction in irradiation time to 10 hrs. and 5 hrs. gave identical yields.

Octane [6]. A mixture of 83 g of triethyl phosphite (previously distilled from sodium) and 73 g of octanethiol in a Pyrex flask fitted with a 24-in. fractionating column was irradiated at a distance of 13 cm with a General Electric 100 Watt S-4 bulb. After 6.25 hrs. irradiation, the mixture was distilled to give 50.3 g (88%) of octane (b.p. 122—124.5°) and 90.9 g of triethyl phosphorothionate (7), b.p.$_{0.5}$ 45°.

c) Photolysis of triarylphosphines

The photolysis of triarylphosphines was studied by GRIFFIN [9] and HORNER [10]. Irradiation of triphenylphosphine (8) in benzene in the absence of oxygen gave about 20% each of diphenylphosphine (9) and biphenyl together with some tetraphenyldiphosphine and terphenyls [9]. The formation of the products may be accounted for by assuming an initial homolysis of the carbon-phosphorus bond:

$$Ar_3P \xrightarrow{h\nu} Ar\cdot + Ar_2P\cdot$$
$$Ar_2P\cdot + C_6H_6 \longrightarrow Ar_2PH + C_6H_5\cdot$$
$$Ar\cdot + C_6H_6 \text{ or } C_6H_5\cdot \longrightarrow Ar-C_6H_5$$
$$2\ Ar\cdot \longrightarrow Ar-Ar$$
$$2\ Ar_2P\cdot \longrightarrow Ar_2P-PAr_2$$

When the irradiation was carried out in benzene-ethanol mixtures [9], 9 and ethyldiphenylphosphine were the major products, these being accompanied by the same side-products as above. Slightly different results were reported by HORNER and DÖRGES [10] in the photolysis of 8 in methanol, phenylphosphine and tetraphenylphosphonium salts being formed in addition to 9.

d) Photochemical synthesis of phosphonium salts

The photolysis of aromatic iodo compounds in the presence of triphenylphosphine constitutes a convenient and apparently general one-step synthesis of the corresponding aryltriphenylphosphonium iodides [11].

Photolysis reactions with triphenylphosphine (8) and the aryl iodides have been carried out at 40—70°, yields and reaction times being as shown in table 37.

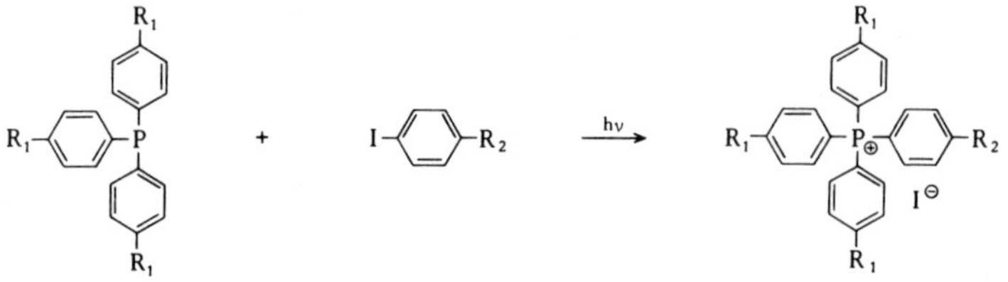

$$R_1\text{-}\langle\rangle\text{-}P(\text{-}\langle\rangle\text{-}R_1)_2 \quad + \quad I\text{-}\langle\rangle\text{-}R_2 \quad \xrightarrow{h\nu} \quad R_1\text{-}\langle\rangle\text{-}P^{\oplus}(\text{-}\langle\rangle\text{-}R_1)_2\text{-}\langle\rangle\text{-}R_2 \; I^{\ominus}$$

8: $R_1 = H$

10: $R_2 = H$
12: $R_2 = CH_3$
14: $R_2 = OCH_3$
16: $R_2 = C_6H_5$
18: $R_2 = OH$

11: $R_1 = R_2 = H$
13: $R_1 = H, R_2 = CH_3$
15: $R_1 = H, R_2 = OCH_3$
17: $R_1 = H, R_2 = C_6H_5$
19: $R_1 = H, R_2 = OH$

20: $R_1 = N(CH_3)_2$

21: $R_2 = N(CH_3)_2$

22: $R_1 = R_2 = N(CH_3)_2$

Table 37 [*11*]

starting material	irr. [hrs.]	product	yield [%]
8 + 10	46	tetraphenylphosphonium iodide (11)	23
8 + 12	346	triphenyl-p-tolylphosphonium iodide (13)	36
8 + 14	249	(4-methoxyphenyl)-triphenylphosphonium iodide (15)	18
8 + 16	137	biphenylyltriphenylphosphonium iodide (17)	9
8 + 18	57	(4-hydroxyphenyl)-triphenylphosphonium iodide (19)	42

The method described above opens an access to tetraarylphosphonium iodides which are obtained only with difficulty (cf. **19** [*11*]) or not at all (cf. **22** [*12*]) by other routes.

The reaction is not restricted to triarylphosphines but can be extended to trialkylphosphines as well, as is demonstrated by the conversion of **23** to **24** [*11*]:

$$(C_4H_9)_3P + C_6H_5I \xrightarrow{h\nu} (C_4H_9)_3(C_6H_5)P^{\oplus}I^{\ominus}$$
$$\quad\; 23 \qquad\quad 10 \qquad\qquad\qquad 24$$

An interesting variation was introduced by PTITSYNA and co-workers [*13*]. When acetone solutions of triphenylphosphine (**8**) and diaryliodonium tetrafluoroborates (cf. **25**) were irradiated, 40—85% yields of aryltriphenylphosphonium tetrafluoroborates (cf. **26**) resulted.

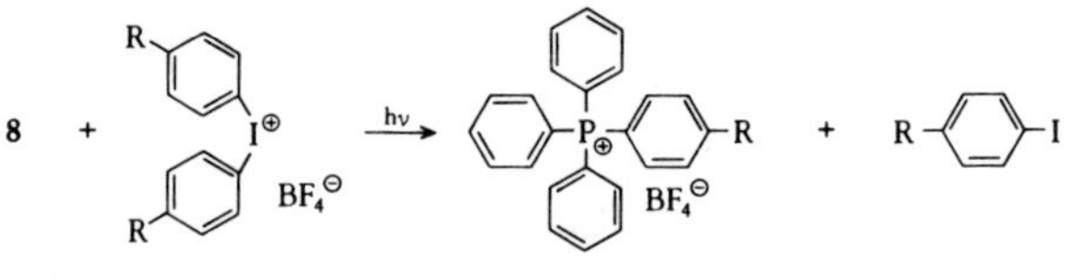

$$8 \; + \; (R\text{-}\langle\rangle)_2 I^{\oplus} \; BF_4^{\ominus} \xrightarrow{h\nu} (C_6H_5)_3P^{\oplus}\text{-}\langle\rangle\text{-}R \; BF_4^{\ominus} \; + \; R\text{-}\langle\rangle\text{-}I$$

25: R = H 26: R = H 10: R = H

Tetraphenylphosphonium iodide (11) [*11*]. A solution of **8** (19 mMoles) and iodobenzene (**10**) (19 mMoles) in chlorobenzene (65 ml) in a Vycor reaction vessel under an atmosphere of dry nitrogen was irradiated for 46 hrs. at 65° with an unfiltered 100 Watt Hanovia mercury resonance lamp. The precipitated product was recrystallized from chloroform-benzene to give 22.5% of **11**, m.p. 323—326°.

Tetrakis-(p-dimethylaminophenyl)-phosphonium iodide (22) [*12*]. A solution, containing 3.91 g of tris-(p-dimethylaminophenyl)-phosphine (**20**) and 2.49 g of 4-iodo-N,N-dimethylaniline (**21**) in 250 ml chlorobenzene, was irradiated under argon with a Philips HPK 125 W mercury high pressure immersion lamp through quartz. After 65 hrs. the clear solution was evaporated in vacuo in a nitrogen atmosphere and the resulting sirup treated with aqueous ethanol to give 0.84 g of tris-(p-dimethylaminophenyl)-phosphine oxide, m.p. 288—290°. On concentration of the filtrate and trituration with hot benzene another 3.49 g of crystals were obtained which consisted largely of the phosphine oxide. Treatment with methanol left 0.22 g of dark violet tetrakis-(p-dimethylaminophenyl)-phosphonium triiodide undissolved, m.p. 239—243°. **22** was finally obtained with $NaHSO_3$; colorless crystals, m.p. 328° (dec.).

Tributylphenylphosphonium iodide (24) [*11*]. Irradiation of a solution of iodobenzene (**10**) and tributylphosphine (**23**) in chlorobenzene for 118 hrs. gave a 6.4% yield of **24**, m.p. 151—152.5°.

Tetraphenylphosphonium tetrafluoroborate (26) [*13*]. A mixture of 0.52 g of **8** and 0.74 g of diphenyliodonium tetrafluoroborate (**25**) was dissolved in 10 ml of absolute acetone. This solution was irradiated for 6 hrs. with a PRK-4 mercury vapor lamp from a distance of 30 cm. A part of the product had separated out during the irradiation (0.48 g) and was filtered off. A second crop of 0.23 g was isolated from the filtrate by means of chromatography the details of which should be taken from the original paper. The total yield of **26** was 0.71 g (85%), m.p. 345°.

e) Photolysis of tetraarylphosphonium salts

The tetraarylphosphonium salts synthesized in one of the foregoing experiments are themselves not stable towards radiation. Thus, when tetraphenylphosphonium chloride was photolyzed in benzene-ethanol [*14*], biphenyl, diphenylphosphine (**9**) and triphenylphosphine (**8**) were obtained besides 33% of unreacted starting material.

Both the photolysis of tetraarylphosphonium salts and the photochemical arylation of triarylphosphines are visualized as proceeding via tetraarylphosphoranyl radicals ($Ar_4P\cdot$) [*9, 10, 14*].

2. Organoarsenic compounds

The photolysis of tris-(trifluoromethyl)-arsine (**27**) to hexafluoroethane (**28**) was carried out by EMELÉUS and collaborators [*15*]. When **27** was irradiated in admixture with iodomethane, besides **28** and trifluoroiodomethane, bis-(trifluoromethyl)-methylarsine (**29**) could be isolated.

$$2 \ (F_3C)_3As \xrightarrow{h\nu} 3 \ F_3C-CF_3 + 2 \ As$$

$$(F_3C)_3As + CH_3I \xrightarrow{h\nu} (F_3C)_2As-CH_3 + F_3CI + F_3C-CF_3$$

$$\quad\quad 27 \quad\quad\quad\quad\quad\quad\quad\quad\quad 29 \quad\quad\quad\quad\quad\quad\quad 28$$

Hexafluoroethane (28) [*15*]. Tris-(trifluoromethyl)-arsine (**27**) (0.738 g) was irradiated with UV light for 16 hrs. in a 120 ml quartz tube, with the lamp situated 1 cm from the tube. A film of arsenic formed on the cooler parts of the tube. The volatile material was composed of 0.078 g of **28** and 0.6 g of **27**.

Bis-(trifluoromethyl)-methylarsine (29) [*15*]. A mixture of **27** (0.65 g) and iodomethane (0.28 g) was irradiated with a UV-lamp for 24 hrs. as above. The reaction products consisted of **28** (0.07 g), trifluoroiodomethane (0.24 g) and 0.42 g of material which contained **27** and **29**. After hydrolytic decomposition of the former with 10% aqueous NaOH the remaining gas was mainly pure **29**.

References

[*1*] G. Sosnovsky: Free Radical Reactions in Preparative Organic Chemistry, pp. 153—192. New York: Macmillan 1964.

[*2*] R. F. Hudson: Structure and Mechanism in Organo-Phosphorus Chemistry. London: Academic Press 1965.

[*3*] F. Ramirez and S. Dershowitz: J. Org. Chem. **22**, 1282 (1957).

[*4*] J. B. Plumb and C. E. Griffin: J. Org. Chem. **28**, 2908 (1963).

[*5*] R. B. LaCount and C. E. Griffin: Tetrahedron Letters **1965**, 3071.

[*6*] F. W. Hoffmann, R. J. Ess, T. C. Simmons and R. S. Hanzel: J. Amer. chem. Soc. **78**, 6414 (1956).

[*7*] C. Walling and R. Rabinowitz: J. Amer. chem. Soc. **81**, 1243 (1959).

[*8*] P. J. Bunyan and J. I. G. Cadogan: J. Chem. Soc. **1962**, 2953.

[*9*] M. L. Kaufman and C. E. Griffin: Tetrahedron Letters **1965**, 769.

[*10*] L. Horner and J. Dörges: Tetrahedron Letters **1965**, 763.

[*11*] J. B. Plumb and C. E. Griffin: J. Org. Chem. **27**, 4711 (1962).

[*12*] G. P. Schiemenz: Chem. Ber. **98**, 65 (1965).

[*13*] O. A. Ptitsyna, M. E. Pudeeva, N. A. Belkevich and O. A. Reutov: Dokl. Akad. Nauk SSSR **163**, 383 (1965); Dokl. Chem. **163**, 671 (1965).

[*14*] C. E. Griffin and M. L. Kaufman: Tetrahedron Letters **1965**, 773.

[*15*] H. J. Emeléus, R. N. Haszeldine and E. G. Walaschewski: J. Chem. Soc. **1953**, 1552.

Chapter 45

Photochemical formation and reactions of organometallic compounds

1. Light-induced formation of organometallic carbonyl compounds from metal carbonyls and organic compounds

The classical method of preparation of these organometallic compounds from the metal carbonyl and organic compounds is the thermal process. An example is the synthesis of butadieneiron tricarbonyl by REIHLEN [1] carried out in a sealed tube (135°, 24 hours).

$$C_4H_6 + Fe(CO)_5 \longrightarrow C_4H_6Fe(CO)_3 + 2\ CO$$

Since about 1958, thermal methods have been extensively replaced by photochemical ones, mostly involving UV-light, with or without simultaneous heating.

For the photochemical preparation of organometallic compounds from metal carbonyls and organic compounds, carbonyls which have been used are $Fe(CO)_5$, $Cr(CO)_6$, $Mo(CO)_6$ and $W(CO)_6$, and as organic reactants: acrylic acid derivatives, acetonitrile, 1,3-butadiene, derivatives of acetylene and of cyclooctatetraene. The significance of amines as organic reactants will be discussed further below.

In a series of cases — and these have facilitated elucidation of the constitution of the reaction products — the organic ligand may be recovered from the organometallic compound.

Thus, (1,2,3,4-tetraphenylbutadiene)-iron tricarbonyl (**2**) produced photochemically from $Fe(CO)_5$ and 1,2,3,4-tetraphenylbutadiene (**1**), liberates this butadiene derivative on heating [2]. In other cases the organic component may not be regained in this way, e.g. duroquinoneiron tricarbonyl formed by reaction of 2-butyne and $Fe(CO)_5$ (cf. p. 461).

STROHMEIER [3, 4] has considered the mechanism of the substitution reactions which are described in this section; the metal carbonyl, by absorption of light, is converted into an excited species which decomposes with loss of a CO group. The fragment thus formed contains a coordination gap which is filled by the organic component of the photoreaction, e. g. an olefine.

The field of photochemical reactions of metal carbonyls with primary, secondary and tertiary amines, phosphines, sulfides and other electron donors with non-bonding electron pairs has been thoroughly reviewed on by STROHMEIER [5]. Table 38 provides some outstanding examples:

Table 38

metal carbonyl	donor	product*	ref.
$Cr(CO)_6$	piperidine	$Cr(CO)_5(pip) + Cr(CO)_4(pip)_2$	[6]
$Cr(CO)_6$	$N(C_2H_5)_3$	$Cr(CO)_5N(C_2H_5)_3$	[6]
$Cr(CO)_6$	2-picoline	$Cr(CO)_5(2\text{-picoline})$	[7]
$Cr(CO)_6$	2,6-lutidine	$Cr(CO)_5(2,6\text{-lutidine})$	[7]
$Cr(CO)_6$	4-chloropyridine	$Cr(CO)_5(4\text{-chloropyridine})$	[7]
$Cr(CO)_6$	quinoline	$Cr(CO)_5(quinoline)$	[7]
$Cr(CO)_6$	isoquinoline	$Cr(CO)_5(isoquinoline)$	[7]
$Cr(CO)_6$	aniline	$Cr(CO)_5(aniline)$	[8]
$Mo(CO)_6$	piperidine	$Mo(CO)_5(pip) + Mo(CO)_4(pip)_2$	[6]
$Mo(CO)_6$	pyridine	$Mo(CO)_5(py) + Mo(CO)_4(py)_2$	[9]
$W(CO)_6$	pyridine	$W(CO)_5(py) + W(CO)_4(py)_2$	[10]
$W(CO)_6$	aniline	$W(CO)_5(aniline)$	[10]
$W(CO)_6$	$N(C_2H_5)_3$	$W(CO)_5N(C_2H_5)_3$	[6]

* (py) = pyridine, (pip) = piperidine

(Phenylcyclooctatetraene)-iron tricarbonyl (4) was obtained [12] by the photochemical reaction of phenylcyclooctatetraene (3) with iron pentacarbonyl under N_2 atmosphere at room temperature for 24 hours. 4 was

isolated by sublimation at 120°/5 mm Hg and purified by recrystallization from methanol. Besides the deep red crystals of 4 (m.p. 65°), the yellow (phenylcyclooctatetraene)-diiron hexacarbonyl 5, m.p. 160—170° under decomposition, was obtained.

References, pp. 470—471

Irradiation of iron pentacarbonyl with terminal olefins leads to the formation of the corresponding olefin iron tetracarbonyl compounds. Thus acrylonitrile formed **6** [*13*] while **7** and **8** were obtained from methyl methacrylate and vinyl acetate respectively [*14*].

$$Fe(CO)_5 + H_2C=CR_1R_2 \xrightarrow[-CO]{h\nu} (OC)_4Fe\text{—}\|\begin{array}{c}R_1\diagdown C \diagup R_2\\CH_2\end{array}$$

6: $R_1 = H$; $R_2 = CN$
7: $R_1 = CH_3$; $R_2 = COOCH_3$
8: $R_1 = H$; $R_2 = OOCCH_3$

Irradiation of iron pentacarbonyl with dimethyl maleate or with dimethyl fumarate resulted in stereospecific addition; thus (dimethyl maleate)-iron tetracarbonyl (m.p. 35—36°) or (dimethyl fumarate)-iron tetracarbonyl (m.p. 133°; dec.) were obtained in 20 % or 51 % yield [*15*].

When a mixture of 2-butyne and $Fe(CO)_5$ is placed in a flask and exposed to sunlight, large orange crystals of duroquinoneiron tricarbonyl (**9**) are formed which decompose at about 50° [*16*]. Strong support for structure **9** was obtained by the treatment of the photoproduct with acids which resulted in 100 % conversion to durohydroquinone and carbon monoxide. Durohydroquinone (**10**) could also be synthesized without isolation of an intermediate complex by merely adding methanolic HCl to the irradiated mixture of 2-butyne and iron pentacarbonyl [*16*].

$$2\ \text{|||} + Fe(CO)_5 \xrightarrow{h\nu} \underset{\mathbf{9}}{\text{[duroquinone]}}\text{---}Fe(CO)_3 \xrightarrow{2HCl} \underset{\mathbf{10}}{\text{[durohydroquinone]}} + FeCl_2 + 3\ CO$$

A complex similar to **9** may conceivably be an intermediate in the REPPE synthesis of hydroquinone from acetylene and iron pentacarbonyl at elevated temperatures [*17*].

Tris-(4-aminopyridine)-molybdenum tricarbonyl (**11**) is formed on UV-irradiation [*18*] of a solution of 4-aminopyridine and $Mo(CO)_6$ in benzene under nitrogen; on account of its low solubility in benzene it is precipitated. If the irradiation is carried out in cyclohexane solution the reaction comes to an end with the replacement of only 2 carbonyls by

$$Mo(CO)_6 + 3\ C_5H_4N\text{-}NH_2 \xrightarrow[C_6H_6]{h\nu} \underset{\mathbf{11}}{(H_2N\text{-}C_5H_4N)_3Mo(CO)_3} + 3\ CO$$

$$Mo(CO)_6 + 2\ C_5H_4N\text{-}NH_2 \xrightarrow[C_6H_{12}]{h\nu} \underset{\mathbf{12}}{(H_2N\text{-}C_5H_4N)_2Mo(CO)_4} + 2\ CO$$

4-aminopyridine, bis-(4-aminopyridine)-molybdenum tetracarbonyl (12) being formed in good yield. For the formation of a (nitrosobenzene)-iron tricarbonyl see p. 469.

Anilinechromium pentacarbonyl [8]. 557.8 mg of $Cr(CO)_6$, 5 ml freshly distilled aniline and 25 ml tetrahydrofuran are irradiated with an Osram HBO 200 mercury high pressure lamp until about 1 equivalent of CO is evolved. After removing the solvent and aniline in a vacuum rotary-evaporator, the yellow residue is dissolved in a little acetone and precipitated with water. The yellow compound resulting from several such precipitations decomposes rapidly in solution in the presence of air. It is therefore best to work in an inert atmosphere. Yield after 3 precipitations: 58% calculated on the basis of $Cr(CO)_6$ used.

Isopreneiron tricarbonyl [11]. A mixture of isoprene (60 ml) and iron pentacarbonyl (60 ml) was irradiated with a sun lamp for 44 hr. The resulting orange solution was concentrated and chromatographed on alumina. Elution with isohexane gave an orange eluate. The eluate was evaporated, but the infra-red spectrum of the remaining oil showed a band at 1640/cm characteristic of an unsaturated impurity. The oil was, therefore, rechromatographed four additional times. Evaporation of the final eluate yielded 1.0 g of a yellow liquid, having no band in its infra-red spectrum at 1640/cm. The yield of pure isopreneiron tricarbonyl was only 1% since the long purification procedure introduced considerable losses.

(1,2,3,4-Tetraphenylbutadiene)-iron tricarbonyl (2) [2]. Tetraphenylbutadiene (1 g) was refluxed with 1.5 g of iron pentacarbonyl in 40 ml of benzene for 15 hrs. whilst being irradiated with UV-light. Recrystallization of the reaction product from methanol and acetone gave the yellow **2**, m.p. 227—231° (dec.).

(Methyl methacrylate)-iron tetracarbonyl (7) [14]. Iron pentacarbonyl (3.099 g) and 100 ml freshly distilled methyl methacrylate were irradiated at 15° with a mercury high pressure lamp Philips HPK 125 W. When 1 mole CO had been liberated excess methyl methacrylate was removed at 25° in vacuo. Distillation in high vacuum yielded 2.27 g (54%) of **7** as orange-red liquid. All operations were performed in an argon atmosphere.

2. Photochemical reactions of organometallic carbonyl compounds

a) Substitution reactions with electron donors

This field has received particular attention from FISCHER and STROHMEIER and their co-workers; table 39 reviews part of the reactions which they have carried out. An extensive survey has been given by FISCHER and WERNER [19].

According to NYHOLM [20] not only the monodentate ligands triphenylphosphine and triphenylarsine react readily with cyclopentadienylmanganese tricarbonyl $[C_5H_5Mn(CO)_3]$ but similar reactions could be effected with the bidentate ligands 1,2-bis-(diphenylphosphino)-ethane (Diphos = $(C_6H_5)_2PCH_2CH_2P(C_6H_5)_2$) and 1,2-bis-(dimethylarsino)-benzene (Diars = o-$C_6H_4[As(CH_3)_2]_2$). With these two compounds, two types of complexes were obtained: irradiation in cyclohexane solution produced complexes in which the ligands act as bridging groups, e. g. [C$_5$H$_5$Mn(CO)$_2$]$_2$(Diphos). Those in which ligand atoms occupy normal chelate positions, e. g. [C$_5$H$_5$Mn(CO)](Diphos), were obtained by irradiation in benzene.

Table 39

organometallic carbonyl cpd.*	donors	products	ref.
C$_5$H$_5$Mn(CO)$_3$	triphenylphosphine	C$_5$H$_5$Mn(CO)$_2$P(C$_6$H$_5$)$_3$ and C$_5$H$_5$Mn(CO) [P(C$_6$H$_5$)$_3$]$_2$	[21]
C$_5$H$_5$Mn(CO)$_3$	ethylene	C$_5$H$_5$Mn(CO)$_2$C$_2$H$_4$	[22]
C$_5$H$_5$Mn(CO)$_3$	1,3-butadiene	C$_5$H$_5$Mn(CO)C$_4$H$_6$	[23]
C$_5$H$_5$Mn(CO)$_3$	1,3-cyclohexadiene	C$_5$H$_5$Mn(CO)$_2$C$_6$H$_8$ and [C$_5$H$_5$Mn(CO)$_2$]$_2$C$_6$H$_8$	[24]
C$_5$H$_5$Mn(CO)$_3$	dimethyl sulfoxide	C$_5$H$_5$Mn(CO)$_2$CH$_3$SOCH$_3$ and C$_5$H$_5$Mn(CO) (CH$_3$SOCH$_3$)$_2$	[25]
C$_5$H$_5$Mn(CO)$_3$	diphenylacetylene	C$_5$H$_5$Mn(CO)$_2$(C$_6$H$_5$C$\equiv$CC$_6$H$_5$)	[26]
C$_5$H$_5$Mn(CO)$_3$	piperazine	C$_5$H$_5$Mn(CO)$_2$C$_4$H$_{10}$N$_2$ and [C$_5$H$_5$Mn(CO)$_2$]$_2$C$_4$H$_{10}$N$_2$ (13)	[27]
C$_5$H$_5$Mn(CO)$_3$	triethylenediamine	C$_5$H$_5$Mn(CO)$_2$C$_6$H$_{12}$N$_2$ and [C$_5$H$_5$Mn(CO)$_2$]$_2$C$_6$H$_{12}$N$_2$ (14)	[27]
C$_5$H$_5$Mn(CO)$_3$	hexamethylene-tetramine	C$_5$H$_5$Mn(CO)$_2$C$_6$H$_{12}$N$_4$	[27]
C$_5$H$_5$Mn(CO)$_3$	triphenylarsine	C$_5$H$_5$Mn(CO)$_2$As(C$_6$H$_5$)$_3$	[20]
C$_5$H$_5$Mn(CO)$_3$	1,2-bis-(dimethyl-arsino)-benzene (Diars)	[C$_5$H$_5$Mn(CO)$_2$]$_2$(Diars)	[20]
C$_5$H$_5$Mn(CO)$_3$	1,2-bis-(diphenyl-phosphino)-ethane (Diphos)	C$_5$H$_5$Mn(CO) (Diphos) and [C$_5$H$_5$Mn(CO)$_2$]$_2$ (Diphos)	[20]
CH$_3$C$_5$H$_4$Mn(CO)$_3$	diphenylacetylene	CH$_3$C$_5$H$_4$Mn(CO)$_2$(C$_6$H$_5$C$\equiv$CC$_6$H$_5$)	[28]
CH$_3$C$_5$H$_4$Mn(CO)$_3$	pyridine	CH$_3$C$_5$H$_4$Mn(CO)$_2$NC$_5$H$_5$	[29]
CH$_3$C$_5$H$_4$Mn(CO)$_3$	piperidine	CH$_3$C$_5$H$_4$Mn(CO)$_2$NC$_5$H$_{11}$	[30]
C$_5$H$_5$V(CO)$_4$	1,3-butadiene	C$_5$H$_5$V(CO)$_2$C$_4$H$_6$	[23]
mesitylene-Cr(CO)$_3$	ethylene	[s-C$_6$H$_3$(CH$_3$)$_3$]Cr(CO)$_2$C$_2$H$_4$	[31]

* C$_5$H$_5$ and CH$_3$C$_5$H$_4$ = cyclopentadienyl and methylcyclopentadienyl

$$\begin{array}{c} \text{CO} \\ | \\ \text{C}_5\text{H}_5\text{Mn}-\text{HN} \\ | \\ \text{CO} \end{array} \begin{array}{c} \text{CH}_2-\text{CH}_2 \\ \diagdown \\ \text{CH}_2-\text{CH}_2 \end{array} \begin{array}{c} \text{CO} \\ | \\ \text{NH}-\text{MnC}_5\text{H}_5 \\ | \\ \text{CO} \end{array}$$

13

$$\begin{array}{c} \text{CO} \\ | \\ \text{C}_5\text{H}_5\text{Mn}-\text{N} \\ | \\ \text{CO} \end{array} \begin{array}{c} \text{CH}_2-\text{CH}_2 \\ \diagup \quad \diagdown \\ -\text{CH}_2-\text{CH}_2- \\ \diagdown \quad \diagup \\ \text{CH}_2-\text{CH}_2 \end{array} \begin{array}{c} \text{CO} \\ | \\ \text{N}-\text{MnC}_5\text{H}_5 \\ | \\ \text{CO} \end{array}$$

14

(Cyclopentadienyl)(pyridine)-manganese dicarbonyl [29]. Cyclopentadienylmanganese tricarbonyl (276.9 mg) dissolved in 25 ml pyridine was irradiated until about 1 equivalent CO had been evolved. After removing the solvent at room temperature under high vacuum, a yellow-brown crystalline residue remained which was taken up in 2 ml methanol under N_2 and transferred to a sublimation apparatus of the cold-finger type. After removing the methanol at room temperature under low vacuum the residue was sublimed at 50° under high vacuum to condense remaining traces of unreacted starting material on the cold-finger. The sublimation was interrupted, the cold-finger cleaned and the bulk of the product sublimed at a bath temperature of 80—90° in high vacuum. This sublimation process was repeated. Yield of resublimed $C_5H_5Mn(CO)_2$-(pyridine) was 157 mg (53% calculated on basis of CO produced).

Related experimental methods have been described by STROHMEIER et al. [10].

(Cyclopentadienyl)(ethylene)-manganese dicarbonyl [22]. A pentane solution of $C_5H_5Mn(CO)_3$ is treated with ethylene at room temperature and normal pressure under N_2 and in the presence of mercury. After 6 hrs. irradiation, the somewhat turbid, yellow-orange solution was filtered and after removal of the solvent, the residue was sublimed at 45°. The more volatile yellow starting material was followed by small quantities of a less volatile orange-red sublimate. This was separated, resublimed and identified as analytically pure (cyclopentadienyl)(ethylene)-manganese dicarbonyl, m.p. 116—118° (dec.). Intensive sunlight caused decomposition with decolorization; the decomposition product no longer melted and was not sublimable.

[Bis-(cyclopentadienylmanganese dicarbonyl)] [1,2-bis-(diphenylphosphino)-ethane] [20]. A solution of 0.63 g of cyclopentadienylmanganese tricarbonyl and 0.2 g of 1,2-bis-(diphenylphosphino)-ethane in cyclohexane (10 ml) was exposed to ultraviolet light for 12 hrs. with occasional shaking. The crystals which separated were filtered off and washed with cyclohexane. The pure yellow $[C_5H_5Mn(CO)_2]_2$(Diphos) crystallized from benzene-light petroleum and was dried in vacuo, m.p. 210—212°; yield 0.43 g.

b) Photochemical decarbonylation of organometallic acyl derivatives to alkyl derivatives

Acyl cyclopentadienyliron dicarbonyl compounds of general formula $RCOFe(CO)_2C_5H_5$ (cf. **15**) fail to undergo thermal decarbonylation. This has hitherto rendered this class of substances of little use in organometallic synthesis. Nevertheless, KING and BISNETTE [32] observed facile decarbonylation when these compounds were subjected to UV-irradiation. Thus, (acetyl)(cyclopentadienyl)-iron dicarbonyl (**15**) on irradiation in hexane solution furnished (cyclopentadienyl)(methyl)-iron dicarbonyl (**16**), and (acryloyl)(cyclopentadienyl)-iron dicarbonyl (**17**) afforded (cyclopentadienyl)(vinyl)-iron dicarbonyl (**18**). The same method allowed also the synthesis of **20** from (benzoyl)(cyclopentadienyl)-iron dicarbonyl (**19**) [32].

$$(C_5H_5)-Fe\underset{CO}{\overset{CO}{\diagdown}}CO-R \quad \xrightarrow[-CO]{h\nu} \quad (C_5H_5)-Fe\underset{CO}{\overset{CO}{\diagdown}}R$$

15: R = CH$_3$ **16:** R = CH$_3$
17: R = CH=CH$_2$ **18:** R = CH=CH$_2$
19: R = C$_6$H$_5$ **20:** R = C$_6$H$_5$

(Cyclopentadienyl)(phenyl)-iron dicarbonyl (20) [*32*]. A solution of 3.0 g of **19** in 25 ml of thiophene-free benzene was irradiated with a 1000 W mercury lamp. The solution was contained in a quartz vessel under nitrogen, 30 cm away from the lamp. After 14 hrs. the solution was filtered and the solvent removed in vacuo. Chromatography of the residue on alumina and elution with pentane gave 0.47 g (17%) of yellow **20**, m.p. 35—36° (after recrystallization from pentane at —78°).

3. Photochemical reactions of organotin compounds

a) Light-induced reaction of hexamethylditin with trifluoroiodomethane

If a mixture of hexamethylditin and excess trifluoroiodomethane, contained in Pyrex tubes, is irradiated with ultraviolet light [*33*], (trifluoromethyl)-trimethyltin (**21**) may be obtained in yields of 85 % and higher. A radical chain mechanism is believed to be operating, probably involving homolytic fission of the tin-tin bond as the first step:

$$(CH_3)_3Sn-Sn(CH_3)_3 \xrightarrow{h\nu} 2\ (CH_3)_3Sn\bullet$$
$$(CH_3)_3Sn\bullet + CF_3I \longrightarrow (CH_3)_3SnCF_3\ (\mathbf{21}) + I$$
$$I + (CH_3)_3Sn-Sn(CH_3)_3 \longrightarrow (CH_3)_3SnI + (CH_3)_3Sn\bullet$$

b) Synthesis of 1,2,3-triphenylazulene by photolysis of an organotin compound

(4-Iodo-1,2,3,4-tetraphenyl-1,3-butadienyl)-dimethyltin iodide (**22**) is readily subject to homolytic cleavage of the carbon-tin bond on photolysis [*34*]. Irradiation in benzene leads to the gradual release of iodine and production of 1,2,3-triphenylazulene (**23**), possibly by way of tetraphenylcyclobutadiene.

1,2,3-Triphenylazulene (23). A solution of 1 g of **22** in 100 ml of dry benzene in a Pyrex flask was irradiated for approximately 8 hours at ambient temperatures by means of a Hanovia Type 7420 light source. The deep iodine color of the solution was removed by washing with thiosulphate and the resulting deep green solution taken to dryness, the residue dissolved in petroleum ether, and chromatographed on neutral alumina of moderate activity. The blue band which appeared was eluted with petroleum ether and the evaporated residue crystallized from dry nitromethane. The resulting blue granules of **23** (0.19 g, 30%) had m.p. 195—200° and after one additional recrystallization melted at 211°.

c) Photosensitized oxygenation of organotin compounds

Unsaturated organotin compounds, e.g. (3-methyl-2-buten-1-yl)-triphenyltin (**24**) and (cyclopentadienyl)-triphenyltin (**25**), were found to be susceptible to sensitized photooxidation (comp. chapter 39). Thus **25** was converted to the epidioxide **26** while **24** was transformed into an oily mixture of allylic hydroperoxides on irradiation in aerated solution and in the presence of sensitizers [*35*].

24   **25**   **26**

(**2,3-Dioxabicyclo[2.2.1]hept-5-enyl)-triphenyltin** (**26**) [*35*]. Irradiation of an acetone solution of **25** containing rose bengal under oxygen at —20° with a sodium lamp afforded a 50% yield of white platelets of **26** which decomposed at about 0°.

4. Photochemical reactions of organomercury compounds

RAZUVAEV and OLDEKOP published a series of papers on the photochemical reactions of diarylmercury compounds in solution. They were able to show that in UV-light, the diarylmercury compounds are converted to the corresponding arylmercury halide, aryl halide, and arene, depending upon the solvent used.

The primary step seems to be the homolysis of the carbon-mercury bond [*36, 37*]:

$$Ar_2Hg \xrightarrow{h\nu} Ar\cdot + ArHg\cdot$$

The radicals formed in this way become stabilized by reaction with the solvent. These reactions may be divided into two groups: the first group concerns reactions in halogenated solvents in which arylmercury halides are formed. The reactions of the second group lead to the formation of elementary mercury if the irradiations are carried out in alcoholic solvents.

References, pp. 470—471

a) Photolysis reactions with formation of arylmercury halides

If diarylmercury compounds are irradiated in carbon tetrachloride, the following reactions occur [38].

$$Ar_2Hg \xrightarrow{UV} ArHg\cdot + Ar\cdot$$
$$ArHg\cdot + CCl_4 \longrightarrow ArHgCl + \cdot CCl_3$$
$$Ar\cdot + CCl_4 \longrightarrow ArCl + \cdot CCl_3$$
$$2 \cdot CCl_3 \longrightarrow Cl_3C-CCl_3$$

To exemplify, the reaction products found in the photolysis of di-o-tolylmercury were: o-chlorotoluene, o-tolylmercury chloride and hexachloroethane [39]:

$$(CH_3C_6H_4)_2Hg + 2\ CCl_4 \xrightarrow{h\nu} CH_3C_6H_4Cl + CH_3C_6H_4HgCl + C_2Cl_6$$

KOTON, ZORINA and OSBERG [40] found that diphenylmercury did not react with bromoform in a sealed tube at 130°. Irradiation with UV-light, however, effected quantitative formation of phenylmercury bromide and benzene [37]. The photochemical reaction of di-o-tolylmercury with chloroform proceeded similarly [39]:

$$(CH_3C_6H_4)_2Hg + 2\ CHCl_3 \xrightarrow{h\nu} C_6H_5CH_3 + CH_3C_6H_4HgCl + \cdot CCl_3 + (\cdot CHCl_2)$$

Phenylmercury bromide [37]. Diphenylmercury (2 g) in bromoform (15 ml) was irradiated with UV-light for 12 hours. Phenylmercury bromide separated out as crystals which after recrystallization from dichloroethane melted at 278°. The filtrate was steam-distilled and a residue (1.4 g) obtained which was identified as phenylmercury bromide. Overall yield of this compound 96% of theory.

b) Photolysis reactions with formation of mercury

UV irradiation of diphenylmercury in isopropanol leads to the formation of benzene, acetone and mercury [37].

$$(C_6H_5)_2Hg + CH_3CHOHCH_3 \xrightarrow{h\nu} 2\ C_6H_6 + Hg + CH_3COCH_3$$

Toluene and formaldehyde (quantitative yield) are formed from di-o-tolylmercury in methanol [39]:

$$(CH_3C_6H_4)_2Hg + CH_3OH \xrightarrow{h\nu} 2\ C_6H_5CH_3 + Hg + CH_2O$$

With glycol monoethyl ether a similar reaction occurs with formation of acetaldehyde:

$$HOCH_2CH_2OC_2H_5 + (CH_3C_6H_4)_2Hg \xrightarrow{h\nu} 2\ CH_3C_6H_5 + Hg + 2\ CH_3CHO$$

Photochemical reaction between diphenylmercury and isopropanol [37]. Diphenylmercury (3 g) in isopropyl alcohol (10 ml) was irradiated for 240 hours with UV-light. Mercury was precipitated; to separate this from diphenylmercury it was washed with hot isopropyl alcohol. Yield of mercury 1.4 g. The filtrate was distilled and treated with water. Benzene (0.6 ml) separated out and was identified by conversion to m-dinitrobenzene. Acetone was detected by formation of its 2,4-dinitrophenylhydrazone.

5. Miscellaneous photochemical reactions of organometallic compounds

a) Light-induced formation of a GRIGNARD reagent from an aliphatic bromide

3-Bromo-3-methyl-1-butyne (**27**) in dry ether does not react with magnesium to form a GRIGNARD compound. However, the reaction may be accomplished in light [41]. Treatment of the GRIGNARD solution with acetaldehyde yields, after working up, 3,3-dimethyl-4-pentyn-2-ol (**28**).

$$\begin{array}{cc}
\text{CH}_3 & \text{CH}_3 \\
| & | \\
\text{Br}-\text{C}-\text{C}\equiv\text{CH} & \text{H}_3\text{C}-\text{CHOH}-\text{C}-\text{C}\equiv\text{CH} \\
| & | \\
\text{CH}_3 & \text{CH}_3 \\
\mathbf{27} & \mathbf{28}
\end{array}$$

(2-Methyl-3-butyn-2-yl)-magnesium bromide [41]. The necessary quantity of magnesium is treated with some $HgCl_2$, covered with some dry ether containing **27** and then irradiated by sunlight from below. This is achieved by means of a metal mirror. The reaction is carried out in a flask fitted with a reflux condenser, a dropping funnel and a stirrer. The onset of the reaction is accompanied by the formation of a white precipitate which disappears more or less rapidly. More ethereal solution of **27** is added to the boiling solution and the irradiation is continued until the magnesium has disappeared.

b) Photodimerization of a metal-complexed olefin

Irradiation of norbornene (bicyclo[2.2.1]hept-2-ene, **29**) in the presence of acetone [42] or acetophenone [43] as sensitizers leads to a distinct mixture of $C_{14}H_{20}$ dimers. In contrast, norbornene, when irradiated in solutions containing cuprous chloride [43] or cuprous bromide [44], afforded the exo,trans,exo isomer of dodecahydro-1,4:5,8-dimethanobiphenylene (pentacyclo[8.2.1.1^{4,7}.0^{2,9}.0^{3,8}]tetradecane, **30**).

The reaction proceeds very smoothly and is remarkably stereospecific. No dimerization occurred either in the dark or on irradiation of **29** in the absence of cuprous salts [44].

Exo,trans,exo-dodecahydro-1,4:5,8-dimethanobiphenylene (30) [44]. A solution of norbornene (**29**) (450 g) in 150 ml anhydrous ether was stirred with cuprous bromide (4 g) under N_2 whilst being irradiated in a tubular reactor for 149 hrs., using a 450 W high-pressure mercury arc and Vycor optics. Distillation afforded 115 g (26%) of **30** as a clear liquid that crystallized. Recrystallization from ethanol gave white needles, m.p. 64—64.5°.

c) Photolysis of aromatic lithium compounds

When phenyllithium in oxygen-free ether solution was irradiated with a 450 W Hanovia high-pressure mercury lamp [45] a 80% yield of biphenyl and metallic lithium was obtained. From 2-naphthyllithium only 2,2'-binaphthyl was generated. Since only small amounts of products resulting from radical attack on the solvent were observed the process is clearly distinguished from the abovementioned photolyses of organomercury compounds.

d) Photolysis of iron pentacarbonyl in nitrobenzene solution

Irradiation of iron pentacarbonyl in nitrobenzene results in liberation of gaseous CO and CO_2 and formation of a bright yellow precipitate which consisted of dimer **31** [46].

(Nitrosobenzene)-iron tricarbonyl dimer (31) [46]. Iron pentacarbonyl (11.7 g) is dissolved in 130 ml of nitrobenzene and, after addition of 3 drops of concentrated HCl, irradiated under argon with a Philips HPK 125 W mercury high-pressure immersion lamp. After 12—15 hrs. approximately 1.3 liters gas are evolved, and 3—4 g of golden material have separated from the dark brown solution. The precipitate after being rinsed with methanol is pure **31**, which decomposes at 120—125°.

References

[1] H. REIHLEN, A. GRUHL, G. VON HESSLING and O. PFRENGLE: Liebigs Ann. Chem. **482**, 161 (1930).
[2] G. N. SCHRAUZER: J. Amer. chem. Soc. **81**, 5307 (1959).
[3] W. STROHMEIER and K. GERLACH: Chem. Ber. **94**, 398 (1961).
[4] W. STROHMEIER and D. VON HOBE: Chem. Ber. **94**, 761 (1961).
[5] W. STROHMEIER: Angew. Chem. **76**, 873 (1964); internat. ed. **3**, 730 (1964).
[6] W. STROHMEIER, K. GERLACH and D. VON HOBE: Chem. Ber. **94**, 164 (1961).
[7] W. STROHMEIER, G. MATTHIAS and D. VON HOBE: Z. Naturforsch. **15b**, 813 (1960).
[8] W. STROHMEIER and K. GERLACH: Z. Naturforsch. **15b**, 413 (1960).
[9] W. STROHMEIER and K. GERLACH: Chem. Ber. **93**, 2087 (1960).
[10] W. STROHMEIER, K. GERLACH and G. MATTHIAS: Z. Naturforsch. **15b**, 621 (1960).
[11] R. B. KING, T. A. MANUEL and F. G. A. STONE: J. Inorg. Nucl. Chem. **16**, 233 (1961).
[12] A. NAKAMURA and N. HAGIHARA: Nippon Kagaku Zasshi **82**, 1387 (1961).
[13] S. F. A. KETTLE and L. E. ORGEL: Chem. and Ind. **1960**, 49.
[14] E. KOERNER VON GUSTORF, M.-J. JUN and G. O. SCHENCK: Z. Naturforsch. **18b**, 503 (1963).
[15] G. O. SCHENCK, E. KOERNER VON GUSTORF and M.-J. JUN: Tetrahedron Letters **1962**, 1059.
[16] H. W. STERNBERG, R. MARKBY and I. WENDER: J. Amer. chem. Soc. **80**, 1009 (1958).
[17] W. REPPE and H. VETTER: Liebigs Ann. Chem. **582**, 133 (1953).
[18] W. STROHMEIER and G. SCHÖNAUER: Chem. Ber. **95**, 1767 (1962).
[19] E. O. FISCHER and H. WERNER: Metall-π-Komplexe mit di- und oligoolefinischen Liganden. Weinheim: Verlag Chemie 1963.
[20] R. S. NYHOLM, S. S. SANDHU and M. H. B. STIDDARD: J. Chem. Soc. **1963**, 5916.
[21] W. STROHMEIER and C. BARBEAU: Z. Naturforsch. **17b**, 848 (1962).
[22] H. P. KÖGLER and E. O. FISCHER: Z. Naturforsch. **15b**, 676 (1960).
[23] E. O. FISCHER, H. P. KÖGLER and P. KUZEL: Chem. Ber. **93**, 3006 (1960).
[24] E. O. FISCHER and M. HERBERHOLD: Z. Naturforsch. **16b**, 841 (1961).
[25] W. STROHMEIER and J. F. GUTTENBERGER: Z. Naturforsch. **18b**, 667 (1963).
[26] W. STROHMEIER and D. VON HOBE: Z. Naturforsch. **16b**, 402 (1961).
[27] W. STROHMEIER and J. F. GUTTENBERGER: Chem. Ber. **96**, 2112 (1963).
[28] W. STROHMEIER, H. LAPORTE and D. VON HOBE: Chem. Ber. **95**, 455 (1962).
[29] W. STROHMEIER and K. GERLACH: Z. Naturforsch. **15b**, 675 (1960).
[30] W. STROHMEIER and J. F. GUTTENBERGER: Z. Naturforsch. **18b**, 80 (1963).
[31] E. O. FISCHER and P. KUZEL: Z. Naturforsch. **16b**, 475 (1961).
[32] R. B. KING and M. B. BISNETTE: J. Organometal. Chem. **2**, 15 (1964).
[33] R. D. CHAMBERS, H. C. CLARK and C. J. WILLIS: Chem. and Ind. **1960**, 76.
[34] H. H. FREEDMAN: J. Org. Chem. **27**, 2298 (1962).
[35] G. O. SCHENCK, E. KOERNER VON GUSTORF and H. KÖLLER: Angew. Chem. **73**, 707 (1961).
[36] G. RAZUVAEV and YU. OLDEKOP: Zhur. obshchei Khim. **19**, 736 (1949); C. A. **43**, 8895 (1949).
[37] G. RAZUVAEV and YU. OLDEKOP: Zhur. obshchei Khim. **19**, 1483 (1949); C. A. **44**, 3450 (1950).
[38] G. A. RAZUVAEV and YU. A. OLDEKOP: Dokl. Akad. Nauk **64**, 77 (1949); C. A. **43**, 4579 (1949).
[39] G. RAZUVAEV and YU. OLDEKOP: Zhur. obshchei Khim. **20**, 181 (1950); Chem. Zentr. **1952**, 350.

[40] M. M. KOTON, T. M. ZORINA and E. G. OSBERG: Zhur. obshchei Khim. **17**, 59 (1947).
[41] Y. PASTERNAK: Comptes rendus **255**, 1750 (1962).
[42] D. SCHARF and F. KORTE: Tetrahedron Letters **1963**, 821.
[43] D. R. ARNOLD, D. J. TRECKER and E. B. WHIPPLE: J. Amer. chem. Soc. **87**, 2596 (1965).
[44] D. J. TRECKER, J. P. HENRY and J. E. MCKEON: J. Amer. chem. Soc. **87**, 3261 (1965).
[45] E. E. VAN TAMELEN, J. I. BRAUMAN and L. E. ELLIS: J. Amer. chem. Soc. **87**, 4964 (1965).
[46] E. KOERNER VON GUSTORF and M.-J. JUN: Z. Naturforsch. **20b**, 521 (1965).

Chapter 46

Light sources and light filters in preparative organic photochemistry

G. O. SCHENCK

Although photochemical experiments are usually easy to perform, a number of theoretical and practical requirements have to be met. Several of the books which have come on the market in recent years contain chapters on the technique of organic photochemistry, and special reference is made in this respect to the monographs of CALVERT and PITTS [1] and MCLAREN and SHUGAR [2]; for earlier literature coping with the elements of photochemical reaction engineering see the bibliography p. 495. At present a comprehensive survey of *all* practical aspects of preparative photochemistry is lacking, however, although partial aspects have been discussed in fair detail by several contributors [1, 2].

In this chapter no attempt will be made to fill this gap but rather to supplement and update the information which may be gained from other sources. Hence, the major part of the first section consists of tables concerning technical data of commercially available gas discharge lamps, which have hitherto not been tabulated in this comprehensive form. Also included are some rather marginal remarks on other artificial light sources and on sunlight. Except for countries with favorable climatic conditions, sunlight as a light source has lost its attractiveness in the course of the last few decades as compared to the days of CIAMICIAN and SILBER (for a tribute to these pioneers in photochemistry, with figures of the early photochemical equipment, see PFAU and HEINDEL [3, 4]).

The second section contains some data on filters, which are intended to supplement the most useful chapters on filters in recent books [1, 2].

For the compilation of the technical data the authors are indebted to F. SCHALLER, Mülheim.

1. Light sources

As stated above, more emphasis will be laid on a survey of commercially available gas discharge lamps rather than on treating all possible light

References, p. 494

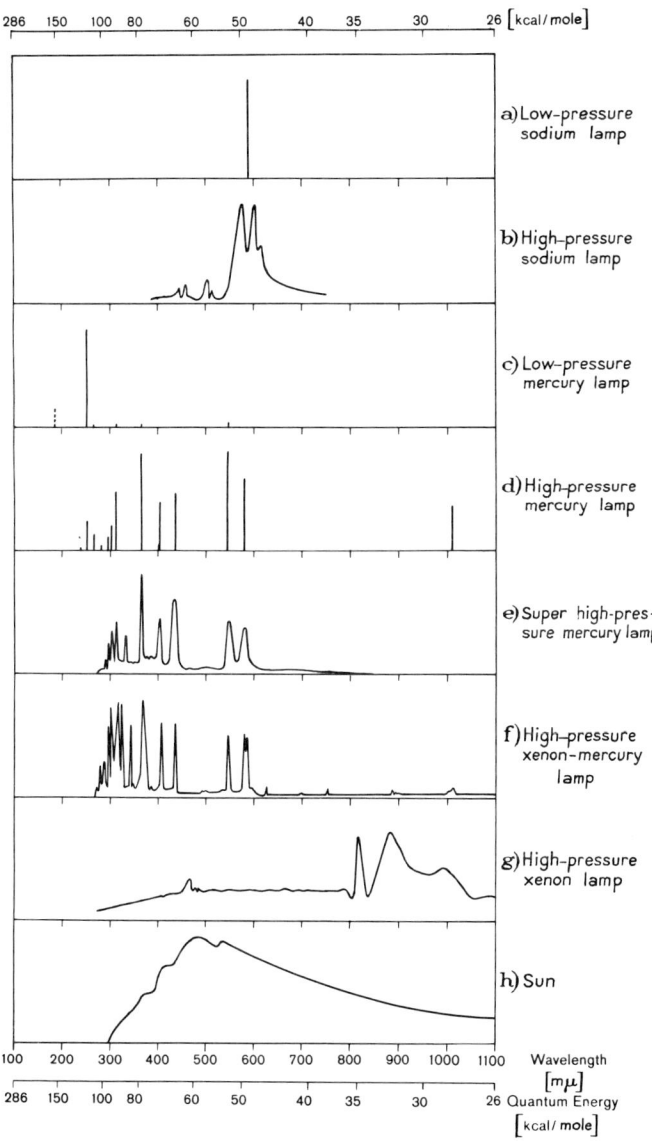

Fig. 1. Relative spectral energy distribution

sources. Hence, in tables 40–46 only closed lamps will be considered the intrinsic properties of which, form, dimensions and starting properties permit mounting within the reaction vessel. Although most of the lamps mentioned below are constructed for illumination, photocopying and sterilization purposes, they may well be used for preparative photochemistry. The main spectral region which has to be considered is the range between 250 and 600 mµ.

In tables 40—46 the most important mechanical, electrical and optical data of some gas discharge and metal vapor discharge lamps are given. The data have been compiled from the information received from the manufacturers and should represent the state of technical development up to 1966. Of course, these lists will not be exhaustive as concerns the number of manufacturers, and many lamps are not included since their technical data were not available.

Figure 1, a—h, shows typical emission spectra of the lamp types most commonly used and of sunlight. Aided by these data the reader should be able to select the lamp most suitable to his problem. One should, however, keep in mind that the spectral energy distribution not only depends on the kind of vapor used, but also on the pressure employed. Generally low pressure lamps show few, sharp lines (resonance lines). With increasing pressure new lines appear at longer wavelengths and an increasing broadening of the lines can be observed. The physical processes in discharge lamps have been considered elsewhere (cf. [5—8]); for the measurement of optical radiation see, inter alia, [1] and [9]. The literature for the period 1924—1958 concerning artificial light sources has been compiled in a Technical Report [10], and recent developments in construction of artificial light sources have been communicated on the Symposium on Solar Radiation Simulation [11].

a) Low-pressure sodium vapor lamps (table 40)

Low-pressure sodium vapor lamps emit nearly monochromatic light of 589 mµ (see fig. 1, a) with a very high yield (up to 125 lm/W).

Most U-shaped sources are permanently connected with a vacuum heat shield. Generally it takes the sodium lamp about 5 to 10 minutes to reach a luminous flux of about 80% of the normal operation luminous flux. In lamps which have not been allowed to cool down to room temperature restarting may be retarded. Choke coils, leak transformers etc. have to be used in the circuit. The average lifetime is about 5000 hours with 1000 starting procedures.

The mounting position of sodium lamps is often critical. Sodium lamps with an energy input of 60 or more Watts can only be started when the lamp is nearly horizontally positioned with a maximum angle of 20°. Nevertheless the larger sodium lamps, which are very interesting for preparative purposes, can be used in vertical immersion vessels if the lamp is started whilst in a horizontal position outside the reaction vessel. When the luminous flux has become constant the lamp is inserted in the immersion well.

References, p. 494

At the end of the irradiation the process is reversed, the current being cut off only when the lamp has been mounted in a horizontal position again [*12*]. The quantum flux given in table 40 has been calculated from the luminous flux data provided by the manufacturers.

b) High-pressure sodium vapor lamps (table 40)

A high-pressure sodium lamp has recently been put on the market by General Electric Co. under the order number LU-400. The lamp is delivered with a ceramic arc-tube in an evacuated glass bulb. The spectrum of this lamp shows several considerably broadened lines (see fig. 1, b). The lamp is designed for any mounting position.

c) Low-pressure mercury lamps (table 41)

Low-pressure mercury lamps (working pressure 0.005—0.1 mm Hg) show emission mainly of the mercury resonance line 253.7 mμ (see fig. 1, c). The fraction of light of longer wavelength strongly depends on the working pressure and varies in the various low-pressure mercury lamps between approximately 10% (NN 30/89, Quarzlampengesellschaft Hanau) and 50% (NK 25/7, Quarzlampengesellschaft Hanau). If quartz is used for the construction of the arc tube, the lamps also show approximately 8% radiation of 185 mμ, whereas lamps with a vycor tube do no emit this wavelength, at which ozone is already formed. The luminous flux of low-pressure mercury lamps strongly depends on the temperature and reaches a maximum at 40—45 °C. The lamps show full emission immediately after turning the current on, and their average lifetime is about 6000—8000 hours.

The quantum flux given in table 41 was calculated from the UV output (W) data for 254 mμ given by the manufacturers.

d) High-pressure mercury lamps (table 42)

High-pressure mercury lamps (working pressure 1—10 atm) show, besides a weak continuum emission, a typical line spectrum (see fig. 1, d). Quite frequently these lamps will also be designated as medium-pressure mercury lamps.

The spectral energy distribution (for examples see tables 47 and 48) depends strongly on the construction characteristics of the lamp and the working pressure. The high pressure of the mercury vapor arises during the first 3—5 minutes of use. Therefore immediately after starting the amperage is twice as high as the stationary working current. During the warm-up the voltage at the lamp rises from 15—30 Volts to the working voltage. Lamps which were not allowed to cool down may start again only after an induction period. According to the type of lamp employed, either a resistance or an inductance coil etc. have to be connected in series with the lamp. The lifetime of high-pressure mercury lamps is about 1000 hours.

High-pressure mercury lamps with halide additives show, besides the mercury lines, the lines of the metals added as their halides. By employing a suitable mixture of metal halides (mostly iodides) a more uniform distribution of energy (resembling daylight) and higher light intensities can be achieved. For examples see table 42.

Much more interesting for photochemical purposes are lamps in which by addition of only one halide certain spectral regions are emphasized. By combination with suitable filters fairly monochromatic light of many different wavelengths can thus be made available. Lamps of this type which are so far only produced to order may be obtained from Quarzlampengesellschaft Hanau. The following metal iodide lamps are available at the moment with the predominating emission range being given in brackets:

TlI (320—380, 530—540 mμ)
NaI (ca. 590 mμ)
CdI_2 (300—360, ca. 640 mμ)
ZnI_2 (ca. 330, ca. 635 mμ)
InI_3 (270—280, 300—330, ca. 410 mμ)
MgI_2 (ca. 280 mμ)

While the normal lifetime of high-pressure mercury lamps with halide additives is 800 hours, that of lamps containing MgI_2 or ZnI_2 is only 400 hours.

e) Super-high-pressure mercury lamps (table 43)

Mercury lamps with working pressures of up to 200 atm are designated here as super-high-pressure lamps, though they are occasionally also termed just high-pressure lamps. They show a much more compact arc than the aforementioned high-pressure mercury lamps of the same output. Their emission spectrum (see fig. 1, e) shows a stronger continuum and broadened mercury lines. Emission below 280 mμ is much weaker than in comparable high-pressure mercury lamps. Super-high-pressure mercury lamps are very important in all cases where light sources of high intensity in the visible region or the near UV are needed. Because of the small dimensions of the arc, the light can be easily focused by optical means. Some super-high-pressure mercury lamps have been constructed for water-cooling, and often they can be obtained already mounted with mirrors or lenses for focusing. Approximately 15 minutes are required until constancy of luminous flux is attained.

References, p. 494

f) Super-high-pressure xenon-mercury lamps (table 44)

Super-high-pressure xenon-mercury lamps also have a very short arc with extremely high brightness. The spectrum (see fig. 1, f) shows mainly the mercury lines with very high intensities down to 290 mμ. The luminous flux is constant within 15 minutes after starting of the arc.

g) High-pressure xenon lamps (tables 45 and 46)

High-pressure xenon lamps (working pressure 1—50 atm) emit a continuum spectrum, which is in the visible region quite similar to that of daylight (see fig. 1, g). The strong emission lines of xenon lie in the infrared region between 800 and 1000 mμ. In the UV the spectral energy distribution corresponds largely to that of extra-terrestrial sunlight, the intensity decreasing with decreasing wavelengths.

High-pressure xenon lamps reach constant working pressure, light intensity etc. immediately after starting. In order to deliver the high voltage pulses of 10—25 kV needed for starting, special starting devices are necessary.

Several firms produce high-pressure xenon lamps with extremely short arcs for projection purposes (see table 45). Hanovia and Duro-Test manufacture short arc lamps with water-cooled anodes for extremely high wattage (up to 20 kW). For preparative purposes the water-cooled types XBF by Osram (cf. table 46) are very well suited since excessive heating of the sample or of parts of the device can be avoided.

h) Sunlight

Though sunlight is in most parts of the world present in abundance its use as a photochemical light source involves certain difficulties, such as low intensities, especially in the short wavelength region, and low reproducibility in intensity due to changes in, for example, the angle of incident radiation during the day and during the seasons of the year. Besides, in insolation experiments precautions have to be taken to exclude thermal reactions which might otherwise be initiated by the infrared radiation which constitutes about 50% of the sun's spectral energy. For the relative spectral energy distribution of sunlight at the earth's surface see fig. 1, h [13].

More information on problems connected with the use of solar energy as a photochemical light source may be gained from the relevant literature (see the bibliography p. 495 for books on solar energy and on atmospheric photochemistry).

i) Incandescent lamps

Incandescent lamps may, for specific purposes, be used as light sources. Since a considerable part of their radiation is emitted in the infrared these lamps may be put into action when visible light as well as heat is required to effect a photochemical reaction. No list of manufacturers and no technical data on incandescent bulbs and on photoflood lamps will be given here.

j) Halogen incandescent lamps

So-called halogen lamps differ from ordinary incandescent lamps in having a small amount of iodine or bromine added to the filling gas, which is contained in a quartz bulb. Such lamps are distinguished by their small dimensions, high efficiency, constant luminous flux and long lifetime. These advantages are achieved by a recycling process, by which the evaporated tungsten is redeposited on the filament. Thus a darkening of the bulb can be avoided. Halogen lamps from different producers, with wattage from 50 to 10,000 Watts, are on the market. For data on such lamps the literature provided by the manufacturers should be consulted.

k) Fluorescent tubes

Various fluorescent tubes may be useful for photochemical work in the visible and near UV region. Because of abundance of the market no data will be given here.

l) Vortex-stabilized plasma lamps

Vortex-stabilized plasma lamps have been produced by the Giannini Scientific Corporation, Santa Ana, Calif. The spectrum of these lamps may be varied within certain limits by variation of filling gas, gas pressure and current intensity. On increase in pressure or current intensity the UV fraction is strongly increased relative to the emission in the visible and IR range. So far, vortex-stabilized plasma lamps can be provided for wattages of 5, 10, 25 and 50 kWatts.

m) Low-pressure gas discharge lamps for the far UV

Below 200 mμ quartz and oxygen begin to absorb light appreciably. Hence, for photochemical experiments in this range the window material has to be lithium fluoride, calcium fluoride or sapphire, and oxygen has to be excluded from the light path.

Besides the low-pressure mercury lamps mentioned above (p. 475) with their weak emission at 185 mμ, the following low-pressure lamps (working pressure 0.1—10 mm Hg) have proved useful:

filling gas	important lines
iodine	206 mμ
xenon	130, 147 mμ
krypton	116, 124 mμ
hydrogen	122 mμ

References, p. 494

Emission of these lamps, which are to our knowledge not yet commercially available, can be sustained without electrodes by the use of microwaves (microwave output 100 W). Thus, contamination from the electrodes can be avoided in the gas filling, and higher light intensities can be achieved. Besides, filling gases may be employed which cannot be used in the presence of metal electrodes, such as chlorine, bromine, or iodine. A quantum flux in the order of $10^{14}-10^{16}$ quanta per second can be achieved; the window size is several square centimeters. For details see McNesby and Okabe [14] and Okabe [15].

n) Flash lamps

In spite of the importance which flash lamps have gained in the studies of fast reactions and in the elucidation of photochemical mechanisms, these lamps do not offer advantages in preparative organic photochemistry. For more detailed information on flash lamp technique the reader is referred to Porter [16] and Calvert and Pitts [1].

o) Lasers

Laser radiations shall also be mentioned only marginally. The laser light is monochromatic, parallel and coherent and offers the possibility of achieving extremely high light intensities by focusing. Because of these properties lasers should be kept in mind as a unique tool for solving special problems. More discussion on the use of lasers in photochemistry will be found elsewhere [17, 18].

Tables

As stated above the following data have been compiled from the manufacturers' literature and no guarantee can be taken for their correctness by these authors. Unless stated to the contrary, the working voltage is always a–c. Arc dimensions cannot be defined as to inner or outer diameter due to lack in information from the manufacturers.

Abbreviations

D	Duro-Test Corp., New York, N.Y. (USA)
G	General Electric Co., Cleveland, Ohio (USA)
Gr	Gräntzel, Physikal. Werkstatt, Karlsruhe (Fed. Rep. Germany)
H	Hanovia Lamp Division, Newark, N.J. (USA)
M	Lampe Mazda, Paris (France)
O	Osram GmbH, Berlin (Fed. Rep. Germany)
P	Philips, Eindhoven (Holland)
PE	PEK Labs. Inc., Sunnyvale, Calif. (USA)
Q	Quarzlampengesellschaft, Hanau (Fed. Rep. Germany)
T	Tungsram, Budapest (Hungary) and Frankfurt/Main (Fed. Rep. Germany)
U	Ushio Kogyo Kaisha Ltd., Tokio (Japan)
W	Westinghouse Lamp Division, Bloomfield, N.J. (USA)

1. voltage of power supply
2. water-cooled tube
3. water-cooled anode
4. diameter of cooling vessel instead of arc tube diameter
5. lamp dimensions incl. heat resistant glass jacket or bulb
6. lamp dimensions incl. blacklight bulb
7. non-ozone-producing type
8. U-shaped arc tube
9. meandered arc tube
10. with reflector for directed radiation
11. with stabilized arc
12. high-pressure mercury lamp with halide additives
13. high-pressure sodium lamp (luminous flux in klm; arc tube diameter instead of lamp diameter)
14. lamp also available with phosphor-covered bulb
15. immersion well system available
16. non-ozone-producing type also available

Table 40. *Sodium vapor lamps* (λ_{max} = 589 mμ)

type	manufacturer	power input W	working voltage V	length over all with socket mm	length of arc mm	lamp diameter mm	quantum flux Nhν/Std.	remarks
SIO 40	M	40		311		53	0.15	8
Na 40 W-2	O	40		310	190	51	0.15	8
SO-X 40 W	P	40	75	303	190	51	0.15	8
SO-I 45 W	P	45	80	257	280	52	0.13	8
SIO 60	M	60		424		53	0.26	8
Na 60 W-2	O	60		424	305	51	0.26	8
SO-X 60 W	P	60	115	417	610	51	0.26	8
SO-I 60 W	P	60	110	310	400	52	0.18	8
SO-I 85 W	P	85	165	424	610	52	0.30	8
SIO 100	M	100		525		61	0.44	8
Na 100 W-2	O	100		525	405	64	0.44	8
SO-X 100 W	P	100	120	518	810	64	0.44	8
SO-I 140 W	P	140	165	525	1000	62	0.50	8
SIO 150	M	150		775		61	0.73	8
Na 150 W-1	O	150		775	645	64	0.73	8
SO-X 150 W	P	150	180	765	1290	64	0.73	8
SIO 200	M	200		1120		66	1.06	8
Na 200 W-2	O	200		1120	965	64	1.06	8
SO-I 200 W	P	200	250	785	1310	62	0.81	8
SO-X 200 W	P	200	265	1110	1930	64	1.06	8
Na 220 W	O	220		1200	1030	45	0.92	
LU 400	G	400		248	100	10	42 klm	13

References, p. 494

Table 41. *Low-pressure mercury lamps* ($\lambda_{max} = 254$ mμ)

type	manufacturer	power input W	working voltage V	length over all with socket mm	length of arc mm	arc tube diameter mm	quantum flux Nhν/Std.	remarks
OZ 4 W	P	4	12	59		34	0.0008	
TUV 4 W	P	4	12	58		34	0.0008	
G4Sll	G	4		57		35	0.0008	
G4T4/1	G	4		146	152	13	0.0038	
NK3/12	Q	4	210	200	120	10	0.0038	
NK 4/4	Q	4	200	116	10	5—10	0.0038	
TUV 6 W	P	6	220	150	75	26	0.0006	
NK 6/20	Q	7	350	220	135	16	0.0069	15
G8T5	G	8		305	216	16	0.0099	
HNS 10	O	10	220	398	296	10	0.029	1
HNS12	O	10	220	221	142	10	0.023	1
93 A-1	H	10	337	410	292	7	0.014	
LO 735 A-7	H	12	12$\simeq$	78		24		
HNS 1350/0.5	O	13	800	502	360	14	0.016	1
NN 15/44	Q	15	58	435	330	15	0.046	
G 15 T 8	G	15		457	356	25	0.027	
86 A-45	H	15	143	136	114	18	0.011	7,8
TUV 15 W	P	15	53	452		26	0.024	
84 A-1	H	16	563	745	628	7	0.031	
TG 16	M	16	60	440	350	25	0.019	
96 A-1	H	17	385	882	743	18	0.033	7
83 A-1	H	20	194	413	273	18	0.024	7
83 A-3	H	20	194	413	273	27	0.020	7
87 A 45	H	20	182	210	267	18	0.033	7,8
HNS 1350/1	O	20	1000	1002	860	14	0.038	1
G 25 T 8	G	25		457	356	25	0.038	
25 W	Gr	25	500	465	510	20		8,16
NK 25/7	Q	27	270	150	70	4—10	0.038	8
NN 30/89	Q	30	108	895	790	15	0.114	15
G 30 T 8	G	30		916	814	25	0.063	
94 A-1	H	30	287	770	628	18	0.056	7
94 A-3	H	30	287	770	628	27	0.050	7
88 A-45	H	30	268	387	623	18	0.079	7,8
TG 30	M	30	107	900	810	25	0.044	
TUV 30 W	P	30	100	909		26	0.061	
G 36 T 6	G	36		915	745	19	0.100	
G 36 T 6/1	G	36		915	745	19	0.075	
HNS 1375/2.0	O	46	1500	2002	1820	14	0.083	1
G 64 T 6	G	65		1625	1475	19	0.137	
100 W	Gr	100	500	485	1100	15		9,16
H 23 KX	G	1500		495	298	29	0.595	

31 Schönberg, Photochemistry

Table 42. *High-pressure mercury lamps*

type	manufacturer	power input W	working voltage V	length over all with socket mm	length of arc mm	arc tube diameter mm	luminous flux klm	remarks
Q 15	Q	15	25	20	4	7	0.5	
Q 25	Q	25	70 ≃	45	5	7	0.9	
St 42	Q	25	41	120	25	7	0.55	11
St 40	Q	45	68 ≃	120	10	7	0.65	11
St 41	Q	45	75 =	120	8	7	1.2	11
MA 80	M	80		156		70	3.3	5,14
HQA 80 W	O	80		156	20	70	3.1	5,14
HP 80 W	P	80		156		70	3.1	5,14
St 75	Q	80	90 =	120	20	10	2.5	11
Q 81	Q	70	72 ≃	125	20	10	2.7	15
H 85 A 3	G	85		143			3	5
SOL 608 A	H	100	100 ≃	156	74	16		15
SH 616 A	H	100	100		43			8
H 38-4 AB	W	100		143	29			5
H 100 A 38	G	100		184			3.6	5,14
H 100 BL 38	G	100		140				5,6
Q 400	Q	120	90	142	30	10	4.2	
MBL/D	M	125	110	115	20	11		11
MA 125	M	125		177		75	5.6	5,14
HQA 125 W	O	125		170	30	75	5.4	5,14
HQV 125 W	O	125		170		75		5,6
HPQ 125 W	P	125	95	88	39	12.5		
HgO 125	T	125	110	135	33	13		
HPK 125 W	P	125	125	95	30	11	4.75	
HP 125 W	P	125		177		75	5.4	5,14
HPR 125 W	P	125	125	222		110	2.8	5,10
HPW 125 W	P	125		177	30	75		5,6
H 39—22 KB	D,W	175		210	51		7.5	5,14
H 175 A 39	G	175		210			7.3	5,14
Q 600	Q	180	125	202	45	13.5	7.0	
S 654 A	H	200	125	235	114	23		15
PRK 2	T	220	70	190	83	17		
Q 300	Q	240	110	240	116	16	4.5	
HgO 250	T	250	115	162	52	19		
H 37—5 KB	D,W	250		210	54		11.8	5,14
UA-2	G	250	92	187	85		6.8	
UV-250 W	P	250	130 =	267	80	21		
HP 250 W	P	250		227		90	11.5	5,14
MA 250	M	250		226		90	12	5,14

References, p. 494

Light sources 483

Table 42 continued

type	manufacturer	power input W	working voltage V	length over all with socket mm	length of arc mm	arc tube diameter mm	luminous flux klm	remarks
HQA 250 W	O	250		226	50	90	11.5	5,14
H 250 A 37	G	250		210			11	5,14
V 300	Q	300	140	445	300	15		
UA-3	G	360	134	263	152		9	
PRK 4	T	375	120	258	147	20		
HgO 400	T	400	125	180	70	21		
H 33—1 CD	D,W	400		288	70		21	5,14
MA 400	M	400		290	120	120	20.5	5,14
HQA 400 W	O	400		292	70	120	20.5	5,14
HQI 400 W	O	400		265	48	65	33.0	5,12,14
HOQ 400 W	P	400	125	422	301	24		
HP 400 W	P	400		290		120	20.5	5,14
HPI 400 W	P	400		283		46	33	5,12,14
H 400 A 33	G	400		287			20.5	5,14
L 679 A	H	450	135	237	114	26		15
Q 700	Q	500	125	346	180	23	12.5	15
A 673 A	H	550	145	237	114	31		15
HOKI 600 W	P	600	600	533	415	35		5
SA 674 A	H	700	150		190			15
H 700 A 35	G	700		364			37	5,14
H 35—18 NA	W	700		370	127			5,14
HOGK 700 W	P	700	190	567	417	24		16
HOG 700 W	P	700	190	567	417	24		7
MA 700	M	700		330		140	37	5,14
PRK 7	T	700	120	322	210	25		
Q 1200	Q	900	230	396	247	23	29	15
MA 1000	M	1000		372		65	52	5,14
H 34—12 GV	D,W	1000		390	127		55	5,14
HQA 1000 W	O	1000		430	120	100	52	5,14
HP/T 1000 W	P	1000		382		65	52	5,14
H 1000 A 34	G	1000		382			55	5,14
UA-11	G	1200	450	560	452		50	
UA-11 B	G	1200	450	566	439		50	5
PQ 1200	Q	1200	480	557	430	30		
HOKI 1200 W	P	1200	550	520	400	35		5
LL 189 A	H	1200	285		305			15
HOKI 1500 W	P	1500	1750	1490	1370	35		5
H 1500 A 23	G	1500		495			81	5
HQA 2000 W	O	2000		430	145	100	125	5,14

31*

Table 42 continued

type	manufacturer	power input W	working voltage V	length over all with socket mm	length of arc mm	arc tube diameter mm	luminous flux klm	remarks
HQI 2000 W	O	2000		430	145	100	190	5,12,14
HPI 2000 W	P	2000		477		100	190	5,12,14
HOKI 2000 W	P	2000	550	710	590	50		5
HOGK 2000 W	P	2000	550	1367	1217	24		16
HOK 2000 W	P	2000	550	780	530	24		
Q 2024.28	Q	2000	240	388	245	23	69	15
Q 2024.100	Q	2000	520	1150	1000	24		15
PQ 1990	Q	2000	550	1365	1180	23		
SS 78 A	H	2100	550		534			15
HOKI 2500 W	P	2500	550	1510	1335	40		5
HOGK 2500 W	P	2500	550	1504	1354	24		16
PQ 2502	Q	2500	550	1475	1338	40		
HOKI 3000	P	3000	1250	1490	1370	35		5
UA-37	G	3000	1500	1330	1220		129	
UA-9	G	3000	535	1450	1270		120	7
Q 3000	Q	3000	650	1465	1300	23	130	
PQ 3018	Q	3000	1500	1487	1350	40		
H 3000 A 9	G	3000		1395			132	5
BMS 47 A	H	3500	925		1220			15
PQ 4026	Q	4000	880	1580	1350	50		
Q 4024.100	Q	4000	1000	1150	1000	24		15
PQ 4029	Q	4500	1100	2007	1800	28		
MSS 77 A	H	4500	960		1070			15
PS 59 A	H	4500	1000		1470			
PIS 57 A	H	5000	1260		1180			
HOKI 5000 W	P	5000	1800	1490	1370	50		5
HST 65 A	H	7500	1700		1500			15
PQ 7500	Q	7500	1650	1355	1215	50		
PQ 9000	Q	9000	1650	1615	1480	50		
Q 10030.150	Q	10000	2200	1650	1500	30		15
HOQ 10 kW	P	10000	3500	1520	1370	24		
Q 20040.150	Q	20000	2200	1650	1500	40		15
HQP 25 KW	O	25000	3500	2110	1800	35		
HOQ 30 kW	P	30000	2500	2000	1800	45	1800	
HQP 40 KW	O	40000	3500	2225	1900	55		
Q 40050.190	Q	40000	3500	2100	1900	50		15
100 KW	D	100000	1350	2100	1800		5200	

References, p. 494

Table 43. *Super-high-pressure mercury lamps*

type	manufacturer	power input W	working voltage V	length over all with socket mm	length of arc mm	arc tube diameter mm	luminous flux klm	remarks
HBO 75 W	O	75	51	122	2.8	30	2.6	
HBO 100 W/2	O	100	20=	90	0.25	10	2.2	
CS 100 W	P	100	20=	85	0.25	9.5	2.0	
PEK 107	PE	100	20=	89	0.35	13	2.2	
PEK 110	PE	100	20=	89	0.35	13	2.2	
HgK 100	Q	100	55	115	2	14	4.0	
CS 150 W	P	150	66≃	140	2	35	7.0	
HBO 200 W/2	O	200	65≃	128	2.2	17	10	
CS 200 W	P	200	57=	122	2.2	17	10	
PEK 202	PE	200	57≃	121	2.5	17	9.5	
PEK 203	PE	200	57≃	121	2.5	17	9.5	
PEK 200−2	PE	200	57≃	104	2.5	17	9.5	
PEK 200−3	PE	200	57≃	104	2.5	17	9.5	
ME/D 250 W	M	250	75≃	141	3.75	50		
SAH 250 A	W	250	37	159	2.5	38	10	
SAH 250 B	W	250	42=	159	2.5	38	10	
HBO 500 W/2	O	500	85≃	170	4.1	28	30	
SP 500 W	P	500	450		12.5		30	2
PEK 500−2	PE	500	77≃	171	2.5	29	24.5	
PEK 500−3	PE	500	77≃	171	2.5	29	24.5	
PEK 500−4	PE	500	77≃	171	2.5	29	24.5	
AH 6−.5	PE	500	350	70	13	2	32.5	2
BH 6−.5	PE	500	350	70	13	2	32.5	
B−H 6	G	900	900	83	25	6	60	
SP 900 W	P	900	750	80	25	7	50	
A−H 6	G	1000	840	83	25	6	65	2
ME/D 1 kW	M	1000	75≃	245	6.5	48	50	
CS 1000 W	P	1000	80	285	4.2	46	50	
SP 1000 W	P	1000	500=		12.5		30	2,10
SAH 1000 A	W	1000	65	240	6.5	51	50	
C-1	PE	1000	700	82	29	1.5	65	2
AH 6−1	PE	1000	700	82	25	2	65	2
BH 6−1	PE	1000	700	82	25	2	65	
A-1	PE	2000	1425	82	29	1	130	2
AH 6−2	PE	2000	1425	108	51	2	130	2
SAH 2500 A	W	2500	65	330	10	70	125	
A-2	PE	4000	2850	104	51	1	260	2

Table 44. *Super-high-pressure mercury-xenon lamps*

type	manufacturer	power input W	working voltage V	length over all with socket mm	length of arc mm	arc tube diameter mm	luminous flux klm	remarks
901 B	H	200	25=	114	2	20	8	
528 B −1/−9	H	1000	72=	190	6	38	40	
537 B	H	1000	65	190	6	38	50	
SAHX 1000 A	W	1000	65	228	6.5	51	50	
929 B	H	2500	50=	317	4	64	120	
SAHX 2500 A	W	2500	65	330	10	70	125	
SAHX 2500 B	W	2500	50=	330	6	70	120	
932 B	H	5000	60=	344	5	86	230	

Table 45. *Xenon short-arc lamps*

type	manufacturer	power input kW	working voltage V	length over all with socket mm	length of arc mm	arc tube diameter mm	luminous flux klm	remarks
X 35	PE	0.035	12=	85	0.35	9	0.7	
X 36	PE	0.035	12=	85	0.35	9	0.7	
X 75	PE	0.075	14=	89	0.38	12	1.4	
X 76	PE	0.075	14=	89	0.38	12	1.4	
XBO 75 W/1	O	0.075	14=	90	0.5	10	0.85	
448 C	H	0.08	16=	84	0.5	10	2.7	
901 C	H	0.15	20=	114	1.9	20	3.2	
510 C 1	H	0.15	21	114	4.0	20	5.0	
XBO 150 W/1	O	0.15	20=	150	2.2	20	3.0	
CSX 150 W	P	0.15	20=	150	2.2	20	3.0	
XBO 150 W	O	0.15	19	150	2.2	20	3.2	
UXL 150	U	0.15	20=	165	2.5	20	3.7	
X 150	PE	0.15	20=	121	1.2	19	3.2	
X 151	PE	0.15	20=	121	1.2	19	3.2	
XE/D 250 W	M	0.25	16	125	3.0	35	5.0	
XBO 250 W	O	0.25	14=	226	1.7	25	4.8	
914 C	H	0.30	17=	135		20		
X 300	PE	0.30	19=	121	1.6	19	6.3	
XBO 450 W/2	O	0.45	18=	175	2.7	29	13	
XBO 450 W	O	0.45	18=	260	2.7	29	13	
CSX 450 W	P	0.45	18=	260	2.7	29	13	
XHP 450	T	0.45	18=	260	2.7	29	13	
959 C	H	0.50	17=	124	0.9	25	18	
XE/D 500 W DC	M	0.50	22=	215	5.5	40	12	
XE/D 500 W AC	M	0.50	20	215	6	40	11	
UXL 500	U	0.50	22=	228	4	30	14	
418 C	H	0.90	31	191	6	36	26	

References, p. 494

Table 45 continued

type	man-ufac-turer	power input kW	working voltage V	length over all with socket mm	length of arc mm	arc tube diam-eter mm	lumi-nous flux klm	re-marks
XBO 900 W	O	0.90	20=	325	3.3	40	30	
CSX 900 W	P	0.90	20=	325	3.3	40	30	
XHP 900	T	0.90	20=	325	3.3	40	30	
XBO 900 W/2	O	0.90	20	239	3.3	40	30	
538 C−1	H	0.95	35=	191	6	38		
UXL 1000	U	1.00	26=	315	4	40	32	
XBO 1600 W	O	1.60	25=	370	4	52	60	
CSX 1600 W	P	1.60	25=	370	4	52	60	
XHP 1600	T	1.60	25=	370	4	52	60	
XE/D 2 kW	M	2.00	25=	315	5.5	65	70	
UXL 2000	U	2.00	30=	365	6	53	70	
UXL 2000 DK	U	2.00	30=	360	6	40	70	
491 C	H	2.20	24=	335	5	56	80	
XBO 2500 W	O	2.50	30=	428	6	57	100	
CSX 2500 W	P	2.50	30=	428	6	57	100	
XHP 2500	T	2.50	30=	428	6	57	100	
UXL 4000 DK	U	4.0	31=	400	7	55	120	
XBO 4000 W	O	4.0	41=	428	7	54	180	
966 C	H	5.0	70	435	10	71	220	
UXL 5000 DK	U	5.0	30=	428	7	60	160	
XE 5000	G	5.0	34=	495	7	89	275	
XBO 6500 W	O	6.5	41=	510	9	60	325	
XE 10000	D	10	40=	725	9	92	550	
XE 20000	D	20	48=	547	12	102	1160	3
L 5020	H	20	45=	760	11	121	1000	3

Table 46. *Xenon long-arc lamps*

type	man-ufac-turer	power input kW	working voltage V	length over all with socket mm	length of arc mm	arc tube diam-eter mm	lumi-nous flux klm	re-marks
XB 1 kW AC	M	1.0	42	300	85	30	22	
XBF 1000 W/1	O	1.0	90	251	50	20	26	2,4
XBF 1000 W/2	O	1.0	90=	251	50	20	23.5	2,4
XO 1500 W	P	1.5	54	380	150	25	31.5	
XBF 2500 W/1	O	2.5	115	277	75	20	77	2,4
XBF 6000 W/1	O	6.0	135	360	110	25	215	2,4
XQO 6000 W	O	6.0	140	890	600	24.5	140	
XO 6000 W	P	6.0	125	855	520	28	140	
XQO 10000 W	O	10.0	140	1200	750	34.5	250	
XO 10000 W	P	10.0	140	1250	750	35	250	
XQO 20000 W	O	20.0	280	1950	1500	34.5	500	

488 Light sources and light filters in preparative organic photochemistry

Table 47. Relative spectral energy distribution of high-pressure mercury lamps of Quarzlampengesellschaft, Hanau

wavelength [nm]	wavenumber [kK]	St 40	St 41	St 42	St 75	Q 25	Q 81	Q 300	Q 400	Q 600	Q 700	Q 1200	Q 2024.28	Q 4024.100	Q 10030.150	Q 20040.15	Q 40050.190
238	42.0	2	2	2	2	1	2	3	2	2	3	3	4			4	(cut off
240	41.7	2	2	2	2	1	2	3	2	2	3	3	4			4	by
248	40.3	7	7	11	7	7	9	10	9	10	11	10	13	16	42	10	glass
254	39.4	68	20	150	30	4	46	81	36	101	59	40	42	58	81	146	sleeve)
265	37.7	24	8	33	17	9	19	26	14	19	24	23	26	33	88	15	—
270	37.0	2	2	3	2	2	2	4	2	3	4	4	3			5	—
280	35.7	7	8	9	6	6	8	9	8	9	11	12	13	10	19	11	—
289	34.6	6	6	7	5.5	6	7	6	7	6	7	9	8	3	5	6	—
297	33.7	17	15	20	15	17	17	16	17	16	17	18	18	14	23	15	2
302	33.1	25	26	28	25	23	30	32	30	30	34	37	39	32	53	24	7
313	31.9	61	63	81	55	59	62	71	60	61	69	63	69	67	81	52	28
334	29.9	7	8	7	7	7	8	7	8	8	7	8	7	4	8	8	7
366	27.3	100	100	100	100	100	100	100	100	100	100	100	100	100	100	100	100
405/408	24.7/24.5	43	50	59	43	39	44	32	42	43	33	46	38	40	57	57	96
436	22.9	61	74	84	67	61	71	61	61	62	69	69	46	68	102	92	154
546	18.3	84	89	103	95	89	92	68	89	91	76	87	58	75	91	123	203
577/579	17.3/17.3	65	63	54	70	53	79	65	87	86	72	78	77	41	81	100	165
factor for computation of light flux in mWatts		6.5	13	4.5	23	7.4	30	45	58	74	120	280	660	1650	4300	9930	12100

References, p. 494

Table 48. Spectral energy distribution of high-pressure mercury lamps of Hanovia, Newark, N.J.

wavelength [Å]		wavenumber [kK]	lamps SOL 608A	SH 616A	S 654A	L 679A	A 673A	SA 674A	LL 189A	SS 78A	BMS 47A	PS 59A	MSS 77A	PIS 57A	HST 65A
13673	(infrared)	7.315	0.65	0.95	1.0	2.6	4.6	4.1	10.15	19.0	34.9	39.0	30.1	26.2	39.3
11287		8.861	0.62	0.85	1.3	3.3	3.8	5.0	6.93	14.2	27.8	30.2	21.9	30.8	46.2
10140		9.865	0.85	1.30	1.8	10.5	12.2	14.6	31.60	55.2	88.8	51.3	78.0	110	165.0
5780	(yellow)	17.30	1.55	1.50	3.4	20.0	23.0	32.1	69.35	120	165	141	175	198	297.0
5461	(green)	18.32	1.35	1.50	3.0	24.5	28.2	34.0	40.52	92.2	156	150	163	193	290.0
4358	(blue)	22.95	1.08	0.84	2.6	20.2	23.3	29.0	53.00	101	151	142	140	144	216.0
4045	(violet)	24.72	0.75	0.51	1.6	11.0	12.7	15.9	24.20	47.3	90.5	94.7	82.8	100	150.0
3660		27.32	1.40	1.82	3.1	25.6	30.1	40.5	97.10	180	268	302	310	295	443.0
3341		29.93	0.13	0.18	0.36	2.4	2.8	3.8	6.93	16.8	20.9	32.3	29.2	31.0	46.6
3130		31.95	1.02	1.30	2.3	13.2	15.0	21.0	50.6	105	153	195	200	180	270.0
3025		33.06	0.41	0.57	0.86	7.2	8.2	11.3	32.9	64.8	80.1	76.3	74.7	78.4	117.6
2967		33.70	0.32	0.30	0.48	4.3	5.0	6.5	15.2	28.2	43.5	44.7	44.3	44.6	66.9
2894		34.55	0.10	0.19	0.20	1.6	1.8	2.3	4.41	8.22	13.8	13.3	15.1	14.0	21.0
2804		35.66	0.12	0.19	0.30	2.4	2.8	3.8	13.9	24.9	31.5	41.7	40.0	42.6	63.9
2753		36.32	0.06	0.08	0.14	0.7	0.8	1.0	4.2	7.5	8.2	13.0	12.3	14.4	21.6
2700		37.04	0.07	0.09	0.14	1.0	1.2	1.3	4.85	8.6	9.8	14.7	13.0	15.2	22.9
2652	(UV)	37.71	0.30	0.47	0.64	4.0	4.6	6.6	27.80	50.2	71.5	89.3	92.2	105	158.0
2571		38.89	0.11	0.19	0.20	1.5	1.8	2.3	6.30	9.30	14.0	18.8	16.4	20.7	31.0
2537		39.42	0.34	0.37	1.10	5.8	5.0	7.3	24.1	53.2	62.8	97.1	87.4	84.7	137.0
2482		40.29	0.10	0.19	0.20	2.3	2.6	3.2	10.15	17.0	26.1	31.9	24.8	27.9	41.8
2400		41.67	0.05	0.12	0.20	1.9	2.2	2.9	7.30	13.6	13.2	22.1	18.7	21.2	31.9
2380		42.02	0.03	0.09	0.12	2.3	2.6	3.2	8.40	15.3	12.6	25.7	22.4	25.5	38.4
2360		42.37	0.02	0.06	0.08	2.3	1.8	2.3	6.20	11.5	10.0	19.9	14.2	19.1	28.7
2320		43.10	0.02	0.02	0.03	1.5	2.4	3.1	7.65	15.5	8.9	24.1	10.3	20.8	31.2
2224		44.98	0.04	0.02	0.03	3.7	4.2	4.7	9.20	21.0	26.5	30.8	21.9	26.3	39.4
Total output Watts			11.49	13.70	25.18	175.8	202.7	261.8	572.9	1099.5	1588.0	1740.9	1737.7	1868.4	2814.4

2. Light filters

The use of monochromatic light or of a selected portion of the spectrum is often necessitated in preparative photochemistry in order to prevent primary photoproducts from undergoing secondary photoreactions. Conventional prism or grating monochromator systems cannot provide monochromatic beams of sufficiently high intensity. But several types of glass and liquid filters, which may be used in combination with the light sources listed in the preceding section, may serve the purpose of selecting narrow lines or portions of the spectrum. In most cases combination of several filters is necessary.

Regarding the spectral properties, two types of filters may be distinguished (not considering heat- or infrared-absorbing filters which cut off the longer wavelength part of the spectrum):

cut-off filters

band-pass filters

which will for convenience not be differentiated further, since the so-called cut-off filters are, strictly speaking, also band-pass filters albeit with extreme band width. When using filters it must be kept in mind that filters heat up when absorbing light and may change their spectral properties, or they may be damaged otherwise by heat or by light. For this reason most filters must be preirradiated and checked spectroscopically.

a) Solid filter materials

Solid filters are most useful for preparative work with the immersion type photoreactor. The walls of the immersion well normally act as such

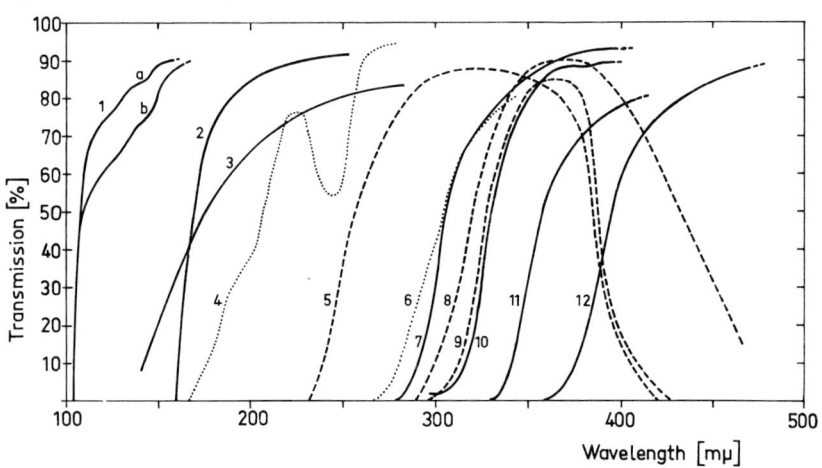

Fig. 2. Transmissions of various solid filter and window materials (cf. table 49)

References, p. 494

filters. Their spectral properties depend on the type of material employed (see fig. 2), and also band-pass filter glass may be used for lamp housings. Immersion wells made from Vycor 7910, Corex 9700 and Pyrex 7740 are now available for lamps from 100 Watts onwards from Hanovia (see table 42). For a survey on solid filter materials see also [8] and [19].

Table 49

No.	material	thickness	ref.
1	lithium fluoride (a: fresh; b: aged)	1—2 mm	[20]
2	Suprasil	10 mm	[21]
3	sapphire	1 mm	cf. [7]
4	fused quartz	10 mm	cf. [7]
5	Corning 9863 glass	3 mm	[22]
6	Solidex	2 mm	
7	Pyrex clear chemical glass	2 mm	cf. [7]
8	Corning 5850 glass	4 mm	[22]
9	Corning 5970 glass	5 mm	[22]
10	window glass	2 mm	
11	Wertheimer Filterglas GWCa	2 mm	
12	Wertheimer Filterglas GWV	2 mm	

b) Liquid filters

A great variety of liquid filter systems have been described in the literature (see, inter alia, BOWEN [23], BASS [24], KASHA [25] and RAPPOLDT [26]). Several transmission curves of liquid filters are shown in fig. 3 as taken from RAPPOLDT [26], and table 50 gives the corresponding

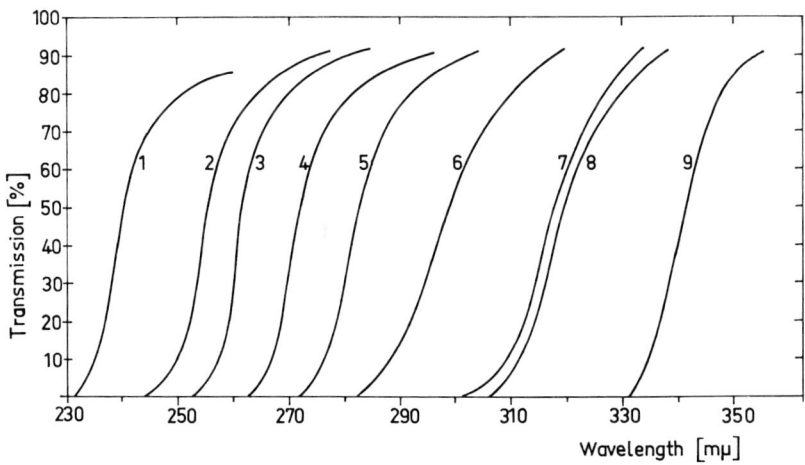

Fig. 3. Transmission of aqueous solutions of some salts (see table 50)

data on filter composition. Some of the filter compositions described in quotations [23—25] are also contained in recent books [2, 18], and CALVERT and PITTS [1] have carried out a very thorough study on light stability, impurity light etc. of a variety of liquid filters recommended by BOWEN, KASHA and others. A particularly useful compilation of prescriptions on filter compositions will be found in LANDOLT-BÖRNSTEIN [8].

When used in immersion type photoreactors, these liquid filters will either be contained in a separate filter jacket of the onion-type reactor or they may serve as coolant proper for the immersion lamp.

Table 50 [26]

No.	NaBr·2H$_2$O	CaCl$_2$	Pb(NO$_3$)$_2$	Hg(NO$_3$)$_2$	Ag$_2$SO$_4$
			solute [g/liter]		
1		500			0.05
2	400				0.024
3	400				0.16
4	400				1.20
5	500				6.40
6	80	375	0.050	0.10	0.30
7	400		0.065	0.10	0.39
8	400			1.00	
9	650		3.00		

c) Interference filters

When using directed light, very pure and intense monochromatic radiation can be obtained by means of interference filters (for a summary on this subject see again [1, 8]). There are two main types of interference filters on the market:

wide band filters
line filters

which differ from each other in the half-peak width. Because of their abundance on the market no details of such filters will be given here and reference is made to the literature provided by the manufacturers (for a selection see table 51).

Table 51

Axler Associates Inc., Corona, N.Y. (USA)
Baird Atomic Inc., Cambridge, Mass. (USA)
Balzers AG, Balzers (Liechtenstein)
Barr and Stroud Inc., Glasgow (United Kingdom)
Bausch and Lomb Optical Co., Rochester, N.Y. (USA)
Jenoptik, Jena (German Dem. Rep.)
Optical Coating Lab. Inc., Santa Rosa, Calif. (USA)
Photovolt Corp., New York, N.Y. (USA)
Schott AG, Mainz (Fed. Rep. Germany)

References, p. 494

Compared with color glass filters and other absorption type filters the interference filters have the advantage of reflecting the undesired spectral portion, so that they are not overheated so easily. When mounting the interference filter attention must be given to the mirror face positioned towards the light source.

In combination with, for example, a 2000 W xenon projector a good selection of spectral bands in the visible region can be attained by means of the interference filters Filtraflex K 1—7 of Balzers (see fig. 4). Filtraflex K 3—7 are now available in special water and heat proof, uncemented

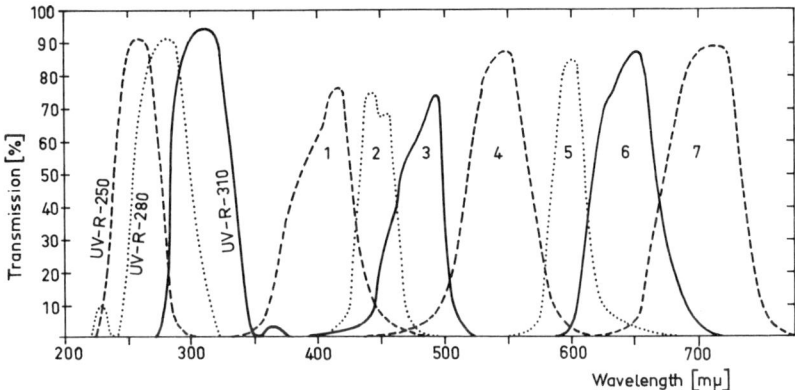

Fig. 4. "Transmission" of interference wide band Filtraflex filters K 1—7 [27] and of reflection interference filters UV-R-250, UV-R-280 and UV-R-310 [28]

design which warrants their mention here. Under direct water cooling these filters are resistant even to the focused full luminous flux of a 2000 Watt high-pressure xenon lamp. Because the Filtraflex K type filters show some transmission in the IR region it is necessary to insert a heat absorbing filter into the light path, preferably behind the interference filter.

d) Reflection interference filters

With usual interference filters lines or bands in the wave length region below 320 mµ can be singled out only with difficulty. In these cases the newly developed reflection interference filters UV-R for wavelengths 250, 280 and 310 mµ (Schott) may be successfully employed (for the working principle of these filters the manufacturer's literature [28] should be consulted). "Transmission" curves for the UV-R filter types available are included in fig. 4 (fourfold reflection at 45°).

References

[1] J. G. CALVERT and J. N. PITTS JR.: Photochemistry. New York: Wiley 1966, esp. pp. 686—814.
[2] A. D. MCLAREN and D. SHUGAR: Photochemistry of Proteins and Nucleic Acids. Oxford: Pergamon 1964, esp. pp. 358—387.
[3] M. PFAU and N. D. HEINDEL: Ann. Chim. [13] **10**, 187 (1965).
[4] N. D. HEINDEL and M. A. PFAU: J. Chem. Educ. **42**, 383 (1965).
[5] R. G. FOWLER: Radiation from Low Pressure Discharges. In: Handbuch der Physik, ed. by S. FLÜGGE, vol. **22**: Gasentladungen II. Berlin — Göttingen — Heidelberg: Springer-Verlag 1956, pp. 209—253.
[6] Quecksilberdampf-Hochdrucklampen. Grundlagen und Anwendungen. Ed. by W. ELENBAAS. Eindhoven: Philips Techn. Bibliothek 1966; English edition: High Pressure Mercury Vapour Lamps and their Applications. Eindhoven: Philips Techn. Bibliothek 1965.
[7] L. R. KOLLER: Ultraviolet Radiation. Second edition. New York: Wiley 1965.
[8] LANDOLT-BÖRNSTEIN: Zahlenwerte und Funktionen aus Physik, Chemie, Astronomie, Geophysik und Technik. Vol. IV: Technik, 3. Teil: Elektrotechnik, Lichttechnik, Röntgentechnik, 6th ed. Berlin — Göttingen — Heidelberg: Springer-Verlag 1957, pp. 881—924: Lichtquellen (E. LAX, P. SCHULZ and A. LOMPE) and pp. 925—962: Lichtfilter (W. GEFFCKEN).
[9] G. BAUER: Measurement of Optical Radiations. London: Focal Pr. 1965; German edition: Strahlungsmessung im optischen Spektralbereich. Braunschweig: Vieweg 1962.
[10] Lamps 1924—1958. Catalog of Technical Reports (CTR) 354 Washington: US Department of Commerce 1958.
[11] Proc. Internat. Symp. Solar Radiation Simulation, Los Angeles 1965. Mt. Prospect: Inst. Environmental Sciences 1965.
[12] G. O. SCHENCK and D. E. DUNLAP: Angew. Chem. **68**, 248 (1956).
[13] R. HERRMANN: Strahlentherapie **76**, 193 (1947).
[14] J. A. MCNESBY and H. OKABE, in: Adv. Photochem., ed. by W. A. NOYES JR., G. S. HAMMOND and J. N. PITTS JR., vol. **3**. New York: Interscience 1964, pp. 157—240.
[15] H. OKABE: J. Opt. Soc. Am. **54**, 478 (1964).
[16] G. PORTER, in: Technique of Organic Chemistry, ed. by A. WEISSBERGER, 2nd ed., vol. **8**, pt. 2: Investigation of Rates and Mechanisms of Reactions, ed. by S. L. FRIESS, E. S. LEWIS and A. WEISSBERGER. New York: Interscience 1963, pp. 1055—1106.
[17] N. J. TURRO: Molecular Photochemistry. New York: Benjamin 1965.
[18] H. H. SELIGER and W. D. MCELROY: Light: Physical and Biological Action. New York: Academic Press 1965, esp. pp. 349—357 (lasers), 358—360 (filters).
[19] K. A. MCCARTHY, S. S. BALLARD and W. L. WOLFE, in: American Institute of Physics Handbook, ed. by D. E. GRAY, 2nd ed. New York: McGraw-Hill 1963, pp. 6/45—6/78.
[20] D. A. PATTERSON and W. H. VAUGHAN: J. Opt. Soc. Am. **53**, 851 (1963).
[21] Form No. 6583 N-EX: Optical Fused Quartz. Hillside: Engelhard Industries Inc., Amersil Quartz Division.
[22] C. W. SILL: Anal. Chem. **33**, 1584 (1961).
[23] E. J. BOWEN: The Chemical Aspects of Light. Second edition. Oxford: Clarendon Press 1949, esp. pp. 276—281.
[24] A. M. BASS: J. Opt. Soc. Am. **38**, 977 (1948).
[25] M. KASHA: J. Opt. Soc. Am. **38**, 929 (1948).
[26] M. P. RAPPOLDT: Thesis Univ. Leiden 1958.
[27] Kurvenblatt No. 246, Breitbandfilter FILTRAFLEX-K, No. 1—7. Balzers: Balzers AG.
[28] Prospekt No. 3703: Interferenz-Reflexionsfilter UV-R-250. Mainz: Schott u. Gen.

A selective bibliography on photochemistry

The following pages are intended to give a comprehensive though by no means exhaustive survey of literature in the field of photochemistry. The first section contains titles of books, or parts thereof, dealing with photochemistry in general. Little or no mention has been made of work on photosynthesis, photobiology, experimental optics and illumination technique. The second section lists some titles of journals and serials, which are (wholly or in part) concerned with photochemistry. The final section contains quotations from reviews and related sources.

More emphasis has been laid on recency than on completeness of the literature covered. Hence, reviews were not scanned before 1945, and occasionally titles of books were included though the respective volumes had not yet appeared.

Books covering photochemical subjects

F. BACHÉR: Chemische Reaktionen organischer Körper im ultravioletten und im Licht der Sonne. In: Handbuch der biologischen Arbeitsmethoden, ed. by E. ABDERHALDEN, Abt. 1, Teil 2/II. Berlin: Urban & Schwarzenberg 1929, pp. 1339–1968.
A. BENRATH: Lehrbuch der Photochemie. Heidelberg: Winter 1912.
A. BERTHOUD: Photochimie. Paris: Doin 1928.
H. F. BLUM: Photodynamic Action and Diseases Caused by Light. New York: Reinhold 1941.
— Carcinogenesis by Ultraviolet Light. Princeton: Univ. Press 1959.
K. F. BONHOEFFER and P. HARTECK: Grundlagen der Photochemie. Dresden: Steinkopff 1933.
E. J. BOWEN: The Chemical Aspects of Light, 2nd ed. Oxford: Clarendon Press 1946.
Comparative Effects of Radiation, ed. by M. BURTON, J. S. KIRBY-SMITH and J. L. MAGEE. New York: Wiley 1960.
J. G. CALVERT and J. N. PITTS JR.: Photochemistry. New York: Wiley 1966.
B. CAPON, M. J. PERKINS and C. W. REES: Organic Reaction Mechanisms 1965, 1966. Chapter 13: Photochemistry. London: Interscience 1966, 1967.
Organic Photochemistry, ed. by O. L. CHAPMAN. Vol. 1. New York: Dekker 1967.
Photochemistry and Reaction Kinetics, ed. by F. S. DAINTON, P. G. ASHMORE and T. M. SUGDEN. Cambridge: Univ. Press 1967.
Solar Energy Research, ed. by F. DANIELS and J. A. DUFFIE. Madison: Univ. of Wisconsin Press 1961.

F. DANIELS: Direct Use of the Sun's Energy. New Haven: Yale Univ. Press 1964.

C. H. DePUY: Molecular Reactions and Photochemistry. Englewood Cliffs: Prentice-Hall 1967.

D. ELAD: Photochemistry of the C—O—C Group. In: The Chemistry of the Ether Linkage, ed. by S. PATAI. London: Interscience 1967, pp. 353—372.

C. ELLIS and A. A. WELLS: The Chemical Action of Ultraviolet Rays. New York: Reinhold 1941.

A. ÉTIENNE: Photo-oxydes d'Acènes. Union Labile de l'Oxygène au Carbone. In: Traité de Chimie Organique, ed. by V. GRIGNARD, G. DUPONT and R. LOQUIN, vol. 17. Paris: Masson 1949, pp. 1299—1332.

Optische Anregung organischer Systeme, ed. by W. FOERST. Weinheim: Verlag Chemie 1966.

T. FÖRSTER: Fluoreszenz organischer Verbindungen. Göttingen: Vandenhoeck & Ruprecht 1951.

Photophysiology, ed. by A. C. GIESE. Vols. 1 and 2. New York: Academic Press 1964.

G. GLOCKLER and S. C. LIND: The Electrochemistry of Gases and other Dielectrics. New York: Wiley 1939.

K. GOLLNICK and G. O. SCHENCK: Oxygen as a Dienophile. In: 1,4-Cycloaddition Reactions, ed. by J. HAMER. New York: Academic Press 1967, pp. 255—344.

The Middle Ultraviolet: Its Science and Technology, Ed. by A. E. S. GREEN. New York: Wiley 1966.

R. O. GRIFFITH and A. McKEOWN: Photo-processes in Gaseous and Liquid Systems. London: Longmans 1929.

Photochemistry in the Liquid and Solid States, ed. by L. J. HEIDT, R. S. LIVINGSTON, E. RABINOWITCH and F. DANIELS. New York: Wiley 1960.

Fluorescence and Phosphorescence Analysis. Principles and Applications. Ed. by D. M. HERCULES. New York: Interscience 1966.

Radiation Biology, ed. by A. HOLLAENDER. Vol. 2: Ultraviolet and Related Radiations, Vol. 3: Visible and Near-visible Light. New York: McGraw-Hill 1955, 1956.

The Chemistry of Ionization and Excitation, ed. by G. R. A. JOHNSON and G. SCHOLES. London: Taylor & Francis 1967.

Luminescence of Organic and Inorganic Materials, ed. by H. P. KALLMANN and G. M. SPRUCH. New York: Wiley 1962.

R. O. KAN: Organic Photochemistry. New York: McGraw-Hill 1966.

G. B. KISTIAKOWSKY: Photochemical Processes (ACS Monogr. Series 43). New York: Chem. Catalog Comp. 1928.

A. F. KLESCHNIN: Die Pflanze und das Licht. Berlin: Akademie-Verlag 1960.

G. KÖGEL: Die Methodik zur Prüfung der Lichtempfindlichkeit der Naturfarbstoffe. In: Handbuch der biologischen Arbeitsmethoden, ed. by E. ABDERHALDEN, Abt. II, Teil 2/II. Berlin: Urban & Schwarzenberg 1931, pp. 1999—2054.

L. R. KOLLER: Ultraviolet Radiation. New York: Wiley 1952 (1st ed.) and 1965 (2nd ed.).

J. KOSAR: Light-sensitive Systems: Chemistry and Application of Nonsilver Halide Photographic Processes. New York: Wiley 1965.

P. LEIGHTON: Photochemistry of Air Pollution. New York: Academic Press 1961.

R. LIVINGSTON: Photochemical Autoxidation. In: Autoxidation and Antioxidants, ed. by W. O. LUNDBERG, vol. 1. New York: Interscience 1961, pp. 249—298.

C. R. MASSON, V. BOEKELHEIDE and W. A. NOYES JR.: Photochemical Reactions. In: Technique of Organic Chemistry, ed. by A. WEISSBERGER, 2nd ed., Vol. 2. New York: Interscience 1956, pp. 257—384.

Symposium on Light and Life, ed. by W. D. McELROY and B. GLASS. Baltimore: Johns Hopkins Press 1961.

A. D. McLaren and D. Shugar: Photochemistry of Proteins and Nucleic Acids. Oxford: Pergamon 1964.
H. Meier: Die Photochemie der organischen Farbstoffe. Berlin: Springer 1963.
H. Meyer and E. O. Seitz: Ultraviolette Strahlen. Berlin: de Gruyter 1949.
D. C. Neckers: Mechanistic Organic Photochemistry. New York: Reinhold 1967.
R. M. Noyes and A. Weller: Photostationary Methods. In: Investigation of rates and mechanisms of reactions, ed. by S. L. Friess, E. S. Lewis and A. Weissberger (Technique of organic chemistry, ed. by A. Weissberger, 2nd ed., vol. 8, pt. 2). New York: Interscience 1963, pp. 845—863.
W. A. Noyes jr. and P. A. Leighton: The Photochemistry of Gases. New York: Reinhold 1941.
— and V. Boekelheide: Photochemical Reactions. In: Technique of Organic Chemistry, ed. by A. Weissberger, vol. 2. New York: Interscience 1948, pp. 79—141.
L. Pincussen: Biologische Lichtwirkungen, ihre physikalischen und chemischen Grundlagen. München: Bergmann 1920.
J. N. Pitts jr. and J. K. S. Wan: Photochemistry of Ketones and Aldehydes. In: The Chemistry of the Carbonyl Group, ed. by S. Patai. London: Interscience 1966, pp. 823—916.
J. Plotnikow: Photochemische Versuchstechnik. Leipzig: Akadem. Verlagsgesellschaft 1912.
— Allgemeine Photochemie. Berlin: de Gruyter 1920 (1st ed.) and 1936 (2nd ed.).
— Photochemische Arbeitsmethoden im Dienste der Biologie. In: Handbuch der biologischen Arbeitsmethoden, ed. by E. Abderhalden, Abt. III, Teil A/2. Berlin: Urban & Schwarzenberg 1930, p. 1653—1912.
G. Porter: Flash Photolysis. In: Investigation of rates and mechanisms of reactions, ed. by S. L. Friess, E. S. Lewis and A. Weissberger (Technique of organic chemistry, ed. by A. Weissberger, 2nd ed., vol. 8, pt. 2). New York: Interscience 1963, pp. 1055—1106.
P. Pringsheim: Fluorescence and Phosphorescence. New York: Interscience 1949.
E. I. Rabinowitch: Photosynthesis and Related Processes. Vol. 1: Chemistry of Photosynthesis, Chemosynthesis and Related Processes in vitro and in vivo. New York: Interscience 1945.
— and R. L. Belford: Spectroscopy and Photochemistry of Uranyl Compounds. New York: Macmillan 1964.
C. Reid: Excited States in Chemistry and Biology. London: Butterworths 1959.
G. K. Rollefson and M. Burton: Photochemistry and the Mechanism of Chemical Reactions. New York: Prentice-Hall 1939.
G. O. Schenck and K. Gollnick: Über die schnellen Teilprozesse photosensibilisierter Substrat-Übertragungen. Untersuchungen über Chemismus und Kinetik der durch Xanthenfarbstoffe photosensibilisierten O_2-Übertragungen. Forschungsberichte des Landes Nordrhein-Westfalen 1256. Köln: Westdt. Verlag 1963.
A. Schönberg: Präparative organische Photochemie. Berlin: Springer 1958.
H. H. Seliger and W. D. McElroy: Light: Physical and Biological Action. New York: Academic Press 1965.
S. E. Sheppard: Lehrbuch der Photochemie. Leipzig: Barth 1916.
D. Shugar: Photochemistry of Nucleic Acids and their Constituents. In: Nucleic Acids, ed. by E. Chargaff and J. N. Davidson, vol. 3. New York: Academic Press 1960, pp. 39—104.
C. J. Spinatelli: Photochemistry Simplified. Chicago: Ziff-Davis 1948.
H. A. Spoehr: Photosynthesis (ACS Monogr. Series 29). New York: Chem. Catalog Comp. 1926.
H. Staude: Photochemie. Mannheim: Bibliogr. Inst. 1962.

E. W. R. STEACIE: Atomic and Free Radical Reactions, 2nd. ed., Vols. 1 and 2 (ACS Monogr. Series 125). New York: Reinhold 1954.

T. SWENSSON: Untersuchungsmethoden biochemisch wichtiger Lichtreaktionen im sichtbaren und ultravioletten Spektralgebiet. In: Handbuch der biologischen Arbeitsmethoden, ed. by E. ABDERHALDEN, Abt. II, Teil 2/I. Berlin: Urban & Schwarzenberg 1928, pp. 1213—1290.

H. von TAPPEINER and A. JODLBAUER: Die sensibilisierende Wirkung fluorescierender Substanzen. Leipzig: Vogel 1907.

A. N. TERENIN: Fotokhimiya Krasitelei i Rodstvennykh Organicheskikh Soedinenii (Photochemistry of Dyes and Related Organic Compounds). Moscow: Izd. Akad. Nauk SSSR 1947; Translation of chapters 5, 7, 8 and 11 available as No. 59—10485/10488 from Kresge-Hooker Scientific Library.

N. J. TURRO: Molecular Photochemistry. New York: Benjamin 1965.

Elementarnye Fotoprotsessy b Molekulakh (Elementary Photoprocesses in Molecules), ed. by F. I. VILESOV et al. Moskva: Izd. Nauka 1966.

F. WEIGERT: Optische Methoden der Chemie. Leipzig: Akadem. Verlagsgesellschaft 1927.

Introduction to the Utilization of Solar Energy, ed. by A. M. ZAREM and D. D. ERWAY. New York: McGraw-Hill 1963.

Reactivity of the Photoexcited Organic Molecule (Proc. 13th Conf. Chemistry Brussels 1965). London: Wiley 1967.

Journals and series in the field of photochemistry

Optika i Spektroskopiya (Journal) **1**. Moskva: Akad. Nauk SSSR 1956.

Optics and Spectroscopy (Journal, transl. of above J.) **6**. Washington: Opt. Soc. America 1959.

Photochemistry and Photobiology (Journal) **1**. London: Pergamon 1962.

Zeitschrift für wissenschaftliche Photographie, Photophysik und Photochemie (Journal) **1**. Leipzig: Barth 1903.

Advances in Photochemistry, ed. by W. A. NOYES JR., G. S. HAMMOND and J. N. PITTS JR., vol. **1**. New York: Wiley 1963.

Reviews on photochemical topics, taken from periodicals and series

YU. A. ARBUZOV: The Diels-Alder Reaction with Molecular Oxygen as Dienophile. Usp. Khim. **34**, 1332 (1965); Russ. Chem. Rev. **34**, 558—573 (1965).

R. G. BENNETT and R. E. KELLOGG: Mechanisms and Rates of Radiationless Energy Transfer. In: Progress in Reaction Kinetics, ed. by G. PORTER, vol. 4. Oxford: Pergamon 1967.

J. M. BONNIER: La Phosphorescence en Chimie Organique. Rev. Inst. Franc. Pétrole Ann. Combust. Liquides **21**, 735—763 (1966).

E. J. BOWEN: Light Absorption and Photochemistry (including Photo-polymerisation and the Effects of Light on Dyes). Quart. Rev. **4**, 236—250 (1950).

G. H. BROWN and W. G. SHAW: Phototropism (Photochromism). Rev. Pure Appl. Chem. **11**, 2—32 (1961).

A. E. CASSANO, P. L. SILVESTON and J. M. SMITH: Photochemical Reaction Engineering. Ind. Eng. Chem. **59** (1), 18—38 (1967).

R. J. CVETANOVIĆ: Mercury Photosensitized Reactions. In: Progress in Reaction Kinetics, ed. by G. PORTER, vol. 2. Oxford: Pergamon 1964, pp. 39—130.

A. S. DAVIS and R. B. CUNDALL: Primary Processes in the Gas-phase Photochemistry of Carbonyl Compounds. In: Progress in Reaction Kinetics, ed. by G. PORTER, vol. 4. Oxford: Pergamon 1967.

W. Davis Jr.: The Gas-phase Photochemical Decomposition of the Simple Aliphatic Ketones. Chem. Rev. **40**, 201—250 (1947).

G. A. Delzenne: Initiations Photochimiques et Transfert d'Energie en Photopolymérisation. Ind. Chim. Belge **30**, 679—692 (1965).

P. de Mayo: Ultraviolet Photochemistry of Simple Unsaturated Systems. In: Advances in Organic Chemistry, ed. by R. A. Raphael, E. C. Taylor and H. Wynberg, vol. 2. New York: Interscience 1960, pp. 367—425.

— and S. T. Reid: Photochemical Rearrangements and Related Transformations. Quart. Rev. **15**, 393—417 (1961).

W. L. Dilling: Intramolecular Photochemical Cycloaddition Reactions of Nonconjugated Olefins. Chem. Rev. **66**, 373—393 (1966).

C. Dufraisse: La Photooxydation. Experientia Suppl. No. **2**, 27—48 (1955).

D. Elad: Some Aspects of Photoalkylation Reactions. Fortschr. Chem. Forsch. **7**, 528—558 (1967).

R. Exelby and R. Grinter: Phototropy (or Photochromism). Chem. Rev. **65**, 247—260 (1965).

E. Fischer: Photochromie und reversible Photoisomerisierung. Fortschr. Chem. Forsch. **7**, 605—641 (1967).

R. M. Hochstrasser and G. B. Porter: Primary Processes in Photo-oxidation. Quart. Rev. **14**, 146—173 (1960).

M. Kasha: Phosphorescence and the Role of the Triplet State in the Electronic Excitation of Complex Molecules. Chem. Rev. **41**, 401—419 (1947).

K. F. Koch: Photochemistry of Tropolones. In: Advances in Alicyclic Chemistry, ed. by H. Hart and G. J. Karabatsos, vol. 1. New York: Academic Press 1966, pp. 257—281.

A. A. Krasnovskii: Reversible Photochemical Reduction of Chlorophyll and its Analogues and Derivatives. Usp. Khim. **29**, 736 (1960); Russ. Chem. Rev. **29**, 344—357 (1960).

V. A. Krongauz: Intermolecular Energy Transfer in Photochemical Reactions in Solutions of Organic Compounds. Usp. Khim. **35**, 1638 (1966); Russ. Chem. Rev. **35**, 678—687 (1966).

S. K. Lower and M. A. El-Sayed: The Triplet State and Molecular Electronic Processes in Organic Molecules. Chem. Rev. **66**, 199—241 (1966).

H. Massey and A. E. Potter: Atmospheric Photochemistry. Roy. Inst. Chem. (London) Lecture Series **1961**, No. 1.

G. R. McMillan and J. G. Calvert: Gas Phase Photo-oxidation. In: Oxidation and Combustion Reviews, ed. by C. F. H. Tipper, vol. 1. Amsterdam: Elsevier 1965, pp. 83—135.

A. Mustafa: Dimerizations in Sunlight. Chem. Rev. **51**, 1—23 (1952).

A. J. C. Nicholson: The Photolysis of the Aliphatic Aldehydes and Ketones. Rev. Pure Appl. Chem. **2**, 174—184 (1952).

W. A. Noyes Jr., G. B. Porter and J. E. Jolley: The Primary Photochemical Process in Simple Ketones. Chem. Rev. **56**, 49—94 (1956).

Organic Photochemistry (Internat. Symp. Strasbourg 1964). Pure Appl. Chem. **9**, 461—621 (1964).

M. Pape: Die Photooximierung gesättigter Kohlenwasserstoffe. Fortschr. Chem. Forsch. **7**, 559—604 (1967).

G. O. Phillips: Photochemistry of Carbohydrates. In: Advances in Carbohydrate Chemistry, ed. by M. L. Wolfrom and R. S. Tipson, vol. 18. New York: Academic Press 1963, pp. 9—59.

Photochemical Processes (Disc. Faraday Soc. Liverpool 1931). Trans. Faraday Soc. **27**, 357—573 (1931).

Photochemical Reactions in Liquids and Gases (Disc. Faraday Soc. Oxford 1925). Trans. Faraday Soc. **21**, 437—656 (1926).

Photochemistry. Kagaku (Kyoto) **18**, 880—1027 (1963); C. A. **60**, 8813 (1964).

Photochemistry in Relation to Textiles (Symposium Harrogate 1949). J. Soc. Dyers Colourists **65**, 585—788 (1949).

The Primary Process in Photochemistry. Ann. N. Y. Acad. Sci. **41**, 169—240 (1941).

C. REID: The Triplet State. Quart. Rev. **12**, 205—299 (1958).

J. SALTIEL: The Mechanisms of some Photochemical Reactions of Organic Molecules. In: Survey of Progress in Chemistry, ed. by A. F. SCOTT, vol. 2. New York: Academic Press 1964, pp 239—328.

K. SCHAFFNER: Photochemische Umwandlungen ausgewählter Naturstoffe. In: Fortschr. Chem. Org. Naturstoffe, ed. by L. ZECHMEISTER, vol 22. Wien: Springer 1964, pp 1—114.

G. O. SCHENCK: Apparate für Lichtreaktionen und ihre Anwendung in der präparativen Photochemie. In: Dechema-Monographien **24**, 105—145 (1955). Weinheim: Verlag Chemie.

A. SCHÖNBERG: The Formation of Peroxides in Light. In: Le Mécanisme de l'Oxydation (8. Conseil Chimie Bruxelles 1950). Bruxelles: Stoops 1950, pp. 217—259; disc. pp. 260—278.

— and A. MUSTAFA: Reactions of Non-enolizable Ketones in Sunlight. Chem. Rev. **40**, 181—200 (1947).

P. SEYBOLD and M. GOUTERMAN: Radiationless Transitions in Gases and Liquids. Chem. Rev. **65**, 413—433 (1965).

J. P. SIMONS: The Reactions of Electronically Excited Molecules in Solution. Quart. Rev. **13**, 3—29 (1959).

R. STEINMETZ: Photochemische Carbocyclo-Additionsreaktionen. Fortschr. Chem. Forsch. **7**, 445—527 (1967).

K. TOKUMARU, A. SUGIMORI, T. AKIYAMA and T. NAKATA: Organic Photochemistry of Carbonyl Compounds. Yuki Gosei Kagaku Kyokai Shi **24**, 1183—1201 (1966).

R. N. WARRENER and J. B. BREMNER: Photochemistry of Unsaturated Systems — a Classification Based on Multicentre Cycloelectronic Redistribution. Rev. Pure Appl. Chem. **16**, 117—173 (1966).

E. L. WEHRY: Photochemical Behaviour of Transition-metal Complexes. Quart. Rev. **21**, 213—230 (1967).

Author Index *

Numbers in brackets are reference numbers

Abell, P. I.   158 [16]
Abramovitch, R. A.   318 [1]
Acker, D. S.   106 [12]
Agnello, E. J.   387 [75]
Akhookh, Y.   366 [13]
Akhtar, M.   15 [10]; 59 [28]; 248, 250 [4]; 369 [24]
Akiyama, T.   500
Aladekomo, J. B.   98 [10]
Alder, K.   4 [22]; 118 [5]; 277 [24]; 375 [16]; 383 [48]
Aldrich, P.   229 [18]; 292 [19]
Alemagna, A.   337 [5]; 337 [6]
Allais, A.   389 [82]; 391 [91]
Allen, C. F. H.   393 [105]; 395, 397 [117]
Allen, R. G.   158 [17]
Altwicker, E. R.   29 [31]
Aly, O. M.   119 [20]
Amiard, G.   15 [9]; 15 [8]; 15 [11]
Amin, J. H.   48 [1]
Anderson, A. W.   263 [11]
Anderson, J. C.   236 [39]; 236, 237 [40]
Anderson, L. C.   398 [128]
Ando, W.   375 [19]
Anet, R.   85 [67]; 85 [68]
Argus, H. J. F.   109 [4]
Anner, G.   369 [20]; 369 [21]; 369 [22]; 369, 370 [25]
Anslow, Jr., W. P.   157 [11]
Aono, I.   415, 417 [9]
Apicella, M.   155 [3]
Appel, R.   427, 429 [7]
Applequist, D. E.   98 [13]; 98 [15]; 348 [23]
Appleyard, J. H.   99 [24]
ApSimon, J. W.   322-324 [12]
Arbuzov, Yu. A.   383, 387, 389, 395 [53]; 498
Arigoni, D.   30, 32 [36]; 31 [44]; 31, 32 [46]; 78, 79 [36]; 226, 227 [10]; 329 [4]

Arndt, F.   295 [4]
Arnold, D. R.   415-417 [8]; 468 [43]
Ashmore, P. G.   495
Asinger, F.   435, 437 [11]; 435 [12]; 435 [13]; 438 [21]
Asker, W.   133, 134 [22]; 398 [131]
Atkinson, J. G.   112, 114 [13]
Attree, G. F.   131 [17]
Aubry, J.   394 [113]
Auerbach, J.   116 [23]
Autrey, R. L.   16 [19]
Awad, W. I.   71, 85, 87 [5]; 119 [17]; 119 [20]; 187, 188 [6]; 187, 188 [7]; 190, 191 [24]; 190 [26]; 191 [28]; 205 [53]
Axelrod, M.   66, 67 [59]; 116 [23]
Aycock, B. F.   158 [16]
Ayer, D. E.   112, 114 [13]
Ayer, W. A.   100 [26]
Ayling, E. E.   269 [13]

Bacchetti, T.   337 [5]; 337 [6]
Bachér, F.   57 [8]; 83 [57]; 495
Bachman, G. L.   275, 276, 278, 284, 285 [10]
Bachmann, W. E.   194, 195 [5]; 205 [20]; 209 [36]; 209 [42]
Bacon, R. G. R.   443 [33]; 443 [34]; 443 [37]; 443, 444 [38]
Baddar, F. G.   129, 131 [13]
Badger, G. M.   53 [19]; 139 [5]; 139, 140 [6]; 139, 140 [7]; 139, 140 [8]; 141, 142 [10]; 187, 188 [17]; 337, 338 [7]; 392 [98]
Badoche, M.   390 [84]
Bäckström, H. L. J.   205 [12]
Baget, J.   393 [108]
Bagli, J. F.   376, 377 [24]
Bailey, P. S.   23, 24 [6]; 57 [16]
Baker, J. W.   4 [19]
Baker, W.   71, 73 [9]
Baker, W. P.   205 [13]
Baldwin, J. E.   18 [26]

* The index part was prepared by O.-A. Neumüller

Ballard, S. S.   491 [19]
Ballester, M.   347, 348 [21]
Bamberger, E.   268, 269 [10]
Banus, J.   363-365 [5]
Barakat, M. Z.   81 [44]; 99 [22]; 119 [13]
Barbeau, C.   463 [21]
Barks, P. A.   8 [49]
Barltrop, J. A.   111 [11]; 424 [27]
Barnard, M.   34, 35 [1]
Bartlett, P. D.   53, 54 [21]; 291 [52]; 416, 420, 421 [7]
Barton, D. H. R.   14 [3]; 16 [19]; 30 [34]; 30 [35]; 30, 32 [39]; 31, 32 [42]; 66 [57]; 223, 224 [15]; 226, 227 [10]; 227 [15]; 227, 228, 230, 234, 235 [16]; 229, 230 [17]; 248 [3]; 250 [12]; 250, 251 [13]; 251 [14]; 253 [19]; 253 [20]; 254 [21]; 319, 320 [3]; 319 [4]; 370 [26]; 270, 271 [27]; 371 [28]; 387 [78]; 442 [32]
BASF,   263, 264 [13]; 264 [21]
Basinski, J. E.   82 [46]; 82, 83 [51]
Bass, A. M.   491, 492 [24]
Bass, J. D.   109, 112 [2]
Bass, R. G.   407 [3]; 431 [18]
Basselier, J. -J.   391, 393 [87]; 394 [114]; 407 [5]; 408 [6]
Bastiansen, O.   347 [22]
Battersby, A. R.   401 [141]
Bauer, G.   321 [10]; 474 [9]
Baumann, P.   15 [14]
Beaton, J. M.   248 [3]; 250 [12]; 250, 251 [13]; 254 [21]
Beavers, L. E.   102 [34]
Becher, J.   51, 52 [14]
Becker, E. I.   394 [115]; 394 [116]
Becker, H. -D.   379, 380 [37]
Beckett, A.   25 [11]
Beckwith, A. J. L.   270, 271 [27]
Beckwith, A. L. J.   371 [28]; 442 [31]
Bedford, C. T.   382 [46]
Beereboom, J. J.   420 [20]
Belford, R. L.   497
Belkevich, N. A.   456, 457 [13]
Bell, A.   393 [105]
Bell, I.   15 [13]
Bell, J. A.   274, 275 [5]
Bellin, J. A.   257 [10]
Bencze, W. L.   195 [9]; 205 [26]
Bendas, H.   201, 202 [14]
Bennett, R. G.   498
Benrath, A.   182, 183 [5]; 495
Benson, S. W.   274, 275 [3]
Berends, W.   91 [89]
Bereza, S.   29 [32]

Bergmann, E.   200, 201 [11]; 205 [25]
Bergmann, E. D.   201, 202 [14]
Bergmann, F.   205, 208 [27]
Bergmann, W.   17 [24]; 46, 47 [24]; 46, 47 [25]; 47 [26]; 152 [3]; 383-385, 387, 389 [51]; 387, 388 [74]
Berlage, F.   401, 402 [143]
Berman, E.   42 [12]
Bernasconi, R.   226, 227 [10]
Bernauer, K.   401, 402 [140]; 401, 402 [142]; 401, 402 [143]
Bernstein, H. I.   84, 85 [60]
Bernstein, S.   358 [47]
Berson, J. A.   268-270 [12]
Berthoud, A.   495
Bessey, O. A.   256 [2]
Bestmann, H. J.   297, 298 [6]; 300 [14]
Beukers, R.   91 [89]
Bickel, A. F.   153 [9]
Biekert, E.   77, 79 [34]; 118, 119, 121, 122, [3]
Bikales, N. M.   394 [116]
Bilow, N.   282, 283 [34]
Binovi, L. J.   358 [47]
Binsch, G.   307-309 [40]
Bird, C. W.   2, 3 [16]; 40 [21]
Birks, J. B.   98 [10]; 99 [24]
Bisnette, M. B.   464, 465 [32]
Bissell, E. R.   421, 422 [22]
Blackburn, G. M.   91 [91]
Bladon, P.   43, 44 [13]; 193 [3]; 386 [72]
Blank, B.   246, 247 [12]
Blanke, E.   226 [5]; 226 [8]
Blattmann, H. -R.   20 [33]
Blum, H. F.   495
Bock, G.   209 [40]
Bockhorn, G. H.   325 [20]
Bodforss, S.   408 [7]
Böck, E.   435 [12]
Böhme, P.   239 [49]
Boekelheide, V.   20 [33]; 496; 497
Böll, W.   281 [30]; 329 [7]
Böll, W. A.   281 [29]
Bohlmann, F.   60, 61 [30]
Boie, I.   450 [57]
Bondy, R.   131-133 [16]
Bonhoeffer, K. F.   495
Bonnett, R.   50, 51 [10]
Bonnier, J. M.   498
Borden, G. W.   5, 6, 9 [28]; 5, 7 [31]; 220 [8]
Bordwell, F. G.   163 [34]; 163, 164 [35]
Borgeaud, P.   149 [5]
Bosshard, H.   30, 32 [36]; 31 [43]; 235 [38]

Bouas-Laurent, H.   99 [21]
Bowden, S. T.   126 [2]
Bowen, E. J.   98 [8]; 390 [83]; 491, 492 [23]; 495; 498
Bowesman, C.   72 [12]
Boykin, Jr., D. W.   44, 45 [16]
Bradow, R. L.   354 [41]; 365 [11]
Bradshaw, J. S.   56, 57 [5]; 97 [1]
Bradsher, C. K.   102 [34]; 142 [13]
Brady, O. L.   63 [43]; 63 [44]
Bram, S.   238 [44]
Brammer, R.   427, 429 [7]
Brand, K.   89, 90 [84]
Brandon, R. L.   8 [44]
Brandt, G. A. R.   439-441 [24]
Brandt, G. R. A.   440 [25]
Brauman, J. I.   469 [45]
Bregman, J.   57, 58 [13]; 84 [63]
Bremer, K.   83 [58]
Bremner, J. B.   500
Brenner, G.   229 [18]; 229 [19]
Breywisch, W.   7, 8 [37]
Brockmann, H.   131, 132, 134 [18]; 132 [19]; 256, 257 [6]; 256 [7]
Brode, W. R.   61 [31]; 62 [34]
Brossi, A.   374, 375 [13]
Brown, B. B.   320 [7]
Brown, E.   268-270 [12]
Brown, E. V.   195 [8]; 205 [24]
Brown, G. B.   52 [16]
Brown, G. H.   42 [8]; 498
Brown, H. C.   233, 234 [31]; 233 [32]; 234 [33]; 342 [4]; 436, 437 [16]
Brown, J. H.   343 [11]
Brown, R. F. C.   323 [15]
Brown, T. L.   98 [13]
Bruce, J. M.   80, 81 [41]; 189 [21]
Bruderer, H.   30, 32 [36]
Brunings, K. J.   387 [75]
Brunken, J.   385, 388 [67]
Bruylants, A.   349 [28]
Bryce-Smith, D.   79, 81 [39]; 93 [104]; 109 [4]; 110 [8]; 114 [17]; 114 [18]; 418 [16]; 418, 419 [17]; 423 [25]
Buchardt, O.   51, 52 [13]; 51, 52 [14]
Buchschacher, P.   26 [18]; 36 [9]
Buchwald, G.   375, 376 [17]
Büchi, G.   1 [3]; 6 [33]; 27 [25]; 30, 32 [36]; 41, 42 [2]; 41, 42 [3]; 58, 60 [18]; 93, 94 [103]; 112, 114 [13]; 415, 417 [2]; 423 [24]
Bunyan, P. J.   454 [8]
Burckhardt, C. A.   195 [9]; 205 [26]
Burckhardt, U.   18 [27]
Burdett, F.   313 [1]

Burgess, E. M.   6 [33]; 27 [25]
Burrell, Jr., E. J.   262 [7]
Burstain, I. G.   428-430 [10]
Burton, M.   495; 497
Busch, M.   380 [39]
Busfield, W. K.   106 [11]
Butenandt, A.   65, 66 [53]; 65 [54]; 65 [55]; 77 [33]; 77, 79 [34]; 118, 191, 121, 122, [3]
Buttery, R. G.   53 [19]; 276, 283 [14]
Buzbee, L. R.   345 [17]
Byers, G. W.   218, 219 [3]; 218, 219 [5]
Bylina, G. S.   399, 400 [134]

Cadogan, J. I. G.   454 [8]
Cainelli, G.   78, 79 [36]
Calas, R.   98 [12]; 98 [14]; 98 [16]; 178 [76]
Calingaert, G.   344 [15]
Calvert, J. G.   472, 474, 479, 492 [1]; 495; 499
Calvin, M.   49, 50 [6]; 49-51 [7]
Cameron, F. K.   98, 101 [6]
Campbell, N.   62 [40]; 129, 131 [14]
Capon, B.   495
Caporale, G.   87, 88 [74]
Cappellina, V.   87, 88 [73]
Capper, N. S.   98 [4]
Caress, E. A.   319, 320 [6]
Cargill, R. L.   2, 3, 10 [6]; 5 [27]; 5 [30]
Carlon, F. E.   251 [14]
Carruthers, W.   136 [30]
Carter, F. L.   329 [6]
Cassano, A. E.   498
Castañer, J.   347, 348 [21]
Castells, J.   386 [70]
Catel, J.   103 [36]; 431 [15]
Cava, M. P.   142 [12]; 221 [11]; 302 [16]; 302 [21]; 306, 308 [38]; 447 [51]; 447, 448 [52]
Cereghetti, M.   26 [18]; 36 [9]; 36, 37 [10]
Cerutti, P.   195, 196 [10]; 196 [11]; 207, 208 [33]; 335-337 [3]; 335, 336 [4]
Chamberlain, N. F.   277 [22]
Chambers, R. D.   465 [33]
Chandra, P.   91 [93]
Chao, O.   302 [18]; 303 [23]
Chao, T. H.   436, 437 [16]
Chapman, O. L.   5, 6, 9 [28]; 5, 7 [31]; 6, 9 [32]; 6, 7, 9 [34]; 8 [45]; 8 [46]; 8 [49]; 28, 30 [29]; 30 [38]; 92 [101]; 92 [102]; 220 [8]; 227, 229 [12]; 227

[13]; 229 [20]; 250 [11]; 266, 267 [2]; 495
Chardonnens, L. 48, 49 [3]
Chaudhuri, N. 276, 283 [14]
Chen, C. S. H. 175, 176 [66]
Chieffi, G. 200 [10]; 200 [13]; 215 [2]; 414, 415 [1]
Chopard-Dit-Jean, L. H. 374, 375 [13]
Christmann, A. 318, 321 [2]; 326 [21]; 326, 327 [22]
Ciamician, G. 1 [1]; 26 [16]; 26 [19]; 26, 27 [21]; 57, 63, 64 [14]; 71, 83, 85 [1]; 76 [27]; 203 [1]; 203 [4]; 203, 205 [6]; 204 [8]; 209 [37]; 225 [3]; 225 [4]; 267-269 [3]; 269 [14]
Cier, A. 157 [13]
Cimbollek, G. 226, 227 [7]
Citarel, L. 446 [47]
Civera, M. 89 [80]
Clar, E. 20 [34]; 132, 134 [21]
Clark, H. C. 465 [33]
Clark, V. M. 50, 51 [10]
Clasen, R. A. 142 [11]
Closs, G. L. 275 [11]; 281 [29]; 281 [30]; 329 [7]
Closs, L. E. 281 [29]
Coates, J. E. 313 [1]
Coe, C. S. 251, 252 [15]
Coffey, R. S. 275, 276, 278, 284, 285 [10]
Coffman, D. D. 421-423 [21]
Cohen, M. D. 83 [52]; 84, 85 [61]
Cohen, S. G. 206 [31]
Cohen, W. D. 194 [6]; 205 [19]; 215 [3]
Colby, T. H. 303, 304 [26]
Coleman, G. H. 242-244 [2]
Collin, P. J. 116 [21]; 334 [2]
Comrie, A. M. 449 [56]
Conca, R. J. 47 [26]
Conway, W. 442 [30]
Cook, A. H. 62, 63 [38]; 298 [10]
Cook, C. D. 29 [31]
Cook, E. W. 421, 422 [23]
Cookson, R. C. 2, 3 [10]; 2, 3 [13]; 2-4 [14]; 2, 3 [15]; 2, 3 [16]; 27 [23]; 27 [24]; 40 [21]; 79-81 [38]; 80 [40]; 218, 219 [4]
Cooper, G. D. 436, 437 [17]
Cope, A. C. 4 [20]; 384, 385 [65]
Corey, E. J. 109, 112 [2]; 244 [8]; 244 [9]; 244, 245 [10]; 376, 383, 389 [27]
Cot, A. 99 [21]
Coulson, C. A. 98 [7]
Coulson, D. R. 418 [15]
Coulter, A. W. 302 [20]

Counsell, R. C. 56, 57 [5]
Courtot, P. 5 [29]; 14 [5]; 103 [37]; 205 [29]; 384, 385 [66]
Covell, J. 64, 65 [50]
Cowan, D. O. 56, 57 [5]
Cowley, B. R. 408 [10]
Cox, D. A. 8 [50]; 79-81 [38]
Coyne, L. M. 66, 67 [59]
Craft, L. 423 [26]
Crawford, E. J. 256 [2]
Criegee, R. 82 [50]; 115 [19]; 212 [52]; 380 [41]; 384, 385, 389, 393 [64]
Cristol, S. J. 2-4 [9]; 165 [38]
Crowley, K. J. 2, 4 [5]; 17 [23]; 19 [29]; 22, 24 [4]
Crumrine, D. 450 [58]
Crundwell, E. 2, 3 [13]; 2-4 [14]; 2, 3 [15]; 2, 3 [16]
Culbertson, T. P. 243, 246 [6]
Cumming, W. M. 53 [17]; 53 [18]
Cundall, R. B. 498
Cuong, N. K. 391, 393 [87]
Curtis, G. G. 302 [19]
Cutts, E. 189 [21]
Cvetanović, R. J. 498

Dahl, E. 169 [49]
Dainton, F. S. 438 [19]; 495
Dale, J. I. 407 [3]; 431 [18]
Dalton, C. 56, 57 [5]
Daniels, F. 350 [32]; 495; 496; 496
Dannenberg, H. 182 [2]
Dannenberg-von Dressler, D. 182 [2]
Dauben, W. G. 2, 3, 10 [6]; 4 [25]; 5 [27]; 5 [30]; 7, 8 [38]; 8 [50]; 10 [54]; 10 [56]; 15 [13]; 15 [14]; 227 [11]; 227, 229 [12]; 229, 230 [22]; 284 [43]; 386 [71]
Davies, A. G. 373, 376 [4]
Davies, R. J. H. 91 [91]
Davies, W. 90 [87]
Davis, A. S. 498
Davis, B. A. 318 [1]
Davis, Jr., W. 25 [15]; 499
Davison, P. S. 129, 131 [14]
Davydova, V. P. 354, 355 [42]
De Boer, C. D. 2, 3, 10 [8]
De Jong, A. W. K. 83 [54]; 85 [66]
De Jonge, J. 316 [7]
De Kock, R. J. 14 [4]; 15 [15]; 15 [17]
De Mayo, P. 3 [18]; 30 [35]; 30, 32 [39]; 42 [4]; 48 [1]; 100 [26]; 103 [35]; 111 [12]; 113 [15]; 113 [16]; 170 [52]; 198, 199, 201 [3]; 199 [4]; 199, 201 [5]; 199 [7]; 229, 230 [17]; 268 [11]; 395-397 [118]; 448 [53]; 448 [54]; 499

De More, W. B.   274, 275 [3]
Dean, P. M.   390 [86]
Deichsel, S.   46 [21]
Deleo, E.   199 [6]
Dellweg, H.   91 [92]; 91 [93]
Delzenne, G. A.   499
Demuynck, R.   391 [91]
Denk, W.   375, 378 [14]; 429 [12]
Denney, D. B.   285 [45]
Denny, R.   428-430 [10]
Depke, F.   383 [54]
DePuy, C. H.   277 [25]; 496
Derible, P.   43, 44 [14]; 43 [15]
Dershowitz, S.   453 [3]
Di Giacomo, A.   262 [6]
Diekmann, J.   161 [25]
Dien, C. -K.   23, 24 [6]; 57 [16]
Dietrich, R.   304 [29]
Dietz, W.   380 [39]
Dijkstra, R.   316 [7]
Dilling, W. L.   499
Dilthey, W.   134, 136 [26]; 239 [47]; 410, 411 [17]; 411 [23]
Dimroth, K.   7, 8 [37]; 46 [20]
Dinjaški, K.   87, 88 [71]
Dinwoodie, A. H.   106 [4]
Dirstine, P. H.   342 [2]
Disselnkötter, H.   277-279, 283 [20]
Djerassi, C.   226, 227 [10]; 409 [12]
Dodson, R. M.   64, 65 [50]
Dörges, J.   321 [10]; 455, 457 [10]
Dörscheln, W.   54 [24]; 55 [25]
Doležal, S.   171 [57]
Dominy, B.   409 [11]
Donaruma, L. G.   262, 263 [5]
Doolittle, R. E.   142 [13]
Dorant, K.   271, 272 [18]
Dorfman, L.   358 [47]
Doumani, T. F.   251, 252 [15]
Doumaux, A. R.   432 [19]
Dowbenko, R.   172, 173 [60]
Drewer, R. J.   139 [5]; 139, 140 [6]; 337, 338 [7]
Druckrey, E.   15 [6]
Dürr, H. G. C.   238 [44]
Duffie, J. A.   495
Dufraisse, C.   196 [13]; 389, 391 [81]; 389 [82]; 390 [84]; 390 [86]; 391, 393 [87]; 391 [88]; 391 [91]; 391 [92]; 391 [93]; 391 [94]; 391 [95]; 392 [97]; 392 [99]; 393 [107]; 393 [108]; 393 [110]; 393 [112]; 394 [113]; 396 [119]; 396 [120]; 396, 397 [123]; 407 [5]; 430, 431 [14]; 499
Dulou, R.   232 [27]; 232, 233 [30]
Dunderdale, J.   364 [9]
Dunitz, J. D.   71, 72 [2]; 71 [10]; 72 [16]

Dunlap, D. E.   384 [61]; 475 [12]
Dunn, F. P.   63 [43]
Dunn, J. H.   343 [10]
Dunston, J. M.   410 [16]
Duschinsky, R.   155 [4]
Dutler, H.   31 [43]; 31 [44]; 235 [38]
Dvoretzky, I.   283 [41]
Dziewoński, K.   71, 72 [3]; 72 [15]

Eastman, R. H.   218 [2]
Eaton, P.   2, 3 [11]
Eaton, P. E.   77, 78 [30]; 112, 113 [14]
Ebel, H. F.   369 [19]
Ebeneder, F.   435, 437 [11]; 435 [12]
Eberhardt, G.   297 [7]
Eberhardt, H. -D.   305, 306, 310 [33]
Ecary, S.   430, 431 [14]
Ecke, G. G.   161 [24]; 345 [17]
Eckell, A.   17 [22]
Eckert, A.   131-133 [16]; 133, 134 [24]; 133 [25]; 149 [8]
Eckoldt, H.   435, 436 [10]; 438 [21]
Edwards, O. E.   322-324 [12]
Ege, G.   330 [8]
Eggert, H.   375, 378 [14]; 429 [12]
Eigen, I.   297 [7]
Eisfeld, W.   377, 387, 388 [31]; 377, 378 [33]
Eistert, B.   191 [29]; 209 [40]; 295 [4]
Eiszner, J. R.   282, 283 [33]
Elad, D.   45 [19]; 160, 161 [23]; 166, 167 [42]; 166-168 [43]; 167 [45]; 167 [46]; 230, 231 [23]; 496; 499
Elenbaas, W.   474 [6]
Elger, F.   268, 269 [10]
El-Assal, L. S.   129, 131 [13]
El-Hewehi, Z.   349 [29]
El-Sayed, M. A.   499
El-Shafei, Z. M.   133 [23]; 401 [138]
Ellis, C.   496
Ellis, L. E.   469 [45]
Elnitskii, A. P.   399, 400 [134]
Eme1éus, H. J.   363-365 [5]; 366, 367 [14]; 439-441 [24]; 440 [25]; 457, 458 [15]
Engler, C.   271, 272 [18]
Englert, L. F.   30 [38]; 229 [20]
English, Jr. , J.   17 [24]; 152 [3]
Erhardt, F.   427 [3]
Erickson, K. L.   17 [22]
Erlenmeyer, H.   357, 358 [46]

Ermolaev, V.   205 [9]
Erway, D. D.   498
Ess, R. J.   454, 455 [6]
Étienne, A.   101 [30]; 101, 102 [31]; 101 [32]; 196 [13]; 389 [80]; 390 [85]; 391 [89]; 391 [94]; 393 [102]; 393 [103]; 393 [106]; 393 [112]; 394 [113]; 396 [119]; 407 [5]; 496
Etter, R. M.   275 [12]
Evanega, G. R.   17 [24]; 152 [3]
Exelby, R.   42 [7]; 499

Failer, G.   77, 79 [34]; 118, 119, 121, 122, [3]
Fainberg, A. H.   362 [2]
Fanta, G. F.   348 [23]
Farid, S.   86 [69]; 88 [78]; 119, 120, 123 [6]; 119, 121, 122, 124 [8]; 119, 121-124 [21]; 419, 420 [18]
Farmer, E. H.   373 [1]; 373 [10]
Farnum, D. G.   285-287 [46]
Faseeh, S. A.   342, 343 [6]
Fateen, A. K.   215-217 [4]
Faugère, J.-G.   98 [14]; 98 [16]
Feer, A.   316 [8]
Feichtinger, H.   435 [14]
Feldkimel-Gorodetsky, M.   38 [17]
Fell, B.   345 [13]
Ferrier, G. S.   53 [18]
Ferrini, P. G.   329 [4]
Fiedler, G.   264 [18]; 264 [19]
Fields, D. B.   421, 422 [22]
Fields, R.   288 [47]; 288, 290 [48]
Fieser, L. F.   329 [5]; 385 [68]
Findley, T. W.   400 [137]
Fine, D. A.   42 [11]
Fisch, M. H.   30 [37]; 229 [21]
Fischer, A.   2, 3, 10 [7]
Fischer, E.   19 [31]; 63 [41]; 127 [10]; 499
Fischer, E. O.   462 [19]; 463, 464 [22]; 463 [23]; 463 [24]; 463 [31]
Fischer, H.   143 [18]; 143, 144 [19]
Flores R, H.   303 [23]
Foerst, W.   496
Foerster, G.   85 [64]
Förster, T.   496
Fokin, A. V.   163 [33]
Fonken, G. J.   7, 8 [38]; 19 [28]; 386 [71]
Foote, C.   430 [13]
Foote, C. S.   375 [19]; 376, 377, 383 [25]; 376, 383 [26]; 428-430 [10]; 428 [11]
Forbes, E. J.   8 [43]; 227, 292 [14]
Foss, R. P.   203, 205 [5]

Foster, D. J.   172 [59]
Fowler, R. G.   474 [5]
Fox, R. E.   42 [10]
Fraenkel, G.   91 [90]
Frainnet, E.   178 [76]
Frampton, V. L.   329 [6]
Francis, W. C.   225 [1]
Frank, I.   143 [14]
Frankel, D. M.   407 [4]
Frankel, J. J.   80 [40]
Frankel, M.   63 [41]
Franzen, V.   169 [49]; 294, 295 [3]
Fray, G. I.   423 [25]
Frederiksen, E.   443, 444 [35]
Freedman, H. H.   465 [34]
Freeman, H. C.   316 [6]
Freeman, J. P.   231 [24]
Frehden, O.   62 [39]
Freidlina, R. Kh.   39 [18]; 39 [19]; 283 [36]
Freudenberger, H.   445 [45]
Frey, H. M.   274, 275, 290 [4]; 290 [50]; 290 [51]; 291 [54]
Freytag, C.   243 [5]
Friedlander, H. N.   35 [6]; 171, 172, 175 [58]
Friedrich, E. C.   98 [15]
Friedrich, W.   65 [55]
Fries, D.   197 [15]; 263, 265 [15]; 263 [16]; 263, 265 [17]
Friese, U.   239 [48]
Fritzsche,   98 [2]
Fritzsche, H.   258 [17]
Fry, A. J.   285 [44]
Fuchs, R.   179 [78]
Fujise, S.-I.   200, 201 [11]
Fulton, J. D.   71, 72 [2]; 71-73 [8]; 72 [13]
Furrer, H.   115 [19]
Furukawa, K.   82, 83 [48]

Gabriel, T.   155 [4]
Gall, R.   357, 358 [46]
Ganguly, A. K.   16 [19]
Ganter, C.   31 [44]; 31 [45]
Garcia-Banús, A.   126, 127 [1]
Gardner, P. D.   8 [44]; 234, 235 [36]
Gassman, P. G.   302 [19]; 303 [25]
Gattermann, L.   445 [44]
Gauss, W.   397 [126]
Gebhardt, A. I.   373 [2]
Geffcken, W.   474, 491, 492 [8]
Gegiou, D.   19 [31]; 127 [10]
Geiseler, G.   438 [21]
Geller, L. E.   248 [3]; 250, 251 [13]
Gembitskii, P. A.   22 [2]

George, M. V.   223, 224 [15]; 442 [32]
Georgian, V.   246, 247 [12]
Gérard, M.   389, 391 [81]
Gerecht, J. F.   373 [2]
Gerlach, K.   460 [3]; 460 [6]; 460, 462 [8]; 460 [9]; 460, 464 [10]; 463, 464 [29]
Geyer, A. M.   179 [77]
Giacalone, A.   199 [6]
Gibbons, C. J.   15 [10]
Gibian, M. J.   205 [16]
Giddings, B.   355, 356 [43]
Giese, A. C.   496
Gilbert, A.   79, 81 [39]; 418 [16]; 418, 419 [17]; 423 [25]
Gilbert, E. E.   434 [4]
Gilman, H.   179 [78]
Gindy, M.   119, 122 [24]; 129, 131 [13]
Ginsburg, D.   79, 81 [105]
Giua, M.   89 [80]
Glass, B.   496
Glick, A. H.   415-417 [8]
Glockler, G.   496
Glos, M.   305, 306, 310 [33]
Glover, D.   78 [37]
Gnoj, O.   253 [19]
Goering, H. L.   158 [16]; 163 [36]
Göth, H.   54 [24]; 55 [25]; 196 [11]; 207, 208 [33]
Göthlich, L.   368 [17]
Goffinet, B.   15 [11]
Gold, E. H.   79, 81 [105]
Golden, J. H.   11 [57]
Goldman, I. M.   1 [3]
Gollnick, K.   375, 376 [17]; 376, 383 [22]; 376, 383 [23]; 377 [32]; 383, 384, 387, 389, 395 [50]; 496; 497
Gomberg, M.   209 [42]
Gondot, L.   157 [13]
Goodman, M.   446 [47]
Goosen, A.   371 [28]
Gordon, J. T.   19 [30]; 127 [3]; 127, 129 [4]; 127 [5]
Gordon, M. P.   155 [4]
Gorodetsky, M.   37, 38 [15]; 37, 38 [16]
Gotthardt, H.   66, 67 [59]
Gould, E. S.   283 [37]
Gouterman, M.   500
Gowenlock, B. G.   252 [16]; 262 [8]
Graf, R.   438 [20]
Gravel, D.   23 [8]; 77 [35]; 194 [4]
Gray, P.   252 [17]
Greeley, R. H.   18 [26]
Green, A. E. S.   496

Greene, F. D.   98, 99 [11]; 98 [19]; 266 [1]
Grellmann, K. -H.   138 [1]; 138 [2]
Greuter, F.   31 [45]
Grewe, R.   8 [41]
Griffin, C. E.   347 [20]; 454, 455 [4]; 454 [5]; 455, 457 [9]; 455-457 [11]; 457 [14]
Griffin, G. W.   64, 65 [49]; 64, 65 [50]; 72 [17]; 82 [46]; 82, 83 [48]; 82, 83 [51]; 193 [1]; 238 [45]
Griffing, M. E.   344 [15]; 344 [16]
Griffith, I.   256 [4]
Griffith, R. O.   496
Grimsley, R. L.   346, 347, 350 [18]
Grinter, R.   42 [7]; 499
Griswold, A. A.   5, 6, 9 [28]; 220 [8]; 266, 267 [2]
Gritter, R. J.   160 [22]
Gross, A.   121 [28]; 295-299 [2]; 318, 321 [2]; 321, 322 [8]; 321, 322 [9]
Grossmann, H.   231-233 [25]
Grovenstein, Jr., E.   109 [5]
Grubb, P. W.   111 [11]
Gründemann, E.   166 [44]
Gruhl, A.   459 [1]
Gumlich, W.   153 [7]
Guryleva, A. A.   177 [74]
Gutsche, C. D.   275, 276, 278, 284, 285 [10]; 278, 279, 284 [26]
Guttenberger, J. F.   463 [25]; 463 [27]; 463 [30]
Guy, R. G.   443 [34]; 443, 444 [38]
Guyot, A.   103 [36]; 431 [15]

Hackmann, E. -A.   191 [29]
Häfliger, O.   374, 375 [13]
Hafez, M. S.   187, 188 [7]
Hafner, K.   325 [19]
Hagemann, H.   16, 17 [21]
Hagihara, N.   460 [12]
Hales, J. L.   106 [10]
Halford, J. O.   398 [128]
Hall, S. N.   343 [10]
Haller, I.   10 [55]
Halwer, M.   257, 258 [15]
Hamer, J.   48 [2]
Hamil, H. F.   234, 235 [36]
Hammond, G. S.   2, 3, 10 [7]; 2, 3, 10 [8]; 4 [24]; 25 [47]; 56 [4]; 56, 57 [5]; 56 [7]; 66, 67 [59]; 74, 75 [21]; 75 [25]; 97 [1]; 203, 205 [5]; 204, 208 [7]; 205 [13]; 276 [16]; 498
Hammond, H. A.   64, 65 [49]
Hanzel, R. S.   454, 455 [6]

Harding, M. J. C.   396 [122]
Harhash, A. H. E.   187 [12]
Harrison, I. T.   59, 61 [20]
Harris, Jr., J. F.   162 [28]; 163, 164 [32]; 164 [37]; 421-423 [21]
Harteck, P.   495
Hartenstein, J. H.   16 [20]; 16, 17 [21]
Hartley, G. S.   62 [36]; 62 [37]
Hartman, R.   409 [15]
Hartmann, I. -M.   72, 73 [18]; 72 [20]
Hartmann, W.   72, 73 [18]; 82, 83 [47]; 110 [9]; 110, 111 [10]; 415, 417 [12]
Hasek, W. R.   90, 91 [88]
Hass, H. B.   342 [3]
Hassel, O.   347 [22]
Hassler, J. C.   283 [38]
Hassner, A.   302 [20]
Haszeldine, R. N.   105, 106 [3]; 106 [4]; 157 [14]; 157, 158, 173, 174 [15]; 170, 173, 174 [54]; 173, 174 [62]; 173, 174 [63]; 174 [64]; 174 [65]; 178 [75;] 179 [77]; 225 [1]; 225 [2]; 234 [34]; 234 [35]; 288 [47]; 288, 290 [48]; 342 [5]; 343 [7]; 347 [20]; 363-365 [5]; 364 [6]; 364, 365 [7]; 364 [8]; 364, 365 [10]; 366 [12]; 366, 367 [14]; 439-441 [24]; 440 [25]; 440 [26]; 441 [28]; 457, 458 [15]
Hatt, H. H.   205 [21]
Hausser, I.   64 [47]; 143, 144 [15]
Haven, Jr., A. C.   4 [20]
Havinga, E.   14 [4]; 15 [12]; 15 [15]; 15 [16]; 15 [17]; 15 [18]
Hawkins, E. G. E.   373, 376 [5]
Hawkins, J. E.   7, 9 [36]
Hayatsu, R.   100 [26]
Hayes, W. K.   229, 230 [22]
Haynes, G. R.   8 [44]
Hebbelynck, M. F.   357 [45]
Heck, G.   209 [40]
Heidt, L. J.   496
Heilbronner, E.   20 [33]
Heindel, N. D.   25 [10]; 472 [3]; 472 [4]
Heinrich, P.   48, 49 [3]
Heiss, H.   310, 311 [41]
Heitzer, H.   205 [17]
Helberger, J. H.   434 [8]
Helferich, B.   119, 122 [22]; 119, 122, 124 [23]; 119, 122 [24]
Hellberg, H.   156 [9]; 156 [10]
Heller, C. A.   42 [11]
Heller, H. G.   129, 131 [14]
Heller, M. S.   36, 37 [11]; 36, 37 [12]

Hemmerich, P.   257, 258 [12]
Hempel, D.   349 [29]
Henderson, A. W.   62 [40]
Henderson, Jr., W. A.   25 [14]; 130, 131 [15]
Henmo, E.   448 [54]
Henrikson, B. W.   348 [23]
Henry, A. J.   72 [14]
Henry, J. P.   468, 469 [44]
Henry, R. A.   42 [11]
Henze, M.   71 [7]
Herberhold, M.   463 [24]
Hercules, D. M.   496
Hered, W.   350 [31]
Herrmann, R.   477 [13]
Hertler, W. R.   244 [8]; 244 [9]; 244, 245 [10]
Herzenstein, A.   367, 368 [15]
Hesp, B.   424 [27]
Hess, D.   99 [20]
Heusler, K.   369 [20]; 369 [21]; 369 [22]; 369 [23]; 369, 370 [25]
Hewett, W. A.   163 [34]; 163, 164 [35]
Hikino, H.   113 [16]
Hill, J.   222, 223 [14]
Hill, R. R.   2-4 [14]
Hilpern, J. W.   71, 73 [9]
Hilpert, S.   271 [19]
Hinderling, R.   357, 358 [46]
Hine, J.   274 [1]
Hinkel, L. E.   269 [13]
Hinman, R. L.   415-417 [8]
Hipps, G. E.   78 [37]
Hirschmann, F.   46, 47 [25]
Hirshberg, Y.   205 [25]; 205, 208 [27]
Hlavka, J. J.   39 [20]; 362 [1]
Hochstrasser, R. M.   499
Hochweber, M.   239 [46]
Hock, H.   373, 374 [7]; 373 [8]; 373 [9]; 374 [11]; 383 [54]
Hodgkins, J. E.   339 [11]
Hodgson, H. H.   316 [5]
Hodson, H. F.   401 [141]
Höft, E.   399 [132]
Hölzle, G.   353 [38]
Höschen, W.   239 [47]
Höver, H.   82 [50]
Hoffmann, F. W.   454, 455 [6]
Hoffmann, H.   283, 284, 286-288, 290, 291 [42]
Hoffmann, R. W.   316, 317 [9]
Hoffmeister, E.   368, 369 [18]
Hofmann, A. W.   242 [4]
Hoganson, E.   266, 267 [2]
Hoganson, E. D.   6, 7, 9 [34]
Hollaender, A.   496

Holmström, B.   257 [10]; 257 [11]; 257 [16]
Homburg, F.   226, 227 [7]; 226 [8]
Hooker, S. C.   147 [2]
Horclois, R.   391 [92]; 391 [93]
Horner, L.   121 [28]; 123 [30]; 283, 284, 286-288, 290, 291 [42]; 294, 301 [1]; 295-299 [2]; 298, 299 [9]; 299, 300 [12]; 302 [17]; 303 [24]; 304 [30]; 313, 314 [2]; 318, 321 [2]; 321, 322 [8]; 321, 322 [9]; 321 [10]; 326 [21]; 326, 327 [22]; 330, 331 [9]; 331, 332 [11]; 455, 457 [10]
Hostettler, H. U.   219 [7]
Howard, E. G.   106, 107 [13]
Hubbert, H. E.   354 [41]; 365 [11]
Huber, H.   260 [1]; 260, 261 [2]
Hudec, J.   2, 3 [10]; 2, 3 [13]; 2-4 [14]; 27 [23]; 79-81 [38]; 80 [40]
Hudson, R. F.   453 [2]
Hübler, M.   8 [39]
Huffman, K. R.   25 [12]; 25 [13]
Hugelshofer, P.   127-129 [8]; 139-141 [4]
Huisgen, R.   135, 136 [29]; 271, 272 [24]; 291 [53]; 307-309 [40]; 315 [4]
Huneck, S.   324 [16]
Hurst, J. J.   28 [26]; 28 [27]
Hutton, T. W.   15 [13]
Huynh, C.   302 [22]
Huyser, E. S.   160 [21]; 175, 176 [67]; 176 [68]; 210, 211 [47]; 355, 356 [43]

IG Farbenindustrie,   135 [27]; 135 [28]; 434 [7]; 438 [22]
Ilinski, M.   80, 81 [42]
Ingold, C. K.   298 [11]
Inhoffen, H. H.   15 [7]; 60 [29]; 60, 61 [30]; 149 [6]
Inman, C. G.   415, 417 [2]
Iriarte, J.   26, 27 [22]; 36, 37 [13]; 222, 223 [14]
Irmscher, K.   15 [7]
Irwin, R. S.   443 [34]; 443 [37]
Ishikawa, M.   52 [15]
Ishimoto, M.   91 [93]
Islam, A. M.   71 [4]; 119 [14]; 119 [20]; 398 [127]
Isler, O.   59, 61 [21]
Ismail, A. F. A.   133, 134 [22]; 133 [23]; 401 [138]
Ivin K. J.   438 [19]
Iwasaki, S.   385, 388 [69]

Jacobs, G.   4 [22]
Jacobs, R. L.   161 [24]
Jacobson, H. W.   106 [12]
Jambor, B.   143 [16]; 143 [17]
James, F. C.   90 [87]
Jamieson, N. C.   139, 140 [7]
Jeger, O.   23, 24 [7]; 23 [8]; 26 [18]; 26, 27 [22]; 30, 32 [36]; 31 [43]; 31 [44]; 31 [45]; 31, 32 [46]; 34, 36 [4]; 36 [9]; 36, 37 [10]; 36, 37 [11]; 36, 37 [12]; 36, 37 [13]; 7 7[35]; 78, 79 [36]; 194 [4]; 222, 223 [14]; 235 [37]; 235 [38]; 329 [4]; 408 [8]; 408, 412 [9]; 412, 413 [24]
Jennen, J.   420 [19]
Jenny, W.   353 [38]
Jensen, E. V.   171-173 [55]
Jerchel, D.   64 [47]; 143, 144 [15]; 143 [18]; 143, 144 [19]
Jerome, J. J.   172 [61]
Jodlbauer, A.   498
Jörgensen, P. F.   62 [39]
Johnson, C. E.   407 [4]
Johnson, C. K.   409 [11]
Johnson, G. R. A.   496
Johnson, H. E.   278, 279, 284 [26]
Jolley, J. E.   499
Jolly, P. W.   199 [4]
Jones, D. G.   298 [10]
Jones, E. R. H.   386 [70]
Jones, J. B.   41 [1]
Jones, J. H.   102 [34]
Jones, J. I.   106 [10]
Jones, W. J.   126 [2]
Jones, Jr., M.   283 [40]
Jorgenson, M. J.   2 [17]; 22, 24 [1]; 22, 24 [3]; 89 [81]; 89 [82]; 415-417 [5]
Joshua, C. P.   139, 140 [8]; 141, 142 [10]
Juday, R. E.   210 [45]
Jun, M. -J.   461, 462 [14]; 461 [15]; 469 [46]
Junghans, K.   132-134 [20]; 200, 201 [12]
Juveland, O. O.   160 [20]; 160 [21]; 165 [40]; 165 [41]

Kabasakalian, P.   248, 249 [5]; 248 [6]; 248, 249 [7]; 249 [8]; 251 [14]; 252, 253 [18]; 253 [19]
Kabitzke, K.   450 [57]
Kägi, D.   31 [45]
Kaiser, E. T.   448, 449 [55]
Kalamazoo Conference,   88 [77]
Kallmann, H. P.   496
Kaluza, F.   270, 271 [17]

Kalvoda, J.   127-129 [8]; 139-141 [4]; 369 [20]; 369 [21]; 369 [22]; 369 [23]; 369, 370 [25]
Kamatani, Y.   57, 58 [17]
Kamel, M.   191 [27]
Kaminsky, L.   50 [11]
Kampmeier, J. A.   368, 369 [18]
Kan, R. O.   100, 101 [28]; 496
Kane, S. S.   233 [32]; 234 [33]
Kaneko, C.   52 [15]
Kaplan, L.   10 [53]
Karabinos, J. V.   297 [8]
Karasch, C.   246, 247 [12]
Karlson, P.   65 [54]
Karlson-Poschmann, L.   77, 79 [34]; 118, 119, 121, 122, [3]
Karrer, P.   257, 258 [14]; 258 [17]; 258 [18]; 268, 270 [7]; 401, 402 [142]; 401, 402 [143]
Kasha, M.   491, 492 [25]; 499
Katritsky, A. R.   54 [22]
Kaufman, M. L.   455, 457 [9]; 457 [14]
Kaufmann, H. P.   23 [5]; 432, 433 [20]; 433 [21]
Kegelman, M. R.   195 [8]; 205 [24]
Kellmann, A.   101 [33]
Kellogg, R. E.   498
Kempf, R.   267, 270 [5]
Kenyon, J.   215, 216 [5]
Kerr, E. R.   344 [15]
Kerwin, J. F.   246, 247 [12]
Kettle, S. F. A.   461 [13]
Kharasch, M. S.   35 [6]; 153 [8]; 169 [48]; 170 [51]; 170, 171 [53]; 171-173 [55]; 171, 172 [56]; 171, 172, 175 [58]; 172 [61]; 233, 234 [31]; 233 [32]; 234 [33]; 349 [30]; 350 [31]; 435 [15]; 436, 437 [16]; 446 [48]
Kharasch, N.   368 [16]; 368 [17]; 434 [1]
Khorlina, M. Ya.   39 [19]; 283 [36]
Kidd, J. M.   440 [26]; 441 [28]
King, J. A.   339 [11]
King, J. F.   448 [53]
King, R. B.   462 [11]; 464, 465 [32]
King, R. W.   5, 7 [31]; 8 [45]; 8 [46]; 92 [102]
Kinkel, K. G.   46, 47 [22]; 384 [58]
Kirby-Smith, J. S.   495
Kirmse, W.   274 [2]; 283, 284, 286-288, 290, 291 [42]; 299, 300 [12]; 302 [17]; 331, 332 [11]; 331 [12]
Kistiakowsky, G. B.   496
Kitzing, R.   16, 17 [21]
Kleiman, J. P.   98 [13]
Klein, E.   376 [28]; 377 [30]; 409 [12]
Klemchuk, P. P.   285 [45]
Kleschnin, A. F.   496
Klessinger, M.   61 [32]
Klinger, H.   186, 187 [1]; 186-188 [2]; 189 [18]; 189 [19]
Kloosterziel, H.   356 [44]
Klosa, J.   338 [9]; 338 [10]
Klose, G.   64, 65 [50]
Klyne, W.   409 [12]
Knaggs, I. E.   82 [49]
Knight, S. A.   27 [23]
Knox, L. H.   276 [18]; 276 [19]; 283 [40]
Knunyants, I. L.   163 [33]
Kobel, H.   156 [8]
Kober, E.   441 [29]
Kobsa, H.   236, 237 [41]
Koch, E.   428 [8]
Koch, K.   227 [11]; 227, 229 [12]
Koch, K. F.   499
Kocsis, K.   329 [4]
Köbner, T.   257, 258 [14]
Koechlin, B. A.   50, 51 [9]
Kögel, A.   496
Kögler, H. P.   463, 464 [22]; 463 [23]
Köller, H.   193 [2]; 207 [32]; 466 [35]
König, C.   325 [19]
König, H.   307-309 [40]
Koerner von Gustorf, E.   461, 462 [14]; 461 [15]; 466 [35]; 469 [46]
Koevoet, A. L.   15 [12]
Kofron, J. T.   423 [24]
Koike, D.   182 [4]
Kolka, A. J.   344 [15]; 344 [16]; 345 [17]
Koller, E.   423 [24]
Koller, G.   71, 73 [6]
Koller, L. R.   474, 491 [7]; 496
Koltzenburg, G.   189 [20]; 189 [22]; 231-233 [25]
Kolvenbach, W.   189 [19]
Konovalova, I. V.   177 [72]; 177, 178 [73]; 177 [74]
Kooyman, E. C.   153 [9]
Kopecky, K. R.   276 [16]
Korn, F.   76 [28]
Kornhauser, A.   91 [93]
Kornis, G.   199 [7]
Korte, F.   74 [22]; 74 [23]; 415, 417 [11]; 468 [42]
Kosar, J.   496
Kost, V. N.   39 [18]; 39 [19]; 283 [36]
Kostin, V. N.   22 [2]
Kotlan, J.   87 [72]
Koton, M. M.   467 [40]
Koyano, K.   63 [45]

Kramer, E.   271, 272 [23]
Krasnovskii, A. A.   499
Krauch, C. H.   75 [24]; 82, 83 [47]; 86 [69]; 86 [70]; 88 [78]; 99 [20]; 110 [7]; 110 [9]; 119, 121, 122, 124 [8]; 168 [47]; 379, 380 [37]; 419, 420 [18]; 444 [39]
Krazinski, H. M.   39 [20]; 362 [1]
Krebaum, L. J.   30, 32 [36]
Kreher, R.   325 [20]
Kröhnke, F.   269 [15]; 272 [25]
Kroke, H.   46 [20]
Krongauz, V. A.   499
Kropp, P. J.   30, 32 [40]; 30, 32 [41]
Kuderna, B. M.   169 [48]
Kuderna, J.   170 [51]
Küster, H.   168 [47]
Kuhls, J.   110 [7]
Kuhn, H. J.   8 [47]; 8, 9 [48]; 65 [51]; 65 [52]; 92 [98]; 92 [99]; 92 [100]
Kuhn, R.   64 [47]; 64 [48]; 143, 144 [15]; 257, 258 [13]
Kuo, C. H.   66 [58]
Kupchan, S. M.   128 [11]
Kurz, P. F.   157 [11]
Kuzel, P.   463 [23]; 463 [31]
Kynaston, W.   106 [10]

Laage, E.   83 [56]
Laber, G.   277 [22]
Lacher, J. R.   105 [2]; 159 [19]
LaCount, R. B.   454 [5]
Läuger, P.   363 [3]
LaFlamme, P.   275 [8]
Lalande, R.   98 [12]; 98 [14]; 99 [21]
Lamchen, M.   50 [11]
Lamola, A. A.   56, 57 [5]
Landquist, J. K.   51, 52 [12]
Landrum, B. F.   421, 422 [23]
Lang, S.   373, 374 [7]; 373 [8]; 373 [9]
Laporte, H.   463 [28]
Larsen, D. W.   163 [36]
Latif, N.   71, 85, 87 [5]; 81 [44]; 99 [22]; 119, 123 [10]; 191, 123 [11]; 119 [13]; 119 [17]; 186, 187 [3]; 191 [28]; 210, 211 [44]; 205 [53]
Laubach, G. D.   387 [75]
Laughlin, R. G.   276, 283 [14]
Lawrence, R. V.   7, 9 [36]; 153, 154 [11]; 387 [76]; 387, 388 [77]
Laws, G. F.   15 [13]; 387 [78]
Lawson, W.   306 [36]
Lax, E.   474, 491, 492 [8]

Le Bel, N. A.   158 [18]
Le Berre, A.   431 [16]
Le Bras, J.   389 [82]
Le Fèvre, R. J. W.   316 [6]
Le Roux, Y.   157 [13]
Lederer, M.   162 [26]
Ledwith, A.   274, 277 [7]
Lee, A.   256 [4]
Leedham, K.   174 [65]
Leermakers, P. A.   204, 208 [7]; 210 [46]; 210 [49]; 218, 219 [3]; 218, 219 [5]; 219 [6]; 276 [16]
Legendre, P.   42 [5]; 59 [19]
Legrand, M.   101 [32]; 393 [104]
Lehfeldt, A.   85 [65]
Lehmann, C.   408, 412 [9]; 412, 413 [24]
Lehmann, G.   338 [8]
Leighton, P.   496
Leighton, P. A.   497
LeMahieu, R.   109, 112 [2]
Lemal, D. M.   285 [44]
Lemke, T. F.   25 [10]
Lengyel, I.   280 [28]
Lenz, G.   266, 267 [2]
Lepage, Y.   390 [85]
LeRosen, A. L.   59 [25]; 59 [26]
Leto, M. F.   175, 176 [66]
Letsinger, R. L.   76 [29]; 205 [11]
Levin, G.   52 [16]
Levin, S. H.   229 [18]; 229 [19]
Levina, R. Ya.   22 [2]
Levine, S. G.   257 [8]
Levinson, A. S.   323 [13]; 323 [14]
Lewis, A.   303 [25]
Lewis, B. B.   246, 247 [12]
Lewis, G. E.   337, 338 [7]; 138, 140 [3]; 139 [5]; 139, 140 [6]; 139, 140 [7]; 139, 140 [8]; 141, 142 [10]
Lewis, R. G.   238 [44]
Liberles, A.   396 [124]; 407 [2]; 431 [17]
Liebermann, C.   80, 81 [42]
Lightner, D. A.   229, 230 [22]
Liisberg, S.   443, 444 [35]
Limaye, D. B.   210 [50]
Lind, S. C.   496
Linden, H.   297 [7]
Lindquist, L. C.   127 [3]
Linebarger, C. E.   98 [5]
Link, W. J.   71, 73 [11]
Linschitz, H.   138 [1]; 138 [2]
Lipinsky, E. S.   415, 417 [2]
Liss, T. A.   384, 385 [65]
Litle, R. L.   302 [16]
Littrell, R.   283 [38]
Liu, R. S. H.   74, 75 [21]
Liu, Y. C.   349 [30]
Livingston, R.   496; 98 [9]

Livingston, R. S.   496
Lodemann, E.   91 [92]; 91 [93]
Lodge, J. E.   93 [104]; 114 [17]; 114 [18]
Löffler, K.   243 [5]
Lohaus, G.   380 [41]
Lohse, C.   51, 52 [14]
Lompe, A.   474, 491, 492 [8]
Lonsdale, K.   82 [49]
Loudon, J. D.   269 [16]
Lower, S. K.   499
Lowry, B. R.   303, 304 [26]
Lowry, O. H.   256 [2]
Loy, M.   25 [13]
Lozeron, H. A.   155 [4]
Lüttke, W.   61 [32]; 262 [8]
Lundberg, W. O.   373 [3]
Lunk, H.   296, 297 [5]
Luther, R.   98 [3]
Lutz, R. E.   23, 24 [6]; 44, 45 [16]; 57 [16]; 407 [3]; 431 [18]
Lwowski, W.   324 [17]; 324, 325 [18]
Lyon, A. M.   292 [57]
Lythgoe, B.   59, 61 [20]

Macaluso, A.   48 [2]
Mackay, D.   330, 331 [10]
MacKellar, F. A.   100 [27]
Mader, F.   196 [12]
Madinaveitia, J.   81 [45]
Magee, J. L.   495
Magnani, A.   246, 247 [12]
Magritte, H.   349 [28]
Mahajan, J. R.   249 [9]; 250 [10]
Maher, J.   381 [44]
Maimind, V. I.   53 [20]
Mallory, F. B.   19 [30]; 127 [3]; 127, 129 [4]; 127 [5]; 127, 128 [6]; 140, 141 [9]
Mangold, D.   221 [11]
Mannsfeld, S.-P.   75 [24]; 82, 83 [47]; 110 [7]; 110 [9]
Mansour, A. K.   191 [9]; 187, 190, 191 [11]; 187, 191 [13]
Mansour, A. K. E.   187 [12]
Mantell, G. J.   446 [48]
Manuel, T. A.   462 [11]
Margerum, J. D.   62 [35]; 416 [6]
Markau, K.   153 [10]
Markby, R.   461 [16]
Marklow, R. J.   178 [75]
Marsden, E.   316 [5]
Marsh, J. K.   98 [4]
Martel, J.   396 [119]; 396 [120]
Martens, T. F.   242-244 [2]
Martin, J. C.   53, 54 [21]
Martin, R. H.   357 [45]
Martin, R. B.   205 [11]

Martin, R. S.   218 [2]
Marx, U. F.   330, 331 [10]
Masamune, S.   300, 301 [15]
Mason, J.   106 [5]; 364 [9]
Massey, H.   499
Masson, C. R.   496
Mateos, J. L.   302 [18]; 303 [23]
Mathieu, J.   302 [22]
Matic, M.   46 [23]
Matthias, G.   149 [7]; 460 [7]; 460, 464 [10]
Mattingly, T. W.   324 [17]
Mattingly, Jr., T. W.   324, 325 [18]
Mattinson, B. J. H.   105, 106 [3]
Mauser, H.   205 [17]
Mayahi, M. F.   349 [27]
Mayer, J. R.   277 [25]
Mayo, F. R.   350 [31]
Mazur, Y.   37, 38 [15]; 37, 38 [16]; 38 [17]
Mazur, R. H.   348 [24]
McBee, E. T.   342 [3]
McCarthy, K. A.   491 [19]
McConnell, D. G.   306, 308 [38]
McCormick, D. B.   258 [19]
McDaniel, R. S.   45 [17]
McDonnell, C. H.   160 [20]
McElroy, W. D.   479 [18]; 496; 497
McGreer, D. E.   45 [17]
McGuire, F. J.   249 [9]; 250 [10]
McHugh, G. P.   63 [44]
McKeon, J. E.   468, 469 [44]
McKeown, A.   496
McLaren, A. D.   91 [96]; 472, 492 [2]; 156 [5]; 497
McLean, M. J.   383-385, 387, 389 [51]
McLeish, W. L.   57 [15]
McMeekin, W.   43, 44 [13]
McMillan, G. R.   499
McNesby, J. A.   479 [14]
McOmie, J. F. W.   71, 73 [9]
Meakins, G. D.   386 [70]
Mecke, R.   256, 257 [6]
Meder, W.   205 [10]
Meerwein, H.   277-279, 283 [20]; 283, 286 [39]
Meier, H.   497
Meinwald, J.   1, 3 [4]; 17 [22]; 302 [19]; 303 [25]
Mellier, M.-T.   393 [107]; 393 [110]; 393, 394 [111]
Meltzer, T. H.   153 [8]
Mendelson, W. L.   377 [34]; 377, 378 [35]; 407 [1]
Merenyi, R.   116 [26]
Mertens, H.-J.   46, 47 [22]; 384 [58]

Merz, H.   123 [30]
Metzger, H.   197 [15]; 261, 262 [3]; 261, 262 [4]; 263 [14]; 263, 265 [15]; 263 [16]; 263, 265 [17]
Metzner, W.   82, 83 [47]; 110 [9]
Meuche, D.   20 [33]
Meyer, H.   131-133 [16]; 149 [8]; 497
Meyer, W. L.   323 [13]; 323 [14]
Meyer-Delius, M.   272 [25]
Meystre, C.   358, 359 [48]; 369 [20]; 369 [21]; 369, 370 [25]
Middleton, W. J.   106 [12]; 106, 107 [13]
Miescher, K.   358, 359 [48]
Migita, M.   205, 207 [22]
Mikhailov, B. M.   49 [4]
Milas, N. A.   157 [11]
Milks, J. E.   410 [19]; 410-412 [20]
Miller, B.   346, 347 [19]; 363 [4]
Miller, L. A.   343 [10]
Miller, R. E.   42 [9]
Miller, W. T.   362 [2]
Milligan, B.   354 [41]; 365 [11]
Mills, J. S.   226, 227 [10]
Minnaard, N. G.   14 [4]
Minor, J. C.   387 [76]
Misani, F.   292 [57]
Mislow, K.   66, 67 [59]
Misrock, S. L.   98, 99 [11]
Mitchell, A.   253 [19]; 254 [21]
Mitchell, M. J.   306, 308 [38]
Mitra, R. B.   109, 112 [2]
Mitsch, R. A.   291 [55]; 291 [56]
Möller, K.   305 [32]; 305, 306, 310 [33]; 305 [34]; 306 [37]; 310, 311 [41]
Mole, T.   275, 278, 281 [13]
Molinet, C.   347, 348 [21]
Molyneux, R. J.   20 [33]
Moore, R. F.   148 [4]; 151, 152 [1]; 182, 183 [1]; 186, 189 [5]
Moore, R. N.   7, 9 [36]
Moore, W. M.   127 [7]; 203, 205 [5]; 205 [13]
Moosmayer, A.   281 [31]
Morduchowitz, A.   37 [14]; 210 [48]
Morgan, D. D.   127 [7]
Morgan, W. H.   269 [13]
Morgan, Jr., L. R.   319, 320 [3]
Moriarty, R. M.   319 [5]; 328, 329 [3]
Moriconi, E. J.   276, 286, 289 [17]; 307, 308 [39]
Morizur, J.-P.   153 [6]
Morkved, E.   448 [53]
Moroz, E.   302 [21]
Morrison, H.   417 [14]
Moss, R. A.   275 [11]

Mostafa, A.   203 [3]; 209 [43]
Moubacher, R.   187 [9]
Moubasher, R.   71, 85, 87 [5]; 81 [44]; 99 [22]; 119, 123 [10]; 119 [13]; 119 [17]; 186, 187 [3]; 191 [28]; 209 [38]; 209 [39]; 209 [41]; 209 [43]; 210, 211 [44]; 205 [53]
Moulines, F.   98 [14]
Moureu, C.   390 [86]
Mousa, G. A.   187, 188 [6]
Mousseron, M.   42 [5]; 42 [6]; 59 [19]; 59 [23]
Mousseron-Canet, M.   42 [5]; 59 [19]
Mowat, J. H.   238 [42]
Mühlmann, R.   131, 132, 134 [18]
Müller, A.   153 [10]
Müller, E.   197 [15]; 260 [1]; 260, 261 [2]; 261, 262 [3]; 261, 262 [4]; 263, 265 [15]; 263 [16]; 263, 265 [17]; 264 [18]; 264 [19]; 264 [20]; 281 [31]; 437, 438 [18]
Müller, M.   387 [73]
Müller, R.   191 [29]
Müller, W.   132, 134 [21]; 384 [62]
Muhr, H.   363 [3]
Mukherjee, T. K.   193 [2]; 207 [32]
Mulcahy, E. N.   119, 122, 124 [23]
Muller, G.   302 [22]
Murawski, D.   245 [11]
Murov, S.   415-417 [5]
Murphy, S.   52 [16]
Murray, J. J.   276, 286, 289 [17]; 307, 308 [39]
Musajo, L.   87, 88 [74]
Mustafa, A.   71 [4]; 76, 83 [26]; 81 [44]; 81 [44]; 90 [86]; 99 [22]; 99 [23]; 107 [14]; 118 [1]; 118, 119, 123 [2]; 119 [9]; 119, 121, 122 [12]; 119 [13]; 191 [13]; 119 [14]; 119 [16]; 119, 121 [18]; 191 [19]; 146-148 [1]; 147, 148 [3]; 187 [10]; 187, 190, 191 [11]; 187 [12]; 187, 191 [13]; 187 [14]; 187 [15]; 187, 188 [16]; 191 [27]; 199 [8]; 200, 201 [9]; 205 [18]; 205-207, 211, 212 [23]; 205, 206 [28]; 211 [51]; 398 [127]; 398, 399 [129]; 398 [130]; 398 [131]; 434 [2]; 445, 446 [43]; 499; 500
Muszkat, K. A.   19 [31]; 127 [10]
Muth, K.   302 [17]; 306, 308 [38]

Nakamura, A.   460 [12]
Nakata, T.   500
Nakazaki, M.   152 [4]; 153 [5]
Nann, B.   23 [8]; 77 [35]; 194 [4]
Napier, D. R.   302 [16]

Narang, S. A.   250 [10]
Naumann, W.   330, 331 [9]
Naylor, M. A.   263 [11]
Neckers, D. C.   210, 211 [47]; 218, 219 [5]; 219 [6]; 497
Nef, J. U.   399 [133]
Neher, C. M.   343 [10]
Nelson, Jr., M. F.   243 [7]
Nesmeyanov, A. N.   39 [18]; 39 [19]; 283 [36]
Neumann, H. -G.   182 [2]
Neumüller, O. -A.   8 [47]; 8, 9 [48]; 65 [51]; 65 [52]; 92 [98]; 92 [100]; 375 [18]; 377 [32]; 378 [36]; 377, 378 [33]
Nichols, G.   242-244 [2]
Nicholson, A. J. C.   499
Nickel, S.   445 [46]
Nickon, A.   249 [9]; 250 [10]; 376, 377 [24]; 377 [34]; 377, 378 [35]; 407 [1]
Nineham, A. W.   64 [46]
Nisikawa, Y.   57, 58 [17]
Noel, F.   446, 447 [50]
Nofre, C.   157 [13]
Nordstrom, J. D.   242 [1]
Norin, T.   409 [12]
Noyes, R. M.   497
Noyes, Jr., W. A.   25 [15]; 496; 497; 498; 499
Noyori, R.   57, 58 [17]; 276, 285, 289 [15]
Nozaki, H.   57, 58 [17]; 276, 285, 289 [15]
Nudenberg, W.   153 [8]; 170 [51]; 349 [30]; 446 [48]
Nussbaum, A. L.   248, 250, 252 [2]; 251 [14]; 253 [20]; 254 [21]
Nussim, M.   415-417 [5]; 418 [15]
Nye, M. J.   218, 219 [4]
Nyholm, R. S.   462, 464 [20]
Nyman, F.   234 [34]; 234 [35]

O'Connell, E. J.   64, 65 [49]; 193 [1]; 238 [45]
Oakes, V.   359 [50]
Ogliaruso, M. A.   394 [115]
Ohloff, G.   34 [3]; 375, 376 [17]; 375 [18]; 377 [30]; 409 [12]
Okabe, H.   479 [14]; 479 [15]
Oldekop, Yu.   466 [36]; 466-468 [37]; 467 [39]
Oldekop, Yu. A.   399, 400 [134]; 467 [38]
Oliveri-Mandalà, E.   199 [6]
Oliveto, E. P.   251 [14]; 253 [19]; 253 [20]; 254 [21]
Olszewski, W. F.   177 [71]

Omran, S. M. A. E.   187 [12]
Omran, S. M. A. R.   215-217 [4]
Opitz, K.   220 [10]
Orgel, L. E.   98 [7]; 461 [13]
Orloff, H. D.   344 [15]; 344 [16]
Orndorff, W. R.   98, 101 [6]
Orton, K. J. P.   313 [1]
Osaki, K.   57, 58 [13]; 84 [63]
Osberg, E. G.   467 [40]
Oster, G.   257 [10]; 446 [47]
Osterroht, C.   450 [57]
Oswald, A. A.   446 [49]; 446, 447 [50]
Oth, J. F. M.   116 [26]
Othman, A. M.   209 [38]
Owings, F. F.   246, 247 [12]

Padeken, H. G.   264 [18]
Padwa, A.   49 [5]; 409 [13]; 409 [14]; 409 [15]; 450 [58]
Pampus, G.   256, 257 [6]
Panico, M. R.   393 [109]
Pape, M.   194 [7]; 205 [10]; 499
Pappas, S. P.   5, 6 [26]; 10 [52]
Paquette, L. A.   6, 7 [35]; 100 [27]; 100, 101 [29]
Parcell, L. J.   106 [6]
Parham, W. E.   90, 91 [88]
Park, J. D.   105 [2]; 159 [19]
Paschalski, C.   72 [15]
Paskovich, D. H.   222 [12]; 289 [49]
Pasternak, Y.   468 [41]
Pasto, D. J.   5, 6, 9 [28]; 6, 9 [32]; 220 [8]
Patai, S.   344 [14]
Paternò, E.   89 [79]; 200 [10]; 200 [13]; 215 [2]; 414, 415 [1]
Patterson, D. A.   491 [20]
Patterson, J. M.   205 [11]
Paudler, W. W.   99, 100 [25]
Pausacker, K. H.   380 [40]
Pavlis, R. R.   353 [40]
Pearce, R. S.   187, 188 [17]
Pearson, E. G.   61 [31]
Pechet, M. M.   248 [3]; 250, 251 [13]
Pedersen, C. J.   161 [25]; 332, 333 [13]
Pelletier, S. W.   57 [15]
Perkin, A. G.   131 [17]
Perkin, Jr., W. H.   306 [35]; 306 [36]
Perkins, M. J.   495
Perold, G. W.   270, 271 [17]
Perret, G.   215 [2]
Perry, C. W.   93, 94 [103]
Peterson, L. I.   82, 83 [51]
Petit, A.   15 [8]; 15 [9]

Petropoulos, C. C.   216 [6]
Petry, R. C.   231 [24]
Petterson, R. C.   64, 65 [50]
Pettit, R.   187, 188 [17]
Pfau, M.   25 [10]; 231 [26]; 232 [27]; 232 [28]; 232, 233 [30]; 472 [3]
Pfau, M. A.   472 [4]
Pfeiffer, P.   271 [22]; 271, 272 [23]
Pfennig, H.   384 [62]
Pfrengle, O.   459 [1]
Pfundt, G.   118, 121, 122 [4]; 119, 120, 123 [6]; 122, 124 [29]; 183, 184 [6]; 183, 184 [9]; 186, 190 [4]
Phillips, G. O.   499
Pilgrim, A.   205 [21]
Pinckard, J. H.   59, 61 [27]
Pincussen, L.   497
Pitt, H. M.   407 [4]
Pitts, Jr., J. N.   25 [47]; 76 [29]; 205 [11]; 416 [6]; 472, 474, 479, 492 [1]; 495; 497; 498
Plotnikow, J.   497
Plumb, J. B.   454, 455 [4]; 455-457 [11]
Pope, R.   99 [24]
Porter, G.   25 [11]; 205, 206 [14]; 205 [15]; 206 [30]; 479 [16]; 497
Porter, G. B.   499
Poschmann, L.   65, 66 [53]; 65 [55]
Postovskii, I. Ya.   255, 257 [1]
Potter, A. E.   499
Potzinger, P.   274 [6]
Pouchot, O.   396, 397 [123]
Praetorius, P.   76 [28]
Prijs, B.   357, 358 [46]
Pringsheim, P.   497
Prinzbach, H.   15 [6]; 16 [20]; 16, 17 [21]
Pschorr, R.   190 [25]
Ptitsyna, O. A.   456, 457 [13]
Pudeeva, M. E.   456, 457 [13]
Pudovik, A. N.   177 [72]; 177, 178 [73]; 177 [74]
Pütter, R.   410, 411 [17]; 411 [23]
Purkis, C. H.   248 [1]

Quimby, W. C.   84, 85 [60]
Quinkert, G.   66 [57]; 220, 221 [9]; 220 [10]; 226 [5]; 226, 227 [7]; 226 [8]; 226 [9]; 227, 228, 230, 234, 235 [16]
Quint, F.   134, 136 [26]

Rabinowitch, E.   496
Rabinowitch, E. I.   497

Rabinowitz, R.   454 [7]
Rabold, G. P.   193 [2]; 207 [32]
Rahman, M.   319 [5]
Ramirez, F.   453 [3]
Ramp, F. L.   4 [20]
Ranjon, A.   396, 397 [123]
Rao, D. V.   109 [5]
Rapalski, G.   71, 72 [3]
Rappen, F.   277-279, 283 [20]
Rappoldt, M. P.   15 [15]; 491, 492 [26]
Rassoul, A. R. A. A.   215, 216 [5]
Rathjen, H.   283, 286 [39]
Rathke, B.   106 [7]; 106 [8]
Ratsimbazafy, R.   431 [16]
Rauen, H. M.   256 [3]
Raymond, E.   400 [136]
Rayner, D. R.   66, 67 [59]
Razuvaev, G.   466 [36]; 466-468 [37]; 467 [39]
Razuvaev, G. A.   647 [38]
Read, A. T.   435 [15]
Readio, P. D.   352 [37]
Reasoner, J.   266, 267 [2]
Recktenwald, G.   205 [11]
Recktenwald, G. W.   76 [29]
Reed, C. F.   434 [5]; 434 [6]
Reeder, E.   50, 51 [9]
Reeder, J. A.   165 [38]
Rees, C. W.   495
Reese, C. B.   236 [39]; 236, 237 [40]
Regitz, M.   209 [40]
Reid, C.   497; 500
Reid, S. T.   100 [26]; 111 [12]; 268 [11]; 395-397 [118]; 499
Reihlen, H.   459 [1]
Reinmuth, O.   170, 171 [53]
Relyea, D. I.   163 [36]
Reppe, W.   263 [12]; 461 [17]
Resnick, C.   201, 202 [14]
Resnick, P.   381 [42]
Reusch, W.   169, 170 [50]; 409 [11]
Reusch, W. H.   16 [19]
Reutov, O. A.   456, 457 [13]
Rheiner, Jr., A.   15 [13]
Rhoads, S. J.   277 [21]
Richards, J. H.   30 [37]; 229 [21]
Richards, T. A.   59 [28]
Richardson, D. B.   283 [41]
Rieche, A.   166 [44]; 379, 380 [38]; 399 [132]
Ried, W.   107, 108 [15]; 271 [20]
Riedl, H.-J.   263 [12]
Rieker, A.   281 [31]
Rigaudy, J.   43, 44 [14]; 43 [15]; 391, 393 [87]
Riiber, C. N.   83 [53]

Rinehart, Jr., K. L. 328 [1]; 328 [2]
Rinker, J. W. 23, 24 [6]; 57 [16]
Rio, G. 396, 397 [123]
Ripley, R. A. 227, 229 [14]
Rivas, C. 25 [9]; 208 [34]
Robb, E. W. 93, 94 [103]; 112, 114 [13]
Robert, E. 165, 166 [39]
Robert, J. 393 [103]
Robert, J.-M. 14 [5]
Roberts, J. D. 342 [2]; 348 [24]
Robinson, B. 401 [139]
Robinson, C. H. 248, 250, 252 [2]; 253 [19]; 253 [20]; 254 [21]
Robinson, R. 306 [35]; 306 [36]
Robinson, T. A. 443 [34]
Robson, R. 111 [11]
Rodighiero, G. 87, 88 [73]; 87, 88 [74]
Roe, A. 354 [41]; 365 [11]
Roedig, A. 296, 297 [5]; 353 [39]
Rogers, D. A. 62 [35]
Rogers, M. T. 98 [15]
Rojahn, W. 376 [28]
Rokach, J. 166-168 [43]; 167 [46]
Rollefson, G. K. 497
Romanelli, M. G. 394 [115]
Rondestvedt, Jr., C. S. 346, 347, 350 [18]
Roosen-Runge, C. 149 [7]
Rose, J. E. 354 [41]; 365 [11]
Rosenthal, D. 423 [24]
Ross, J. 373 [2]
Rubin, M. B. 78 [37]; 183 [7]; 183, 184 [8]; 184 [10]; 208, 212 [35]
Rucker, J. T. 343 [12]
Rudy, H. 257 [9]; 257, 258 [13]
Rüegg, R. 59, 61 [21]
Rummert, G. 60, 61 [30]
Rupert, C. S. 91 [95]
Rupp, E. 153 [7]
Russell, D. W. 268, 270 [8]
Russell, G. A. 342 [4]; 348, 349 [25]; 348 [26]
Russell, J. R. 177 [71]
Rust, F. F. 162-164 [27]; 162 [29]; 162, 163 [30]; 163 [31]; 176, 177 [69]; 177 [70]
Rutgers, J. G. 416 [6]
Rutschmann, J. 156 [8]
Rydon, H. N. 359 [50]
Ryf, H. 31 [44]

Sachs, D. H. 103 [37]; 205 [29]
Sachs, F. 267, 270 [5]; 271 [19]
Sage, M. 171, 172 [56]
Sakata, M. K. 218 [2]

Sakla, A. B. 366 [13]
Sakurai, H. 415, 417 [9]; 415, 417 [10]
Salamon, M. 264 [18]; 264 [20]
Salomon, H. 257, 258 [14]; 258 [17]; 258 [18]
Saltiel, J. 28, 30 [30]; 56 [4]; 56, 57 [5]; 56 [6]; 120 [7]; 500
Sanders, G. M. 15 [18]
Sandhu, S. S. 462, 464 [20]
Sandros, K. 205 [12]
Šantavý, F. 8 [42]
Sargent, M. V. 19, 20 [32]; 127, 128 [9]
Sasin, R. 177 [71]
Sasse, W. H. F. 116 [21]
Sattar, A. B. M. A. 448 [53]; 448 [54]
Saunders, Jr., W. H. 319, 320 [6]
Sauvage, G. 391 [95]; 391, 392 [96]; 392 [97]; 392 [100]
Savitz, M. L. 127 [3]
Sax, K. J. 358 [47]
Scanlan, J. T. 400 [137]
Schaafsma, Y. 153 [9]
Schach von Wittenau, M. 382 [47]; 420 [20]
Schäfer, F. 297 [7]
Schaffner, K. 23, 24 [7]; 23 [8]; 26 [18]; 26, 27 [22]; 30 [33]; 31 [44]; 31 [45]; 34, 36 [4]; 36 [9]; 36, 37 [10]; 36, 37 [11]; 36, 37 [12;] 36, 37 [13]; 56 [1]; 66 [56]; 77 [35]; 127-129 [8]; 139-141 [4]; 194 [4]; 222, 223 [13]; 222, 223 [14]; 235 [37]; 408 [8]; 408, 412 [9]; 412, 413 [24]; 500
Schaller, F. 226 [6]
Schaltegger, H. 351 [33]; 351 [34]; 443 [36]
Scharf, D. 74 [23]; 415, 417 [11]; 468 [42]
Scharf, H.-D. 74 [22]
Scheffer, A. 411 [22]
Scheindlin, S. 256 [4]
Schenck, G. O. 2, 3 [12]; 8 [47]; 8, 9 [48]; 46, 47 [22]; 65 [51]; 65 [52]; 72, 73 [18]; 72 [19]; 75 [24]; 82, 83 [47]; 86 [69]; 86 [70]; 92 [98]; 92 [100]; 109, 110 [1]; 110 [6]; 110 [7]; 110 [9]; 110, 111 [10]; 116 [22]; 118, 121, 122 [4]; 119, 121, 122, 124 [8]; 119, 122 [15]; 121 [25]; 121 [26]; 121 [27]; 151 [2]; 182 [3]; 183, 184 [9]; 186, 190 [4]; 189 [20]; 189 [22]; 194 [7]; 203 [2]; 205 [10]; 226 [6]; 231-233 [25]; 232 [29]; 274 [6]; 277, 279 [23]; 278,

279 [27]; 375, 378 [14]; 375, 376, 381, 383, 396 [15]; 375, 376 [17]; 375 [18]; 376, 381 [20]; 376, 383 [21]; 376, 383 [22]; 376, 383 [23]; 376, 378 [29]; 377 [30]; 377 [32]; 377, 378 [33]; 378 [36]; 379, 380 [37]; 381 [43]; 383 [49]; 383, 384, 387, 389, 395 [50]; 383 [55]; 383 [56]; 383 [57]; 384 [58]; 384 [59]; 384 [60]; 384 [61]; 384 [62]; 384 [63]; 415, 417 [12]; 419, 420 [18]; 427 [4]; 427 [5]; 427, 429 [6]; 427, 429 [7]; 428 [8]; 428-430 [10]; 429 [12]; 430 [13]; 439 [23]; 444 [39]; 444 [40]; 445 [42]; 461, 462 [14]; 461 [15]; 466 [35]; 475 [12]; 496; 497; 500
Scherb, H.   435 [13]
Schiedt, U.   77, 79 [34]; 118, 119, 121, 122, [3]
Schiemenz, G. P.   456, 457 [12]
Schlatmann, J. L. M. A.   15 [16]
Schlenk, W.   281 [32]; 367, 368 [15]
Schlessinger, R. H.   142 [12]; 447 [51]; 447, 448 [52]
Schlientz, W.   156 [7]
Schlittler, E.   258 [17]; 258 [18]
Schlubach, H. H.   169 [49]
Schmid, H.   54 [24]; 195, 196 [10]; 196 [11]; 207, 208 [33]; 239 [46]; 335-337 [3]; 335, 336 [4]; 401, 402 [142]; 401, 402 [143]
Schmidlin, J.   126, 127 [1]
Schmidt, E. W.   437, 438 [18]
Schmidt, G. M. J.   57, 58 [13]; 83 [52]; 84, 85 [61]; 84 [62]; 84 [63]
Schmidt, U.   153 [10]; 450 [57]
Schmidt, W.   396 [125]; 435, 437 [11]
Schmidt-Thomée, G. A.   121 [27]; 439 [23]
Schmitz, E.   166 [44]; 245 [11]; 379, 380 [38]
Schmitz, P.   277 [24]
Schneider, R. A.   1, 3 [4]
Schnider, O.   374, 375 [13]
Schocken, V.   445 [41]
Schöllner, R.   374, 375 [12]
Schönauer, G.   461 [18]
Schönberg, A.   71, 85, 87 [5]; 81 [44]; 88 [75]; 88 [76]; 98 [18]; 99 [22]; 99 [23]; 106, 107 [9]; 107 [14]; 118 [1]; 118, 119, 123 [2]; 119, 123 [10]; 119, 123 [11]; 119, 121, 122 [12]; 119 [13]; 119 [17]; 119, 121 [18]; 132-134 [20]; 133, 134 [22]; 146-148 [1]; 147, 148 [3]; 153 [7]; 186, 187 [3]; 187,
188 [6]; 187 [9]; 187 [14]; 187, 188 [16]; 189 [23]; 190, 191 [24]; 190 [26]; 191 [28]; 199 [8]; 200, 201 [9]; 200, 201 [12]; 203 [3]; 205 [53]; 205 [18]; 205-207, 211, 212 [23]; 209 [39]; 209 [41]; 210, 211 [44]; 215-217 [4]; 239 [48]; 383, 387, 389 [52]; 398, 399 [129]; 398 [130]; 440, 441 [27]; 445, 446 [43]; 445 [46]; 497; 500
Schöpf, C.   338 [8]
Schöpp, K.   258 [17]; 258 [18]
Scholes, G.   496
Scholl, M.-J.   394 [114]
Schomburg, G.   75 [24]
Schorta, R.   23 [8]; 77 [35]; 194 [4]
Schrauzer, G. N.   459, 462 [2]
Schreiber, E. C.   387 [75]
Schreier, E.   156 [8]
Schröder, G.   116, 117 [24]; 116 [25]; 116 [26]
Schroeter, S.   375, 376 [17]; 375 [18]
Schroeter, S. H.   427, 429 [7]
Schütz, O.   445 [46]
Schuller, W. H.   7, 9 [36]; 153, 154 [11]; 387 [76]; 387, 388 [77]
Schulte-Elte, K. H.   34 [3]
Schulte-Elte, K.-H.   376, 378 [29]; 379, 380 [37]; 387-389 [79]; 428-430 [9]; 428-430 [10]
Schulte-Frohlinde, D.   426, 427 [1]; 427 [3]
Schultze, H.   399 [132]
Schulz, P.   474, 491, 492 [8]
Schulze, H.   445 [44]
Schulze-Buschoff, H.   383 [56]
Schumacher, H.-J.   434, 436 [9]
Schumacher, M.   118 [5]; 383 [48]
Schuster, D. I.   116 [23]
Schwall, H.   209 [40]
Schwenke, W.   300 [14]
Schwieter, U.   59, 61 [21]
Scott, A. I.   382 [46]
Searles, S.   416 [6]
Searles, Jr., S.   142 [11]; 415, 416 [4]
See, L. B.   442 [31]
Seebach, D.   115 [20]
Seese, W. S.   302 [20]
Seffl, R. J.   105 [2]
Seitz, E. O.   497
Seliger, H. H.   479 [18]; 497
Selzer, H.   191 [29]
Selzer, R.   91 [93]
Sen, S. C.   187 [8]
Sen Gupta, A. K.   23 [5]; 432, 433 [20]; 433 [21]

Serebryakov, E. P.   370 [26]
Sernagiotto, E.   1 [2]
Setlow, J. K.   91 [94]
Setser, D. W.   283 [38]
Sevchenko, A. N.   399, 400 [134]
Seybold, P.   500
Shafiq, M.   30 [35]; 30, 32 [39]; 229, 230 [17]
Shalaby, A. F. A. M.   119 [9]; 187, 190, 191 [11]
Shand, A. J.   426 [2]
Shannon, J. S.   334 [1]
Sharkey, W. H.   106 [12]; 106, 107 [13]
Shaw, W. G.   42 [8]; 498
Shechter, H.   71, 73 [11]; 89, 90 [85]
Sheehan, J. C.   215-217 [1]; 280 [28]
Shemyakin, M. M.   53 [20]
Sheppard, S. E.   497
Sherman, G. M.   138 [1]
Shima, K.   198 [2]; 415, 417 [9]; 415, 417 [10]
Shinzawa, K.   49 [8]
Shugar, D.   54 [22]; 54 [23]; 91 [96]; 156 [5]; 472, 492 [2]; 497; 497
Siddiqui, M. N.   206 [31]
Sidky, M. M.   189 [23]
Sieber, W.   316, 317 [9]
Siegert, F. W.   176 [68]
Sieglitz, A.   239 [49]
Siemer, H.   60 [29]
Silber, P.   1 [1]; 26 [16]; 26 [19]; 26, 27 [21]; 57, 63, 64 [14]; 71, 83, 85 [1]; 76 [27]; 203 [1]; 203 [4]; 203, 205 [6]; 204 [8]; 209 [37]; 225 [3]; 225 [4]; 267-269 [3]; 269 [14]
Silberman, H.   334 [1]; 334 [2]
Sill, C. W.   491 [22]
Silveston, P. L.   498
Simkin, R. D.   410, 411 [21]
Simmons, M. C.   283 [41]
Simmons, T. C.   454, 455 [6]
Simon, M. I.   381 [45]
Simons, J. P.   500
Simonsen, J.   30 [34]
Sina, A.   88 [75]; 88 [76]; 119, 123 [10]; 186, 187 [3]; 210, 211 [44]
Singer, L. A.   416, 420, 421 [7]
Sircar, A. C.   187 [8]
Sisido, K.   276, 285, 289 [15]
Skau, E. L.   46, 47 [24]; 46, 47 [25]; 387, 388 [74]
Skell, P. S.   158 [17]; 275 [9]; 275 [12]; 352 [37]; 353 [40]
Skladnev, A. A.   163 [33]

Skovronek, H. S.   275 [12]
Slade, R. E.   342 [8]
Slates, H. L.   66 [58]
Slomp, G.   100 [27]; 100, 101 [29]
Smith, H. G.   8 [45]; 8 [46]; 8 [49]; 92 [101]; 92 [102]
Smith, J. M.   498
Smith, K. C.   156 [6]
Smith, P. A. S.   262 [9]; 320 [7]
Smith, S. L.   227, 229 [12]
Snell, R. L.   2-4 [9]
Sobhy, M. E. D.   398 [131]
Söll, H.   375 [16]
Soliman, G.   215, 216 [5]
Sonnenberg, J.   396 [121]
Sonntag, F. I.   57, 58 [13]; 84, 85 [61]; 84 [63]
Sosnovsky, G.   155 [1]; 263 [10]; 341 [1]; 434 [3]; 453 [1]
Späth, E.   71, 73 [6]
Speers, L.   292 [57]
Speier, J. L.   351 [35]
Spietschka, E.   121 [28]; 294, 301 [1]; 295-299 [2]; 298, 299 [9]; 303 [24]; 321, 322 [8]
Spinatelli, C. J.   497
Splitter, J. S.   49, 50 [6]; 49-51 [7]
Spoehr, H. A.   497
Sproesser, U.   205 [17]
Spruch, G. M.   496
Srinivasan, R.   10 [55]; 14 [1]; 14, 18 [2]; 18 [25]; 26 [17]; 26 [20]; 218 [1]; 417 [13]
Stacey, F. W.   160 [20]; 160 [21]; 162 [28]; 163, 164 [32]; 165 [40]
Staehelin, A.   101, 102 [31]; 391 [89]; 393 [106]
Standke, O.   189 [18]
Starr, J. E.   218 [2]
Starratt, A. N.   319 [4]
Staude, H.   497
Staudinger, H.   29 [32]; 299 [13]; 445 [45]
Stauff, J.   434, 436 [9]
Steacie, E. W. R.   498
Steel, J. K.   53 [17]
Steele, B. R.   157 [14]; 173, 174 [62]
Stefani, A. P.   159 [19]
Steinberger, F. K.   57, 58 [12]; 83 [59]
Steinmetz, R.   2, 3 [12]; 109, 110 [1]; 109 [3]; 110 [6]; 110, 111 [10]; 161 [22]; 232 [29]; 278, 279 [27]; 415, 417 [12]; 500
Stephenson, A.   106, 107 [9]
Stephenson, E. F. M.   205 [21]
Steppan, H.   304 [29]

Stermitz, F. R.   127 [7]
Sternbach, L. H.   50, 51 [9]
Sternberg, H. W.   461 [16]
Sternhell, S.   334 [1]; 334 [2]
Stevens, I. D. R.   290 [51]
Stewart, H. N. M.   136 [30]
Stiddard, M. H. B.   462, 464 [20]
Stiles, A. R.   176, 177 [69]; 177 [70]
Stiles, M.   18 [27]
Still, I. W. J.   89 [83]
Stobbe, H.   57, 58 [12]; 83 [55]; 83 [58]; 83 [59]; 85 [65]; 129 [12]
Stockmann, H.   155, 160 [2]
Stöhr, H.   313, 314 [2]
Stoermer, R.   57 [11]; 63 [42]; 83 [56]; 85 [64]; 155, 160 [2]; 165, 166 [39]; 322 [11]
Stoessl, A.   199, 201 [5]; 448 [53]; 448 [54]
Stoll, A.   156 [7]
Stolpp, T.   440, 441 [27]
Stone, B. R.   155 [3]
Stone, F. G. A.   462 [11]
Stothers, J. B.   42 [4]; 100 [26]; 170 [52]; 198, 199, 201 [3]
Stright, P. L.   90, 91 [88]
Strohmeier, W.   460 [3]; 460 [4]; 460 [5]; 460 [6]; 460 [7]; 460, 462 [8]; 460 [9]; 460, 464 [10]; 461 [18]; 463 [21]; 463 [25]; 463 [26]; 463 [27]; 463 [28]; 463, 464 [29]; 463 [30]
Struve, H.   8 [40]
Subbarow, Y.   358 [47]
Subrahmanyam, G.   218, 219 [4]
Süs, O.   304, 306, 309, 310 [27]; 304, 305 [28]; 304 [29]; 305 [32]; 305, 306, 310 [33]; 305 [34]; 306 [37]; 310, 311 [41]; 314, 315 [3]
Sugden, T. M.   495
Sugimori, A.   500
Sugowdz, G.   116 [21]; 334 [2]
Suida, H.   267 [4]
Suppan, P.   205, 206 [14]; 205 [15]; 206 [30]
Susemihl, W.   374 [11]
Sutton, D. A.   46 [23]; 373 [1]; 373 [10]
Swensson, T.   498
Swern, D.   177 [71]; 400 [137]

Tadros, W.   366 [13]
Takeshita, H.   113 [15]
Taliaferro, J. D.   175, 176 [67]
Tanaka, I.   49 [8]; 63 [45]
Tanasescu, I.   267 [6]; 268 [9]; 271 [21]
Tanner, D. W.   98 [8]
Tarbell, D. S.   442 [30]
Taub, D.   66 [58]
Tautz, W.   155 [4]
Taylor, D.   62 [40]
Taylor, E. C.   99, 100 [25]; 100, 101 [28]
Taylor, J. W.   109 [5]
Taylor, M. J.   106 [11]
Taylor, R. P.   205 [11]
Taylor, W.   98 [7]
Taylor, W. C.   31, 32 [42]; 376, 383, 389 [27]
Templeton, W.   170 [52]; 198, 199, 201 [3]
Tennant, G.   269 [16]
Terenin, A.   205 [9]
Terenin, A. N.   498
Ter-Sarkisyan, G. S.   49 [4]
Thaler, W.   351, 352 [36]
Theilacker, W.   396 [125]
Thelen, P. J.   243 [7]
Thiele, J.   305 [31]
Thiessen, W. E.   227 [11]
Thoai, N.   45 [18]
Thompson, H. W.   248 [1]
Thomson, R. H.   426 [2]
Throndsen, H. P.   78, 79 [36]
Tiefenthaler, H.   54 [24]; 55 [25]
Tiers, G. V. D.   71, 73 [11]; 89, 90 [85]
Timmons, C. J.   19, 20 [32]; 127, 128 [9]
Tobler, E.   172 [59]
Todd, A.   50, 51 [10]
Toki, S.   415, 417 [10]
Tokumaru, K.   500
Tomaschek, R.   133 [25]
Tomoeda, M.   223, 224 [15]; 442 [32]
Toromanoff, E.   196 [13]; 196 [14]
Townley, E.   251 [14]; 253 [19]
Townley, E. R.   248, 249 [5]; 248 [6]; 248, 249 [7]; 249 [8]; 252, 253 [18]
Träger, L.   91 [93]
Traylor, T. G.   291 [52]
Trecker, D. J.   35, 36 [7]; 35, 36 [8]; 105 [1]; 468 [43]; 468, 469 [44]
Trefilova, L. F.   255, 257 [1]
Treibs, W.   77 [31]; 77 [32]; 374, 375 [12]
Tröster, H.   239 [49]
Trotman, J.   252 [16]
Trumbull, E. R.   4 [20]
Tseng, C.-Y.   408 [10]
Tsuda, K.   385, 388 [69]
Tsutsumi, S.   198 [2]
Türck, G.   91 [93]

Turro, N. J.   2, 3, 10 [7];
  2, 3, 10 [8]; 4 [24]; 56, 57 [5];
  56 [7]; 74, 75 [21]; 218, 219 [3];
  218, 219 [5]; 219 [6]; 479 [17];
  498
Turro, Jr., N. J.   75 [25]

Überwasser, H.   369, 370 [25]
Uhler, R. O.   89, 90 [85]
Ullman, E. F.   25 [12]; 25 [13]; 25
  [14]; 130, 131 [15]; 410 [18]; 410
  [19]; 410-412 [20]
Umezawa, B.   250 [10]
Undheim, K.   359 [50]
Urry, W. H.   35, 36 [7]; 35, 36 [8];
  105 [1]; 160 [20]; 160 [21]; 165
  [40]; 165 [41]; 169 [48]; 170, 171
  [53]; 171-173 [55]; 172 [61]; 282,
  283 [33]; 282, 283 [34]; 282, 283
  [35]
Urscheler, H.   15 [13]
Utzinger, E. C.   31 [44]; 31, 32
  [46]

Valade, J.   178 [76]
Valentine, D.   75 [25]
Van Auken, T. V.   328 [1]; 328 [2]
Van de Vloed, H.   277-279, 283 [20]
Van der Beek, P. A. A.   399 [135]
Van der Kuip, G.   15 [17]
Van der Linden, T.   342 [9]
Van Driel, H.   451 [59]
Van Meter, J. P.   447 [51]
Van Tamelen, E. E.   5, 6 [26];
  9 [51]; 10 [52]; 229 [18]; 229
  [19]; 469 [45]
Van Vunakis, H.   381 [45]
VanAllan, J. A.   395, 397 [117]
Vaughan, W. E.   162-164 [27]; 162
  [29]; 162, 163 [30]; 163 [31]; 176,
  177 [69]; 177 [70]
Vaughan, W. H.   491 [20]
Veber, D. F.   72 [17]
Veeger, C.   257, 258 [12]
Vellturo, A. F.   82 [46];
  82, 83 [48]
Velluz, L.   15 [8]; 15 [9]; 15 [11];
  391 [88]; 391 [88]; 392 [99]; 392
  [99]; 393 [101]
Ver Nooy, C. D.   346, 347, 350 [18]
Verloop, A.   15 [12]; 15 [17]
Vesley, G. F.   210 [46]; 210 [49];
  218, 219 [5]; 219 [6]
Vetter, H.   461 [17]
Vickery, B.   110 [8]
Vidal, F.   344 [13]
Vilesov, F. I.   498

Vilkas, M.   232 [27]; 232, 233 [30]
Vinje, M. G.   45 [17]
Vladimirtsev, I. F.   255, 257 [1]
Voegtli, W.   363 [3]
Vogel, E.   4 [23]
Vogt, I.   269 [15]
Vogt, V.   56, 57 [5]
Von Bünau, G.   194 [7]; 274 [6]
Von E. Doering, W.   275 [8]; 275,
  278, 281 [13]; 276, 283 [14]; 276
  [18]; 276 [19]; 27 7[22]; 277 [25];
  283 [40]
Von Gross, E.   119, 122 [22]
Von Gustorf, E. K.   461, 462 [14];
  461 [15]; 466 [35]; 469 [46]
Von Halban, H.   239 [46]
Von Hessling, G.   459 [1]
Von Hobe, D.   460 [4]; 460 [6]; 460
  [7]; 463 [26]; 463 [28]
Von Meyer, A.   182, 183 [5]
Von Planta, C.   59, 61 [21]
Von Rintelen, H.   277-279, 283 [20]
Von Schickh, O.   263 [12]; 263 [14]
Von Tappeiner, H.   498
Von Wilucki, I.   86 [70]
Von Wittenau, M. S.   382 [47]; 420
  [20]
Vonderwahl, R.   277 [22]
Voronkov, M. G.   354, 355 [42]

Wachholtz, F.   57, 58 [10]
Wacker, A.   91 [92]; 91 [93]
Wagner-Jauregg, T.   257, 258 [13]
Walaschewski, E. G.   457, 458 [15]
Wald, G.   59 [22]
Waldmann, H.   256 [3]
Wallace, T. J.   160 [22]
Walling, C.   205 [16]; 346, 347
  [19]; 349 [27]; 363 [4]; 454 [7]
Walter, H. C.   106 [12]
Walter, M.   374, 375 [13]
Wan, J. K. S.   497
Wang, S. Y.   91 [97]; 155 [3]
Wani, M. C.   257 [8]
Warburg, E.   56 [2]
Warburg, O.   445 [41]
Warren, P. C.   210 [49]
Warrener, R. N.   500
Warszawski, R.   235 [37]
Wasserman, H. H.   396 [124]; 407
  [2]; 431 [17]; 432 [19]
Waters, W. A.   148 [4]; 151, 152
  [1]; 157 [12]; 182, 183 [1]; 186,
  189 [5]; 330, 331 [10]
Wawzonek, S.   242 [1]; 243, 246 [6];
  243 [7]
Wayne, R.   253 [19]
Weaver, S. D.   298 [11]

Weber, K.-H.   304 [30]
Weber, P.   342 [3]
Weedon, B. C. L.   59 [28]
Wegemund, B.   226 [5]; 226, 227 [7]
Wehrli, H.   23, 24 [7]; 23 [8]; 26 [18]; 36 [9]; 36, 37 [10]; 36, 37 [11]; 36, 37 [12]; 66 [56]; 77 [35]; 194 [4]; 408 [8]; 408, 412 [9]
Wehry, E. L.   500
Weigert, F.   98 [3]; 498
Weil, L.   381 [44]
Weinberg, K.   31 [44]; 31, 32 [46]
Weinlich, J.   220 [10]
Weisbuch, F.   45 [18]
Weiss, J.   98 [7]
Weiss, K.   193 [2]; 207 [32]
Weissermel, K.   162 [26]
Weissman, L.   72 [16]
Weitz, E.   411 [22]
Weitz, H. M.   64 [48]
Weizmann, C.   205 [25]
Weizmann, M.   344 [14]
Weller, A.   497
Wells, A. A.   496
Welstead, Jr., W. J.   407 [3]; 431 [18]
Wender, I.   461 [16]
Wendler, N. L.   66 [58]
Wenger, R.   23, 24 [7]
Werner, H.   283, 286 [39]; 462 [19]
Werner, V.   426, 427 [1]
Wessely, F.   78, 88 [71]; 87 [72]
Wettstein, A.   358, 359 [48]; 369 [20]; 369 [21]; 369 [22]; 369, 370 [25]
Wexler, H.   373 [6]
Wexler, S.   375 [19]; 376, 377, 383 [25]; 376, 383 [26]; 428-430 [10]; 428 [11]
Weygand, F.   143 [14]; 297, 298 [6]; 297 [7]; 300 [14]
Whalley, E.   106 [11]
Whipple, E. B.   468 [43]
White, D. M.   396 [121]
White, E. H.   396 [122]
White, R. C.   106 [6]
Whitear, B. R. D.   27 [23]
Whitham, G. H.   28 [26]; 28 [27]
Wiberg, K. B.   303, 304 [26]
Wieland, H.   4 [21]; 369, 370 [25]
Wieland, P.   369 [20]; 369 [22];
Wiemann, J.   45 [18]
Wiersdorff, W. W.   220 [10]
Wierzchowski, K. L.   54 [22]; 54 [23]
Wilk, M.   107, 108 [15]; 271 [20]
Wilkinson, F.   25 [47]
Willard, J.   350 [32]

Willemart, A.   98 [17]
Willemart, M. A.   391 [90]
Willey, F. G.   10 [56]; 284 [43]
Williams, A.   252 [17]
Williams, I. A.   43, 44 [13]; 193 [3]
Williams, R. B.   277 [22]
Williams, R. O.   2, 3 [10]
Williams, R. W. J.   386 [70]
Willis, C. J.   465 [33]
Wilson, H. R.   218, 219 [5]
Wilson, J. W.   408 [10]
Wilson, R. M.   215-217 [1]
Wilt, J. W.   282, 283 [35]
Wilzbach, K. E.   10 [53]
Windaus, A.   7, 8 [37]; 149 [5]; 149 [7]; 385, 388 [67]
Winey, D. A.   35, 36 [8]
Winkler, B.   5, 7 [31]
Winnick, C.   391 [94]
Wipke, W. T.   10 [54]
Wirth, H.   381 [43]; 444 [40]; 445 [42]
Wirtz, R.   383 [57]
Wislicenus, J.   57, 58 [9]
Wittig, G.   344 [13]; 369 [19]; 397 [126]
Wolf, K.   46 [20]
Wolf, W.   368 [16]
Wolfe, W. L.   491 [19]
Wolfe, Jr., J. R.   98, 99 [11]
Wolff, A.   65 [54]; 77 [33]
Wolff, M. E.   242, 243 [3]; 246, 247 [12]
Wolff, R. E.   226, 227 [10]
Wolfrom, M. L.   297 [8]
Wolgast, R.   72 [19]
Wolinsky, J.   229 [18]; 229 [19]
Wolovsky, R.   63 [41]
Wood, C. S.   19 [30]; 127 [3]; 127, 129 [4]; 127 [5]; 127, 128 [6]; 140, 141 [9]
Wood, G. W.   384, 385 [65]
Wood, H. C. S.   257, 258 [12]
Woodworth, R. C.   275 [9]
Wormser, H. C.   128 [11]
Wright, J. B.   238 [43]; 358 [49]
Wuesthoff, M. T.   428-430 [10]
Wulf, W.   8 [41]
Wulfers, T. F.   448, 449 [55]
Wulff, D. L.   91 [90]
Wyatt, P.   2, 3, 10 [8]
Wylde, J.   42 [5]
Wyman, G. M.   56, 57 [3]; 61 [31]; 61 [33]; 62 [34]; 62 [35]
Wynberg, H.   176 [68]; 451 [59]

Yamada, S.   52 [15]
Yang, C. S.   258 [19]

Yang, D.-D. H.   34 [2]; 34, 36 [5]; 37 [14]; 198, 210 [1]
Yang, N. C.   22, 24 [1]; 22, 24 [3]; 25 [9]; 34, 35 [1]; 34 [2]; 34, 36 [5]; 37 [14]; 41, 42 [2]; 41, 42 [3]; 58, 60 [18]; 198, 210 [1]; 208 [34]; 210 [48]; 415, 416 [3]; 415-417 [5]; 418 [15]
Yao, H. C.   381 [42]
Yates, P.   2, 3 [11]; 2 [17]; 89 [81]; 89 [82]; 89 [83]; 285-287 [46]; 410 [16]
Yip, R. W.   42 [4]; 103 [35]; 111 [12]
Yogev, A.   37, 38 [15]
Yorke, W.   72 [13]
Yost, W. L.   195 [9]; 205 [26]
Young, S. T.   98 [13]
Youssefyeh, R. D.   45 [19]; 160, 161 [23]
Yuan, E. P.   254 [21]
Yudis, M. D.   248 [6]

Zaher, H. A. A.   187, 191 [13]
Zahler, W. D.   315 [4]
Zanker, V.   196 [12]
Zarem, A. M.   498
Zavarin, E.   80, 81 [43]
Zayed, S. M. A. D.   187, 188 [16]
Zechmeister, L.   59, 60 [24]; 59 [25]; 59 [26]; 59, 61 [27]; 62 [39]
Zehender, F.   257, 258 [14]
Zenhäusern, A. F.   61 [33]
Ziegler, H.   119, 122, 124 [23]; 277, 279 [23]; 384 [60]
Ziegler, I.   256 [5]
Ziegler, K.   383 [55]; 384 [59]
Zimmerman, H. E.   28, 30 [28]; 222 [12]; 238 [44]; 289 [49]; 408 [10]; 410, 411 [21]; 423 [26]
Zorina, T. M.   467 [40]
Zwitkowits, P.   183 [7]; 183, 184 [8]; 208, 212 [35]
Zyatkov, I. P.   399, 400 [134]

Reaction Index *

acetals
dehydrogenation of acetals to esters 238
isomerization of nitro acetals to nitroso esters 268, 269
sensitization in isomerization of cyclic acetals to esters 45, 46
acetylene compounds
acetylene compounds (alkynes)
unsaturated compounds *
addition of aldehydes to acetylene compounds with formation of ketones 169
addition of formamides to acetylene compounds 167
addition of halogens to acetylene compounds 342
addition of hydrogen halides to acetylene compounds with formation of organic halides 158
addition of organic halides to acetylene compounds 172, 174
addition of organosilicon compounds to olefins and acetylene compounds 178, 179
complexation with nitrogen compounds, nitroso compounds, sulfoxides, phosphines, arsines, acetylene compounds and olefins as electron donors 459-464, 469
cycloaddition of acetylene compounds to aromatic compounds 115, 116
cycloaddition of acetylene compounds to aromatic compounds followed by isomerization of the adducts to 8-membered rings 114
cycloaddition of carbenes to acetylene compounds with formation of cyclopropene derivatives 281
cycloaddition of carbonyl compounds to acetylene compounds with formation of oxete derivatives 423
cycloaddition of α,β-unsaturated carbonyl compounds to acetylene compounds with formation of cyclobutene derivatives 115
cyclodimerization of acetylene compounds followed by isomerization 92-94
dimerization of aryl substituted acetylene compounds 92-94
sensitization in addition reactions with olefins and acetylene compounds 155, 160-163, 166-168, 176, 177
acid chlorides
replacement of hydrogen by the chloro-formyl group with formation of acid chlorides 233, 234
acids
acids see also keto acids
acids see also unsaturated acids
addition of water to ketenes with formation of acids 227, 229, 295-297, 301-306, 309, 310
hydrolysis of phenols with formation of acids 239, 240
isomerization of epidioxides to pseudo acids 428-430
isomerization of nitro aldehydes to nitroso acids 267-270
isomerization of organic nitrites to hydroxamic acids 252, 253
oxidation of organic halides with formation of acids 225

* Items marked with an asterisk should be consulted as additional keywords

photolysis of cyclic ketones to acids   1, 26, 28, 225-230
photolysis of diazonium salts in inorganic acids with formation of organic halides or phenols   314, 315

addition
addition of hydrogen peroxide to C=C bonds with formation of glycols see MILAS reaction
addition of oxygen to 1,3-dienes with formation of epidioxides see peroxidation
addition of oxygen see also hydroperoxidation and peroxidation
addition see also insertion, alkylation and arylation
addition of alcohols to C=C bonds with formation of alcohols   159, 160, 195, 196, 233
addition of alcohols to C=C bonds with formation of ethers   8-11, 159, 160, 193, 194, 432, 433
addition of alcohols to epidioxides with formation of alkoxy hydroperoxides   428-430
addition of alcohols to ketenes with formation of esters   219, 238, 239, 295-297, 301, 303, 304
addition of alcohols to ketones with formation of glycols   207, 208
addition of alcohols to SCHIFF bases with formation of oxazolidine derivatives   335-337
addition of alcohols to unsaturated acids followed by dehydration to lactones   231-233
addition of aldehydes to acetylene compounds with formation of ketones   169
addition of aldehydes to methylene groups with formation of alcohols   201
addition of aldehydes to olefins with formation of ketones   169, 188, 201
addition of amines to ketenes with formation of amides   228-230, 297
addition of ammonia, amines and formamides to C=C bonds   165-168
addition of aromatic halides to phosphines with formation of phosphonium salts   455-457
addition of carbenes to oxygen   291
addition of cyclic ethers to C=C bonds   160, 161
addition of diazo compounds to C=C bonds with formation of pyrazoline derivatives   288, 292
addition of formamides to acetylene compounds   167
addition of halogens to acetylene compounds   342
addition of halogens to aromatic compounds with formation of organic halides   342-345
addition of halogens to C=C bonds with formation of organic halides   341, 350, 353
addition of hydrogen halides to acetylene compounds with formation of organic halides   158
addition of hydrogen halides to olefins with formation of organic halides   157-159
addition of hydrogen sulfide, thiols and thiocarboxylic acids to C=C bonds   162-164
addition of hypohalites to olefins with formation of ethers   162
addition of ketenes to thiols with formation of esters of thiocarboxylic acids   297, 298
addition of keto-carbenes to organic halides   307, 308
addition of ketones to alcohols with formation of pinacols   203
addition of ketones to methylene groups with formation of carbinols   198-202
addition of ketones to olefins   169, 170
addition of magnesium to organic halides with formation of organomagnesium compounds   468
addition of maleic anhydride derivatives to olefins   168
addition of mercury to organic halides with formation of organomercury compounds   366, 367
addition of nitrosyl chloride to olefins   159
addition of organic halides to acetylene compounds   172, 174

addition of organic halides to C=C bonds   170-174
addition of organic halides to 1,3-dienes   171, 172, 175, 176
addition of organogermanium compounds to olefins   179
addition of organosilicon compounds to olefins and acetylene compounds   178, 179
addition of phosphines to C=C bonds   176
addition of phosphonates and phosphonothioates to C=C bonds   177, 178
addition of phosphonates to 1,4-quinones   453, 454
addition of 1,2-quinone imines to aldehydes with formation of amides   190, 191
addition of 1,2-quinone oximes to aldehydes   191
addition of 1,2-quinones to aldehydes with formation of esters   186-188
addition of 1,4-quinones to aldehydes with formation of esters   189
addition of 1,4-quinones to aldehydes with formation of ketones   188, 189
addition of 1,2-quinones to aralkyl compounds with formation of
   alcohols   182-184
addition of 1,2-quinones to ethers   184
addition of 1,4-quinones to hydroaromatic compounds or aralkyl compounds
   with formation of ethers   151, 182
addition of sulfenyl chlorides or sulfonyl chlorides to olefins   164, 165
addition of sulfur compounds to C=C bonds with formation of sulfides and
   thiols   162-164
addition of sulfur dioxide and chlorine or oxygen with formation of
   sulfonyl chlorides or sulfonic acids   434-439
addition of water following rearrangement of cross-conjugated
   cyclohexadienones   30-32
addition of water to C=C bonds with formation of alcohols   155, 156
addition of water to ketenes with formation of acids   227, 229, 295-297,
   301-306, 309, 310
addition of water to keto-carbenes with formation of lactones   306-308
addition of water to pyrimidine derivatives   155, 156
addition reactions with C=C bonds   155-179
sensitization in addition reactions of alcohols to unsaturated acids   231-233
sensitization in addition reactions with olefins and acetylene compounds   155,
   160-163, 166-168, 176, 177
alcohols
alcohols see also carbinols
addition of alcohols to C=C bonds with formation of alcohols   159, 160, 195, 196,
   233
addition of alcohols to C=C bonds with formation of ethers   8-11, 159, 160, 193,
   194, 432, 433
addition of alcohols to epidioxides with formation of alkoxy
   hydroperoxides   428-430
addition of alcohols to ketenes with formation of esters   219, 238, 239, 295-297,
   301, 303, 304
addition of alcohols to ketones with formation of glycols   207, 208
addition of alcohols to SCHIFF bases with formation of oxazolidine
   derivatives   335-337
addition of alcohols to unsaturated acids followed by dehydration to
   lactones   231-233
addition of aldehydes to methylene groups with formation of alcohols   201
addition of ketones to alcohols with formation of pinacols   203
addition of 1,2-quinones to aralkyl compounds with formation of
   alcohols   182-184
addition of water to C=C bonds with formation of alcohols   155, 156
cyclization of aryl substituted amines involving incorporation of
   alcohols   335, 336
cyclization of SCHIFF bases to heterocyclic nitrogen compounds involving
   incorporation of alcohols   334, 335
dihydrodimerization with alcohols as hydrogen donors   203-211
hydroperoxidation of alcohols   379

photolysis of acyl azides in alcohols with formation of carbamates   321, 322, 325, 326
photolysis of organomercury compounds in alcohols with formation of mercury and carbonyl compounds   467, 468
photolysis of 1,4-quinone diazides in alcohols with formation of alkoxy phenols   310, 311
photolysis of sulfonyl azides in alcohols, sulfoxides and sulfides   326, 327
photolysis of α,β-unsaturated ketones in alcohols with formation of esters   227, 238
photolysis of unsaturated sultones in alcohols with formation of sulfonates   448
reduction of C=O bonds with hydrogen donors involving formation of alcohols   113, 194, 195, 204, 208
reduction of diazonium salts with alcohols as hydrogen donors   313, 314
reduction with alcohols, ethers or hydrocarbons as hydrogen donors   193-197
sensitization in addition reactions of alcohols to unsaturated acids   231-233
sensitization in cyclization reactions of nitrogen compounds with alcohols   335, 336

aldehydes
aldehydes see also unsaturated aldehydes
carbonyl compounds *
addition of aldehydes to acetylene compounds with formation of ketones   169
addition of aldehydes to methylene groups with formation of alcohols   201
addition of aldehydes to olefins with formation of ketones   169, 188, 201
addition of 1,2-quinone imines to aldehydes with formation of amides   190, 191
addition of 1,2-quinone oximes to aldehydes   191
addition of 1,2-quinones to aldehydes with formation of esters   186-188
addition of 1,4-quinones to aldehydes with formation of esters   189
addition of 1,4-quinones to aldehydes with formation of ketones   188, 189
cycloaddition of aldehydes to olefins with formation of oxetane derivatives   414-417, 420-423
decarbonylation of aldehydes   222, 223
dihydrodimerization of aldehydes with formation of pinacols   203, 204
dimerization of aldehydes   105, 107, 108
isomerization of aldehydes   27
isomerization of furan derivatives to aldehydes   45
isomerization of nitro aldehydes to nitroso acids   267-270
oxidation of aldehydes with formation of acyl peroxides   399, 400

alkenes
alkenes see olefins

alkylation
addition *
alkylation see also insertion reactions of carbenes or keto-carbenes

alkynes
alkynes see acetylene compounds

amidation
formation of amides *
amidation   166-168
amidation of aromatic compounds with formamides   230, 231
sensitization in amidation reactions   166-168, 230, 231

amides
amides see also formamides
formation of amides see also amidation
addition of amines to ketenes with formation of amides   228-230, 297
addition of 1,2-quinone imines to aldehydes with formation of amides   190, 191
isomerization of nitrones to amides or lactams   48-52
isomerization of oximes to amides   48
reduction of acyl azides with hydrogen donors involving formation of amides   321-325

amines
addition of amines to ketenes with formation of amides 228-230, 297
addition of ammonia, amines and formamides to C=C bonds 165-168
cyclization of aryl substituted amines involving incorporation of
    alcohols 335, 336
dealkylation of N-alkyl amines 255-257
dehydrocyclization of aryl substituted amines to carbazole derivatives 138
oxidation of amines 381
reduction of azides to amines 319, 320
reduction of nitro compounds to amines 270, 271
ammonia
addition of ammonia, amines and formamides to C=C bonds 165-168
anthracene derivatives
dimerization of anthracene derivatives 97-99, 101, 102, 266
aralkyl compounds
aralkyl compounds (compounds containing methylene groups)
aromatic compounds *
addition of 1,2-quinones to aralkyl compounds with formation of
    alcohols 182-184
addition of 1,4-quinones to hydroaromatic compounds or aralkyl compounds
    with formation of ethers 151, 182
hydroperoxidation of aralkyl compounds 146, 373
thiocyanation of aralkyl compounds 443, 444
arenes
arenes see aromatic compounds and fused polycyclic hydrocarbons
ARNDT-EISTERT reaction
ARNDT-EISTERT reaction 295, 296
aromatic compounds
aromatic compounds (arenes)
aromatic compounds see also fused hydrocarbons and heterocyclic compounds
aromatic compounds see also hydroaromatic compounds and aralkyl compounds
addition of halogens to aromatic compounds with formation of organic
    halides 342-345
amidation of aromatic compounds with formamides 230, 231
cycloaddition of acetylene compounds to aromatic compounds 115, 116
cycloaddition of acetylene compounds to aromatic compounds followed by
    isomerization of the adducts to 8-membered rings 114
cycloaddition of carbenes to aromatic compounds involving ring
    enlargement 276-279
cycloaddition of olefins to aromatic compounds 111, 112
cycloaddition of α,β-unsaturated carbonyl compounds to aromatic
    compounds 109, 110, 140
dehydrocyclization of aromatic compounds to fluorene derivatives 126, 127
dehydrocyclization of aromatic compounds to fused polycyclic
    hydrocarbons 130-136
dehydrocyclization of aryl substituted olefins to aromatic compounds 127-131, 136
dimerization of aromatic compounds with formation of 8-membered rings 11, 93,
    97-102, 390
valence tautomerization of aromatic compounds 10
aromatization
aromatization 272, 229, 234, 235
aromatization involving dehydrogenation by hydrogen acceptors 127-136, 138-
    144, 152-154
arsines
organoarsenic compounds *
complexation with nitrogen compounds, nitroso compounds, sulfoxides,
    phosphines, arsines, acetylene compounds and olefins as electron
    donors 459-464, 469
photolysis of arsines 457, 458

arylation
addition *
arylation   285, 286, 368, 455
azides
photolysis of acyl azides in alcohols with formation of carbamates   321, 322, 325, 326
photolysis of acyl azides with formation of isocyanates   321-324
photolysis of azides with formation of imenes   318-327
photolysis of sulfonyl azides in alcohols, sulfoxides and sulfides   326, 327
reduction of acyl azides with hydrogen donors involving formation of amides   321-325
reduction of azides to amines   319, 320
sensitization in the photolysis of azides   321, 325
azines
C=N bonds *
cleavage of azines by ketones involving formation of nitriles   339
aziridine derivatives
cycloaddition of acyl imenes to C=C bonds with formation of aziridine derivatives   325
cycloaddition of carbenes to C=N bonds with formation of aziridine derivatives   280
azo compounds
N=N bonds *
cycloaddition of ketenes to azo compounds   298, 299
dehydrocyclization of aryl substituted azo compounds or SCHIFF bases to aromatic heterocyclic nitrogen compounds   138-142, 337, 338
dimerization of imenes with formation of azo compounds   321
photolysis of diaroyl azo compounds with formation of aromatic 1,2-diketones   330, 331
photolysis of diazonium salts with formation of azo compounds   316
photolysis of 1,2-quinone diazides with formation of azo compounds   309, 310
photolysis of 1,4-quinone diimine dioxides with formation of azo compounds   332, 333
stereoisomerization of azo compounds   62-64
azoxy compounds
isomerization of azoxy compounds   53

BARTON reaction
BARTON reaction (isomerization of organic nitrites to hydroxy oximes)
BARTON reaction   248-251
bicyclo[3.2.0]heptene derivatives
bicyclo[3.2.0]heptene derivatives see cyclobutene derivatives
bicyclo[3.1.0]hexene derivatives
bicyclo[3.1.0]hexene derivatives see cyclopropane derivatives
bromine
bromine, chlorine and iodine see halogens

C-C bonds
insertion of carbenes into C-C bonds   281
C-H bonds
halogenolysis of C-H bonds   342-344, 347-349, 351, 352, 354-359, 365
insertion of acyl imenes into C-H bonds with formation of carbamates   324, 325
insertion of carbenes into C-H bonds   278-280, 283-286
insertion of imenes into aromatic C-H bonds   319
insertion of keto-carbenes into aromatic C-H bonds   311
C-O bonds
insertion of carbenes into C-O bonds   279, 280, 282

C=C bonds
addition of hydrogen peroxide to C=C bonds with formation of glycols see MILAS reaction
compounds containing C=C bonds see also olefins
unsaturated compounds *
addition of alcohols to C=C bonds with formation of alcohols   159, 160, 195, 196, 233
addition of alcohols to C=C bonds with formation of ethers   8-11, 159, 160, 193, 194, 432, 433
addition of ammonia, amines and formamides to C=C bonds   165-168
addition of cyclic ethers to C=C bonds   160, 161
addition of diazo compounds to C=C bonds with formation of pyrazoline derivatives   288, 292
addition of halogens to C=C bonds with formation of organic halides   341, 350, 353
addition of hydrogen sulfide, thiols and thiocarboxylic acids to C=C bonds   162-164
addition of organic halides to C=C bonds   170-174
addition of phosphines to C=C bonds   176
addition of phosphonates and phosphonothioates to C=C bonds   177, 178
addition of sulfur compounds to C=C bonds with formation of sulfides and thiols   162-164
addition of water to C=C bonds with formation of alcohols   155, 156
addition reactions with C=C bonds   155-179
cycloaddition of acyl imenes to C=C bonds with formation of aziridine derivatives   325
cycloaddition of carbenes to C=C bonds with formation of cyclopropane derivatives   275-279, 288, 300, 301
cycloaddition of thioketones to C=C bonds with formation of thietane derivatives   448, 449
oxidative cleavage of C=C bonds   239, 401
reduction of C=C bonds with hydrogen donors   193, 194
stereoisomerization at C=C bonds   56-62

C=N bonds
C=N bonds see also azines, imines and SCHIFF bases
cycloaddition of carbenes to C=N bonds with formation of aziridine derivatives   280
cycloaddition of ketenes to C=N bonds   299, 300
reduction of C=N bonds   257
reduction of C=N bonds with hydrogen donors   195, 196
stereoisomerization at C=N bonds   63, 64

C=O bonds
cycloaddition of carbenes to C=N bonds with formation of epoxy compounds   279, [280
reduction of C=O bonds with hydrogen donors involving formation of alcohols   113, 194, 195, 204, 208

C3O-cycloadducts
C3O-cycloadducts see oxetane derivatives

C4O2-cycloadducts
C4O2-cycloadducts see dioxin derivatives

C4-cycloadducts
C4-cycloadducts see cyclobutane derivatives

C8-cycloadducts
C8-cycloadducts see 8-membered rings

cage compounds
cyclodimerization with formation of cage compounds   79-81, 89
intramolecular cycloaddition with formation of cage compounds   2-4

carbamates
insertion of acyl imenes into C-H bonds with formation of carbamates   324, 325
photolysis of acyl azides in alcohols with formation of carbamates   321, 322, 325, 326

carbazole derivatives
heterocyclic nitrogen compounds *
dehydrocyclization of aryl substituted amines to carbazole derivatives   138
intramolecular cyclization of imenes with formation of carbazole
    derivatives   320
carbenes
alkylation *
carbenes see also keto-carbenes
addition of carbenes to oxygen   291
addition of carbenes to organic halides   282, 283
cycloaddition of carbenes to acetylene compounds with formation of
    cyclopropene derivatives   281
cycloaddition of carbenes to aromatic compounds involving ring
    enlargement   276-279
cycloaddition of carbenes to C=C bonds with formation of cyclopropane
    derivatives   275-279, 288, 300, 301
cycloaddition of carbenes to C=N bonds with formation of aziridine
    derivatives   280
cycloaddition of carbenes to C=O bonds with formation of epoxy compounds   279, 280
cycloaddition of carbenes to furan derivatives or thiophene derivatives   278, 279
dimerization of carbenes with formation of olefins   288, 289, 291
dimerization of thio carbenes   331, 332
insertion of carbenes into C-C bonds   281
insertion of carbenes into C-H bonds   278-280, 283-286
insertion of carbenes into C-O bonds   279, 280, 282
insertion of carbenes into N-H bonds   287, 288
insertion of carbenes into O-H bonds with formation of ethers   286, 287
isomerization of carbenes with formation of olefins   288-290
photolysis of diazo compounds with formation of carbenes   274-291
photolysis of thiadiazole derivatives with formation of thio carbenes   331, 332
reduction of carbenes with hydrogen donors   287, 300
carbinols
alcohols *
addition of ketones to methylene groups with formation of carbinols   198-202
photolysis of carbinols   199
carbonyl compounds
carbonyl compounds see also aldehydes and ketones
PATERNO-BUECHI reaction *
cycloaddition of carbonyl compounds to acetylene compounds with formation of
    oxete derivatives   423
photolysis of organomercury compounds in alcohols with formation of mercury
    and carbonyl compounds   467, 468
replacement of carbon monoxide in metal carbonyl compounds by electron
    donors with formation of organometallic carbonyl compounds   459-462, 469
chlorine
bromine, chlorine and iodine see halogens
addition of sulfur dioxide and chlorine or oxygen with formation of
    sulfonyl chlorides or sulfonic acids   434-439
replacement of chlorine in nitrogen-halogen compounds by hydrogen   242
chloronitroso compounds
nitroso compounds *
reduction of chloronitroso compounds to oximes   197
CIAMICIAN addition
intramolecular cycloaddition *
CIAMICIAN addition   1-4, 10, 11, 27
cis-trans isomerization
stereoisomerization *

Reaction Index

cleavage
photolysis *
cleavage of azines by ketones involving formation of nitriles 339
cleavage of pinacols by ketones 206, 207, 212
cleavage of pinacols by 1,4-quinones 211, 212
oxidative cleavage of C=C bonds 239, 401
ring cleavage 14-17, 26, 27, 218-221, 225-230, 239, 348, 448
complexation
complexation in cyclodimerization reactions 468, 469
complexation with nitrogen compounds, nitroso compounds, sulfoxides, phosphines, arsines, acetylene compounds and olefins as electron donors 459-464, 469
CURTIUS rearrangement
CURTIUS rearrangement 321-324
cyanation
cyanation (replacement of hydrogen by the cyano group)
cyanation of aliphatic compounds or ethers with formation of nitriles 260, 261
cyclization
cyclization following dehydrohalogenation of N-chloroamines see HOFMANN-LOEFFLER reaction
cyclization of unsaturated compounds involving dehydrogenation see dehydrocyclization
isomerization *
cyclization 15-20, 25, 27, 28, 34-37, 252, 257, 278, 284, 285, 315, 322, 330, 338
cyclization of aryl substituted amines involving incorporation of alcohols 335, 336
cyclization of olefins to cyclic peroxides involving incorporation of oxygen 397, 398
cyclization of SCHIFF bases to heterocyclic nitrogen compounds involving incorporation of alcohols 334, 335
cyclization of α,β-unsaturated carbonyl compounds 27, 28
intramolecular cyclization of acyl imenes with formation of lactams 322-324
intramolecular cyclization of aromatic nitro compounds to indole derivatives 271, 272
intramolecular cyclization of imenes with formation of carbazole derivatives 320
intramolecular cyclization with formation of furan derivatives 41, 215-217, 426, 427
intramolecular cyclization with formation of pyran derivatives 41, 42
isomerization of ketones by intramolecular cyclization to cyclobutanols 34-37, 409
isomerization of unsaturated ketones by intramolecular cyclization to cyclohexenols 37
photolysis of nitrogen-halogen compounds followed by cyclization to heterocyclic nitrogen compounds 242-246
sensitization in cyclization reactions of nitrogen compounds with alcohols 335, 336
cycloaddition
reversal of cycloaddition see retro-cycloaddition
intramolecular cycloaddition see also CIAMICIAN addition
PATERNO-BUECHI reaction *
cycloaddition of acetylene compounds to aromatic compounds 115, 116
cycloaddition of acetylene compounds to aromatic compounds followed by isomerization of the adducts to 8-membered rings 114
cycloaddition of acyl imenes to C=C bonds with formation of aziridine derivatives 325
cycloaddition of aldehydes to olefins with formation of oxetane derivatives 414-417, 420-423
cycloaddition of carbenes to acetylene compounds with formation of cyclopropene derivatives 281

cycloaddition of carbenes to aromatic compounds involving ring
  enlargement 276-279
cycloaddition of carbenes to C=C bonds with formation of cyclopropane
  derivatives 275-279, 288, 300, 301
cycloaddition of carbenes to C=N bonds with formation of aziridine
  derivatives 280
cycloaddition of carbenes to C=O bonds with formation of epoxy compounds 279,
  280
cycloaddition of carbenes to furan derivatives or thiophene derivatives 278, 279
cycloaddition of carbonyl compounds to acetylene compounds with formation of
  oxete derivatives 423
cycloaddition of 1,2-dicarbonyl compounds to unsaturated compounds with
  formation of dioxin derivatives 118-124, 419
cycloaddition of 1,2-dicarbonyl compounds to unsaturated compounds with
  formation of oxetane derivatives 121, 122, 124, 419, 420
cycloaddition of ketenes to azo compounds 298, 299
cycloaddition of ketenes to C=N bonds 299, 300
cycloaddition of ketones to olefins with formation of oxetane
  derivatives 414-417, 419-423
cycloaddition of olefins to aromatic compounds 111, 112
cycloaddition of 1,4-quinones to 1,3-dienes with formation of pyran
  derivatives 424
cycloaddition of 1,4-quinones to olefins with formation of oxetane
  derivatives 418, 419
cycloaddition of 1,4-quinones to olefins 418, 419
cycloaddition of sulfur dioxide to 1,2-quinones 439
cycloaddition of thioketones to C=C bonds with formation of thietane
  derivatives 448, 449
cycloaddition of α,β-unsaturated carbonyl compounds to acetylene compounds
  with formation of cyclobutene derivatives 115
cycloaddition of α,β-unsaturated carbonyl compounds to aromatic
  compounds 109, 110, 114
cycloaddition of α,β-unsaturated carbonyl compounds to olefins 110-113
cycloaddition of unsaturated compounds with formation of cyclobutane
  derivatives 109-117
intramolecular cycloaddition 27
intramolecular cycloaddition of unsaturated ketones to oxetane
  derivatives 417, 418
intramolecular cycloaddition with formation of cage compounds 2-4
intramolecular cycloaddition with formation of cyclobutane derivatives 1-4, 10, 11
sensitization in cycloaddition reactions 110, 111, 115
sensitization in intramolecular cycloaddition reactions 2-4

cyclobutane derivatives
cyclobutane derivatives (C4-cycloadducts)
fragmentation of cyclobutane derivatives see also retro-cycloaddition
cycloaddition of unsaturated compounds with formation of cyclobutane
  derivatives 109-117
cyclodimerization of unsaturated compounds with formation of cyclobutane
  derivatives 70-92
intramolecular cycloaddition with formation of cyclobutane derivatives 1-4, 10, 11
stereoisomerization of cyclobutane derivatives 65

cyclobutanols
isomerization of ketones by intramolecular cyclization to cyclobutanols 34-37, 409

cyclobutene derivatives
cycloaddition of α,β-unsaturated carbonyl compounds to acetylene compounds
  with formation of cyclobutene derivatives 115
valence tautomerization of cyclic 1,3-dienes with formation of cyclobutene
  derivatives 4-9, 27, 28, 227

cyclodimerization
dimerization *
complexation in cyclodimerization reactions 468, 469
cyclodimerization 115, 262, 280, 316, 317
cyclodimerization of acetylene compounds followed by isomerization 92-94
cyclodimerization of aryl substituted olefins 70-73, 89, 90
cyclodimerization of 1,3-dienes 74, 75
cyclodimerization of olefins 74, 468, 469
cyclodimerization of pyran derivatives 89
cyclodimerization of pyrimidine derivatives 91
cyclodimerization of 1,4-quinones 79-81
cyclodimerization of 1,4-quinones with formation of oxetane derivatives 79, 80
cyclodimerization of SCHIFF bases 141, 142
cyclodimerization of thiocarbonyl compounds to dithietane derivatives 106, 107, 449, 450
cyclodimerization of α,β-unsaturated acids 23, 82-85
cyclodimerization of unsaturated compounds with formation of cyclobutane derivatives 70-92
cyclodimerization of α,β-unsaturated ketones 75-81, 89, 91, 92
cyclodimerization of unsaturated lactones 85-88
cyclodimerization of unsaturated sulfur compounds 90, 91
cyclodimerization with formation of cage compounds 79-81, 89
sensitization in cyclodimerization reactions 23, 72-76, 82, 83, 86, 468
cyclohexadienones
addition of water following rearrangement of cross-conjugated cyclohexadienones 30-32
isomerization of cross-conjugated cyclohexadienones 28-32, 235
isomerization of linear-conjugated cyclohexadienones to unsaturated ketones 227, 229
cyclohexenols
isomerization of unsaturated ketones by intramolecular cyclization to cyclohexenols 37
cyclooctatetraene derivatives
photolysis of cyclooctatetraene derivatives 409, 410
cyclopropane derivatives
cycloaddition of carbenes to C=C bonds with formation of cyclopropane derivatives 275-279, 288, 300, 301
photolysis of pyrazoline derivatives with formation of cyclopropane derivatives and olefins 328, 329
reduction of cyclopropane derivatives 65
stereoisomerization of cyclopropane derivatives 64, 65
valence tautomerization of 1,3,5-trienes with formation of cyclopropane derivatives 15-17, 19
cyclopropene derivatives
cycloaddition of carbenes to acetylene compounds with formation of cyclopropene derivatives 281
photolysis of pyrazole derivatives with formation of cyclopropene derivatives 281, 329

dealkylation
dealkylation (elimination of alkyl groups)
photolysis *
dealkylation 253, 254
dealkylation of N-alkyl amines 255-257
dealkylation of N-alkyl heterocyclic nitrogen compounds 257, 258
sensitization in dealkylation reactions 256, 257
decarbonylation
decarbonylation 210, 394, 395, 459-465, 469
decarbonylation of acyl xanthates 223, 224

decarbonylation of aldehydes   222, 223
decarbonylation of aromatic ketones   220, 221
decarbonylation of cyclic ketones involving ring contraction   218, 219, 221
decarbonylation of cyclic ketones with formation of olefins   218-220
decarbonylation of ketenes   222
decarboxylation
decarboxylation   370, 409, 410, 469
photolysis of oxadiazoline derivatives involving decarboxylation   337
dehalogenation
dehalogenation   133, 134, 158, 368
dehalogenation followed by dimerization   366
dehydrocyclization
dehydrocyclization   (cyclization of unsaturated compounds involving dehydrogenation)
dehydrocyclization   19, 25, 63
dehydrocyclization of aromatic compounds to fluorene derivatives   126, 127
dehydrocyclization of aromatic compounds to fused polycyclic hydrocarbons   130-136
dehydrocyclization of aromatic heterocyclic sulfur compounds   136
dehydrocyclization of aryl substituted amines to carbazole derivatives   138
dehydrocyclization of aryl substituted azo compounds or SCHIFF bases to aromatic heterocyclic nitrogen compounds   138-142, 337, 338
dehydrocyclization of aryl substituted olefins to aromatic compounds   127-131, 136
dehydrocyclization of aryl substituted pyridinium or tetrazolium salts to aromatic heterocyclic nitrogen compounds   142-144
dehydrocyclization with formation of fused heterocyclic hydrocarbons   25, 130, 132-136, 138-144
dehydrocyclization with oxygen , iodine or quinones as hydrogen acceptors   126-136, 138, 141, 142
dehydrodimerization
dehydrodimerization   (dimerization involving dehydrogenation)
dehydrodimerization   38, 146-149, 153, 183, 199-201, 230
dehydrodimerization with oxygen , quinones or dyes as hydrogen acceptors   146-149
sensitization in dehydrodimerization reactions   147-149
dehydrogenation
dehydrogenation   (elimination of hydrogen)
cyclization of unsaturated compounds involving dehydrogenation see dehydrocyclization
dimerization involving dehydrogenation see dehydrodimerization
aromatization involving dehydrogenation by hydrogen acceptors   127-136, 138-144, 152-154
dehydrogenation of acetals to esters   238
dehydrogenation of hydroaromatic compounds with quinones , disulfides or dyes as hydrogen acceptors   151-154
sensitization in dehydrogenation reactions   153, 154
dehydrohalogenation
cyclization following dehydrohalogenation of N-chloroamines see HOFMANN-LOEFFLER reaction
dehydrohalogenation   128, 134-136
deoxygenation
deoxygenation   (elimination of one oxygen atom)
deoxygenation   430-432
deoxygenation of nitrones   52
depolymerization
depolymerization   78, 85, 91, 92
desulfuration
desulfuration   (elimination of sulfur)
desulfuration   442, 454, 455

Reaction Index

desulfuration of disulfides to sulfides   439, 440
desulfuration of thiirane derivatives and dithietane derivatives with
   formation of olefins   449, 450
diazirine derivatives
photolysis of diazirine derivatives   290, 291
diazo compounds
addition of diazo compounds to C=C bonds with formation of pyrazoline
   derivatives   288, 292
photolysis of diazo compounds with formation of carbenes   274-291
sensitization in photolysis reactions of diazo compounds   276
diazo ketones
cyclic diazo ketones see also quinone diazides
ketones *
photolysis of cyclic diazo ketones involving ring contraction   302-306, 309, 310
photolysis of diazo ketones with formation of keto-carbenes   121, 294-311
diazonium salts
photolysis of diazonium salts   313-315
photolysis of diazonium salts in inorganic acids with formation of organic
   halides or phenols   314, 315
photolysis of diazonium salts with formation of azo compounds   316
reduction of diazonium salts with alcohols as hydrogen donors   313, 314
dicarbonyl compounds
isomerization of unsaturated nitro compounds to oximes of dicarbonyl
   compounds   266, 267
1,2-dicarbonyl compounds
1,2-dicarbonyl compounds see also 1,2-quinones , 1,2-diketones and 1,2,3-
   triketones
cycloaddition of 1,2-dicarbonyl compounds to unsaturated compounds with
   formation of dioxin derivatives   118-124, 419
cycloaddition of 1,2-dicarbonyl compounds to unsaturated compounds with
   formation of oxetane derivatives   121, 122, 124, 419, 420
DIELS-ALDER reaction
DIELS-ALDER reaction   75, 118, 123, 424, 432
DIELS-ALDER reaction with oxygen   382, 383, 428
intramolecular DIELS-ALDER reaction   4
1,2-dienes
isomerization of 1,3-dienes to 1,2-dienes   14, 15, 22-24
1,3-dienes
addition of oxygen to 1,3-dienes with formation of epidioxides see
   peroxidation
addition of organic halides to 1,3-dienes   171, 172, 175, 176
cycloaddition of 1,4-quinones to 1,3-dienes with formation of pyran
   derivatives   424
cyclodimerization of 1,3-dienes   74, 75
isomerization of cyclic 1,3-dienes to 1,3,5-trienes   14-17
isomerization of 1,3-dienes to 1,2-dienes   14, 15, 22-24
isomerization of 1,3,5-trienes to cyclic 1,3-dienes   18-20, 127-133, 139, 141, 142
peroxidation of cyclic 1,3-dienes   382-388, 394, 407, 408
valence tautomerization of aliphatic 1,3-dienes   10
valence tautomerization of cyclic 1,3-dienes with formation of cyclobutene
   derivatives   4-9, 27, 28, 227
1,5-dienes
isomerization of cyclic 1,5-dienes   18
dihydrodimerization
dihydrodimerization (dimerization involving reduction)
dihydrodimerization   195, 196, 215-217, 283, 284, 290
dihydrodimerization of aldehydes with formation of pinacols   203, 204
dihydrodimerization of keto acids with formation of glycols   210, 211
dihydrodimerization of ketones with formation of pinacols   203-209

dihydrodimerization with alcohols as hydrogen donors 203-211
1,2-diketones
1,2-dicarbonyl compounds *
photolysis of diaroyl azo compounds with formation of aromatic 1,2-diketones 330, 331
photolysis of 1,2-diketones 231
1,3-diketones
isomerization of epoxy ketones to 1,3-diketones or 1,4-diketones 408, 409, 412, 413
1,4-diketones
isomerization of epoxy ketones to 1,3-diketones or 1,4-diketones 408, 409, 412, 413
dimerization
dimerization involving dehydrogenation see dehydrodimerization
dimerization involving reduction see dihydrodimerization
dimerization see also cyclodimerization
dehalogenation followed by dimerization 366
dimerization 126, 212, 335
dimerization of aldehydes 105, 107, 108
dimerization of anthracene derivatives 97-99, 101, 102, 266
dimerization of aromatic compounds with formation of 8-membered rings 11, 93 97-102, 390
dimerization of aryl substituted acetylene compounds 92-94
dimerization of carbenes with formation of olefins 288, 289, 291
dimerization of furan derivatives 103
dimerization of heterocyclic nitrogen compounds with formation of 8-membered rings 99-102
dimerization of heterocyclic oxygen compounds with formation of 8-membered rings 102, 103
dimerization of imenes with formation of azo compounds 321
dimerization of nitroso compounds 105-107, 248, 249, 252, 261-263, 269
dimerization of organic halides 105, 106
dimerization of pyran derivatives 102, 103
dimerization of pyridine derivatives 99-101
dimerization of thio carbenes 331, 332
dioxin derivatives
dioxin derivatives (C4O2-cycloadducts)
cycloaddition of 1,2-dicarbonyl compounds to unsaturated compounds with formation of dioxin derivatives 118-124, 419
disproportionation
disproportionation 134, 135, 139, 141, 143, 199, 226, 242, 252, 338
disulfides
sulfur compounds *
dehydrogenation of hydroaromatic compounds with quinones, disulfides or dyes as hydrogen acceptors 151-154
desulfuration of disulfides to sulfides 439, 440
halogenation of disulfides with formation of sulfenyl chlorides 441
insertion of mercury into disulfides 441
photolysis of disulfides 152, 153, 439-441
dithietane derivatives
cyclodimerization of thiocarbonyl compounds to dithietane derivatives 106, 107, 449, 450
desulfuration of thiirane derivatives and dithietane derivatives with formation of olefins 449, 450
dyes
dehydrodimerization with oxygen, quinones or dyes as hydrogen acceptors 146-149
dehydrogenation of hydroaromatic compounds with quinones, disulfides or dyes as hydrogen acceptors 151-154

electron donors
complexation with nitrogen compounds, nitroso compounds, sulfoxides, phosphines, arsines, acetylene compounds and olefins as electron donors   459-464, 469
replacement of carbon monoxide in metal carbonyl compounds by electron donors with formation of organometallic carbonyl compounds   459-462, 469
replacement of carbon monoxide in organometallic carbonyl compounds by electron donors   462-464
elimination
elimination of alkyl groups see dealkylation
elimination of hydrogen see dehydrogenation
elimination of one oxygen atom see deoxygenation
elimination of sulfur see desulfuration
fragmentation *
NORRISH type II process *
photolysis *
elimination   35-37, 139, 347, 359, 429, 465
elimination of oxygen from epidioxides   383, 385, 390, 391
elimination of sulfur dioxide from cyclic sulfones   447, 448
elimination reactions with nitro compounds   266, 271
elimination reactions with organic nitrites   251-254
isomerization of saturated ketones via elimination   25, 26
endoperoxides
endoperoxides and cyclic peroxides see epidioxides
ene-synthesis
ene-synthesis   168, 375, 381, 382, 396
enol
rearrangement of enol esters   37, 38
enolization
enolization   22, 24-26, 42, 59, 130, 208
epidioxides
epidioxides (endoperoxides and cyclic peroxides)
addition of oxygen to 1,3-dienes with formation of epidioxides see peroxidation
addition of alcohols to epidioxides with formation of alkoxy hydroperoxides   428-430
elimination of oxygen from epidioxides   383, 385, 390, 391
isomerization of epidioxides to epoxy compounds   46, 47, 384, 387, 407, 408, 431
isomerization of epidioxides to furan derivatives   394
isomerization of epidioxides to pseudo acids   428-430
epimerization
stereoisomerization *
epimerization   65-67, 249, 251
epoxy compounds
cycloaddition of carbenes to C=O bonds with formation of epoxy compounds   279, 280
isomerization of epidioxides to epoxy compounds   46, 47, 384, 387, 407, 408, 431
oxidation of olefins with formation of epoxy compounds   378, 407
epoxy ketones
ketones *
isomerization of cyclic epoxy ketones to pyran derivatives   409-411
isomerization of epoxy ketones to 1,3-diketones or 1,4-diketones   408, 409, 412, 413
valence tautomerization of epoxy ketones with formation of pyrylium compounds   410, 411
esters
isomerization of aryl esters with formation of acyl phenols see FRIES rearrangement
addition of alcohols to ketenes with formation of esters   219, 238, 239, 295-297, 301, 303, 304

addition of ketenes to thiols with formation of esters of thiocarboxylic
  acids 297, 298
addition of 1,2-quinones to aldehydes with formation of esters 186-188
addition of 1,4-quinones to aldehydes with formation of esters 189
allylic rearrangement of esters 234, 235
dehydrogenation of acetals to esters 238
isomerization of nitro acetals to nitroso esters 268, 269
isomerization of unsaturated esters 234-237
photolysis of esters 38
photolysis of α,β-unsaturated ketones in alcohols with formation of
  esters 227, 238
rearrangement of enol esters 37, 38
replacement of acyloxy groups in esters by hydrogen 235
sensitization in isomerization of cyclic acetals to esters 45, 46
ethers
addition of alcohols to C=C bonds with formation of ethers 8-11, 159, 160, 193,
  194, 432, 433
addition of cyclic ethers to C=C bonds 160, 161
addition of hypohalites to olefins with formation of ethers 162
addition of 1,2-quinones to ethers 184
addition of 1,4-quinones to hydroaromatic compounds or aralkyl compounds
  with formation of ethers 151, 182
cyanation of aliphatic compounds or ethers with formation of nitriles 260, 261
hydroperoxidation of ethers 379, 380
insertion of carbenes into O-H bonds with formation of ethers 286, 287
insertion of keto-carbenes into O-H bonds with formation of ethers 286, 307,
  308, 310, 311
photolysis of organic hypohalites with formation of ethers 369, 370
reduction with alcohols, ethers or hydrocarbons as hydrogen donors 193-197

fluorene derivatives
dehydrocyclization of aromatic compounds to fluorene derivatives 126, 127
formamides
amides *
addition of ammonia, amines and formamides to C=C bonds 165-168
addition of formamides to acetylene compounds 167
amidation of aromatic compounds with formamides 230, 231
formazans
stereoisomerization of oximes, nitrones and formazans 63, 64
fragmentation
fragmentation of cyclobutane derivatives see also retro-cycloaddition
fragmentation see also elimination
photolysis *
FRIES rearrangement
FRIES rearrangement (isomerization of aryl esters with formation of acyl
  phenols)
FRIES rearrangement 236, 237
furan derivatives
heterocyclic oxygen compounds *
cycloaddition of carbenes to furan derivatives or thiophene derivatives 278, 279
dimerization of furan derivatives 103
intramolecular cyclization with formation of furan derivatives 41, 215-217,
  426, 427
isomerization of epidioxides to furan derivatives 394
isomerization of furan derivatives to aldehydes 45
isomerization of furan derivatives to ketones 44, 45
oxidation of furan derivatives to unsaturated lactones 239
peroxidation of furan derivatives 427-432
valence tautomerization of furan derivatives 44

fused heterocyclic hydrocarbons
dehydrocyclization with formation of fused heterocyclic hydrocarbons   25, 130, 132-136, 138-144
fused polycyclic hydrocarbons
fused polycyclic hydrocarbons (arenes)
dehydrocyclization of aromatic compounds to fused polycyclic hydrocarbons   130-136
isomerization of fused polycyclic hydrocarbons   20, 21
peroxidation of fused polycyclic hydrocarbons   389-394

glycols
addition of hydrogen peroxide to C=C bonds with formation of glycols see MILAS reaction
glycols see also pinacols
addition of alcohols to ketones with formation of glycols   207, 208
dihydrodimerization of keto acids with formation of glycols   210, 211

halides
halides (halogen compounds)
halides see also nitrogen-halogen compounds
addition of aromatic halides to phosphines with formation of phosphonium salts   455-457
addition of halogens to aromatic compounds with formation of organic halides   342-345
addition of halogens to C=C bonds with formation of organic halides   341, 350, 353
addition of hydrogen halides to acetylene compounds with formation of organic halides   158
addition of hydrogen halides to olefins with formation of organic halides   157-159
addition of keto-carbenes to organic halides   307, 308
addition of magnesium to organic halides with formation of organomagnesium compounds   468
addition of mercury to organic halides with formation of organomercury compounds   366, 367
addition of organic halides to acetylene compounds   172, 174
addition of organic halides to C=C bonds   170-174
addition of organic halides to 1,3-dienes   171, 172, 175, 176
addition of carbenes to organic halides   282, 283
dimerization of organic halides   105, 106
halogenolysis of organic halides   347, 348, 354, 362-365
isomerization of organic halides   39, 362
oxidation of organic halides   225, 234
oxidation of organic halides with formation of acids   225
photolysis of diazonium salts in inorganic acids with formation of organic halides or phenols   314, 315
photolysis of organic halides   128, 133-136, 354-359, 362-371
photolysis of organomercury compounds in organic halides   466, 467
halogenation
halogenation   238, 341-359, 434, 435, 437
halogenation of disulfides with formation of sulfenyl chlorides   441
halogen compounds
halogen compounds see halides
halogenolysis
halogenolysis (replacement of substituents by halogens)
halogenolysis of C-H bonds   342-344, 347-349, 351, 352, 354-359, 365
halogenolysis of nitroso compounds   347
halogenolysis of organic halides   347, 348, 354, 362-365
halogenolysis of sulfonyl chlorides   346, 347, 350

halogens
halogens (bromine, chlorine and iodine)
replacement of substituents by halogens see halogenolysis
addition of halogens to acetylene compounds 342
addition of halogens to aromatic compounds with formation of organic halides 342-345
addition of halogens to C=C bonds with formation of organic halides 341, 350, 353
replacement of halogens by hydrogen 362-364
heterocyclic nitrogen compounds
heterocyclic nitrogen compounds see also pyridine derivatives and carbazole derivatives
heterocyclic nitrogen compounds see also pyrimidine derivatives and pyrazole derivatives
cyclization of SCHIFF bases to heterocyclic nitrogen compounds involving incorporation of alcohols 334, 335
dealkylation of N-alkyl heterocyclic nitrogen compounds 257, 258
dehydrocyclization of aryl substituted azo compounds or SCHIFF bases to aromatic heterocyclic nitrogen compounds 138-142, 337, 338
dehydrocyclization of aryl substituted pyridinium or tetrazolium salts to aromatic heterocyclic nitrogen compounds 142-144
dimerization of heterocyclic nitrogen compounds with formation of 8-membered rings 99-102
oxidation of heterocyclic nitrogen compounds 381, 395-397, 399, 401, 402
photolysis of nitrogen-halogen compounds followed by cyclization to heterocyclic nitrogen compounds 242-246
reduction of aromatic heterocyclic nitrogen compounds 195, 196
heterocyclic oxygen compounds
heterocyclic oxygen compounds see also furan derivatives and pyran derivatives
dimerization of heterocyclic oxygen compounds with formation of 8-membered rings 102, 103
isomerization of heterocyclic oxygen compounds 44-47
isomerization of ketones to heterocyclic oxygen compounds 41-44
oxidation of heterocyclic oxygen compounds with formation of peroxides 398, 399
heterocyclic sulfur compounds
dehydrocyclization of aromatic heterocyclic sulfur compounds 136
oxidation of heterocyclic sulfur compounds with formation of peroxides 398, 399
HOFMANN-LOEFFLER reaction
HOFMANN-LOEFFLER reaction (cyclization following dehydrohalogenation of N-chloroamines)
HOFMANN-LOEFFLER reaction 242-246
hydroaromatic compounds
hydroaromatic compounds (compounds containing methylene groups) aromatic compounds *
addition of 1,4-quinones to hydroaromatic compounds or aralkyl compounds with formation of ethers 151, 182
dehydrogenation of hydroaromatic compounds with quinones, disulfides or dyes as hydrogen acceptors 151-154
hydroperoxidation of hydroaromatic compounds 373-375, 380, 382
hydrocarbons
reduction with alcohols, ethers or hydrocarbons as hydrogen donors 193-197
hydrogen
elimination of hydrogen see dehydrogenation
replacement of halogens by hydrogen 362-364
hydrogen acceptors
aromatization involving dehydrogenation by hydrogen acceptors 127-136, 138-144, 152-154

dehydrocyclization with oxygen, iodine or quinones as hydrogen
  acceptors 126-136, 138, 141, 142
dehydrodimerization with oxygen, quinones or dyes as hydrogen
  acceptors 146-149
dehydrogenation of hydroaromatic compounds with quinones, disulfides or
  dyes as hydrogen acceptors 151-154
hydrogenation
hydrogenation see reduction of C=C bonds
hydrogen donors
dihydrodimerization with alcohols as hydrogen donors 203-211
reduction of acyl azides with hydrogen donors involving formation of
  amides 321-325
reduction of carbenes with hydrogen donors 287, 300
reduction of C=C bonds with hydrogen donors 193, 194
reduction of C=N bonds with hydrogen donors 195, 196
reduction of C=O bonds with hydrogen donors involving formation of
  alcohols 113, 194, 195, 204, 208
reduction of diazonium salts with alcohols as hydrogen donors 313, 314
reduction of quinones by hydrogen donors 151, 152
reduction with alcohols, ethers or hydrocarbons as hydrogen donors 193-197
hydrogen sulfide
addition of hydrogen sulfide, thiols and thiocarboxylic acids to C=C
  bonds 162-164
hydrolysis
hydrolysis see also addition of water
hydrolysis 225-230, 239, 240, 287
hydrolysis of nitro sulfenyl chlorides to amino sulfonic acids 270, 271
hydrolysis of phenols with formation of acids 239, 240
hydrolysis of unsaturated ketones to unsaturated acids 227-230, 238, 239
hydroperoxidation
hydroperoxidation (replacement of hydrogen by the hydroperoxy group)
addition of oxygen *
hydroperoxidation and peroxidation of organotin compounds 466
hydroperoxidation of alcohols 379
hydroperoxidation of aralkyl compounds 146, 373
hydroperoxidation of ethers 379, 380
hydroperoxidation of hydroaromatic compounds 373-375, 380, 382
hydroperoxidation of nitrogen compounds 380, 389, 396, 397
hydroperoxidation of olefins 373, 375-378, 382, 387
hydroperoxidation of phenols 382
sensitization in hydroperoxidation reactions 375-382, 466
hydroperoxides
oxidation involving formation of hydroperoxides see hydroperoxidation
addition of alcohols to epidioxides with formation of alkoxy
  hydroperoxides 428-430
oxidation of thiols and olefins with formation of hydroperoxides of
  sulfides 446, 447

imenes
cycloaddition of acyl imenes to C=C bonds with formation of aziridine
  derivatives 325
dimerization of imenes with formation of azo compounds 321
insertion of acyl imenes into C-H bonds with formation of carbamates 324, 325
insertion of imenes into aromatic C-H bonds 319
intramolecular cyclization of acyl imenes with formation of lactams 322-324
intramolecular cyclization of imenes with formation of carbazole
  derivatives 320
isomerization of imenes to imines 318-320
photolysis of azides with formation of imenes 318-327

imines
C=N bonds *
imines see also 1,2-quinone imines
isomerization of imenes to imines   318-320
incorporation
cyclization of aryl substituted amines involving incorporation of alcohols   335, 336
cyclization of olefins to cyclic peroxides involving incorporation of oxygen   397, 398
cyclization of SCHIFF bases to heterocyclic nitrogen compounds involving incorporation of alcohols   334, 335
indole derivatives
intramolecular cyclization of aromatic nitro compounds to indole derivatives   271, 272
insertion
addition *
alkylation *
insertion of acyl imenes into C-H bonds with formation of carbamates   324, 325
insertion of carbenes into C-C bonds   281
insertion of carbenes into C-H bonds   278-280, 283-286
insertion of carbenes into C-O bonds   279, 280, 282
insertion of carbenes into N-H bonds   287, 288
insertion of carbenes into O-H bonds with formation of ethers   286, 287
insertion of imenes into aromatic C-H bonds   319
insertion of keto-carbenes into aromatic C-H bonds   311
insertion of keto-carbenes into O-H bonds with formation of ethers   286, 307, 308, 310, 311
insertion of mercury into disulfides   441
iodine
bromine, chlorine and iodine see halogens
dehydrocyclization with oxygen, iodine or quinones as hydrogen acceptors   126-136, 138, 141, 142
isocyanates
photolysis of acyl azides with formation of isocyanates   321-324
isomerization
isomerization of aryl esters with formation of acyl phenols see FRIES rearrangement
isomerization of organic nitrites to hydroxy oximes see BARTON reaction
isomerization see also intramolecular cyclization, rearrangement, stereoisomerization, valence tautomerization
cycloaddition of acetylene compounds to aromatic compounds followed by isomerization of the adducts to 8-membered rings   114
cyclodimerization of acetylene compounds followed by isomerization   92-94
isomerization of aldehydes   27
isomerization of aromatic nitro compounds to nitroso compounds   267-270
isomerization of aryl substituted thiophene derivatives   451
isomerization of azoxy compounds   53
isomerization of carbenes with formation of olefins   288-290
isomerization of cross-conjugated cyclohexadienones   28-32, 235
isomerization of cyclic 1,5-dienes   18
isomerization of cyclic 1,3-dienes to 1,3,5-trienes   14-17
isomerization of cyclic epoxy ketones to pyran derivatives   409-411
isomerization of cyclic ketones to ketenes   226, 227, 229
isomerization of cyclic ketones to unsaturated lactones   219
isomerization of 1,3-dienes to 1,2-dienes   14, 15, 22-24
isomerization of epidioxides to epoxy compounds   46, 47, 384, 387, 407, 408, 431
isomerization of epidioxides to furan derivatives   394
isomerization of epidioxides to pseudo acids   428-430
isomerization of epoxy ketones to 1,3-diketones or 1,4-diketones   408, 409, 412, 413

isomerization of furan derivatives to aldehydes 45
isomerization of furan derivatives to ketones 44, 45
isomerization of fused polycyclic hydrocarbons 20, 21
isomerization of heterocyclic oxygen compounds 44-47
isomerization of imenes to imines 318-320
isomerization of keto-carbenes to α,β-unsaturated ketones 294, 295
isomerization of ketones 22-33
isomerization of ketones by intramolecular cyclization to cyclobutanols 34-37, 409
isomerization of ketones to heterocyclic oxygen compounds 41-44
isomerization of linear-conjugated cyclohexadienones to unsaturated ketenes 227, 229
isomerization of nitro acetals to nitroso esters 268, 269
isomerization of nitro aldehydes to nitroso acids 267-270
isomerization of nitrogen-halogen compounds 53, 54, 246, 247
isomerization of nitrones to amides or lactams 48-52
isomerization of nitrones to oxaziridine derivatives 49-52, 63
isomerization of nitroso compounds to oximes 248, 252, 262, 263
isomerization of organic halides 39, 362
isomerization of organic nitrites to hydroxamic acids 252, 253
isomerization of oxete derivatives to ketones 423
isomerization of oximes to amides 48
isomerization of phosphites to phosphonates 454
isomerization of pyran derivatives 42, 46, 89
isomerization of saturated cyclic ketones to unsaturated aldehydes 26, 27
isomerization of saturated ketones via elimination 25, 26
isomerization of 1,3,5-trienes to cyclic 1,3-dienes 18-20, 127-133, 139, 141, 142
isomerization of α,β-unsaturated carbonyl compounds to β,γ-unsaturated carbonyl compounds 22-25
isomerization of unsaturated compounds 1-69
isomerization of unsaturated esters 234-237
isomerization of unsaturated ketones by intramolecular cyclization to cyclohexenols 37
isomerization of α,β-unsaturated ketones to unsaturated lactones 239
isomerization of unsaturated nitro compounds to oximes of dicarbonyl compounds 266, 267
sensitization in isomerization of cyclic acetals to esters 45, 46
sensitization in isomerization reactions 18, 30

ketenes
addition of alcohols to ketenes with formation of esters 219, 238, 239, 295-297, 301, 303, 304
addition of amines to ketenes with formation of amides 228-230, 297
addition of ketenes to thiols with formation of esters of thiocarboxylic acids 297, 298
addition of water to ketenes with formation of acids 227, 229, 295-297, 301-306, 309, 310
cycloaddition of ketenes to azo compounds 298, 299
cycloaddition of ketenes to C=N bonds 299, 300
decarbonylation of ketenes 222
isomerization of cyclic ketones to ketenes 226, 227, 229
isomerization of linear-conjugated cyclohexadienones to unsaturated ketenes 227, 229
rearrangement of keto-carbenes to ketenes 121, 294-299, 301-306, 309, 310
keto acids
acids *
ketones *
dihydrodimerization of keto acids with formation of glycols 210, 211

keto-carbenes
alkylation *
carbenes *
addition of keto-carbenes to organic halides   307, 308
addition of water to keto-carbenes with formation of lactones   306-308
insertion of keto-carbenes into aromatic C-H bonds   311
insertion of keto-carbenes into O-H bonds with formation of ethers   286, 307, 308, 310, 311
isomerization of keto-carbenes to α,β-unsaturated ketones   294, 295
photolysis of diazo ketones with formation of keto-carbenes   121, 294-311
rearrangement of keto-carbenes to ketones   121, 294-299, 301-306, 309, 310

ketones
carbonyl compounds *
ketones see also diazo ketones, epoxy ketones, keto acids and unsaturated ketones
addition of alcohols to ketones with formation of glycols   207, 208
addition of aldehydes to acetylene compounds with formation of ketones   169
addition of ketones to alcohols with formation of pinacols   203
addition of ketones to methylene groups with formation of carbinols   198-202
addition of ketones to olefins   169, 170
addition of 1,4-quinones to aldehydes with formation of ketones   188, 189
cleavage of azines by ketones involving formation of nitriles   339
cleavage of pinacols by ketones   206, 207, 212
cycloaddition of ketones to olefins with formation of oxetane derivatives   414-417, 419-423
decarbonylation of aromatic ketones   220, 221
decarbonylation of cyclic ketones involving ring contraction   218, 219, 221
decarbonylation of cyclic ketones with formation of olefins   218-220
dihydrodimerization of ketones with formation of pinacols   203-209
isomerization of cyclic ketones to ketenes   226, 227, 229
isomerization of cyclic ketones to unsaturated lactones   219
isomerization of furan derivatives to ketones   44, 45
isomerization of ketones   22-33
isomerization of ketones by intramolecular cyclization to cyclobutanols   34-37, 409
isomerization of ketones to heterocyclic oxygen compounds   41-44
isomerization of oxete derivatives to ketones   423
isomerization of saturated cyclic ketones to unsaturated aldehydes   26, 27
isomerization of saturated ketones via elimination   25, 26
oxidation of thioketones to ketones   445, 446
photolysis of aralkyl ketones   215-217
photolysis of cyclic ketones to acids   1, 26, 28, 225-230

lactams
intramolecular cyclization of acyl imenes with formation of lactams   322-324
isomerization of nitrones to amides or lactams   48-52

lactones
lactones see also unsaturated lactones
addition of alcohols to unsaturated acids followed by dehydration to lactones   231-233
addition of water to keto-carbenes with formation of lactones   306-308
photolysis of nitrogen-halogen compounds with formation of lactones   370, 371

light filters
light filters   490-493

light sources
light sources   472-489
mercury lamps as light sources   473, 475-477, 481-486, 488, 489
sodium lamps as light sources   473-475, 480
xenon lamps as light sources   473, 477, 486, 487

8-membered rings
cycloaddition of acetylene compounds to aromatic compounds followed by isomerization of the adducts to 8-membered rings 114
dimerization of aromatic compounds with formation of 8-membered rings 11, 93, 97-102, 390
dimerization of heterocyclic nitrogen compounds with formation of 8-membered rings 99-102
dimerization of heterocyclic oxygen compounds with formation of 8-membered rings 102, 103
mercury
addition of mercury to organic halides with formation of organomercury compounds 366, 367
insertion of mercury into disulfides 441
mercury lamps
mercury lamps as light sources 473, 475-477, 481-486, 488, 489
methylene groups
compounds containing methylene groups see hydroaromatic compounds and aralkyl compounds
MILAS reaction
MILAS reaction (addition of hydrogen peroxide to C=C bonds with formation of glycols)
MILAS reaction 157

N-H bonds
insertion of carbenes into N-H bonds 287, 288
N-oxides
N-oxides see also nitrones
N=N bonds
N=N bonds see also azo compounds
stereoisomerization at N=N bonds 62-64, 252
nitriles
cleavage of azines by ketones involving formation of nitriles 339
cyanation of aliphatic compounds or ethers with formation of nitriles 260, 261
nitrites
isomerization of organic nitrites to hydroxy oximes see BARTON reaction
elimination reactions with organic nitrites 251-254
isomerization of organic nitrites to hydroxamic acids 252, 253
photolysis of organic nitrites 248-254
nitro compounds
elimination reactions with nitro compounds 266, 271
intramolecular cyclization of aromatic nitro compounds to indole derivatives 271, 272
isomerization of aromatic nitro compounds to nitroso compounds 267-270
isomerization of unsaturated nitro compounds to oximes of dicarbonyl compounds 266, 267
reduction of nitro compounds to amines 270, 271
nitrogen compounds
nitrogen compounds, oxygen compounds and sulfur compounds see also heterocyclic compounds
complexation with nitrogen compounds, nitroso compounds, sulfoxides, phosphines, arsines, acetylene compounds and olefins as electron donors 459-464, 469
hydroperoxidation of nitrogen compounds 380, 389, 396, 397
sensitization in cyclization reactions of nitrogen compounds with alcohols 335, 336
nitrogen-halogen compounds
halides *
isomerization of nitrogen-halogen compounds 53, 54, 246, 247

photolysis of nitrogen-halogen compounds followed by cyclization to
   heterocyclic nitrogen compounds   242-246
photolysis of nitrogen-halogen compounds with formation of lactones   370, 371
replacement of chlorine in nitrogen-halogen compounds by hydrogen   242
nitrones
N-oxides *
deoxygenation of nitrones   52
isomerization of nitrones to amides or lactams   48-52
isomerization of nitrones to oxaziridine derivatives   49-52, 63
stereoisomerization of oximes, nitrones and formazans   63, 64
nitrosation
nitrosation (replacement of hydrogen by the nitroso group)
nitrosation   261-263
nitroso compounds
nitroso compounds see also chloronitroso compounds
complexation with nitrogen compounds, nitroso compounds, sulfoxides,
   phosphines, arsines, acetylene compounds and olefins as electron
   donors   459-464, 469
dimerization of nitroso compounds   105-107, 248, 249, 252, 261-263, 269
halogenolysis of nitroso compounds   347
isomerization of aromatic nitro compounds to nitroso compounds   267-270
isomerization of nitroso compounds to oximes   248, 252, 262, 263
nitrosyl chloride
addition of nitrosyl chloride to olefins   159
photolysis of nitrosyl chloride   262-264
NORRISH type II process
NORRISH type II process see also elimination
NORRISH type II process   25, 26, 36, 37

O-H bonds
insertion of carbenes into O-H bonds with formation of ethers   286, 287
insertion of keto-carbenes into O-H bonds with formation of ethers   286, 307,
   308, 310, 311
olefins
olefins (alkenes)
compounds containing C=C bonds *
PATERNO-BUECHI reaction *
unsaturated compounds *
addition of aldehydes to olefins with formation of ketones   169, 188, 201
addition of hydrogen halides to olefins with formation of organic
   halides   157-159
addition of hypohalites to olefins with formation of ethers   162
addition of ketones to olefins   169, 170
addition of maleic anhydride derivatives to olefins   168
addition of nitrosyl chloride to olefins   159
addition of organogermanium compounds to olefins   179
addition of organosilicon compounds to olefins and acetylene compounds   178, 179
addition of sulfenyl chlorides or sulfonyl chlorides to olefins   164, 165
complexation with nitrogen compounds, nitroso compounds, sulfoxides,
   phosphines, arsines, acetylene compounds and olefins as electron
   donors   459-464, 469
cyclization of olefins to cyclic peroxides involving incorporation of
   oxygen   397, 398
cycloaddition of aldehydes to olefins with formation of oxetane
   derivatives   414-417, 420-423
cycloaddition of ketones to olefins with formation of oxetane
   derivatives   414-417, 419-423
cycloaddition of olefins to aromatic compounds   111, 112
cycloaddition of 1,4-quinones to olefins   418, 419

Reaction Index 547

cycloaddition of 1,4-quinones to olefins with formation of oxetane
   derivatives 418, 419
cycloaddition of α,β-unsaturated carbonyl compounds to olefins 110-113
cyclodimerization of aryl substituted olefins 70-73, 89, 90
cyclodimerization of olefins 74, 468, 469
decarbonylation of cyclic ketones with formation of olefins 218-220
dehydrocyclization of aryl substituted olefins to aromatic compounds 127-131, 136
desulfuration of thiirane derivatives and dithietane derivatives with
   formation of olefins 449, 450
dimerization of carbenes with formation of olefins 288, 289, 291
hydroperoxidation of olefins 373, 375-378, 382, 387
isomerization of carbenes with formation of olefins 288-290
oxidation of olefins with formation of epoxy compounds 378, 407
oxidation of thiols and olefins with formation of hydroperoxides of
   sulfides 446, 447
photolysis of pyrazoline derivatives with formation of cyclopropane
   derivatives and olefins 328, 329
sensitization in addition reactions with olefins and acetylene compounds 155,
   160-163, 166-168, 176, 177

organoarsenic compounds
organoarsenic compounds see also arsines
organogermanium compounds
addition of organogermanium compounds to olefins 179
organolithium compounds
photolysis of organolithium compounds 469
organomagnesium compounds
addition of magnesium to organic halides with formation of organomagnesium
   compounds 468
organomercury compounds
addition of mercury to organic halides with formation of organomercury
   compounds 366, 367
photolysis of organomercury compounds 369
photolysis of organomercury compounds in alcohols with formation of mercury
   and carbonyl compounds 467, 468
photolysis of organomercury compounds in organic halides 466, 467
organometallic carbonyl compounds
replacement of carbon monoxide in metal carbonyl compounds by electron
   donors with formation of organometallic carbonyl compounds 459-462, 469
replacement of carbon monoxide in organometallic carbonyl compounds by
   electron donors 462-464
organophosphorus compounds
organophosphorus compounds see also phosphines, phosphonates and
   phosphonothioates
organosilicon compounds
addition of organosilicon compounds to olefins and acetylene compounds 178, 179
organotin compounds
hydroperoxidation and peroxidation of organotin compounds 466
photolysis of organotin compounds 465, 466
oxadiazoline derivatives
photolysis of oxadiazoline derivatives involving decarboxylation 337
oxaziridine derivatives
isomerization of nitrones to oxaziridine derivatives 49-52, 63
oxazolidine derivatives
addition of alcohols to SCHIFF bases with formation of oxazolidine
   derivatives 335-337
oxetane derivatives
oxetane derivatives (C3O-cycloadducts)

PATERNO-BUECHI reaction *
cycloaddition of aldehydes to olefins with formation of oxetane
 derivatives   414-417, 420-423
cycloaddition of 1,2-dicarbonyl compounds to unsaturated compounds with
 formation of oxetane derivatives   121, 122, 124, 419, 420
cycloaddition of ketones to olefins with formation of oxetane
 derivatives   414-417, 419-423
cycloaddition of 1,4-quinones to olefins with formation of oxetane
 derivatives   418, 419
cyclodimerization of 1,4-quinones with formation of oxetane derivatives   79, 80
intramolecular cycloaddition of unsaturated ketones to oxetane
 derivatives   417, 418
oxete derivatives
cycloaddition of carbonyl compounds to acetylene compounds with formation of
 oxete derivatives   423
isomerization of oxete derivatives to ketones   423
oxidation
oxidation involving formation of hydroperoxides see hydroperoxidation
oxidation of aldehydes with formation of acyl peroxides   399, 400
oxidation of amines   381
oxidation of cyclopentadienones with formation of α,β-unsaturated
 ketones   394, 395, 397
oxidation of furan derivatives to unsaturated lactones   239
oxidation of heterocyclic nitrogen compounds   381, 395-397, 399, 401, 402
oxidation of heterocyclic oxygen compounds with formation of peroxides   398, 399
oxidation of heterocyclic sulfur compounds with formation of peroxides   398, 399
oxidation of olefins with formation of epoxy compounds   378, 407
oxidation of organic halides   225, 234
oxidation of organic halides with formation of acids   225
oxidation of phosphites to phosphates   454, 455
oxidation of sulfoxides to sulfones   444
oxidation of thiocarbonyl compounds to sulfinic acids   444, 445
oxidation of thioketones to ketones   445, 446
oxidation of thiols and olefins with formation of hydroperoxides of
 sulfides   446, 447
sensitization in the oxidation of sulfur compounds   444, 445
oximation
oximation   262-265
oximes
isomerization of organic nitrites to hydroxy oximes see BARTON reaction
oximes see also 1,2-quinone oximes
isomerization of nitroso compounds to oximes   248, 252, 262, 263
isomerization of oximes to amides   48
isomerization of unsaturated nitro compounds to oximes of dicarbonyl
 compounds   266, 267
reduction of chloronitroso compounds to oximes   197
stereoisomerization of oximes, nitrones and formazans   63, 64

oxygen
addition of oxygen to 1,3-dienes with formation of epidioxides see
 peroxidation
elimination of one oxygen atom see deoxygenation
addition of oxygen see also hydroperoxidation and peroxidation
addition of carbenes to oxygen   291
addition of sulfur dioxide and chlorine or oxygen with formation of
 sulfonyl chlorides or sulfonic acids   434-439
cyclization of olefins to cyclic peroxides involving incorporation of
 oxygen   397, 398
dehydrocyclization with oxygen, iodine or quinones as hydrogen
 acceptors   126-136, 138, 141, 142

Reaction Index

dehydrodimerization with oxygen, quinones or dyes as hydrogen
  acceptors 146-149
DIELS-ALDER reaction with oxygen 382, 383, 428
elimination of oxygen from epidioxides 383, 385, 390, 391
oxygen compounds
nitrogen compounds, oxygen compounds and sulfur compounds see also
  heterocyclic compounds

PATERNO-BUECHI reaction
PATERNO-BUECHI reaction see also cycloaddition of carbonyl compounds to
  olefins
intramolecular PATERNO-BUECHI reaction 43, 417, 418
PATERNO-BUECHI reaction 414-420
peroxidation
peroxidation (addition of oxygen to 1,3-dienes with formation of
  epidioxides)
addition of oxygen *
hydroperoxidation and peroxidation of organotin compounds 466
peroxidation of cyclic 1,3-dienes 382-388, 394, 407, 408
peroxidation of furan derivatives 427-432
peroxidation of fused polycyclic hydrocarbons 389-394
sensitization in peroxidation reactions 47, 382-390, 408, 428-432, 466
peroxides
endoperoxides and cyclic peroxides see epidioxides
cyclization of olefins to cyclic peroxides involving incorporation of
  oxygen 397, 398
oxidation of aldehydes with formation of acyl peroxides 399, 400
oxidation of heterocyclic oxygen compounds with formation of peroxides 398, 399
oxidation of heterocyclic sulfur compounds with formation of peroxides 398, 399
phenols
isomerization of aryl esters with formation of acyl phenols see FRIES
  rearrangement
hydrolysis of phenols with formation of acids 239, 240
hydroperoxidation of phenols 382
photolysis of diazonium salts in inorganic acids with formation of organic
  halides or phenols 314, 315
photolysis of 1,4-quinone diazides in alcohols with formation of alkoxy
  phenols 310, 311
phosphates
oxidation of phosphites to phosphates 454, 455
phosphines
organophosphorus compounds *
addition of aromatic halides to phosphines with formation of phosphonium
  salts 455-457
addition of phosphines to C=C bonds 176
complexation with nitrogen compounds, nitroso compounds, sulfoxides,
  phosphines, arsines, acetylene compounds and olefins as electron
  donors 459-464, 469
photolysis of phosphites and phosphines 454, 455
phosphites
phosphites see also phosphonates
isomerization of phosphites to phosphonates 454
oxidation of phosphites to phosphates 454, 455
photolysis of phosphites and phosphines 454, 455
phosphonates
organophosphorus compounds *
phosphites *
addition of phosphonates and phosphonothioates to C=C bonds 177, 178

addition of phosphonates to 1,4-quinones   453, 454
isomerization of phosphites to phosphonates   454
phosphonium salts
addition of aromatic halides to phosphines with formation of phosphonium salts   455-457
photolysis of phosphonium salts   457
phosphonothioates
organophosphorus compounds *
addition of phosphonates and phosphonothioates to C=C bonds   177, 178
photochromism
photochromism   42, 410, 411
photolysis
photolysis of 1,2-quinone diazides involving ring contraction see SUES reaction
photolysis see also cleavage, dealkylation, elimination, fragmentation
photolysis of acyl azides in alcohols with formation of carbamates   321, 322, 325, 326
photolysis of acyl azides with formation of isocyanates   321-324
photolysis of aralkyl ketones   215-217
photolysis of arsines   457, 458
photolysis of azides with formation of imenes   318-327
photolysis of carbinols   199
photolysis of cyclic diazo ketones involving ring contraction   302-306, 309, 310
photolysis of cyclic ketones to acids   1, 26, 28, 225-230
photolysis of cyclooctatetraene derivatives   409, 410
photolysis of diaroyl azo compounds with formation of aromatic 1,2-diketones   330, 331
photolysis of diazirine derivatives   290, 291
photolysis of diazo compounds with formation of carbenes   274-291
photolysis of diazo ketones with formation of keto-carbenes   121, 294-311
photolysis of diazonium salts   313-315
photolysis of diazonium salts in inorganic acids with formation of organic halides or phenols   314, 315
photolysis of diazonium salts with formation of azo compounds   316
photolysis of 1,2-diketones   231
photolysis of disulfides   152, 153, 439-441
photolysis of esters   38
photolysis of nitrogen-halogen compounds followed by cyclization to heterocyclic nitrogen compounds   242-246
photolysis of nitrogen-halogen compounds with formation of lactones   370, 371
photolysis of nitrosyl chloride   262-264
photolysis of organic halides   128, 133-136, 354-359, 362-371
photolysis of organic hypohalites with formation of ethers   369, 370
photolysis of organic nitrites   248-254
photolysis of organolithium compounds   469
photolysis of organomercury compounds   369
photolysis of organomercury compounds in alcohols with formation of mercury and carbonyl compounds   467, 468
photolysis of organomercury compounds in organic halides   466, 467
photolysis of organotin compounds   465, 466
photolysis of oxadiazoline derivatives involving decarboxylation   337
photolysis of phosphites and phosphines   454, 455
photolysis of phosphonium salts   457
photolysis of pyrazole derivatives with formation of cyclopropene derivatives   281, 329
photolysis of pyrazoline derivatives with formation of cyclopropane derivatives and olefins   328, 329
photolysis of 1,4-quinone diazides in alcohols with formation of alkoxy phenols   310, 311
photolysis of 1,2-quinone diazides with formation of azo compounds   309, 310

photolysis of 1,4-quinone diimine dioxides with formation of azo
    compounds   332, 333
photolysis of sulfides   216, 217, 440, 441
photolysis of sulfonyl azides in alcohols, sulfoxides and sulfides   326, 327
photolysis of thiadiazine derivatives   316, 317
photolysis of thiadiazole derivatives with formation of thio carbenes   331, 332
photolysis of thioketones to sulfides   449
photolysis of α,β-unsaturated ketones in alcohols with formation of
    esters   227, 238
photolysis of unsaturated sultones in alcohols with formation of sulfonates   448
photolysis of xanthates   442
sensitization in photolysis reactions of diazo compounds   276
sensitization in the photolysis of azides   321, 325
pinacolization
pinacolization   198, 200, 203-211, 215, 339
pinacols
glycols *
addition of ketones to alcohols with formation of pinacols   203
cleavage of pinacols by ketones   206, 207, 212
cleavage of pinacols by 1,4-quinones   211, 212
dihydrodimerization of aldehydes with formation of pinacols   203, 204
dihydrodimerization of ketones with formation of pinacols   203-209
dihydrodimerization of 1,2,3-triketones with formation of pinacols   209
polymerization
polymerization   79, 153, 310, 427
pyran derivatives
heterocyclic oxygen compounds *
cycloaddition of 1,4-quinones to 1,3-dienes with formation of pyran
    derivatives   424
cyclodimerization of pyran derivatives   89
dimerization of pyran derivatives   102, 103
intramolecular cyclization with formation of pyran derivatives   41, 42
isomerization of cyclic epoxy ketones to pyran derivatives   409-411
isomerization of pyran derivatives   42, 46, 89
pyrazole derivatives
heterocyclic nitrogen compounds *
photolysis of pyrazole derivatives with formation of cyclopropene
    derivatives   281, 329
rearrangement of pyrimidine derivatives or pyrazole derivatives   54, 55
pyrazoline derivatives
addition of diazo compounds to C=C bonds with formation of pyrazoline
    derivatives   288, 292
photolysis of pyrazoline derivatives with formation of cyclopropane
    derivatives and olefins   328, 329
pyridine derivatives
heterocyclic nitrogen compounds *
dimerization of pyridine derivatives   99-101
pyridinium or tetrazolium salts
dehydrocyclization of aryl substituted pyridinium or tetrazolium salts to
    aromatic heterocyclic nitrogen compounds   142-144
pyrimidine derivatives
heterocyclic nitrogen compounds *
addition of water to pyrimidine derivatives   155, 156
cyclodimerization of pyrimidine derivatives   91
rearrangement of pyrimidine derivatives or pyrazole derivatives   54, 55
pyrylium compounds
valence tautomerization of epoxy ketones with formation of pyrylium
    compounds   410, 411

quinone diazides
cyclic diazo ketones *
1,2-quinone diazides
photolysis of 1,2-quinone diazides involving ring contraction see SUES reaction
photolysis of 1,2-quinone diazides with formation of azo compounds   309, 310
1,4-quinone diazides
photolysis of 1,4-quinone diazides in alcohols with formation of alkoxy
  phenols   310, 311
1,4-quinone diimine dioxides
photolysis of 1,4-quinone diimine dioxides with formation of azo
  compounds   332, 333
1,2-quinone imines
addition of 1,2-quinone imines to aldehydes with formation of amides   190, 191
1,2-quinone oximes
addition of 1,2-quinone oximes to aldehydes   191
quinones
dehydrocyclization with oxygen, iodine or quinones as hydrogen
  acceptors   126-136, 138, 141, 142
dehydrodimerization with oxygen, quinones or dyes as hydrogen
  acceptors   146-149
dehydrogenation of hydroaromatic compounds with quinones, disulfides or
  dyes as hydrogen acceptors   151-154
reduction of quinones by hydrogen donors   151, 152
1,2-quinones
1,2-dicarbonyl compounds *
addition of 1,2-quinones to aldehydes with formation of esters   186-188
addition of 1,2-quinones to aralkyl compounds with formation of
  alcohols   182-184
addition of 1,2-quinones to ethers   184
cycloaddition of sulfur dioxide to 1,2-quinones   439
1,4-quinones
addition of phosphonates to 1,4-quinones   453, 454
addition of 1,4-quinones to aldehydes with formation of esters   189
addition of 1,4-quinones to aldehydes with formation of ketones   188, 189
addition of 1,4-quinones to hydroaromatic compounds or aralkyl compounds
  with formation of ethers   151, 182
cleavage of pinacols by 1,4-quinones   211, 212
cycloaddition of 1,4-quinones to 1,3-dienes with formation of pyran
  derivatives   424
cycloaddition of 1,4-quinones to olefins   418, 419
cycloaddition of 1,4-quinones to olefins with formation of oxetane
  derivatives   418, 419
cyclodimerization of 1,4-quinones   79-81
cyclodimerization of 1,4-quinones with formation of oxetane derivatives   79, 80

racemization
racemization   66, 67
rearrangement
isomerization *
rearrangement see also VON AUWERS rearrangement, FRIES rearrangement,
  CURTIUS rearrangement
addition of water following rearrangement of cross-conjugated
  cyclohexadienones   30-32
allylic rearrangement of esters   234, 235
rearrangement of enol esters   37, 38
rearrangement of keto-carbenes to ketenes   121, 294-299, 301-306, 309, 310
rearrangement of pyrimidine derivatives or pyrazole derivatives   54, 55

reduction
dimerization involving reduction see dihydrodimerization
reduction of acyl azides with hydrogen donors involving formation of amides 321-325
reduction of aromatic heterocyclic nitrogen compounds 195, 196
reduction of azides to amines 319, 320
reduction of carbenes with hydrogen donors 287, 300
reduction of C=C bonds with hydrogen donors 193, 194
reduction of chloronitroso compounds to oximes 197
reduction of C=N bonds 257
reduction of C=N bonds with hydrogen donors 195, 196
reduction of C=O bonds with hydrogen donors involving formation of alcohols 113, 194, 195, 204, 208
reduction of cyclopropane derivatives 65
reduction of diazonium salts with alcohols as hydrogen donors 313, 314
reduction of nitro compounds to amines 270, 271
reduction of quinones by hydrogen donors 151, 152
reduction with alcohols, ethers or hydrocarbons as hydrogen donors 193-197
sensitization in reduction reactions 65, 193

replacement
replacement (substitution)
replacement of hydrogen by the cyano group see cyanation
replacement of hydrogen by the hydroperoxy group see hydroperoxidation
replacement of hydrogen by the nitroso group see nitrosation
replacement of hydrogen with formation of sulfonyl chlorides see sulfochlorination
replacement of hydrogen with formation of thiocyanates see thiocyanation
replacement of substituents by halogens see halogenolysis
replacement of acyloxy groups in esters by hydrogen 235
replacement of carbon monoxide in metal carbonyl compounds by electron donors with formation of organometallic carbonyl compounds 459-462, 469
replacement of carbon monoxide in organometallic carbonyl compounds by electron donors 462-464
replacement of chlorine in nitrogen-halogen compounds by hydrogen 242
replacement of halogens by hydrogen 362-364
replacement of hydrogen by the carboxamide group 230, 231
replacement of hydrogen by the chloro-formyl group with formation of acid chlorides 233, 234
replacement of hydrogen with formation of sulfenyl chlorides 437, 438

retro-cycloaddition
retro-cycloaddition (reversal of cycloaddition)
fragmentation of cyclobutane derivatives *
retro-cycloaddition 35, 112, 116, 117, 141, 142, 219, 416, 421, 423, 449

ring cleavage
ring cleavage see cleavage

ring contraction
photolysis of 1,2-quinone diazides involving ring contraction see SUES reaction
decarbonylation of cyclic ketones involving ring contraction 218, 219, 221
photolysis of cyclic diazo ketones involving ring contraction 302-306, 309, 310
ring contraction 447

ring enlargement
cycloaddition of carbenes to aromatic compounds involving ring enlargement 276-279
ring enlargement 325

SCHIFF bases
C=N bonds *
addition of alcohols to SCHIFF bases with formation of oxazolidine derivatives   335-337
cyclization of SCHIFF bases to heterocyclic nitrogen compounds involving incorporation of alcohols   334, 335
cyclodimerization of SCHIFF bases   141, 142
dehydrocyclization of aryl substituted azo compounds or SCHIFF bases to aromatic heterocyclic nitrogen compounds   138-142, 337, 338
sensitization
sensitization see also individual sensitizers in the sensitizer index
sensitization in addition reactions of alcohols to unsaturated acids   231-233
sensitization in addition reactions with olefins and acetylene compounds   155, 160-163, 166-168, 176, 177
sensitization in amidation reactions   166-168, 230, 231
sensitization in cyclization reactions of nitrogen compounds with alcohols   335, 336
sensitization in cycloaddition reactions   110, 111, 115
sensitization in cyclodimerization reactions   23, 72-76, 82, 83, 86, 468
sensitization in dealkylation reactions   256, 257
sensitization in dehydrodimerization reactions   147-149
sensitization in dehydrogenation reactions   153, 154
sensitization in hydroperoxidation reactions   375-382, 466
sensitization in intramolecular cycloaddition reactions   2-4
sensitization in isomerization of cyclic acetals to esters   45, 46
sensitization in isomerization reactions   18, 30
sensitization in peroxidation reactions   47, 382-390, 408, 428-432, 466
sensitization in photolysis reactions of diazo compounds   276
sensitization in reduction reactions   65, 193
sensitization in stereoisomerization reactions   56-58, 63, 65
sensitization in the oxidation of sulfur compounds   444, 445
sensitization in the photolysis of azides   321, 325
sodium lamps
sodium lamps as light sources   473-475, 480
stereoisomerization
isomerization *
stereoisomerization see also cis-trans isomerization of unsaturated compounds
stereoisomerization see also epimerization
sensitization in stereoisomerization reactions   56-58, 63, 65
stereoisomerization   22-24, 70, 127, 141, 420
stereoisomerization at C=C bonds   56-62
stereoisomerization at C=N bonds   63, 64
stereoisomerization at N=N bonds   62-64, 252
stereoisomerization of azo compounds   62-64
stereoisomerization of cyclobutane derivatives   65
stereoisomerization of cyclopropane derivatives   64, 65
stereoisomerization of oximes, nitrones and formazans   63, 64
stereoisomerization of α,β-unsaturated carbonyl compounds   58-62
substitution
substitution see replacement
SUES reaction
SUES reaction (photolysis of 1,2-quinone diazides involving ring contraction)
SUES reaction   304-306, 309, 310
sulfenyl chlorides
addition of sulfenyl chlorides or sulfonyl chlorides to olefins   164, 165
halogenation of disulfides with formation of sulfenyl chlorides   441
hydrolysis of nitro sulfenyl chlorides to amino sulfonic acids   270, 271
replacement of hydrogen with formation of sulfenyl chlorides   437, 438

sulfides
sulfur compounds *
addition of sulfur compounds to C=C bonds with formation of sulfides and
    thiols   162-164
desulfuration of disulfides to sulfides   439, 440
oxidation of thiols and olefins with formation of hydroperoxides of
    sulfides   446, 447
photolysis of sulfides   216, 217, 440, 441
photolysis of sulfonyl azides in alcohols, sulfoxides and sulfides
    326, 327
photolysis of thioketones to sulfides   449
sulfinic acids
oxidation of thiocarbonyl compounds to sulfinic acids   444, 445
sulfochlorination
sulfochlorination (replacement of hydrogen with formation of sulfonyl
    chlorides)
sulfochlorination   434-437
sulfones
elimination of sulfur dioxide from cyclic sulfones   447, 448
oxidation of sulfoxides to sulfones   444
sulfonic acids
addition of sulfur dioxide and chlorine or oxygen with formation of
    sulfonyl chlorides or sulfonic acids   434-439
hydrolysis of nitro sulfenyl chlorides to amino sulfonic acids   270, 271
sulfonyl chlorides
replacement of hydrogen with formation of sulfonyl chlorides see
    sulfochlorination
addition of sulfenyl chlorides or sulfonyl chlorides to olefins   164, 165
addition of sulfur dioxide and chlorine or oxygen with formation of
    sulfonyl chlorides or sulfonic acids   434-439
halogenolysis of sulfonyl chlorides   346, 347, 350
sulfoxidation
sulfoxidation   438, 439
sulfoxides
complexation with nitrogen compounds, nitroso compounds, sulfoxides,
    phosphines, arsines, acetylene compounds and olefins as electron
    donors   459-464, 469
oxidation of sulfoxides to sulfones   444
photolysis of sulfonyl azides in alcohols, sulfoxides and sulfides   326, 327
sulfur
elimination of sulfur see desulfuration
sulfur compounds
nitrogen compounds, oxygen compounds and sulfur compounds see also
    heterocyclic compounds
sulfur compounds see also sulfides, disulfides and xanthates
sulfur compounds see also unsaturated sulfur compounds
addition of sulfur compounds to C=C bonds with formation of sulfides and
    thiols   162-164
sensitization in the oxidation of sulfur compounds   444, 445
sulfur dioxide
addition of sulfur dioxide and chlorine or oxygen with formation of
    sulfonyl chlorides or sulfonic acids   434-439
cycloaddition of sulfur dioxide to 1,2-quinones   439
elimination of sulfur dioxide from cyclic sulfones   447, 448
sultones
photolysis of unsaturated sultones in alcohols with formation of sulfonates   448

thiadiazine derivatives
photolysis of thiadiazine derivatives   316, 317
thiadiazole derivatives
photolysis of thiadiazole derivatives with formation of thio carbenes   331, 332
thietane derivatives
cycloaddition of thioketones to C=C bonds with formation of thietane derivatives   448, 449
thiirane derivatives
desulfuration of thiirane derivatives and dithietane derivatives with formation of olefins   449, 450
thiocarbonyl compounds
thioketones *
cyclodimerization of thiocarbonyl compounds to dithietane derivatives   106, 107, 449, 450
oxidation of thiocarbonyl compounds to sulfinic acids   444, 445
thiocarboxylic acids
addition of hydrogen sulfide, thiols and thiocarboxylic acids to C=C bonds   162-164
addition of ketenes to thiols with formation of esters of thiocarboxylic acids   297, 298
thiocyanates
replacement of hydrogen with formation of thiocyanates see thiocyanation
thiocyanation
thiocyanation (replacement of hydrogen with formation of thiocyanates)
thiocyanation of aralkyl compounds   443, 444
thioketones
thioketones see also thiocarbonyl compounds
cycloaddition of thioketones to C=C bonds with formation of thietane derivatives   448, 449
oxidation of thioketones to ketones   445, 446
photolysis of thioketones to sulfides   449
thiols
addition of hydrogen sulfide, thiols and thiocarboxylic acids to C=C bonds   162-164
addition of ketenes to thiols with formation of esters of thiocarboxylic acids   297, 298
addition of sulfur compounds to C=C bonds with formation of sulfides and thiols   162-164
oxidation of thiols and olefins with formation of hydroperoxides of sulfides   446, 447
thiophene derivatives
cycloaddition of carbenes to furan derivatives or thiophene derivatives   278, 279
isomerization of aryl substituted thiophene derivatives   451
1,3,5-trienes
isomerization of cyclic 1,3-dienes to 1,3,5-trienes   14-17
isomerization of 1,3,5-trienes to cyclic 1,3-dienes   18-20, 127-133, 139, 141, 142
valence tautomerization of 1,3,5-trienes with formation of cyclopropane derivatives   15-17, 19
1,2,3-triketones
1,2-dicarbonyl compounds *
dihydrodimerization of 1,2,3-triketones with formation of pinacols   209

unsaturated acids
acids *
addition of alcohols to unsaturated acids followed by dehydration to lactones   231-233
hydrolysis of unsaturated ketones to unsaturated acids   227-230, 238, 239
sensitization in addition reactions of alcohols to unsaturated acids   231-233

Reaction Index 557

α,β-unsaturated acids
cyclodimerization of α,β-unsaturated acids   23, 82-85
unsaturated aldehydes
aldehydes *
isomerization of saturated cyclic ketones to unsaturated aldehydes   26, 27
α,β-unsaturated carbonyl compounds
cyclization of α,β-unsaturated carbonyl compounds   27, 28
cycloaddition of α,β-unsaturated carbonyl compounds to acetylene compounds with formation of cyclobutene derivatives   115
cycloaddition of α,β-unsaturated carbonyl compounds to aromatic compounds   109, 110, 114
cycloaddition of α,β-unsaturated carbonyl compounds to olefins   110-113
isomerization of α,β-unsaturated carbonyl compounds to β,γ-unsaturated carbonyl compounds   22-25
stereoisomerization of α,β-unsaturated carbonyl compounds   58-62
β,γ-unsaturated carbonyl compounds
isomerization of α,β-unsaturated carbonyl compounds to β,γ-unsaturated carbonyl compounds   22-25
unsaturated compounds
cyclization of unsaturated compounds involving dehydrogenation see dehydrocyclization
stereoisomerization *
unsaturated compounds see also olefins, C=C bonds and acetylene compounds
cycloaddition of 1,2-dicarbonyl compounds to unsaturated compounds with formation of dioxin derivatives   118-124, 419
cycloaddition of 1,2-dicarbonyl compounds to unsaturated compounds with formation of oxetane derivatives   121, 122, 124, 419, 420
cycloaddition of unsaturated compounds with formation of cyclobutane derivatives   109-117
cyclodimerization of unsaturated compounds with formation of cyclobutane derivatives   70-92
isomerization of unsaturated compounds   1-69
unsaturated ketones
ketones *
hydrolysis of unsaturated ketones to unsaturated acids   227-230, 238, 239
intramolecular cycloaddition of unsaturated ketones to oxetane derivatives   417, 418
isomerization of unsaturated ketones by intramolecular cyclization to cyclohexenols   37
α,β-unsaturated ketones
cyclodimerization of α,β-unsaturated ketones   75-81, 89, 91, 92
isomerization of keto-carbenes to α,β-unsaturated ketones   294, 295
isomerization of α,β-unsaturated ketones to unsaturated lactones   239
oxidation of cyclopentadienones with formation of α,β-unsaturated ketones   394, 395, 397
photolysis of α,β-unsaturated ketones in alcohols with formation of esters   227, 238
valence tautomerization of cyclic α,β-unsaturated ketones   23
unsaturated lactones
lactones *
cyclodimerization of unsaturated lactones   85-88
isomerization of cyclic ketones to unsaturated lactones   219
isomerization of α,β-unsaturated ketones to unsaturated lactones   239
oxidation of furan derivatives to unsaturated lactones   239
unsaturated sulfur compounds
sulfur compounds *
cyclodimerization of unsaturated sulfur compounds   90, 91

valence tautomerization
isomerization *
valence tautomerization of aliphatic 1,3-dienes   10
valence tautomerization of aromatic compounds   10
valence tautomerization of cyclic 1,3-dienes with formation of cyclobutene derivatives   4-9, 27, 28, 227
valence tautomerization of cyclic α,β-unsaturated ketones   23
valence tautomerization of epoxy ketones with formation of pyrylium compounds   410, 411
valence tautomerization of furan derivatives   44
valence tautomerization of 1,3,5-trienes with formation of cyclopropane derivatives   15-17, 19
VON AUWERS rearrangement
VON AUWERS rearrangement   40

water
hydrolysis *
addition of water following rearrangement of cross-conjugated cyclohexadienones   30-32
addition of water to C=C bonds with formation of alcohols   155, 156
addition of water to ketenes with formation of acids   227, 229, 295-297, 301-306, 309, 310
addition of water to keto-carbenes with formation of lactones   306-308
addition of water to pyrimidine derivatives   155, 156

xanthates
sulfur compounds *
decarbonylation of acyl xanthates   223, 224
photolysis of xanthates   442
xenon lamps
xenon lamps as light sources   473, 477, 486, 487

Sensitizer Index

1'-acetonaphthone as sensitizer 57
2'-acetonaphthone as sensitizer 75
acetone as sensitizer 3,45,58,74,160-163,166-168,176-178,230,231,335,336,468
acetophenone as sensitizer 57,58,447,468
benzil as sensitizer 110
benzophenone as sensitizer 23,30,57,65,75,82,83,110,111,115,168,193,232,233,276,321,335,379,380,447
benzophenone, 4,4'-bis-(dimethylamino)-, as sensitizer 74
benzo[a]pyrene as sensitizer 382
biacetyl as sensitizer 57
bromine as sensitizer 56-58
2,3-butanedione as sensitizer 57
β-carotene as sensitizer 86
chlorine as sensitizer 234
chlorophyll as sensitizer 381,383
cyclododecanone as sensitizer 58

8H,16H-dibenzo[a,j]perylene-8,16-dione as sensitizer 397,408
dinaphtho[2,3-b:2',3'-d]thiophene as sensitizer 429
eosin as sensitizer 63,388,397,401,429
eosin blue as sensitizer 388
eosin sodium salt as sensitizer 385
erythrosin B as sensitizer 153,154,387,388
fluoren-9-one as sensitizer 57
hematoporphyrin as sensitizer 377,378
iodine as sensitizer 59-61,63
mercury as sensitizer 18
methylene blue as sensitizer 378,381,384,397,401,431
protoporphyrin methyl ester as sensitizer 378
riboflavine as sensitizer 256,257
rose bengal as sensitizer 377,378,384,389,430,444,445,466
rubrene as sensitizer 381,408
triphenylene as sensitizer 58
uranyl chloride as sensitizer 76

Compound Index

10(5→4)-abeo-androstane-3,5-dione, 17β-acetoxy- 412
———, 17β-acetoxy-4-methyl- 412
10(5→4)-abeo-androst-1-ene-3,5-dione, 17β-acetoxy- 412,413
abietic acid, 6,14-epidioxy-18-hydroperoxy-9,14-dihydro- 387-389
acenaphthene 199,201
acenaphthene, 1-benzoyl- 199
1,2-acenaphthenediol, 1,2-diphenyl- 211
acenaphthenequinone 187
14,15-acenaphtheno[12](2,5)cyclopentadienophan-17-one 395,397
acenaphth[1,2-c][1,2]oxathiin 7,7-dioxide, 9,10-diphenyl- 448
acenaphthylene 71-73
acenaphthylene, 1,2-diisobutyryl- 395
2-acenaphthylenesulfonate, methyl 1-(α-benzoyl-α-phenylmethyl)- 448
[14](1,2)acenaphthylenophane-1,14-dione 395,397
acetaldehyde 169,186,188-191,300,321,322,334,365,415-417,421,422,467
acetaldehyde, methoxy- 279,280
———, trifluoro- 422
acetals 45
acetamide, N,N-difluoro- 231
———, phenyl- 231
———, m-tolyl- 231
———, o-tolyl- 231
———, p-tolyl- 231
acetanilide, N-methyl-2-phenyl- 297,298
acetate, cis-2-chlorocyclohexyl thio- 163
———, ethyl diazo- 279
———, ethyl diphenyl- 296
———, ethyl phenyl- 45
———, O-ethyl thio- 450
———, ethyl tribromo- 175,176
———, methyl (benzo[a]pyren-6-ylthio)- 442
———, methyl benzoyldiazo- 294
———, methyl chloro- 283
———, methyl diazo- 275,278,281-283
———, methyl dibromo- 283
———, methyl phenyl- 322
———, methyl trichloro- 282
———, cis-2-methylcyclohexyl thio- 163
———, trans-2-methylcyclohexyl thio- 163
———, 2-methyl-3-pentyl thio- 164
———, vinyl 461
acetic acid 286,287,436,438
acetic acid, benzoyl- 338
———, diphenyl- 296
———, mercapto- 442
———, phenyl- 200
———, thio- 163,164,442
acetic acid γ-lactone, (2,5-dihydroxyphenyl)-diphenyl- 29
acetic anhydride 132,134,399,400,438
acetoacetate, ethyl 2-diazo-4,4,4-trifluoro- 300
———, ethyl 4,4,4-trifluoro- 300
1'-acetonaphthone 415,421
2'-acetonaphthone 415
acetone 169,170,196,198,199,201,203,204,206,207,212,252,279,300,307,308,420,421,467
acetone, acetyl- 37,38
———, benzylidene- 295
———, diazo- 298,300
———, dibenzylidene- 76,77
———, 1,3-dichlorotetrafluoro- 422,423
———, hexafluoro- 421,422
———, methoxy- 279,280
acetone p-bromophenylhydrazone 381

Compound Index

acetonitrile 307—309,459
acetonitrile, (2-methoxyethoxy)- 260,261
acetophenone 65,205,414,415,423
acetophenone, 2-(1-acenaphthenyl)-2-hydroxy-2-phenyl- 199,201
——, benzylidene-2'-nitro- 271,272
——, 2-chloro-2-phenyl- 216,217
——, 2-diazo- 297
——, 2,3-dihydroxy- 236
——, 2',5'-dihydroxy- 189
——, 3,4-dihydroxy- 236
——, 2,2-diphenyl- 216,217
——, 2-methyl- 24,25
——, 2-(4-oxo-2,5-cyclohexadienylidene)-2-phenyl- 423,424
——, 2-phenyl-2-(phenylthio)- 216,217
acetophenone hydrochloride, 2-dimethylamino-2-phenyl- 216,217
acetyl azide, phenyl- 321
acetyl chloride, trichloro- 407
——, trifluoro- 234
acetyl fluoride, chlorodifluorothio- 106
——, trifluorothio- 106
acetylacetone 113
acetylacetone see also acetone, acetyl-, 37,38
acetylene 114,115,163,174,461
acetylene, (2,4-dinitrophenyl)-phenyl- 271,272
——, diphenyl- 92—94,115,116,122,220,221,409,410,423,463
——, (2-nitrophenyl)-phenyl- 271,272
——, phenyl- 92,93,163,172
acetylene compounds 342
acetylene derivatives 459
acetylenedicarboxylate, diethyl 167
——, dimethyl 24,25,41
acetylenedicarboxylates, dialkyl 167
acetylenedicarboxylic acid 232
acetylenes 423
acids, γ-hydroxy 231
——, α-keto- 358
——, α,β-unsaturated 82—85
acridan 195,196

acridan, 9-(1-hydroxyethyl)- 195,196
——, 10-methyl- 195,196
——, 10-methyl-9-(1-hydroxyethyl)- 195,196
9-acridanthione, 10-phenyl- 445
acridine 101,195,196
9-acridinecarboxanilide 48,49
acridinium chloride, 10-methyl- 195,196
acrylates 163
acrylic acid, cis-3-benzoyl-2-methyl- 23,24,57
——, trans-3-benzoyl-2-methyl- 23,24,57
acrylic acid derivatives 459
acrylic acids, β-aroyl- 23,24
acrylonitrile 173,174,31,461
acrylophenone, 2,3-diphenyl- 119,120
actinomycin C2 256,257
actinomycin C2, N-methyl- 256,257
actinomycins 256,257
acyl derivatives, organometallic 464
acyl hypoiodites 370
adamantane 264
2-adamantanone oxime 264
adduct, 10-methyleneanthrone-oxygen 2:1- 398
adenine 52
adenine 1-oxide 52
adipamide 166,168
adipic acid 370
adipic acid diazide 321
alcohols 159,160,294—297,300,301,310,311,313,314
aldehydes 169,201,202,203,399,400,414—417,421—423
aldehydes, nitroso- 249,252
aldehydes, α-keto-, dialkyl acetals 358
aldosterone 21-acetate 250
aldoximes, aryl- 48
aldrin 447
C-alkaloid-A 401,402
C-alkaloid-E 402
alkaloids, ergot 156,157
alkanes, chloronitroso- 261—263
alkanesulfonyl chloride, amino- 435
alkanethiols 454
2-alkanones 34,35
alkyl hypoiodites 369,370

Compound Index

allantoin, 1,3-dimethyl- 381,382
allenes 23
alloocimene 19
alloxan 209
alloxantin 209
alloxazine 258
alloxazine, 7,8-dimethyl- 257,258
allyl alcohol 157,163,176
allyl chloride 342
amides 48,321,322,324
amides, N,N-difluoro- 231
——, N-iodo- 370,371
amines 165,459,460
amines, N-halogenated 242-247
amino acids, N-(2,4-dinitrophenyl)- 268
ammonia 165
β-amyrone 226
androsta-3,5-diene, 3,17β-diacetoxy- 38
androsta-1,4-diene-3,11,20-trione, 21-acetoxy-17α-hydroxy- 31,32
androsta-1,4-dien-3-one, 17β-acetoxy- 31,235
——, 17β-acetoxy-2,4-dimethyl- 31
——, 17β-acetoxy-2-methyl- 31,32
——, 17β-hydroxy-4-methyl- 31,32
5α-androstan-17β-ol nitrite, 3α-acetoxy- 253
5α-androstan-3-one, 17β-acetoxy- 194
——, 17β-acetoxy-4α,5-epoxy- 412
——, 17β-acetoxy-4α,5-epoxy-4β-methyl- 412
5α-androstan-17-one, 3β-acetoxy-5-chloro-6β-hydroxy- 369,370
——, 3β-acetoxy-5-chloro-6β,19-oxido- 369,370
——, 3α-hydroxy- 65,66
5α,13α-androstan-17-one, 3α-hydroxy- 65,66
5β-androstan-3-one, 17β-acetoxy-4β,5-epoxy- 412
——, 17β-acetoxy-4β,5-epoxy-4α-methyl- 412
androstan-17-one, 16-diazo-3β-hydroxy- 302,303
androst-5-en-19-al, 3,3:17,17-bis-(ethylenedioxy)- 222,223
androst-5-en-19-al-19-d, 3,3:17,17-bis-(ethylenedioxy)- 222
androst-5-ene, 3,3-ethylenedioxy-17β-formyl- 253
androst-5-ene-16α-carboxylic acid, 3β-hydroxy- 302

androst-5-ene-16β-carboxylic acid, 3β-hydroxy- 302
androst-4-ene-3,17-dione 253,254
androst-4-en-3-one 77,79
androst-4-en-3-one, 17β-acetoxy- 194
——, 17β-acetoxy-6β-acetyl- 38
——, 17β-hydroxy- 23
androst-5-en-3-one, 17β-acetoxy-4-acetyl- 38
androst-5-en-17-one, 16-diazo-3β-hydroxy- 302
——, 3β-methoxy- 226
——, 3β-methoxy-16,16-dimethyl- 226
5α-androst-1-en-3-one, 17β-acetoxy-4α,5-epoxy- 412
5β-androst-1-en-3-one, 17β-acetoxy-4β,5-epoxy- 412, 413
10α-androst-4-en-3-one, 17β-hydroxy- 22-24
10α-androst-5-en-3-one, 17β-hydroxy- 22-24
androst-4-en-3-one dimer 77,79
androsterone 65,66
angelate, methyl 328
angelic acid 57
angelicin dimer 87
angelicin, 5-methoxy-, see isobergapten
anilides 48-50
aniline 165,166,228,229,322,338,460,462
——, N-benzylidene- 300
——, N,N-benzylidene- 141,142
——, N-butyl- 319
——, 2-chloro-N,N-dimethyl- 314
——, N,N-dibenzoyl- 49,50
——, N-(4-dimethylaminobenzylidene)- 141,142
——, N,N-dimethyl-4-nitroso- 49
——, N-diphenylmethylene- 141,299
——, N-(diphenylmethylene)- 319,320
——, 4-iodo-N,N-dimethyl- 456,457
——, N-methyl- 297,298,315
——, N-(2-nitrobenzylidene)- 267,270
——, 4-nitro-2-nitroso- 268,270
anisaldehyde 204,421
p-anisaldehyde 187,188,190,191,338
anisaldehyde azine 339
p-anisil 330
anisole 184,313,368

Compound Index

anisole, 4-azido- 321
——, 4-iodo- 456
anthracene 97-99,152,389
anthracene, 9-amino- 99
——, 9-cyano- 98
——, 9,10-dihydro- 149,152
——, 9,10-dihydroperoxy-9,10-diphenyl- 389
——, 1,4-dimethoxy-9,10-diphenyl- 391,393
——, 1,4-dimethoxy-9,10-di-(2-pyridyl)- 390
——, 9,10-dimethyl- 98
——, 9,10-diphenyl- 98,393
——, 9,10-epidioxy-9,10-dihydro- 389,391
——, 9,10-epidioxy-9,10-dihydro-9-methyl- 389,391
——, 9,10-epidioxy-9,10-dimethoxy- 389,391
——, 1,4-epidioxy-1,4-dimethoxy-9,10-diphenyl- 391,393
——, 9,10-epidioxy-9,10-diphenyl- 389,393
——, 9,10-epidioxy-2-methoxy-9,10-diphenyl- 391
——, 9-nitro- 266,267
——, 9-phenyl- 98
anthracene dimer 97-99
anthracene dimer, 9-amino- 99
——, bromo- 98
——, chloro- 98
——, 9-cyano- 98
——, ethyl- 98
——, methyl- 98
——, 9-nitro- 266
anthracene peroxide 389,391
9,10[9',10']anthracenoanthracene, 9,10,11,16-tetrahydro- 97-99
[2+2](9,10)anthracenophane 11
[2+2](9,10)anthracenophane photoisomer 11
9-anthraldehyde 98,99,187,415,416
9-anthraldehyde dimer 98,99
9-anthranol 98
anthraquinone 148,266,398,401
anthraquinone, 1-selenocyanato- 353
anthraquinone oxime 266,267
anthraquinone-1-selenenyl bromide 353
9-anthroic acid 98
9-anthroic acid dimer 98

9-anthrol, 4a,10-dihydro- 25
——, 9,10-dihydro-9-phenyl-10-(thioxanthen-9-ylidene)- 132,133
anthrone, 10-benzylidene- 71,119,120
——, 10-(fluoren-9-ylidene)- 401
——, 10-methylene- 119,120,398
——, 10-thioxanthen-9-ylidene- 132,133
——, 10-xanthen-9-ylidene- 132-134
9-anthrone 25,146,147
9-anthrone, 10-diphenylmethylene- 132,134
aromadendrene 152,153
arsenic 458
arsine, bis-(trifluoromethyl)-methyl- 457,458
——, triphenyl- 462,463
——, tris-(trifluoromethyl)- 457,458
aryl diazonium salts 313-316
aryl diazosulfonates 316
ascaridole 46,47,383,384
atisine 323
7-azabicyclo[2·2·1]heptane, 7-methyl- 244
7-azabicyclo[4·1·0]heptane, 7-carboethoxy 324,325
7-azabicyclo[4·1·0]heptane-7-carboxylate, ethyl 324,325
2-azabicyclo[3·2·0]hept-6-en-3-one, 1,2,4,6-tetramethyl- 6,7
——, 1,4,6-trimethyl- 6,7,9
2-azabicyclo[3·2·1]octan-3-one, d,1-2-hydroxy-1,8,8-trimethyl- 252,253
17α-aza-D-homo-5α-androstan-17-one, 3α-acetoxy-17a-hydroxy- 253
6-azaindole 306
6-azaindole-3-carboxylic acid 305,306
7-aza-4(3H)-quinolone, 3-diazo- 305,306
1-azatricyclo[4·2·1·0³,⁷]nonane, 3,7-dimethyl- 244
3-azatricyclo[3·3·1³,⁶·0]nonane, 1,5-dimethyl- 244
azenes 318
1H-azepine-1-carboxylate, ethyl 325
1H-azepin-2-one, 2,3-dihydro-1,3,5,7-tetramethyl- 6,7
——, 2,3-dihydro-3,5,7-trimethyl- 6,7,9
2H-azepin-2-one, hexahydro- 262,263
2-azetidinone, 3-methyl-1,4-diphenyl- 300
——, 1,3,3,4,4-pentaphenyl- 299,300

Compound Index

2-azetidinones 299
azibenzil 121,294–296,298,299
azides 318–327
azides, acyl 321–324
-----, alkyl 319
-----, aryl 320,321
-----, sulfonyl 326,327
-----, triarylmethyl 319
-----, triterpenoid acyl 324
azines 339
aziridinone, 1,3-triphenyl- 280
2H-azirine, 3-phenyl- 318
azo compounds 309,310,321,332,333
azo compounds, diaroyl 330,331
azo dyes 309
azobenzene 138,140–142,298,299,332,333
azobenzene, 3-amino- 139,140
-----, 4-amino- 139,140
-----, 4-bromo-2-hydroxy- 53
-----, 4,4'-dimethoxy- 332
-----, 4,4'-diphenyl- 321
-----, 2-hydroxy- 53
-----, 4-hydroxy- 53
-----, 4-methoxy- 332
-----, 2-methyl- 139
-----, 3-methyl- 139
-----, 4-methyl- 139,337,338
-----, 2,4,6-trimethyl- 139,140
cis-azobenzene 62,63
cis-azobenzene, 4-methyl- 62
trans-azobenzene 62,63,139,141,142
azobenzene-4-carboxylic acid 139,140
2,2'-azobenzenedicarboxylic acid 107,108
azobenzenes, acetyl- 140
-----, chloro- 140
-----, iodo- 140
-----, nitro- 140
azodi-p-anisoyl 330
azodibenzoyl 298,330
azodibenzoyl, 2,2'-dichloro- 330
-----, 4,4'-dichloro- 330,331
azodicarboxylate, diethyl 298,299

azomethines 299
1,1'-azonaphthalene 140
cis-1,2'-azonaphthalene 62,63
2,2'-azonaphthalene 140
2,2'-azonaphthalene, 1-hydroxy- 53
cis-2,2'-azonaphthalene 62,63
cis-2,2'-azopyridine 62
α,α'-azotoluene 298,299
azoxy compounds 53
azoxybenzene 53
O,N-azoxybenzene, 4-bromo- 53
2,2'-azoxynaphthalene 53
azulene 277
azulene, 4,8-dimethyl- 277
-----, 1-phenyl- 93,94
-----, 1,2,3-triphenyl- 93,94,465,466
2-azulenecarboxylic acid, 2,4,5,6,7,8-hexahydro- 304,305

benz[a]acridine, 12-methyl- 49
benz[b]acridine 101,102
benz[b]acridine dimer 101,102
benz[a]acridine-12-carboxanilide, 4'-dimethylamino- 49
benzaldehyde 186,187,189–191,204,215,216,338,399,400,414,415,417,423
benzaldehyde, p-chloro- 187,190
-----, 2-chloro- 190
-----, 4-chloro- 400
-----, 4-iodo- 368
-----, 4-methoxy- 421
-----, 2-nitro- 267–269
-----, 3-nitro- 267
-----, 4-nitro- 201,202,267
-----, 2-nitroso- 107,108
-----, 4-phenyl- 368
benzaldehyde acetals, 2-nitro- 268,269
benzaldehyde azine 339
benzaldehyde azine, 2-chloro- 339
-----, 4-chloro- 339
-----, 4-dimethylamino- 339
-----, 4-nitro- 339

Compound Index

benzaldehyde O-benzyloxime, 4-nitro- 63
benzaldehyde diethyl acetal, 2-nitro- 268,269
benzaldehyde diethyl acetal, 4-nitro- 269
benzaldehyde hydrazone, 2-(2-phenylethyl)- 279
benzaldehyde O-methyloxime, 3-nitro- 63
——, 4-nitro- 63
anti-benzaldehyde oxime, 4-nitro- 63,64
syn-benzaldehyde oxime 48
syn-benzaldehyde oxime, 4-nitro- 63,64
benzaldehyde p-toluenesulfonylhydrazone 289
benzamide 48,231,321,322
benzamidine, N,N'-dibenzoyl- 396
benzanilide 49,322
benzanilide, 2-nitroso- 267,270
benz[a]anthracene 99
benz[a]anthracene, 7,12-epidioxy-7,12-diphenyl- 393
benz[a]anthracene dimer 99
benzene 109,110,114-117,116,117,231,276-279,285,286,
290,311,319,325,341-345,368,369,455,467,468
benzene, 1-arsenoso-2-nitro- 268,270
——, 1,2-bis-(dimethylarsino)- 462,463
——, bromo- 354,363
——, 1-bromo-2-chloro- 363
——, 1-bromo-3-chloro- 363
——, 1-bromo-4-chloro- 363
——, 1-bromo-4-nitro- 363
——, (1-bromo-3,3,3-trichloro-2-methylpropyl)- 171,
172
——, tert•butyl- 435
——, chloro- 110,314,344,346,354,365
——, (1-chloroethyl)- 355,356
——, o-dibenzoyl- 205
——, 1,2-dibenzoyl- 396,430
——, 1,4-dibromo- 363
——, (1,2-dibromo-2-chloroethyl)- 350
——, 1,2-dichloro- 344,363
——, 1,3-dichloro- 363
——, 1,4-dichloro- 346,347,363
——, 1,2-diiodo- 368,369
——, 1,4-diiodo- 370
——, 2,4-dinitro-1-selenocyanato- 353
——, ethyl- 355,356,443
——, (2-ethylbutyl)- 284
——, hexachloro- 347,348,355
——, hexaphenyl- 93
——, iodo- 354,365,368,370,456,457
——, 1-iodo-4-nitro- 368
——, isopropyl- 443
——, (2-methylpentyl)- 284
——, nitro- 131,133
——, pentaphenyl- 152
——, (1,3,3-tetrabromopropyl)- 171,172
——, 1,2,3,5-tetraphenyl- 46
——, (1,2,2-tribromoethyl)- 350
——, 1,2,4-tri-tert•butyl- 10
——, (1,2,2-trichloroethyl)- 346,347
——, 1,2,4-trimethyl- 182
benzene, nitro-, see nitrobenzene
benzenearsonic acid, 2-nitroso- 268,270
benzenediazonium chloride, 4-chloro- 313,314
——, 2-dimethylamino- 314
——, 4-dimethylamino- 314
——, 4-methoxy- 313
——, 4-nitro- 313
benzenediazonium sulfate, 4-anilino- 314,315
benzenes, chlorine substituted 344
benzeneselenenyl bromide, 2,4-dinitro- 353
benzenesulfenic acid, 2,4-dinitro- 270,271
benzenesulfenyl chloride, 2,4-dinitro- 270,271
benzenesulfonamide, N,N-dibutyl- 243
benzenesulfonate, methyl 346
benzenesulfonic acid, 2-(acetoxyamino)-4-nitro- 270,
271
——, 2-amino-4-nitro- 270,271
benzenesulfonyl chloride 346
benzenesulfonyl chloride, 4-bromo- 346,347
benzenethiol 216
benzenethiol, 4-chloro- 446,447
benzhydrol 153,194,195,203,210,238
benzhydrol, 4-chloro- 194
——, 4-chloro-4'-methyl- 194
——, 4,4'-dimethyl- 194
——, 4-methoxy- 194
——, 4-methyl- 194

Compound Index

——, 4-phenyl- 194
benzhydrols 194
benzidines 139
benzil 119–121,199,201,231,419,420
——, 4,4'-dichloro- 330,331
——, 4,4'-dimethoxy- 330
benzil visnagin 1:1 adduct 419,420
benzimidazole, 7-methyl- 54
——, 2-phenyl- 337
2-benzimidazolecarboxylate, butyl 337
——, cyclohexyl 337
——, ethyl 337
——, isopropyl 337
benzimidazoles 54
benzoate, 4-tert-butyl-2-chlorophenyl 236
——, 4-tert-butylphenyl 236,237
——, 4-tert-butylphenyl 4-amino- 237
——, 4-tert-butylphenyl 4-tert-butyl- 237
——, 4-tert-butylphenyl 4-chloro- 237
——, 4-tert-butylphenyl 4-cyano- 237
——, 4-tert-butylphenyl 4-nitro- 237
——, ethyl 2-nitroso- 267–269
——, methyl 2-nitroso- 268
——, phenyl 236,237
benzoates, alkyl 2-nitroso- 268
benzo[c]cinnoline 138,139
benzo[c]cinnoline, 3-amino- 139–141
——, 2,4-dimethyl- 139,140
——, 1-methyl- 139
——, 2-methyl- 139,337,338
——, 3-methyl- 139
——, 4-methyl- 139
——, 1,2,4-trimethyl- 139,140
benzo[c]cinnoline-2-carboxylate, methyl 139,140,142
benzo[c]cinnoline-2-carboxylic acid 140
benzo[c]cinnolinecarboxylic acids 140
benzo[c]cinnolines, acetyl- 140
——, chloro- 140
——, iodo- 140
——, nitro- 140
benzocyclobutene, 4,6-dimethyl-1-mesityl- 289
——, cis-1,2-diphenyl- 221

——, trans-1,2-diphenyl- 221
benzocyclobutene-1-carboxylic acid 302,303
benzocyclobutene-1-carboxylic acid, 2,2-diphenyl- 306,307
5H-benzocyclohepten-2-one, 3-diazo-2,3,6,7,8,9-hexahydro- 304,305
3H-1,4-benzodiazepine, 7-chloro-4,5-epoxy-4,5-dihydro-2-methylamino-5-phenyl- 50,51
3H-1,4-benzodiazepine 4-oxide, 7-chloro-2-methylamino-5-phenyl- 50,51
1,4-benzodioxin, 5,6,7,8-tetrachloro-2,3-dihydro-2,3-diphenyl- 123
——, 5,6,7,8-tetrachloro-2,3-dihydro-2-oxo-3,3-diphenyl- 121
——, 5,6,7,8-tetrachloro-2,3-dihydro-2-phenyl- 123
1,3-benzodioxole, 4,5,6,7-tetrachloro-2,2-dimethyl- 307,308
benzo-1,4-dithiin, 2-nitro- 90,91
benzo[1,2,3-kl:6,5,4-k'l']dithioxanthene 133
benzo[1,2,3-kl:6,5,4-k'l']dixanthene 133,134
11H-benzo[b]fluorene, 5,10-diphenyl- 129–131
11H-benzo[a]fluorene-11-carboxylic acid 304
benzofuran 419
benzofuran, 2,3-diphenyl- 119,120
——, 2,4,5,6,7,7a-hexahydro-7a-hydroperoxy-7a-methoxy-3,6-dimethyl- 430
——, 5-methoxy-2-(3-methoxyphenyl)- 216,217
——, 7-methoxy-2-(3-methoxyphenyl)- 216,217
——, 2-phenyl- 215–217
——, 4,5,6,7-tetrahydro-3,6-dimethyl- 430
2(3H)-benzofuranone, 5-hydroxy-3,3-diphenyl- 29
benzofuran-2(3H)-one, 3-phenyl- 146
benzofuran-2-one, 2,4,5,6,7,7a-hexahydro-7a-hydroxy-3,6-dimethyl- 430
benzofuroxan, 4-phenyl- 320
benzohydroxamate, ethyl 321,322
benzoic acid 370
benzoic acid, 2,2'-azodi- 107,108
——, 2-benzoyl- 210,211
——, 2-bromo- 363
——, 3-bromo- 363
——, 2-formyl- 211

Compound Index

-----, 4-formyl-2-nitroso- 267
-----, 2-nitroso- 267-269
benzoic acids, phenylazo- 140
benzoin acetate 216,217
benzoin acetate, 3,3'-dimethoxy- 216,217
benzo[g]isoquinoline, 5,10-epidioxy-5,10-diphenyl- 393
benzo[b]naphtho[1,2-d]thiophene 136
benzonitril, 2-chloro- 339
-----, 4-chloro- 339
-----, 4-dimethylamino- 339
-----, 4-methoxy- 339
-----, 4-nitro- 339
benzonitrile 111,112,114,339
benzonitrile, 2-N-methylamino- 55
benzonitrile oxide 262
16H-benzo[4,5]phenaleno[1,2,3-kl]thioxanthen-16-ol, 16-Phenyl- 132,133
16H-benzo[4,5]phenaleno[1,2,3-kl]thioxanthen-16-one 132,133
16H-benzo[4,5]phenaleno[1,2,3-kl]xanthen-16-one 134
benzo[a]phenanthro[1,10,9-jkl]acridinium chloride, 10-Phenyl- 135
benzo[a]phenanthro[1,10,9-jkl]xanthylium perchlorate 134-136
benzo[a]phenanthro[1,10,9-jkl]xanthylium perchlorate, 3-methyl- 135,136
benzo[a]phenanthro[1,10,9-jkl]xanthylium picrate 136
benzo[a]phenanthroxanthylium salts 134-136
benzo[a]phenazine-5,6-diol 187,188
benzo[a]phenazine-5,6-diol diacetate 187,188
benzo[a]phenazine-5,6-dione 187,188
benzo[a]phenazin-5(7H)-one, 6-benzoyloxy- 187,188
-----, 6-hydroxy- 188
-----, 6-(4-methoxybenzoyloxy)- 187,188
benzophenone 24,148,194,195,200,201,203-207,209,232, 291,339,414,415,417,445,446
benzophenone, 2-amino- 206,315
-----, 4-amino- 206,415
-----, 4'-amino-5-tert•butyl-2-hydroxy- 237
-----, 2-benzyl- 24,25,208

-----, 4,4'-bis-(dimethylamino)- 194
-----, 4,4'-bis-(methylthio)- 205-207
-----, 5-tert•butyl-3-chloro-2-hydroxy- 236
-----, 5-tert•butyl-4-chloro-2-hydroxy- 237
-----, 5-tert•butyl-4'-cyano-2-hydroxy- 237
-----, 5-tert•butyl-2-hydroxy- 236,237
-----, 5-tert•butyl-2-hydroxy-4'-nitro- 237
-----, 4-chloro- 194,415
-----, 3-chloro-4-hydroxy- 236
-----, 4-chloro-4'-methyl- 194
-----, 2-diazo- 315
-----, 4',5-di-tert•butyl-2-hydroxy- 237
-----, 4,4'-dichloro- 205,207
-----, 2,5-dihydroxy- 189
-----, 4,4'-dimethoxy- 205,207,446
-----, 2,6-dimethyl- 24,25
-----, 4,4'-dimethyl- 194,415
-----, 4-dimethylamino- 206
-----, 2,2',4,4',6,6'-hexamethyl- 222
-----, 2-hydroxy- 236,237
-----, 4-hydroxy- 206,236,237
-----, 4-methyl- 194
-----, 2-methyl- 24,25,208
-----, 4-methyl- 194,415
-----, 4-(methylthio)- 205-207
-----, 2-phenyl- 205
-----, 3-phenyl- 205
-----, 4-phenyl- 194,205
benzophenone anil 299
benzophenone as filter, 2,2'-dihydroxy- 416
benzophenone azine 280,283,287,290
benzophenone 2,4-dinitrophenylhydrazone, 2-hydroxy- 237
benzophenone C-oxide 291
benzophenone peroxide, dimeric 291
benzophenones, 2-alkyl- 24,25
benzopinacol 153,200,201,203,204,206,209,211,339
benzopinacol see also 1,2-ethanediol, tetraphenyl-,
benzo[g]pteridine-2,4(1H,3H)-dione 258
1H-2-benzopyran, 1,3-diphenyl- 119,120
2H-1-benzopyran, 2,2-dimethyl- 119,120
5H-1-benzopyran, 6,7,8,8a-tetrahydro-2,5,5,8a-tetramethyl- 41,42

568   Compound Index

4H-1-benzopyran-Δ⁴,α-acetic acid γ-lactone,
    α-(2-benzoyloxyphenyl)-3-hydroxy-2-phenyl-   239
3H-2-benzopyran-3-one, 1,4-dihydro-1,1-diphenyl-   306,
    307
11H-[2]benzopyrano[3,4-b]phenanthro[9,10-e][1,4]dioxin,
    9a,15b-dihydro-11-oxo-9a-phenyl-   122,124
benzo[a]pyren-6-acetate, methyl   442
benzo[a]pyren-6-acetic acid   442
benzo[a]pyrene   442
benzo[a]pyrene-6-thiol acetate   442
2-benzopyrylium-4-oxide, 1,3-diphenyl-   411
----, 3-methyl-1-phenyl-   411
benzo[g]quinazoline   101
benzo[g]quinoline, 5,10-epidioxy-5,10-diphenyl-   393
benzo[f]quinoline, 2-butyl-3-pentyl-   334
----, 2-butyl-3-phenyl-   334
----, 1,3-diphenyl-   135,136
----, 2-isopropyl-3-phenyl-   334
----, 3-(4-methoxyphenyl)-   334
----, 3-phenyl-   334
benzo[g]quinoline   101
benzo[h]quinoline, 5,10-epidioxy-5,10-diphenyl-   393
benzo[h]quinoline-5,6-dione   119,120,187
benzo[h]quinoline-5,6-dione imine   190
benzo[h]quinoline-5,6-dione oxime   191
benzo[h]quinolin-5-ol, 6-benzamido-   190
benzo[b]quinolizinium bromide   102
benzo[b]quinolizinium dimer dibromide   102
benzo[a]quinolizinium perchlorate   142
benzo[a]quinolizinium perchlorate, 10-chloro-   142
    134,189,211,212,332,418,423,424
F-benzoquinone, tetrabromo-   119,120
----, tetrachloro-   119-121,123,187,439
----, 3,4,5,6-tetrachloro-   184
1,4-benzoquinone   79,81,148
1,4-benzoquinone, 2,3-dimethyl-   80,81
----, 2,5-dimethyl-   79,80
----, 2,6-dimethyl-   80
o-benzoquinone diazide, 3,4,5,6-tetrachloro-   307-309
p-benzoquinone diazide   310,311
F-benzoquinone diimine N,N'-dioxide, N,N'-bis-(4-
    methoxyphenyl)-   332

----, N,N'-diphenyl-   332,333
F-benzoquinone diimine N,N'-dioxides   332
1,4-benzoquinone dimer, 2,5-dimethyl-   79,80
----, 2,6-dimethyl-   80
F-benzoquinone imine N-oxide, N-phenyl-   332,333
p-benzoquinone imine N-oxides   332
1,4-benzoquinones, methyl substituted   79-81
benzo[f]quinoxaline-5,6-dione, 3-phenyl-   187
benzo[c]tetrazolo[2,3-a]cinnolin-4-ium chloride,
    2-phenyl-   143,144
(benzo[c]tetrazolo[2,3-a]cinnolin-4-ium nitrate),
    7,7'-methylenebis-   143,144
----, 2,2'-p-phenylenebis-   143,144
benzo[c]tetrazolo[2,3-a]cinnolinium salts   143,144
benzo[b]thiophene, 3-β-styryl-   136
benzo[b]thiophene 1,1-dioxide   70,90
benzo[b]thiophene 1,1-dioxide, 3-bromo-   90
----, 3,4-dimethyl-   90
----, 3-methyl-   90
benzo[b]thiophene 1,1-dioxide dimer   90
benzo[b]thiophen-3(2H)-one, 2-phenyl-   146
benzo[1,2,3-kl]thioxantheno[6,5,4-kl']xanthene   133
12H-benzo[b]xanthen-12-one, 5a,6-dihydro-11-hydroxy-6-
    phenyl-   130
----, 11-hydroxy-6-phenyl-   25,130,131
----, 11-phenyl-   130,131
1,2-benzoxathiin 2,2-dioxide, 4-methyl-   448
1,3-benzoxathiole, 4,5,6,7-tetrachloro-2-phenylimino-
    307,308
benzoxazole, 4,5,6,7-tetrachloro-2-methyl-   307-309
2-benzoxazolinone   322
benzyl azide   321,322
benzyl alcohol   203
benzyl bromide   351
benzyl bromide, p-bromo-   344
benzyl chloride   262,357
benzyl chloride, p-bromo-   356
benzylamine   338
benzylamine, N-phenyl-   141,142
benzyne   368,369
bergapten   87,88
bergapten dimer   87,88

Compound Index 569

biacetyl 169,199,231
9,9'-biacridan 101,195,196
9,9'-biacridan, 10,10'-dimethyl- 195,196
10,10'-bi-9-anthrol 131-133
10,10'-bi-9-anthrol diacetate 132,134
Δ¹⁰,¹⁰'-bianthrone 131-134
Δ¹⁰,¹⁰'-bianthrone, 1,1',8,8'-tetrachloro- 133
10,10'-bianthrone 146-148,266
1,1'-bianthryl, 9,10:9',10'-bis-epidioxy-9,9',10,10'-
  tetraphenyl- 391,392
—, 9,9',10,10'-tetraphenyl- 391,392
9,9'-bianthryl, 10,10'-diphenyl- 392
(3,3'-bibenzofuran)-2,2'(3H,3'H)-dione, 3,3'-diphenyl-
  146,147
Δ⁴,⁴'-bi-4H-1-benzopyran, 3,3'-epoxy-2,2'-diphenyl-
  239
2,2'-bi-p-benzoquinone, 5,5'-dibromo- 426,427
—, 5,5'-dichloro- 426,427
—, 5,5'-dimethyl- 426,427
—, 5,5'-diphenyl- 426,427
(2,2'-bibenzo[b]thiophene)-3,3'(2H,2'H)-dione,
  2,2'-diphenyl- 146,147
bibenzyl 216,220,221
bibenzyl, 2-(diazomethyl)- 278,279,284
—, 4,4'-dimethyl- 148,183
bicyclobutane 10
bicyclo[5·3·0]deca-1(10),7-diene-9-carboxylic acid
  304,305
bicyclo[6·2·0]decan-10-one, 1-hydroxy- 35,36,116
bicyclo[6·4·0]dodecane-2,6-dione, 4,4-dimethyl- 113
cis-bicyclo[8·2·0]dodecan-1-ol 34
trans-bicyclo[8·2·0]dodecan-1-ol 34
bicyclo[2·2·1]heptadiene 2,3,10,328,329,418
bicyclo[3·2·0]hepta-2,6-diene 4,5
bicyclo[4·1·0]hepta-2,4-diene 276-278
bicyclo[4·1·0]hepta-2,4-diene-7-carboxylate, ethyl
  277,279
—, methyl 384
bicyclo[2·2·1]heptadiene-2,3-dicarboxylic acid 2-4
bicyclo[3·2·0]hepta-3,6-dien-2-one, 5-methoxy- 4,6,9
bicyclo[2·2·1]heptane 264
bicyclo[2·2·1]heptane, exo-2-acetonyl- 169,170

—, exo-2-bromo- 158
—, 3-endo-bromo-2-exo-(trichloromethyl)- 171,172
—, endo-3-chloro-exo-2-p-toluenesulfonyl- 165
—, exo-cis-2,3-dibromo- 158,159
—, trans-2,3-dibromo- 158,159
bicyclo[4·1·0]heptane 275,276
bicyclo[4·1·0]heptane, 7-phenyl- 275,276
bicyclo[2·2·1]heptane derivatives see also norbornane
  derivatives
bicyclo[2·2·1]heptan-2-one, 3-diazo- 303,304
—, 3-diazo-1,7,7-trimethyl- 303
bicyclo[2·2·1]heptan-2-one oxime 264
bicyclo[2·2·1]heptane 415,417
bicyclo[2·2·1]hept-2-ene 74,165,169-172,411,412,468,
  469
bicyclo[2·2·1]hept-2-ene, 2-bromo- 158
bicyclo[3·2·0]hept-6-ene 4,5,9
bicyclo[3·2·0]hept-6-ene, 1,5-diphenyl- 4,5
bicyclo[3·2·0]hept-6-en-2-ol, 5-methoxy- 4,5
bicyclo[3·2·0]hept-6-en-3-ol, 6-methoxy- 4,5
bicyclo[2·2·1]hept-2-en-7-one, 1,5,5-trimethyl- 27,28
bicyclo[2·2·1]hept-5-en-2-one 116
bicyclo[2·2·1]hept-5-en-2-one, 2,7,7-trimethyl- 28
bicyclo[3·1·1]hept-2-en-6-one 116
bicyclo[3·2·0]hept-2-en-7-one 116
bicyclo[3·2·0]hept-6-en-2-one, 1,7-dimethyl- 115
—, 6,7-dimethyl- 115
—, 5-methoxy- 4,6
—, 1,4,4-trimethyl- 4,6,27,28
bicyclo[2·2·0]hexa-2,5-diene 6
bicyclo[2·2·0]hexa-2,5-diene, 1,2,5-tri-tert-butyl- 10
bicyclo[2·1·1]hexane-5-carboxaldehyde, 1,6,6-trimethyl-
  27
bicyclo[2·1·1]hexane-5-carboxylate, methyl 303,304
bicyclo[2·1·1]hexane-5-carboxylic acid, 1,6,6-trimethyl-
  303
bicyclo[2·1·1]hex-2-ene, 5-isopropyl-2-methyl- 17
bicyclo[3·1·0]hex-2-ene, 6-isopropyl-2-methyl- 17
—, 1,2,3,5,6-pentaphenyl- 17
—, 3,4,6,6-tetramethyl- 19
bicyclo[3·1·0]hex-2-ene-3-carboxylate, ethyl 15
bicyclo[2·2·1]hex-2-ene-1,4-dicarboxylate, dimethyl
  2,3-dimethyl- 115

bicyclo[3.1.0]hex-2-ene-1,5-dicarboxylate, dimethyl 16,17
bicyclo[3.1.0]hex-3-ene-1,3-dicarboxylate, dimethyl 16
bicyclo[2.2.0]hex-5-ene-2,3-dicarboxylic anhydride 5,6
bicyclo[3.1.0]hex-2-enes 15-17,19
bi-1-cyclohexen-1-yl 388,389
bi-2-cyclohexen-1-yl 170,198,199,201,325
bicyclohexylidene 122,219
bicyclo[4.2.0]octa-2,4-diene, 1-cyano-7,8-trimethyl- 111,112
bicyclo[4.2.0]octa-2,7-diene 5,7
bicyclo[2.2.2]octane 264
cis-bicyclo[3.3.0]octane 264
cis-bicyclo[3.3.0]octane, exo-6-chloro-exo-2-(trichloromethyl) 172,173
bicyclo[4.2.0]octane-7,8-dicarboxylates, dimethyl 111
bicyclo[4.2.0]octane-7,8-dicarboxylic anhydride 111
bicyclo[4.2.0]octan-2-one, 7-benzyloxy- 112
-----, 7,7-dimethyl- 112
-----, 8,8-dimethyl- 112
bicyclo[2.2.2]octan-2-one oxime 264
cis-bicyclo[3.3.0]octan-2-one oxime 264
cis-bicyclo[3.3.0]octan-3-one oxime 264
bicyclo[4.2.0]oct-7-ene 4,5
bicyclo[5.1.0]oct-3-ene 18
bicyclo[2.2.2]oct-2-ene-5,6-dione, 1,2,3,4-tetrachloro-7-phenyl- 123
bidesyl 215-217
(7,7'-biergosta-5,8,(22-triene)-3β,3β'-diol 149
(4,4'-biflavylene, 3,3'-epoxy- 239
Δ⁹,⁹'-bifluorene 440,441
9,9'-bifluorene 148,183,290
9,9'-bifluorene, 9,9'-bis-(phenylthio)- 440,441
-----, 9,9'-dihydroxy- 206,207
-----, 9,9'-diphenyl- 126,127,367,368
2,2'-biindan-1,1',3,3'-tetrone, 2,2'-dihydroxy- 209
1,1'-biisoquinoline, 1,1',2,2',3,3',4,4'-octahydro-1,1'-dimethyl- 335
(Δ²,²'(¹H,¹'H)-binaphthalene)-1,1'-dione, 4,4'-dimethyl- 427

(2,2'-binaphthalene)-1,1',4,4'-tetrone, 1,1',4,4'-tetrahydro- 426,427
-----, 1,1',4,4'-tetrahydro-3,3'-dihydroxy- 147
1,1'-bi-1-naphthol, cis-1,1',2,2',3,3',4,4'-octahydro- 208
-----, trans-1,1',2,2',3,3',4,4'-octahydro- 208
1,1'-binaphthyl 128
2,2'-binaphthyl 469
2,2'-biphenalene-1,1',3,3'-tetrone, 2,2'-dihydroxy- 209
biphenyl 276,455,457,469
biphenyl, 2-azido- 320
-----, 4-azido- 321
-----, 2-azido-3,5-dibromo- 320
-----, 2-azido-3-nitro- 320
-----, 2,2'-bis-(4-methylphenylacetyl)- 208,212
-----, 2,2'-dibenzoyl- 212
-----, 2-hydroxy- 368
-----, 4-hydroxy- 311,368
-----, 4-hydroxy-2'-methyl- 311
-----, 4-hydroxy-4'-methyl- 311
-----, 2-iodo- 369
-----, 4-iodo- 456
-----, 2-methoxy- 368
-----, 4-nitro- 368
biphenyl-2d, 2'-(diazomethyl)- 285
4-biphenylamine, N-phenyl- 311
biphenylene-1,4-dione, 1,4,4a,8b-tetrahydro-5,8-dihydroxy-2,3,6,7-tetramethyl- 80,81
3,3'-biphthalide 211
3,3'-biphthalide, 3,3'-diphenyl- 210,211
5α-bisnorcholanoyl xanthate, O-ethyl 3β-acetoxy-11-oxo- 223,224
2,2'-bithiophene 451
2,3'-bithiophene 451
3,3'-bithiophene 451
Δ⁹,⁹'-bithioxanthene 133
9,9'-bithioxanthene 146-148
Δ⁹,⁹'-bixanthene 133,134
9,9'-bixanthene 148,183,184
9,9'-bixanthene, 9,9'-dihydroxy- 206,207,211
d,l-borneol nitrite 252,253

Compound Index

bromine 238,341,342,344,349-353,358,359,362,363
bromine isotope 82 363
bromoform 467
N-bromosuccinimide see succinimide, N-bromo-,
BUECHNER acids 277
bullvalene 116,117
butadiene, 2,3-dimethyl- 74
1,2-butadiene 14
1,3-butadiene 10,14,74,175,176,342,424,459,463
1,3-butadiene, 2-tert·butyl- 176
-----, 2,3-di-tert·butyl- 176
-----, 2,3-dimethyl- 281,418,424
-----, 1,4,4-tetrakis-(4-methoxyphenyl)- 366
-----, 1,2,3,4-tetraphenyl- 459,462
butanal 105
butanal, 4-nitroso- 249
butane 435
butane, 1-azido- 319
-----, 1-bromo- 352
-----, 3-bromo-1,1,1-trichloro- 171
-----, 2-bromo-2,3,3-trimethyl- 342
-----, 1-chloro- 352
-----, 1-chloro-2,3-dimethyl- 348,349,356
-----, 2-chloro-2,3-dimethyl- 349,356
-----, 1,4-diiodo- 370
-----, 2,3-diiodo- 353,354
-----, 2,3-dimethyl- 348,356
-----, 2,2,3,3,4,4-heptafluorodiazo- 288,289
-----, nonafluoro-2-iodo- 173,174,364,365
-----, nonafluoro-1-nitroso- 364,365
-----, nonafluoro-2-nitroso- 348
-----, 1,2,4-trichloro-2-(chloromethyl)- 366
-----, 1,2,4-trichlorohexafluoro-4-iodo- 225
-----, 1,3,4-trichlorohexafluoro-1-iodo- 342
-----, 2,2,3-trimethyl- 203
2,3-butanediol, 2,3-dimethyl- 211
2,3-butanediol, 2,3-diphenyl- 294,295
1,3-butanedione, 2-diazo-1-phenyl- 408
-----, 2-methyl-1-phenyl- 38
-----, 1-phenyl- 193,409
1,4-butanedione, 1,4-diphenyl- 193,409
-----, 2,3-epoxy-1,2,3,4-tetraphenyl- 431
-----, 1,2,3,4-tetraphenyl- 215-217,450
2,3-butanedione 169,199,231
2,3-butanedione 363
butane-1,4-disulfonyl chloride 435
butanedisulfonyl chlorides 435
butanes, bromochloro- 352
-----, dibromo- 352
butanesulfonyl chlorides 435
1-butanethiol 164
butanol 205
tert·butanol 325,326
1-butanol, 4-nitroso-4-phenyl- 248
2-butanol 210,211,232,379
2-butanol, 2-hydroperoxy- 379
1-butanone, 3,4-epoxy-1,4-diphenyl- 35
-----, trans-3,4-epoxy-1,4-diphenyl- 409
2-butanone 210,211,279
2-butanone, 3-(2-cyclohexen-1-yl)-3-hydroxy- 199
-----, 3-diazo-4-phenyl- 295
-----, 4-(6,6-dimethyl-2-methylenecyclohexylidene)- 42
-----, 4-hydroxy-4-(2-nitrophenyl)- 271
-----, 3,3,4,4-pentachloro-1-diazo- 296,297
-----, 4-(2,6-trimethyl-2-cyclohexenylidene)- 59
butatriene, tetraphenyl- 89,90
1-butene 163,176
1-butene, 4-chloro- 348
-----, 4-chloro-2-(chloromethyl)- 348
-----, 2H-heptafluoro- 288-290
-----, hexafluoro-4,4-diiodo- 105
-----, pentafluoro-1-iodo- 174
-----, 3,3,4,4-pentafluoro-1-iodo- 174
2-butene, 1-diazo-2,3-dimethyl- 281
-----, 2,3-dichloro-1,1,1,4,4-hexafluoro- 342
-----, 2,3-diethoxy- 450
-----, 2,3-dimethyl- 218-220,218,219,376,378,415,417
-----, 1,1,1,4,4-hexafluoro- 288
-----, 2-methyl- 111,112,122,124,414-417
-----, 2-phenyl- 119,120
cis-2-butene 275,278,353,415,448,449
trans-2-butene 275,278,353,415,448,449
3-butene, 1-hydroperoxy-2,3-dimethyl- 376,378
1,4-butenedione, cis-1,4-diphenyl- 193

572   Compound Index

2-butene-1,4-dione, 2,3-dibenzoyl-1,4-diphenyl-   239
——, cis-1,4-diphenyl-   238,239
——, 1,2,3,4-tetraphenyl-   394,395,431
——, 1,2,4-triphenyl-   238,239
3-butenoate, ethyl 2,4-diphenyl-4-phenoxy-   238,239
——, ethyl 4-phenoxy-4-phenyl-   57
2-butenoic acid, 2-methyl-   238
3-butenoic acid, 2,4-diphenyl-4-phenoxy-   296,297
——, 3,4,4-trichloro-   238
3-butenoic acid γ-lactone, 3-benzoyl-4-hydroxy-2-(α-phenoxybenzylidene)-4-phenyl-   239
3-buten-2-ol acetate, 1-phenyl-   171,172
3-buten-1-one, 2-hydroxy-1,3-diphenyl-   408,409
——, 1,2,3,4-tetraphenyl-   44
——, 1,2,4,4-tetraphenyl-   44
3-buten-2-one, 4-phenyl-   295
——, 3,4,4-trichloro-1-diazo-   296,297
tert-butyl hypochlorite   162
tert-butyl nitrite   251
1-butyl nitrite, 4-phenyl-   248
butylamine   435
butylamine, N-butyl-4-chloro-   246
——, N-butylidene-   242
——, N,N-dibromo-1-propyl-   245,246
1-butylamine   319
butylidenimine   319
butylphosphonate, dibutyl   454
1-butyne, 3-bromo-3-methyl-   468
2-butyne   115,281,459,461
2-butyne, hexafluoro-   342
butyraldehyde   105,169
butyraldehyde, 3-methyl-   268
——, 4-nitroso-   249
butyraldehyde diethyl acetal, 3,3-dimethyl-2-oxo-   238
butyrate, ethyl 3,3-dimethyl-2-oxo-   238
——, methyl 2,2,3-trimethyl-   219
butyric acid, 3-amino-   165
——, 3-anilino-   165
——, 3,4-dichloropentafluoro-   225
——, 3-ethoxy-   160
——, 3-hydroxy-   155
——, 3,3'-iminodi-   165

——, 3-methoxy-   160
——, 3,3,4,4-pentachloro-   296
butyric anilide, 3-anilino-   165,166
butyronitrile, 4,4,4-trifluoro-2-iodo-   174
butyryl fluoride, heptafluoro-   422

cafestol   432
cafestol, 1-dehydro-   432
cafestol 17-acetate, 3,19-dihydro-3,19-dimethoxy-   432,433
cage compounds   2-4,79-81,89
calciferol   15
C-calebassine   401,402
camphene   163,284,285
camphidine, N-chloro-   244
α-campholanaldehyde, d,l-nitroso-   252
α-campholenaldehyde   26
α-campholenaldehydes   252,253
camphor, diazo-   303
d,l-camphor   26
camphor p-toluenesulfonylhydrazone   284,285
camphoric anhydride   303
cantharidine   383,384
caproate, cyclohexyl   170
caproic acid   225,226
caproic acid-2-d   226
ε-caprolactam   262,263
carbamate, tert-butyl N-tert-butoxy-   321
——, diethyl tetramethylenedi-   321
——, ethyl N-tert-butoxy-   325
——, ethyl 1-cyclohexenyl-   325
——, ethyl 2-cyclohexenyl-   325
——, ethyl 3-cyclohexenyl-   325
——, ethyl cyclohexyl-   324
——, ethyl N-ethoxy-   325
——, ethyl N-isopropoxy-   325
——, ethyl N-methoxy-   325
——, ethyl phenyl-   322
——, methyl benzyl-   321,322
carbamate, ethyl-, see urethan
carbamates, ethyl N-alkoxy-   325
carbanilate, ethyl   321,322
carbazole   138,320

Compound Index

carbazole, 3-chloro- 314,315,314,315
-----, 1,3-dibromo- 320
-----, dihydro- 138
-----, 2,7-dinitro- 320
-----, 1,2,3,4,9a-hexahydro-4a-methyl- 195,196
-----, 9-methyl- 138
-----, 9-phenyl- 138
1H-carbazole, 2,3,4,4a-tetrahydro-4a-methyl- 336,337
4aH-carbazole, 1,2,3,4-tetrahydro-4a-methyl- 195,196
carbazole chloride, 3-diazo- 314,315
carbazole-9a-methanol, 1,2,3,4,4a,9a-hexahydro-4a,9-dimethyl- 336
-----, 1,2,3,4,4a,9a-hexahydro-4a-methyl- 336
1H-carbazolium iodide, 2,3,4,4a-tetrahydro-4a,9-dimethyl- 336
carbazolium salts 195
-----, 9-dimethyl- 336
carbene 274-276,279-284
carbene, carboethoxy- 277
-----, carbomethoxy- 275,281,283
-----, difluoro- 289
-----, dimethyl- 290
-----, diphenyl- 275,280,283,284,286,287,290
-----, methyl- 290
-----, phenyl- 275,276,284,289
-----, trifluoromethyl- 288,290
carbenes 294,295,300,304,305,307-309,318,331
carbenes, carboalkoxy- 277,278
carbinols 198-202
carbinols, dialkyl-(cyclohexenyl)- 198,199
carbon dioxide 268,337,469
carbon monoxide 218-223,369,394,395,459-462,464,465,469
carbon tetrabromide 170-173
carbon tetrachloride 170-173,282,283,307,467
carbonate, diethyl 449
-----, 0,S-diethyl thio- 449
carbonyl compounds, metal 146
carbonyl compounds, metal 459-465,469
carbonyl fluoride 225
carbostyryl 51,52
carbostyryl, 1-methyl- 52
-----, 3-methyl- 52

carbostyryls, methyl- 51,52
all-trans-β-carotene 60,61
15-cis-β-carotene 60,61
carotenoids 59-61
carveol 376
carvone 1,3,27,376
carvonecamphor 1,3,27
α-caryophyllene alcohol, hydroxyimino- 249,250
α-caryophyllene alcohol nitrite 249,250
α-cedrene 168
α-cedrene dimer 168
β-cedrene-5-succinic anhydride, α,β-dimethyl- 168
-----, β-methyl- 168
celiobial, hexaacetyl- 122
chalcone, 2'-nitro- 271,272
chalcone oxides 408
chelates 463
chloranil 151,152,182,189,418,419,453
chlorine 261-265,291,341-349,351,355,362-365,407,434-437,441
chloroform 282,283,355,467
chola-4,23-dien-3-one, 22-bromo-24,24-diphenyl- 358
-----, 24,24-diphenyl- 358,359
cholanthrene, 3-methyl- 99
cholanthrene dimer, 3-methyl- 99
chola-4,20(22),23-trien-3-one, 24,24-diphenyl- 358,359
2,4-cholestadiene 47
cholesta-2,4-diene 387,388
cholesta-3,5-diene 10,11
cholesta-5,7-dien-3β-ol benzoate 358
cholesta-4,6-dien-3-one 78,79
cholesta-4,6-dien-3-one dimer 78,79
5α-cholestane-3,6-dione, 4α,5-methylene- 329
5α-cholestane-3,6-dione, 4α,5-pyrazoline 329
5α-cholestan-6β-ol, 3β-acetoxy-19-hydroxyimino- 250
5α-cholestan-6β-ol nitrite, 3β-acetoxy 250
5α-cholestan-2-one, 4α,5-epoxy- 47
5α-cholestan-3-one, 4α,5-epoxy- 378,379
cholest-5-ene, 3β-benzoyloxy- 351,358
-----, 3β-tosyloxy- 351
5α-cholest-3-ene, 2α,5-epidioxy- 47,388

Compound Index

cholest-4-ene-3,6-dione, 4-methyl- 329
cholest-4-en-3β-ol 377,379
cholest-5-en-3β-ol 377,378,443,444
cholest-5-en-3β-ol, 7α-hydroperoxy- 443,444
-----, 7-thiocyanato- 377
5α-cholest-6-en-3β-ol 377
5α-cholest-6-en-3β-ol, 5-hydroperoxy- 377,378
cholest-5-en-3β-ol benzoate 351,358
cholest-5-en-3β-ol benzoate, 7β-bromo- 351,358
cholest-5-en-3β-ol esters, 7β-bromo- 443
cholest-5-en-3β-ol tosylate 351
cholest-5-en-3β-ol tosylate, 7β-bromo- 351
cholest-4-en-3-one 378,379
cholest-4-en-7-one, 3β-acetoxy- 234,235
cholest-5-en-7-one, 3β-acetoxy- 234,235
5β-cholest-3-en-7-one, 5-acetoxy- 234,235
cholesterol 377,378,443,444
cholesteryl benzoate 351,358
cholesteryl tosylate 351
chromium dicarbonyl, (ethylene)(mesitylene)- 463
chromium hexacarbonyl 459,460,462
chromium pentacarbonyl, aniline- 460
-----, (4-chloropyridine)- 460
-----, isoquinoline- 460
-----, (2,6-lutidine)- 460
-----, (2-picoline)- 460
-----, piperidine- 460
-----, quinoline- 460
-----, (triethylamine)- 460
chromium tetracarbonyl, bis-(piperidine)- 463
chromium tricarbonyl, mesitylene- 25,130,131
chromone, 3-benzoyl-2-benzyl- 463
-----, 2-phenyl-4-thio- 445
chrysanthenone 28
chrysene 128
5,6-chrysenedione 119,120,187,439
5(6H)-chrysenone, 6-diazo- 304
cinnamaldehyde 186-188,338
cinnamaldehyde, 2-nitro- 269
cinnamate, ethyl 171,172
cinnamic acid, 2-nitro- 271
cis-cinnamic acid 57,58,83-85

cis-cinnamic acid, 4-methoxy- 57,58
trans-cinnamic acid 57,58,83-85
trans-cinnamic acid, 4-methoxy- 57,58
cinnamic acid dimer, 2-hydroxy- 85
cis,trans-tricyclo[6•4•0•0²,⁷]dodeca-3,11-diene 75
citral 27
citronellol 377
colchicine 8,9,92
complexes 462
conanine 244,245
conessine, dihydro- 244,245
coumarin 70,85,86,343
coumarin, 3-chloro- 343,344
-----, 3-phenyl- 85
coumarin dimer, 3-phenyl- 85
coumarin dimers 85,86
o-cresol 235
crotonaldehyde 169
crotonate, methyl 3,4,4,4-tetrachloro- 296,297
crotonic acid 155,157,160,165,166,231,232
crotonic acid, 3,4,4,4-tetrachloro- 296,297
-----, 3,4,4-trichloro- 296,297
-----, 4,4,4-trifluoro- 173,174
crotonic anilide 165
crotonitrile, 3-amino-2-(1-iminoethyl)- 54,55
cumene 373,443
cumene, α-isothiocyanato- 443
-----, α-thiocyanato- 443
cuprous bromide 468
cuprous chloride 18,468
C-curarine-I 402
cyanamide 445
cyanogen 365,366
cyanogen chloride 260,261
cyanuric chloride 260
1β,5-cyclo-5β,10α-androstan-2-one, 17β-hydroxy- 23
1β,5-cyclo-5β,10α-androst-3-en-2-one, 17β-hydroxy-1,3-dimethyl- 31
-----, 17β-hydroxy-1-methyl- 31,32
cyclobuta[1,2-b:3,4-b']bis[1,4]benzodithiin, dinitro- 90,91

Compound Index

cyclobuta[1,2-c:3,4-c']bis[1]benzopyran-6,12-dione,
  6,6a,6b,12,12a,12b-hexahydro- 85,86
cyclobuta[1,2-c:4,3-c']bis[1]benzopyran-6,7-dione,
  6,6a,6b,7,12b,12c-hexahydro- 85,86
cyclobut[a]acenaphthylene, 6b,7,8,8a-tetrahydro-7,8-
  dihydroxy-7,8-diphenyl- 199
cyclobuta[1,2-a:3,4-a']diacenaphthylene, cis-
  6b,6c,12b,12c-tetrahydro- 72,73
-----, trans-6b,6c,12b,12c-tetrahydro- 72,73
cyclobutadiene, tetraphenyl- 93,465
cyclobuta[1,2-d:4,3-d']dipyrimidine-2,4,5,7-tetrone,
  dodecahydro-4,4b-dimethyl- 91
cyclobuta[b]naphthalene, trans-1,2-dihydro-1,2-diphenyl-
  447
cyclobutane 264
cyclobutane, 1,2-dicinnamoyl-3,4-diphenyl- 76,77
-----, 1,3-dicinnamoyl-2,4-diphenyl- 76
-----, cis-1,2-divinyl- 74
-----, trans-1,2-divinyl- 74
-----, cis,trans,cis-1,2,3,4-tetracyano- 82,83
-----, cis,trans,cis-1,2,3,4-tetrakis-(4-amidinophenyl)-
  72,73
-----, tetrakis-(diphenylmethylene)- 89,90
-----, cis,trans,cis-1,2,3,4-tetraphenyl- 71-73
-----, trans,trans,trans-1,2,3,4-tetraphenyl- 71,73
cyclobutane derivatives 70-92
cyclobutane-1,2-dicarboxylic anhydride, 3,3,4-trichloro-
  1,2-dimethyl- 111
1,3-cyclobutanedione, tetramethyl- 218
1,2,3,4-cyclobutanetetracarboxylate, tetramethyl 82,
  83
1,2,3,4-cyclobutanetetracarboxylic dianhydride 82
1,2,3,4-cyclobutanetetracarboxylic dianhydride,
  dimethyl- 82
-----, tetramethyl- 82,83
cyclobutanetetracarboxylic 1,2:3,4-dianhydride,
  tetramethyl- 168
1,2,3,4-cyclobutanetetracarboxylic 1,2:3,4-diimide 82
1,2,3,4-cyclobutanetetracarboxylic 1,2:3,4-diimide,
  N,N'-dicyclohexyl- 82
-----, N,N'-diphenyl- 82
cyclobutanol, 1-methyl-2-vinyl- 37

1-cyclobutanol, 2,3-epoxy-cis-1,2-diphenyl- 35,409
-----, 1-methyl-2-propyl- 34,35
cyclobutanols 34-37
cyclobutanols, 1-acyl- 35
cyclobutanone, 2-butyl-3-ethyl-2-hydroxy- 35,36
-----, hexafluoro- 422
-----, hexamethyl- 219,220
-----, 2-hydroxy-2-methyl- 35
-----, 2-hydroxy-3-methyl-2-propyl- 35
cyclobutanone oxime 264
cyclobutanones, 2-hydroxy- 35,36
cyclobuta[l]phenanthrene, trans-1,2-dihydro-1,2-diphenyl-
  221
cyclobuta[l]phenanthrene-1,2-dicarboxylic anhydride,
  1,2,2a,10b-tetrahydro- 110
cyclobutene 10
cyclobutene derivatives 4-10
cyclobutene-1,2-dicarboxylate, dimethyl 115
cyclobutyl nitrite 249
cycloamphidine 244
3,5-cyclo-5α-cholestane, 6β-ethoxy- 10,11
cyclodecane 264
1,2-cyclodecanedione 35,36
cyclodecanone 34,35
cyclodecanone oxime 264
cyclododecane 263
cyclododecanol 34
cyclododecanone 34
cyclododecanone oxime 263
all-trans-1,5,9-cyclododecatriene 57,58
cis,cis,trans-1,5,9-cyclododecatriene 57,58
cis,trans,trans-1,5,9-cyclododecatriene 57,58
3-cyclododecen-1-one 34
1,3-cycloheptadiene 4,5,9,384,385
1,3-cycloheptadiene, 1,4-diphenyl- 4,5,384,385
2,4-cycloheptadien-1-ol, 5-methoxy- 4,5
3,5-cycloheptadien-1-ol, 5-methoxy- 4,5
2,4-cycloheptadien-1-one, 5-methoxy- 4,6
-----, 2,6,6-trimethyl- 4,6
3,5-cycloheptadienone, 2-methyl- 220
3,5-cycloheptadien-1-one 220
6aH-cyclohepta[a]naphthalene, 5,6-dihydro- 278,279

576 Compound Index

cycloheptane 264,437,438
cycloheptanesulfenyl chloride 438
cycloheptanone oxime 264
cycloheptatriene 276-279,325
1,3,5-cycloheptatriene 4,5
1,3-cycloheptatriene, 7-phenyl- 276
cycloheptatrienecarboxylic acids 277
2,4,6-cycloheptatrien-1-one, 4-methoxy- 4,6,9
cyclohexadec[a]acenaphthylene-7,20-dione, tetradecahydro- 395,397
1,3-cyclohexadiene 14,18,19,75,171,175,384,463
1,3-cyclohexadiene, 5-benzylidene-6-(α-hydroxybenzylidene)- 24,25,208
-----, trans-5,6-dimethyl- 19
-----, 5-(α-hydroxybenzylidene)-6-methylene- 24,25, 208
-----, 1,2,3,4,5-pentaphenyl- 17,152
-----, 1,5,5,6-tetramethyl- 19
2,5-cyclohexadiene-Δ¹,α-acetic acid, 4-dichloromethyl-4-methyl- 40
2,5-cyclohexadiene-1-acetic acid lactone, 1-hydroxy-4-oxo-α,α-diphenyl- 29
1,3-cyclohexadiene-1-carboxylate, ethyl 15
1,3-cyclohexadiene-1,4-dicarboxylate, dimethyl 16,17
-----, dimethyl 2,3-dimethyl- 115
trans-3,5-cyclohexadiene-1,2-dicarboxylic acid 14,15
3,5-cyclohexadiene-1,2-dicarboxylic anhydride, 1,2-dimethyl- 383,384
1,5-cyclohexadienes 14-20
1,5-cyclohexadiene-1-sulfonic acid, 3-diazo-4-oxo- 309,310
2,5-cyclohexadien-1-imine, 4-diazo-N-phenyl- 311
2,4-cyclohexadien-1-one, 6-acetoxy-2,6-dimethyl- 228
-----, 6-acetoxy-4,6-dimethyl- 228
-----, 6-acetoxy-6-methyl- 228,230,235
-----, 6-acetoxy-2,4,6-trimethyl- 228,229,234,235
-----, 6,6-diacetoxy-4-methyl- 228,230
-----, 6-diazo- 305
-----, 6,6-dimethyl- 228
-----, 2,3,4,5-tetrachloro-6-diazo- 307-309
cyclohexa-2,5-dien-1-one, 4-(4-aminophenyl)-4-methyl- 337,338
2,5-cyclohexadien-1-one, 4-(α-benzoylbenzylidene)- 423,424
-----, 4-diazo- 310,311
-----, 4-diazo-2,6-dichloro- 311
-----, 2,6-di-tert•butyl-4-hydroxy-4-phenyl- 29
-----, 4-(α-phenylbenzylidene)- 29
cyclohexadienones, cross-conjugated 28-32
2,4-cyclohexadien-1-ones 227-230,234,235,448
2,5-cyclohexadien-1-ones 28-32
cyclohexane 196,197,198,206,233,234,260-265,264,283, 284,286,290,319,320,324,350,351,435,437-439
cyclohexane, acetonyl-2-acetyl- 113
-----, azido- 319,320
-----, trans-1,2-bis-(2-hydroxy-2-propyl)- 198,199
-----, bromo- 351,352
-----, cis-1-bromo-4-tert•butyl- 352
-----, trans-1-bromo-4-tert•butyl- 352
-----, chloro- 262,435
-----, 1-chloro-2-methyl-1-nitroso- 197
-----, 1-chloro-1-nitro- 261,265
-----, 1-chloro-1-nitroso- 197,261,262
-----, cyano- 260
-----, cis-1,2-dibromo- 158
-----, trans-1,2-dibromo- 352
-----, cis-1-trans-2-dibromo-4-tert•butyl- 352
-----, 1,4-di-tert•butyl- 264
-----, 1,4-dimethyl- 264
-----, 1,2,3,4,5,6-hexachloro- 343
-----, iodo- 370
-----, methyl- 198,233,350,435
-----, nitroso- 262
-----, 1,3,5-trimethyl- 264
cyclohexane dimer, nitroso- 261-263
cyclohexanecarbamate, ethyl 324
cyclohexanecarboxamide 167
cyclohexanecarboxylic acid 370
cyclohexanecarboxylic acid chloride 233,234
1,3-cyclohexanedione, 5,5-dimethyl- 113
cyclohexanes, dibromo- 352
-----, hexachloro- 341-345
cyclohexanesulfenyl chloride 438
cyclohexanesulfonic acid 438,439

Compound Index   577

cyclohexanesulfonyl acetyl peroxide   438
cyclohexanesulfonyl chloride   435
1,3,5-cyclohexanetrione, hexamethyl-   219,220
cyclohexanone   26,170,210,218,225,226,307,308,320
cyclohexanone, 2-acetyl-   37,38
------, 2-acetyl-cis-2,5-dimethyl-   409
------, 2-acetyl-trans-2,5-dimethyl-   409
------, 2-acetyl-2-methyl-   38
------, 2-cyclohexyl-   170
------, 2-(2-cyclohexylideneethylidene)-   59,61
------, 2-methyl-   38,225
------, 2-(β-methylallyl)-   112
------, 3-(β-methylallyl)-   112
------, 2-methyl-2-(1-methyl-2-oxocyclohex-1-yl)-   38
------, 2-octyl-   170
cyclohexanone oxime   261-265
cyclohexanone oxime, 2,5-di-tert-butyl-   264
------, 2,5-dimethyl-   264
------, 2-methyl-   197
------, 2,4,6-trimethyl-   264
cyclohexene   111,113,163,167,170,171,176,198,199,201,
276,290,324,325,418
cyclohexene, 1-acetyl-   22
------, 3-acetyl-   22
------, 1-bromo-   158
------, 3-bromo-4-(trichloromethyl)-   172,175
------, 6-bromo-3-(trichloromethyl)-   172,175
------, 1-chloro-   163
------, 1-methyl-   163,276
------, 3-methyl-   276
------, 4-methyl-   276
------, 1,3,4,5,6-pentachloro-   344
------, 3,4,5,6-tetrachloro-   344
------, 4-vinyl-   74
1-cyclohexene, 1,2-dimethyl-   373,374
2-cyclohexene, 1,2-dimethyl-1-hydroperoxy-   373,374
------, 2,3-dimethyl-1-hydroperoxy-   373,374
2-cyclohexene-methanol, α,α-dimethyl-   198,199,201
------, α-ethyl-   201
3-cyclohexen-1-ol, 1-methyl-   37
1-cyclohexen-1-ol acetate   37,38
1-cyclohexen-1-ol acetate, 2-methyl-   37,38

1-cyclohexen-1-ol benzoate   38
1-cyclohexenols, 1-alkyl-   37
2-cyclohexenone   112
2-cyclohexen-1-one, 3,5-dimethyl-   77
cyclohexyl nitrate   262
cyclohexylamine   228-230,319,320
cyclohexylamine, N-chloro-N-methyl-   244
cyclones   394,395,397
cyclononane   264
cyclononanone oxime   264
1,3-cyclooctadiene   4,5
1,5-cyclooctadiene   7,18,418
cis,cis-1,5-cyclooctadiene   172,173
1,5-cyclooctadienes   18
cyclooctane   264
cyclooctanone   35,36,116
cyclooctanone oxime   2626,264,265
cyclooctatetraene   114,115,116
cyclooctatetraene, 1-cyano-2,3-dibutyl-   114
------, 1-cyano-2,3-diethyl-   114
------, phenyl-   93,460
------, 1,2,4,7-tetraphenyl-   409,410
1,3,5,7-cyclooctatetraene, 1,3,5,7-tetramethyl-   102,
103
cyclooctatetraene derivatives   459
cyclooctatetraene dimer   116,117
cyclooctatetraenecarboxylate, methyl   114
1,3,5-cyclooctatriene   5,7
cyclooctene   418,419
cyclooctene chloranil 2:1 adduct   418,419
8H-cyclopent[a]acenaphthylen-8-one, 7,9-diisopropyl-
395
cyclopentadiene   75,116,171,384,394
cyclopentadiene, 5-cyclohexyl-   284,286
------, 5-diazo-   284,286
------, 1,4-diphenyl-   384,394
cyclopentadiene-5-carboxylic acid   305
cyclopentadiene-2-sulfonic acid, 5-carboxy-   309,310
2,4-cyclopentadien-1-cl, pentaphenyl-   394
cyclopentadienone, tetraphenyl-   368,369,394,395
cyclopentadiencnes   394,395,397
9H-cyclopenta[1,2-b:4,3-b']dipyridine   200,201

37   Schönberg, Photochemistry

578   Compound Index

9H-cyclopenta[1,2-b:4,3-b']dipyridine, 9-oxo-   200,201
cyclopentane   233,264,437,438
cyclopentane, acetonyl-2-acetyl-   113
——, 1,2:3,4-diepoxy-5-methylene-1,2,3,4-tetraphenyl-   407,408
cyclopentaneacetaldehyde, 2,2-dimethyl-3-methylene-   252,253
——, 2,2,3-trimethyl-3-nitroso-   252
1-cyclopentanecarboxaldehyde, 2-isopropenyl-5-methyl-   27
1,2-cyclopentanedione   442,443
1,3-cyclopentanedione, hexamethyl-   219,220
cyclopentanesulfenyl chloride   438
cyclopentanone   218
cyclopentanone oxime   264
2H-cyclopenta[l]phenanthren-2-one, 1,3-dihydro-1,3-diphenyl-   221
cyclopentene   74,111-113,171
3-cyclopentene, 4-acetyl-3,3,4-trimethyl-   26
2-cyclopentene-1-acetaldehyde, 2,2,3-trimethyl-   227
2-cyclopentene-1-acetate, methyl 4-oxo-   227,229
2-cyclopentene-1-acetic acid, 4-oxo-   77,78
cyclopentenone   77,78
2-cyclopentenone   112,113,115
2-cyclopenten-1-one, 4-benzoyl-2,5-di-tert-butyl-   29
——, 4,5-epoxy-3,4-diphenyl-   409,410
——, 4,5-epoxy-2,3,4,5-tetraphenyl-   410
——, 4,5-epoxy-2,4,5-triphenyl-   410
2-cyclopentenone dimer   115
cyclopentyl nitrite   249
11β,19-cyclo-5α-pregnan-1α-ol, 3,3:20,20-bis-(ethylenedioxy)-   37
18,20-cyclopregn-5-en-20-ol, 3β-acetoxy-   36,37
——, 3,3-ethylenedioxy-   36,37
cyclopropane   342
cyclopropane, 1-acetyl-1-methyl-   45
——, 1-benzoyl-1,2-triphenyl-   44,45
——, 1,1-bis(chloromethyl)-   348
——, chloro-   342
——, (chloromethyl)-   348
——, cis-1,2-dibenzoyl-   64
——, trans-1,2-dibenzoyl-   64,65
——, cis-1,2-dimethyl-   275
——, trans-1,2-dimethyl-   275
——, cis-1,2-diphenyl-   64,65
——, trans-1,2-diphenyl-   64,65
——, hexafluoro-   291
——, hexamethyl-   219,220
——, 1-hydroxy-1-methoxy-2,2,3,3-tetramethyl-   219
——, methyl-   348
——, 1,1,2,-trifluoro-3-(trifluoromethyl)-   288
cyclopropanecarboxaldehyde, cis-3-methyl-trans-2-propenyl-   45
——, trans-3-methyl-trans-2-propenyl-   45
cyclopropanecarboxylate, methyl cis-1,2-dimethyl-   328
——, methyl trans-1,2-dimethyl-   328
——, methyl trans-1,3-dimethyl-   275,278
cis-cyclopropanecarboxylate, methyl cis-2,3-dimethyl-   275,278
trans-cyclopropanecarboxylate, methyl cis-2,3-dimethyl-   275,278
cyclopropenes, vinyl-   175,176
cyclopropanone, tetramethyl-   218,219
cyclopropene, 3-(diazoacetyl)-1,2-diphenyl-   300,301
——, 1,2-dimethyl-   281
——, 1,3-trimethyl-   281,329,330
cyclopropenecarboxylate, methyl 1,2-dimethyl-   281
cyclopropenone, diphenyl-   220,221
11,19-cyclosteroids   37
16,20-cyclosteroids   36,37
cycloundecane   264
cycloundecanone oxime   264

cis-9-decalol   34,35
5,6-decanedione   35,36
decanoate, ethyl   45
2-decanol, 2-methyl-   160
2-decanone   168,169
2-decanone, 6-benzylidene-   423
2,4,6,8-decatetraene, 3-cyano-2-methyl-   112
1-decene   166
cis-dec-2-ene-4,6,8-triyn-1-ol   41
trans-dec-2-ene-4,6,8-triyn-1-ol   41
5-decyne   114,423
dehydroabietic acid   153,154

dehydroergosterol 386
dehydrogriseofulvin 66
9-cis-3-dehydroretinoic acid 59,60
9-cis,13-cis-3-dehydroretinoic acid 59,60
deoxybenzoin 205,216
deoxybenzoin derivatives 215-217
deoxyribonucleic acid 91
des-D-[pregna-5,13(18)diene, 3,3-ethylenedioxy-14β-vinyl- 37
desyl chloride 216,217
desyl compounds 215-217
deuterium 222
deuterium oxide 226
dianthracene 97-99,101,102,149
diars see benzene, 1,2-bis-(dimethylarsino)-,
1,4-diazabicyclo[2.2.2]octane 463
1,8-diazafluorene 200,201
1,8-diazafluoren-9-ol, 9-(1,8-diazafluoren-9-yl)- 200,201
1,8-diazafluoren-9-one 200,201
6,7-diazatetracyclo[3.2.1.0²·⁴.0²·⁶]non-6-ene 328,329
8,9-diazatetracyclo[4.3.0.0²·⁴.0³·⁷]cn-8-ene 328,329
3,7-diazatricyclo[4.2.2.2²·⁵]dodeca-9,11-diene-4,8-dione 100,101
3,7-diazatricyclo[4.2.2.2²·⁵]dodeca-9,11-diene-4,8-dione, 3,7-bis-(2-hydroxyethyl)- 100
-----, 1,5-dimethyl- 100
-----, 3,7-dimethyl- 100
-----, 9,11-dimethyl- 100
-----, 2,3,6,7-tetramethyl- 100
-----, 3,7,9,11-tetramethyl- 100
3,7-diazatricyclo[4.2.2.2²·⁵]dodeca-3,7,9,11-tetraene, 4,8-bis-(benzylamino)- 100
-----, 4,8-diamino- 100,101
-----, 4,8-diamino-10,12-dichloro- 100
-----, 4,8-diamino-1,5-dimethyl- 100
-----, 4,8-diamino-2,6-dimethyl- 100
1,2-diazetidine, 3,4-bis-(4-dimethylaminophenyl)-1,2-diphenyl- 141,142
1,2-diazetidin-3-one, 1,2-dibenzcyl-4,4-diphenyl- 298
-----, 1,2-dibenzyl-4,4-diphenyl- 298,299
-----, 1,2,4-tetraphenyl- 298,299

-----, 1,2,4-triphenyl- 298,299
1,2-diazetidin-3-ones 298,299
diazirine 291
diazirine, difluoro- 291
-----, dimethyl- 290
diazoacetate, methyl 275,278,281-283
diazoketones 294-303
diazomethane 274-279,281-283,286,291,325
diazomethane, benzoyl- 298
-----, dimesityl- 289
-----, diphenyl- 275,280,283,286-288,290,291
-----, methyl- 290
-----, phenyl- 275,284,289
diazonium salts 313-316
diazosulfonates, aryl 316
dibenza,i]acridine 101
dibenz[b,h]acridine 101
dibenz[b,h]acridine, 8,13-epidioxy- 393
dibenz[a,j]acridinium chloride, 7,14-diphenyl- 135
11a,11b,12-octahydro- 81
dibenzo[c,h]cinnoline 140
dibenzamide, N-phenyl- 49,50
8H-dibenz[a,de]anthracen-8-one, 13-phenyl- 132,134
dibenzo[b,h]biphenylene-5,6,11,12-tetrone, 5,5a,5b,6,11, 11a,11b,12-octahydro- 81
dibenzo[a,e]cyclooctadiene 220,221
dibenzo[a,e]cyclooctadiene, 3,4-diphenyl- 221
dibenzo[a,e]cyclooctatetraene, 5,6-diphenyl- 18
-----, 5,11-diphenyl- 18
dibenzo-c-dioxin, 1,2,3,4,4a,6a,7,8,9,10-decahydro- 388,389
dibenzofuran, 1,2,3,4,4a,5a,6,7,8,9-decahydro-4a-hydroperoxy-5a-methoxy- 429,430
-----, 1,2,3,4,6,7,8,9-octahydro- 429
dibenzofuran-1,4-dione, 8-hydroxy-3,7-dimethyl- 426
dibenzo[a,o]erylene, 7,16-diphenyl- 392
dibenzo[a,o]erylene-7,16-diol diacetate 132
dibenzo[a,o]erylene-7,16-dione 131-133
dibenzo[a,o]erylene-7,16-dione, 1,6,8,11,12,15-hexachloro- 133,134
8H,16H-dibenzo[a,j]erylene-8,16-dione 394
8H,16H-dibenzo[a,j]erylene-8,16-dione, 4b,12b-epidioxy- 393,394

Compound Index

7H,16H-dibenzo[a,o]perylene-7,16-diyl, 3a,11b-epidioxy-7,16-diphenyl- 392
10H,13H-ditenzo[a,o]perylene-10,13-diyl, 3a,11b-epidioxy-7,16-diphenyl- 392
dibenzo[c,i]phenanthridine 142
dibenzo[a,j]xanthene, 14-(2-chlorophenyl)- 134-136
----, 14-phenyl- 134,135
14H-dibenzo[a,j]xanthene, 14-methylene- 119,120
----, 14-phenyl- 398
dibenzo[a,j]xanthylium perchlorate, 14-(2-chlorophenyl)- 134-136
----, 14-phenyl- 134,135
----, 14-p-tolyl- 134-136
dibenzoxanthylium salts 134-136
dibutylamine 242,243
dibutylamine, N-bromo- 243
----, N-chloro- 242,243,246
1,2-dicarbonyl compounds 419,420
3,5:4,6-dicyclo-5α-chclestane 10,11
endo-dicyclopentadiene 2,3,75
exo-dicyclopentadiene 75,446
dienol acetates 37,38
5,12,11-diepoxydibenzo[a,e]cyclooctadiene, 5,6,11,12-tetrahydro-5,6,11,12-tetraphenyl- 103
diethylamine 287,288,381
diethylamine, N-(diphenylmethyl)- 287,288
C-dihydrctoxiferine-I 401,402
diimide, bis-(2-chlorobenzoyl)- 330,331
----, bis-(4-chlorobenzoyl)- 330
----, di-p-anisoyl- 330
----, dibenzoyl- 330
diiron hexacarbonyl, (phenylcyclooctatetraene)- 460
β-diketones 113
1,2-diketones 35,36,231
1,3-diketones 408,409,412,413
dimedone 113
1,4:5,8-dimethanobiphenylene, exo,trans,exo-dodecahydro- 74,468,469
1,4:5,8-dimethanonaphthalene, 5,6,7,8,9,9-hexachloro-2-(4-chlorophenylthio)-1,2,3,4,4a,5,8,8a-octahydro-3-hydroperoxy- 447

dinaphtho[1,8-bc:1',8'-fg][1,5]dithiocine 7,7,14,14-tetroxide 316,317
dinaphtho[1,2-b:2'3'-d]furan, 7-hydroxy-5,12-dimethyl- 427
dinaphtho[1,2-b:2',3'-d]furan-7,12-dione, 5-hydroxy-426,427
dinaphtho[1,2-b:2,3'-d]furan-7-cne, 7,12-dihydro-5,12-diethyl- 427
dinitrogen tetroxide 291
2,3-dioxa-7-azaticyclo[2.2.1]hept-5-ene 395,396
2,3-dioxabicyclo[2.2.1]hept-5-ene 384
2,3-dioxaticyclo[2.2.1]hept-5-ene, 1,4-diphenyl- 384
----, 7-methylene-1,4,5,6-tetraphenyl- 407,408
2,7-dioxaticyclo[3.2.0]hept-3-ene, 6,6-diphenyl- 415-417
----, 6-ethyl- 415,416
----, 6-phenyl- 415-417
2,3-dioxabicyclo[2.2.1]hept-5-en-7-ol, 1,4,5,6,7-pentaphenyl- 394
6,7-dioxaticyclo[3.2.2]non-8-ene 384,385
6,7-dioxaticyclo[3.2.2]non-8-ene, 1,5-diphenyl- 384,385
2,3-dioxaticyclo[2.2.2]oct-5-ene 384
2,3-dioxabicyclo[2.2.2]oct-7-ene-5,6-dicarboxylic anhydride, 5,6-dimethyl- 383,384
2,3-dioxa-5,7-diazabicyclo[2.2.1]hept-5-ene, 1,4,6-triphenyl- 396
dioxane 160,184
1,2-dioxane, 3,3,6,6-tetrakis-(4-methoxyphenyl)- 397
1,3-dioxane, 2-heptyl- 45
----, 2-phenethyl- 45
1,4-dioxane 184,260,379,380
1,4-dioxane, 2-cyano- 260
----, 2-hydroperoxy- 379,380
----, 2-(2,3,4,5-tetrachloro-6-hydroxyphenoxy)- 184
3,9-dioxapentacyclo[6.4.0.0²·⁷.a⁴·¹¹¹.0⁵·¹⁰]dodecane-6,12-dione, 2,4,8,10-tetramethyl- 2,3,89
2,7-dioxatricyclo[6.2.0.0³·⁶]decane-1,6-diol, 3,8-dimethyl- 35
3,7-dioxatricyclo[4.2.2.2⁵]dodeca-9,11-diene-4,8-dione, 2,6,9,11-tetramethyl- 102
3,9-dioxatricyclo[6.4.0.0²·⁷]dodeca-4,10-diene-6,12-dione, 2,4,8,10-tetramethyl- 2,3

Compound Index

3,11-dioxatricyclo[6·4·0·0²·⁷]dodeca-5,9-diene-4,12-
  dione, 2,6,8,10-tetramethyl-   102,103
3,11-dioxatricyclo[4·1·0·0²·⁴]heptane, 5-methylene-
  1,2,4,6-tetraphenyl-   407,408
6,7-dioxatricyclo[3·2·2·0²·⁴]non-8-ene-3-carboxylate,
  methyl   384
1,4-dioxins, dihydro-   419
1,3-dioxolane, 2-benzyl-   45
----, 2,4-dimethoxy-   279,280
----, 2-heptyl-   45
----, 2-nonyl-   45
----, 2-pentyl-   45
----, 2-phenethyl-   45
1,3-dioxole derivatives   186
diphenylamine   138
diphenylamine, 2-amino-5-methyl-   338
----, 4-bromo-   314,315
----, 4-chloro-   314,315
----, 4-ethoxy-   311
----, N-methyl-   138
----, 4-phenyl-   311
diphenylmethane   373
diphos see ethane, 1,2-bis-(diphenylphosphino)-,
diphosphine, tetraphenyl-   455
diselenide, bis-(2,4-dinitrophenyl)   353
dispiro[5·1·5·1]tetradecane-7,14-dione   219
disulfide, 2-benzothiazolyl   152,153
----, bis-(2-chloro-1,1,2-trifluoroethyl)   163
----, bis-(heptafluoropropyl)   441
----, bis-(trifluoromethyl)   164,439-441
----, dicyclohexyl   437
----, diphenyl   152,153
----, isoamyl   152
disulfides   152,153
dithiafulvenes   331
1,3-dithietane, 2,4-bis-(chlorodifluoromethyl)-2,4-
  difluoro-   106,107
----, 2,4-bis-(trifluoromethyl)-2,4-difluoro-   106,107
----, 2,2,4,4-tetrachloro-   450
dithietanes   106
1,4-dithiin, tetraphenyl-   331,332
1,4-dithiins   331,332

1,3-dithiole, 2-benzylidene-4-phenyl-   331,332
----, 4,5-diphenyl-2-diphenylmethylene-   331,332
1,3-dithioles, 2-methylene-   331,332
dodecane   435
dodecanediamide   166
1-dodecanethiol   446
durohydroquinone   461
dyes   146
dyes, azo   309
trans-dypnone oxide   408,409

enol acetates   37,38
enol benzoates   38
eosin   149
epi-α-caryophyllene alcohol nitrite   249,250
epidioxide, hetero-cerodianthrone   393,394
----, rubrene   390
trans-epidioxide, phellandrene   384
epidioxides   46,47,382-397,407
epidioxides, acene   389-394
7,12-epithiopleiadene 13,13-dioxide, 7,12-dihydro-   447,448
epoxides   399,400,407-413
5,8-epoxy-5H-benzocycloheptene-6,7-dicarboxylate,
  dimethyl 8,9-dihydro-8-methyl-9-oxo-5-phenyl-   411
----, dimethyl 8,9-dihydro-9-oxo-5,8-diphenyl-   411
5,11-epoxy-1,4-methano-5H-dibenzo[a,d]cyclohepten-10-one,
  1,4,4a,10,11,11a-hexahydro-5,11-diphenyl-   411,412
Δ⁸-ergolene-8-carboxylic acid, 6-methyl-   156
5α-ergosta-6,22-diene, 3β-acetoxy-5,8α-epidioxy-   385,388
5α-ergosta-7,14-diene, 3β-acetoxy-   387,388
5α-ergosta-8(14),22-diene, 3β-acetoxy-7,15-epidioxy-   387
5α-ergosta-6,22-dien-3β-ol, 5,8α-epidioxy-   385,388
ergosta-5,7,9(11),22-tetraene, 3β-acetoxy-   385,388
ergosta-6,8(14),9(11),22-tetraene, 3β-acetoxy-   387
ergosta-5,7,9(11),22-tetraen-3β-ol   386
ergosta-3,5,8,22-tetraen-7-one   385,388
5α-ergosta-6,8,22-triene, 3β-acetoxy-11α,14α-epidioxy-   386,387
5α-ergosta-7,14,22-triene, 3β-acetoxy-   387
ergosta-5,7,22-trien-3β-ol   385,386,388

Compound Index

5α-ergosta-6,9(11),22-trien-3β-ol, 5,8α-epidioxy- 386
9β-ergosta-5,7,22-trien-3β-ol 385,386
ergosta-5,7,22-trien-3β-ol see also ergosterol 15, 17, 19
ergosta-5,8,22-trien-7-one, 3β-acetoxy- 385,388
ergosterol 8,15,17,19,148,149,153,385,386,388
9β-ergosterol 7,8,385,388
ergosterol E3 acetate 387
ergot alkaloids 156,157
ergotamine 156
ergotaminine 149
erythrosin 45
esters 45
estra-1,4-dien-3-one, 10β,17β-diacetoxy- 235
estradiol 17β-acetate 235,236
estra-1,3,5(10)-trien-3-ol, 17β-acetoxy- 235,236
13α-estra-1,3,5(10)-trien-17-one, 3-hydroxy- 65,66
estr-5-ene, 3,3:17,17-bis-(ethylenedioxy)- 222,223
estr-5-ene-10-d, 3,3:17,17-bis-(ethylenedioxy)- 222
estrone 65,66
ethane 401
ethane, 1-azido-1,1-diphenyl- 319
—————, 1,2-bis-(4-biphenylyl)-1,1,2,2-tetraphenyl- 367
—————, 1,1-bis-(4-bromophenyl)-2,2,2-trichloro- 363
—————, 1,1-bis-(4-chlorophenyl)-1,2,2,2-tetrachloro- 363
—————, 1,2-bis-(diphenylphosphino)- 462-464
—————, 1-bromo-1,1,2-trichloro- 282
—————, 1-bromo-1,1,2-trifluoro- 157
—————, 1-bromo-1,2,2-trifluoro- 157
—————, 1-tert-butoxy-2,2-dichloro-1,1,2-trifluoro- 162
—————, 2-chlorotetrafluoro-1-nitroso- 364
    dibenzoyl- 193
—————, 1,2-dibromo-2-chloro-1-phenyl- 350
    dibromotetrachloro- 350
—————, 1,1-dichloro-2,2,2-trifluoro- 234
—————, 1,2-dichlorotrifluoro-1-nitroso- 347
—————, 1,2-dimethoxy- 260,261
—————, 1,2-diphenyl- 216

—————, 1,2-di-p-tolyl- 183
—————, 1,2-epoxymethoxy- 279,280
    eroxytetrachloro- 407
—————, hexachloro- 283,407,467
—————, hexafluoro- 365,457,458
—————, hexaphenyl- 126,281,282
—————, iodo- 366
    pentafluoroiodo- 366,367
    pentafluoronitroso- 364
—————, tetrafluoro-1,2-diiodo- 292
—————, 1,1,2-tetrafluoro-1-(trichlorosilyl)- 178, 179
—————, 1,1,2,2-tetrakis-(4-methoxyphenyl)- 200,201
—————, 1,1,2,2-tetraphenyl- 148,183,216,217,220,221, 28,284,288,290
—————, 1,2,2-tribromo-1-phenyl- 350
—————, 1,1,2-trichloro-2,2-difluoro-1-nitroso- 159
—————, 1,1,2-trichloro-2-phenyl- 346,347
—————, 1,1,2-trichlorotrifluoro- 347
—————, 2,2,2-trifluorodiazo- 288
1,2-ethanediol, 1,1-bis-(4-methoxyphenyl)- 207
—————, 1,2-bis-(4-methoxyphenyl)- 204,207,208
—————, 1,2-bis-(4-methylthiophenyl)-1,2-diphenyl- 206,207
—————, 1,2-di-1-naphthyl- 204
—————, 1,2-diphenyl- 204,414,417
—————, d,l-1,2-diphenyl- 210
—————, tetrakis-(4-chlorophenyl)- 211
—————, tetrakis-(4-methoxyphenyl)- 207,211
—————, tetrakis-(4-methylthiophenyl)- 206,207
—————, tetraphenyl- 203,204,206,209,211
—————, tetra-p-tolyl- 211
    triphenyl- 203
1,2-ethanediol dihydrochloride,
1,2-bis-(4-dimethylaminophenyl)-1,2-diphenyl- 206
1,2-ethanediol, tetraphenyl-, see also benzopinacol
1,4,2,3-ethanediylidenenaphthalene, 1,2,3,4-tetrahydro-2,3-dimethyl-9,10-diphenyl- 116
—————, 1,2,3,4-tetrahydro-9,10-diphenyl- 115,116
ethanes, tetraaryl- 146,147
ethanesulfonic acid, 2-amino- 435
ethanesulfonyl chloride, 2-amino- 435

Compound Index 583

-----, 1,2-dibromo-2-phenyl- 350
1,1,2,2-ethanetetracarboxylate, tetramethyl 167
1,1,2,2-ethanetetracarboxylic acid 167
ethanethiol 297,298
ethanethiol, 2-chloro- 164
-----, 2-chloro-1,1,2-trifluoro- 163
-----, 1,1,2-trifluoro-2-methoxy- 163,164
ethanol 160,193,196,204,205,207,209,210,232,238,268,
  286,294,296,300,307,308,311,321,322,325,334,363,
  364,429
ethanol, 2,2-bis-(4-methoxyphenyl)-1,1-diphenyl- 200,
  201
-----, 2-ethoxy- 467
-----, 1,1,2,2-tetraphenyl- 200
1,4-ethenobenzo[3,4]cyclobuta[1,2-a]cyclopropa[c]
  cycloheptene, 1,1a,4,4a,4b,8a,8b,8c-octahydro- 116,
  117
ether, benzhydryl methyl 286,287
-----, benzyl vinyl 112
-----, diallyl 163
-----, diethyl 196,197,260,261,283
-----, divinyl 163
-----, ethyl hydroquinone 311
-----, ethyl isopropyl 283
-----, ethyl propyl 283
-----, ethyl vinyl 119,120,122,124
-----, glycol monoethyl 467
-----, methyl trifluorovinyl 163,164
-----, methyl α-tropolone 227
-----, methyl γ-tropolone 4,6,9
ethers 159-161,260
ethers, α-cyano- 260
ethyl mercaptan 297,298
ethylamine 435
ethylene 163,173,290,401,421,422,463,464
ethylene, 1,1-bis-(4-bromophenyl)-2,2-dichloro- 397
-----, 1,1-bis-(4-methoxyphenyl)- 363
-----, 1,1-bis-(4-methoxyphenyl)-2-bromo- 366
-----, chlorotrifluoro- 162,163,347
-----, cis-dibenzoyl- 193,238,239
-----, 1,2-dichloro- 415
-----, dichlorodifluoro- 422

-----, 1,1-dichloro-2,2-difluoro- 159
-----, 1,1-difluoro- 164,422
-----, 1,1-diphenyl- 119,120,189,307,415,449
-----, 1,2-di-4-Pyridyl- 119,120
-----, tetrabenzoyl- 239
-----, tetrachloro- 350,407
-----, tetracyano- 161
-----, tetrafluoro- 163,174,178,179,421,422
-----, tetrafluoro- 291
-----, tetramesityl- 222,289
-----, tetramethyl- 218,219,376,378
-----, 1,1,2-trichloro- 111
-----, trifluoro- 157,288,290
-----, trifluoroiodo- 105
-----, trifluoronitroso- 347
-----, triphenyl- 118-120
ethylenediamine, N,N'-diphenyl- 335,336
eucarvone 4,6,27,28,115

ferric chloride 128,129,139,140
1-fluoranthenecarboxylic acid, 3-hydroxy- 239
3-fluoranthenol 239,240
fluorene 147,183,284,344
fluorene, 2-bromo- 344
-----, 9-bromo- 344
-----, 9-chloro-9-phenyl- 367,368
-----, 9-cyclohexyl- 284,290
-----, 9-diazo- 284
-----, 9-diphenylmethyl- 284
-----, 1,2,3,4,4a,9a-hexahydro- 374,375
-----, 9-isopropylidene- 416,417,421
-----, 9-phenyl- 367,368,398
fluorene-9-dl 285
fluorene-9-acetic acid, 1-carboxy- 239,240
fluorene-9-carboxylic acid 304
fluorene-9-malonic acid, 1-carboxy- 239
fluorene-9-one 205-207,212,315,401,415-417,421
folic acid 255,256
formaldehyde 401,467
formamide 166-168,230,231
formamide, N-tert-butyl- 167
-----, N,N-diphenyl- 49

584 Compound Index

formate, tert-butyl azido— 325,326
——, ethyl 279,280
——, ethyl azido— 324,325
——, methyl 279,280
formazan, 3-ethyl-1,5-diphenyl— 64
——, 1,3,5-triphenyl— 64,143
formazans 63,64
fulgide, 6,7-bis-(2-methoxyphenyl)— 129,131
——, 6,7-bis-(4-methoxyphenyl)— 129
——, 6,7-diphenyl— 129
——, 6,7-di-p-tolyl— 129
fulvene, 1,2,3,4-tetraphenyl— 407,408
fumarate, diethyl 167
——, dimethyl 57,58,82,83,461
fumarates, dialkyl 167
fumaric acid 56-58,82,157,231,232
fumaronitrile 82,83,420,421
2-furaldehyde 187,429
2-furaldehyde, 4,5-dimethyl— 89
furan 110,119,120,219,278,415,417,428,429
furan, 3,4-bis-(4-bromophenyl)-2,5-epidioxy-2,5-dihydro-2,5-diphenyl— 431
——, 2-cyanotetrahydro— 260
——, 3-cyanotetrahydro— 260
——, 2:3:4,5-diepoxytetrahydro-2,3,4,5-tetraphenyl— 431
——, 2,3-dihydro— 278
——, 2,5-dihydro-2,5-dimethoxy-2,3,4,5-tetraphenyl— 431
——, 2,3-dihydro-4,5-dimethyl— 45
——, 2,5-dihydro-5-hydroperoxy-2-methoxy-2,5-dimethyl— 428
——, 2,5-dihydro-5-hydroperoxy-2-methoxy-2-methyl— 429
——, 2,3-dihydro-3-methyl-2-propenyl— 45
——, 2,3-dihydro-2,2,4,5-tetraphenyl— 44,45
——, 2,5-dihydro-2,3,5,5-tetraphenyl— 44
——, 2,5-dimethyl— 428,429
——, 2,5-diphenyl— 119,120
——, 2-(2,4-hexadiynylidene)-2,5-dihydro— 41
——, 2-hydroperoxytetrahydro— 379,380
——, 2-methyl— 427,429
——, tetrahydro— 160,161,184,260,283,379,380
——, tetrahydro-2-methyl— 283
——, tetrahydro-3-methyl— 283
——, tetrahydro-2-octyl— 161
——, tetraphenyl— 431
furan derivatives 426-433
furan peroxide dimer, 2,5-dimethyl— 428,429
2-furanol, 5-benzoyl-2,5-dihydro-2,3,4,5-tetraphenyl— 394
2-furanol cyclic acetal, 5-benzcyl-2,5-dihydro-2,3,4,5-tetraphenyl— 394
furan-5-one, 2-ethoxy-2,5-dihydro— 429
furan-5-ones, 2-alkyl-2,5-dihydro-2-hydroxy— 428,429
[2·2](2,5)furanophane 432
2-furansuccinic anhydride, tetrahydro— 161
furfural 429
furfural see also 2-furaldehyde
2H-furo[2,3-c][1]benzopyran-2-one, 1-(2-benzoyloxyphenyl)-4-phenyl— 239
5H-furo[3,2-g]-1-benzopyran-5-one, 4-benzoyloxy-9-hydroxy-7-methyl— 189
furo[2,3-c:5,4-c']bis[1]benzopyran, 6,8-diphenyl— 239
furochromones 419
furocoumarins 87,88,419
furoxan, diphenyl— 262

galactal, triacetyl— 122
gammexane 341-343
germane, octyltriphenyl— 179
——, tetraphenyl— 179
——, triphenyl— 179
L-glucal, 3,4,6-triacetyl— 119,122,124
D-glucopyranose 3,4,6-triacetate, 1,2-0-9',10'-phenanthrylene— 122,124
glutamic acid, N-(p-aminobenzoyl)— 255,256
glutaraldehyde 338
glyceric acid, 2-methyl— 210
glycerol 157
glycol monoethyl ether 467
glyoxal 231
glyoxal diethyl acetal, phenyl— 238
glyoxylate, cyclohexyl phenyl— 210

Compound Index

-----, ethyl phenyl- 210,211,238
glyoxylic acid, phenyl- 210,211
guaiazulene 152,153
guaiene 152,153
α-gurjunene 15

halides 39,40
harmyrin 306
hecogenin acetate 43,44
helianthrene, meso-diphenyl- 392
helianthrone 131-133
helianthrone, 1,4,5,8,10,15-hexachloro-
   133,134
α-heptacyclene 72,73
β-heptacyclene 72,73
heptadecane, 11-acetoxy-1-iodo- 370
2,5-heptadienoic acid, 6-acetoxy-4-methyl- 228
3,5-heptadienoic acid, 6-acetoxy- 228,230
-----, 6-acetoxy-2-methyl- 228
-----, 6,6-diacetoxy-4-methyl- 228,230
-----, 6-methyl- 228
2,5-heptadienoic acid cyclohexylamide, 6-acetoxy-2,4-dimethyl- 229
-----, 6-acetoxy-4-methyl- 228,229
3,5-heptadienoic acid cyclohexylamide, 6-acetoxy- 228-230
heptanal 169
heptanamide 166
heptanamide, 2-methyl- 167
4-heptanamine, N,N-dibromo- 245,246
heptane 233,261,435,437
heptane, 4-amino- 245,246
-----, chloronitroso- 261
-----, 4-chloro-4-nitroso- 197
heptane dimers, nitroso- 261
4-heptanone 198
1,3,5-heptatriene 220
1-heptene 166,167
3-heptene 167
5-hepten-2-one 417,418
5-hepten-2-one, 6-methyl- 418
6-hepten-2-one 37

hexacyclo[5·4·1·0²·⁶·0⁴·¹¹·0⁵·⁹·0¹⁰·¹²]dodecane,
   1,2,3,4,11-hexachloro- 2,3
1,3-hexadiene, 5,5,6,6-pentafluoro-1-iodo- 174
1,4-hexadiene, 2-isopropyl- 218
3,5-hexadienoate, ethyl 2-methylene- 15
3,4-hexadienoic acid 14,22-24
hexamethylenetetramine 463
hexanamide 324
hexanamide, 2-ethyl- 167
hexane 363,364
hexane, dodecafluoro-2-iodo-4-(trifluoromethyl)- 173,174
-----, 1-phenyl- 284
hexanedioate, dimethyl 2,5-bis-(diazo)-3,4-dioxo- 301
hexanoate, ethyl 45
1-hexanol 334
2-hexanol acetate, 4-bromo-6,6,6-trichloro- 171,172
hexanoyl azide 324
1,2,4-hexatriene 14
1,3,5-hexatriene 14,18,19,220
1,3,5-hexatrienes 14-20
2,5-hexatrienoate, ethyl 2-methyl- 15
5-hexenal 26
hexene 364
1-hexene 160,166,449
1-hexene, 1,1-diphenyl- 449
3-hexene, 6-bromo-1,1,1-trichloro-3-methyl- 175,176
her-3-enedioate, dimethyl 2,5-dimethylene- 16
trans-4-hexenoate, ethyl 2,2,6-tribromo- 175,176
4-hexenoic acid γ-lactone, 4-hydroxy-2,2,3,5-pentamethyl- 219,220
3-hexen-2-one, 5-methyl- 22,24
-----, 4,5,5-trimethyl- 22,24
4-hexen-2-one, 5-methyl- 22,24
5-hexen-2-one 417,418
hexyl alcohol 334
1-hexyne 169
3-hexyne 114
hydantoin, 1,3-dibromo-5,5-dimethyl- 359
hydrazine, 1,2-bis-(4-chlorobenzoyl)- 331
-----, tetrafluoro- 231
hydrazobenzene, 4-methyl- 338

586 Compound Index

hydrazobenzenes 139
hydrazones 380,381
hydrazyl, 2,2-diphenyl-1-picryl- 153
hydrindantin 209
hydrobenzoin 204,414,417
hydrobromic acid 314,315
hydrochloric acid 314,315
hydrocinnamate, ethyl 45,46
——, ethyl β-bromo-α-(trichloromethyl)- 171,172
——, propyl 45
hydrogen bromide 157-159
hydrogen chloride 134,263-265,345,346,356,367,368
hydrogen peroxide 131,157,379,399
hydrogen sulfide 162-164
hydroperoxide, 1-benzyl-1-isochromanyl 380
——, cholesterol 377,378
——, cumyl 373,374
——, α,α-dimethylbenzyl 373,374
——, 1,4-dioxan-2-yl 379,380
——, diphenylmethyl 373,374
——, 2,3,4,5,6,7-hexahydro-4a(1H)-naphthyl 376,378
——, 1-indanyl 373,374
——, 1-isochromanyl 379,380
——, p-methylbenzyl 373,374
——, myrtenyl 375,378
——, trans-pinocarveyl 379,380
——, tetrahydrofuryl 375
——, verbenyl 373-382,396,397,407
hydroperoxides 373-382,396,397,407
hydroperoxides, organotin 466
hydroquinone 212,461
hydroquinone, tetrachloro- 151,182
hydroquinone monobenzoate, tetrachloro- 189
hydroxamic acids 252,253
hydroxylamine, N,N-bis-(trifluoromethyl)-O-nitroso- 105,106
——, O-methyl-N-(p-toluenesulfonyl)- 326
hypericin 132
hypiodites, acyl 370
——, alkyl 369,370

imene, carboethoxy 324

imenes 318,321,322,324,325
8H-imidazo[5,1-a:4,3-a']diisoquinoline, 5,6,10,11,15b,15c-hexahydro-15b,15c-dimethyl- 335
imidazole 54,396
imidazole, 2,4,5-triphenyl- 396
4H-imidazole, 4-hydroperoxy-2,4,5-triphenyl- 335-337
imidazolidine, d,l-1,3-dimethyl-4,5-diphenyl- 335-337
——, meso-1,3-dimethyl-4,5-diphenyl- 335-337
——, 1,3-diphenyl- 318
imidogens 277,373,374
indan 277,373,374
indan, 2-benzylidene-1-(diphenylmethylene)- 129-131
——, 1-bromo-2-(trichloromethyl)- 171,172
——, 4,7-dimethyl- 277
——, 1-hydroperoxy- 373,374
——, 2-phenyl- 278,279,284
1,3-indandione, 2,2-diphenyl- 43,44
——, 2-methyl-2-phenyl- 43
1-indanone, 2-diazo- 302
——, 2-diazo-3,3-diphenyl- 306,307
——, 2,3-epoxy-2,3-diphenyl- 410-412
——, 2,3-epoxy-2-methyl-3-phenyl- 411
2-indanone 220,221
2-indanone, 1-diazo-3,3-diphenyl- 306-308
——, 1,3-diphenyl- 221
5(6H)-indanone, 6-diazo- 305
1,2,3-indantrione 119,120,209
1H-indazole, 1-methyl- 55
——, 7-methyl- 54
1H-indazoles 54
1H-indazoles, 2-alkyl- 55
indene 119,120,122,124,171,419
indene-1-carboxylic acid 304,306
indene-1,2-dicarboxylate, dimethyl 3-phenyl 309,310
indene-3-sulfonic acid, 1-carboxy- 309,310
indeno[1,2,3-fg]naphthacene, 4b,9-epidioxy-9,10-diphenyl- 393
10H-indeno[1,2-b]phenanthro[9,10-e][1,4]dioxin, 9a,14b-dihydro- 122,124
indigo see indigotin
indigotin 61,271,272

Compound Index 587

indigotin, 1,1'-diacetyl- 61
-----, 5,5'-dibromo- 61
indole, 1-acetyl- 52
3H-indole, 3,3-dimethyl-2-phenyl- 335,336
indole-3-carboxylic acid 305,306
3H-indole-2-carboxylic acid 1-oxide, 3-hydroxy- 271
indoline, 2-tert•butyl-2-hydroxy-1,3,3-trimethyl- 401
-----, 2-hydroxy-1,3,3-trimethyl- 401
-----, 2-hydroxy-1,3,3-trimethyl-2-phenyl- 401
2-indolinemethanol, 1,3-dimethyl-2-phenyl- 336
indolizine, octahydro- 242
iodine 127-129,131,136,141,142,291,344,353,364,368-371,465
iodine chloride 354,364,365
iodonium tetrafluoroborate, diphenyl- 456,457
iodonium tetrafluoroborates, diaryl- 456
cis-α-ionone 58-60
trans-α-ionone 41,58-60
cis-β-ionone 41,42
trans-β-ionone 41,42
iron dicarbonyl, (acetyl)(cyclopentadienyl)- 464,465
-----, (acryloyl)(cyclopentadienyl)- 464,465
-----, (benzoyl)(cyclopentadienyl)- 464,465
-----, (cyclopentadienyl)(methyl)- 464,465
-----, (cyclopentadienyl)(phenyl)- 464,465
-----, (cyclopentadienyl)(vinyl)- 464,465
iron dicarbonyl compounds, (acyl)(cyclopentadienyl)- 464
iron hexacarbonyl, (phenylcyclooctatetraene)di- 460
iron pentacarbonyl 459-462,469
iron tetracarbonyl, (acetonitrile)- 461
-----, (dimethyl fumarate)- 461
-----, (dimethyl maleate)- 461
-----, (methyl methacrylate)- 461,462
-----, (vinyl acetate)- 459
iron tricarbonyl, butadiene- 459,461
-----, duroquinone- 462
-----, isoprene- 462,469
-----, (nitrosobenzene)- 460
-----, (phenylcyclooctatetraene)- 460
-----, (1,2,3,4-tetraphenyl-1,3-butadiene)- 459,462
iron tricarbonyl dimer, (nitrosobenzene)- 469

isatogen, 6-nitro-2-phenyl- 271,272
-----, 2-phenyl- 271,272
isoalloxazine, 10-(2,3-dihydroxypropyl)- 258
-----, 7,8-dimethyl-10-(5-hydroxypentyl)- 258
-----, 10-(2-hydroxyethyl)- 258
-----, 10-methyl- 258
-----, 7,8,10-trimethyl- 257,258
isoascaridole 46,47
isobenzofuran, 1,3-diphenyl- 103,396,430,431
-----, 1,3-epidioxy-1,3-dihydro-1,3-diphenyl- 430,431
isobenzofuran dimer, 1,3-diphenyl- 103
isobergapten 87,88
isobergapten dimer 87,88
isobidesyl 217
d,l-isoborneol nitrite 252
isobutane 350,435
isobutene 401
isobutylamine 381,435
isobutylene 112
isobutyrate, methyl 219
isochromane 379,380
isochromane, 1-benzyl- 380
-----, 1-benzyl-1-hydroperoxy- 379,380
-----, 1-hydroperoxy- 379,380
-----, 1-hydroperoxy-1-methyl- 379,380
-----, 1-methyl- 379,380
3-isochromanone, 1,1-diphenyl- 306-308
isocolchicine 8,9
isocoumarin 419
isocoumarin, 3-methyl-4-phenyl- 411
isocoumarin, 3-phenyl- 87,119,120,122,124,419,420
isocoumarin dimer, 3-phenyl- 87
isocrotonic anilide 165
isocyanate, phenyl 321,322
isocyanates 321-323
isodrin 2,3
isoguanine 52
isoguanine 1-oxide 52
isoindigo, 1,1'-dimethyl- 289
isoindole, 1,3-epidioxy-1,2,3-triphenyl- 396
-----, 1,2,3-triphenyl- 396
isolysergic acid 156

588 Compound Index

isooctane 233
isopentyl alcohol 334
isophotosantonic lactone 30,32
isopimpinellin 87
isoprene 74,424,462
isopropanol 203-212,379
isopropanol see 2-propanol
isopropenyl acetate 37,38
isopropenyl benzoate 38
isopropylamine 287
isopropylamine, N-(diphenylmethyl)- 287
isoprocalciferol 7,8,385,386
isoquinoline 460
isoquinoline, N-aryl-1,2,3,4-tetrahydro- 399
-----, 3,4-dihydro-1-methyl- 335
-----, 1,2,3,4-tetrahydro-2-p-tolyl- 399
isothiocyanates 443
isovaleraldehyde 189

jonones 23

kahweol 432
kahweol 17-acetate 432,433
ketene 35,116
ketene, (carbomethoxy)-phenyl- 294
-----, dimesityl- 222
-----, dimethyl- 219
-----, diphenyl- 119-121,294,295,298
-----, thio- 331,332
ketenecarboxylate, methyl phenyl- 294
ketenes 294-299,301,302,304,309,310
ketenes, butadienyl- 227
ketenimine, dimethyl-N-(2-cyano-2-propyl)- 301
ketipate, dimethyl 2,5-bis-(diazo)- 420,421
ketone, dibenzyl 220,221
-----, dimesityl 222
-----, di-1-naphthyl 205
-----, methyl 1-methylcyclopropyl 45
-----, methyl 2-naphthyl 208
-----, methyl 2-pyridyl 205
-----, methyl 3-pyridyl 205
-----, methyl 4-pyridyl 205

-----, 1-naphthyl 5-tert-butyl-2-hydroxyphenyl 237
-----, 1-naphthyl phenyl 194,205,415
-----, 2-naphthyl phenyl 415
-----, phenyl 2-pyridyl 205
-----, phenyl 3-pyridyl 205
-----, phenyl 4-pyridyl 195,205
-----, phenyl 1,2,2-triphenylcyclopropyl 44,45
MICHLER's ketone see benzophenone, 4,4'-bis-(dimethylamino)-
ketones 146,169,414-423
ketones, dialkyl 198,199
-----, diazo- 294-304
-----, epoxy- 407-413
-----, thio- 445
-----, α,β-unsaturated 75-81
khellinhydroquinone monobenzoate 189
khellinone 119,120
khellinquinone 189

lactams 51,52,322-324
γ-lactones 43,44,231,370,371
5α-lanostan-3-one 226,227
lead tetraacetate 369-371
d,l-leucine, N-(2,4-dinitrophenyl)- 268,270
levopimaric acid 7,9,153,154,387
(+)-limonene 178,375,376
lithium 469
lithium, 2-naphthyl- 469
-----, phenyl- 469
lobelanine 338
lophine 396
luniandrosterone 65,66
lunichrome 257,258
α-lumicolchicine 8,9,65,92
β-lumicolchicine 8,9,65,92
γ-lumicolchicine 8,9,65,92
lumi-dihydrotoxiferine-I 402
luniergotamines 156,157
luniestrone 65,66
luniflavine 257,258
lumihecogenin acetate 43,44
lumiisocolchicine 8,9

Compound Index 589

lumiisocolchicine-methanol adduct 8,9
lumiprednisone acetate 31,32
lumisantonin 30,32,229,230
5β-lumista-6,22-diene, 3β-acetoxy-5,8β-epidioxy- 386
5β-lumista-5,7,9(11),22-tetraene, 3β-acetoxy- 386
5α-lumista-6,9(11),22-triene, 3β-acetoxy-5,8α-epidioxy- 386
5β-lumista-6,9(11),22-triene, 3β-acetoxy-5,8β-epidioxy- 386
lumista-5,7,22-trien-3β-ol 385,386
9α-lumista-5,7,22-trien-3β-ol 385,386
lumista-5,7,22-trien-3β-ol see also lumisterol
lumisterol 8,15,19,385,386 15,19
9α-lumisterol 7,8,385,386
lumisteryl acetate β-epidioxide 386
2,6-lutidine 460
lysergic acid 156

magnesium 468
magnesium bromide, (2-methyl-3-butyn-2-yl)- 468
malealdehydic acid pseudo ethyl ester 429
maleate, diethyl 167
——, dimethyl 57,58,461
maleates, alkyl 110
——, dialkyl 167
maleic acid 56-58,82,157,231-233
maleic anhydride 82,109-111,114,161,345
——, dimethyl- 82,110,111,168
——, methyl- 82,168
——, phenyl- 345
maleimide 396
maleimide, N-cyclohexyl- 82
——, dimethyl- 82
——, 2,3-diphenyl- 19,20
——, N-phenyl- 82
maleimides 110
——, N-phenyl- 110
maleonitrile 420,421
mandelate, ethyl 210
manganese carbonyl, [bis-(triphenylphosphine)] (cyclopentadienyl)- 463
——, (1,3-butadiene)(cyclopentadienyl)- 463
——, (cyclopentadienyl)[bis-(dimethyl sulfoxide)]- 463

——, (cyclopentadienyl)(diphos)- 463,463
manganese dicarbonyl, (1,3-cyclohexadiene)(cyclopentadienyl)- 463
——, (cyclopentadienyl)(dimethyl sulfoxide)- 463
——, (cyclopentadienyl)(diphenylacetylene)- 463
——, (cyclopentadienyl)(ethylene)- 463,464
——, (cyclopentadienyl)(hexamethylenetetramine)- 463
——, (cyclopentadienyl)(piperazine)- 463
——, (cyclopentadienyl)(pyridine)- 464
——, (cyclopentadienyl)(triethylenediamine)- 463
——, (cyclopentadienyl)(triphenylarsine)- 463
——, (cyclopentadienyl)(triphenylphosphine)- 463
——, (diphenylacetylene)(methylcyclopentadienyl)- 463
——, (methylcyclopentadienyl)(piperidine) 463
——, (methylcyclopentadienyl)(pyridine) 463
manganese dicarbonyl], (1,3-cyclohexadiene)[bis-(cyclopentadienyl- 463
manganese dicarbonyl](diars), [bis-(cyclopentadienyl- 463
manganese dicarbonyl](diphos), [bis-(cyclopentadienyl- 463,463
manganese dicarbonyl](piperazine), [bis-(cyclopentadienyl- 463
manganese dicarbonyl](triethylenediamine), [bis-(cyclopentadienyl- 463,464
manganese tricarbonyl, cyclopentadienyl- 462-464
——, (methylcyclopentadienyl)- 463
maticarianol, trans-dehydro- 41
p-mentha-1,5-diene 14,17
p-mentha-1(7),8-dien-2-ol 376
p-mentha-2,8-dien-1-ol 376
p-menthane, 2,9-bis-(trichlorosilyl)- 178
——, 1,2:3,4-diepoxy- 46,47
p-menthan-3-one, 4,8-epoxy- 409
p-menth-1-ene, 3,6-epidioxy- 384
——, 9-(trichlorosilyl)- 178
p-menth-8-ene, 1,2-epoxy- 376
menthofuran 430
menthofuran, 3,9-dihydro-9-hydroperoxy-3-methoxy- 430
——, 3,9-dihydro-3-hydroxy-9-oxo- 430

Compound Index

d,l-menthol, 10-nitroso- 248,249
menthone 26,27,225,226
d,l-menthyl nitrite 248,249
mercuric sulfide 441
mercury 363-367,441,466-468
——, bis-(trifluoromethylthio)- 441
——, diphenyl- 467,468
——, di-o-tolyl- 467
mercury bromide, phenyl- 467
mercury chloride, o-tolyl- 467
mercury compounds, diaryl- 466-468
——, organo- 466-468
mercury halides, aryl- 466,467
mercury iodide, (2-iodophenyl)- 369
——, (trifluoromethyl)- 367
mesitol, 3-acetoxy- 228,229,234,235
mesonaphthodianthrone 131-134
[2·2]metacyclophane 1,9-diene, 8,16-dimethyl- 20
metal carbonyl compounds 459-465,469
methacrylate, methyl 461,462
methacrylic acid 232
methane 401
methane, azidotriphenyl- 319,320
——, 4-biphenylylchlorodiphenyl- 367
——, 4-biphenylyldiphenyl- 367
——, bis-(4-methoxyphenyl)- 200,201
——, bromodifluoronitroso- 364
——, bromotrichloro- 170-172,175,176,282,283
——, bromotriphenyl- 398
——, chlorodifluoronitro- 291
——, chlorodifluoronitroso- 364
——, cyclohexylphenyl- 284
——, dichlorodifluoro- 291
——, difluorodiiodo- 291,292
——, difluorodinitro- 291
——, diphenyl- 147,148,183,193,200
——, di-p-tolyl- 200
——, iodo- 366,457,458
——, nitro- 153
——, (2-nitrophenyl)-diphenyl- 267
——, nitroso- 252
——, trifluoro- 363-365
——, trifluoroiodo- 170,173,174,363,364,365-367,457,458,465
——, trifluoronitroso- 105,106,364
——, triphenyl- 126,127,398
methane, diazo-, see diazomethane
methane dimers, nitroso- 252
methanes, diaryl- 146,147
methanesulfenyl chloride, trichloro- 356
——, trifluoro- 164
methanesulfinic acid, aminoimino- 444,445
methanesulfonyl chloride, trichloro- 355,356
——, (trimethylsilyl)- 436,437
methanethiol 163,176
7,20-methanocyclohexade[a]acenaphthylen-21-one, dodecahydro- 395,397
4,7-methanoindene, 5-(4-chlorophenylthio)-3a,4,5,6,7,7a-hexahydro-6-hydroperoxy- 446
methanol 8,9,160,205,207,208,210,219,238,268,286,287,296,297,301,303,304,307,308,313,314,321,322,325,326,335-337,429-433,448,467
methanol, (2-nitrosophenyl)-diphenyl- 267
1,4-methanonaphthalene-5,8-dione, 1,4,4a,8a-tetrahydro-2-4
3,5,6-methenocyclopentapyrazole, 3,3a,4,5,6,6a-hexahydro-328,329
methyl, triphenyl- 126,127,281,282
——, tris-(4-biphenylyl)- 126
methylamine, N-benzylidene- 335,336
methylamine hydrochloride 338
methylene blue 134
methylene, see also carbene 274
molybdenum hexacarbonyl 459-461
molybdenum pentacarbonyl, piperidine- 460
——, pyridine- 460
molybdenum tetracarbonyl, bis-(4-aminopyridine)- 461,462
——, bis-(piperidine)- 460
——, bis-(pyridine)- 460
molybdenum tricarbonyl, tris-(4-aminopyridine)- 461
morphinane, (+)-3-methoxy-N-methyl- 374,375
——, (−)-3-methoxy-N-methyl-10-oxo- 374,375
mustard gas 163,164
myrcene 2,3

Compound Index

naphthacene, 5,12-dichloro-6,11-diphenyl- 391
-----, 5,12-dichloro-5,12-epidioxy-6,11-diphenyl- 391, 392
-----, 5,12-dichloro-6,11-epidioxy-6,11-diphenyl- 391, 392
-----, 5,12-epidioxy-5,12-dihydro- 391,392
-----, 5,12-epidioxy-5,12-diphenyl- 391,392
-----, 5,12-epidioxy-5,6,11,12-tetraphenyl- 390
1-naphthaldehyde 204,415,421
1-naphthaldehyde, 2-methoxy- 187,190
2-naphthaldehyde 415
naphthalene 110,115,115,116,152,231
naphthalene, 2,3-bis-(bromomethyl)- 357
-----, 1-bromo- 97
-----, 2-bromo- 97
-----, trans-decahydro-4a-isocyanato-1,1-dimethyl- 323
-----, 1,4-dibromo-2,3-dimethyl- 357
-----, 1,2-dibromo-1,2,3,4-tetrahydro- 151,152
-----, 1,2-dihydro- 116,356,357
-----, 2,3-dimethyl- 151,152
-----, 1-methoxy- 97
-----, 2-methoxy- 97
-----, 1-methyl- 97,356
-----, 2-methyl- 97,233,356
-----, 1,2,3,4,5,6,7,8-octahydro- 376,378
-----, 1-phenyl- 93,94
-----, trans-1-styryl- 128
-----, 2-styryl- 119,120
-----, 1,2,3,4-tetrahydro- 151,152,182
-----, 1,2,3,4-tetrahydro-1-(tetrachloro-4-hydroxyphenoxy)- 151,182
-----, 1,2,3,4-tetraphenyl- 368,369
-----, 1,2,3-triphenyl- 93,94
naphthalene dimer, 2-methoxy- 97
naphthalene, 1,2,3,4-tetrahydro-, see also tetralin
1-naphthaleneazobenzene 140
4a-naphthalenecarboxamide, trans-decahydro-1,1-dimethyl- 323
4a-naphthalenecarboxylic acid azide, trans-decahydro-1,1-dimethyl- 323
4a-naphthalenecarboxylic acid lactam, 6-amino-trans-decahydro-1,1-dimethyl- 323

-----, 1-aminomethyl-trans-decahydro-1-methyl- 323
2,3-naphthalenedicarboxylate, dimethyl 1,4-dihydro-1-hydroxy-1-phenyl- 24,25
naphthalene-2,3-dicarboxylic anhydride, 5-methoxy-1-(2-methoxyphenyl)- 129,131
-----, 7-methoxy-1-(4-methoxyphenyl)- 129
-----, 7-methoxy-1-p-tolyl- 129
-----, 1-phenyl- 129
naphthalene-1,2-diol, 3-acetyl-4-cyano- 188
-----, 4-cyano-3-propionyl- 188
1,4-naphthalenediol 151
1-naphthalenemethanol 204
1-naphthalenemethanol, 1,2,3,4-tetrahydro-1-hydroxy- 208
2-naphthalenemethanol, α-methyl- 208
1-naphthalenesulfonic acid, 4-diazo-3,4-dihydro-3-oxo- 309,310
1,4[1',4']naphthalenonaphthalene, 5,8,9,12-tetrahydro-2,14-dimethyl- 97
-----, 5,8,9,12-tetrahydro-2,15-dimethyl- 97
1(2H)-naphthalenone, 3,4-dihydro- 205,208
1(5H)-naphthalenone, 6,7,8,8a-tetrahydro-3,8a-dimethyl- 30,32
2(1H)-naphthalenone, 1-diazo- 304,306
2(4aH)-naphthalenone, 5,6,7,8-tetrahydro-3,4a-dimethyl- 30,32
naphthalide, benzylidene- 119,120
1-naphthamide 231
1-naphthoate, 4-tert•butylphenyl 237
-----, methyl 6,7,8-trimethoxy- 227,229
naphthodianthrene, meso-diphenyl- 392
meso-naphthodianthrone 131-134
1-naphthol 97,362
1-naphthol, 2-(2-naphthylazo)- 53
2-naphthol 97,316
2-naphthol, 1-(2-bromo-p-tolylazo)- 316
-----, 5,6,7,8-tetrahydro-1,4-dimethyl- 34,35
4a-naphthol, cis-decahydro- 30
2-naphthol dibenzoate, 1-amino- 191
1,2-naphthoquinone 439
1,2-naphthoquinone, 4-cyano- 187,188
-----, 3,4-dichloro- 119,120

Compound Index

-----, 3-nitro- 439
1,4-naphthoquinone 81,151,189
1,4-naphthoquinone, 2-anilino-3-bromo- 255
-----, 2-anilino-3-chloro- 255,257
-----, 3-bromo-2-N-ethylanilino- 255
-----, 3-chloro-2-N-ethylanilino- 255,257
-----, 3-chloro-2-(N-phenylacetamido)- 255
-----, 2,3-dimethyl- 81
-----, 2-hydroxy- 147
-----, 2-methyl- 81
-----, 4a,5,8,8a-tetrahydro-6,7-dimethyl- 424
2,1-naphthoquinone diazide 304
1,4-naphthoquinone dimer, 2-methyl- 81
2,1-naphthoquinone imine, N-benzoyl- 191
1,2-naphthoquinones, 4-aryloxy- 187
naphtho[1,8-de][1,2,3]thiadiazine 1,1-dioxide 316,317
naphthothiam blue 316
naphtho[1,8-bc]thiete 1,1-dioxide 316,317
naphtho[2,3-c]thiophene 2,2-dioxide, 1,3-dihydro- 447
-----, 1,3-dihydro-1,3-diphenyl- 447
naphtho[3,2,1-kl]thioxanthene, 9,13b-epidioxy-9-phenyl- 393
2H-naphtho[1,2-d]triazole-4,5-dione, 2-phenyl- 187
2H-naphtho[1,2-d]triazole-4,5-dione oxime, 2-phenyl- 191
naphtho[3,2,1-kl]xanthene, 9,13b-epidioxy-9-phenyl- 393
1-naphthylamine 97,338
1-naphthylamine, N-(1-naphthylmethylene)- 142
2-naphthylamine 97
2-naphthylamine, N-benzylidene- 334
-----, N-(4-methoxybenzylidene)- 334
1,8-naphthyridine-7-carboxylic acid, 3-diazo-3,4-dihydro-4-oxo- 306
1,6-naphthyridin-4(3H)-one, 3-diazo- 306
1,7-naphthyridin-4(3H)-one, 3-diazo- 305,306
neoabietic acid 387,388
neoprednisone acetate 31,32
nicotine 381
nitramide, bis-(trifluoromethyl)- 106
nitrenes 318
nitriles 260

nitrites 248-254
nitrobenzene 469
nitrogen 274-291,294-311,313-322
nitrogen oxide 261-263,265,266,364,365
nitrone, α-acridin-9-yl-N-phenyl- 48,49
-----, α-benz[a]acridin-12-yl-N-(4-dimethylaminophenyl)- 49
-----, α-benzoyl-α,N-diphenyl- 49,50
-----, N-tert-butyl-α-(4-nitrophenyl)- 50,51
-----, α-cyano-α,N-diphenyl- 63
-----, α-(4-dimethylaminophenyl)-N-(3-nitrophenyl)- 50
-----, N,α-diphenyl- 49
nitrones 48-51,63
nitroso compounds 248,249,261-263,267-270
nitrosyl chloride 159,263-265,291
nonanamide 166,167
nonane, 3-bromo-1,1,1-trichloro- 171
-----, 1,1,1,3-tetrabromo- 171,173
-----, 1,1,1,3-tetrachloro- 171
2-nonanol, 2-methyl- 168
2-nonanone 34
2-nonene, 3-bromo-1,1,1-trichloro- 172
2-nonyne 172
D-norandrostane-16-carboxylic acid, 3β-hydroxy- 302,303
norascaridole 384,385
norbornadiene 2,3,10,328,329
norbornane, exo-2-bromo- 158
-----, exo-cis-2,3-dibromo- 158,159
-----, trans-2,3-dibromo- 158,159
norbornane derivatives see also bicyclo[2·2·1]heptane derivatives
norbornene 74,165,169-172,415,417,468,469
2-norbornene, 2-bromo- 158
norcamphor, dehydro- 116
-----, diazo- 303,304
norcaradiene 276-278
2,4-norcaradiene-7-carboxylate, ethyl 277,279
norcaradiene-7-carboxylate, methyl 384
norcarane 275,276
A-norcholest-5-ene, 3α-ethoxymethyl- 10,11

Compound Index

A-nor-B-homoandrostane-3,6-dione, 17β-acetoxy-5-methyl- 412
A-nor-B-homosteroids 31
19-norpregna-1(10),3-diene-2,11,20-trione, 21-acetoxy-17α-hydroxy-5-methyl- 31,32
C-nor-5α-pregnan-11-one, 3,3:20,20-bis-(ethylenedioxy)- 26,27
19-norpregna-1,3,5(10)-triene-11,20-dione, 21-acetoxy-1,17α-dihydroxy-4-methyl- 31,32
17-nor-13,17-secoandrosta-5,13(18),15-triene, 3,3-ethylenedioxy- 37
C-nor-11,13-seco-5α-pregn-13-en-11-al, 3,3:20,20-bis-(ethylenedioxy) 26,27
D-norsteroids 302,303
nortricyclanone p-toluenesulfonylhydrazone sodium salt 285
nyctanthic acid, dihydro- 226

octadecanamide, 9-carbamoyl- 167
----, 10-carbamoyl- 167
octadecanoic acid, 9,10-epoxy- 400
9-octadecen-1-ol 400
3,6-octadienoic acid, 3,7-dimethyl- 28
Δ1(9)-octalin, 10-hydroperoxy- 376,378
Δ9-octalin 376,378
octanamide 166,167
octanamide, N-tert-butyl- 167
----, 7-oxo- 168
octane, 1-azido- 319
----, 1,2,4,5,7,8-hexachlorododecafluoro- 366
2,7-octanedione, 1,8-bis-(diazo)- 301
3,6-octanedione, 4-butyl- 169
4,5-octanedione 35
octanethiol 163,455
1-octanimine 319
octanoate, ethyl 45
----, propyl 45
1-octanol, 4-nitroso- 248,249
2-octanol 160,232
2-octanone 34,35
4-octanone, 5-hydroxy- 105

1,3,5,7-octatetraene isomers, 1,8-diphenyl- 59-61
1,3,5-octatriene, 3,7-dimethyl- 14
2,4,6-octatriene 19
2,4,6-octatriene, 2,6-dimethyl- 19
2,4,6-octatrienedioic acid 15
5-octenal, 3,7-dimethyl- 26,27
1-octene 160,161,165,166,169-171,173,177
4-octene, trans-4H,5H-tetradecafluoro- 288,289
1-octene-3,8-diol, 2,6-dimethyl- 377
3-octene-2,8-diol, 2,6-dimethyl- 377
2-octen-4-one, 3-butyl-2-phenyl- 423
octyl nitrite 248,249
1-octylamine 319
1-octylamine, N-phenyl- 319
1-octyne 172
2-octyne 172
oleamide 167
oleate, methyl 373,400
olefins, fluoro- 421
oleic acid 400
organoarsenic compounds 457,458
organomercury compounds 466-468
organometallic compounds 459-469
organophosphorus compounds 453-457
organosilicon compounds 354,355,436,437
organotin compounds 465,466
organotin hydroperoxides 466
6-oxa-1-azabicyclo[3.1.0]hexane, 2,2-dimethyl- 50,51
----, 2,2,5-trimethyl- 51
2-oxabicyclo[3.2.0]hept-3-ene-6,7-dicarboxylic anhydride, 6,7-dimethyl- 110
2-oxabicyclo[2.2.0]hexane, 1,3-dimethyl- 417,418
----, 1-methyl- 417,418
----, 1,3,3-trimethyl- 418
5-oxabicyclo[2.1.1]hexane, 1,6,6-trimethyl- 418
2-oxabicyclo[3.1.0]hexane-6-carboxylate, methyl 278
2-oxabicyclo[3.1.0]hex-3-ene-6-carboxylate, ethyl 278
6-oxabicyclo[3.1.0]hex-2-en-2-one, 4,5-diphenyl- 409,410
----, 1,3,4,5-tetraphenyl- 410,411
----, 1,3,5-triphenyl- 410

Compound Index

8-oxabicyclo[3.2.1]oct-6-en-3-one, 2,2,4,4-tetramethyl- 219
oxadiaziridines 53
1,3,4-oxadiazole derivatives 331
Δ²-1,2,4-oxadiazoline-3-carboxylate, butyl 5-oxo-4-phenyl- 337
——, cyclohexyl 5-oxo-4-phenyl- 337
——, ethyl 5-oxo-4-phenyl- 337
——, isopropyl 5-oxo-4-phenyl- 337
Δ²-1,2,4-oxadiazolin-5-one, 3,4-diphenyl- 337
oxalyl chloride 233,234
oxalyl-bis-diazoacetate, dimethyl 301
oxamide 167,230
oxapenenone 132-134
1-oxaspiro[3.3]heptane, nonafluoro-3-(trifluoromethyl)- 422
1-oxaspiro[5.5]undeca-3,7,10-trien-9-one, 3,4-dimethyl- 424
13-oxatetracyclo[7.3.1¹·¹¹.0⁰·⁹]tridec-2-ene-7,10-dione 432
13-oxatetracyclo[8.2.1⁵·⁸¹¹.0⁶·¹⁰]tridec-11-ene-4,7-dione 432
1,2-oxathiin dioxides 448
1,2-oxathiolane 2,2-dioxide, 5-oxo- 436,437
3-oxatricyclo[4.2.1.0²·⁵]nonane, 4,4-diphenyl- 415-417
oxaziridine, 2-tert-butyl-3-(4-nitrophenyl)- 50,51
——, 3-cyano-2,3-diphenyl- 63
——, 2,3-diphenyl- 49
oxaziridines 48-52
oxazirino[2,3-d][1,4]benzodiazepine, 8-chloro-3,9b-dihydro-4-methylamino-9b-phenyl- 50,51
oxazole derivatives 190,191
oxazolidine derivatives 335
2-oxazolidinone, 5,5-dimethyl- 325,326
1H,3H,4H-oxazolo[4,3-k]carbazole, 5,6,7,7a-tetrahydro-7a-methyl- 336,337
1H,3H-oxazolo[3,4-a]indole, 9,9a-dihydro-9,9-dimethyl-9a-phenyl- 335,336
2H-oxazolo[4',5':3,4]naphtho[1,2-d]triazole, 5-(4-methoxyphenyl)-2-phenyl- 191,192
5H-oxazolo[4,5-b]phenoxazine-4,6-dicarboxylate, dimethyl 9,11-dimethyl- 256,257

11aH-oxazolo[4,5-b]phenoxazine-4,6-dicarboxylate, dimethyl 2,3-dihydro-11a-hydroxy-2,2,9,11-tetramethyl- 256,257
oxetane 414
oxetane, 2-(9-anthryl)-3,3,4-trimethyl- 415,416
——, 2,2-bis-(chlorodifluoromethyl)-3,4,4-trifluoro-3-(trifluoromethyl)- 422,423
——, 2,2-bis-(trifluoromethyl)- 422
——, 2-(4-chlorophenyl)-3,3-dimethyl-2-phenyl- 415, 416
——, 2-[N-(2-cyano-2-propyl)-imino]-3,3-dimethyl-4,4-diphenyl- 421
——, 3-[N-(2-cyano-2-propyl)-imino]-4,4-dimethyl-2,2-diphenyl- 421
——, 3-[N-(2-cyano-2-propyl)-imino]-2-(4-methoxyphenyl)-4,4-dimethyl- 421
——, 3,3-dichloro-2,2-difluoro-4-methyl- 422
——, 3,3-difluoro-2,2-bis-(trifluoromethyl)- 415,416
——, 4,4-difluoro-2,2-bis-(trifluoromethyl)- 415,416
——, 3,3-dimethyl-2,2-diphenyl- 415,416
——, 3,3-dimethyl-2,2-di-p-tolyl- 415,416
——, 3,3-dimethyl-2-phenyl-2-p-tolyl- 415,416
——, 3-isopropylidene-2,3,4,4-tetramethyl- 219,220
——, 3-methyl-2,2-diphenyl- 415,416
——, 2-methyl-3-phenyl- 415-417
——, 2,2,3,3-tetrafluoro-4-methyl- 422
——, 2,3,4,4-tetrafluoro-3-(trifluoromethyl)-2-(heptafluoropropyl)- 422
——, 3,3,4,4-tetramethyl-2,2-diphenyl- 415-417
——, 2,3,4-trimethyl-2-phenyl- 415
——, 3,4,4-trifluoro-2,3-bis-(trifluoromethyl)- 422
——, 3,4,4-trifluoro-2-(1,1,2,2,3,3,4,4-octafluorobutyl)-3-(trifluoromethyl)- 422
——, 3,4,4-trifluoro-2,2,3-tris-(trifluoromethyl)- 422
——, 3,3,4-trimethyl-2,2-diphenyl- 415
——, 3,3,4-trimethyl-2-(2-naphthyl)-2-phenyl- 415
——, 3,3,4-trimethyl-4-phenyl- 414,415,417
——, 2,3,4-trimethyl-4-phenyl- 414,415,417
2,3-oxetanedinitrile, 4,4-dimethyl- 420,421

Compound Index 595

oxetanes 43,414-423
oxetanes, imino- 420,421
-----, α-keto- 121,122,124,419
-----, polyfluoro- 421
-----, spiro- 418-420
3-oxetanimine, N-(2-cyano-2-propyl)-2-(4-methoxyphenyl)-4,4-dimethyl- 421
oxetes 423
4H,6H-oxeto[3',2':4,5]furo[3,2-g]-1-benzopyran-4-one, 6-benzoyl-5b,7a-dihydro-5-methoxy-2-methyl-6-phenyl- 419,420
N-oxides 51-53
oxime, benzaldehyde 48
oxime O-ethers 63
oximes 63,64,197,248,249
oxindole, 3-chloro-3-(trichloromethyl)-1-methyl- 307
-----, 3-diazo-1-methyl- 276,286,289,307,308
-----, 1,3-diphenyl- 280
-----, 3-ethoxy-1-methyl- 286,307,308
-----, 1,3,3-trimethyl- 401
oxygen 127,128,130,131,133-136,138,141,142,146,147, 183,184,196,225,234,239,257,291,334-336,350,373-402, 407,408,427-432,438,442,444-447,454,455,466

palustric acid 153,387
paraconic acid, 2-ethyl-2-methyl- 232
-----, 2-hexyl-2-methyl- 232
-----, 2-methyl- 232
pentacene, 6,13-dihydro-6-methylene- 21
-----, 5,14-epidioxy-5,7,12,14-tetraphenyl- 393
-----, 6-methyl- 21
pentacene dimer 99
pentacyclo[5·3·0·0²·⁶·0³·⁹·0⁵·⁸]decane 2,3
pentacyclo[5·3·0·0²·⁶·0³·⁹·0⁵·⁸]decan-4-one 2,3
pentacyclo[5·3·0·0²·⁶·0³·⁹·0⁵·⁸]decan-4-one, 2,3,5,6-tetrachloro- 2,3
pentacyclo[12·8·0·0³·¹²·0⁴·¹¹·0¹⁵·²²]docosane-2,13-dione, 1,3,12,14-tetrachloro- 418,419
pentacyclo[6·4·0·0²·⁷·0⁴·¹¹·0⁵·¹⁰]dodecane-3,6,9,12-tetrone 79,81
pentacyclo[6·4·0·0²·⁷·0⁴·¹¹·0⁵·¹⁰]dodecane-3,6,9,12-tetrone, 1,2,7,8-tetramethyl- 80,81

pentacyclo[9·3·2·0²·⁹·0³·⁸·0¹⁰·¹²]hexadeca-4,6,13,15-tetraene 116,117
pentacyclo[6·4·0¹·³·6·¹⁹·¹²·0⁰·²⁴·0²·⁷]tetradecane 468,469
pentacyclo[6·4·0¹·³·6·¹⁹·¹¹²·0⁰·²·¹⁷]tetradecane 74
pentacyclo[8·2·0¹·⁴·¹⁷·0²·⁹·0³·⁸]tetradecane 74,468,469
pentacyclo[6·2·1·0²·⁷·0⁴·¹⁰·0⁵·⁹]undecane-3,6-dione 2,3
1,3-pentadiene 74
1,3-pentadienes 56,57
2,4-pentadienoic acid, 5-phenyl- 342,343
3-pentadienone, 1,5-diphenyl- 76,77
pentaerythritol, bis-(2-nitrobenzylidene)- 268
-----, (α-hydroxy-2-nitrosobenzylidene)-(2'-nitrobenzylidene)- 268
2-pentalenecarboxylic acid, 2,4,5-tetrahydro- 305
pentanal, 5-nitroso- 249
pentanamide, 2-propyl- 167
pentane 233,284
pentane, 2-bromo-2,3,4,4-tetramethyl- 349,350
-----, 2,2,4,4-tetramethyl- 349
-----, 1,1,1-trifluoro-5-iodo- 173
1,5-pentanedione, 1,5-diphenyl- 65
2,3-pentanedione 35
2,4-pentanedione 37,38,113
pentanoate, methyl 5-carbamoyl 166
2-pentanol 379
2-pentanol, 2-hydroperoxy- 379
1-pentanol acetate, 3-bromo-5,5-trichloro-1-phenyl- 171,172
2-pentanone 34,279
2-pentanone, 3,4-epoxy-4-phenyl- 408
3-pentanone 198,279
2-pentenal, 4-oxo- 429
4-pentenamide 166,168
2-pentene, 2-methyl- 163,164
2-penten-4-imine, 2-amino-3-cyano- 54,55
4-pentenoate, methyl 166
4-pentenoic acid, 2,3-dichloro-5-phenyl- 342,343
2-pentenoic acid γ-lactone, 4-hydroxy-4-methyl- 414,417
3-penten-1-ol, 3-methyl-1-phenyl- 233

Compound Index

-----, 4-methyl-1-phenyl- 414,417
4-penten-2-ol acetate 171,172
3-Penten-2-one, 3-methyl- 45
4-penten-2-one, 4-tert-butyl- 22
-----, 3-methyl- 45
4-pentyn-2-ol, 3,3-dimethyl- 468
perbenzoic acid 399,400
Perion 263
Peroxide, acetyl 4-chlorobenzoyl 400
-----, acetyl cyclohexanesulfonyl 438
-----, acetyl p-toluoyl 400
-----, anthracene 389,391
-----, bis-(1-hydroperoxy-1-methylethyl) 379
-----, bis-(2-hydroperoxy-2-propyl) 379
-----, bis-(9-phenyl-9-xanthenyl) 399
-----, bis-(1,2,3,4-tetrahydro-2-p-tolyl-1-isoquinolyl) 399
-----, rubrene 383
-----, triphenylmethyl 398
Peroxides, acyl 427
-----, transannular 399,400
α-phellandrene 382-397
1,2,3-Phenalendrene 14,17,384
1-Phenalenone 119,120,209
1-Phenalenone, 2,3-dihydro- 193,207
Phenalen-1-yl, 1-hydroxy- 207
Phenanthrene 109,110,127,129,132,133,139
Phenanthrene, 3-bromo- 128,129
-----, 1-carbomethoxy-10-nitro- 128
-----, 9,10-dicyano-4a,4b-dihydro- 19,20
-----, 9,10-dicyano-9,10-dihydro- 19,20,127
-----, 4a,4b-dihydro- 19,20,127,128,132
-----, 4a,4b-dihydro-1,3,4a,4b,6,8-hexamethyl- 19,127
-----, trans,anti,cis-perhydro-1β-isocyanato-1α,4aβ-dimethyl- 322-324
-----, 9,10-sulfonyldioxy- 439
phenanthrene-1β-carboxylic acid azide, trans-1,2,3,4,4a,9,10,10a-octahydro-6-methoxy-1α,4aβ-dimethyl- 323
-----, trans,anti,cis-perhydro-1α,4aβ-dimethyl- 324

phenanthrene-1β-carboxylic acid lactam, 4aβ-aminomethyl-trans-1,2,3,4,4a,9,10,10a-octahydro-6-methoxy-1α-methyl- 323
-----, 4aβ-aminomethyl-trans,anti,cis-perhydro-1α-methyl- 322-324
9,10-phenanthrenedicarboximide, 9,10-dihydro- 19,20
9,10-Phenanthrenediol 151,152,182,183,186
9,10-Phenanthrenediol, trans-9,10-bis-(4-methylbenzyl)- 212
-----, 9,10-diphenyl- 211
9,10-phenanthrenequinone 118-124,131,148,151,152,182-184,186-188,419,439
9,10-phenanthrenequinone, 2-nitro- 439
-----, 3-nitro- 439
-----, 4-nitro- 439
9,10-phenanthrenequinone imine 190,191
9,10-phenanthrenequinone imine, 7-isopropyl-1-methyl- 190
9,10-Phenanthrenequinone oxime 191
phenanthrenes 19,20
Phenanthrenes, substituted 127-129
Phenanthridine 141,142
Phenanthridine, dihydro- 141
-----, 9-dimethylamino- 141,142
-----, 6-phenyl- 141
Phenanthridizinium perchlorate 142
Phenanthridizinium perchlorate, 10-chloro- 142
Phenanthridizinium salts 142
Phenanthro[9,10-d]-1,3,2-dioxathiole 2,2-dioxide 439
Phenanthro[9,10-f][1,4]dioxin, 2,3-dihydro-2,3-diphenyl- 118,120,123,124
-----, 2,3-dihydro-2-oxo-3,3-diphenyl- 121
-----, 2,3-dihydro-2,2,3-trimethyl- 122,124
-----, 2,3-diphenyl- 122
-----, 2-ethoxy-2,3-dihydro- 122,124
Phenanthro[9',10':5,6]-1,4-dioxino[2,3-b]phenanthro[9,10-e][1,4]dioxin, 9a,19a-dihydro- 123,124
9-Phenanthrol, 10-acetamido- 190,191
-----, 10-amino- 190
-----, 10-tenzamido- 190
-----, 10-(1,4-dioxan-2-yloxy)- 184
-----, 10-(2-hydroxybenzoyloxy)- 188

Compound Index 597

—————, 10-(4-methylbenzyloxy)- 183
—————, 10-(phenoxymethoxy)- 184
—————, 10-(tetrahydro-2-furyloxy)- 184
9(10H)-phenanthrone, 10-diazo- 304
—————, 10-hydroxy-10-(2-methylbenzyl)- 183
—————, 10-hydroxy-10-(4-methylbenzyl)- 183,184
—————, 10-hydroxy-10-(9-xanthenyl)- 183,184
9(10H)-phenanthrones, 10-alkylaryl-10-hydroxy- 183
phenanthro[9,10-d]oxazole, 2-(4-methoxyphenyl)- 191
—————, 2-phenyl- 190,191
phenanthro[9,10-d]-4-oxazoline, 2-hydroxy-2-methyl- 190
phenanthro[1,10,9,8-opgra]perylene, 7,14-diphenyl- 392
phenanthro[1,10,9,8-opgra]perylene-7,14-dione 131-134
phenanthro[1,10,9,8-opgra]perylene-7,14-dione, 1,6,8,13-tetrachloro- 133,134
11H-phenanthro[9,10-b]pyrano[2,3-e][1,4]dioxin, 11-acetoxymethyl-12,13-diacetoxy-9a,12,13,13a-tetrahydro- 122,124
phenanthro[9,10,1-def]quinoline, 2-phenyl- 135,136
9-phenanthrylamine, 10-acetoxy- 190
phenazine 196
phenazine, 5,10-dihydro- 196
phenanthrenequinone-acetylene 2:1 adduct 123,124
phenol 236
phenol, 4-anilino- 314,315
—————, 4-benzoyloxy-2,3,5,6-tetrachloro- 189
—————, 2,6-dichloro-4-pyridyl- 311
—————, 2-dimethylamino- 315
—————, 4-dimethylamino- 314
—————, 4-ethoxy- 311
—————, 2-iodo- 368
—————, 4-iodo- 368,456
—————, 4-phenyl- 311
—————, 2-phenylazo- 53
—————, 3,4,5,6-tetrachloro-2-(1,4-dioxan-2-yloxy)- 184
—————, 3,4,5,6-tetrachloro-2-methoxy- 307,308
—————, 2,3,5,6-tetrachloro-4-(4-methylbenzyloxy)- 182
—————, 2,3,5,6-tetrachloro-4-(1,2,3,4-tetrahydro-1-naphthyloxy)- 151,182

phenols 316
Phenols, p-alkoxy- 310,311
—————, nitroso- 191
3H-phenoxazine-1,9-dicarboxylate, dimethyl 2-amino-4,6-dimethyl-3-oxo- 256,257
—————, dimethyl 4,6-dimethyl-2-methylamino-3-oxo- 256,257
—————, dimethyl 4,6-dimethyl-3-oxo-2-(2-propylamino)- 256,257
phenyl isocyanate 280,321,322
phenyl isothiocyanate 307,308
phosgene 407
Phosphate, 4-hydroxy-2,3,5,6-tetrachlorophenyl dimethyl 453,454
—————, tributyl 454,455
—————, triethyl 454
—————, triisopropyl 454
—————, trimethyl 454
phosphates, trialkyl 454
Phosphine 176,177
Phosphine, butyl- 176,177
—————, dibutyl- 176,177
—————, diphenyl- 455,457
—————, ethyldiphenyl- 455
—————, phenyl- 177,456,457
—————, tributyl- 455-457,462,463
—————, triphenyl- 456,457
—————, tris-(4-dimethylaminophenyl)- 457
phosphine oxide, tris-(4-dimethylaminophenyl)- 460
phosphines 460
phosphines, trialkyl- 456
—————, triaryl- 455,456
phosphite, tributyl 454,455
—————, triethyl 454,455
—————, triisopropyl 454
—————, trimethyl 454
phosphites, dialkyl 177
—————, trialkyl 454
phosphonate, dibutyl 177,454
—————, dibutyl octyl- 177
—————, dimethyl 453
phosphonates, dialkyl 177,453

598 Compound Index

phosphonium chloride, tetraphenyl- 457
phosphonium iodide, biphenyltriphenyl- 456
——, (4-hydroxyphenyl)-triphenyl- 456
——, (4-methoxyphenyl)-triphenyl- 456
——, tetrakis-(4-dimethylaminophenyl)- 456,457
——, tetraphenyl- 456,457
——, tributylphenyl- 456,457
——, triphenyl-p-tolyl- 456
phosphonium salts, aryltriphenyl- 455-457
——, tetraaryl- 455
phosphonium tetrafluoroborate, tetraphenyl- 456,457
phosphonium tetrafluoroborates, aryltriphenyl- 456
phosphonium triiodide, tetrakis-(4-dimethylaminophenyl)- 457
phosphonothioate, dibutyl 177,178
——, dibutyl octyl- 177,178
phosphonothioates, dialkyl 177,178
——, dialkyl alkyl- 177,178
phosphoranyl radicals, tetraaryl- 457
phosphorothionate, triethyl 454,455
phothecogenin acetate 43,44
photoisopyrocalciferol 7,8
photolevopimaric acid 7,9
photopyrocalciferol 7,8
photosantonic acid 30,229,230
photosantoninic acid 30
phthalaldehyde 107,187
phthalic anhydride, 1,2-dihydro- 5,6
phthalide, benzylidene- 71
——, 3-benzylidene- 119,120
——, 3-ethylidene- 119,120
——, (2-formyl-α-hydroxybenzyl)- 107
——, cis-3-α-methylbenzylidene- 43
——, trans-3-α-methylbenzylidene- 43
——, 3-α-phenylbenzylidene- 43,44
2-picoline 460
pimpinellin 87,88
pimpinellin dimer 87,88
pinacol 203,211
pinacol hexahydrate 203
pinane 264

3-pinanone oxime 264
4-pinanone oxime 264
α-pinene 375,378
β-pinene 2,3,375,378
pinocamphone oxime 264
piperazine 463
piperidine 165,381,460,463
piperidine, 1-bromo-2-propyl- 242
——, 2-octyl- 165
piperitone 77
piperitone dimers 77
piperonal 190
piperylene 74
piperylenes 56,57
pleiadene dimer 447,448
podocarpic acid azide, O-methyl 323
polyenes 58-61
poly(trifluoromethyl)-methylene 288
precalciferol 15,19,59
prednisone acetate 31,32
pregna-5,16-dien-20-one, 3β-acetoxy- 193,194
——, 3β-acetoxy-16-methyl- 193,194
5α-pregnane, 3β-dimethylamino-20α-(N-chloro-N-methylamino)- 244,245
——, 3β-dimethylamino-20α-methylamino- 245
5α-pregnane-20-carboxamide, 3β-acetoxy-11-oxo- 371
5α-pregnane-20-carboxylic acid lactone, 3β-acetoxy-16β-hydroxy-11-oxo- 371
pregnanes, 11-oxo- 37
——, 20-oxo- 26,36,37
5α-pregnane-20-yl xanthate, O-ethyl 3β-acetoxy-11-oxo- 223,224
5α-pregnan-11-one, 3β-acetoxy- 223
——, 3,3;20,20-bis-(ethylenedioxy)- 37
——, 20α-(N-chloro-N-methylamino)-3β-hydroxy- 247
——, 18-chloro-20α-methylamino-3β-trifluoroacetoxy- 246,247
5β-pregnan-11-one, 3α-acetoxy-20β-hydroxy- 251
——, 3α-acetoxy-20β-hydroxy-18-hydroxyimino- 250,251
5β-pregnan-11-one nitrite, 3α-acetoxy-20β-hydroxy- 250,251

Compound Index 599

pregn-4-en-18-al, 21-acetoxy-11β-hydroxy-3,20-dioxo- 250
pregn-4-ene-3,20-dione 358
pregn-4-ene-3,20-dione, 21-acetoxy-11β-hydroxy-18-hydroxyimino- 250
pregn-4-ene-3,20-dione nitrite, 17α-hydroxy- 253,254
pregn-4-ene-3,20-dione 11-nitrite, 21-acetoxy-11β-hydroxy- 250
pregn-5-en-20β-ol nitrite, 3,3:21,21-bis-(ethylenedioxy)- 253
pregn-5-en-3-one, 20-hydroxy-18-hydroxyimino- 250,251
pregn-5-en-20-one, 3β-acetoxy- 36,37,193,194
-----, 3β-acetoxy-16α-(1-hydroxyethyl)- 193,194
-----, 3β-acetoxy-16α-methyl- 194
-----, 3,3-ethylenedioxy- 26,36
pregn-4-en-3-one nitrite, 20-hydroxy- 250,251
pregn-4-en-3-one 20-nitrite, 17α,20α-dihydroxy- 253, 254
-----, 17α,20β-dihydroxy- 253,254
prismane 10
progesterone 358
propane 163,171,342,435,437
propane, 2-azido-2-phenyl- 319
-----, 2-bromo-2-(chloromethyl)-1,3-dichloro- 282
-----, 1-bromo-1,1,2,3,3-hexafluoro- 157,158
-----, 2-(bromomethyl)-2-(chloromethyl)-1,3-dichloro- 282
-----, 2-(4-bromophenylazo)-2-hydroperoxy- 381
-----, 2-bromo-1,2,3-trichloro- 282
-----, 1-tert-butoxy-2-chloro-1,1,2,3,3-hexafluoro- 162
-----, 2-chloro- 342
-----, 3-chloro-1,1,1-trifluoro- 364,365
-----, 3-chloro-1,1,1-trifluoro-3-iodo- 174,365
-----, 1,2-dibromohexafluoro- 158
-----, 1,3-dichloro-2,2-bis-(chloromethyl)- 282,283
-----, 1,3-dichloro-2-(chloromethyl)-2-methyl- 282
-----, 1,2-dichloro-1,1,3,3-pentafluoro-2-methyl- 343
-----, 3,3-dichloro-1,1,1-trifluoro- 365
-----, 1,2-epoxy-2-methyl- 279
-----, heptafluoro-1-iodo- 173,225
-----, heptafluoro-1-nitroso- 364
-----, 1,1,1,3,3,3-hexaphenyl- 281,282
-----, 2-isothiocyanato-2-phenyl- 443
-----, 2-phenyl-2-thiocyanato- 443
-----, 1,2,2-trichloropentafluoro- 362
-----, 3,3,3-trichloro-1,1,1-trifluoro- 365
-----, 1,1,1-trifluoro-3-iodo- 173,364,365
1,3-propanedione, 1,3-diphenyl- 408
1,2-propanedione 1-oxime, 1-phenyl- 266,267
propane-1,3-disulfonyl chloride ,437
1-propanesulfenyl chloride, heptafluoro- 441,442
1-propanesulfonyl chloride 435,437
2-propanesulfonyl chloride 435,437
2-propanol 160,193-196,203-212,232,233,238,300,313, 379,467,468
2-propanol, 2-hydroperoxy- 379
1-propanone, 1-cyclohexyl- 201
2-propanone, 1,3-diphenyl- 220,221
-----, 1-hydroxyimino-1-phenyl- 266,267
1-(1,2,2-trimethylcyclopropyl)- 22,24
propene, cis-1-bromo- 290,415
propene, 3-bromopentafluoro- 158
-----, 2-bromo-1,3-trichloro- 362
-----, 2-bromo-1,3-trichloro- 39
-----, 2-bromo-3,3-trichloro- 39
-----, 3-bromo-1,1,2-trichloro- 39
-----, 2-cyclopropyl- 175,176
-----, hexafluoro- 157-159,173,174,422,423
-----, 2-methyl- 415
-----, 1,1,3,3,3-pentafluoro-2-methyl- 343
-----, perfluoro- 162
-----, 3,3,3-trifluoro- 292
-----, 3,3,3-trifluoro-2-methyl- 292
1-propene, 1,1-diphenyl- 119,120,449
-----, 3,3,3-trifluoro-1-iodo- 174
1-propene-1,3-dione, 2-methyl-3-phenyl- 294,295
1-propenesulfonate, methyl 2-(2-hydroxyphenyl)- 448
1-propen-2-ol acetate 37,38
1-propen-2-ol benzoate 38
propiolate, methyl 114
Propiolic acid 232,233

600   Compound Index

propionaldehyde   169,188,201,415
propionate, ethyl 2-benzoyl-   294
——, ethyl thio-   297,298
——, methyl 2-bromo-   282
——, methyl 3-bromo-2-methyl-   282
——, methyl 3-bromo-2,3-trichloro-   282,283
——, methyl 3-chloro-2,2-bis-(chloromethyl)-   282
——, methyl 2,3,3-trichloro-   282,283
propionic acid   436,437
propionic acid, 3-benzoyl-2-methylene-   23,24
——, 2-chloro-   437
——, 3,3-dichloro-2-p-tolyl-   40
——, 3-hydroxy-2,3,3-triphenyl-   200
propionic acid anhydride, 3-sulfo-   436,437
propionitrile, 2,3-dimethyl-   260,261
——, 2-ethoxy-   260,261
——, 2-isocyanato-2-methyl-   421
propionyl isocyanate, 3-bromo-   54
propiophenone, 2,3-epoxy-3-methyl-3-phenyl-   408,409
——, 2,3-epoxy-3-phenyl-   408
propylamine   381,435
propyne   158
pseudoindoxyl, 2,2-diphenyl-   280
psoralen   87
psoralen dimer   87
psoralen, 5-methoxy-, see bergapten
psoralen, 8-methoxy-, see xanthotoxin
6-pteridinecarboxaldehyde, 2-amino-4-hydroxy-   255,256
6-pteridinecarboxylic acid, 2-amino-4-hydroxy-   255,256
pteroylglutamic acid   255,256
pulegone α-oxide   409
pulegone β-oxide   409
purine, 6-amino-2-hydroxy-   52
purine 1-oxide, 6-amino-2-hydroxy-   52
purpurogallin, tetra-C-methyl-   227,229
purpurogallin tetramethyl ether   227,229
pyran, cis-4-methyl-2-(2-methyl-1-propenyl)-tetrahydro-   377
——, trans-4-methyl-2-(2-methyl-1-propenyl)-tetrahydro-   377

2H-pyran, 2-benzyl-2,4,6-triphenyl-   46
4H-pyran, 4-benzyl-2,4,6-triphenyl-   46
2H-pyran-2-one, 4,6-dimethyl-   102
——, 4,5-diphenyl-   409,410
——, tetraphenyl-   394,395,410-412
——, 3,5,6-triphenyl-   410
4H-pyran-4-one, 2,6-dimethyl-   89
pyrazole   54
pyrazole, 4-acetoxy-3-benzoyl-5-phenyl-   286,287
——, 3-benzoyl-4-diazo-5-phenyl-   285-287
——, 3-benzoyl-5-diazo-4-phenyl-   285-287
——, 3-benzoyl-4,5-diphenyl-   285,286
——, 3-benzoyl-4-hydroxy-5-phenyl-   287
——, 3-benzoyl-4-phenyl-   287
3H-pyrazole, 3,3,5-trimethyl-   281,329,330
3H-Pyrazole-4,5-dicarboxylate, dimethyl 3,3-diphenyl-   330
3H-Pyrazoles   329,330
2-Pyrazoline, 5-methyl-5-(trifluoromethyl)-   292
——, 5-(trifluoromethyl)-   292
——, 3,4,5-tris-(trifluoromethyl)-   288
1-Pyrazoline-3-carboxylate, methyl cis-3,4-dimethyl   328
——, methyl trans-3,4-dimethyl   328
pyrazolines, steroid   329
1-Pyrazolines   329
2-Pyrazolines   329
Pyrene, trans-10b,10c-dihydro-10b,10c-dimethyl-   20
Pyridazine, 1,2-dibenzoylhexahydro-3,5-dioxo-4,4,6,6-tetraphenyl-   298
Pyridazine-1,2-dicarboxylate, diethyl hexahydro-3,5-dioxo-4,4,6,6-tetraphenyl-   298,299
Pyridine   271,272,311,460,463,464
Pyridine, 2-acetyl-   205
——, 3-acetyl-   205
——, 4-acetyl-   205
——, 2-amino-   100,101
——, 4-amino-   461,462
——, 2-amino-5-chloro-   100
——, 2-amino-3-methyl-   100
——, 2-amino-6-methyl-   100
——, 2-amino-5-nitro-   100

Compound Index   601

——, 2-benzoyl-   205
——, 3-benzoyl-   205
——, 4-benzoyl-   205
——, 2-benzylamino-   100
——, 4-chloro-   460
——, 3,5-diacetyl-1,4-dihydro-2,6-dimethyl-4-(2-
     nitrophenyl)-   268-270
——, 3,5-diacetyl-1,4-dihydro-2,6-dimethyl-4-(4-
     nitrophenyl)-   269
——, 3,5-diacetyl-2,6-dimethyl-4-(2-nitrosophenyl)-
                                                        268-270
——, 2,5-diamino-   100
——, 2,4-dichloro-3-cyano-6-β-styryl-   71,73
——, 2-styryl-   119,120
Pyridine dimer, 2-amino-   100,101
——, 2,4-dichloro-3-cyano-6-β-styryl-   73
Pyridine-3,3'-azo-2',5'-dimethylpyrrole, 4-hydroxy-2,6-
     dimethyl-   310
Pyridine-1-carboxylate, ethyl 1,4-dihydro-4-thioxo-
                                                        449,450
Pyridine-3-carboxylate, ethyl 5-acetyl-1,4-dihydro-2,6-
     dimethyl-4-(2-nitrophenyl)-   269
5-Pyridinecarboxylic acid, 2-amino-   100
Pyridine-3,5-dicarboxylate, diethyl 1,4-dihydro-2,6-
     dimethyl-4-(2-nitrophenyl)-   269
4-pyridinemethanol, α-phenyl-   195
Pyridinium bromide, 4'-chlorostyryl-   142
——, 1-styryl-   142
Pyridinium salts, styryl-   142
4-Pyridinol, 2,6-dimethyl-3-(2,5-dimethylpyrrol-3-
     ylazo)-   310
2(1H)-Pyridone   99-101
2(1H)-Pyridone, 3-chloro-   100
——, 6-chloro-   100
——, 1,4-dimethyl-   100
——, 1,6-dimethyl-   100
——, 1-(2-hydroxyethyl)-   99,100
——, 1-methyl-   100
——, 3-methyl-   100
——, 4-methyl-   100
2(3H)-Pyridone, 3-diazo-   305
4(3H)-Pyridone, 3-diazo-2,6-dimethyl-   310

2(1H)-pyridone dimer   99-101
pyrimidine, 4-amino-2,6-dimethyl-   54,55
pyrimidines   155,156
pyrocalciferol   7,8,385,386
Pyrocatechol   236
pyrocatechol monoacetate   236
Pyromucic acid   429
α-Pyronene   19
pyrrole   395-397
Pyrrole, N-methyl-   396
——, 2,3,5-tetraphenyl-   382,396,397
2H-Pyrrole, 2-hydroperoxy-2,3,4,5-tetraphenyl-   396,
                                                        397
Pyrrole-2-carboxylic acid   305
Pyrrole-3-carboxylic acid, 2,5-dimethyl-   310
Pyrrolidine   319,381
pyrrolidine, 1-butyl-   243
Pyrrolidines   243
2-Pyrrolidinone, 5,5-dimethyl-   50,51
1-Pyrroline, 3,3,-tricyano-2-(tricyanovinyl)-4-(2-
     tetrahydrofuryl)-5-imino-   161
1-Pyrroline 1-oxide, 5,5-dimethyl-   50,51
——, 2,5,-trimethyl-   50,51
3-Pyrrolin-2-one, 5-hydroxy-   395-397
Pyrrolizidine   245,246
1H-Pyrrolo[2,3-c]pyridine   306
1H-Pyrrolo[2,3-c]pyridine-3-carboxylic acid   305,306
Pyruvaldehyde diethyl acetal   238
Pyruvaldehyde diethyl acetal, bromo-   238
Pyruvate, alkyl diazo-   299
——, ethyl   238
pyruvic acid   210
Pyrylium-3-oxide, 2,4,5,6-tetraphenyl-   410
——, 2,4,6-triphenyl-   410

quadricyclene   2,3,10,328,329
quinaldehyde   187
quinaldine 1-oxide   52
quinazoline, 4-amino-2-benzamido-6-methyl-   359
——, 2,4-dibenzamido-6-(bromomethyl)-   359
——, 2,4-dibenzamido-6-methyl-   359
quinazoline dibenzoate, 2,4-diamino-6-methyl-   359

Compound Index

13,17-seco-13α-androst-5-en-17-oic acid, 3β-methoxy- 226
9,10-secoergosta-5(10),6,8,22-tetraen-3β-ol 15,19
9,10-secoergosta-5,7,10(19),22-tetraen-3β-ol see also calciferol 15
3,4-seco-5α-lanostan-3-oic acid 226,227
13,17-secopregna-5,13(18)-dien-20-cne, 3,3-ethylenedioxy- 26,37
1,10-seco-3,5(10)-santadien-8α,12-olide, 1-carboxy- 229,230
12,13-seco-5α,22α,25D-spirost-13-en-12-al, 3β-acetoxy- 43,44
13,17-secosteroids 26
sensitizers 2,3,56-61,63,65,72-76,82,83,86,110,111, 115,148,149,153,155,160-163,166-168,176-178,193,230- 233,276,277,321,325,335,336,375-397,401,402,408,427- 432,444,445,447,466
serine ethyl ester, C-benzyl- 201,202
——, 2-(benzyloxymethyl)-3-(4-nitrophenyl)- 201, 202
silane, bis-(1,1,2,2-tetrafluoroethyl)-dimethyl- 179
——, (bromomethyl)-chlorodimethyl- 351
——, chlorotrimethyl- 351
——, dichlorodimethyl- 354
——, dimethyl- 179
——, (1,1,2,2,3,3,4,4-octafluorobutyl)-dimethyl- 179
——, (1,1,2,2-tetrafluoroethyl)-dimethyl- 179
——, tetramethyl- 436,437
——, trichloro- 178,179
——, trichloro-(chloromethyl)- 354,355
——, trichloro-(chlorophenyl)- 354,355
——, trichloro-(dichloromethyl)- 354,355
——, trichloromethyl- 354,355
——, trichlorophenyl- 354,355
——, trichloro-(1,1,2,2-tetrafluoroethyl)- 178,179
——, trichloro-(trichloromethyl)- 354,355
silanes, dialkyl- 179
sorbate, methyl 23
sorbic acid 14,22-24
spiro[1,3-benzodioxole-2,1''-cyclohexane], 4,5,6,7- tetrachloro- 307,308

quinhydrone 189,212
quinodimethane, α,α,α',α'-tetracyano- 161
quinoline 460
quinoline, 8-(bromomethyl)- 357-359
——, 8-(dibromomethyl)- 357-359
——, 8-methyl- 357,358
——, 2-β-styryl- 71
quinoline 1-oxide 51,52
quinoline 1-oxides, methyl- 51
4(3H)-quinolone, 3-diazo- 305,306
quinomethane, diphenyl- 29
quinone diazides, p-imino- 310,311
o-quinone diazides 304-310
p-quinone diazides 310,311
quinone oximes 191
quinones 146,453
p-quinones 418,419,423,424
1,2-quinones 439
1,4-quinones 79-81
quinoxaline 1,4-dioxide 51,52
quinoxaline 1-oxide 51
2-quinoxalinol 51,52
2-quinoxalinol 4-oxide 51,52

radicals, benzenesulfenyl 153
——, diphenylhydroxymethyl 206,207
——, 1-hydroxyphenalen-1-yl 207
——, sulfenyl 153
——, tetraarylphosphoranyl 457
——, thiyl 153,440,442
resin acids 153,154
retinol 59
riboflavine 257,258
rose oxides 377
rubrene 390

salicylaldehyde 187,188
salicyloyl azide 322
2,4-santadien-8α,12-olide, 1-oxo- 229,230
santonin 30-32
SCHIFF bases 141,142,299,334,335

Compound Index

spiro[2H-1-benzopyran-2,2'-indoline], 1'',3'',3''-trimethyl-
  6-nitro- 42
spiro[bicyclo[4·1·0]heptane-7,3'-indolin]-2'-one,
  1'-methyl- 276
spiro[bicyclo[3·1·0]hex-3-ene-6,1'-cyclopentan]-2-one,
  1,4-dimethyl- 30
spiro[2,5-cyclohexadiene-1,10'-[9']-oxabicyclo[6·2·0]
  decan]-4-one 418
spiro[2,5-cyclohexadiene-1,10'-[9']-oxabicyclo[6·2·0]
  decan]-4-one, 2,3,5,6-tetrachloro- 418,419
spiro[2,5-cyclohexadiene-1,8-[7]oxabicyclo[4·2·0]oct-3-
  ene]-2',4,5'-trione, 2,3',5,6'-tetramethyl- 79,80
spiro[cyclopropane-1,3'-indolin]-2-one, 1'-methyl-2,2-
  diphenyl- 307
spiro[cyclopropene-1,9'-fluorene], 2,3-dicarbomethoxy-
  330
spiro[cyclopropene-1,9'-fluorene]-2,3-dicarboxylate,
  dimethyl 330
spiro[4·5]deca-6,9-dien-8-one, 6,9-dimethyl- 30,32
spiro[fluorene-9,2'-oxetane], 3',3',4'-trimethyl- 415-
  417,421
spiro[fluorene-9,3'-[3H]-Pyrazole], 4',5'-
  dicarbomethoxy- 330
spiro[fluorene-9,3'-[3H]-Pyrazole]-4',5'-dicarboxylate,
  dimethyl 330
spiro[fluoren-9,2'-oxetane], 4'-[N-(2-cyano-2-propyl)-
  imino]-3',3'-dimethyl- 421
spiro[indeno[2,1-b]oxete-2(2aH),9'(10'H)-phenanthren]-
  10'-one, 7,7a-dihydro- 419
spiro[3H-naphtho[2,1-b]pyran-3,9'-thioxanthene] 119,
  120
spirooxetanes 418-420
spiro[2H-oxeto[2,3-b]benzofuran-2,9'(10'H)-phenanthren]-
  10'-one, 2a,7a-dihydro- 419
spiro[1H,4H-oxeto[2,3-c][2]benzofuran-1,9'(10'H)-
  phenanthrene]-1',10'-dione, 2a,8b-dihydro- 419
  -----, 2a,8b-dihydro-2a-phenyl- 419
spiropentane 347,348
spiropentane, chloro- 347,348
spiropyrans 42
5α,22α,25D-spirostan, 3β-acetoxy-12α,14α-epoxy- 43,
  44

5α,22α,25D-spirostan-12-one, 3β-acetoxy- 43,44
spiro[5·5]undecane 264
spiro[5·5]undecanone oximes 264
stannane, tributyl- 204,208
stearamide 371
stearic acid, 12-acetoxy- 370
γ-stearolactone 371
steroids, 11,19-cyclo- 37
  -----, 18,20-cyclo- 36,37
  -----, 17-oxo- 65,66
stilbamidine 70-73
2-stilbazole 119,120
stilbene 19,20,70,71,73,118-120,123,127,129,132,133,
  139,415
stilbene, α-benzoyl-α'-benzoyloxy- 431
  -----, 2,2'-bis-(2,4,6-trimethylbenzyl)-3,3',5,5'-
  tetramethyl 289
  -----, 4-bromo- 128
  -----, 2-carbomethoxy-2'-iodo-α-nitro- 128
  -----, 2-carbomethoxy-α-nitro- 128
  -----, α-chloro- 119,120,122
  -----, dibenzoyl- 431
  -----, cis-dibenzoyl- 394,395
  -----, trans-dibenzoyl- 394,395
  -----, α,α'-dicyano- 127
  -----, trans-α,α'-dicyano- 19,20
  -----, 2,2',4,4',6,6'-hexamethyl- 19,127
cis-stilbene 120,123,124
cis-stilbene, 4,4'-bis-(dimethylamino)- 141,142
trans-stilbene 120,123,124,289
trans-4,4'-stilbenedicarboxamidine 70-73
stilbenes 56
stilbenes, substituted 127-129
styrene 118-120,123,163,169,171,172,415,417,446,
  448
styrene, α-azido- 318
  -----, β-methyl- 171,172
  -----, β-methyl-β-nitro- 266,267
styrene-β-sulfonate, methyl 2-hydroxy-α-methyl-
  346,347,350
styrene-β-sulfonyl chloride 301
suberate, diethyl 301
suberic acid 167
succinamide, pentyl-

Compound Index

succinate, diethyl carbamoyl- 167
——, diethyl 1,2-dicarbamoyl- 167
succinic acid dilactone, 2,3-bis-(1-hydroxy-1-methylethyl)- 232
succinic acid monolactone, 2,3-bis-(1-hydroxy-1-methylethyl)- 232
succinic anhydride, bis-(2-methoxybenzylidene)- 129,131
——, bis-(4-methoxybenzylidene)- 129
——, bis-(p-methylbenzylidene)- 129
2-chloro-3-(2,3,4,5,6-pentachlorocyclohexyl)- 345
——, 2-chloro-3-phenyl- 345,346
——, dibenzylidene- 129
——, 2,3-dichloro- 345
succinimide, N-bromo- 53,54,238,357-359
——, N-chloro- 357
succinonitrile, tetramethyl- 421
sudan G 46
sulfates, cyclic 439
sulfenyl chlorides 164,437
sulfenyl radicals 153
sulfide, bis-(2-chloroethyl) 163,164
——, bis-(2-chloro-1,1,2-trifluoroethyl) 163
——, bis-(trichloromethyl) 356
——, bis-(1,1,2-trifluoro-2-methoxyethyl) 163,164
——, bis-(trifluoromethyl) 439,440
——, 2-chloro-1,1-difluoroethyl trifluoromethyl 164
——, 2-chloro-2,2-difluoroethyl trifluoromethyl 164
——, (2-cyclopropyl)-1-propyl methyl 176
——, dibutyl 164
——, dimethyl 327
——, di-4-pyridyl 449
——, methyl 2-methyl-2-penten-1-yl 176
sulfides 327,460
sulfides, alkyl β-hydroperoxyalkyl 446,447
——, aryl desyl 216
sulfimide, S,S-dimethyl-N-(p-toluenesulfonyl)- 327
sulfimides, N-sulfonyl 438
sulfinic acids 438

sulfone, diethyl 444
——, dimethyl 444
——, diphenyl 346
sulfones 165
sulfones, cyclic 447
sulfonic acids 438,439
sulfonyl chlorides 165,434-437
sulfonylhydrazones 284,285,289
sulfoxide, (±)-t-butyl methyl 67
——, diethyl 444
——, dimethyl 326
——, (±)-methyl p-tolyl 444,463
——, (±)-1-naphthyl p-tolyl 67
sulfoxides 66,67,326
sulfoxides, alkyl β-hydroxyalkyl 446,447
sulfoximide, S,S-dimethyl-N-(p-toluenesulfonyl)- 326,327
sulfoximides, N-sulfonyl 326,327
sulfur 440,446,450
sulfur compounds, organic 434-451
——, unsaturated 90,91
sulfur dichloride 437
sulfur dioxide 346,355,434-439,445-447
sulfuric acid 445
sulfuryl chloride 354,355,434-437
sultones, unsaturated 448
suprasterol-II 15,17

tachysterol 15,59
tartaric acid, dimethyl- 210
——, diphenyl- 210,211
tartrate, diethyl diphenyl- 210,211
taurine 435
terebic acid 232,233
terephthalaldehyde, nitro- 267
p-terphenyl 409,410
terphenyls 455
α-terrinene 383,384
testosterone 23
10α-testosterone 22-24

Compound Index 605

testosterone acetate 194
testosterone ester 77
testosterone ester dimers 77
tetrabenzo[2.2]paracyclophane 11
tetrabenzo[2.2]paracyclophane photoisomer 11
tetracene dimer 99
tetracycline, anhydro-6-demethyl- 39
—, 7-bromo-6-demethyl-6-deoxy- 39,362
—, 11a-bromo-6-demethyl-6-deoxy- 39,362
—, N-tert•butyl-7-chloroanhydro- 382
—, N-tert•butyl-7-chloro-6-deoxy-6-
  hydroperoxydehydro- 382
—, 7-chloroanhydro- 382
—, 7-chloro-6-deoxy-6-hydroperoxydehydro- 382
—, 6-demethyl-6-deoxy- 362
—, 9,N-di-tert•butyl-7-chloroanhydro- 382
—, 9,N-di-tert•butyl-7-chloro-6-deoxy-6-
  hydroperoxydehydro- 382
tetracyclo[2.2.1.0²·⁶.0³·⁵]heptane 2,3,10,328,329
tetracyclo[3.2.0.0²·⁴.0⁴·⁶]heptane 2,3,10
tetracyclo[2.2.1.0²·⁶.0³·⁵]heptane-2,3-dicarboxylic acid 2-4
tetracyclo[3.2.0.0²·⁷.0⁴·⁶]heptane-1,5-dicarboxylic acid 2-4
tetracyclo[2.2.0.0²·⁶.0³·⁵]hexane, 1,2,5-tri-tert•butyl- 10
tetracyclone 368,369,394,395
tetralin 151,152,182,374
α-tetralone 205,208
2,4,8,10-tetraoxaspiro[5.5]undecane, 3,9-bis-(2-
  nitrophenyl)- 268
2,4,8,10-tetraoxaspiro[5.5]undecan-3-ol, 9-(2-
  nitrophenyl)-3-(2-nitrosophenyl)- 268
2H-tetrazolium bromide, 2-(1-naphthyl)-3,5-diphenyl- 143
—, 2-(2-naphthyl)-3,5-diphenyl- 143
2H-tetrazolium chloride, 2-(4-biphenylyl)-3,5-diphenyl- 143
—, 2,3-bis-(4-nitrophenyl)-5-phenyl- 143
—, 2-(4-nitrophenyl)-3,5-diphenyl- 143,144
—, 2,3,5-triphenyl- 143,144
—, 2,3,5-tris-(4-methoxyphenyl)- 143

2H-tetrazolium chloride), 5,5'-p-phenylenebis-(2,3-
  diphenyl- 143,144
2H-tetrazolium nitrate], 3,3'-(methylenedi-p-phenylene)-
  bis-[2,5-diphenyl- 143,144
tetrazolium salts, triphenyl- 143,144
1,2,4,5-tetroxane, 3,3,6,6-tetraphenyl- 291
theophylline 381
2-thiabicyclo[3.2.0]hept-3-ene-6,7-dicarboxylic
  anhydride, 6,7-dimethyl- 110
2-thiabicyclo[3.1.0]hex-3-ene-6-carboxylate, ethyl
  278,279
1,2,3-thiadiazine S,S-dioxide 316,317
1,2,3-thiadiazole, 4,5-diphenyl- 331,332
—, 4-phenyl- 331,332
1,2,3-thiadiazoles 331,332
cis-thianthrene 5,10-dioxide 67
trans-thianthrene 5,10-dioxide 67
thietanes 416,449
thiirane, trans-2,3-dibenzoyl-2,3-diphenyl- 450
—, 2,3-diethoxy-2,3-dimethyl- 450
thiiranes 450
thiobenzophenone 445,446,448,449
thiobenzophenone, 4,4'-bis-(dimethylamino)- 445
—, 4,4'-dimethoxy- 445,446
thiocyanates 443
thiocyanogen 443,444
thiocyanogen chloride 443,444
thioethers 439
4-thioflavone 445
cis-thioindigo 61,62
trans-thioindigo 61,62
thioketene 331,332
thioketones 445
thiols 162,163,446,447
thiophene 110,278,279
thiophene, 2-phenyl- 451
—, 3-phenyl- 451
thiophene 1,1-dioxide, 4-bromotetrahydro-3-
  (trichloromethyl)- 171,172
—, 2,5-dihydro- 171
thiophenol 152
thiophosgene 106,107

4H-thiopyran-4-thione, 2,6-diphenyl- 445
thiourea 444,445
thioxanthene 146-148
——, 9-benzylidene- 119,120
——, 9-(1-naphthyl)- 398
——, 9-phenyl- 398
thioxanthene-9-thione 445
thioxanthen-9-ol, 9-(xanthen-9-yl)- 199
thioxanthen-9-one 199
Δ⁹,⁹'-thioxanthylidenexanthene 133
thiyl radicals 153,440
thujone 218
thymine 91
thymine dimer 91
thymoquinone 80,81
thymoquinone dimer 80,81
tiglate, methyl 328
tiglic acid 57
tin, (cyclopentadienyl)-triphenyl- 466
——, (2,3-dioxabicyclo[2.2.1]hept-5-enyl)-triphenyl-
  466
——, hexamethyldi- 465
——, (3-methyl-2-buten-1-yl)-triphenyl- 466
——, (trifluoromethyl)-trimethyl- 465
tin compounds, organo- 465,466
tin hydroperoxides, organo- 466
tin iodide, (4-iodo-1,2,3,4-tetraphenyl-1,3-butadienyl)-
  dimethyl- 465
——, trimethyl- 465
tolan 423
tolan, 2,4-dinitro- 271,272
tolan, 2-nitro- 271,272
tolan see also acetylene, diphenyl-
p-tolualdehyde 191,400
toluene 110,196,231,233,261,262,276,290,311,351,356,
  357,443,467
toluene, p-bromo- 344,355,356
——, o-chloro- 467
——, p-chloro- 233
——, α-chloro- 262
——, 4-iodo- 456
——, octachloro- 347,348

toluene dimer, α-nitroso- 261
p-toluenediazonium chloride 313
p-toluenediazosulfonate, sodium 3-bromo- 316
p-toluenesulfonyl azide 326,327
p-toluenesulfonyl chloride 165
p-toluidine 165
toxiferine-I 401,402
trans-tricyclo[5·3·0·0²·⁶]decane 74
tricyclene 284,285
trans-tricyclo[5·3·0·0²·⁶]deca-3,9-diene 75
tricyclo[5·2·1·0²·⁶]deca-4,8-dien-3-one 2,3
tricyclo[5·2·1·0²·⁶]deca-4,8-dien-3-one, 2,4,5,6-
  tetrachloro- 2,3
tricyclo[3·3·1·1³·⁷]decane 264
cis,trans-tricyclo[5·3·0·0²·⁶]decane-3,8-dione 77,
  78,115
cis,trans,cis-tricyclo[5·3·0·0²·⁶]decane-3,10-dione
  77,78
cis,trans,cis-tricyclo[5·3·0·0²·⁶]decan-3-one
  112,113
tricyclo[3·3·1·1³·⁷]decan-2-one oxime 264
tricyclo[3·3·2·0²·⁸]deca-3,6,9-triene 116,117
tricyclo[4·2·2·0²·⁵]dec-9-ene-3,4,7,8-tetracarboxylic
  3,4:7,8-dianhydride 109,110,114
tricyclo[4·4·0·0²·⁶]dec-4-en-3-one, 2,4-dimethyl- 30
exo-tricyclo[6·2·2·0²·⁷]dodeca-3,9-diene 75
cis,cis,cis-tricyclo[6·4·0·0²·⁷]dodeca-3,11-diene 75
cis-tricyclo[6·4·0·0²·⁷]dodeca-4,10-diene-6,9,12-
  tetrone, 4,5,10,11-tetramethyl- 80,81
tricyclo[6·4·0·0²·⁶]dodecane-2,6-diol, 4,4-dimethyl-
  113
tricyclo[7·3·0·0²·⁷]dodecane-1,9-diol, 11,11-dimethyl-
  113
tricyclo[6·4·0·0²·⁷]dodecane-3,9-dione, 1,5,7,11-
  tetramethyl- 77
tricyclo[6·4·0·0²·⁷]dodecane-3,12-dione, 5,7,8,10-
  tetramethyl- 77
tricyclo[2·2·1·0²·⁵]heptane, 1,7,7-trimethyl- 284,285
tricyclo[2·2·1·0³·⁵]heptan-2-one, 1,7,7-trimethyl- 303
tricyclo[3·3·0·0²·⁶]octane 18
tricyclo[4·2·0·0²·⁵]octane-1,2,5,6-tetracarboxylate,
  tetramethyl 115

Compound Index

tricyclo[3.3.0.0²·⁷]octan-3-one, 1,2-dimethyl- 1,3
tricyclo[3.3.0.0²·⁸]oct-3-ene 7
tricyclo[2.1.0.0²·⁵]pentan-3-one, 1,5-diphenyl- 300, 301
tricyclo[6.2.1.0²·⁷]undeca-4,9-diene-3,6-dione 2,3
triethylamine 381,460
triethylenediamine 463
trimethylene oxide 414
2,3,7-trioxatricyclo[2.2.1]hept-5-ene, 5,6-bis-(4-bromophenyl)-1,4-diphenyl- 431
—, 1,4-dimethyl- 428
3,5,7-trioxatricyclo[4.1.0.0²·⁴]heptane, 1,2,4,6-tetraphenyl- 431
triphenylamine 138
triphenylmethyl 126,127
trisulfide, bis-(trifluoromethyl) 440
1,3,5-trithiane 349
1,3,5-trithiane 1,1,3,3,5,5-hexaoxide, 2,2,4,4,6,6-hexachloro- 349
1,2,4-trithiolane, 3,3,5,5-tetraphenyl- 445,446
α-tropolone 227,229
α-tropolone methyl ether 227
γ-tropolone methyl ether 4,6,9
α-truxillic acid 76,83–85
β-truxinic acid 83–85
δ-truxinic acid 76
tungsten hexacarbonyl 459,460
tungsten pentacarbonyl, aniline- 460
—, pyridine- 460
—, (triethylamine)- 460
tungsten tetracarbonyl, bis-(pyridine)- 460

undecanamide 166
undecanoate, methyl 11-carbamoyl- 166
—, tributyl 11-phosphono- 177,178
undecanoates, trialkyl phosphono- 177,178
10-undecenamide 166
10-undecenoate, butyl 177
—, methyl 166
10-undecenoates, alkyl 177
uracil, 1,3-dimethyl- 155,156
—, 5-fluoro- 155

—, 5-fluoro-6-hydroxy- 155
—, 6-hydroxy-1,3-dimethyl- 155,156
urea, diphenyl- 322
urethan 322,324,325
uretidinedione, 1,3-diphenyl- 280
urotropine 463
ursa-9(11),12-dien-28-oate, methyl 3β-acetoxy- 16
ursa-8(26),9(11),12-trien-28-oate, methyl 3β-acetoxy- 16
ursa-8,11,13-trien-28-oate, methyl 3β-acetoxy- 16
usnic acid 66

valeraldehyde, 2,2,3,4,4,5,5-octafluoro- 422
valeric acid, 2,3,4,5-tetrachloro-5-phenyl- 342
valeric acid γ-lactone, 4-hydroxy-2,4-dimethyl- 232
—, 4-hydroxy-3,4-dimethyl- 232
—, 4-hydroxy-2,4-dimethyl-2-(diphenylhydroxymethyl)- 232
—, 4-hydroxy-3-(1-hydroxy-1-methylethyl)-4-methyl- 233
vanadium dicarbonyl, (1,3-butadiene)(cyclopentadienyl)- 463
vanadium tetracarbonyl, cyclopentadienyl- 463
verbanone oxime 28
verbenone 264
vinyl acetate 461
vinyl chloride 119,123,124,163,164,174
visnagin 419,420
vitamin A 59
vitamin D2 15

water 28,30–32,155,156,225–227,229,230,234,235,270, 271,287,295–297,301–310,336,363,364,379

xanthate, O,O-diethyl S,S-glutaryldi- 442
—, O-ethyl 3β-acetoxy-11-oxo-5α-bisnorcholanoyl 223,224
—, O-ethyl 3β-acetoxy-11-oxo-5α-pregnane-20-yl 223,224
—, O-ethyl S-benzyl 223,224
—, O-ethyl S-phenylacetyl 223,224

608   Compound Index

xanthates, S-acyl   223,224,442
xanthene   147,183,184,199-201
xanthene, 9-benzyl-   398,399
-----, 9-benzylidene-   119,120
-----, 4-chloro-9-phenyl-   398
-----, 1,2,3,4,4a,9a-hexahydro-4a-methoxy-   374,375
-----, 9-(4-methoxyphenyl)-   398
-----, 9-(1-naphthyl)-   398
-----, 9-phenyl-   398
-----, 9-m-tolyl-   398
-----, 9-o-tolyl-   398
-----, 9-p-tolyl-   398
9-xanthenemethanol, α,α-diphenyl-   199,200
xanthene-9-thione   445
xanthen-9-ol, 9-(xanthen-9-yl)-   199-201
xanthen-9-one   148,199,201,205-207,212,415
xanthotoxin   87,88
xanthotoxin dimer   87,88
xylal, diacetyl-   122
m-xylene   231,233,443,444
o-xylene   110,182,183,231
p-xylene   148,182,183,231,373
p-xylene, decachloro-   347
-----, α,α,α',α'-tetracyano-   161
-----, α,α,α',α'-tetracyano-α-(2-tetrahydrofuryl)-
                161